A DICTIONARY OF
MECHANICAL ENGINEERING TERMS

Originally Compiled By

J. G. HORNER, A.M.I.M.E.

NINTH EDITION REVISED AND ENLARGED BY

G. K. GRAHAME-WHITE

M.I.R.T.E., A.M.S.E., M.Inst.P.S., F.R.S.A.

PART 1: MORE MODERN TERMS

PART 2: BASIC TERMINOLOGY

Oxford & IBH Publishing Co. Pvt. Ltd.
New Delhi
(A Unit of CBS Publishers & Distributors Pvt Ltd)

CBS

Dedicated to Education

CBS Publishers & Distributors Pvt Ltd

New Delhi • Bengaluru • Chennai • Kochi • Kolkata • Mumbai
Hyderabad • Jharkhand • Nagpur • Patna • Pune • Uttarakhand

Dictionary of
Mechanical Engineering Terms, Ninth Edition

ISBN-13: 978-81-204-1729-8
ISBN-10: 81-204-1729-1

Reprint: 2018

OXFORD & IBH

New Delhi
(A Unit of CBS Publishers & Distributors Pvt Ltd)

CBS Publishers & Distributors Pvt Ltd
204 FIE, Patparganj Industrial Area, Delhi 110 092
E-mail: delhi@cbspd.com, cbspubs@airtelmail.in

Ph: 4934 4934   Fax: 4934 4935   Website: www.cbspd.com
                                                     e-mail: publishing@cbspd.com;
                                                     publicity@cbspd.com

Branches

- **Bengaluru:** Seema House 2975, 17th Cross, K.R. Road, Banasankari 2nd Stage, Bengaluru 560 070, Karnataka
  Ph: +91-80-26771678/79   Fax: +91-80-26771680   e-mail: bangalore@cbspd.com
- **Chennai:** No. 7, Subbaraya Street, Shenoy Nagar, Chennai 600 030, Tamil Nadu
  Ph: +91-44-26680620, 26681266   Fax: +91-44-42032115   e-mail: chennai@cbspd.com
- **Kochi:** Ashana House, 39/1904, AM Thomas Road, Valanjambalam, Ernakulam 682 016, Kochi, Kerala
  Ph: +91-484-4059061-65,67   Fax: +91-484-4059065   e-mail: kochi@cbspd.com
- **Kolkata:** No. 6/B, Ground Floor, Rameswar Shaw Road, Kolkata-700014 (West Bengal), India
  Ph: +91-33-2289-1126, 2289-1127, 2289-1128   e-mail: kolkata@cbspd.com
- **Mumbai:** 83-C, Dr E Moses Road, Worli, Mumbai-400018, Maharashtra
  Ph: +91-22-24902340/41   Fax: +91-22-24902342   e-mail: mumbai@cbspd.com

Representatives

Hyderabad	0-9885175004	**Jharkhand**	0-9811541605	**Nagpur**	0-9021734563
Patna	0-9334159340	**Pune**	0-9623451994	**Uttarakhand**	0-9716462459

Printed at Chaman Interprises, Daryaganj, Delhi, India

PREFACE TO THE NINTH EDITION

WHEN the late Mr. J. G. Horner compiled the original edition of this work, he aimed at producing a comprehensive dictionary of the general and traditional terms used by draughtsmen, pattern-makers, moulders, smiths, boiler-makers, fitters, turners, erectors and engineering storekeepers.

The result was more than a dictionary. It might best be described as a condensed encyclopaedia of mechanical engineering practice, with the practical aspects as strongly represented as the theoretical (no doubt as a result of the twenty-seven years of his life which the author had spent on the shop floor).

Engineering, however, does not stand still. Continual progress has introduced many terms unknown to an older generation of craftsmen. The mechanical engineer to-day must be something of a metallurgist; his work often overlaps that of the electrical and electronic engineer; radiography, radiology and ultrasonics play important parts in the inspection and test departments; and an acquaintance with chemistry and physics will often be found essential.

It has, therefore, as always, been difficult to know where in this new edition to draw the dividing line between terms associated primarily with mechanical engineering and those which would be more appropriately found in dictionaries devoted to other branches of science and technology. The latest editor of Mr. Horner's work can only hope that the inevitably tentative balance he has attempted to strike will be regarded as a reasonable one.

The principal improvement in this new edition has been the addition of some 190 new terms to Part I of the *Dictionary*. Earlier entries in this Part have again been revised; and asterisks have been added to those terms appearing in Part I which are further (or alternatively) defined in Part II.

Part II itself (save for the correction of one " howler " that had escaped detection through eight previous editions) has been left unaltered. The motive here was frankly economic. The

PREFACE

printing surface for this Part is held in the form of flat metal
plates. The alteration to these plates involves cutting out the
offending line or paragraph from the plate, resetting it in type,
converting this into plate form, and soldering in the new text
into the place left by the old. This is an expensive process
and the Publishers felt that the admitted inconvenience of
having the more recent terms used in mechanical engineering
listed separately from the more general and traditional terms
would be more widely acceptable than an otherwise inevitably
drastic raising of the price of the *Dictionary* as a whole.

Other additions to the Ninth Edition comprise a number of
useful *conversion factors* and *tables of equivalents* and a selected
list of the more important *British Standards publications* relating to mechanical engineering.

PART 1

DICTIONARY OF MODERN TERMS

USED IN

MECHANICAL ENGINEERING

A

Abrasion Test.—Test for abrasion by means of a scratch on the smooth surface of the material tested. By comparing the results on different materials by reference to the hardness scale (q.v.) the relative hardness is found ; or a better test is by means of the Sclerometer, which is an instrument consisting of a diamond at one end of a lever attached to a vertical pillar. This diamond is loaded, and on rotating the pillar by hand it makes a scratch. The weight in grammes required to produce a scratch of standard depth gives a measure of the hardness. See Abrasion.*

Abrasive.—Mineral employed in sharpening, grinding or polishing operations. Natural abrasives include diamond dust, corundum and emery ; corundum and emery consist mainly of aluminium oxide. Artificial abrasives are based either on silicon carbide (e.g. " Carborundum "), aluminium oxide ("Aloxite," "Alundum," etc.) and diamond dust. (See Bond, Grit.)

Abscissa.—Usually the horizontal distance of a point from the origin of co-ordinates ; the *ordinate* being the corresponding vertical distance. Together they define the position of a point with respect to the axes, which may be oblique, or polar (q.v.).

Absorption Dynamometer.—A dynamometer, which, like the Prony or the rope-brake dynamometer, measures the work done. (See Dynamometer, Basic Terminology.)

Accelerated Circulation.—Relates to the circulation of water in a steam boiler, specifically in water tube types. The principle of the Galloway tubes and cross tubes is amplified into the nests of tubes which connect the water chamber below with the steam drum above.

Acceleration.—Denotes the rate at which the velocity of a moving body increases. It is uniform when the velocity increases by equal amounts in equal intervals of time ; variable when it does not. The units employed in calculation are the F.P.S. (foot pound second) system, and the C.G.S. (centimetre gramme second) system.

Accelerator.—Pedal in a motor vehicle which controls the power and speed, by acting on the throttle valve and thereby regulating the mixture of air and petrol entering the engine. Also, a machine for imparting very high speeds to nuclear particles by means of electric or magnetic forces. There are two main types of particle accelerator, the linear accelerator and the cyclotron.

Accumulator.*—See Pt. 2, Basic Terminology : (*a*) An *hydraulic* accumulator consists of a heavily weighted tank, which is raised by means of a pump, thereby storing up energy to be utilised as required, by its descent.

(*b*) A *heat* accumulator is a large container, in which a mass of material is placed which can absorb the heat of the steam from the engine exhaust of a steam-turbine and return it as required. It functions in a similar way to the Regenerator of a blast furnace.

(*c*) An *electric* accumulator consists of storage cells which are charged by a dynamo.

Acetylene Gas.—Results from the action of water on calcium carbide. The acetylene flame has a high illuminating value, being from ten to fifteen times that of coal gas. But its chief value to engineers is in combination with oxygen for use in a blowpipe for cutting metal plates, and welding.

Acme Thread.—Screw thread of American origin, the section of which is a mean between the square and vee threads. Used extensively for feed screws. The flanks have an inclined angle of 29°.

Actinometer.—Instrument for measuring the intensity of the solar rays.

Actual Size.—Measured size of a dimension or part.

Adapter.—Appliance by means of which objects of different sizes will interchange on a spindle or other fitting.

Addendum.*—The radial distance between the pitch circle of a gear wheel and the tip of a tooth. In the British Standard form of tooth it is equal to $0 \cdot 3183 \times$ circular pitch.

Addendum Circle.—In designing wheel teeth, the circle that passes through the tips of the teeth.

Adit.—Horizontal passage into a mine, etc.

Advance.—In an internal combustion engine cylinder, *advance* occurs when the timing of the ignition is altered to cause the spark in the cylinder to pass at an earlier point. Only done when running at high speed to allow the explosion time to take effect when the piston has reached the top of its stroke. See also Angular Advance.*

Æolotropic.—A material is so described when the stresses in it vary in different directions. Synonymous with anisotropic, and opposite to isotropic.

Aerial Transportation.—Effected by means of carriers or hoists running on suspended cableways of wire rope and actuated by steam or electricity. Largely employed in mining districts, in the coaling of ships off shore, and similar duties where a rail or roadway is impracticable.

Aero-thread Insert.—Threaded insert made from phosphor-bronze, carbon steel or stainless steel helical wire, and used in parts made of magnesium and other light alloys in conjunction with high-tensile steel studs or bolts. A truncated-vee thread form is used on the external thread of the insert, and a shallow arc-section thread on the screw or stud. Aero-thread inserts are widely used on aero engines, including sparking plug bushings.

Afterburner or **Reheater.**—A combustion chamber in turbine-type engines for aircraft into which fuel is injected for the purpose of reheating the hot exhaust gases, thereby developing increased thrust. The complete assembly usually consists of the combustion chamber, flame holder, injection nozzle, and exhaust ejector. The unit is either mounted within or attached to the exhaust tail pipe of the engine.

After-Burning.—A jet engine functions by sucking in air from the front of the engine and compressing it in a flame tube, into which fuel is then injected and ignited. The rapidly-expanding air rushes rearwards, driving the turbines to suck in more air and also pushing the aircraft forward. The power of the engine can be almost doubled by injecting more fuel behind the turbine blades and " reburning " the air. This process is called *after-burning* or *reheat*. See also Reheating.*

Age-hardening.—The change in strength and hardness which occurs in many aluminium alloys during a period of about four days after heat-treatment. The metal is normalised by heating it to 400–500° C. and quenching in water. During the age-hardening process the copper and magnesium in the alloy, as the case may be, separate out of the solid solution and are precipitated around the crystal boundaries, increasing the strength and hardness of the metal. Some alloys do not age-harden at room temperatures, but must be maintained at a temperature of 160° C. for about 18 hours. To avoid the necessity for frequently normalising light alloy sheets and rivets which must be maintained in a ductile state during production work, age-hardening is delayed by storing the material in refrigerators immediately after heat treatment. The beginning of age-hardening can thus be delayed from the usual 1–2 hours or 100 hours or more. See also Hardening.*

Aggregate.—The coarse material which is used with sand and cement to form concrete. (See Pt. 2, Basic Terminology.) It may consist of almost any hard and fairly clean material, such as broken stone or brick, slag, shingle, etc.

Air-chuck.—Pneumatically operated chuck used to grip work in a machine tool. A mechanical safety device is generally incorporated to prevent the chuck releasing the work if the air pressure should fail.

Air Cleaner.—Filtering device which extracts dust and grit from the air entering any mechanism.

Air Gap.—The space between the armature and the magnets of an electric motor.

Air Hardening.—The hardening of high speed steels in a blast of air.

Air Hoist.—A light suspended hoist, operated by compressed air lifting a piston enclosed in a cylinder.

Air Lift.—Machine by which water is lifted from deep wells by the aid of compressed air.

Air-lift Pump.—Pump which acts by forcing air from an air-compressor down a small pipe, which is passed down a well or bore-hole from which water is to be raised. The air pipe is bent upward at its lower end so as to discharge into the rising main. The air bubbles thus formed reduce the density of the mixture as compared with that of the water in the well or bore-hole, and in consequence of the difference in weight between the aerated water in the rising main and that outside it discharge takes place, the amount depending upon the depth to which the main pipe is immersed below the level of the water surface in the boring or sump. The simplicity of the apparatus required recommends the process, especially for temporary purposes. The hotter the liquid pumped the greater the efficiency.

Air Lock.—(1) Chamber in a caisson working with compressed air, within which the men remain both when descending and returning, in order to become accustomed to the differences in the pressures of the external air, and that in the caisson. (2) Accumulation of air in a pipe system, which diminishes the flow of water in a pipe and may completely stop it by collecting at the summit of a bend.

Air Valve.—Valve attached to the summit of a pipe line to discharge automatically any air which accumulates there so as to prevent an air-lock.

Alclad.—The resistance to corrosion of high-strength heat-treated alloys, such as those used in aircraft construction, is considerably increased by a coating of pure aluminium. Such alloys form the Alclad group and consist of a core of strong alloy, generally 17S, 24S or 26S, having a thin layer of high-purity aluminium on both sides. The coating of pure aluminium is itself very resistant to corrosion, and in addition exerts an electrolytic protective action to any exposed surfaces of the alloy core, such as to sheared edges, scratches, or abrasions.

Alignment Chart.—Graphical means of obtaining quickly the results of numerical formulæ. Also, a chart showing means of testing, and the tolerances for machine tools.

All-geared Drive.—Applied to the growing practice of operating speeds and feeds of machine tools through nests of gears instead of by belts. The gears are enclosed in boxes, which usually form oil wells. The term " all-geared head " is applied to a lathe headstock so fitted.

Allowance.—Prescribed difference between the high limit for a shaft and the low limit for a hole in order to provide a certain class of fit (q.v.). The allowance may be either positive or negative according to whether a clearance fit (q.v.) or an interference fit (q.v.) is required.

Alpha-Beta Brass.—Brasses suitable for hot-working, at temperatures between 600–800° C., including casting, extruding, hot-rolling and hot-stamping, usually containing 58–60% copper. Other elements, such as nickel, tin, iron, aluminium and manganese, are often added to modern brasses to improve their properties. So-called manganese bronze (q.v.) is one of these alloys.

Alpha Brass.—Alloys suitable for cold-working, wire drawing and tube making containing not less than 63% copper. Although they harden during cold working, alpha brasses can be readily softened by annealing at 600° C. Cartridge brass (70% copper and 30% zinc) is a typical example of this group.

Alpha Iron.—The phase of iron present in iron or steel below the critical range.

Alternating Current.—Current of electricity which is reversed from 25 to 60 times per second, with change of pressure.

Alternator.—Electric generator supplying an alternating current.

Aluminium* (see Pt. 2, Basic Terminology) is now produced at a comparatively low cost, both in the pure state and alloyed with other metals. It is not found pure in nature, but occurs as an oxide, and a double fluoride of aluminium and sodium. Bauxite is the mineral from which aluminium is extracted. Aluminium is melted in crucibles without a flux at a temperature only a little higher than its melting point, which is 1,210° F. Large quantities are melted in a reverbertory furnace. Its shrinkage is about the same as that of gun-metal, and large risers and good venting are therefore necessary. It can be forged hot or cold. It is filed best with single cut files. In addition to a wide range of mechanical and structural uses, it is also employed for some electrical transmission lines. Its specific gravity ranges from 2·6 to 2·7. A cubic foot weighs about 159 lb.

Aluminium Alloys.—The value of these is greater than that of the pure metal. They are termed "light" or "heavy" according as they contain a large or a small percentage respectively of aluminium. Many have trade names. They are largely used for castings and sheets, etc., where a combination of great strength with lightness is required.

Aluminium Bronze.—An alloy of copper with aluminium, in various and widely differing proportions. A tensile strength as high as 57·9 tons per inch has been obtained with this material. Four grades are prepared, the strongest containing 10% of aluminium, the weakest 2¼%. The tensile strength of the first is 30 tons per square inch, with an elongation of 22% in 4 in.; that of the last 20 tons, and 40% elongation, the other grades occupying intermediate positions.

Amalgamation.—The separation of gold and silver from their ores, by taking advantage of the property which mercury possesses of forming an amalgam with those metals.

American Cloth. Sometimes used with red lead and oil, to form the steam joints of flanges, in a similar manner to gauze wire, or asbestos sheet.

American Pipe Thread.—A standard American screw thread specified in two types, tapered and straight. Tapered threads based on the Briggs standard, having a taper of $\frac{1}{16}$ in. per ft., are most usually employed; straight or parallel threads are, however, specified for pressure-tight joints on oil and fuel pipe fittings, grease cups, free-fitting and loose-fitting mechanical joints, and loose-fitting hose couplings.

American Screw Thread.—An American standard thread, previously known as United States Thread or Sellers thread. Having a thread angle of 60°, with flat roots and crests, the thread is specified in the following series : coarse-thread series; fine thread series; 8-pitch series; 12-pitch series and 16-pitch series.

American Standard System of Limits.—Classification of fits adopted in the United States of America. Under this system fits are defined as loose, free, medium, snug, wringing, tight, medium force and heavy shrink fits.

American Turbine.—Mixed flow turbine which is a modification of the Francis turbine, in which the vanes are twisted so as to discharge parallel to the axis of rotation. Higher speed and power are obtained under low heads by reducing the diameter of the runner and increasing its capacity by making the buckets deeper and reducing the number of vanes.

Ammeter.—Instrument designed to show by direct reading the number of amperes of electric current passing in a circuit. Abbreviation of ampere-meter. There are many varieties.

Ammonia.—NH_3 is used in refrigerating apparatus, the work of the ammonia being to absorb the heat in its passage from the liquid to the gaseous state. A pound of ammonia absorbs 555·5 B.Th. units in passing from one to the other condition at zero F.

Amorphous.—A term applied to a non-crystalline metallic surface produced by polishing, when the crystal structure has been completely broken down and the molecules are irregularly distributed, as in the liquid state. Experimental evidence supports the view that the amorphous film may be regarded as an under-cooled liquid metal. See also Beilby Layer.

Ampere.—The unit of electrical current strength. A pressure of one volt will force one ampere through a resistance of one ohm.

Ampere-hour.—The quantity of electricity delivered by a current of one ampere in one hour.

Ampere-meter.—Same as Ammeter (see above).

Amplitude.—The amplitude of an oscillation is the maximum distance it moves from its central position.

Aneroid Barometer.—This type of barometer consists of a cylindrical metal box, the lid of which is a thin corrugated metal plate. It is exhausted of air, and the atmospheric pressure acting on the thin elastic lid bends it, and by a delicate system of levers registers the movement by a pointer moving over a graduated scale which gives the height in feet above mean sea level.

Angle Bending Machines.—Used for dealing with other sections besides angles. They embrace two types, one being that in which the bars are bent to various curvatures by means of rolls, the other in which they are squeezed into angular forms in presses. See also Angle* and Angle Iron.*

Angle Bevelling Machine.—One in which angles are set between rollers to acute or obtuse sections. The angle may be uniform, or be varied from one end to the other.

Angle Cleat.—Piece of steel plate bent to an angle (usually a right angle), and used to connect cross-beams, etc., to the main beams of a structure or to the stanchions.

Angle Cutter.—Milling cutter used for milling flutes on taps, reamers, spiral mills, and other cutters.

Angle Gauge.—A gauge used for testing or finding angles. It usually consists of a number of metal blades or leaves, the ends of which are ground to an angle. Its use is limited, but it can often be used instead of a protractor. See also Angle* and Protractor.*

Angle Wheels or Spiral Gears.—Form of screw gearing (q.v. in Pt. 2, Basic Terminology) in which the teeth form portions of many helices, and by means of which motion is transmitted between shafts which cross each other, but are not parallel. In these gears, the friction between the teeth is wholly that of sliding, hence their wear, and the mechanical loss occasioned is excessive. Skew bevels (q.v. in Pt. 2, Basic Terminology) were formerly sometimes used in preference to angle wheels, because they develop less friction, but they are rarely employed now.

Angular Hole Drilling.—Method of form cutting by which holes of square, hexagonal, and other sections are produced by a chisel-like tool. Attached to a spindle moving under the control of the form, it works the shape predetermined.

Angular Momentum.—Equivalent to moment of momentum. See also Angular Velocity.*

Angus Smith's Process.—Process employed for protecting water-pipes from corrosion. The pipes are heated to 154° F., and immersed in a bath of pitch, to which a little oil is usually added, the temperature of the bath being not less than 98° F.

Anisotropic.—Not isotropic. A material whose elastic properties vary in different directions. Same as æolotropic (q.v.).

Anode.—The conductor by which a current of electricity enters a liquid or a gas.

Anodic Treatment.—The formation of a protective oxide layer on the surface of aluminium and aluminium alloys by suspending the work in a bath of sulphuric acid or chromic acid and passing an electric current from the part, which forms the positive electrode or anode, through the electrolyte to a suitable cathode. With chromic acid a stainless steel or carbon cathode is used, with a bath temperature of 40–45° C., a current of 3–4 amp./sq. ft., and a voltage rising from zero to 40 volts in the first 15 min., maintained at 40 volts for 35 min., a further rise to 50 volts within 5 min., and a steady voltage of 50 volts for 35 min. With sulphuric acid as an electrolyte a lead anode is used with the bath at room temperature and a steady current of 10–20 amps. at about 40 volts for 30 min. Anodic films may be coloured by aniline dyes, giving a lasting wear-resisting and attractive finish.

Anodising.—See Anodic Treatment.

Antenna.—The group of insulated aerial wires for transmitting or receiving radio communication waves. Synonymous with *Aerial*.

Antielastic Bending.—Transverse bending in a beam associated with the longitudinal bending caused by the load, but of opposite curvature, which results from the lateral contraction of the longitudinally extended filaments and the lateral expansion of the longitudinally compressed filaments, in accordance with Poisson's Ratio (q.v.). The effect is neglected in the practical theory of beams.

Appliance.—An aid or adjunct to a machine tool, or to a shop operation. Templets, jigs and fixtures are the most striking examples.

Apron* (see Pt. 2, Basic Terminology).—Also, a covering of timber, stone or metal, to protect a surface from the action of water flowing over it.

Aqueduct.—The principal modern examples are those by which the water of lakes is conveyed to great cities, largely through steel pipes, siphons, tunnels, and cut and cover.

Arboring.—Signifies the shouldering back of a flat bearing face, to receive the washers and nuts of attachment bolts. It is done by means of a broad facing cutter wedged transversely in a boring bar, or arbor. Where fillets or radii intersect flange faces, and where the surface of a casting is from any cause uneven, arboring is usually and properly resorted to. See also Arbor* and Arbor Chuck.*

Arc Crater.—In welding, the depression left in the weld metal where the arc is broken.

Arc Lamp.—An electric lamp in which, a gap being made in an electric circuit between two carbon rods, a spark or arc of gaseous matter passes across, carrying with it a minute quantity of the carbon from the positive side of the gap.

Arc Welding.—Method of welding in which the metal is fused by the heat of an electric arc. The process can be considered under four headings : metallic arc welding ; carbon-arc welding ; argon or helium shielded arc welding ; and atomic hydrogen welding. Each process is described under its respective heading. See also Technotherm-Rakos Welding, Weibel Welding Process.

Archimedes' Principle.—The principle discovered by Archimedes (a Greek mathematician, 287–212 B.C.) that, when a body is weighed in a fluid, it appears to lose weight by an amount equal to the weight of the fluid it displaces.

Ardoloy.—Sintered carbide alloy used in the production of cutting tools. (See Sintered Carbide Tools.)

Argonarc Welding.—See Shielded Arc Welding.

Armature.—The revolving member in a dynamo, comprising an iron case on a shaft, carrying on its outer surface copper conductors which cut through the magnetic lines of the field, and receive by induction the currents generated in the field magnets.

Armour Plates.—These are slabbed and rolled from cast ingots of alloy steel, and subsequently hardened. Plates are planed and drilled before hardening, and afterwards corrected by grinding. Th presses for slabbing, and the rolls are the largest used in steel mills, the ingots often weighing 70 tons.

Armoured Cables.—Electric conductors protected with lead, and usually supplemented with other materials.

Armoured Concrete.—A term sometimes applied to reinforced, or ferro-concrete.

Armoured Hose.—Rubber or other hose which is covered with steel wire wound spirally round the outside, thus protecting it from cuts, abrasion or other damage.

Arsenical Copper.—Arsenic in amounts up to approximately 0·5% is added to copper with the primary intention of obtaining slightly increased strength and toughness, and most copper products made in Great Britain, other than electrical equipment, are made from arsenical copper. The best known arsenical copper conforms to B.S.1035-1040 and has an arsenic content of 0·3-0·5%. Further specifications are in existence for the production of arsenical copper for purposes such as railway materials. See also Arsenic.*

Artesian Wells.—These are bored through successive strata with augers and chisels, and lined with tubes, smaller in diameter as the depth increases. The debris is removed with a shell, being a pipe fitted with a valve which allows the borings to enter, but prevents their escape. For boring very hard rocks a drill set with black diamonds is sometimes used.

Assembling.—Signifies the union of the parts which comprise a piece of mechanism, without any correction by hand fitting. To ensure this, all parts must be perfectly interchangeable, to ensure which result they must be tooled to gauge within very fine limits of tolerance.

Assembly Drawing.—A drawing which shows the relative arrangement of various parts when they are fitted together to form an assembly. The majority of drawings used in an engineering workshop refer to the manufacture of one part only of an assembly, e.g., for the manufacture of a lathe chuck, one drawing will show details for the body, another for the jaws, another for the scrolls, and so on. The assembly drawing will show all these parts gathered together in their correct position to form the chuck. See also Drawing.*

Asymmetric Section.—One which is not symmetrical with respect to an axis.

Atomic Energy.—Term applied in a general sense to the energy obtained when the atoms of an element are split as a result of bombardment by high-speed atomic particles. Nuclear fission (q.v.) is accompanied by the generation of a large amount of heat, which can be transferred to a steam turbo-generator or can be employed to heat a gas for use in a gas turbine. The fission also causes intense radiation which produces radioactive metals and chemicals having many applications in industry, medicine and research.

Atomic Hydrogen Welding.—Method of welding in which a single-phase, alternating-current arc is struck between the tips of two tungsten electrodes which project from tubular nozzles through which hydrogen or cracked ammonia gas is fed. The electrical energy of the arc, combined with the catalytic effect of the electrodes, dissociates the molecular hydrogen into its atomic state. The recombination of the hydrogen atoms outside the arc results in the liberation of a very much greater heat than is obtainable from a gas welding flame. A further advantage is that hydrogen is an active reducing agent and thus prevents contamination and oxidation of the weld metal.

Atomic Space Lattice.—See Lattice or Space Lattice.

Atomiser.—A nozzle by means of which liquid fuel is reduced to a fine spray which readily ignites. The means by which the fuel is atomised are steam or air pressure, or by mechanical arrangements. Used also for disinfecting, cooling, perfuming, etc.

Audit System.—A term used to describe a system under which the quantity and quality of each day's output from a factory is accurately assessed and reported to a senior executive.

Auger.*—A tool used for boring holes in wood or in the earth. An earth auger, or ground auger, is often used to speed up the laying of foundations for bridges, buildings and other constructional projects. Since drilling does not vibrate the soil system, foundations can be drilled very close to other buildings which could be damaged by normal pile-driving operations. Another advantage is that the noise of augering is far less than that caused by a pile driver. See also Pile* and Pile Driver.*

Austenite.—Solid solution of carbon or iron carbide, Fe_3C, in gamma iron. All steels above their critical range consist of this solid solution. Its composition may vary in carbon content from a trace to 1·75%. It may also contain alloys such as manganese, nickel, chromium, molybdenum, tungsten, etc. Austenite is retained in certain alloy steels by air- or oil-cooling from above their critical ranges. It is not retained in any appreciable quantity in plain carbon steels even on fast cooling by a water-quench.

Auto-collimator.—See Optical Tooling.

Autographic Diagram.—Diagram made on squared paper by the machine to which it is attached and used in connection with engine performances, steam indicators, in testing machines, furnaces, electrical apparatus, etc.

Automatic Arc Welding Machine.—A class of welding machine in which the arc is struck and maintained at the correct voltage by automatic regulation of the feed of the electrode wire. The machine may move along the work on a motor-driven carriage, as in welding deck plates in shipbuilding, or the work may be moved longitudinally or rotated beneath the electrode carrier. On some machines a finely divided slag-forming flux is deposited ahead of the electrode to shield the molten metal from the atmosphere; the "melt," as this flux is termed, is a highly resistant conductor of the welding current and thus increases the heat generated by the passage of the current.

Automatic Gauging Machine.—Where a 100% check on all dimensions on all parts produced is necessary, an automatic, multi-dimensional gauging machine is used. As many as twelve dimensions can be checked and the results rendered visible by signal lights on a screen on which an enlarged view of the part is shown, the lights being related to the surfaces which are being measured. The machine may be arranged to check parts for ovality, eccentricity, taper, squareness, and other conditions, in addition to straightforward measurements, and may check and sort into appropriate bins simple parts at speeds up to as high as 10,000 components per hour.

Automatic Welding Machine.—A development of spot or seam resistance welding machines (q.v.) in which a succession of spots is made automatically at speeds of from 100 to 200 spots per minute as the work is fed through the machine. In automatic seam welders (q.v.) the work passes through the rollers at 6–13 ft. per min. See also Automatic Arc Welding Machine.

Automatic Work.—The development of automatic tools and mechanisms has been rapidly extending, partly with the object of lessening labour costs and charges, partly as a necessity imposed by the growth of the interchangeable system of manufacture. It includes both machine types and details. Gear-cutting and grinding are examples of the first ; stops, trips, and reverses of the second. These occur in many machine tools of common types. In the automatics proper, there are the cams, chucks, magazine and wire feeds, self-releasing dies, etc. See also Automatic.*

Automation.—Term first used to describe the installation, on a large scale, of automatic machines which performed successive operations, including transference of components between the machines, which had previously been carried out manually. The term now has a much wider significance, being applied to any process where the physical or mental functions of an operator are replaced by automatic machines and computers. (See Electronic Control of Machine Tools.)

Automobile.—A mechanically propelled vehicle running on common roads. Steam, electricity, alcohol, and petroleum spirit are the agencies employed, the last-named vastly predominating. The manufacture of the engines and gears for these vehicles has occasioned the design of a large number of new and modified machine tools, and has greatly influenced shop methods.

Axial Turbine.—One in which the water passes through the wheel parallel to the axis of rotation. (See Jonval turbine.)

Axle, Front.—A steel beam in a motor-car or other vehicle to carry the front wheels. In the former the stub axles which carry the wheels are hinged to king pins for steering. The wheels steer together by arms coupled by a track rod. See also Axle.*

Axle-grinding Machine.—One in which new or worn railway axles are ground, the wheels being on their axle. The latter is centred between two heads at opposite ends of a bed, and two grinding wheels on carriages in front are rotated and traversed along the journals.

Axle Lathe.—A lathe usually double, that is, having a loose poppet at each end of the bed, between which the axle is centred. It is rotated through a central headstock with double drivers, through gears with speed changes. Two slide-rests deal with the axle journals.

B

Babbitt.—The high proportion of tin renders the alloy described in Pt. 2, Basic Terminology somewhat expensive. A cheaper babbitt which is effective for less heavily loaded bearings contains : 60% tin (minimum), 2·25–3·75% copper, 9·5–11·5% antimony, 26% lead (maximum), 0–08% bismuth (maximum). Other alloys are marketed under trade names such as Glacier Metal, Hoyt Metal, Magnolia Metal, etc. See also Babbitt's Metal.*

Back-fire.—This takes place in internal combustion engines, when ignition occurs before the piston has reached the end of its stroke ; in which case the explosion may oppose the turning moment on the shaft. It is therefore dangerous to anyone attempting to start the engine by hand, and may cause a breakage to the shaft. See also Back Firing.*

Backing Belt.—An extra belt sometimes fitted to a screw-cutting lathe in order to reverse the direction of rotation, and so run the rest back without withdrawing the clasp nut. It avoids the difficulty of catching threads of odd pitches. But other arrangements are preferable. See also Reversing Motion.*

Backing-off Lathe or Relieving Lathe.—One in which the teeth of milling cutters and taps are backed-off by a to-and-fro movement of a cutting tool on the slide rest.

Backing Sand.—A foundry term for the sand which has been knocked out of moulds, and is re-used behind the facing sand of a new mould ; usually forms the bulk of the sand in the mould.

Back Rest.—Toolpost attached to the back of a lathe cross-slide. When taking heavy cuts the tool is mounted on the back rest or back toolpost, so that the reaction of the cut is taken by the bottom half of the head stock bearing, affording greater rigidity. A steady carried on the back rest is also used in cylindrical grinding with the work between centres to take the thrust of the grinding wheel. See also Back Steady Rest.*

Back Stick.—Wooden stick used to support the rear flange of a thin metal blank during the initial stages of spinning metal in the lathe. By taking the thrust of the metal spinning tool, the back stick prevents the metal from buckling or tearing.

Backwater Curve.—This is the curve which the surface of a canal or river assumes, when an obstruction such as a submerged weir is constructed across it.

Bacteria.—These have a special interest to the engineer, because they are the agents in the modern treatment of sewage. The anaerobic bacteria, which do not require oxygen, disintegrate the matter, liquefying it, and breaking it up into nitrogen and ammonia chiefly, after which the aerobic bacteria, which require oxygen, produce a pure effluent in filter beds.

Bag Work.—Relates to the deposition of concrete in a liquid mass enclosed in a bag of jute, holding 100 tons, less or more.

Balance.*—A lever turning on a pivot or fulcrum, used for the weighing of objects. See also Steel Yard* and Fulcrum.*

Baling Press.—An hydraulic press in which loose materials are compressed into a small space for export or storage.

Ball Bearings.—These have largely displaced common bearings not only in the cycles and motor vehicles, for which they were first adopted, but also in engineers' machinery. It is usual now to see races of ball bearings fitted to take the end thrust of lathe spindles, of drill spindles, of the radial arms of drilling machines, of the knees of milling machines and in innumerable other machine parts, and parts of engines, etc. The balls are made in special automatic machines, turned, hardened and ground, and sorted automatically according to size. Several distinct designs of bearings have been developed.

Ballistic Pendulum.—A pendulum for ascertaining the kinetic energy in a projectile. It consists of a large block of wood attached to the end of a strong iron stem and swings about a horizontal axis at its upper end. When the projectile strikes the block and imbeds itself in it, the height to which it rises measures the kinetic energy imparted, and from this the velocity and force of the blow may be deduced.

Ball Mill.—A rotating cylinder usually, within which a number of loose balls of iron or steel crush hard materials, as slag, ores, emery, bones, glass, etc.

Ball Turning.—Is done by means of a special rest moved by worm gears in a circular path, or by means of a form tool.

Band Conveyor.—Conveyor consisting of a power-driven fabric, leather or metal band.

Banding Press.—A machine by means of which copper bands are tightened around projectiles.

Bank.—To bank an aeroplane is to tilt it laterally.

Barba's Law.—This law states that geometrically similar bars deform similarly. Important in testing.

Bar Gauge.—Gauge used in place of a plug gauge (q.v.) to check the dimensions of plain holes having diameters greater than about $4\frac{1}{2}$ in.

Barlow's Formula.—Formula due to Barlow which attempts to calculate the stresses in a thick cylinder when subjected to internal or external pressure, or both combined. It does not appear to agree with experimental results as well as the formula of Lamé, which is now therefore preferred. (See Lamé's formula.)

Barometer* (see Pt. 2, Basic Terminology).—Also see Aneroid and Fortin's barometer.

Barrage.—The damming of the waters of a stream for irrigation or industrial purposes.

Barrel Wheel.—The large wheel which is keyed upon the same shaft as the lifting barrel of a crane, and by which the latter is directly revolved. It is the last one in the train of gearing.

Barring.—(1) The turning round of an engine flywheel with an iron bar, to get the engine over dead centres (q.v. in Pt. 2, Basic Terminology) in readiness to start. Points of leverage are afforded by fixed pins, suitably placed adjacent, and a few holes are drilled in the flywheel to take the point of the bar. (2) Also the initial turning of a large engine by a smaller engine. (3) Poking away lumps of fuel from the immediate vicinity of the tuyère holes of a cupola, an iron bar being thrust through the sight hole for the purpose, in order to make a free passage for the blast. See also Barring Engine.*

Base.—In an electrical or mechanical sense, the structural foundation on which an item is to be permanently assembled. It differs from a base plate inasmuch as it is an integral part of the item for which it is a foundation. See also Base Plate.*

Basic Size.—Size of a dimension or part in relation to which all limits of variation are determined.

" Basis " Brass.—Brass containing 61·5% (min.) to 64% (max.) of copper ; it may be either an alpha-beta brass or an alpha brass (q.v.), according to its precise composition and treatment. Suitable for a wide range of uses, including those where hardness is essential, or where good cold-working is required.

Batten Plates.—Flat steel plates connecting the rolled steel sections which form the uprights of a stanchion, to prevent secondary buckling between the points of support. See also Battens.*

Battery.—A combination of elements for the production of electrical action when immersed in acids. The term is principally applied to voltaic batteries, of which there is a large variety, such as Daniel's, Bunsen's, Grove's, the Bichromate, the Leclanché, Smee's, etc., among the commonest. See also Secondary Battery. See also Battery of Boilers.*

Beam Micrometer.—A beam caliper which has a micrometer screw in the sliding head.

Beam Theory.—The simplified theory of beams, used in practice, is due to Bernoulli. It neglects the effects of shearing stresses and transverse bending (see Antielastic bending) and is therefore only an approximation to the more exact theory. It is, however, sufficiently accurate, in general, for practical purposes provided the beam has a vertical plane of symmetry, that the resultant load on it acts in this plane, and that the limit of elasticity of its material is not exceeded.

Bearing Plate.—Bearing plate is riveted to the bottom flange of a girder whenever the area of the flange surface resting on a wall or pier is insufficient to keep the bearing stress of the supporting material within its safe working limit.

Beetle.—Heavy wooden rammer such as paviors use for setting paving blocks. See also Beetle Head.*

Beilby Layer.—Surface layer of amorphous (q.v.) metal formed by polishing. A Beilby layer is also formed, for example, during the running-in of working surfaces of an engine provided that adequate lubrication is maintained.

Bel.—A unit used in electrical engineering. It is ten times the size of the more frequently-used decibel (q.v.).

Belfast Sand.—A red, fine-grain moulding sand, possessing good bonding and moderate refractory properties. Used as a facing sand or for making green-sand moulds in casting aluminium alloys, brass and light iron castings ; it is often blended with rock sand.

Belt Conveyor.—Wide belt running over pulleys, transporting materials laid and carried on it.

Belt Polisher.—Machine in which a belt charged with abrasive is carried around pulleys. Also termed a Linisher.

Belt Sander.—Sandpapering machine in which a belt charged with abrasive is moved rapidly by rotating pulleys for finishing woodwork.

Bench Block.—A metal block the working surface of which is drilled and grooved, used by engineers for driving pins or dowels into round or flat work. It is usually of round or square section, but sometimes in the shape of a V-Block so that a round bar may be supported in it for drilling purposes. See also Bench.*

Bendix Gear.—A Bendix pinion mounted on a screw thread which draws it into mesh with the teeth on the flywheel rim. When the engine starts the pinion runs back along the thread and is thrown out of mesh.

Benzene or Benzol (C_6H_6).—A by-product of gas manufacture, often used as a fuel for internal combustion engines. It dissolves indiarubber and is therefore useful in tyre manufacture and for tyre repairs, etc.

Berm.—A horizontal surface, as for a pathway. In canals and railways it forms the level top of the embankment.

Bernoulli's Theorem.—The fundamental theorem of hydraulics, which states that the total energy of a flowing fluid consists of its potential, kinetic and pressure energies, the sum-total being constant.

Bevel Gear Planer.—Several machines of this type are now made. They are either of the form or the generating type. (See Generating Gear-cutting Machines.) In the form machines the planer arm that carries the cutting tool is controlled by an enlarged copy of the tooth—the form—which is usually three times larger than the tooth, in order to reduce any error present. The wheel blank is so mounted in relation to the cutting tool that the latter always cuts in a plane which, if pro-longed, would terminate in the apex of the pitch cone. The blank is rotated by suitable mechanism between each cut. In some machines two tools reciprocate and cut alternately on opposite sides of one tooth. The principal objection to the use of a form is the possibility of in-accuracy in the form itself. This, however, can be but slight, and it is obviated in some cases by shaping the form in a machine of the generating type. See also Bevel Gearing.*

Birfurcated Rivet.—Also called *Split Rivet* (q.v.). A rivet with a split shank or two prongs which are opened out and tapped down for the ·purpose of fastening together sheets of light material. See also Rivet.*

Bilge.—(1) The bottom of a ship on each side of the keel. (2) The swelled part of barrel. See also Bilge Pipes.*

Binder Pulley.—A pulley the sole function of which is to bind or tighten a belt or cord on its driving and driven pulleys, when, owing to extension or shrinkage of the belt or cord, the tension becomes variable in amount, or where the pulley centres are very close. The pulley is made adjustable.

Black Light Crack Detector.—See Zyglo Crack Detector.

Blanking Die.—One in which sheet metal is punched out or blanked preparatory to subsequent treatment. See also Blank.*

Blast Engine.—Blast or blowing engines are now largely driven by the waste gases from the furnaces, after they have been cleaned to get rid of dust and tar. See also Blast Furnace.*

Blast-furnace Gas.—The waste gases from a blast-furnace are now utilised for driving large gas-engines. The gases taken from the main are forced by a blower through a scrubber into a holder regulator and thence to the engine. The small hydrogen content enables very high compression pressures to be used without danger of pre-ignition. See also Blast Furnace.*

Blast-furnace Hoist.—Usually an inclined hoist electrically operated by which ore, coke and limestone are taken to the top of the furnace.

Blind Cylinder.—One case with a solid end. See also Blank End* and Blank Flange.*

Blown Castings.—A foundry term used to describe castings in which bubbles or blowholes have been caused by gases, steam, etc., generated when the mould is cast, finding their way into the molten metal.

Blue-Heat.—A range of temperature between about 243° and 315° C. corresponding with a blue-black colour, during which steel becomes injuriously affected by work done upon it. Bending a strip once only while at blue-heat alters the material for the worse, notwithstanding that it may be bent cold with relatively little injury.   See also Blueing.*

Bob.*—A lap (q.v. in Pt. 2, Basic Terminology).

Bobbing.—A term sometimes applied to the lapping (see Lap in Pt. 2, Basic Terminology) of metals, the lead or copper lap being termed a bob.

Body Flange.—When a pattern flange is fitted temporarily over the body or outside of a pipe, for convenience of sliding it along into any position for stopping off odd lengths, it is called a body flange, to distinguish it from the flanges which are fitted in grooves permanently, at the ends of pipe patterns.

Boiler Tap.—Threading tap designed to tap holes for the reception of boiler stays.   The tap may be a one-piece tool, or may consist of two shell taps secured by keys to a long shaft.   See also Boiler Stays.*

Bollman Truss.—A favourite type of truss in America for small-span bridges.   The main booms are supported at regular intervals by vertical struts, each of which is braced by tension rods attached to the extremities of the truss at the abutments.

Bolt.—A bolt is a headed and externally threaded mechanical device formed from a pin, rod or wire and designed for insertion through holes in assembled parts to mate with a nut and is normally intended to be tightened or released by turning that nut.

Bolt Forcer.—Small hydraulic machine for forcing bolts out of couplings, etc., by the pressure of a ram.

Bond.—In grinding wheel terminology the term refers to the matrix of non-abrasive material used to bind or hold the abrasive grains together in the form of a wheel or other shaped article.   Bond imparts mechanical strength and " grade " or hardness, which can be varied at will to produce desired characteristics for any given grinding operation.   Several different types of bond are employed, dependent on the particular requirements.   See Vitrified Bond, Silicate Bond, Resinoid Bond, Rubber Bond, Shellac Bond.

Bonnet* (see Pt. 2, Basic Terminology).—Also, in a motor-car, the metal shield that covers the engine.

Booster.—An auxiliary dynamo, by which the pressure on feeders in a distributing system is maintained or increased.

Borda's Mouthpiece.—This consists of a thin tube projecting into a tank containing water.   When the water discharges under the head in the tank, it forms a contracted vein at the sharp entrance to the tube, and the quantity delivered is, theoretically, only half that due to the head.   Actually the coefficient of discharge is about 0·52.

Boring.—Strictly, the term boring refers to the cutting out of holes of large size, the term drilling relating to those of small diameter, but the term is somewhat vague, and loosely applied, since we speak of boring bits and boring lathes, which operate on small as well as large holes.   See also Borings.*

Boron Carbide.—Extremely hard compound, obtained from boron trioxide and coke, used as an abrasive for cutting tools.

Boss Tools.—Moulders' sleekers used for smoothing the sides and bottom faces of boss moulds.   The body of the tool is of a hollow semicircular cross section, and has a flat horizontal set-off piece at the bottom.   It is attached to a long vertical handle, or stalk, above.

Bottom Clearance.—The distance between the points of the teeth of one gear wheel and the roots of another which are in mutual engagement, the clearance being given to prevent interference and friction between points and roots. See also Bottoming* and Clearance.*

Bought-Out.—A term used to describe items of a specialist nature which cannot be economically produced by a principal manufacturer and which can be more cheaply bought from an outside firm specializing in the manufacture of such components (e.g. ball bearings, shaft seals, gears, etc.).

Bouncing-Pin.—See Knockmeter and Highest Useful Compression Ratio.

Box Angle Plate.—Cast iron box in the faces of which slots are machined. Each surface is also machined flat and true, so that the box forms a rigid, accurate mounting for work during drilling, milling or planing operations.

Box Jig.—A jig (q.v.) in the form of a box, in which work to be drilled is located. Hardened bushes are inserted in the sides of the jig to guide the drills.

Box Tools.—Combinations of two or more separate tools secured in a box which is attached to the faces of a lathe turret. They are of great value, since they often include centring, drilling, turning, or knurling, with vee or roller steadies.

Boyden's Diffuser.—Fixed annular casing surrounding the wheel of a centrifugal pump and fitting closely to the outer periphery of the moving vanes. The sides of the casing diverge gradually, which causes a loss of velocity head accompanied by a gain of pressure head, resulting in a small gain of efficiency. Not much used at present, due to increase of size and weight.

Brachistochrone.—The plane of quickest descent for a body starting at the top of a circle and meeting it again.

Bramah Press.—(See Hydraulic Press, Basic Terminology.) See also Bramah's Press.*

Brass.*—The word " brass " is commonly used as a general term for the whole range of copper-zinc alloys, although it is also often confined to those which contain about 55–80% of copper. The alloys of compositions of from 80–95% of copper and 5–20% of zinc are generally called " gilding metals " because of their golden colour. The terms " Dutch metal," "red brass," " pinchbeck," " tombac," " Prince's metal," and many others are also used for certain alloys included in this class. The various brasses are characterised by the ease with which they can be cast, extruded, forged, rolled, pressed, drawn, machined or otherwise fabricated. These alloys also show marked resistance to atmospheric and marine corrosion ; they are used in machinery, equipment and instruments which are exposed to adverse conditions, or which depend for their functioning on freedom from rust. (See also Pt. 2, Basic Terminology.)

Breather.—Ventilator on the crankcase of an internal combustion engine.

Breathing Space.—The area allowed for elastic movements between the furnace flues and the gusset stays, or the tubes in horizontal boilers.

Bressummer.—A beam of wood, iron or stone, supporting a wall over a door or other opening. A kind of lintel.

Bricks (see Pt. 2, Basic Terminology).—These are common bricks, facing bricks, paviours, blue bricks, stock bricks, malms, wirecut, pressed bricks, glazed, terra-cotta, etc. See also Brick.*

Brickwork Flues.—(1) The external flues in a Lancashire or Cornish boiler, forming split draught, or wheel draught, as distinguished from the internal flues in the boiler itself. (2) Chequering (q.v. in Pt. 2, Basic Terminology).

Bright Annealing.—See Close Annealing.

Brimstone.—A term commonly given to sulphur which has been fused into sticks or blocks. See also Sulphur.*

Brinell Test.—Hardness test in which a hardened steel ball, 10 mm. in diameter, is forced into the surface of the specimen with a load of 3,000, 1,000 or 500 kg., the area of the impression being related to the Brinell Number. The test may be used for soft metals and alloys and toughened steels. It is not suitable for extremely hard tool steels or case-hardened work owing to the deformation of the steel ball caused by the resistance of a hard specimen. If case-hardened work is tested, a false figure may be obtained, owing to the case being crushed into the core under the heavy load. In preparing samples for test, the surface should be prepared by filing, and finished with rough emery cloth. Deep file marks tend to spoil the ball impression.

Briquette.—A test-piece of cement and sand of British Standard Specification dimensions, consisting of one part by weight of Portland cement to three parts by weight of standard sand mixed with water to specification.

Briquette Fuel.—Fuel compressed into blocks of various shapes for use in hand-fired boilers, and for domestic use in the form of ovoids.

British Association (B.A.) Thread.—Thread form with an angle of 47½° and rounded crests and roots. Sometimes called the Swiss or Thury Thread, being based on a Swiss standard, the B.A. thread is used for small diameters below ¼ in. The thread sizes are specified by numbers ranging from 0 to 25 ; the larger the number the smaller the diameter.

British Standard System of Limits.—System of limits covered by B.S.164 and recommended by the British Standards Institution (see British Standards). The system is on a hole basis (see Limits System). A unilateral system is recommended, but bilateral tolerances are included in the specification. Tolerances in the metric system are issued as a supplement to B.S.164, while tolerances for workshop and inspection gauges for controlling work made to the British Standards System are given in B.S.969.

British Standards.—Standard specifications for materials, manufactured articles, methods of testing, codes of practice, terms, definitions, etc., issued by the British Standards Institution after consultation with all interested Government Departments, Industrial Committees, etc. These standards are usually referred to as B.S.709, B.S.499, Pt. 2, etc.

British Thermal Unit or B.Th.U.—The quantity of heat required to raise the temperature of 1 lb. of water at or near 32° F. through 1° F.

Brittle-lacquer Technique.—Method of qualitative stress analysis (q.v.) in which the part to be tested is coated with a brittle lacquer which cracks when the part is subjected to stress. The lacquer is sensitive to temperature and humidity and is therefore supplied in a wide range of grades to cover all conditions likely to be encountered. The location, direction and magnitude of tensile strains can be revealed by applying the lacquer while the part is unstressed. Compression strains may be determined by allowing the coating to attain a neutral condition while the component is subjected to maximum load.

Broaching Machine.—One built on the design of the keyway cutting machine, but used for finishing square and polygonal holes. The machine may also be of the surface-broaching type. (The term

" broaching " formerly applied to reamering is no longer used in that sense.)   See also Broaching.*

Broad Cutting.—Taking finishing cuts, of considerable width, with square-nosed tools, off the surfaces of turned or planed work.

Bronze.*—Term covering a range of copper-tin alloys used for bearing and structural purposes. British bearing bronzes are usually phosphor-bronzes, i.e., copper-tin alloys containing 5-20% tin, 0-20% lead, with a small percentage of phosphorus, and the remainder copper. In some cases 2-6% zinc is added in lieu of phosphorus, when the alloys are known as gun-metals (q.v.). In addition, there are special brasses, aluminium-bronzes (90% copper, 10% aluminium anprox.), silicon-bronzes (96% copper, 4% silicon approx.), berylliu -copper (97·5% copper, 2·5% beryllium approx.) and other copper alloys which are sometimes used as bearing metals.

Bronze Welding.—Method of welding in which metals are united by melting bronze or brass filler rods into the joint.

Brush.—In electric current generators and motors, the pieces of copper or carbon blocks that bear against the cylindrical surface of the commutator.

Bucket-piston.—Piston provided with one or more valves which permit of the fluid in the cylinder passing in *one* direction only.

Buckle.*—Usually applied to the clip that binds together the plates of laminated springs and attaches them to the axle. Also called a U-bolt.

Buckstaves.—The cast-iron plates which form the outer casings of rever-beratory furnaces, and which are supported by the bolts passing from one side to the other.

Buffing.—Polishing process by means of emery or other powder, usually applied to a high-speed rotating mop. (See also Pt. 2, Basic Terminology.) See also Buff Wheel.*

Bulkhead.—On ships, etc., a partition which divides the interior into separate compartments. Also a long face of a wharf parallel to the stream.

Bulldozer.—(1) Heavy type of horizontal forging press of American design used in the railway and waggon shops.   (2) Heavy motor-driven vehicle mounted on caterpillar tracks and provided at the front with a steel blade, used to level rough ground, etc.

Bulk Modulus or **Modulus of Cubic Compressibility.**—The ratio of the compressive stress to the compressive strain in a solid material subjected to a uniform stress all over.   See also Modulus.

Bull Metal.—Alloy used for hydraulic and marine work, and for gun fittings, both in the form of bar and of castings. The toughness and ductility of the alloy increase with high temperatures, so that at 204° C. the strength of rolled bull metal is equal to that of mild steel.

Bull Wheel.—The large wheel which engages with and drives the rack of a planing machine.

Buna.—A term used in the plastics industry to describe a type of syn-thetic rubber which has been manufactured from butadiene and sodium.

Bus Bar.—A copper conductor used in electric lighting or power stations to receive the current from the dynamos, the distributing leads being connected to the bus wires or bars.

Butterfly Throttle.—An elliptical plate pivoted on its centre so that it throttles the steam or gas passing into the cylinder, or closes the induction pipe altogether, as required.   See also Butterfly Valve.*

Butt Welding.—Method of resistance welding (q.v.) in which the parts to be joined are clamped in the copper jaws of a welding machine which is connected to a source of low-voltage, high-amperage current. The ends of the parts are brought into light contact while current is passed through them to heat the metal, and the weld is completed by applying increased pressure.

By-pass.—A passage through which a gas or a liquid is allowed to flow as an alternative to its ordinary channel.

By-product.—A secondary product thrown out in the course of a manufacture, and, unless utilised, a waste product.

C

Cabinet Leg.—An extension of the hollow or boxed design of framing to the legs of lathes and other machine tools. The interior is utilised as a cupboard or cabinet for gears or tools, and sometimes as an oil tank.

Cableway.—Cable of wire rope carried on standards at terminal stations and made the means of conveying a carriage or buckets over the intermediate space.

Cage.*—In a travelling crane, the frame within which the attendant sits and from which he controls the movements.

Calciner.—Specifically, a reverberatory furnace (q.v. in Pt. 2, Basic Terminology) having a hearth of large area nearly as broad as long, and used in the roasting of the ores of copper. See also Calcining Furnace.*

Calcium Carbide (CaC_2).—Fused compound of lime and hard coal. In combination with water gives off acetylene gas. See also Calcium.*

Calibration.—Ascertaining the amount of variation from absolute accuracy in a scientific instrument.

Caliper Gauge.—Solid gauge of horseshoe type, or a modified form of caliper rule with vernier readings. See also Caliper* and Caliper Rule.*

Calorimeter.*—Instrument used to ascertain the amount of heat in a body —in the case of fuel, its calorific value. Fuel is burned in a tube enclosed in a vessel of water, and the comparison of the original and the final temperature of the latter, with allowances for radiation, etc., gives the calorific value. In the calorimeter for determining the dryness of steam, the steam is admitted to a closed vessel containing cold water and the final temperature is the basis of the calculations made to determine the percentage of water in the steam.

Calorizing or **Calorising.**—A process of rendering the surface of steel or iron resistant to oxidation or rusting by the application to it of an aluminium coating. The coating is deposited by spraying the iron or steel object with a mixture of aluminium powder and alumina, in a temperature of 800° to 1000° C. See also Corrosion.*

Camels.—A kind of barge or hollow floating vessel, which when filled with water to partially sink it, is fastened, one on each side of a sunken ship, so that when the water is pumped out of them they rise and lift the ship, so that she can be floated into shallower water.

Camming.—Laying out the cam strips at their proper angles and positions on the cam drum of an automatic for actuating the tool slides.

Canadian Standards Association.—No electrical equipment in Canada may, by law, be sold unless it has been approved by the Canadian Standards Association. Any machine or appliance which incorporates electrical components of any kind is regarded as " electrical equipment ", even though such components may form only a part of the complete machine. C.S.A. approval is therefore required even for such equipment as machine tools, printing presses or bottle-washing machines. All gas or oil-burning equipment must also be C.S.A.-approved. See also British Standards.

Candle Power.—The standard of illumination, being that of a sperm candle burning at the rate of 120 grains per hour.

Cantilever Bridge.—A bridge, the shore spans of which are carried wholly from the abutments, being, therefore, self-supporting. See also Cantilever.*

Cantilever Crane.—One in which the arm or jib hangs out from the supporting member, and is counterbalanced. See also Cantilever.*

Capacitor.—Formerly also called a Condenser (q.v.).* An electrical component, consisting of two or more electrodes separated by a layer of insulating material, whose primary purpose is to act as a temporary store of electrical energy when it is connected to a source of voltage.

Capacity* (see Pt. 2, Basic Terminology).—(1) The volume of the cylinder(s) and combustion chamber(s) of an internal combustion engine. (2) The storage capacity in the case of electrical secondary batteries; the quantity of electrical current which they can supply when charged without undue exhaustion. It is expressed in ampere-hours.

Cape Chisel.—A term sometimes applied to a cross-cut chisel (q.v. in Pt. 2, Basic Terminology).

Capillarity* (see Pt. 2, Basic Terminology).—The reaction between liquid surfaces of different kinds, or between liquid and solid surfaces due to surface tension.

Capstan Tool Rest.*—This appliance affords one of the most striking examples of development in later machine shop practice. From a small adjunct to an ordinary lathe it has grown into the most important fitting of many special lathes. The larger capstans or turrets are fitted with elaborate boxes of tools for turning, boring, facing, threading, knurling and cutting off. Some are hollow to permit long rods to pass clear through without excessive overhang of the tool. Instead of half a dozen tools, some turrets are now capable of carrying sixteen to eighteen. Some also carry a supplementary turret or cut-off slide. The largest turrets are rotated on ball races, frequently by power. The movement of the tools is arrested automatically at the right moment by stops, and by a knock-out arrangement, after which the turret is rotated to bring the next tool into action. The movements of the turret synchronise with the automatic feeding of the stock when doing bar work. Often also movements of a cross slide carrying one or more tools synchronise with those of the turret, its action also being controlled by stops. Rotary oil pumps and pipes and spreaders are fixed to all modern capstan lathes to flood the tools with lubricant, a feature which conduces to heavy and easy cutting. The development of the capstan lathe also includes the tooling of castings of fair size, and often these are placed in the machine automatically from a feeding magazine.

Carboloy.—Sintered carbide alloy used in the production of cutting tools. See Sintered Carbide tools.

Carbon Arc Welding.—Method of welding in which an arc is struck between a carbon electrode and the work, the electrode and work being connected to a low-voltage direct current supply. A small pool of molten metal is formed on the surface of the work, into which additional metal is fed from a filler rod, in a somewhat similar manner to oxy-acetylene welding (q.v.). Carbon arc welding tends to form a somewhat brittle and porous weld.

Carbon Residue.—The amount of coke or carbon which remains after a lubricating oil has been heated, in the absence of air, to a temperature of about 550° C. under carefully standardised conditions. The amount of carbon remaining after the test is indicative of the quality of the oil —the lower the weight of carbon residue, the more stable the oil under variable temperature conditions. See also Oils.*

Carbon Steel.—Denotes common steel, or that which is produced by the union of a definite quantity of carbon with iron. This term is necessary, since alloys possessing steel-like qualities are now made by a union of iron with manganese, silicon, vanadium, chromium, tungsten and other elements, hence termed alloy steels.

Carbon Tetrachloride.—A heavy, colourless liquid having a pleasant smell which forms the active constituent of many portable fire-extinguishers. Prepared from chloroform or carbon disulphide and chlorine, it is non-inflammable. When scattered over a burning mass, it has the property of evolving heavy vapours which cut off the supply of oxygen, and so extinguish the flames. See also Carbon.*

Carborundum.—An abrasive material, the hardness of which is second only to the diamond, and which is used considerably where emery and corundum had previously been employed. It is manufactured in electric furnaces from a mixture of coke, sand, sawdust and salt.

Carburettor.—An appliance in which an inflammable gas, such as petrol, is mixed with air to form an explosive mixture to operate internal combustion engines.

Carburising.—See Case Hardening (Basic Terminology) and Cyaniding (Basic Terminology). See also Carburization.*

Cardan Shaft.—A term sometimes used for any shaft which transmits power, as in a motor or the propeller shaft of a ship.

Card System.—A method of indexing and tabulating facts and figures for ready reference. Its great advantage is that it is perfectly elastic and expansive, which book records are not. See also Cards.*

Cartridge Brass.—70/30, or cartridge brass, combines optimum ductility with good strength. It is therefore largely used for articles requiring high ductility to facilitate manufacture, with high strength in the finished product, including the making of cartridge cases and other complex shapes which are deep drawn, and also for automobile radiator shells and similar articles of deep or unsymmetrical pressed shape.

Castigliano's Theorem.—A well-known theorem in mathematics applicable to the determination of the stresses in a statically indeterminate structure. It states that the differential coefficient of the total strain-energy in a structure with respect to any load upon it is equal to the displacement of its point of application in the direction of this load.

Cast Iron* (see Pt. 2, Basic Terminology).—The tensile strength of cast iron seldom averages more than seven or eight tons per square inch, but its compressive strength varies from 35 to 40 tons. In rough calculations the weight of a cubic inch of iron is usually taken at 0·263 lb., and of a cubic foot at 450 lb.; but this is necessarily approximate, since grey, white, and mottled irons have different sp. gravities. Grey iron melts at from 1,600° C. to 1,700° C., and white iron at from 1,400° C. to 1,500° C.

Castle Nuts.—These nuts are provided with vertical slots in the upper portion of their height, through which a split pin may be passed to lock the nut in position.

Cathetometer.—Instrument used in physical measurements to determine accurately the height of an object, which is sighted in a telescope, a scale and vernier then being observed by an independent magnifying system.

Cathode or Kathode.—That part of a galvanic battery by which the electric current leaves a substance through which it has passed, or the surface at which the current passes out of the electrolyte; the negative pole.

Cathode-ray Oscillograph.—See Cathode-ray Tube.

Cathode-ray Tube.—Glass tube in which a narrow beam of electrons is directed on to a screen coated with a substance which fluoresces when struck by the electrons. A spot of light is thus formed which may be moved in any direction over the screen by deflecting the electron beam. This is done by means of two electrostatic and/or magnetic fields at right angles to one another. Cathode-ray tubes are extensively used in laboratory work to delineate wave forms, engine efficiency curves, etc., and in electronic instruments and comparators (q.v.).

Cathodic Protection.—Method of protecting buried and submersed steel objects by ensuring that the structure to be protected at all times remains cathodic in a controlled electrical circuit. Anodes, made from a metal which is much more reactive than that to be protected, are buried or immersed close to the object, zinc or magnesium being commonly used. In some cases the anode is connected to the positive terminal of a supply of electric current, the object to the negative terminal.

Caustic Embrittlement.—A form of corrosion in boilers caused when the caustic constituent in water attacks the grain boundaries of the metal forming the plates. It generally occurs at the seams of riveted joints. See Corrosion.*

Cavitation or Separation is the partial vacuum sometimes caused by the separation of the water flowing along a pipe or rotating in a forced vortex at high speed.

Cementation Furnace.—The furnace used for the conversion of bar iron into blister steel. It is an oblong, arched chamber of fire-brick, supplied with small chimneys leading into a tall stack. The chamber contains a fireplace, running longitudinally between two oblong troughs or pots, in which the bars are converted into steel. These may measure 12 ft. 0 in. long, by 3 ft. 0 in. wide, by 3 ft. 0 in. deep. The pots rest upon and are flanked by a series of bearers and divisions of masonry, which form channels for the diffusion of the heat and flame from the fireplace. There is a tap hole in one end of each pot to receive the proof or trial bars. The furnaces are built in ranges of five or six. See also Cementation.*

Cementite.—A hard constituent of high carbon steel, having the formula Fe_3C. May occur in globular, massive or network form.

Central Station.—The power station in which electricity is generated for distribution through a surrounding area.

Centre Casting.—Pouring a mould at the centre, which is often preferable to pouring at the edges. Also pouring steel ingots from a crane, around which the moulds are arranged in a circle.

Centre of Buoyancy.—Centre of gravity of the volume of a liquid displaced by a body floating in it.

Centre of Flotation.—Centroid of the surface of flotation, that is, of the area of intersection of the water surface with a body floating in it.

Centre of Pressure.—The point at which the resultant of a system of forces cuts a given plane.

Centrifugal Casting.—Pouring moulds while in rapid rotation to drive the heavy metal outwards, leaving the lighter matters to rise into a central head.

Centrifugal Sand Mixer.—Denotes a large group, in which the sand is disintegrated and mixed by concentric rows of prongs rotating at a high speed.

Centrifuge.—A machine by which moisture is extracted from bodies by rotation at a high speed. Also termed a hydro-extractor.

Centrifuse.—A process in which molten cast-iron is spun into a hot steel brake-drum, the centrifugal force assisting the fusing of the two metals. The process is sometimes used in the manufacture of brake drums in order to combine the desirable friction characteristics of cast-iron with the superior strength of steel. See also Cast Iron, Central Forces* and Steel Facing.*

Centrode.—The locus of the virtual or instantaneous centre of a body as the body changes its position.

Centroid.—A term to be preferred to *centre of gravity* when dealing with areas, which, of course, have no weight or gravity. See also Centre of Gravity.*

Ceramic.—Term applied to any form of pottery and used, in the engineering sense, to describe materials, such as porcelain and mixtures of aluminium oxide with silica and refractory oxides, which are used either as electrical insulators or as tips for cutting tools. (See Ceramic Tool.)

Ceramic Tool.—Type of cutting tool that can be used for machining almost all metals, including titanium and vanadium and also to machine abrasive materials, such as carbon, graphite, asbestos, and plastics which are difficult to machine with ordinary metal turning tools. An example is Sintox, which has a hardness approaching that of a diamond, combined with a compressive strength of 100 tons per sq. in. and good resistance to vibration (unlike some other ceramics). Normally, ceramic tools must be dressed with diamond wheels but it is possible to use green grit wheels provided that only light pressure is exerted. The development of excessive heat will crack or break the tips.

Cetane Number.—A measure of the ignition value of a fuel oil. The cetane number of a fuel oil is the percentage of cetane in a mixture of cetane and 1-methylnaphthalene which has the same ignition quality.

Chain Feed.*—Many machine tools have speeds and feeds transmitted through pitch chains in preference to using belts.

Chain Conveyor.—On slat and cross bar chain conveyors a pair of power-driven chains move along each side of a trough. Attached to the chains at intervals are slats or cross bars which span the trough and thus push the goods ahead of them. Overhead chain conveyors are provided with suspended platforms or hooks to carry parts from one part of a factory or warehouse to another.

Chain Gearing.—Chain gearing is used for transmitting power to machines by means of chain-wheels and sprocket-wheels. Projections on the wheel fit into cavities in the chain, or vice versa, so that the two are constrained to move together without slipping. The two principal forms of chain are Renold's bush roller chain and Renold's inverted tooth chain.

Chain Saw.—An endless chain, having chisel or saw-like teeth on its links, used for cutting mortices in wood in a machine.

Change Point.—(1) In a workable machine the relative motions of all its parts must be completely constrained. If, however, in any portion a want of constraint occurs—that is, if through lack of proper coercion it is possible for that portion to remain still, or to take some other motion than that intended, that position is termed a *change point*. Constraint is usually brought upon change points by duplication of the mechanism, so that one portion is in complete constraint whilst the other is passing its change point. An example occurs in pairs of coupling-rods or links, with their cranks placed at 90° of angle with each other. (2) Critical Point (q.v.).

Change Valve.—A valve used on a hydraulic crane, for directing the water pressure into two, or three, lifting cylinders at will, either singly or in combination, in order to obtain different degrees of power adapted to the load to be lifted. It is operated by a slide valve passing over ports.

Change Wheels.*—The practice of changing sets of gears by hand for every screw of a different pitch is giving place to that of employing a nest of gears, by which the movement of a lever putting various sets into operation, effects changes in a second or two. This practice has followed the Hendey-Norton device, and it is now applied to a considerable number of lathes, and also to many other machines for effecting feed and speed changes. Another growing practice is that of connecting the feed rod of the lathe with the same gears that actuate the lead screw, to drive either one as required.

Chapmanising.—Surface-hardening process resembling nitriding (q.v.), in which iron and steel parts are heated in a liquid salt bath at about 815° C. Gaseous nitrogen, said to be formed by the electrical decomposition of ammonia, forms the hardening agent in the bath. The advantage claimed for this process is that the use of special steels, as required for normal nitriding, is unnecessary.

Characteristic Curves for Turbines.—There is a large variation in the efficiency of a turbine when the gate and speed are varied ; for small gate openings and low speeds the efficiency is very low. To obtain the conditions for maximum efficiency a diagram is plotted showing the efficiencies for all conditions, and from this diagram the conditions for maximum efficiency can be seen. The points which enable the curves to be drawn are found by testing the turbine for various gate openings and at various speeds. Such curves are known as characteristic curves for the turbines.

Charging Machine.—Crane built in special designs for charging open hearth, or reheating steel furnaces. They are of low type, running on rails on the ground, or of overhead type similar to overhead cranes, except for the details of the charging bar. These machines are operated electrically. See also Charge* and Charging.*

Chassis.—The mechanical part of a vehicle, as distinct from the body.

Chock.—Any piece of material used for filling up a chance hole.

Choke Tube.—In the carburettor of an internal combustion engine, air is drawn into the engine cylinder through a *choke tube* causing a depression which results in petrol issuing from a jet fed by the float chamber.

Chrome-Nickel Steel.—A specially strong, tough, hard steel used in cases of heavy loads. Its tensile strength varies from 42 to 130 tons per square inch, with an elongation of 30–10% and a reduction of area from 66% to 35%.

Chromium (Symbol Cr).—Metallic element used in small quantities to form a chromium-steel alloy. An addition of 5% increases the capacity of steel to resist shock. It is also used for chromium plating, to resist oxidation.

Choking of Vents.—See Vents.

Chordal Pitch.—When a number of holes are equally spaced round a circle, the distance from the centre of one hole to the centre of an adjacent hole, measured in a straight line, is known as the chordal pitch.

Chucking Lathe.—A face lathe without a tailstock. Face work only is done upon this lathe, and often only a portion of that work, such as turning a rim, or boring a hole, etc., the rest being completed in another machine.   See also Chuck and Chucking.*

Cippoletti Weir or Trapezoidal Notch.—A weir having the form of a trapezoid, the sides sloping upward from the bottom, so as to counteract the diminution of discharge due to end contractions as the depth increases.

Circle Cutter.—A tool (sometimes also called a Washer Cutter or a Tank Cutter) consisting of a pivot and of an arm or arms to which a cutter is attached. It is used for cutting out gaskets, etc., or for cutting disks from tanks, by rotating the arm about the pivot. The shank of the pivot is usually held in the jaws of an engineer's drill brace. See Cutter Pin Drill.*

Clapeyron's Theorem.—This theorem, due to Clapeyron, is commonly known as the *Theorem of Three Moments*. It gives the relation between any three unknown bending moments at the extremities of any two consecutive spans of a continuous girder, and thereby leads to their determination.

Clearance Fit.—Condition in which there is a positive allowance (q.v.) between the largest permissible shaft diameter and the smallest permissible hole diameter. See also Clearance.*

Cleat.—Piece of bent steel plate used for connecting two parts of a steel structure by means of rivets. (See Angle Cleat.)

Close Annealing.—Usually carried out by heating the work in a closed container in the presence of a non-oxidising atmosphere, followed by slow cooling in a similar atmosphere, to produce a surface free from the oxides which form in an open furnace. Also termed bright annealing.

cm.p.s.—Centimetres per second.

Coal-cutting Machinery.—Four designs of machines are now commonly used for cutting coal, employing respectively a bar, a rotating disc, a chain, and a pick operating percussively. These machines are operated by electricity, or compressed air.

Coalite.—Fuel obtained from the low-temperature (500° to 600° C.) distillation of coal. The residual coke contains approximately 10% of volatile matter, and is convenient for domestic fires, being easily ignited and giving off the greater portion of its heat by radiation and without smoke.

Coal Tip.—Machine which hoists a waggon-load of coal and discharges it through a shoot.

Cod.—This term is often used in the foundry, to denote a green sand core, a mass of sand upon a drawback, or any loose portion of green sand—as "a cod of sand." The term is also sometimes used to denote a small tongue, or filling-up piece, such as a glut.

Cofferdam.—A wall of piles, between or within which excavation is done for foundations.

Coil Clutch.—One in which the grip is effected by an elastic coil of steel the friction of which is made to hold the periphery of a drum by pressure applied in different ways.

Cold Blow.—Results when the percentage of foreign elements, such as silicon and phosphorus, undergoing oxidation in a Bessemer converter, is not sufficient to maintain the high temperature required for the successful conduct of the process. (See Bessemer Skulls in Basic Terminology.)

Cold Soldering.—Process of amalgamation of metallic surfaces by the aid of mercury. A hard amalgam is made of five or six parts of pure silver, three or four parts of tin, and 3–5% of bismuth. This alloy is melted and cast into ingots, the ingots reduced to fine filings, and these filings mixed, when required, with enough mercury to form a stiff paste, which hardens in about an hour.

Cold Welding.—A process of uniting metals solely by pressure, based on the fact that ductile metals can be made to "flow" if sufficient pressure is applied. The process is usually applied to aluminium and alloys of aluminium, to copper, cadmium, lead and nickel, including combinations of these metals; e.g., the welding of copper to aluminium. Intermittent and continuous welds are made by applying pressure to the mating surfaces, which must be absolutely clean, by dies of the required shape; alternatively, the prepared parts may be passed between rollers. A significant development of pressure welding is the use of intense vibratory energy produced by magnetostriction (q.v.) which results in a solid-state metallurgical bond, without fusion. This method is particularly useful in joining very thin sheets to heavy sections.

Collecting Rings.—Metallic rings fixed to a crane post and insulated, with a revolving portion carrying brushes of carbon or gun-metal making spring contact.

Collector.—A vessel or drum fixed on the upper part of a water-tube boiler into which the steam generated in the tubes ascends, and whence dry steam is taken to the engines. Also an appliance for picking up electric currents from conductors.

Combination Grits.—In certain grinding applications, for example in grinding crankshafts, it is desirable to use combinations of several grit sizes, which are referred to as combination grits. They are shown by adding numerals to the grit symbol, e.g. 301, 463, 367, 1201. In order of increasing "fineness" the numbers used are 1–3–5–7–9, thus No. 7 combination would contain a finer mixture of grits than No. 5 and so on. No. 9 is seldom used. See also Grit.*

Combination Square.—A tool used by engineers for marking-off purposes. It generally comprises a flat metal blade or ruler, a square head, a centre square, a protractor head and a scriber. Sometimes spirit levels are also fitted into the square and protractor heads. To ensure continued accuracy, the combination square needs to be treated as a precision instrument and kept in its case when not in use. See also Combination Caliper.*

Combustion Chamber* (see Pt. 2, Basic Terminology).—Also, in a motor vehicle, the space above the cylinder enclosed by a detachable head, in which the mixture of air and petrol is burnt.

Coming-down.—Boiler crowns are said to "come down" when the plates are bulged downwards into the firebox through overheating, due to the accumulation of deposit thereon.

Commercial Fastener.—A commercial fastener is a fastener manufactured to published standards and stocked by manufacturers or distributors. The material, dimensions and finish of commercial fasteners conform to the quality levels generally recognized by manufacturers and users as commercial quality.

Commutator.—Cylinder formed of alternate sections of conducting and non-conducting materials parallel to the axis of the cylinder. It is fixed to the shaft of the generator or motor and rotates with it. Its object is to convert alternating into continuous current.

Comparator.—(1) Gauging instrument which provides the person controlling a manufacturing process with immediate information concerning the parts which are being produced. In most instances a pointer moving over a graduated scale indicates the actual size of the part and its relation to the specified limits (q.v.). A mechanically operated comparator consists of a fixed and a moving anvil, the latter operating the pointer through a magnifying linkage. See also Electronic Comparator and Pneumatic Gauge. (2) An instrument for comparing accurately two objects of very nearly equal lengths, e.g. a standard metre with its copy. It consists of two microscopes fixed on two stone pillars, which are sighted upon the extremities of the pieces to be compared, the cross-wires of the microscopes being capable of fine adjustment.

Compensating Cylinders.—Cylinders are used for large pumps, as in the Worthington pumping engines, intended to equalise the pressure throughout the stroke of the engine by storing energy during the first portion of the forward stroke and restoring it during the remainder of the stroke. The engine and pump are in line, and the compensating cylinders lie between them on either side of it. They are provided with plungers turning on gudgeons connected with the piston-rod. These plungers move in cylinders fixed to the frame and oscillate about their centres. The cylinders contain water which is in communication with an accumulator under air pressure. During the first half of the engine stroke the plungers are forced into the compensating cylinders, thereby compressing the air in the accumulator, the work done being restored by the compressed air during the remainder of the stroke.

Compensating Lever.—An equal-armed lever, which is used to obtain a mean between unequal pressures. In the locomotive, the lever is a rigid bar, pivoted at its centre, and the ends of its equal arms are attached to the springs at front and back, thus equalising the loads on their wheels, and so rendering the engine less liable to jerks and shocks when running over rough ways. The reason is obvious, since for a lever with equal arms to be in equilibrium, the initial forces acting thereon must immediately become equalised. See also Compound Oils.*

Compound Belts.—Two or more belts which run freely and independently on each other.

Compounded Oil.—(1) A lubricating oil which may be blended from different grades of mineral and vegetable oils, and which is compounded with certain chemical elements in order to improve its film strength, reduce "gumming," sludge and carbon formation, etc. (2) A cutting fluid (q.v.) produced by blending a fatty and a mineral oil, producing an oil which combines the advantages of both its constituents. Particularly suitable for low-speed, heavy cuts on tough materials.

Compound Girder.—One which is built up of rolled sections with plates.

Compound Locomotive.—One in which the steam passes from a high into a low pressure cylinder or cylinders before exhausting into the air. There are several systems in use, in which the number and disposition of the cylinders varies.

Compound Pendulum.—Every actual pendulum is compound; even a small mass attached to a thin thread being only a close approximation to the ideal simple pendulum. (See Simple Pendulum.)

Compound Slide Rest.—One which has sliding and surfacing, and swivelling motions.

Compound Stanchion.—A compound stanchion consists of two or more stanchions connected by single or double latticing or by batten plates (q.v.).

Compressed Air.—Used extensively in mining, in pneumatic tools, in moulding machines, in sand blasting, and in some machine tools. The normal pressure is usually 80 lb. per square inch.

Compression Ratio.—In an internal combustion engine, the ratio of the total volume enclosed in the cylinder at outer dead-centre of the piston to the volume at the end of the compression stroke. Can also be expressed as " The ratio of the combustion chamber, plus the piston-swept volume, to the combustion chamber volume. " See also Compression.*

Conductor.—Either bare conductors, or insulated cables, the former generally being understood. They must be insulated, and the current conveyed by them is taken up by collectors.

Con Rod.—Colloquial abbreviation of *Connecting-rod.* (See Pt. 2, Basic Terminology.) See also Connecting Rod.*

Conservation of Momentum.—Important principle of mechanics which states that if two or more masses impinge on one another, the total momentum of the system remains unchanged, the momenta lost by some of the bodies being gained by the others.

Contact Breaker.—Device for opening and closing an electrical circuit in quick succession.

Continuity of Flow.—The law of continuity for an incompressible fluid states that when flow takes place in a closed channel of varying section, the product of velocity by area of cross-section is constant for all sections.

Continuous Gauge.—Gauge which enables readings to be taken on work which is moving longitudinally or which is revolving. Continuous gauges may be of the dial type, mechanically operated by a plunger, or may be electrically operated, movement of the gauging plunger or roller being converted into an electrical signal which is amplified and shown on a dial calibrated in 0·001 in. or 0·0001 in.

Continuous Girder.—A continuous beam or girder is one which is continuous over its supports, and the stresses in which are therefore statically indeterminate. (See Clapeyron's Theorem above.)

Controller.—Form of switch for starting and controlling the speed of electric motors.

Controlling Force.—In the case of a pendulum governor, the controlling force is the radial force required at each ball to maintain the balls at a given distance from the axis of rotation when the governor is stationary.

Conveyor.—Means of moving parts or materials from place to place with the minimum of handling and manual effort. See Band Conveyor, Chain Conveyor, Pneumatic Conveyor, Roller Conveyor, Vibrating Conveyor.

Coolant.—(1) A fluid (e.g., water, ethylene glycol or a mixture of the two) used in the cooling system of a liquid-cooled internal combustion engine; (2) Fluid used during machining to cool and lubricate the cutting or grinding tool and the work. (See Cutting Fluid.)

Cooling System.—To reduce the heat generated in internal combustion engines, air is allowed to flow through the radiator, so as to abstract the heat from the circulating water ; or by forcing the water through the cylinder jackets with a pump ; or, again by using natural circulation with convection currents. (See Thermo-syphon System.) See also Cooling Surface.*

Cooling Tower.—Structure in which the heated circulating water of a condensing plant is brought into contact with large volumes of air.

Copper-Aluminium Alloys.—Copper-aluminium alloys possess good casting

2 +

characteristics. The tendency to hot-shortness and high casting shrinkage which is associated with the group, decreases with increasing copper content. They combine excellent mechanical properties with very good machineability.

Copper-Lead Bearing Alloys.—Alloys having the approximate composition, 70% copper and 30% lead. These are of modern development and show advantages over "white metal" bearings, being particularly suitable for high speeds with comparatively heavy loads.

Copper Smoke.—The gases which escape from the hearth of the reverberatory furnace in which the calcination of copper ores is performed. They consist of sulphurous acid, with a little of the vapour of sulphuric acid, arsenious acid, and hydrofluoric acid gas.

Core.—Also, the part of a nuclear reactor which contains the atomic fuel, the moderator and the control rods.

Core or Kern.—The area of a cross-section of a structure within which the resultant of the loads acting on it must lie, in order that the stress on the section may not change sign. See Core.*

Core-making Machines.—These are either push-through machines, in which the core is pushed along in the longitudinal axis of a cylindrical box of metal. Or the cores are squeezed between the halves of boxes between a presser head and table by the action of hand or hydraulic power. See also Core Maker.*

Core Vents.—The methods of venting cores do not differ materially from those employed for moulds (see Venting in Basic Terminology). Cores are vented with the wire, with ashes, with strings or ropes, and with rods, and the sand employed is also, when dried, of a porous character. Core vents require to be well secured where they fit into their prints, or into one another, or where they abut against the mould. Wet loam is often daubed around to secure these, and dried with a "devil," or open fire.

Corundum.*—Oxide of aluminium, next in hardness to diamond and used as an abrasive. Sometimes termed white sapphire. Small amounts of impurities produce the ruby and sapphire gemstones.

Coulomb.—The unit employed in the measurement of electrical charge. A coulomb is the number of electrons passing a given point in a circuit when a current of one ampere flows for one second.

Counter or Revolution Counter or Engine Counter.—Automatic recorder for registering the number of revolutions made by an engine. The motion of the engine is transmitted to the counter, either from a reciprocating part of the engine by a pawl and ratchet-wheel, or from a revolving part by means of an endless screw. An index-hand registers the number of revolutions.

Counterbore* (see Pt. 2, Basic Terminology).—Signifies also to bore a larger hole concentric with another and smaller hole, the hole thus bored forming a recessed or shouldered portion to the other, to receive the head of a cheese-headed screw, or of a shouldered pin. A pin drill is commonly used in counterboring. Sometimes the pin drill itself is termed a counterbore. The term is also sometimes used to signify the simple enlargement of holes already drilled with a smaller drill, not in the sense of reaming, but of enlarging with a pin drill. The enlarged portion of the bore of a cylinder into which the covers are checked is called the counterbore.

Counterforts.—Vertical projections of masonry or reinforced concrete projecting from the back of a wall at intervals with the intention of strengthening it, though except in the case of reinforcement, usually failing to do so, because the counterforts break away from the wall.

Countersunk Rivet.—A rivet, the head of which is shaped to fit into a countersunk hole, so that it lies flush with the face of the plate or bar in which it fits. It is employed in cases where a flush face is necessary, in order to permit of the attachment of other pieces of work. See also Countersink* and Countersunk.*

Coursed Masonry.—Masonry in which the stones are rectangular.

Crack Detection.—The detection of cracks and flaws in iron, steel and non-ferrous materials. In the past it was usual to section or to test to destruction representative samples from a batch of work. Specialised products, however, such as large turbine wheels and rotors, could not be tested, while even the percentage sampling system did not preclude the possibility of flaws being present in the remaining articles in a batch. Modern practice, therefore, is to subject important items to individual tests by means of X-ray or Gamma-ray inspection (see Radiology) or by the use of magnetic, fluorescent, electronic or supersonic crack detection equipment. See Electronic Crack Detector, Magnetic Crack Detector, Supersonic Flaw Detector, and Zyglo Crack Detector.

Cramped Joint.—Brazed joint made in copper and tin sheets, by a notching and lapping of the edges.

Crampoon.—A pair of hooked nippers for raising stones, timber, etc., by a crane.

Crankcase.—The casing which encloses the crankshaft of an internal combustion engine. See also Crank Shaft.*

Crank Hoop.—Hoop of wrought iron or of steel shrunk around the webs of locomotive cranks, to prevent accident in case of sudden fracture of the web.

Crankshaft Lathe.—Very massive design of lathe for machining marine cranks. It usually carries four tool rests and a heavily geared head-stock, and a tailstock, the spindle of which is gear-operated.

Creep.—The elongation or shrinking of a metal. The term is usually (but not always) applied to the changes that take place when the metal is subjected to tensile or compressive stress at high temperatures. The creep strength of metals is of particular importance in such applications as the blades of gas turbines, which are subjected to very high temperatures and centrifugal forces. Creep takes place in three stages: an initial, decelerating rate, termed primary creep; an approximately steady rate, termed secondary creep; and a final accelerating stage, in tension, or a series of accelerations and decelerations in compression, both known as tertiary creep. When the tertiary stage begins, the useful life of the metal is virtually exhausted. See also Creeping.*

Critical Eccentricity.—The distance of the point in which the resultant load on a structure cuts any cross-section, from the centroid of the section, when the stress on the remote edge is zero.

Critical Load.—The critical load on a pillar or stanchion is the load which first causes it to yield by buckling.

Critical Point.—If a piece of carbon steel is heated gradually, at a certain point it will absorb heat without showing rise in temperature. This is the critical point at which the carbon changes its state, and is known as the point of decalescence. When it cools slowly from a temperature above this point, the metal will actually increase in temperature, in spite of its surroundings being colder. This is termed the point of *recalescence*. Until steel reaches its point of *decalescence* it retains its magnetic properties. This point is from 100° to 200° F. above the recalescent point.

Critical Range.—There are several critical points (q.v.) in steel: for example three in low carbon steel between 690° and 900° C. The range of temperature over which these points occur is known as the critical range.

Critical Velocity.—For low velocities the motion of a fluid is a steady stream-line flow, but when a certain speed is reached, eddy currents arise and the flow is said to be *turbulent*. The velocity at which the flow first changes from steady to turbulent is known as the critical velocity.

Crowd Weight.—The total weight concentrated on a definite area of a floor, of men, or other units, crowded closely.

Crowned Pulley.—A crowned pulley is one whose surface width is convex, the object of curving it being to prevent lateral movement of the belt which drives it.

Crucible Furnace.—Any furnace in which crucibles are heated. Many are fired with gas. In many coke-fired furnaces the waste heat is utilised in the preliminary heating up of a crucible. Numbers of furnaces tilt. Batteries of furnaces are built up of individual units. See also Crucible.*

Cup Drum.—Term sometimes used to denote that form of sheave-wheel (q.v. in Dict. of General Terms) the rim of which is recessed out for the individual links of a chain.

Current Meter.—Instrument used for measuring the velocity of running water. It consists of a helical propeller running in very smooth bearings, the speed of the current being determined by a revolution counter. An electric current is passed to the propeller shaft from a battery above water and a commutator fixed to the shaft makes and breaks circuit at each revolution, the revolutions being registered on the counter.

Curving of Castings.—When curvings are badly proportioned, that is, when there are light and heavy flanges or ribs in close proximity, the cooling does not take place regularly. One portion becomes "drawn" by the other, becoming curved in consequence. Usually, the section having heaviest metal goes concave lengthwise, but this is a general rule only, since in some cases the opposite condition obtains. A knowledge of the amount and manner of curving can only be gathered by experience, and then but approximately. Patterns are cambered in the reverse direction. See also Curving of Patterns.*

Cut-and-Cover.—Work which is partly open cut and partly tunnelling.

Cutanit.—Sintered carbide alloy used in the production of cutting tools. See Sintered Carbide Tools.

Cut-in.—To connect into circuit any electric appliance, mechanism or conductor.

Cutting Fluid.—An oil or other lubricant primarily used to lubricate and cool a cutting tool, and in addition to wash away chips and swarf, reduce the power necessary to drive a machine tool, improve the finish on the work, and prevent corrosion and rusting. Fluids used include soluble oils, mineral oils (lard oil), compound oils, sulphurised oils, sulpho-chlorinated oils, soap suds and turpentine or paraffin.

Cutting Oil.—See Cutting Fluid.

Cut-out.—The reverse of cut-in (see above).

Cutter-stock.—Head or holder in which a cutting blade is fastened when in use.

Cut-water.—(1) The forward edge of the stem or prow of a vessel; (2) The edge of a *starling* (q.v.) facing upstream, to divide the water on each side of the pier, which it protects.

Cyaniding.—The superficial hardening of steel by heating it in contact with a cyanide salt, followed by quenching.

Cylinder Gate.—Outside or inside cylinder ring which is raised or lowered in order to regulate the flow of water into a turbine.

Cylinder Grinding Machine.—One in which usually the cylinder is fixed, and the grinding is done by a planet spindle.

Cylinder Ratio.—Ratio between the capacities of the cylinders in compound engines.

D

D'Alembert's Principle.—Mathematical artifice for reducing problems of rigid dynamics to the familiar laws of static equilibrium. The difficulty to be overcome consists in the fact that a rigid body consists of particles which are not free to move as in the case of a single particle because their motions are restrained by their adherence to adjacent particles.

Dalton's Laws.—(1) The pressure exerted by, and the quantity of, a vapour which saturates a given space, are the same for the same temperature, whether this space is filled by a gas or is a vacuum. (2) The pressure exerted by a mixture of a gas and a vapour or of two vapours, or of two gases, is equal to the sum of the pressures which each would exert if occupying the same space alone.

Damped Vibration.—When the vibrations of a body are retarded by friction, or by the resistance of a liquid, the amplitude gradually diminishes, until the body is brought to rest. The vibrations are then said to be *damped*.

Damper.*—Frictional device used to check vibration; usually applied to long crankshafts to damp their elastic oscillation under variable torque.

Dash Pot.*—Device employed to damp the movement of a mechanism. The movement of a piston within a cylinder is restricted by the fluid friction of air or a liquid, the flow of which from one side of the piston to the other is controlled by a calibrated orifice. A one-way valve is often incorporated in the assembly to give a differential damping action. See also Pt. 2, Basic Terminology.

Deacon Meter.—Meter for determining the varying flow of water through a pipe and thereby detecting leakage.

Dead Beat.—An instrument which reaches its reading quickly is said to be *dead beat*. Ordinarily the pointer, if moving quickly, would swing to and fro a number of times before coming to rest if not prevented by damping to be dead beat.

Dead-bright.—Practically equivalent to burnishing. It means that machined surfaces shall be finished with dead smooth files and oil until all tool marks are obliterated, the grain closed up, and a polished face imparted. Such a surface keeps bright for many years. The term is used in specifications,

Dead-hole.—A hole which is not a thoroughfare hole; that is, one that is bored only for a certain distance into a piece of metal. This term is used chiefly in connection with the use of taps, the plug and bottoming taps being made expressly for tapping dead-holes, which could not be reached with taper taps.

Deceleration.—The rate of reduction or retardation in the speed of a moving object. It is usually measured in feet per second per second. See also Acceleration.

Decibel.—The unit of power-level difference in electrical engineering. The tenth part of a bel. It is not a measure of the actual intensity of a given sound but an expression of the ratio between the relative intensity of two sounds.

Dedendum.—The depth of a wheel tooth below the pitch circle.

Degradation of Energy.—Since in every transformation of energy from one form to another, some of the energy is converted into heat, the total quantity of *available* energy in the universe is continually diminishing, because, since the process is not reversible, the energy will ultimately be in the form of heat at a uniform level of temperature. This process is known as the *degradation* or *dissipation* of energy.

Degreasing.—The removal of oil or grease from metal parts. Two methods are in common use: (1) Liquid degreasing, in which the work is immersed in a suitable solvent, such as petrol, paraffin or liquid trichlorethylene; or in a bath containing a hot concentrated solution of caustic soda or a proprietary degreasing chemical, followed by thorough rinsing in hot water. Caustic solutions are unsuitable for some metals and alloys, e.g., aluminium and its alloys. (2) Vapour degreasing, carried out in a cabinet filled with trichlorethylene vapour which is obtained by heating liquid trichlorethylene in the base of the cabinet. The vapour condenses on the work and drains back through a filter tray to the sump or may be collected in a trough and diverted to an external container. Solvent which is mixed with oil may be reclaimed by distillation.

Delta Metal.—An alloy of copper and zinc, with a small quantity of iron. It is prepared in various grades, both cast and forged. It is adapted for screw propellers because it has good resisting power to corrosion. It is also suitable for small-gear wheels, being stamped hot in suitable dies, and for pump-work and ornamental fittings.

Demagnetiser.—Device used to remove residual magnetism from parts which have been held on a magnetic chuck (q.v.) or which have been subjected to magnetic crack detection (q.v.). Rotating cutters may also become magnetised. In the platen type of demagnetiser the parts are placed on platens which are influenced by powerful electromagnets energised by an alternating current which produces rapidly changing polarity in the magnets. In the aperture type, the parts are passed through an aperture surrounded by electromagnets which are similarly energised by an alternating current.

Denaturise.—To render alcohol undrinkable, so that it may be used with economy for industrial purposes.

Depth Gauge.—Gauge used by wood and metal workers for testing the depths of holes and recessed portions. The sliding portion is usually graduated to serve as a rule, and a vernier, or a micrometer is included in the finer types.

Desaxe Engine.—Reciprocating engine mechanism in which the line of stroke, when produced, does not pass through the axis of the crankshaft. During the forward stroke the obliquity of the connecting rod is diminished, the pressure on the guide-bars is reduced and the turning moment is slightly more uniform. The advantage is somewhat reduced on the return stroke, but on the whole a small gain remains.

Designer.—In engineering, a person whose responsibility covers an entire project from its first conception to the issue of detailed instructions for operation. See also Design* and Prototype.

Destructor Furnace.—One designed for the burning of town's refuse, by which steam is raised for the generation of electrical energy.

Detent.—A term sometimes used to describe any device which locks, unlocks, detains or checks a movement. An example is the hand-brake locking mechanism on a motor vehicle. See also Paul or Pawl* and Ratchet.*

Detonation.—Unduly rapid combustion of fuel in an internal combustion engine, producing a knocking sound, roughness and loss of power.

Dial Gauge.*—Also termed Dial Test Indicator (abbrev. D.T.I.). Gauge in which a pointer, moving over a graduated scale, is actuated through magnifying linkage by a spring-loaded plunger or finger. The pointer is set to zero by the use of a reference gauge or with the gauge plunger in contact with the work, and the gauge will then indicate any departure, plus or minus, from this basic setting. The gauge can thus be used to detect ovality in cylinders, bushes, on shafts, etc., lack of truth on plane surfaces, end-float on bearings, and so on. May be mounted on a scribing block or clamped to any suitable support ; may also be provided with a handle and two-point anvil for use as a cylinder gauge. C.f. Comparator.

Dial Test Indicator.—Often abbreviated to D.T.I. See Dial Gauge.

Diamond.—Allotropic form of carbon which in its various forms is the hardest substance known. (See Hardness Scale.) In engineering, diamonds are widely used to tip tools (see Diamond Tool) and, in powder form, as abrasives. (See Diamond Dust, Diamond Wheel.)

Diamond Dust.—Diamond dust or powder is used as an abrasive for charging laps, of various types, when it is mixed with oil or grease to a paste-like consistency. Mixtures of sintered materials and diamond dust, which are more economical than the pure dust, have also been developed.

Diamond Pyramid Number or Vickers Pyramid Numeral.—See Vickers Hardness Test.

Diamond Riveting.—Arrangement of the rivets in a joint connecting two flat bars to end, or a flat bar to a gusset plate ; the arrangement consists of beginning with one rivet and increasing by one in each successive row until the required number is reached, with a symmetrical arrangement about the junction of the joint.

Diamond Tool.—(1) Diamond-tipped tool used for turning, boring, reaming and facing. Diamond-tipped tools consist of shaped diamonds, ground and polished as required, and mounted in suitable tool holders. In reamers two shaped diamonds are used. Although used for boring, turning, reaming and finishing ferrous large work, such as calender rolls for paper-making machines, diamond tools are more generally used on plastics, compressed fibre, rubber and on non-ferrous alloys such as aluminium, brass, bronze, copper, etc. (2) Diamond-set tool used in truing grinding wheels. The diamond is usually set in the end of a soft steel rod, and may be held by peening or brazing, the diamond being pushed into the molten brazing metal in a hole in the end of the rod, which usually has a round shank to allow it to be clamped in varying positions in order to equalise the wear on the diamond. See also Diamond Point.*

Diamond Wheel.—Abrasive wheel composed of diamond particles bonded into a resin or metallic matrix. Small wheels often contain the diamond abrasive throughout the structure but larger wheels usually have only a relatively thin surface coating of diamond-impregnated bond on the working surface, a numeral indicating the thickness in thirty-seconds of an inch. The concentration of diamond particles is denoted by a letter, A being a low concentration and C a high percentage.

Diatomite.—A form of silica, similar in appearance to chalk but lighter in weight and more porous, used as an insulator against heat, cold or sound. In the form of a powder or in bricks, it is used in furnaces, ovens, boilers, etc., and it is a constituent part of most fireproof cements.

Die Casting.—Casting produced in a permanent steel mould, instead of in a sand mould. Metals used for die casting must have a lower melting point than those generally employed for sand casting. Zinc-base alloys, with melting points in the neighbourhood of 400° C., are generally favoured ; aluminium alloys are also die-cast, but the pouring temperature of 700° C. is apt eventually to cause minute cracks on the surfaces of the dies. In gravity die casting the metal flows into the die by gravity alone ; the process is unsuitable for thin sections, due to the metal freezing before the die is filled. In pressure die casting, the metal is forced into the mould at high speed by a plunger which may exert a pressure of 1,500 lb. per sq. in. Dense castings, free from blow-holes or porosity, result. See also Die.*

Die Forging.*—The forging of smiths' work in dies is largely practised in the case of repetitive work. Dies are single, or double. If the top portion of the forging is plain, a single bottom die suffices ; if otherwise, a portion of the forging is formed in a top die. The forging takes place under the drop hammer, or the steam or pneumatic hammer, or in the forging press. Care has to be taken in cutting off the exact quantity of metal required for the forging, in order that on the one hand the forging shall finish clean and sharp everywhere, and that on the other there shall be as little superfluous fin as possible. The practice varies with the type of hammer or press used, and with the dimensions of the forgings made. A quantity of pressed plated work is also done with or without welding. The introduction of mild steel into the smithy has had the effect of increasing the range of die forging because, being practically destitute of fibre, shapes can be stamped out of solid plate or bar that would when made in iron have involved preliminary bending, welding, etc. (See Drop Forgings.)

Dielectric.—A non-conductor of electricity.

Dielectric Heating.—Process used to heat so-called electrically non-conducting materials, which usually also possess low thermal conductivity. Heating is produced by placing the material between two metal plates to which a high voltage, high-frequency alternating current is applied. The high-frequency current causes molecular disturbances which result in the generation of heat. The process is widely used in the plastics and rubber industries and for curing thermo-setting glues used in the manufacture of furniture, plywood and chip-board. The uniform heating obtained is also an advantage in drying yarns and fibres, powders, chemicals, tobacco and similar substances, without the risk of surface overheating.

Die Nut.—Square nut of hardened steel and of the usual thickness, but having three or four grooves cut in it, used for cleaning the threads of studs which have been damaged, the nut being run down and back again.

Diesel Engine.—Internal combustion engine of high efficiency, in which a high temperature is obtained by high compression of the air. A metered quantity of oil fuel is injected and is ignited by the heat thus developed. Power is regulated by metering the quantity of fuel supplied and not, as in a petrol engine, by throttling the air intake.

Diesel Index.—A term often used as an indication of the ignition quality of a diesel fuel, alternatively to the cetane number (q.v.)—from which, however, it may differ widely. The index is calculated with reference to the aniline point and specific gravity of the fuel.

Differential Accumulator.—Type of accumulator arranged to work a
2*

machine such as a hydraulic riveter, where the force exerted at the beginning of the operation is comparatively small, but which at the end is large. The capacity being small, the load descends rapidly and with increasing speed, and as it approaches the end of its stroke, the kinetic energy acquired in falling is converted rapidly into pressure energy in the water, thereby causing an enormous force to bear on the object, such as a rivet head, upon which it bears.

Differential Arc Lamp.—One in which the carbons are fed forward, as consumed, by the action of solenoids.

Differential Principle.—The mechanical principle which has for its basis the difference between two members connected rigidly together, by the application of which in various machines a very large mechanical gain is obtained with a small expenditure of power. It denotes in general the compounding of two or more distinct motions in such a way that uniform variations of speed shall be imparted to certain pieces of mechanism relatively to one another. See also Differential Motion.*

Diffuser.—See Boyden's Diffuser.

Dip Rod.—A simple depth gauge to show the level of a fluid in any container.

Direct Casting.—The pouring of heavy castings from the blast furnace, thus saving the cost of remelting. Tunnel segments, keel blocks, tubbing, and kentledge are now often cast thus.

Direct Current.—A current from a primary battery flows in one direction. In a dynamo the current generated alternates, but it may be converted into direct current by a commutator (q.v.).

Disc Feed.—A feed used on many machine tools, obtained by driving a large disc at a constant speed, from which another disc at right angles is driven at variable rates by the regulation of its distance from the centre.

Disc Friction.—The friction on a rotating disc, as in the case of steam and water turbines rotating at high velocities in a casing.

Disc Grinder.—Grinding machine in which steel discs have sheets of emery or other abrasive cemented on the face. The work is presented to the wheels on canting tables. A disc sander has sheets of glass paper for use in woodworking.

Disintegrator.—Machine for breaking up or granulating materials. Two discs carry teeth or prongs on their faces. One is usually fixed, the other revolves at a high speed, and the materials are reduced to powder between them.

Displacement Can.—See Eureka Can.

Displacement Diagram.—Graphical diagram which enables the displacements of the joints of a framed structure under its load to be determined more readily and completely than by calculation.

Distributor.—In the ignition system of an internal combustion engine, the spark occurs across the points of a sparking plug screwed into each cylinder, and these are connected by insulated high-tension leads to the *distributor*, which delivers high-voltage current to each plug in turn.

Doble Vane.—A vane which consists of a double cup, with central cutwater, which splits the jet impinging on the vanes of a Pelton wheel and turns it through an angle of about 160°, so as to clear the wheel at exit. Named after its inventor.

Dolly Bolt.—Bolt passing through the laminated pieces that form the springs of a vehicle, and which is continued into the axle, to prevent the U-bolts from slipping.

Dot Punch.—Also called a Prick Punch. Similar to a centre punch except that it has a longer and finer point, it is chiefly used in setting-out, e.g., for dotting the intersection of lines indicating a curve and so on. The long fine point does not obscure the user's vision. See also Puch.*

Double Shear Steel.*—Made in the same way as ordinary shear steel, but the process is repeated two or more times.

Downdraught Carburettor.—An inverted arrangement which feeds the explosive mixture downwards into the inlet manifold instead of discharging it upwards.

Dowson Gas.—Gas made by passing a mixture of superheated steam and air through incandescent fuel.   Although about four volumes of Dowson gas are required to do the same work in a gas engine as one volume of town gas, yet because Dowson gas is produced so cheaply it is economical to use for heating purposes for furnace work of all kinds and for gas engines.

Draft Tube or Draught Tube.—(See Suction Tube.)

Drag Link.*—In a motor vehicle, the link which acts on the track rod to move the steering arms.

Draw.—The *draw* on a pile is the batter given to one side of it to cause it to press closely against the adjacent pile, while driving.

Drawing Dies.—Dies used in power presses for pressing sheet metal into cup-like shapes.   In deep drawing more than one operation is necessary.

Drawing Number.—Ideally, every engineering drawing should be allocated a drawing serial number to facilitate reference and identification. The drawing should always be identified by this serial number, and should not be re-drawn and given another serial number unless it has been so modified or altered as to be no longer interchangeable with the previous design.   An individual part or item, though it may be used in several different models, types or sizes of mechanisms, should always retain its original drawing serial number and should be detailed on one drawing only.   The drawing number should in principle be located in the bottom right-hand corner of the sheet, but it may be repeated in other corners of the drawing to ensure that it is visible when the drawing is filed away.   A register should be maintained for the systematic allocation of drawing numbers.   See also Item Number.

Drip.—Small channel cut under the lower projecting edge of a coping to cause any water reaching it to fall off and prevent it from reaching the wall.

Drop Arm.—In a motor vehicle, the arm actuated by the steering wheel, which in its turn acts on the drag link (see above).

Drop Forgings.—Term applied to forgings stamped in dies under the drop hammer.   It is employed rather loosely to signify any class of light stamped forgings, irrespective of the particular hammer used, in either steel, iron, copper, or aluminium.   The dies used may be of cast iron, or more accurately and permanently cut in steel.   Frequently an intricate forging can be made thus in a single heat within a minute, that would require several heats and skilled labour at the forge.   Drop forgings are also all alike when made in the same dies.

Drop Hammer.—Heavy hammer used for forging.   It slides between vertical guides, and is suspended by a belt or rope that passes over and is attached to a pulley of large diameter overhead, or to a board between pulleys, and to the opposite end of which gear is attached, by which the weight is lifted.   The Brett patent lifters are now largely used for operating drop hammers.

Drop Worm.—A worm which can be dropped out of engagement with its wheel, either by hand or automatically. It is used on many lathes, milling and other machines.

Dryness Fraction.—The proportion of dry steam in a mixture of steam and water; that is, wet steam. See also Dry Steam.*

Dry Puddling.*—White or refined iron (q.v.) only is suited for dry puddling. The principal oxidising agent is the atmosphere, which plays over the surface of the pasty metal on the furnace bed, but the action is assisted by hammer scale (q.v.) added to the charge, and the oxide of iron forming the bed. The pasty metal is rabbled over the bed to assist the oxidising action, the carbon being removed as carbonic oxide, while the silicon, sulphur, phosphorus, manganese, mostly enter the tap cinder in the form of oxides. As the metal becomes purified its fusibility decreases.

Dual Ignition.—Duplicated system of ignition serving two sparking plugs per cylinder.

Dumb Irons.—The forgings usually riveted to the ends of the side members of the frame of the chassis of a motor-car and forming an extension, so as to act as a bracket for carrying the ends of the springs at the back or front.

Duralumin.—A metal alloy of aluminium, copper, manganese and magnesium, containing over 90% of aluminium, and possessed of high tenacity and great hardness. It can be rolled, forged and drawn when hot or cold.

Dust Exhausting.—Exhaust fans and pipe systems are used in grinding shops and in wood-working establishments for withdrawing dust and chips.

Dynamite.—An explosive of reddish-brown colour prepared from Kieselguhr, a silicious earth, with nitroglycerine. One kilo. develops by its explosion 7,250,000 footpounds of energy in $\frac{1}{30000}$ of a second.

Dynamo.—Machine driven by steam or water power for converting the mechanical energy expended in driving it, into electrical energy in the form of current. The dynamo may be an alternator, that is, one which generates an alternating current, or a continuous current dynamo. An alternating current may be converted into a continuous current or *vice versa* by means of a rectifier or a converter.

Dyne.—The C.G.S. unit of force. It is the force which imparts an acceleration of one centimetre per second every second to a mass of one gramme.

E

Eccentricity.—When the resultant of the loads acting on a structure cuts any cross-section or the cross-section produced at a distance from its centroid, the point where it cuts it (the Load-point) is said to be that much eccentric, and causes bending moment, the amount of which depends upon the magnitude of the eccentricity.   See also Eccentric.*

Elastic Clutch.—A type which in various forms has come into extensive use in consequence of the high speeds of electric and other motors. The elasticity is an element of safety not possessed by rigid types.

Electrical Resistance Welding.—Method of welding, employing the heat caused by the resistance to the passage of electric current across the junction of the parts to be welded, combined with pressure to force the parts into contact. See Butt Welding, Flash Welding, Percussion Welding, Projection Welding, Seam Welding, Spot Welding, Stitch Welding.

Electric Crane.—Crane driven by an electric motor or motors, independent of the mechanism of the crane.   The current is conveyed along copper conductors from the generating set, collectors sliding along the rods and conveying the current to the motors on the crane.   The modern type of electric crane is that in which a separate motor is fitted for each motion of the crane, one, or two for lifting, another for travelling, and a third, in the case of an overhead crane, for traversing the crab crossways, each rated for its duty, hence these are commonly termed three-motor travellers.   When an auxiliary hoist is fitted for light loads, it will also have its own motor.

Electric Crucible.—Crucible used for melting difficultly-fusible substances or for reducing ores, etc., by the electric arc produced within it.

Electric Cutting.—The electric arc will cut any metal, ferrous or non-ferrous, and is particularly useful for cutting cast iron.   The voltage of the cutting arc varies from 25 to 45 volts and the current from 800 to 1,200 amp.

Electric Drive.—Method of driving shafting, machine and other tools. A length of shafting may be driven by its own motor, and thence drive the machine tools of that particular section by belting from the short length of shafting.   In the second method the motor is attached to or built into the machine.

Electric Furnace.—One in which high temperatures are obtained by the formation of an arc between positive and negative carbon electrodes. See also Induction Heating.

Electric Lighting.—Illumination by arc, incandescent or fluorescent lamps.

Electric Motor.—An electrical machine which has its field magnet system and the brushes of its armature excited by electricity brought from a dynamo or battery.   Motors are either shunt wound, series, or compound wound, each suited to different conditions.

Electrode Holder.—Device provided with an insulating handle, used to hold the electrode during metallic arc welding (q.v.) and to convey the welding current to the electrode.

Electro-deposition.—Term generally used to describe the electro-chemical deposition of relatively heavy coatings of metal, as used for engineering and similar purposes, e.g. the deposition of chromium or nickel on steel parts in order to build up worn areas or to provide a hard wearing

surface. Electro-deposition differs from electroplating (q.v.) not only in the greater thickness of metal deposited, but also in the fact that one or more undercoats of a different metal such as copper or nickel are not usually applied, the need for maximum adhesion necessitating a sound bond between the deposited metal and the base metal. To ensure this the surface of the part is frequently roughened by anodic etching to provide a satisfactory key.

Electro-erosion.—Metal may be removed by electrical erosion in four ways: (1) Electrolytic erosion, in which a direct current is passed through an electrode and the workpiece, both surrounded by a conducting fluid, termed the electrolyte. (2) Electro-arcing, in which an arc is maintained between an electrode and the workpiece, the latter being cooled by the passage of a coolant which also washes away the eroded particles. (3) Electro-sparking, in which a succession of sparks is generated by the discharge of current stored in capacitors. (4) Electro-abrasion, a punch or drill being operated at an ultrasonic frequency. (See Ultrasonic Drilling.)

Electroformed Moulds.—The economical production of dies of the most intricate pattern for die casting and plastics production is rendered possible by an electroforming process which consists essentially of accurate machining or carving, in plastics, of a master to which a high degree of finish is given. The master is metallised and a thick shell of hard nickel (500 D.P.N.) is electroformed on it. The shell is levelled at the back with a deposit of copper and is then removed from the master and given a mirror finish; after machining the back, it is mounted in a steel bolster. The main advantage of this process lies in the relative ease of cutting, finishing and checking a positive master as compared with a negative cavity in steel.

Electrolytic Corrosion.—Corrosion which occurs when two metals having dissimilar electrical potentials are placed in contact or in close proximity in the presence of an electrolyte. The moisture of the air is sufficient to form an electrolyte, while the presence of salts in solution (as in sea water) will aggravate the effect.

Electrometallurgy.—A generic term covering all the various electrical processes by means of which metal can be worked, prepared or surface-finished. Such processes include, among others, electro-deposition (q.v.), electro-refining, operations in electric furnaces, electro-steeling and electro-galvanising. See also Electronics, Electroplating, and Electro-polishing.

Electromotive Force.—The force which causes an electric current to flow between two points of differing electrical charge when the points are joined by an electrical conductor. The unit of measurement for e.m.f. is the volt—a volt being the difference of potential which will cause a current of one ampere to flow through a resistance of one ohm.

Electron Diffraction Analysis.—The use of the diffraction of X-rays from the atoms in the lattice of crystalline material to produce a pattern of diffracted X-ray beams, determined by the arrangement of the atoms. Crystals of a given substance in the same condition always yield the same characteristic pattern; consequently an unknown substance may be identified by comparing its X-ray crystal diffraction pattern with the patterns given by known substances. Examination by the method of crystal analysis leaves the specimen examined untouched and unaffected. Although it is therefore often of great usefulness when only very small amounts of material are available for examination, its uses also include the examination of corrosion products in roller bearings, the detection of directional properties in metal sheet intended for pressing, the presence of residual stress in a metal, and the difference between coarse-grain and fine-grain structures, revealed as differences in the character of the rings. Thus the method forms a reliable means of studying the degree of cold work of a specimen, the effect of fatigue and the changes during heat treatment. Another method of diffraction analysis is the use of so-called monochromatic X-rays emitted by a cobalt target in an X-ray tube (see X-rays). In these methods the diffraction pattern is studied—not the shadow picture of the internal structure of the material.

Electronic Comparator.—Instrument used in production engineering inspection for the following purposes : material and component inspection for chemical composition and hardness; detection of cracks and flaws of all types, on or below the surface ; inspection of plating and coatings, including verification of adhesion ; surface finish comparison ; stress-strain analysis ; dimensional comparison with micro-inch sensitivity. The instrument produces electrical impulses of a definite frequency which are passed to a detector or adaptor. When this is applied to a standard or reference component, the frequency of the impulses is modulated and this modulation is revealed on a meter which can be adjusted to give a full-scale deflection. Any subsequent samples which are identical with the standard component will not alter the meter reading but a divergence from standard, in the form of chemical composition, physical condition, or a crack or flaw, will cause a deflection of the needle which varies according to the nature of the fault.

Electronic Computer.—A computer in which electronic circuits are used to solve mathematical problems. Analogue computers work from varying physical magnitudes, e.g., speeds of rotating parts, light intensities, voltages, etc.; a slide rule is a form of analogue computer. Digital computers work from pulses representing actual numbers and alphabetical symbols, the abacus or bead frame being a primitive example of this type.

Electronic Control of Machine Tools.—Method of controlling any type of machine tool in which the movements of the cutter or cutters are determined by data fed to an electronic control unit by a punched tape or a magnetic tape. Alternatively, the control unit may be governed from a desk carrying a set of dials on which ordinate dimensions can be set up manually by the operator. In a third form, the electronic control is operated by a tracer which follows a template or pattern. An important application of magnetostriction (q.v.) is the use of a nickel bar surrounded by a solenoid winding to give an inching movement in feeding work to a grinding wheel. By clamping the bar at each end alternately and switching the current on and off, increments of a few millionth of an inch are obtainable. (See Automation, Electronic Computer, Magnetostriction.)

Electronic Crack Detector.—See Electronic Comparator.

Electronics.—In engineering, the term is applied to the science of employing electric current at radio frequencies in inspection and flaw detection, to heat materials and to control processes. See Dielectric Heating, Electronic Comparator, Ignitron, Induction Heating, Thyratron.

Electronic Timer.—Electronic control, embodying thyratron and/or ignitron valves (q.v.) used to time various processes—for example to control the heating programme on automatic welding machines.

Electroplating.—The deposition of one metal on another by electrolytic action.

Electropolishing.—Polishing process for metals and alloys in which the article to be polished forms the anode of an electrolytic bath, as in anodic treatment (q.v.). The composition of the electrolyte, however, is such that the anodic film dissolves at a slightly greater rate than its rate of formation. Projecting portions of the surface of the metal are dissolved more rapidly than the general surface, so that a smooth, polished surface is obtained. The process is employed to prepare samples of metals for microscopic examination, since it has the advantage of producing a surface which is free from a Beilby layer (q.v.), and is also applied commercially to the polishing of aluminium alloys and stainless steels.

Electrostatic Crack Detector.—Apparatus for detecting cracks in non-conducting materials having glazed surfaces, such as ceramics, glass, etc. The surface is treated with an electrically-conducting fluid which remains in any cracks after the surface has been wiped clean, forming electrical conducting paths. If an electric charge is applied to the article and the surface is dusted with magnetic powder, the particles will collect along the lines of the cracks.

Electrostatic Precipitator.—Device consisting of alternate parallel rows of negatively-charged wires and earthed collecting plates, used to remove dust or ash from air, gas or smoke flowing through the precipitator. A direct current having a potential of 15–80 kV is usually applied to the wires.

Element.—A term used in work measurement to describe a part of a task or work-cycle which has been selected for observation and analysis. See also Time and Motion Study.

Eureka Can or **Displacement Can.**—A vessel fitted with a spout which is used to determine the volume of any solid of irregular shape. The object is immersed in fluid in the can, and the resulting liquid displacement is measured. See also Archimedes' Principle and Volume.*

Ellipse of Stress.—If the stress-intensity at any point be plotted in magnitude and direction from a pole, the locus of the extremity of the radius vector, as the stress changes its direction, will describe an ellipse, of which the principal stresses, that is the greatest and least stresses, are its semi-major and semi-minor axes. This ellipse is known as Rankine's ellipse of stress, and from it the stresses on any plane may be found, when the principal stresses are known. See also Ellipse.*

Ellipsoid of Stress.—The ellipsoid which corresponds to the ellipse of stress above, when considering stresses in three-dimensional space.

Elliptic Chuck.—Chuck used in a lathe which enables pieces of material to be turned, whose cross-sections are to be elliptical.

Elliptic Trammels.—Mechanism for drawing ellipses, consisting of a fixed plate having two straight grooves cut in it at right angles to each other. In these grooves, a bar carrying the tracer slides, and can be adjusted to alter the form of the ellipse.

Elliptical Gears.—Gear wheels, the pitch peripheries of which are elliptical in outline. In these, the centres of the shafts being placed in the intersections of the conjugate and transverse diameters, there are four rates of revolution, two maximum and equal, two minimum and equal, coinciding with the major and minor diameters. These are, therefore, variable gears (q.v.). Elliptical gears, the shafts of which are in the focal centres, are used occasionally for producing a quick return (q.v. in

Pt. 2, Basic Terminology) motion. The shaft of the driven wheel must then be movable in space.

Elmarid.—Sintered carbide alloy used in the production of cutting tools. See Sintered Carbide Tools.

Elongation.*—A term used to indicate the amount by which, in a tensile test, a piece of metal will stretch before it is fractured. It is expressed as a percentage of the original length of the test piece, the length of which must always be stated. (Thus a test piece might have " an elongation of 18 per cent on 2 inches ".) See also Tensometer.

Enclosed Motor.—One in which the moving parts are enclosed in a casing as a protection from dust. Ventilating grids are usually provided in the casing.

Engineer's Tapers.—See Taper-Gauge.

Entropy.—A " thermo-dynamic function." A property of a body which remains constant so long as there is no communication of heat to the body, but which increases when heat enters, and diminishes as heat leaves the body. In the words of Professor Ripper, " Entropy is length on a diagram whose height is absolute temperature, and whose area is energy in heat units."

Epoch.—If the time of a body moving with simple harmonic motion is measured, not from the extremity of its path but from some other point, the distance of this point from the extremity is known as the *epoch.*

Equilateral Triangle.—See Triangle.*

Erg.—On the C.G.S. system, the unit of work is the work done when a force of one dyne is exerted through a distance of one centimetre, and is called an *erg.*

Escutcheon.—The outside movable plate that protects the keyhole of a lock from dust.

Etching.—The preparation of highly polished metal surfaces in order to examine the crystal structure or composition under the microscope. The process consists of immersing the metal in a chemical solution—usually alcohol and nitric acid—until the polished surface appears slightly dulled. The dilute acid solution attacks the grain boundaries so that minute channels outline the structure.

Euler's Formula.—This formula, due to Euler, gives the relaxation between the critical or buckling load on an " ideal strut " and its flexural rigidity (EI) ; whereby an ideal strut is to be understood one which is perfectly straight, of uniform cross-section, whose material is homogeneous and isotropic, and which is axially loaded. This formula forms the mathematical basis of the numerous more practical formulæ that have been devised for allowing, as far as possible, for the deviations which necessarily arise from the ideal conditions, such as Rankine's, Gordon's, Moncrieff's, etc., etc.

Eureka Can or **Displacement Can.**—A vessel fitted with a spout which is used to determine the volume of any solid or irregular shape. The object is immersed in fluid in the can, and the resulting liquid displacement is measured. See also Archimedes' Principle and Volume.*

Eutectic.—Mixture of substances which has a minimum melting point.

Evaporative Cooling.—Cooling system which takes advantage of the latent heat of evaporation, by allowing the cooling fluid to boil and then condensing and returning it to the cylinder jacket.

Evaporator.—Appliance in which the auxiliary feed water for steam engines is evaporated, the water being taken from the condenser. See also Evaporation.*

Excavators.—These may be classified as: (1) Power shovels or navvies; (2) Power drag-lines; (3) Grabbing cranes or clam shells; (4) Dredger excavators. See also Excavator* and Grab.*

Exhaust Injector.—Form of injector by means of which exhaust steam at the atmospheric pressure is made to force water into a boiler against steam of high pressure. The arrangement of the mechanism is designed to cause the exhaust steam to come into contact with the cold water in a combining cone, and produce condensation, and a partial vacuum, into which more steam and water rushes, improving the vacuum, until sufficient velocity and force are generated to overcome the back-pressure due to the steam in the boiler. See also Exhaust Steam.*

Expanded Steel.—Steel sheet which is pierced and opened out in a machine, yielding lozenge-shaped spaces. Often embedded in concrete.

Explosive Rivet.—A rivet the end of which has been bored out to take an explosive charge or compound. When the charge is exploded, by percussion or by heat, the rivet expands and opens out sufficiently to form a head. See also Rivet Tail and Riveting.*

Express Boiler.—Water-tube boiler having the tubes arranged for very rapid steam raising.

Extensometer.—Instrument which is attached to a test piece for the purpose of measuring accurately the small strains under load. It is carried on the gauge-points of the piece tested. There are many varieties.

Extrusion.—Process for producing long lengths of metal of constant cross-section. A round billet of the metal (usually brass or an aluminium alloy) in hot, plastic condition, is forced out by a ram through one or more dies in the end of a container of the same cylindrical shape as the billet. Extruded sections form the basis of low-cost fabrication methods for small articles whose design permits them to be cut out from the shaped bars.

F

Fabricated.—This term, applied to a manufactured article, generally implies that the article is constructed of two or more different pieces of metal which have been permanently joined together by welding, brazing, soldering or riveting.

Factor of Safety* (see Dict. of General Terms).—If the calculation is one determining the load on a structure, the load corresponding with the ultimate strength will be *divided* by the factor of safety in order to give the safe working load. The factor of safety may also be determined in reference to the *elastic strength* of the structure instead of by its ultimate strength. This, *divided* by the factor of safety, will give the safe working load. Factors of safety will range from about three in iron structures, subject to dead loads only, to ten in structures in the same material subject to live loads, and will reach twenty or more in the case of masonry.

Fall.*—The rope used with pulleys in hoisting. (See also Pt. 2, Basic Terminology.)

False-work.—The scaffold, centre, or other temporary supports for a structure while being built.

Farad.—The basic unit of electrical capacitance. A capacitor—formerly also referred to as a condenser—has a capacitance of one farad when a charging current of one ampere flowing for one second causes a change of one volt in the potential difference between its plates.

Fascines.—Bundles of twigs and small branches, laid cross-wise, for forming foundations on soft ground.

Fastener.—A fastener is a mechanical device designed specifically to hold, join, couple, assemble or maintain equilibrium of single or multiple components. The resulting assembly may function dynamically or statically as a primary or secondary component of a mechanism or structure. Based on the application intended, a fastener receives varying degrees of built-in precision and engineering capability, insuring adequate, sound service under any imaginable pre-established environmental condition.

Fastening.—Fastening is the act or process by which single or multiple components are held, joined, coupled, assembled or otherwise maintained in equilibrium.

Fatigue Allowance.—Allowance made in time study (q.v.) for the fatigue caused by the physical and mental demands of the work, the monotony or danger of the job, the working conditions, the duration of the working periods and similar factors. The allowance will vary considerably; for example, 10% for assembly work with the operator seated at a well-designed work place, as compared with 50% for a fitter carrying out rough heavy filing.

Fatty Oils.—Fatty oils, especially lard oil, are among the most generally used " straight " cutting oils (q.v.) when there is no shortage of raw materials; they have a better spreading action than straight mineral oils, but may become rancid and form gummy residues. See also Oils.*

Feather Edge.—A keen edge, or, as the workmen say, an edge going off, or tapering off, to nothing.

Feed Gear.—The mechanism, usually automatic, by which the extent of the operations of cutting tools used in machines is governed. Combinations of gearing, cams, feed screws, ratchet-wheels, etc., are variously employed; with the result of imparting a precise and definite amount of forward or downward movement, either to the tool or to the work, immediately after each cutting stroke. See also Feed.*

Feeler Gauge.—Also called a *Thickness Gauge*. The gauge consists of a number of leaves of thin tempered steel, each of which has been accurately ground to a different thickness. Every leaf is marked with a number representing its thickness either in thousandths of an inch or in millimetres. The leaves are held in a metal case for protection. The purpose of the gauge is to provide a means of measuring the clearances between two surfaces, e.g., in testing the fit of joints, in gauging the clearance between a piston and its cylinder wall, or in setting a milling cutter.

Felloe or Felly.—The parts which form the circular rim of a wheel, into which the outer ends of the spokes fit, the whole being often surrounded by a steel tyre, shrunk on.

Ferret.—Jagged piece of iron driven into brickwork, masonry, etc., to secure a hold.

Ferrite.—Iron containing little or no carbon. Low carbon steel, viewed under the microscope, shows regions of ferrite and regions of pearlite (q.v.). Alpha iron is ferrite. An unalloyed carbon-free iron in the annealed condition consists of polyhedral grains of ferrite.

Ferro-concrete or Reinforced Concrete.—A combination of concrete and steel, in which the steel is used to assist the concrete, particularly where it is liable to be put in tension or shear.'

Ferrous.—A term used to describe all metallic alloys in whose composition there is a large proportion of the element iron. See also Ferrite, Ferrous Oxide* and Iron.*

Field-Magnets.—Electromagnets by which the magnetic field of force is produced in any electrically operated mechanism.

Field Rivets.—Rivets which are driven at site, in order to connect large portions of a structure that have been assembled in the shop. So called to distinguish them from shop-driven rivets.

Figures of Equivalent Resistance.—Since the moment of resistance of an element of a beam section is equal to its area multiplied by the stress on it and by its distance from the neutral axis ; and since the stress on it is proportional to its distance from the neutral axis, it follows that if the element be moved parallel to the neutral axis and its width diminished in proportion as its distance from the neutral axis diminishes, the stress on it will be the same as that of the element most distant from the neutral axis, and the new section thus obtained will have a uniform stress on it, and the resultant stress will act through its centroid. Such a figure is known as a figure of equivalent resistance, because its moment of resistance is the same as that of the actual section it replaces. The maximum stress multiplied by the area of this figure gives the resultant force for either side of the neutral axis, and multiplying this by the distance between the centroids of the two areas gives the moment of resistance required. This graphical method is useful for awkward cross-sections.

Filament Lamps.—Lamps which consist of metal or carbon filaments enclosed in glass bulbs exhausted of air, which become incandescent due to their electrical resistance to the current passing through them. The Osram, Mazda and other lamps have tungsten filaments which are of pure drawn tungsten wire.

Filler Metal.—Material for addition to a fusion weld (q.v.).

Fillet Gauge.—See Radius Gauge.

Fin.*—Projection on the sides of rolled steel sections, formed by the squeezing out of the metal between the flat bodies of the rolls on each side of the groove, owing to insufficient allowance having been made for the spread of the material.

Fire Point.—A term used in the testing of oils. After flash point is reached, heating is continued until the oil ignites and continues to burn for a period of at least five seconds when a flame is applied to its surface. The temperature of the oil when this occurs is known as its fire point. See Flash Point.*

Firth Hardometer.—Hardness tester in which a diamond indentor may be used with alternative loads of 10, 30 or 120 kg. for hardened steels, or a steel ball for soft materials.

Fit.—Relationship between two mating parts which determines the amount of clearance or interference present when they are assembled together. See Clearance Fit, Force Fit, Interference Fit, Transition Fit.

Fixture.—Attachment to a machine tool, which holds pieces of work in definite positions, but does not guide the tools.

Flame Cutting.—See Oxy-acetylene Cutting.

Flame Hardening.—See Oxy-acetylene Flame Hardening.

Flame Planing.—See Oxy-acetylene Cutting.

Flame Washing.—A term used in welding and brazing to describe a process in which, after the building up of splines, gear teeth, etc., undue rugosities in the metal are smoothed by the surface-melting of the deposit. The object is to save time in the final dressing when the work is cold.

Flash Boiler.—Rapidly steaming boiler in which the steam is generated in coils of small tubes. Steam pressures will range to 1,000 lb. Oil burners are used. The boilers are safe, the steam is superheated, scale seldom forms, and the space occupied is small. The steam is supplied as required and on the instant.

Flashings.—Broad strips of zinc, lead, etc., having one edge inserted into the joints of the brickwork or masonry, an inch or two above a roof, etc., and flattened down, to prevent rain leaking through the joint between the roof and the brick chimney, etc., which projects above it.

Flash Welding.—Method of electrical resistance welding (q.v.) in which the parts are clamped in the jaws of an electrical welding machine. Their edges are brought into contact and are then separated in order to strike an arc, the heat of which fuses the metal. The parts are then forced into contact while still incandescent.

Flow Diagram.—A term used in Method Study to describe a scale diagram or model of the working area under consideration. It shows the location of various activities, the nature of these activities, and the routes taken by workers and equipment in the performance of them.

Flow Process Chart.—A term used in Method Study to describe a process chart analysing a given procedure. On the chart symbols are used to provide full details of what every worker does, what happens to material, and how equipment is used.

Fluctuation of Energy, Coefficient of.—The ratio of the change of energy due to change of speed, to the work done per cycle, in the case of an engine flywheel.

Fluctuation of Speed, Coefficient of.—The ratio of the fluctuation of speed to the mean speed per cycle in the case of a flywheel. The fluctuation of energy is twice the fluctuation speed.

Fluid Coupling.—Device that transmits a rotary drive through the medium of oil or other liquid. A simple coupling consists of two cupped members, having internal radial vanes, enclosed in an oil-filled casing. When the driving member or impellor is rotated the oil in the vanes is flung outwards and also in the direction of rotation, impinging on the vanes of the driven member or turbine which is thus rotated by the reaction of the oil streams, provided that the speed of the impellor is sufficiently high. At low speeds considerable "slip" takes place between the two members but this is reduced to approximately $1\frac{1}{2}$ per cent at and above the "coupling" speed. In addition to acting as an hydraulic clutch the fluid coupling can also function as a torque converter (q.v.) by the addition of one or more stationary reaction members.

Fluid Friction.—The laws of fluid friction, above the critical velocity, as determined by Froude, are as follows : (1) The frictional resistance varies approximately as the square of the velocity ; (2) It varies with the nature of the surface ; (3) The resistance per square foot of surface decreases as the length of the body increases, but is practically constant for long lengths ; (4) It varies with the density of the fluid.

Fluorescent Screen.—Viewing screen used in radiology (q.v.), which becomes luminous or fluoresces when struck by X-rays ; consists of a sheet of cardboard coated with barium platino-cyanide, calcium tungstate or zinc sulphide.

Focal Points.—Points at which the shearing forces due to a moving load are liable to change sign as the load traverses a girder. Any panels of a braced girder which include the focal points must either be counterbraced or the diagonal subject to tension must be designed to resist compression.

Follsain Process.—A method by which ferrous surfaces are treated to resist oxidation at high temperatures, or attack by corrosive acids. The material to be treated is packed with a mixture of powdered aluminium, chromium chloride and a catalyst, and is then subjected to a maintained temperature of 1000° C. for a period of three to four hours.

Force Fit.—A fit between a shaft and its hole which requires considerable pressure to make one go in the other, as that of a lever, a screw, or a hydraulic ram. See Interference Fit.

Forced Lubrication.—The supply of oil under pressure to the bearings of engines and to machine tools.

Fork-lift Truck.—Form of motor-powered stacking truck (q.v.) provided with forked carrying arms instead of a platform.

Form Tools.—Tools which are shaped to the same profiles as the articles they have to turn. They are used chiefly on the cross slides of turret lathes.

Forms.—Timber or metal work intended to serve as moulds or supports in or upon which concrete may be applied. The forms are removed when the concrete has hardened.

Fortin's Barometer.—Form of barometer arranged to enable the level of the mercury in the cistern to be brought to a fixed mark by making the bottom of the cistern flexible. Before taking a reading on the vertical scale, the surface of the mercury is adjusted until it exactly touches the point of an ivory pin fixed to the top of the cistern.

Fottinger's Hydraulic Transmitter.—Transmission gear consisting of two centrifugal pump wheels keyed to the primary shaft, which is driven by a turbine. On the secondary shaft or propeller shaft three impulse reaction turbine wheels are mounted, the after pair for ahead driving and the forward one for astern driving.

Fourier's Theorem.—States that any periodic motion may be analysed into a number of harmonic motions, the periods of which are commensurate.

Fourneyron Turbine.—Outward flow reaction turbine. Was the first highly efficient turbine but was difficult to govern and is now practically obsolete .

Foxtail.—Thin wedge inserted into a slit at the lower end of a pin, so that as the pin is driven down, the wedge enters it and causes it to swell, and thereby to hold more firmly.

Frame.—(1) Any articulated structure built up of tension and compression members ; (2) a chassis structure for motor vehicles consisting of channel-section side members connected at intervals by cross members. See also Framing.*

Francis Turbine.—Inward-flow reaction turbine having several important advantages and also to a large extent self-governing. It can lift to about 500 ft. Is in very general use.

Frazing.—Removing the fin from forged nuts and bolts in a frazing machine.

Free Air.—Air at atmospheric pressure.

Free-cutting Steel.—Steel to which has been added sulphur, phosphorus, lead and/or other elements for the purpose of inducing in the metal a degree of brittleness to improve its machining properties. See also Steel.*

Freewheel.—Automatic over-running device or mechanical one-way clutch.

Fretting Corrosion.—A type of corrosion which occurs at the contact area when two metal surfaces under load are subjected to vibratory and/or oscillatory motion. It results in surface decolorization, pitting, and the accumulation of a brown powdery deposit, principally iron oxide. The fatigue strength of a component can be impaired by this type of corrosion. See also Corrosion.*

Frequency.—The number of oscillations per minute or second. The reciprocal of the periodic time.

Friction Axis.—In any link of a turning pair, when motion occurs, the line of thrust or pull on it will not coincide with the axis of the link, due to the effect of friction. The actual line of thrust or pull on the link is known as the *friction axis*. Its displacement from the axis of the link affects the torque transmitted.

Friction Back Gear.—Device in which the back gear of a lathe is put into and out of engagement by means of a friction clutch within or next the cone pulleys, by means of a lever while the lathe is running. See also Friction Gearing.*

Full Nut.—A nut whose height is equal to the diameter of its bolt.

Funicular Polygon or **Link-Polygon.**—An imaginary chain of links which would be in equilibrium under a given system of loads acting on a structure. The line of pressure for the structure is a link-polygon.

Funicular Railway.—A railway for mounting very steep grades by means of rack and pinion rails, or by means of clutches which grasp the sides of the rails.

Fuse.—A safety device in electrical circuits, consisting of a fine wire which fuses if the current flow becomes too great. See also Fusibility.*

Fusion Face.—A surface to be welded by a fusion-welding process (q.v.).

Fusion Welding.—Method of uniting metals in which the metal is melted along the joint or seam, with or without the addition of metal from a filler rod. See Atomic Hydrogen Welding, Carbon Arc Welding, Metallic Arc Welding, Oxy-acetylene Welding, Shielded Arc Welding, Thermit Welding.

G

G.A.—Grate area.

Gamma Iron.—When pure iron is heated, a thermal critical point (q.v.) occurs at 900° C., marking an internal rearrangement of the atoms. Here a change occurs from alpha iron to gamma iron. Gamma iron is formed in steels when heating through the critical range (q.v.), the temperature of formation depending on the composition of the steel.

Gamma Rays.—Radiations of shorter wavelength than X-rays (q.v.), used to examine heavy sections of steel and other dense metals; copper alloys up to 6 in. and steel up to 12 in. in thickness may be penetrated. Gamma rays are given off by a minute quantity of radium or mesothorium sealed into a glass sphere approx. 0·04 in. dia. See Radiology.

Gang Mills.—Large milling cutters built up of several separate smaller cutters either to form a long parallel mill, or more often, profiled forms.

Gang Tool.—A bolster or tool holder on which a number of similar tools or cutters are mounted, used in lathes and planing machines, and in press-tool engineering. See also Capstan Tool Rest.*

Gap Gauge.—(1) Gauge, in the form of a pair of gauging anvils held in a rigid frame, used to check the dimensions of shafts, external threads and other parts. Gap gauges may be of the solid or adjustable type, those for screw threads being adjustable. (2) A form of plate gauge used by smiths, made by cutting a series of notches of different definite sizes in the edge of a piece of plate. These notches are slipped over those portions of a flat bar which are being forged to a required size.

Gas Furnace.—A regenerative furnace (q.v.) or specifically a reheating furnace (q.v.) in which gaseous instead of solid fuel is used. The gas may be made in a portion of the furnace itself, or in a separate producer. These are essentially reverberatory furnaces, either with or without regenerators.

Gas Engine* (see Pt. 2, Basic Terminology).—The latest development in gas engines is in an immense increase in dimensions and power, due to the utilisation of the waste heat from blast furnaces. Modern gas engines may be classified as follows : (1) Engines igniting at constant pressure, with previous compression of the charge, as in the Diesel engine (q.v.) ; (2) Engines igniting at constant volume, with previous compression. The great majority of gas engines work on the four-stroke cycle, suction, compression, working stroke and exhaust, the cylinders being usually single-acting.

Gas-engine Starters.—Large gas engines are usually started by means of compressed air pumped by the engine into a reservoir when running ; or gas and air may be pumped into the cylinder by a small engine and ignited just as the larger engine has passed its in-dead-centre.

Gas Turbine.—Power unit consisting essentially of one or more combustion chambers in which a fuel, generally liquid, is burnt in combination with air which is supplied by a compressor. The gases generated are led to a turbine wheel in which they expand and do work. The compressor may be driven directly from the turbine shaft, or may be operated by a second turbine through which the exhaust gases pass. In order to obtain improved efficiency, various arrangements of two-stage turbines, compressors and heat exchangers are used. See also Steam Turbine and Turbine, Water.*

Gas Welding.—See Oxy-acetylene Welding.

Gauge.*—Measuring instrument used to control the dimensions of work during manufacture. Gauges may be divided into two classes, workshop gauges and inspection gauges. Workshop gauges are used during production and for inspection of the parts by the works inspection staff before they are handed over to the final inspection department or purchaser, who may carry out final inspection. Inspection gauges are used only for final inspection and should be so dimensioned that they will not reject any parts which lie between the prescribed limits. Solid gauges are of the GO or NOT GO type. Specialised gauging instruments may also be used, which are usually of the comparator type. See Automatic Gauging Machine, Comparator, Continuous Gauge, Dial Gauge, Gap Gauge, Optical Flat, Optical Gauge, Plug Gauge, Pneumatic Gauge, Slip Gauge, Thread Gauge, Toolmaker's Microscope, Workshop Microscope.

Gravity.*—Any unsupported body in the vicinity of Earth tends to fall vertically towards the centre of the Earth with a constant or uniform acceleration. This tendency to fall is caused by a force known as the force of gravitation. According to Newton's theory, any piece of matter attracts any other piece of matter with a force which is directly proportional to the product of their masses, and inversely proportional to the square of the distance between them. By virtue of its mass, the Earth attracts every object in its neighbourhood. This attractive force of the Earth is called gravity. The acceleration it causes is called the acceleration of gravity. See also Acceleration.

Graphite Paint.—A mixture of oil and graphite used to protect iron and steel structures against atmospheric corrosion. See also Graphite.*

Gear Box.—An arrangement which provides a series of speed ratios between the input and output shafts by means of toothed gears mounted on a mainshaft and a layshaft. The gears are slid into mesh, and the power is conveyed to the layshaft and back again to the mainshaft, with reduction of speed. To permit sliding, the mainshaft gears are mounted on serrations called *splines* (q.v.).

Gear Cutting.—Classes of machines used in gear cutting are the bevel gear planers (q.v.) and the generating gear-cutting machines (q.v.). Worm wheel hobbing machines (q.v.) and spur gear hobbing machines have also come to the fore. See also Gear Cutters.*

Generating Gear-Cutting Machines.—In these a form or templet is not used, but the mechanism of the machine, in some cases including the shapes of the tools used, produces theoretically correct curves for wheels of various pitches and sizes. These machines are still in course of development, and their use will extend by reason of the demand for perfectly formed wheel teeth for high-class machinery.

Generators.—A current generator is any apparatus for maintaining an electric current, whether mechanical, as in the case of a dynamo (this being the most common use of the term); thermal, as in a thermo-electric battery; or chemical, as in a voltaic battery.

German Silver.—Another name for nickel-silver. See also Nickel.*

Gilding Metals.—Copper-zinc alpha alloys which contain 80–95% copper, used for various decorative purposes, such as for shop fronts, etc., where, when treated to yield a rich brown colour, they are frequently misnamed " bronze." To reveal the golden colour of the gilding metals for jewellers' work and similar applications, the surface is pickled in nitric acid, or polished, to remove the copper-coloured skin which is formed during heat treatment.

Girard Turbines.—These may be either axial-flow or radial-flow. In either case the turbine is of the impulsive type, that is to say it works entirely by the reaction on the vanes due to the change of momentum of the water in passing through them. The wheel works under atmospheric pressure, the vanes being supplied with ventilating holes. The guide vanes occupy only opposite quadrants of the ring which carries them, and the water supply is regulated by rotating this ring so as to cut off completely the flow through the vanes which it covers. These turbines are usually of the axial-type.

Grade.*—Grade, or hardness, of a grinding wheel is a measure of the strength with which the abrasive grains are held in position by the bond; it should not be confused with the hardness of the abrasive grains themselves. In use a grinding wheel must wear away in order to maintain good cutting properties, therefore the ideal wheel for any particular operation is one in which the bond strength or grade has been so selected that while excessive wear is avoided, blunting of the cutting face does not occur.

Grain Growth.—At temperatures above 650° C., some crystals in iron begin to absorb adjacent crystals, and this results in an increase in the average grain size. This action is termed grain growth and tends to weaken the metal. Subsequent heating and cooling to produce a structure of small crystal grains is termed annealing. See also **Annealing.***

Gravity Feed.—A method of transferring material, especially in liquid or powder form, from one point to another without the aid of pumps or power-assisted conveyors. To achieve gravity feed, the supply source must be situated at a higher level than the discharge point. See also Gravity* and Gravity Wheel.*

Grease Gun.—Hand-operated pump for forcing grease through a nipple into a bearing for the purpose of lubrication.

Green Sand Moulding.—All common work, which constitutes by far the largest proportion done, is made in green sand. The sand mixtures used are weak and friable by comparison with dry sand mixtures, and green sand moulds are therefore not adapted to stand the enormous liquid pressure of heavy casts. Sometimes a compromise is made by drying the surface only (see Skin Drying). Also, the sand being still damp at the time of pouring, more and careful venting is necessary than in moulds made of dry sand and loam; and there is a slightly greater risk of waster castings. Thin, narrow, weak corners, and sections of sand, are also more liable to become washed away in green than in dry sands. Drying increases the cost of the work. See also Green Sand.*

Grid System or Zoning.—Sometimes called the Marginal Grid Reference System. A system in which the margins of a large drawing are numbered and/or lettered in such a way that a particular dimension or feature on the drawing can be readily located.

Grinding Wheels.—These are made of other abrasives (q.v.) besides emery, carborundum being the principal. The size of the grains of which a wheel is composed determines the " grit " (q.v.) or its degree of coarseness, the binding agent determines the " grade," or the degree of hardness. About 90 grades are made, while the sections are varied, and the methods of mounting are numerous. See also Grinding.*

Grit.*—The grit or grain of a grinding wheel refers to the size of the particles or abrasives used. The sizes are indicated by standard numbers corresponding to the number of meshes in the screen through which they will pass. For example a 36-grit will pass through a screen having 36 meshes to the linear inch.

Groin.—Arch formed by two segmental arches or vaults intersecting at right angles ; also a kind of pier built from the shore outwards into the sea, to intercept shingle or gravel.

Grommet.—A circular moulding with or without a central hole, having two flanges and a groove on its periphery. The groove helps the moulding to seat itself firmly and to remain in position when it is inserted in a hole or clamp. Most grommets are moulded from plastics, rubber or some other resilient material, and are often used to support, protect and/or insulate wires or cables passing through them. A grommet without a central hole is called a blind grommet. See also Grommet Washer.*

Grounds.—A term used in North Britain to denote the wood blocking or framing upon which a piece of mechanism is bolted—a hand winch, for example.

Gudgeon Pin.—Also termed Wrist Pin or Piston Pin. Pin connecting a piston to a connecting rod or piston rod of an engine. See also Pt. 2, Basic Terminology. See also Gudgeon.*

Guide Screw Stock.—A form of die stock (q.v. in Pt. 2, Basic Terminology). The dies are divided into three portions, one becoming the guide, the other two the actual cutters. The cutters are placed in radial slots, and are very narrow, so that they operate to the best advantage, in opposition to the rigid guide piece.

Gutermuth Valve.—Valve which consists of a single piece of sheet metal, either of steel or gun-metal, bent into a spiral except at one end for a short length, which remains flat and serves as the flap of the valve. It gives results much superior to the ordinary ring or mushroom valve.

Gutta-percha.*—Coagulated latex obtained from various types of rubber trees. Has good dielectric properties and is waterproof, rendering it an effective electrical insulator. (See Pt. 2, Basic Terminology.)

Gyroscope.—Rotating wheel able to turn freely about any axis. Such an apparatus tends to maintain itself in the plane in which it moves, and is applied in practice to stabilise the motion of a number of different mechanisms and instruments. If a torque is applied to a gyroscope tending to alter the plane of spin, it turns about an axis at right angles to that about which the torque is applied, until the plane and direction of spin of the wheel coincide with the plane and direction of the torque. This effect is known as *precession.*

Gyroscopic Stabiliser.—An arrangement for giving automatic stability. See Gyroscope.

Gyrostat.—A gyroscope (q.v.) suspended so that its freedom to rotate about one particular axis is partly or wholly removed.

H

Hack Sawing Machine.—Machine which grips work, reciprocates and feeds the blade, relieving it on the backward stroke. See also Hacksaw.*

Hand Ramming.—Ramming (q.v.) performed by hand rammers, as distinguished from that effected by means of a presser (q.v.) in a machine. Ramming is done by hand in many moulding machines, but others are fitted with a pressing plate. See also Hand Rammer.*

Hardening.*—The heating of a steel to a temperature above its critical range (q.v.), followed by rapid cooling by quenching in oil or water or by an air blast in order to retard the transformation from austenite (q.v.) to pearlite (q.v.). See also Babbitt's Metal.*

Hardening Furnace.—The old practice of heating articles to be hardened in an open fire has been abandoned when quantities are being dealt with. The furnaces, usually oil or gas fired, are provided with pyrometers for the measurement of temperatures.

Hardenite.—Steel which has been suddenly quenched at a high temperature, and which contains the maximum of carbon.

Hard Facing.—The deposition of a surface layer of harder metal on a softer base metal to provide improved qualities, such as toughness, hardness, or resistance to heat, corrosion or abrasion. Hard facing is also used to build up worn surfaces, at the same time restoring the original qualities or providing improved resistance to wear or deterioration. The metal is usually deposited by fusion welding but the term can also be applied when the surface is built up by electrolytic plating.

Hard Iron.*—Cast iron comparatively low in silicon, such as solidifies with a high proportion of combined carbon, and on that account possesses good wearing hardness and strength making it suitable for cylinders and other cases subject to hard wear.

Hardness Scale.—The relative hardness of materials which differ greatly in this respect, is sometimes tested roughly by making a scratch upon the smooth surface of the piece tested, with the sharp point of a mineral selected for comparison of the results. The materials chosen for the purpose form the hardness scale. According to Mohs' scale they are, in ascending order of hardness: (1) Talc; (2) Gypsum; (3) Calcite; (4) Fluorite (Fluorspar); (5) Apatite; (6) Orthoclase (Feldspar); (7) Quartz; (8) Topaz; (9) Corundum (Sapphire); (10) Diamond. Hardness varies on different faces of a crystal and in some cases in different directions on any one face.

Hardness Tests.—The most widely used methods for testing the hardness of metals are those known as indentation tests, in which a hardened steel ball is employed for annealed, normalised or toughened metals and a diamond indentor for hardened tool steels and case-hardened work. See Brinell Test, Vickers Hardness Test, Rockwell Hardness Test, Firth Hardometer and Sceleroscope. Abrasion tests are also used as a measure of hardness: see Abrasion Test, Hardness Scale.

Hatching.—In engineering drawing, the practice of shading by means of lightly-drawn unbroken lines, usually equally-spaced at an angle of 45° to the horizontal, to indicate broken or cut surfaces. The lines are also known as Sectioning Lines.

Headers.*—Vessels, generally of a sinuous form, into which the end of the tubes in water-tube boilers enter and discharge their steam.

Heading Machine.—Power press in which bars are upset to produce the heads of bolts, rivets and spikes. See also Heading Tool.*

Head-race.—The channel which supplies the forebay in a turbine plant.

Heater.*—A cubical lump of wrought iron or steel made red or white hot and laid upon a portion of plated work which requires to be slightly set or bent, and which cannot be taken to the fire. The practice is objectionable though unavoidable in many instances. Also a lump of cast iron used in the foundry for skin-drying moulds. It is made red hot and suspended in the mould or laid above it on suitable supports.

Heat Treatment.—A succession of heating and cooling cycles applied to a metal or alloy in order to obtain the desired properties, such as hardness, ductility, grain size, etc. See Annealing, Hardening, Solution Treatment, Sub-zero Heat Treatment.

Heliarc Welding.—See Shielded Arc Welding.

Helical Gear Cutting.—Single helical gears are cut as are spiral gears with single rotary cutters, or with hobs. Double helicals are cut in the same way if the teeth are staggered and divided at the apex. If they are continuous they are cut with end mills. Double helical bevels are cut with end mills. The spiral curves are obtained through change gears. Continuous double helical gears may be generated on a Sunderland gear planer. See also Helical Gear.*

Henry.—The basic unit of measure for electrical inductance. A coil has an inductance of one henry when a change of current at the rate of one ampere per second causes a back-e.m.f. of one volt.

Heterogeneous.—A term used in metallurgy to indicate that the composition of a piece of metal is not uniform, but varies in quality and crystal structure. The opposite of Homogeneous.*

Highest Useful Compression Ratio.—The highest compression ratio at which a fuel can be used in an efficiently-designed internal combustion engine without detonation occurring at any mixture strength, and with any ignition timing. The ratio is ascertained by testing the fuel in a variable compression engine with the aid of a knockmeter. See Knockmeter.

High-speed Tool Steels.—Alloy steels, variously compounded with chromium, tungsten, molybdenum, silicon, etc., which are hardened at a white heat, and employed to most advantage at temperatures at which the cuttings turn blue and smoke. They are used for tools in lathe and planer, for milling cutters, drills, etc.

Hip- or Hipped Roof.—One that slopes four ways, thus forming angles called *hips.*

Hobbing.—Cutting the threads of worm wheels, dies, or chasers with a hob or master tap in a lathe. Also cutting the teeth of worm, spur and spiral gears with hobs in machines designed specially for this work. See also Hob.*

Hobbing Machines.—In these, spur and spiral gears as well as worm wheels are cut by a hob, the same hob serving for each type of gear by altering the angle of inclination of the thread. These though recent have largely taken the place of machines using single rotary cutters. The blank rotates constantly as the hob is fed, until the teeth are finished all round. The term Hobbing is also applied to a special process of die-sinking under pressure.

Hodograph.—Curve drawn through the extremities of the vectors radiating from a pole, and which represent the velocities of a body at successive intervals of time. It greatly simplifies the study of curvilinear motion, especially when irregular. The velocity along the hodograph gives the acceleration of a body along its path.

Hog-back Girders.—These girders have a curved upper boom, the lower boom being usually horizontal, and the depth increases towards the centre. The moment of resistance to bending is thereby kept more nearly uniform than when the booms are parallel, with a saving of material and weight. (See Fish-belly girders in Basic Terminology.)

Holding-plates or Anchors.—Broad plates of iron sunk into the ground and generally surrounded by masonry, for resisting the pull of the cables of a suspension bridge and for similar purposes.

Hole-grinding Machines.—A large group, in which the grinding spindle is supported plumb with the axis of the work being ground instead of being carried by the head of a wheel used for external grinding. In most machines the spindle is horizontal, in a few it is disposed vertically. The growth of these machines is due mainly to the increasing practice of hardening bushes and bearings generally.

Honing.—Process for finishing the bores of engine cylinders, etc., by rubbing them with abrasive stones. The honing machine resembles a vertical drilling machine. The stones are carried in a holder attached to the spindle of the machine, and radial pressure is applied either by springs or by a positive feed. The spindle is also given a reciprocating as well as a rotary movement. See also Hone.*

Hook Gauge.—Instrument used in measuring the flow of water over a weir and in similar cases. To determine the level of the water with precision, a hook is provided which slides on a fixed scale, and is adjusted until the point of the hook just touches the water surface. The height of the hook is then referred to a bench-mark by means of a level.

Hooke's Law.—Extension is proportional to load, which must be qualified by adding, within the limit of proportionality of the material.

Horizontal.—A term used to indicate the position of anything which is on a level with, or parallel to, the horizon, or at right angles to the vertical or to a plumb line. Thus, a horizontal plane. See also Parallel Lines,* Plan Section,* Inclined Plane* and Perpendicular.*

Horses.—The sloping timbers which carry the steps of a staircase.

Hot Tears.—A term used in foundries to describe small cracks, either on the surface or in the body of the metal, which may result when the metal has been prevented from shrinking while it is cooling in the mould. When the stresses set up in this way are severe, the casting may be fractured, or even broken, when the mould is opened.

House-to-House System.—Step-down transformers for electric transmission may be arranged on the *house-to-house* or *distributed* system, or on the *sub-station* system. In the former each building is fitted with one or more step-down transformers; in the latter, the transformers are placed in suitably situated sub-stations, and a number of buildings are supplied by one transformer, or from a number, as the case may be.

Humpage's Gear.—An epicyclic train of wheels frequently used in lathes, capstans and electrical machinery, for reducing the speed of a shaft to the lower speed of one it drives.

Humphrey Pump.—See Internal combustion pump.

Hunting.—With a fully isochronous governor any small increase of speed causes the sleeve to move to its extreme position for a minimum opening of the valve. The driving torque is now too small and the speed of the machine falls below normal, so that the sleeve moves to the full open position, causing the speed to rise above normal. The speed therefore rises and falls continuously. This effect is known as *hunting*.

Hydraulic Bolt Forcer.—Machine for forcing coupling bolts, pins, etc., in and out of machines. It consists of a hollow steel sliding ram, the ends of which project through the front and back ends of the cylinder in which the ram moves. Inserted in the hollow ram is a steel drift which passes through the centre of the ram, having a head on one end and a shoe for the bolt at the other.

Hydraulic Brake.—(1) This brake in its simplest form consists of a cylinder fitted with a piston and rod, and fitted with some liquid, usually oil, water or glycerine. The two ends of the cylinder are connected either by small holes in the piston or otherwise. The energy of a heavy body such as a moving train, or of a gun during its recoil, is then absorbed without the use of spring buffers. (2) The term is also applied to braking systems in which brake shoes or discs are actuated by pistons working in hydraulic cylinders.

Hydraulic Dynamometer.—Mechanical device for absorbing and measuring the energy developed by an engine or motor. It consists of a double disc fixed to the power shaft and carrying on its outer face a series of narrow pockets inclined at 45° to the axis of the shaft. A casing carrying similar discs inclined in the opposite direction surrounds the disc and carries a graduated lever which carries the brake load. Water is supplied to the pockets and the water being thrown outward by centrifugal action is projected forward into stationary pockets in the casing, from which it enters other moving pockets, and so on. The change in angular momentum produced is measured by the resisting movement of the brake load.

Hydraulic Efficiency.—The theoretical hydraulic efficiency of a turbine is the ratio between the work done by the turbine per pound of water and the available head. The actual efficiency is, however, much less than this, because of the lost energy in the supply pipe, in the wheel and suction tube. The hydraulic efficiency of a centrifugal pump is the ratio of the actual lift or manometric head to the head generated by the pump.

Hydraulicising.—This process consists in using a jet of water under high pressure to disintegrate the soil and wash it away. The water, discharged at a very high velocity through nozzles of about six inches diameter mounted on ball-and-socket joints, is played on the face of earth to be removed, and the debris is carried away by sluices.

Hydraulic Joint.—Form of joint used to connect piping which is subject to water pressure. The flanges are checked into one another by a male and female fitting, terminating in bevelled edges or shoulders, and a ring of indiarubber, screwed up and pressed between the bevelled faces, renders the joint water-tight.

Hydraulic Mean Depth.—Ratio of the cross-sectional area of a stream to the length of the wetted perimeter.

Hydraulic Mean Radius.—Equivalent to hydraulic mean depth when referring to pipe flow.

Hydraulic Mining.—The breaking down of auriferous soil by a jet of water directed against it under a head of several hundred feet.

Hydrogen Embrittlement.—Embrittlement of steel caused by the absorption of hydrogen during pickling and some plating processes. Ageing for 24 hours or heating at 200° C. will often correct this fault. See also Hydrogen.*

Hydrometer.—Instrument used to measure the density of liquids. Employed to test the density of petrol and heavier oils and to test the specific gravity of accumulator electrolyte.

Hydrostatic Weighing Machine.—One in which a cylinder of oil is pressed by a piston, from the rod of which a load is suspended, and which being displaced gives corresponding readings on the dial of a pressure gauge.

Hygrometer.—An instrument for measuring the relative humidity of the air. The usual type is known as a " wet-and-dry bulb " thermometer. See also Thermometer.*

Hypoid Gear.—Form of spiral bevel gear in which the axis of the pinion is above or below that of the gear.

Hysteresis.—(1) The resistance of iron and steel to magnetisation—its magnetic inertia, or the lagging of the magnetic flux behind the magnetic force producing it. (2) Mechanical hysteresis is the lagging in the relation of strain to stress, as when a tie-bar is alternately loaded and unloaded, the length under any intermediate load being a little greater during unloading than during loading, causing a loop on the stress-strain diagrams. This slight imperfection in elasticity is negligible in practice.

I

Ice.—Water in the solid state. Pure water freezes and becomes ice at 0° C. (32° F.), at normal atmospheric pressure; but impurities mixed with the water lower this freezing point. Melting ice absorbs more heat than does any other solid. See also Freezing Point.*

Idle Time.—Refers to time lost when for various reasons machine tools are out of use, vehicles or vessels unable to travel, etc.

Ignition System.—In petrol-driven engines, the arrangement used to ignite the mixture of air and petrol. The electric current, generated by a magneto or by a battery and coil, is carried by high-tension leads to the distributor (q.v.) which delivers the current to each sparking plug in turn. In each case the instant of sparking is governed by the contact breaker, and the timing can be altered by moving it in relation to its drive. See also Ignition* and Igniter.*

Ignitron Valve.—Valve used in automatic timing controls where short timing periods are necessary and where a contactor or relay is unsuitable. The ignitron valve is essentially a mercury arc rectifier fitted with an igniter or control electrode. To render the ignitron conductive the igniter circuit is controlled by thyratron valves (q.v.). When the igniter is in circuit, a hot spot is created in the mercury pool cathode, starting the conductive period and allowing an arc to form between the anode and cathode. When the igniter circuit is broken, the valve continues to conduct until the current wave approaches zero, when the arc is extinguished. The valve then ceases to conduct until the igniter circuit is again energised.

Impact Extrusion.—A method of shaping material, usually metal, by squeezing it out between the wall of a die and a punch, so producing a thin-walled, hollow container. Tooth-paste tubes and such-like items are generally manufactured by this process.

Impact Testing.—The failure of materials used in high-speed machinery under repeated impulses, when such have shown themselves satisfactory as regards strength and elongation in a static tensile test, has led to the use of special impact testing machines of various kinds, the most usual consisting of a pendulum having a tup with a striking edge at its centre of percussion. The test pieces are 2 ins. long, $\frac{3}{4}$ in. thick, and $\frac{3}{8}$ in. broad, with a V nick of 60° in each. They are held in a vice, and the hammer head swings as a pendulum from a definite height, strikes the test piece, fractures or bends it, and rises on the other side to a height which is recorded by a pointer moving over a scale. The difference in height of the initial and final positions of the hammer multiplied by the weight of the hammer, measures the energy absorbed by the blow. The scale is graduated directly in foot pounds of work done. See also Impact.*

Imperfect Frame.—In the case of a framed structure, an *imperfect frame* is one which has less than the requisite number of members required to maintain equilibrium apart from the rigidity of its joints. The stresses in them cannot therefore be found. (See Perfect Frame.)

Impost.—The upper part of a pier from which an arch springs.

Impulse Turbine.—A turbine which acts by the reaction due to the change of momentum of the water in passing round the curved vanes of the runner. It works under atmospheric pressure as in the case of the Pelton wheel and the Girard turbine in contradistinction to a pressure turbine, which is submerged, or if not submerged is fitted with a suction tube which discharges under water.

Inching.—Adjusting the slide of a machine tool by small increments.

Indeterminate Frame.—An indeterminate frame is one which has redundant members, so that the stresses in it cannot be found by purely statical analysis, but require the elasticity of its members to be considered. It is said to be statically indeterminate of the first, second, etc., degree, according to the number of redundant members it contains, for each of which an elastic equation is required.

Index Centres.—The head, and the tail stock between which work is carried to be pitched or indexed. They are plain, or universal in type, and are used on milling and gear-cutting machines.

Indexing.—Dividing the circle for pitching, for purposes of milling, fluting and gear cutting. When a spiral movement is imparted to the work through a lead screw and change gears, and a swivel table, the indexing is universal in character.

Indicator.*—Device for magnifying minute measurements when testing machines, and in machining operations. (See Dial Gauge.)

Indium.—A silvery-white metal, more malleable than lead and resistant to atmospheric corrosion at normal temperatures, which is used in the manufacture of bearings to improve their strength and hardness. Alloys containing indium offer increased resistance to corrosion, and possess good anti-friction properties. Indium is also used in electroplating.

Induced Draught.—Artificial draught produced by means of an exhausting fan located in the base of a chimney.

Induction Coil or Ruhmkorff.—Consists of a core of soft iron wire comprising a comparatively few turns of fairly thick insulated copper wire. This constitutes the *primary* coil, which is covered with insulating material such as paraffin wax, and round this is wound the *secondary* coil, which consists of a very large number of turns of fine well-insulated wire. The ends of the secondary coil are connected to a spark gap such as that between the electrodes of the sparking plug of an internal combustion engine. The object of an induction coil is to produce in the secondary circuit an induced electromotive force (e.m.f.) higher than that in the primary circuit when the current flowing through the primary winding is interrupted, so that the current jumps the air gap and produces a spark.

Induction Heating.—Method of heating electrically conductive materials or materials enclosed in conductive containers, used for melting, heat treatment, soldering, brazing and sintering, in addition to other specialised applications. The work is placed within a coil which carries a high-frequency alternating current. The alternating flux generated in the work quickly raises the temperature to the required figure.

Induction Valve.—The valve by which the flow of gases from the carburettor to the cylinder of the motor is timed and regulated ; usually operated from the camshaft.

Influence Line.—A line diagram which shows how some specified effect or *influence* undergoes change when a load (usually taken as a unit-load) traverses a girder or other structure. Thus we may have an influence line for shearing force, for bending moment, for the horizontal thrust of an arch, or other effects.

Ingot Iron.—Mild steel, low in carbon, which has been prepared by the open hearth, or the Bessemer processes; a term employed to distinguish it from weld iron or wrought iron. See also Ingot.*

Ingot Saw.—Hot iron saw specially constructed for sawing hot ingots. These saws are made several feet in diameter, for cutting ingots of several inches square.

Ingot Tilter.—Machine by which ingots are tilted or turned over through an angle of 90° between each pass of the rolling mill.

Injection.*—Admission of fuel under pressure to the combustion chamber of a diesel engine, or to the induction manifold or individual cylinder of a petrol engine. (See also Pt. 2, Basic Terminology).

Injection Moulding.—A process in which softened thermoplastic material is forced from a heated cylinder and injected into a mould.

Injector.*—Spring-loaded valve through which fuel is injected into the combustion chamber of a diesel or petrol engine. (See also Pt. 2, Basic Terminology).

Injector Hydrant.—Since the chief loss of energy in the ordinary jet pump (q.v.) is due to impact on collision of the two jets, it was suggested that this loss might be reduced by diminishing the velocity of the high-pressure water, or increasing that of the low-pressure water, in stages. High-pressure water from the hydraulic power supply mains, therefore, is injected into the low-pressure town main water, giving a larger discharge and lifting it to a much greater height. This hydrant is specially adapted to fire-extinguishing purposes, whenever high-pressure water is available.

Inspection.*—A term used in method study to indicate that a check has been or is to be made for quality and/or quantity. The requirement for inspection will be recorded on a chart by means of an "activity" symbol. See also Symbol* and Time and Motion Study.

Instantaneous Centre.*—See Virtual Centre.

Insulation.—The covering of electrical conductors with materials which offer a very high resistance to the passage of electricity ; or the isolation of conductors from the earth.

Insulator.—Any non-conducting substance ; synonymous with dielectric.

Intensifier.—Device frequently employed in place of the hydraulic accumulator, for converting a low water-pressure into a higher. The water at low pressure operates a piston in a large cylinder, which in turn operates a ram of smaller diameter in a smaller cylinder. The areas of the two cylinders are proportional to the difference in the low and high pressures required.

Intensifying Screen.—In direct exposures to X-rays, less than 1% of the radiation affects the photographic film ; intensifying screens are therefore utilised to increase the photographic effect. A fluorescent intensifying screen has the property of emitting visible, actinic rays from areas which are activated by X-rays. These rays are capable of affecting a photographic film held in close contact with the fluorescent screen, but there is necessarily some diffusion of details of the X-ray image. Where no such diffusion can be tolerated, a useful, but smaller, increase in exposure effect can be obtained by the use of lead-foil intensifying screens, provided the kilovoltage is high enough. (See Radiology.)

Interchangeable Gears.—Gears, the teeth of which are so designed that any others of any number of teeth, if on the same pitch, will mesh together correctly. This can only be secured by designing the teeth on a common basis. The Odontograph scale of Professor Willis effects this result by taking a generating circle of the same diameter as the radius of the smallest wheel of a set (of 11 or 12 teeth) and constructing all the teeth on that basis. In any case in interchangeable gears, the roots of the small pinions will be weak and those of the larger wheels excessively strong. Pinions having a diameter smaller than the one which is taken as the base, will have flanks undercut. In involutes interchangeability is secured by a constant angle of pressure $14\frac{1}{2}°$, 18°, 22°, etc. The higher the angle the less is the undercut in small pinions.

Interference Fit.—Condition in which there is a negative allowance (q.v.) causing interference between the largest permissible hole and the smallest permissible shaft, the shaft being larger than the hole. (See Force Fit).

Intermediate Gear.—Gear which is interposed between power applied and the work done, for the purpose of obtaining advantage.

Intermittent Jet.—A type of ram-jet engine in which the pressure of air in the intake forces open a valve. Air entering the combustion chamber is sprayed with fuel, and ignited as soon as enough is present to support combustion. The pressure set up by combustion then forces the intake valve to close, and the hot expanding gases escape at high speed through the exhaust pipe. As soon as pressure in the combustion chamber is again lower than that outside, the ram effect forces open the intake valve again, and the operation is repeated. By this means, the compression due to forward speed is increased by pressure waves within the unit.

Internal Combustion or Humphrey Pump.—Pump which acts by the explosion of a mixture of gas and air above the surface of the water in a closed vessel having an outlet valve. The pressure developed by the explosion forces the water through the outlet valve, which then closes, and a new supply enters the combustion chamber and the cycle is repeated. Such pumps are particularly suitable up to lifts of 150 ft. The explosions occur at the rate of about 15–30 times per min.

Invar.—Nickel-steel alloy containing 36% of nickel. It has a negligible amount of expansion with change of temperature and is used for standard measures of length, surveying tapes, precision levels and staffs for same and in other cases where expansion and contraction is an important source of error.

Invert.—An inverted arch frequently built under openings, in order to distribute the pressure more evenly over the foundation.

Investment Casting.—A precision casting process employing the " lost wax " technique. Although used for many years in dental mechanics and for casting small statuary, investment casting has only comparatively recently been applied to mechanical engineering ; it is valuable in casting complex shapes in materials which are difficult to machine. A wax pattern is cast in a split master die, prepared from a master pattern ; wax gates and risers are separately cast and joined to the pattern. A finely powdered refractory material is applied to the wax pattern, which is then "invested" by coating it with further refractory material, mixed with a binder and water to a creamy paste. When the investment has been dried in an oven, at a temperature below the melting point of the wax, the wax is melted out and the resulting mould is fired to harden it. The molten metal is then poured into the cavity of the hot mould. A recent development is the use of mercury as a pattern material. (See Mercast Process.)

Isochronous Governor.—In cases where the speed of rotation of a machine is required to be constant, some arrangement is required by which the *height* of governor is kept the same for all positions of the sleeve. Such a governor is said to be *isochronous*. See also Isochronous.*

Isometric Projection.—Method of drawing sometimes used by architects and engineers, which enables objects to be drawn in false perspective. All horizontal edges of planes are drawn at 30° to the base of the paper, and all vertical edges at 90° to it. The result combines the plan and elevation in one and the same view in a simple manner and to scale.

Isotropic.—A material is said to be isotropic when it is equally elastic in all directions. Some solids, such as crystals, exhibit different physical properties in different directions, and these are said to be *œolotropic* or *anisotropic* (q.v.).

Item.—A term used to describe any separate article or part. It may be either a complete article such as a lathe chuck or chuck key, or a component of an assembly (e.g., a piston ring or a piston for an air compressor). See also Item Number, Chuck,* Key,* Ring,* Piston* and Piston Ring.*

Item Number.—Parts on an assembly drawing (q.v.) are indicated by an item number, not by the drawing number. The item number is usually enclosed in a circle outside the outline of the drawing, with a pointer or leader line connecting it to its detail. In the Item column of the drawing, a series of numbered items are shown, and these numbers correspond with the numbers shown against every part on the drawing. (See Drawing Number.)

Jack Rafters or **Common Rafters.**—These are small rafters laid on the purlins of a roof, to support shingling laths, etc.

Jacob's Ladder.—Either (1) A rope ladder fitted with wooden treads or steps; or (2) An endless chain to which buckets or containers are fitted for use as an elevator or carrier. See also Endless Chain.*

Jambs.—The sides of an opening through a wall, etc., as in a door, window or fireplace jambs.

Jamb-linings.—The facing of woodwork with which jambs are covered and hidden.

Jenkins' Bend Test.—A test in which the ductility of a flat bar, usually of steel, is ascertained by subjecting it to an alternating bend test through 180°. The test is usually performed on a bending machine which has a constant radius of bend. See also Bending Machine* and Test Specimens.*

Jet.*—Calibrated tube or drilled plug used to meter the flow of liquid or air. See also Pt. 2, Basic Terminology.

Jet Propulsion.—A method of propulsion (usually for an aircraft) by re-action. The power unit obtains oxygen from the air, and a fuel turbine jet discharges hot gas through a tail pipe and nozzle. The thrust so obtained propels the vehicle or aircraft.

Jet Pump.—Pump used for lifting water against a head. Water from a high-pressure supply is led through a converging passage, its pressure diminishing as its velocity increases, in accordance with Bernoulli's law (q.v.). It then discharges into the delivery pipe through a diverging passage. At issue from the nozzle, the pressure is below that of the atmosphere and water from the suction pipe rushes in to join the high-pressure jet, the mixture being carried forward into the discharge pipe. This type of pump is adapted to continuous pumping and for drainage operations, where a fair pressure supply is available and where the quantity of water to be lifted and the working head are small. The principle of action is the same as that of the steam injector for forcing the feed water into a boiler against the steam pressure.

Jetty.—A pier, mound or mole projecting into the water, as for a wharf, pier, etc.

Jig.—An appliance which locates and holds a piece of work, and guides the tools which operate on it. A *fixture* holds, without controlling the tools.

Jigging.—The practice of tooling work held in jigs, adopted in the inter-changeable system, when articles are tooled in quantities.

Jockey Pulleys.—Used to change the direction of a belt drive, when the planes of the driving and driven pulleys are inclined to one another.

Jockey Weight.—Weight which is slid along a lever, in a weighing or testing machine, for purposes of precise adjustment.

Joggled (see Dict. of General Terms).—Also, stiffeners are said to be *joggled* or *crimped* when they are bent over the flange angles of a web-plate girder, in order to avoid the use of packing strips. See also Joggle.*

Johannson Gauge Blocks.—See Slip Gauge.

Joists* (see Pt. 2, Basic Terminology).—*Binding* joists are for carrying *common* joists. The common joists are then called *bridging* joists.

Ceiling joists are small ones under roof trusses or girders, for sustaining the plastered ceiling only.

Jolt-Ramming Machine.—A machine used in a foundry to pack sand in a moulding box. Pattern plate, moulding box and sand are together lifted automatically and then allowed to fall on to a rigid anvil, the process being repeated (usually by pneumatic operation) some 300 times a minute. The sudden arrest of the downward movement of the sand effectively rams it round the pattern. See also Moulding Machine.*

Jolt-Squeeze Machine.—A machine similar in action to a jolt-ramming machine (q.v.), but possessing additional mechanism which lifts the moulding box after every cycle of jolting action and presses it against a squeeze-plate under a pressure of several tons, to reinforce the ramming process.

Jominy Test.—A test to show how a steel will harden at different cooling rates between about 500° C. per second and 2° C. per second. The test consists of heating a specimen of steel, and then water-quenching one end of it under controlled conditions. The hardness at varying distances from the quenched end is then measured.

Jonval Turbine.—This type of hydraulic turbine is axial flow, and the water from the guide vanes discharges into the radial vanes of the runner in a vertical direction. The simplest of these turbines consists of one ring of horizontal moving vanes, but a later type has several of these rings on the same axis, one above another, the power being regulated by closing one or more rings completely by means of a *register gate* (q.v.).

Jump Weld.—A butt-welded joint, formed by bringing the ends of a bar together, and jumping them up on the anvil, or with the hammer.

K

Kaplan Water Turbine.—A water turbine of the propeller type in which the pitch of the blades may be varied in relation to the load, thereby increasing efficiency over a large load-range. See also Turbine, Water.*

Kater's Pendulum, Captain.—A bronze bar having two knife-edges, one fixed rigidly to it and the other movable. The latter can be adjusted so that the bar oscillates about it in exactly the same time as it does about the fixed knife-edge. The distance between the knife-edges is then the length of the *ideal* simple pendulum (q.v.). It is used to determine the acceleration due to gravity (*g*) at any place.

Keeper.*—A bar of soft iron used to connect the poles of a magnet, in order to prevent it from losing its magnetism.

Keepers.—The pieces of metal or wood which keep a sliding bolt in its place, and guide it in sliding.

Keller Furnace.—An early type of electric furnace for smelting iron. Heat is generated by passing an electric current through the charge sufficient to cause arcing between the charge and electrodes inserted for the purpose into the furnace. See also Charge,* Smelting* and Electric Furnace.

Kelvin Scale or **Absolute Scale** of temperature has no negative temperatures. Thus the lowest temperature that can be reached (the absolute zero of temperature) is recorded as 0° on the Kelvin scale; while it is $-273 \cdot 1°$ on the Celsius or Centigrade scale, and $-459 \cdot 4°$ on the Fahrenheit scale. See also Absolute Zero* and Absolute Temperature.*

Kelvin's Law.—Law deduced by Lord Kelvin to determine the most economical size of copper wire for transmission of power, but was afterwards modified to read : " the most economical area of conductor is that for which the annual cost of energy wasted is equal to the annual interest on that portion of the capital outlay which is proportional to the weight of the conductors used."

Kennedy Water Meter.—Positive type of meter, the volume of water flowing being measured by continually filling and emptying a cylinder of known volume. It works automatically and the discharge is automatically registered.

Kern.—See Core.

Key Gauges.—Plate gauges, both male and female, used for checking the width of keys and key seatings, the one being notched to embrace the sides of keys of a given size, the fellow one fitting closely within this notch, and used to check the corresponding width of key-way. See also Key.*

Key-seat Clamps.—Clamps made of high-grade steel which have been designed for the purpose of transforming an ordinary steel rule into a key-seat rule (q.v.) or box square. Generally used in pairs and capable of being attached or detached from a rule instantly, they form a good substitute for a more expensive tool.

Key-seat Rule.—A rule similar in shape to a box square or angle bar, whose edges or flanges are graduated for measuring. It is used for drawing lines parallel to the axis of round bars and tubing, usually in

3*

the marking-out of key-ways on shafting. It is normally made of high-carbon steel, hardened and tempered.

Key-way Seating Machines.—These are either of the vertical type, with the advantage that the wheel to be key grooved is laid upon a level table, or they are horizontal. The key groove is produced by a cutter held in a bar that is reciprocated. These machines offer several advantages over the practice of key-way cutting in the slotting-machine, both in regard to economy in time, and of uniformity in shape, neither are their utilities limited to key seating.

Kibble.—The bucket used for raising earth, stone, etc., from shafts or mines.

Kilowatt.—A unit of power equal to 1,000 watts. It is equivalent to 1,000 joules per second, or 737 foot-pounds per second, or approximately 1·34 horse-power.

Kilowatt-hour.—The standard unit of electrical energy, representing the energy derived from the consumption of 1,000 watts of power in one hour. Commonly abbreviated to kWh. The term can also be used to express the work done in one hour at a rate of one kilowatt.

Kinematic Chain.—A closed kinematic chain is composed of a series of links, so paired that the movement of each link is absolutely constrained in relation to all the others in the chain. When a single link of a kinematic chain is fixed, a mechanism is produced.

Kinematic Design.—Method of design which ensures that all components co-join in such a manner that the motion or constraint of a prescribed point follows or maintains a defined and intended path or position. The principle is often followed in design of precision instruments. See also Kinematics.*

Kinematic Elements.—Signifies certain profiles or forms, or methods of constraint imparted to bodies in mutual engagement or connection, by means of which their relative motions are absolutely constrained in all possible positions of the bodies. These elements must of necessity always occur in pairs, and they are the fundamental elements in all mechanisms. The most important are the turning pairs, the sliding pairs and the twisting pairs. The type of the first is the revolving pin with collars, that of the second the sliding block or guide block, that of the third the common screw and nut.

Kinematic Links.—Bodies which are essential and elementary parts of all mechanisms, as rods, levers, etc., which are connected together by two or more kinematic elements (q.v.) by which their relative movements are rigidly constrained.

Kinetics.—Dynamics is divided into (a) *Kinematics*, which discusses problems of pure motion without regard to the masses moved or to the forces which produce the motion ; (b) *Kinetics*, in which the masses moved and the forces acting must be taken into account. See also Kinetic Energy.*

Kite Mark.—A registered certification trade mark owned by the British Standards Institution. Manufacturers may be licensed to use the mark on their products if they agree to follow a system of inspection, sampling and testing which conforms to the requirements of the relevant British Standard. See also Marks.*

Knife Tool.—A lathe finishing tool, which is used for side cutting, and made in right- and left-hand forms.

Knock Off Joint.—A joint used in the rods of deep-well pumps. The jointed ends of the rods are enlarged to a square section and scarfed and notched to fit against one another, and are confined by a clasp or bridle embracing them. The joint is tapered lengthwise, and the hole in the clasp is tapered to correspond, so that the tendency is always for the clasp to tighten round the joint.

Knocking* (see Pt. 2, Basic Terminology).—Audible hammering sound produced by loose parts or by detonation in the cylinder of an engine.

Knockmeter.—An electrical meter used in conjunction with an apparatus known as a " bouncing-pin " for measuring the intensity of detonation of petroleum fuels. It consists of a heater-coil or resistance, a thermo-couple, a graduated scale and a recording needle or pointer. The " bouncing-pin " assembly consists of a plain metal tube or body screwed vertically into the cylinder head of a variable compression engine, at approximately the point where detonation occurs. At the lower end of the tube is a thin steel diaphragm on which rests a steel rod or pin about $\frac{1}{4}$ in. diameter. This pin protrudes slightly from the upper end of the tube. A leaf spring presses on top of the pin, and immediately above it is a second leaf spring. The two springs are set a pre-determined distance apart, and each has an electrical contact in circuit with an electrical generator and the knockmeter itself. Nor-mally, no current flows in the circuit; but when detonation occurs, the resulting increase of gas pressure causes the diaphragm to be deflected, the degree of deflection depending on the intensity of the detonation. A large enough deflection causes the pin to bounce up and cause the lower contact point to touch the upper one, thus completing the circuit. The more intense the detonation, the longer the circuit is closed and the more current flows. The heating effect produced by this current in the resistance in the knockmeter is proportional to the value of the current and the time for which it flows, and so proportional to the intensity of detonation. The thermocouple circuit is arranged so that the record-ing pointer registers the degree of knock or detonation against a graduated scale. See also Highest Useful Compression Ratio.

Knurling.—Generally now used instead of " milling " when applied to the edge of the head of an adjusting screw, etc., as a knurled head, or a knurled edge. See also Knurling Tool.*

Krypton.—A rare gas present in the atmosphere (from which it is ob-tained by liquefaction) used in certain types of gas-filled electric lamps to retard the burning rate of the filaments and to withstand vibration.

L

Labyrinth Packing or Seal.—Consists of a series of grooves cut in the piston of an engine, etc., so that any escape of steam through the clearance space between the piston and cylinder expands in succession into these grooves with loss of velocity and gain of pressure at each groove, so that the total difference of pressure between the two sides of the piston is reduced and the leakage thereby diminished.

Lamé's Formula.—This formula, due to Lamé, aims at calculating the stresses in thick cylinders of hydraulic presses, wire-wound guns, shrunk-on rings, etc. It is to be preferred to Barlow's formula (q.v.).

Lamina.—A general term covering (in its most commonly used sense) any object or pattern which has been cut from a flat sheet of material of uniform thickness, such as cardboard, plywood, steel, etc. See also Lamellar* and Laminated.*

Lattice or Space Lattice.—Regular arrangement of the atoms forming structural units of the crystals of a solid metal or alloy. (See also Pt. 2, Basic Terminology.) See also Lattice.*

Laundry Machinery.—An immense group designed for washing, drying and ironing.

Lay.—A term used in connection with surface roughness or finish to indicate the direction required for toolmarks and/or scratches. The lay, usually unimportant though sometimes required for functional reasons, is indicated on engineering drawings by a note or by other suitable means such as arrowheads or hatching.

Lead Tempering.—Method of tempering sometimes adopted for armour plate. The advantage is uniformity of texture. For heating many classes of cutlery preparatory to hardening, a bath of melted lead or of an alloy of lead with tin is used. The advantage is that the temperature of such a bath is uniform, and therefore more reliable than the test of colour. By varying the proportions of lead and tin the temperature of the bath can be varied within a range of more than 100° C.

Leaf Spring (sometimes also called *Carriage Spring*).—A combination of two or more strips or leaves of specially-tempered resilient metal, clamped or bolted together in such a way as to be capable of acting independently of one another, designed to be mounted between the axle and frame of railway rolling-stock and other vehicles for the purpose of minimising road shocks. See Spring* also Laminated Spring.*

Least Angle of Traction.—The least angle which a force acting upon a body will make with the normal on the plane on which it tests, in order that the body shall commence to slide against friction. It is equal to the friction angle, and the coefficient of friction for the two surfaces in contact is the tangent of this angle.

Least Work, Principle of.—This important principle states that whenever possible, Nature always effects its purpose with the least expenditure of energy. For example, in a framed structure which is statically indeterminate, the internal stresses will so adjust themselves that the total strain-energy in the members is the least possible. In order to find the stress in a redundant member, therefore, we find the differential coefficient of the total strain-energy with respect to the force in this member and equate the result to zero, in accordance with the method of maxima and minima values.

Leftward Welding.—Method of welding in which the blowpipe is used in the downhand position and is moved from right to left along the seam with a slight zig-zag or weaving motion, the welding wire being moved progressively ahead of the blowpipe.

Leslie's Cube.—A cubical metal container the vertical sides of which have different surfaces. It is used in conjunction with a thermopile, a galvanometer, a thermometer, and water to show that the radiating power of a surface depends not only on the temperature but also on the nature of the surface. See also Thermometer.*

Lever Crank Chains.—Kinematic links which are connected by pairs of mechanical elements in such a way that, one pair being fixed, all the others have movement or motion of rotation. Illustrations occur in the eccentric and crank.

Licker.—An attachment which picks up a drop of oil from a fixed lubricator and drops it on an oil supply of a moving part.

Lifting Cylinder.—The cylinder of an hydraulic crane, which is used for lifting the load, as distinguished from the turning cylinder (q.v.). There are in the largest cranes three such cylinders, by means of which, when worked separately or in unison, different degrees of power can be exerted.

Lifting Magnet.—Electromagnet suspended from a crane hook, which picks up masses of iron and steel when current is switched on through a flexible cable.

Lighter.—A scow, raft or other vessel, used for unloading vessels out from the shore.

Light Railways.—Any railways of gauge narrower than the standard, mostly from 18 in. to 30 in. gauge. They are used for transit in new countries, in contractors' work, and largely for industrial purposes, in engineers' and other shops.

Limit Gauge.—When a piece of work has to be tooled with a view to interchangeability, a limit of variation is permitted on each side of the correct dimensions, and gauges are made to these limits, and used to test the work by. The limits will vary in fineness with the nature of the work, and may range from $\frac{1}{100}$ inch to $\frac{1}{5000}$ inch. The two dimensions are frequently combined in one gauge, which is stamped at one end " Go," and at the other " Not go." See Gauge.

Limit of Proportionality.—Equivalent to Elastic Limit (q.v. Pt. 2, Basic Terminology).

Limits.—Limits of size for a dimension or part ; the two extreme permissible sizes for that dimension.

Limits of Tolerance.—The difference between the two limits of size—usually abbreviated to Limits (q.v.)—and the basic size (q.v.) of a dimension. Each limit of tolerance should be associated with its appropriate sign, i.e. plus for oversize and minus for undersize. The difference between the limits of tolerance is equal to the tolerance (q.v.) for that dimension.

Limit System.—Method of classifying limits (q.v.). A limit system is said to be on a *hole basis* when the hole is the constant member and varying fits (q.v.) are obtained by varying the size of the shaft ; and on a *shaft basis* when the shaft is the constant member and the fits are obtained by varying the size of the hole. A *unilateral* system is one in which the lower limit of the hole or the upper limit of the shaft is equal to its basic size ; the limits are in one direction only, minus or plus. A limit system is termed *bilateral* when the limits for the basic member are disposed above and below the basic size for that member. **See also** British Standard System of Limits and Newall System of Limits.

Lincoln Milling Machine.—A horizontal spindle machine, in which the spindle has capacity for vertical adjustment over a table the height of which does not vary.

Linde Welding.—Method of butt welding mild steel pipes, in which the weld is made at a temperature below the normal melting point of the steel. This is achieved by using an oxy-acetylene flame fed with an excess of acetylene, which first de-oxidises and then carburises the surface of the steel at the joint. The additional carbon in the iron lowers its melting point, enabling a satisfactory weld to be obtained.

Line Inspection.—A system of inspection (also known as *Patrol Inspection*) in which an inspector visits a machine or a work-bench at frequent intervals of time, and checks components while work is actually in progress. The system has the advantage that defective work can be detected before much of it has been produced, and that faults can often be rectified before the set-up itself is dismantled. See also Inspection.*

Line of Pressure.—The imaginary line which shows the course of the resultant of the external loads acting on a structure. See also Line of Pressures.*

Line of Resistance.—The imaginary line which indicates the course of the resultant of the internal stresses throughout a structure. If it should coincide with the Line of Pressure (see above) the cross-sections will be subject to uniform stress; but if not, there will be bending moments causing variable stresses according to the amount of eccentricity.

Link Belting.—Belting composed of a number of short links cut from leather, arranged parallel and retained in positions by pins which permit the links to pivot freely, and to bend around small pulleys and transmit power easily between those which are situated at a short distance apart. Being more flexible than continuous belting, the necessity for running a tight belt is avoided, with all the losses of power, and strains on belt and bearings that result from that practice.

Link Grinding Machine.—A special design for grinding the curves of the slot links of valve gears. The spindles are of planet type and the links are moved about a centre adjustable for radius. Both horizontal and vertical types are built.

Lintel.—Horizontal beam across an opening in a wall, as seen in windows, doors, etc.; when of wider span and supporting heavy brickwork or masonry, it is known as a bressummer.

Lip Drill.—A common drill, the cutting faces of which are slightly hollowed out backwards, immediately above the cutting edges, in order to give a slight amount of front rake to the tool. A lip drill cuts faster than an ordinary flat drill, but loses its edge quicker, necessitating frequent re-grinding.

Liquid.—A fluid which does not expand, but remains in the bottom of a containing vessel when pressure is removed. Liquids are practically incompressible, differing in this respect from gases. Liquids expand with heat, though to a much smaller extent than gases.

Liquid Fuels.—Mineral hydrocarbons, chiefly petroleum oils. They can only be used in the form of a fine spray with air.

List.—Rim of tin formed along the bottom edges of tin plate during the process of tinning, due to the accumulation of the superfluous tin which has drained down. See also List Pot.*

Little Giant Turbine.—A double water turbine having two tiers of buckets, that is to say, two turbines keyed to the same shaft, one above the other, and both running in the same casing, the upper tier discharging at the top and the lower one under the bottom of the case. It is governed by a cylinder gate (q.v.).

Live Rail.—Insulated rail which carries the positive current for the supply of electric cars, which, passing through the motor, returns through the wheels to the running rails through which current returns.

Load-Extension Diagram.—Graphical representation of the relation between the extension of a test piece and the load which causes it. Similar to a stress-strain diagram.

Load Spreader.—A support or material used to distribute the weight of a concentrated load over a larger area so as to avoid exceeding the permissible floor-loading stress. Any material can be used as a support, provided that it will sustain its share of the load and that its weight is not so great as itself to add to the stress on the floor.

Local Extension.—The extension which occurs on the short length including fracture, of a test piece, and forming the greater part of the total extension, which includes the elastic extension.

Locks.—*Mortice locks* are those which are entirely embedded within the thickness of the door ; *Rim locks* are those which are screwed against the face of the door. Locks are either *right* or *left*.

Log Frame Saw.—Sawing machine which carries a number of saw blades in a swing frame reciprocated vertically for breaking down logs into deals or planks or boards.

Long Distance Transmission.—This is a result of the introduction of electricity into engineers' work. Specifically it applies to the transmission of electricity from the source of power to the locality where it is to be utilised. Its earliest application is the water power in certain districts in America, Switzerland, Germany, Sweden, etc., of which electricity is generated at the falls and then transmitted from the dynamos at the power station to motors and lamps at a distance. Current is now being transmitted to over 200 miles. The power can be used for lighting, driving, including tramcars, etc. Long distance transmission on the North-East coast of England, where large water-powers are not available, is in operation by producing cheap gas in the coalfields from slack and from waste furnace gases, driving large gas-engines thereby, and generating electricity for distribution.

Lost Wax Process.—See Investment Casting.

Louvre.—A kind of vertical window, frequently placed at the tops of roofs, etc., and provided with horizontal slats, which permit ventilation and exclude rain. The term is also applied to ventilation slots in cabinets and other enclosures for electrical equipment, power units and engines.

Lubrication* (see Pt. 2, Basic Terminology).—There are two principal methods of lubricating bearings. In one the oil flows into them at ordinary atmospheric pressure, while in the other it is forced into them under pressure. In the first system *needle* lubricators, *siphon* lubricators, *pad* lubrication, *bath* lubrication, *ring* lubrication, *splash* lubrication are used. In internal combustion engines and in other cases where the bearings carry heavy pressures, oil is usually forced under pressure by a small pump to all the bearings and moving parts that require lubrication.

Lubricator* (see Pt. 2, Basic Terminology).—Also a device for feeding lubricants to machines. Often designed so that the oil supply can be seen passing drop by drop through sight-feed glasses.

Luder's Lines.—If a test piece of mild steel be highly polished before testing in tension, a series of lines, inclined at about 60° to the axis in both directions, are clearly seen after the yield point is reached. These lines, called Luder's Lines, show that molecular slip is taking place in the direction of the shear stresses.

M

Machine Tapper.—A very useful appliance by means of which holes are tapped more quickly and accurately than by hand. It is fitted to the socket of a drilling machine, and thus revolved. By means of an internal spring the driving pressure on the tap is released before it becomes so excessive as to cause risk of fracture of the tap—the socket slipping round merely, without operating. The tap has to be run back by reversing gear attached to the drilling machine, but specially designed tapping machines have an internal reversing gear which operates when the pressure is released. One exception is the Pawsons.

Machine-working Taps.—Taps (see Pt. 2, Basic Terminology) made for use in a screwing machine, and employed chiefly for tapping nuts. They are longer in the shank than hand-working taps, so that several nuts may be threaded at one time. There are also *taper taps*, and *thoroughfare taps*. See also Machine Tap.*

Macnaughting.—The term formerly applied in the North of England to the compounding of steam-engines, after Macnaught, who, in 1845, patented an engine in which a high-pressure cylinder was incorporated with an ordinary condensing beam-engine.

Macrostructure.—A term used in metallurgy to describe the general arrangement of crystals in a metal, as seen by the naked eye or at low magnification. Also, the general distribution of impurities in a mass of metal after the surface of the metal has been etched.

Magazine Feed.—Any mechanism by which pieces of work are fed to an automatic machine singly as by means of shoots, or rotating discs. It does for single separate pieces what the wire or bar feed does for articles cut successively from a rod or bar.

Magnalium.—An alloy of aluminium with magnesium in various proportions.

Magnesium.—A metal lighter than aluminium. Used for light alloys.

Magnesium-Aluminium Alloys.—Light casting alloys which are specifically lighter than unalloyed aluminium. Their physical characteristics are good, and resistance to corrosion is even higher than that of the silicon-aluminium alloys. From the production stand-point, their ready machineability makes them attractive. As opposed, however, to the silicon-aluminium group, the magnesium-aluminium alloys do not possess such good foundry characteristics. With magnesium contents above 1%, there is a notable increase in the formation of dross during melting. More especially in heavy sections, these alloys tend, when cast in green sand, to develop a dark outer skin, which is commonly referred to as " burning," but which appears to be due primarily to intergranular shrinkage. The magnesium-aluminium alloys machine well, take a high polish and are readily anodised.

Magnetic Block.—Laminated block, consisting of a number of high-permeability steel plates separated by non-magnetic material, and used in conjunction with a magnetic chuck (q.v.) when machining parts having irregular surfaces or projections which necessitate raising them above the surface of the magnetic chuck. The laminated construction of the blocks allows the magnetic flux to pass from the poles of the chuck to the work, whereas plain steel blocks would, in most cases, " short-circuit " the flux. Laminated blocks or adaptor plates are also used to distribute the flux evenly over the face of the chuck when it is necessary to hold a large number of small parts, such as washers.

Magnetic Brake.—A brake which depends for its action on the cessation of the flow of an electric current. The brake goes on automatically through the action of weights or springs, when the current ceases to flow, and it is lifted by the passing of the current through solenoids. It is therefore automatically safe, and is used for hoisting machinery.

Magnetic Chuck.—Magnetic device for holding work during grinding, machining or marking-out. Magnetic chucks may be of the electro-magnetic type, in which the poles are energised by coils through which an electric current is passed. Non-electric types embody nickel-aluminium alloy permanent magnets which can be moved by a lever in relation to the pole-pieces in the face of the chuck, thus enabling the chuck to be energised or de-energised at will.

Magnetic Coolant Separator.—Device which magnetically attracts and removes from the coolant used in machine tools, the magnetic swarf and most of the entrained abrasive particles, thus permitting repeated use of the cleaned liquid coolant.

Magnetic Crack Detector.—Apparatus used to detect cracks present on, or close to, the surfaces of articles made of magnetic materials. The principle of operation is that if a magnetic field passes through a magnetic material, the field will be distorted in the neighbourhood of a fault ; if the part is immersed in a fluid containing finely divided magnetic particles, or the detector fluid is poured or sprayed over the component, the particles will become aligned along the path of the crack or fault, thus rendering it visible. Longitudinal crack detectors pass a heavy magnetising current through the part in order to reveal cracks which are approximately parallel to the axis of the direction of current flow. Transverse crack detectors are provided with magnetising coils of heavy cable and are used to reveal cracks which lie in directions approximately at right angles to the axis of the part. Universal crack detectors subject the part to magnetisation in both directions simultaneously. Portable fluid magnetic detectors take the form of transparent cells containing magnetic fluid which are applied to the surface of an article which has previously been magnetised.

Magnetic Field.—The space through which the magnetic lines of force, between the poles of a magnet, pass. The field of magnetic influence.

Magnetic Separator.—Apparatus for separating magnetic substances from mixtures by means of electro-magnets, such as iron filings from brass, or iron particles from clay in porcelain manufacture. See also Magneting.*

Magnetic Sine Table.—See Magnetic Chuck.

Magnetic Sorting Bridge.—Apparatus consisting of two coils having similar electrical characteristics, connected to an electronic unit incorporating a cathode ray tube. A reference sample is inserted in the left-hand coil to provide a standard against which test pieces, which are inserted into the right-hand coil, can be checked, any variation from standard being shown by deflection of the trace on the cathode ray tube.

Magneto.—A compact form of dynamo fitted to some internal combustion engines, so that when driven by the engine it shall supply electric current to ignite the charges in the cylinders at the right instant.

Magnetostriction.—Change in dimension which occurs in certain ferro-magnetic materials when placed in a magnetic field. Although four magnetostriction effects can be observed, the Joule effect, or change in length with magnetisation, is generally accepted as the most important and is employed in a number of industrial processes. The basic principle is to magnetize a nickel bar by passing a current from an ultrasonic generator through a coil of wire, causing the bar to constrict in length at each cycle, the greatest change per unit length being about 30 parts in one million. The vibration of the bar can be used in a number of ways. (See Electronic Control of Machine Tools, Ultrasonic Cleaning, Ultrasonic Drilling, Ultrasonic Inspection, Ultrasonic Soldering, Ultrasconics.)

Maintenance Schedule.—See Servicing Schedule.

Manderla's Method.—The mathematical treatment of the stresses in a framed structure, allowing for the rigidity of the joints.

Manganese Steel.—A compound of manganese with iron. Though manganese is present in all steel, the above term is given specifically to compounds containing more than 7% or ranging from 7% to over 22%. The properties of manganese steel are, extreme hardness, toughness, and ductility. The effect of water hardening upon it is the reverse of that upon carbon steel, increasing the ductility as well as the tensile strength. See also Manganese.*

Manifolds.—In an internal combustion engine, pipes with branches connected to the cylinder ports, etc., which are used both to convey the mixture from the carburettor, and to carry the exhaust gases to the exhaust pipe and silencer.

Manometer.—A pressure gauge.

Manometric Efficiency.—The manometric efficiency of a centrifugal pump is the ratio of the head obtained from the pump per pound of water to the theoretical head, when losses due to disc friction, mechanical friction and slip are neglected.

Man Power.—Roughly, the tenth of a horse-power or 3,300 foot pounds per minute.

Mantles.—The fabric of which Welsbach gas mantles are made is either ramie thread or artificial silk impregnated with thorium oxide and cerium oxide. Other mantles are practically the same.

Marginal Grid Reference System.—See Grid System or Zoning.

Martensite.—Supersaturated solid solution of carbon in alpha iron, the essential constituent of fully hardened steel and the hardest that can be obtained by quenching. Under the microscope martensite appears as a light-etching, acicular structure.

Marking Machine.—One used for rolling or stamping graduations on machine elements as distinct from engraving machines.

Mattock.—A kind of pick with broad edges for digging.

Maxwell's Needle.—An apparatus for determining the torsion modulus (C) by means of a horizontal bar or " needle " suspended at its centre from a wire and set in oscillation. The period being found, the value of the torsion modulus may be deduced. A refined apparatus of this kind was used by Clerk Maxwell for the study of the viscosity of gases.

Mechanical Advantage.—The mechanical advantage of a machine is the ratio of the resistance overcome to the effort expended.

Mechanical Draught.—A means of increasing the draught of boilers either by an exhaust fan to draw the products of combustion through the flues or by blowing air into the closed stokehold or ashpit.

Mechanical Equivalent of Heat.—The amount of mechanical work required to raise 1 pound of water 1° F. and is equal to 778 foot pounds.

Mercast Process.—Method of investment casting (q.v.) in which a hardened steel pattern die, forming a negative of the finished part, is filled with liquid mercury, which is then frozen by immersing the die in dry ice and acetone. The frozen mercury pattern is withdrawn from the die and dipped into a suspension of a refractory material in a solvent which will evaporate at the temperature of the frozen mercury. Further layers are built up, using a coarser material than the inner surface layer, before the mercury pattern is melted out. As the refractory mould is fragile, it is backed up with sand or shot before the casting metal is poured.

Mesh Weld.—Group of spot welds (q.v.) made by a resistance spot welding process in which overlapping rods, wires or strips are welded together by pressure exerted by electrodes of relatively large area.

Metacentre.—The metacentre of a floating body is the point in which the resultant force due to the buoyancy of the liquid cuts the displaced position of the axis of the body, which is vertical when at rest.

Metacentric Height.—The distance GM between the centre of gravity (G) of a floating body and the metacentre (M), for an infinitely small displacement from the position of rest. As long as the metacentre lies above the centre of gravity, that is, as long as the metacentric height is positive, the body has a *righting* couple tending to bring the body back to its position of rest, but when the metacentric height is negative there is an *upsetting* couple ; and when it is zero the body is in equilibrium in all positions. The metacentric height is of great importance in discussing the stability of ships.

Metal Mixer.—A vessel in which various grades of iron are received from a blast furnace or Bessemer Cupola, to be mixed previous to treatment in the Converter.

Metal Spraying.—A method for the protection of iron and steel against corrosion. Atomised particles of molten zinc or aluminium are directed from a gun or pistol on to the surfaces to be protected. The particles spread on impact, and good covering power is obtained. The relatively low temperature at which the sprayed metal reaches the surface also makes it possible to use this method for the coating of paper, textiles, wood, and unglazed metal.

Metallic Arc Welding.—Fusion welding method in which the metal at the joint or seam is melted by an electric arc struck between a welding rod and the work. The welding rod, usually coated with flux, is clamped in an insulated holder and is connected to a source of direct current, in the form of either d.c. mains, a motor generator or a rectifier and transformer for use on a.c. mains. Electric arc welding can also be carried out by using alternating current fed from a simple transformer. In order to strike the arc about 60 volts is required, whereas during welding 25 volts is sufficient for most purposes. The current varies between 10 and 500 amp., depending on the type of work. The temperature of the arc is approximately 3,600° F., causing a small pool of molten metal to form on the work. The tip of the electrode is also fused, and from this small globules of metal are forced across the arc and deposited in the pool by electrical attraction, thus enabling welding to be carried out in an overhead position if required.

Metallic Packings.—Loose blocks held in contact with rods in stuffing boxes, by means of springs, with or without the aid of the steam pressure. There are numerous ways in which the blocks are arranged. See also Metallic Packing.*

Metallography.—The science of the constitution of metals, especially the study of their internal structure under the microscope. Microscopic observation can tell an expert much about the likely effect on a given piece of metal or alloy of (e.g.) heat-treatment or the imposition of varying degrees of strain, or about the causes of a recent fracture. See also Microstructure of Metals.

Metallisation.—See Metal Spraying.

Method of Sections.—This method is used to determine the forces acting in the members of a framed structure under load, by taking an imaginary section through three bars in which the forces are unknown and including that one for which the stress is sought. Then by taking moments about the point in which the other two bars intersect (produced if necessary), an equation is found involving only one unknown quantity.

Metrology.—The science or technique of measurement, covering all measuring instruments from the common rule to electronic instruments which can measure to an accuracy of 0·00001 inch. See also Measurement.*

Michel Bearing.—A bearing in which the parts are split up and pivoted, to enable them to take up the best position for efficient lubrication. These bearings will permit of a pressure of 5,000 pounds per square inch of bearing surface, as against 50 or 60 pounds per square inch in ordinary bearings. Used also for thrust blocks.

Micro-inch.—One millionth part (0·000001) of an inch.

Micron.—One millionth part of a metre. Therefore, one thousandth part of a millimetre. Approximately equal to one twenty-five thousandth part of an inch.

Microphone.—Instrument which includes a contact of variable electric resistance, such as is affected by slight vibrations of the air. It is used as part of a circuit including a telephone and current generator. As the contact varies, the resistance of the circuit and the current intensity change, so that sounds are emitted by the telephone corresponding to these changes produced by the voice at the microphone. (See Telephone.)

Microstructure of Metals.—A branch of research by means of which much has been learned of the effects of heat treatment, of alloying, of strain, of cleavage, crystallisation, hardness, annealing, etc.

Minimum Structure.—This is the least allowance for vehicles which have to pass through arches, tunnels, etc.

Mitre-sill.—The sill against which the lock gates of a canal shut.

Modification.—A general term covering any change in the design of an item. The purpose of modification is either to improve the efficiency of the product in use or to facilitate its manufacture.

Modular Ratio (Symbol *m*).—In reinforced concrete calculations the ratio of the stretch-modulus for steel to the stretch-modulus for concrete. It varies very much with the stress, but is usually taken as 15 in practice, which is a mean value for the range of stress occurring in ordinary cases.

Module.—The equivalent of diametral pitch. It is also that pitch expressed as a metric unit, being the pitch diameter in millimetres divided by the number of teeth in a gear.

Modulus.*—A measure or coefficient; as the modulus of linear elasticity (E); the modulus of shear (C); the bulk modulus (K); modulus of rupture, and so on.

Modulus of Rigidity.—The multiplier or co-efficient expressing the relation between the intensity of shear stress and the amount of shear strain.

Mohs' Scale.—Scale introduced by Mohs to measure the hardness of minerals. (See Hardness Scale.)

Mole.—The registered trade name of a special type of wrench which can either be used as an ordinary adjustable wrench or can be locked on to a job where it will stay locked, thus leaving the operator with both hands free to perform other tasks. It has a spring-loaded quick-release lever which is retained close against a lower handle when in the locked position. When locked, the mole has great gripping power. See also Spanner* and Wrench.*

Molybdenum (Symbol Mo).—A rare metal chiefly used as a steel alloy.

Momental Ellipse.—If the maximum and minimum radii of gyration for a cross-section are plotted at right angles to one another from a pole, so as to form the principal semi-axes of an ellipse, this ellipse is known as the *momental ellipse* for the cross-section, and from it the radius of gyration may be at once found about *any* other axis.

Momentum Grade.—The kinetic energy of a train can be used to overcome resistances greater than those on the ruling gradient on a railway by approaching the up-grade with the momentum acquired on the down-grade. Such gradients, steeper than the ruling gradient, which can be overcome in this way are known as *momentum grades*.

Mond Gas.—A producer gas obtained by the conversion of bituminous fuels into power and heating gas, with the recovery of nitrogen in the form of sulphate of ammonia.

Morse Tapers.—Standard tapers from 0 to 7 for fitting the shanks of drills and other tools to machine spindles.

Motion Plate.—Plate of cast steel, or of rolled steel, or iron plate, which is bolted transversely between the longitudinal frame-plates of a loco-motive, to carry those ends of the motion-bars farthest from the cylinder. The connecting-rods and the valve-rods pass through openings in this plate on their way to the crank axle.

Motion Study.—Scientific method of analysing and recording on special charts the essential movements which comprise any given operation. A form of motion study shorthand is generally used, consisting of symbols known as *Therbligs* (q.v.). The charts may be used as a basis for the construction of new motion patterns, besides providing data for the reorganisation of the workplace and improvement of jigs, fixtures, tools, etc., in order to speed up the work and reduce operator fatigue. In advanced motion study, motion picture cameras may be employed or " still " pictures may be taken to record the paths of lights attached to the operator's wrists, hands or fingers. From these records wire models of the paths followed by the operator's hands may be constructed in order that the movements may be studied and regrouped to form more effective motion patterns.

Movable Centre.—A centre of motion which is not fixed in space, but which revolves around a fixed centre (q.v. in Pt. 2, Basic Terminology).

Muff.—A joining tube driven into the ends of two adjoining pipes.

Muffler.—Another term for a silencer.

Mullion.—A vertical bar separating the compartments of a window. The horizontal bars are called transomes (see Pt. 2, Basic Terminology).

Multiple Drill.—Besides the drills of this class used for plating and boiler making, there are other more recent types of machines. One of these is used for drilling holes in the flanges of pipes and cylinders, the drills being adjustable in circles of different diameters, to drill all the holes in a flange simultaneously. Similar machines are used to drill the holes in retaining rings of the wooden wheel centres of railway waggons and carriages, and for many other applications. Sensitive drills are also fitted with multiple spindles for drilling holes in line, or arranged in various patterns. See also Multiple Drilling Machine.*

Multiple Moulding.—System of moulding some classes of light work in which the top and bottom parts of moulds are formed on opposite faces of the same body of sand in one box part. Several of these being superimposed the metal runs from top to bottom of the series, filling all the moulds in its ascent.

Multi-stage Pump.—Centrifugal pumps with single impeller will lift to about 150 feet; for larger lifts a second impeller is mounted on the same shaft, so that the discharge from the first is passed into the second, and if necessary, from the second into a third and so on. The total lift is theoretically proportional to the number of pumps on the shaft. Such pumps are described as one-stage, two-stage, three-stage, etc., as the case may be, or in general as *multi-stage* pumps.

Mushet Steel.—A self-hardening alloy steel, which contains a large percentage of tungsten, besides high carbon. It is cooled in air, is made in six " tempers," and is characterised by great endurance.

Mutilated Gears or **Segmental Gears.**—Term commonly applied to a class of gears not often employed, but which are extremely useful in some kinds of automatic mechanisms. In these the teeth, instead of being continuous, occupy a portion only of the pitch peripheries. Their action is therefore either absolutely intermittent, ceasing altogether for an indefinite period, or partially intermittent, ceasing to operate on one portion of the driven mechanism, but operating instead upon another.

N

Needle Nozzle.—Nozzle having a needle which tapers gradually to a point and which can be moved axially to and fro within the pipe that surrounds it. The water flowing through the annular space between the needle and the discharge pipe of a Pelton wheel can be regulated by moving the needle in the direction of the jet, so as partly or wholly to fill the opening of the orifice. It is automatically adjusted by an ordinary ball governor acting on a relay mechanism (q.v.).

Needle-roller Bearing.—See Roller Bearing.

Neutral Flame.—In gas welding, a flame to which the gases are supplied in such proportion as to produce perfect combustion. The neutral flame displays the white cone sharply defined and as large as possible. Oxygen and acetylene give a neutral flame when they are present at the torch in approximately equal volumes.

Neutral Point.—The point in an influence diagram (q.v.) at which, if a load be placed, it will produce no effect on the particular influence under consideration. See also Neutral Axis.*

Newall System of Limits.—System of limits, widely used in Great Britain, which has a hole basis (see Limit System). Two classes of tolerance (q.v.) are allowed for holes or bushes, Class A being the finer grade and Class B being used as inspection limits. Modern machine tools are capable of working to Class A limits.

Newel.—The open space surrounded by a stairway.

Newel-Post.—A vertical post sometimes used for supporting the outer ends of a flight of steps. Also the large baluster often placed at the foot of a stairway.

N-girder.—(See Pratt Girder.)

Nickel-silver.—Term covering a range of copper-nickel-zinc alloys, usually consisting of about 60% copper with 10–30% nickel, remainder zinc. The alloys have an attractive silver colour, are especially resistant to corrosion and possess comparatively high tensile strength, ranging from about 25 tons/sq. in. (annealed) to 45 tons/sq. in. (hard drawn).

Nickel Steel.—Nickel added to steel increases its strength and raises its elastic limit.

Nicrosilal.—The registered trade name for an alloy cast-iron which combines the austenite-forming properties of nickel with the heat-resisting qualities of high-silicon cast-iron and a high resistance to corrosion. It contains about 18% nickel, 4% to 5% silicon, 2% to 4% chromium, with traces of manganese, sulphur and phosphorus of less than 1% each.

Nitrading.—See Nitriding.

Nitralloy Steel.—See Nitriding.

Nitration.—See Nitriding.

Nitriding.—Also termed nitrogen hardening, nitration or nitrading. Method of producing a hard surface on steel, in which the part is heated to a temperature of about 500° C. for 40–90 hrs. in an atmosphere of ammonia gas. The gas dissociates, part of the atomic nitrogen which is formed passing into solid solution with the outer layers of the steel as iron nitride. A " case " of intense hardness (900–1000 Brinell) is produced without the necessity for further heat treatment. The process can be applied successfully only to certain classes of alloy steels known collectively as Nitralloy steel. (See also Chapmanising.)

Nitrogen Hardening.—See Nitriding.

Nomag.—The registered trade name for a non-magnetic, high-resistance, austenitic cast-iron developed specially for the electrical industry. It contains about 11% nickel, 7% manganese, and 1·5% silicon.

Nominal Size.—The size by which a dimension or part is referred to as a matter of convenience.

Non-magnetic Alloys.—These alloys of iron or steel are not affected by magnetism, as in the case of manganese-steel, which usually contains from 10% to 15% of manganese and 1% of carbon. It is very hard and tough and will outwear chilled cast iron many times over. It is used for watch-springs to prevent them becoming magnetised, and for the discs of magnetic hoists, because the smallest particles of iron and steel will not cling to the discs after the current is turned off.

Normal Temperature and Pressure.—The standard to which reference is made when it is desired to state the volume of gas at a definite temperature and pressure. Equal to the pressure exerted by a column of mercury 76 cm. high at a temperature of 0° C. Air which exerts this pressure at 0° C. is said to be "at normal temperature and pressure", or "at standard temperature and pressure".

Normalising.—This process consists in heating steel to about its critical temperature (900° C.) and allowing it to cool slowly. The process is almost the same as annealing, but is used in preference when it is wished to relieve abnormal internal stresses due to cold working fatigue.

Nosing.—The slight projection often given to the front edge of the tread of a step; usually rounded.

Nuclear Fission.—The "splitting" of an atom by bombarding the nucleus with high-speed atomic particles, termed neutrons, which may be accelerated by a special device such as a cyclotron or which may be liberated by other atoms of an unstable element. The nucleus consists of protons and neutrons, which are held closely together. Around it revolve electrons in orbits. The nature of an element is determined by the number of protons and electrons in its atom. The number of neutrons may vary, forming what are termed different isotopes of the same elements. Heavy elements, in which the atomic mass exceeds about 210, can be more readily broken up by the impact of a free neutron; at present, only Uranium and Thorium are known to be sufficiently unstable to serve as fuel in a nuclear reactor. When a neutron strikes the nucleus of a Uranium 235 atom, the atom splits into two unequal parts. This fission involves the destruction of some of the material of the atom which is changed into energy, with the liberation of a large amount of heat and very penetrating radiation. Both these products of fission are used in various applications of atomic energy (q.v.).

Nut.—A nut is a perforated block (usually of metal) possessing an internal, or female, screw thread, intended for use on an external, or male, screw thread such as a bolt for the purpose of tightening or holding two or more bodies in definite relative positions.

Nut-lock Plate.—Circular lock plate used to hold on the follower plate nuts of a steam-engine piston. The lock plate is put on and the nuts tightened; a cold chisel is then used to bend the plate up against the side of the nut to prevent it from moving. See also Nut Lock* and Lock Nut.*

O

Oakum.—A loose, fibrous material made from hemp or jute fibre and usually impregnated with oil or pine tar. It is used as a packing, for stopping leaks, and for caulking seams.

Odd-Side Moulding.—A method which is much in use among brass moulders. An odd-side is a false mould, corresponding with the dummy mould prepared in the first stage of ordinary moulding, done by turning over. That is, the patterns are arranged in a box of sand or plaster with their joint faces properly made, and often with the ingates in position, and from this odd-side the bottom part of the mould is rammed. Alternatively two odd-sides may be used, one for the bottom, the other for the top part. The advantage is that the trouble of arranging the patterns and making the sand joints and ingates for each separate mould is saved, since an odd-side will do duty for many scores of mouldings, the actual number depending on the care with which it is made and the permanence of its material.

Ohm.—The unit of electrical resistance, or that which would allow an electromotive force of one volt to pass a current of one ampere. Also termed the B.A. Unit, or British Association Unit. 1,000,000 ohms = 1 megohm.

Ohmmeter.—Instrument for testing cables, reading to ohms, or megohms.

Ohm's Law.—The fundamental law expressing the relation between electromotive force (E), current (I), and resistance (R), viz. $E = IR$.

Oil Bath.—A reservoir of oil, in which revolving parts subject to excessive friction are partially or wholly immersed, to provide them with constant lubrication. Worm gearing, when run at a high speed, is often immersed in an oil bath. So also are nearly all change speed and feed nests of gears.

Oil Engine.—A form of internal combustion engine, differing from the gas engine in the form in which the fuel is introduced. It is vaporised within the cylinder and mixed with air either before or after its introduction to form an explosive mixture. Heavy crude oils are used, and the engines work on the Otto Cycle. See Diesel Engine.

Oil Filter.—Used for purifying oil, by passing it through a fine wire mesh, a stack of metal discs, a felt pad or a cartridge containing a chemically treated powder in order to remove impurities, such as carbon, dust, etc.

Oil Sand.—A silica sand, used in foundries for coremaking, which has been bonded with an organic binder. It often contains a proportion of linseed oil or other drying oils.

Oldham's Coupling.—Mechanism for connecting two parallel shafts whose axes are slightly out of line.

Omtimeter (Optimeter).—High-precision measuring instrument, made in both horizontal and vertical forms, in which the measuring scale, which is optically magnified over 1,000 times, is graduated in 0·00005-in. divisions so that readings to 0·00001 in. may be estimated when observed through the eyepiece, or when projected on a screen in the projection attachment.

Open Circuit.—An electric circuit is said to be *open* when no current passes ; e.g. when a switch is open the current ceases to flow.

Open. Link Chain.—Chain in which the links are not supported across their centres with a stay or stud. Nearly all the chain used by engineers is of the open-link type, and of the close or short-linked class. The proof strain is about double the safe working load. Chains should be frequently lubricated with oil, applied with a brush. After they have been in use for several months they become crystalline and brittle; hence the reason for occasional annealing, which is performed by heating them over a wood fire, or in a stove, and then allowing them to cool down gradually.

Opening Die.—A form of screwing die used in many screw machines and turret lathes which opens and clears the thread on coming to the end. The turret and die are then run back, so saving the trouble and possible injury to the thread due to the reversal of solid dies. The spindle also runs continuously in one direction. Most threads of medium and large size are cut with opening dies on screw machines.

Open-spindle Headstock.—The spindle of a capstan lathe which is pierced through crosswise sufficiently to allow the operator to insert his hand and manipulate a bolt or pin which is being dealt with.

‡ **Optical Flat.**—Precision-ground, polished disc of crown or optical glass or quartz, used to measure errors in the flatness of gauge anvils or precision-finished surfaces. When the optical flat is placed on the work the presence of air between the flat and the surface of the work, due to irregularities in the latter, causes interference between the light reflected from the work and that reflected from the optical flat. This produces rainbow-coloured lines known as interference bands or Newton's Rings. With good-quality optical flats, each band represents twelve millionths of an inch. For precise inspection the bands are viewed through filters producing alternate light and dark bands of one colour wave, allowing measurements to be made to ten millionths of an inch. Straight bands indicate a plane surface; bands which curve away from the point of contact, a concave surface; and those which curve around the point of contact, a convex surface.

Optical Gauge.—Gauge which enables the profiles of screw threads, cutters, and similar parts to be magnified and examined with great accuracy. See Toolmaker's Microscope, Projection Gauge, Workshop Microscope.

Optical Square.—Pentagonal prism, often termed a pentaprism, having two silvered reflecting surfaces, which bends a ray of light entering one face through exactly 90 deg.

Optical Tooling.—Precision method of checking the accuracy of certain types of work, particularly where large dimensions are involved, such as the alignment of marine engine and propeller shaft bearings, locomotive frames, and other large structures, in addition to checking precision components such as dividing heads. The minimum distance between datum points is in the region of 1-2 ft. The method makes use of an alignment telescope, used in conjunction with a collimator and various accessories such as optical squares. The auto-collimator is an example of this equipment. When necessary the rays of light between the telescope and the reflector can be turned through 90° by using an optical square, thus enabling the angular relationship between horizontal and vertical surfaces to be accurately checked.

Ordinate.—The co-ordinate, usually vertical, of a point, which, together with the abscissa (q.v.), defines its position relative to a fixed pair of axes. See also Ordinates.*

Orientation.—The determination of the position of an instrument, building or other body in a given direction, as when a plane-tube is oriented or fixed in the direction required.

Osmium Lamp.—An incandescent lamp having a filament made of osmium.

Otto Cycle.—The cycle of events—compression, ignition, explosion, exhaust—which takes place in an internal-combustion engine operating on the " 4 stroke or Otto cycle ", so named after the German engineer, Dr. Nicholas Otto, who first applied the principle in his gas engine. The four events in the cycle occur at every half-revolution of the crankshaft (i.e., on every stroke of the piston).

Out of Wind (pronounced *wynd*).—Perfectly straight or flat.

Oval Chuck.—A compound chuck in which the amount of eccentricity is controlled by a worm wheel and tangent screw.

Oval Hole Cutting.—This is done in the boiler shop by a machine in which a motion like that of the oval chuck is imparted to the cutter.

Overhaul.—When a machine runs backward under the load it carries it is said to *overhaul*. In the case of a machine which refuses to overhaul, without special means provided to prevent it, as in the case of Weston's differential pulley-block, and in any case where more than 50% of the work done in overcoming the resistance is lost in friction, the machine will support its load unaided. (See also Pt. 2, Basic Terminology.) See also Overhauling.*

Overhead Tracks.—Single trolley tracks hung from a ceiling or roof, and carrying trolleys of hand or electrical types. Practically the whole of a floor area can be covered by a suitable disposition of these tracks. See also Overhead Traveller* and Shop Traveller.*

Overhead Valves.—Valves in an internal combustion engine which are mounted with their stems upward in the cylinder head. They may be operated either by push rods and rockers or from an overhead camshaft.

Overstressing.—The deliberate stressing of material beyond its elastic limit. The method is often used, and is quite safe. Chains, for example, railway coupling links and crane hooks of standard design are overstressed at what is called the proof load. See Proof Load.*

Oxy-acetylene Cutting.—Method of cutting iron or steel plates with the aid of an oxy-acetylene flame. The method, also termed flame cutting, consists of heating the metal until red hot. The acetylene is then turned off and a high-pressure jet of pure oxygen is directed on to the heated spot, resulting in the metal being burnt away. By moving the nozzle slowly, a cut of any desired pattern and inclination to the vertical can be produced, of no greater width than that produced by a saw. For complicated shapes the torch is built into a machine and is guided by a master template. For long cuts the torch is mounted on a portable motor-driven machine adapted for straight-line or curved-edge cutting, "flame planing," bevelling and similar preparation of plate for welding. See also Powder Cutting Process.

Oxy-acetylene Flame Hardening.—Method of hardening iron-carbon alloys by rapid and intensive local heating by oxy-acetylene flames to above the upper critical point (q.v.), followed by immediate quenching. The method of arranging the blowpipe and quenching jet varies with the type of work. The work may be stationary and the burner and jets move in relation to it, or vice versa. Alternatively the burner and jets may move longitudinally while the work revolves. Again, the revolving work may be heated by a stationary burner which is then extinguished or withdrawn and the quenching spray applied. The appropriate method may be applied to the hardening of the gear wheel teeth, gear rings, tyres, axles, cams and camshafts, rollers and spindles, crankpins and journals, valve stem ends and tappets, machine tool beds and other applications.

Oxy-acetylene Welding.—Method of welding in which an oxy-acetylene flame is used to melt the edges of the parts to be joined. In most cases additional metal is fed into the joint from a filler rod of suitable composition. See Leftward Welding, Linde Welding, Rightward Welding, Vertical Welding.

Oxygen-Free Copper.—Also known as " O.F.H.C. Copper " or " Oxygen-free high conductivity copper ". Pure cathode copper is melted in intimate contact with charcoal and then cast, the whole process taking place in an inert atmosphere. It has high electrical conductivity, and greatly improved rolling and drawing properties compared with phos-phorised copper. Distinguished from tough-pitch (q.v.) by the absence of oxygen.

P

Painting Machine.—One which paints objects, however large or intricate, by spraying the pigment, contained in a tank, using compressed air at 25 lb. pressure or more. See also Paint.*

Pallet.—Specially designed platform on which goods are placed to enable them to be handled by a fork-lift truck (q.v.). The pallet is generally of wooden construction and of sufficient height to allow the forks of the truck to pass under it ; it may accomodate one package or may form a base for a stack of packages. See also Paint.*

Paramagnetic.—Substances which are attracted by magnets such as iron, nickel and cobalt, as opposed to diamagnetic substances which are repelled. These last are very few and have no practical value.

Parent Metal.—A term used to denote the metal of which the parts to be welded, brazed or soldered are made. In the metallic arc welding process, the parent metal itself constitutes one of the electrodes. See also Metallic Arc Welding.*

Parge.—To make the inside of a flue smooth by plastering it.

Pascal's Law.—The law states that whatever pressure is exerted upon fluid enclosed in a vessel is transmitted equally and undiminished to every surface of the fluid in all directions. The principle of the law finds practical application in (e.g.) the hydraulic press and hydraulic brakes. See also Hydraulic Brake; Bramah's Press.*

Pasting.—Securing together the halves or sections of cores, previously rammed in separate portions, with clay water or with flour.

Patter.—A term used in automobile engineering to describe a kind of up-and-down or see-saw movement of the front wheels of a motor vehicle when the vehicle is in motion. Patter is caused by oscillation of the front axle across its centre. See also Centre of Oscillation.*

Pavior.—Paving stone. Also a rammer for driving paving stones.

pdl.—Poundal.

Pearlite.—Formation in steel consisting of alternate plates or layers of ferrite and cementite. Pearlite has a constant composition for a given class of steel ; the carbon content of pearlite in plain carbon steels, for example, is 0.85%. The thickness and distance between the alternate lamellae are largely governed by the rate at which the material is cooled, slow cooling producing coarse pearlite and quicker cooling fine pearlite.

Pediment.—The triangular surface formed by the vertical termination of a roof, consisting of two sloping sides and the line joining the eaves.

Pelton Wheel.—Water wheel of small diameter which is rotated by means of a jet of water directed under great pressure into the buckets.

Pendulum* (see Pt. 2, Basic Terminology).—Also used for clocks in combination with an escapement. The length of the pendulum is adjustable to correct the time of a beat. (See Simple and Compound Pendulum.)

Pendulum Saw.—Circular saw for light cross cutting, the mandrel of which is suspended by a frame from trunnion bearings above.

Pentaprism.—See Optical Square.

Percussion Welding.—Modification of the projection welding process (q.v.). The parts are brought into light contact in order to heat the metal by the temperature rise caused by the resistance to the flow of current across the joint. A sudden heavy pressure is then applied to forge the parts together. Generally used for welding studs, tubes, etc., to plates.

Perfect Frame.—A perfect frame is one which has just sufficient members to render the stresses in them statically determinable.

Periodic Time.—The recurring interval in which a complex cycle of operations is performed, as in the case of an oscillating pendulum or spring.

Petrol Engine.—An engine in which the vapour of petroleum oil (either petrol or paraffin) is utilised as the fuel, being mixed and sprayed with a current of air, and ignited. There is a large field yet for the employment of these engines, because, in addition to their economy, they have the advantage that, the fuel supply being independent of the engine, no fixed pipe connections are necessary as in the case of gas and of steam engines. Petrol or paraffin engines are used for pumping, for driving dynamos, for working rock-drills and air-compressors, for driving launches and motor vehicles. See also Petroleum Engine.*

Petroleum.—A rock oil, found extensively in Europe, Asia, the Far East and in North and South America. It occurs chiefly in the Silurian and the lower Tertiary formations. It is employed in various forms for lubrication, for illumination, and as fuel. The preparations of petroleum for lubricating have the advantage over oils of animal and vegetable origin, that they do not absorb oxygen from the air, which causes spontaneous combustion. As heavy oils of this kind do not thicken, and retain their properties when heated, they are valuable for heavy bearings, and engine cylinders. Petroleum is utilised for prime movers (see Petrol Engine) and also for liquid fuel.

Petter Oil-engine.—Small vertical engines working on the two-stroke cycle, suitable for farm work, small machinery, electric lighting, etc.

Phase.—In wave motion, oscillating motion or similar periodic phenomena the interval of time passed from the time a particle has moved through the middle point of its course.

Phosphorised Copper.—Deoxidised (q.v.) copper in which the oxygen has been removed by the use of phosphorus. It has enhanced drawing and rolling properties, but reduced electrical conductivity. It also forms high-density castings, accompanied by considerable shrinkage. See also Phosphorus.*

Photo-conductive Cell.—Photo-electric cell (q.v.) which uses the property of selenium by varying its resistance to electric current when illuminated. A thin film of selenium is deposited on a metal grid which forms a high-resistance path between two conductors. When the cell is illuminated, the resistance of the grid falls and the flow of current increases.

Photo-elasticity.—Method of determining the location and direction of stress distribution in a material. A model of the part is made in transparent plastic and is viewed, while appropriately stressed, by passing a beam of polarised light through it. The stressed areas are revealed by coloured patterns caused by optical interference between the two components of the polarised light passing through the models. The most highly stressed areas are those in which the interference bands are closest together.

Photo-electric Cell.—Also termed Photocell. Device for converting light into electrical energy, used in engineering in counting, control, safety devices and fire-detection equipment. In general, the light-sensitive cell receives light from a constant exciting lamp and triggers a relay circuit when the light is interrupted. Examples of its use are the counting of components or packages on a conveyor; switching off a machine which is incorrectly loaded or when the light beam is interrupted by the hands or body of an operator; or, in fire-detection equipment, the operation of the sprinkler installation if the beam is obscured by smoke. Photo-electric cells may also be used to switch on lights automatically when the level of illumination falls below a predetermined degree. See Photo-conductive Cell, Photo-emissive Cell, Photo-voltaic Cell.

Photo-emissive Cell.—Photo-electric cell (q.v.) which contains a curved metal plate coated with an alkali-metal such as caesium, rubidium, potassium, antimony-caesium or silver-caesium, which emits electrons in proportion to the degree of illumination falling on it. The cell also contains an anode wire. When a voltage of 20–80 volts is applied across the anode and the cathode plate, the flow of electrons forms a conducting path which, as stated above, will vary with the intensity of illumination, thus producing a variation in the current passed by the cell.

Photometer.—An apparatus for measuring the intensity of light, such as Bunsen's, Rumford's, and others.

Photo-voltaic Cell.—Photo-electric cell (q.v.), sometimes termed a photronic cell, dry-disc cell or barrier-layer cell, containing a plate coated with selenium, or, in some cases, copper oxide, over which is deposited a thin, transparent metallic layer. The metal backing is connected to one terminal of the cell and the transparent metal layer to the other terminal. When light falls on the cell, electrons are liberated by the selenium and a potential difference is set up between the terminals of the cell. A current of from several micro-amperes up to one or two milliamperes is generated, according to the size of the cell, and this current is used to operate a relay.

Pickering Governor.—One in which the balls are connected to the centres of cambered springs of flat steel. As speed increases the ends of the springs are pulled inwards and reduce the opening of the throttle valve.

Pierre-perdue.—Random stone or rough stone thrown into the water to make a foundation, and allowed to find its own slope.

Pile Sawing Machine.—Saw operated by hand or power for cutting off the heads of piles level after they have been driven. See also Pile.*

Pilot.—Anything which precedes and guides a fitting or tool.

Pilot Wheel.—A hand wheel with cross handles.

Pinning.*—Clogging of file teeth by embedded material. (See also Pt. 2, Basic Terminology.) See also Pin* and Pinny.*

Pintle.—A vertical projecting pin, like that often placed at the tops of crane-posts, and over which the holding rings at the tops of the wooden guys fit. Also such as are used for the hinges of rudders or of window shutters to turn around.

Pipe Bending Machine.—Machine for bending pipes and tubes between rolls operated by a lever. See also Pipe Bending.*

Pipe Calipers.—Calipers having long arms, with suitable graduations for measuring the thicknesses of pipes several feet inwards from the ends.

Piping.—The drawing down or hollowing out in a conical depression of an ingot of cast steel, due to the shrinkage of the metal. In cast iron it would be termed a "draw." (See Feeding, in Basic Terminology.) See also Pipe.*

Pitchometer.—An instrument by which the pitch of a propeller is measured in place.

Pitot Tube.—An instrument for measuring the flow of air or water, especially in pipes. It consists essentially of a tube of fine bore bent at right angles at its lower end and tapering to a fine opening. When held so that this opening faces the current, the rise of water in the tube measures the velocity head and pressure head, and when the opening is turned through 180° so as to be parallel to the current, the rise in the tube again measures the combined head. The velocity head due to the current alone is the difference of the two heads, and may be calculated from the formula $V = C\sqrt{h_1 - h_2}$ where h_1 and h_2 are the two heads, and C is a coefficient whose value is 2·16 approximately.

Plain Machine.—Milling machine in which the table is not capable of swivelling around a vertical axis.

Planet Spindle.—The spindle of a grinding wheel, which, while rotating, also travels bodily in a circular path.

Planimeter.—Instrument by means of which areas of irregular outlines can be obtained without mensuration.

Plano-milling Machine.—A heavy design, built on the same general outlines of the common planer, and dealing generally with the same class of work that is done on the latter. The spindles carried in bearings on the cross rail may be either horizontal or vertical. Also horizontal spindles for end mills are often carried on one or both of the housings.

Plastic.—(1) Materials are said to be plastic when they have no tendency to return to their original shape after the action of external forces operating upon them is withdrawn. Clay, putty, etc., are plastic substances. The term is employed specifically, to distinguish such substances from those which are termed elastic. (2) Generic name for organic substances, mostly of synthetic or semi-synthetic origin, which become plastic under heat and pressure, allowing them to be moulded, cast, extruded, etc. Plastics are termed *thermoplastic* when they soften whenever heated and *thermosetting* when the application of heat during moulding renders them hard and resistant to softening under subsequent applications of heat. Chief among the plastics are derivatives of cellulose and casein, in addition to natural substances such as shellac and bitumen. (See Synthetic Resins.)

Plastic Metal.—A white alloy, having a low coefficient of friction, and anti-corrosive properties which render it adaptable for high-speed bearings, and for use under water. It can be employed for fan bearings, locomotive axle-boxes, glands, and for coating propeller-blades. It contracts very little in cooling, and fuses at a low temperature.

Plenum Process.—The opposite of an exhausting or vacuum process. It is that employed in pressure systems of ventilation, and in the supply of compressed air to caissons.

Plinth.—The square lowest member of the base of a column.

Plug Gauge.*—Gauge used to check the dimensions of holes or internal threads. The plug may be plain or screwed, parallel or tapered, and fitted with a handle. When the gauging portion and the handle are manufactured separately the gauge is termed a renewable-end type. When the GO and NOT GO dimensions are incorporated on a single plug the gauge is termed a progressive gauge.

Plumb Rule.—Rule or straightedge used to test the perpendicular accuracy of a supposed vertical surface. It carries a plumb line against the centre of its longitudinal axis, and has a recess, cut near its lower end, for the plumb bob. See also Plumb Line.*

Plumbing.—A term used in engineering to cover any piping or tubing fitted to a machine or to a structure. See also Pipe.*

Plums or Fillers.—Large masses of stone embedded in mass concrete for the sake of economy.

4+

Plywood.—Plywood consists of two or three plys or layers of thin wood sheets adhering together and placed with the grain alternately at right angles. See also Ply.*

Pneumatic Conveyor.—Conveyor consisting of a tube through which carriers are moved at high speed by vacuum or air pressure generated by an exhauster or compressor. In more elaborate systems carriers are sorted and diverted into the correct tubes automatically by relays operated by switches which are energised by a series of metal bands on the body of each carrier.

Pneumatic Gauge.—Extremely sensitive gauge in which the rate of flow of air from small jets in the gauging member is determined by the distance of the jets from the surfaces to be measured. The jets may be incorporated in a plug, ring or snap type of gauge (q.v.). The rate of flow from the jets affects the air pressure exerted on a column of liquid in a glass tube which is backed by a scale graduated in the required units, thus enabling semi-skilled personnel to carry out checks to a very high degree of accuracy. Variations in dimensions may be magnified up to 9,000 times. A further advantage is that no wear occurs on the gauge surfaces as there is no contact with the work.

Pneumatic Hoists.—Light hoists by which loads are lifted directly by the movement of a piston in a long cylinder which is suspended over the work. These are now largely used in foundries, machine shops, etc. They are fixed or portable, in the latter case being fitted with a little trolley to run along an overhead track, or along the jibs of crane frames or of travelling gantries. Compressed air is brought to them by flexible hose, and the valves are operated by dependent chains and levers.

Pneumatic Tools.—Compressed air is used for operating many small tools which were formerly driven by hand or by water-power, or by running cords or ropes, the application being of especial value where portability is required. The compressed air is brought in flexible rubber armoured tube to the tools, and the latter are either held in the workman's hands or in a frame, as in the case of some portable drills and tapping machines, etc. The principal tools which are used thus are drills, tappers, riveters, caulking hammers, chipping chisels, beading tools, wood-boring augers, foundry rammers, etc.

Pneumatic Transmission of Power.—Transmission of power through pipes by means of compressed air may be carried out by means of pipes from which the air is partially exhausted. The system is used for light parcels, in shops and elsewhere. On a larger scale power is transmitted, as in Paris, by means of compressed air through large pipes, the velocity of flow being as much as 50 feet per second. As the air is easily available in any quantity and easily disposed of, the system is quite successful, and for tunnelling operations it has the important advantage of ventilating the tunnel. Considerable power may be transmitted on account of the high velocity obtainable without undue friction.

Point Chuck.—The dead centre, or point centre, attached to the headstock of a lathe, for turning work which is pivoted between centres. It is either fitted with a coned end into the hollow of the mandrel nose, or it is screwed over the nose, like the face, and cup chucks. In the latter case it is also usually slotted through for the driver (q.v. in Pt. 2, Basic Terminology). The latter method is adopted on small lathes, the former on those of best construction. Also called driver chuck, and running centre chuck.

Poisson's Ratio.—The ratio of lateral to longitudinal strain. Its value for metals is about 0·3.

Polar Co-ordinates.—The determination of the position of a point by stating its distance from the origin of co-ordinates and the angle its radius vector makes with the initial line. Convenient for curves of spiral form, such as the logarithmic spiral and similar curves.

Pole-plate.—Longitudinal timber resting on the ends of the tie-beams of roofs, and for supporting the feet of the common rafters, when such are used.

Poling Boards.—Short boards used for timbering trenches.

Polyphase Currents.—Alternating currents which differ in phase by equal increments of a complete cycle of alternation. Usually two, three, or six phases are adopted.

Poncelot Wheel.—An undershot water wheel, which is carefully designed to avoid impact at entrance of the water to the vanes and having sufficient blade depth to allow the water to rise until it comes to rest under gravity, and by discharging backward at outlet it loses as little kinetic energy as possible.

Pop.—A term used in North Britain to denote a small boss which receives a set screw.

Portable Drill.—Type of small hand-drilling machine, used chiefly for drilling holes in parts and materials of all types *in situ*. The machine is held in the hands, and may be operated with an endless cord, from a pulley overhead, the cord passing over guide pulleys, the distances of which are adjustable; or electrically; or by compressed air.

Portable Machine Tools.—Representative of a growing practice in consequence of the increase in the dimensions of castings and forgings, many of which are too massive to be taken to machines. The portable tools are brought to their work and often attached to it, or more frequently to a floor plate, to which the work is bolted down. The tools are operated by compressed air or electricity, and include drills, planers, shapers, keyseaters, and milling and grinding machines.

Portal Crane.—A jib crane of rotating type mounted on a gantry, which permits of the passage of railway or road waggons below between the legs.

Potash Hardening.—A shop term, which distinguishes mere surface hardening, effected by the contact of potash at a red heat with wrought iron, from that effected in a case-hardening pot, in prolonged contact with animal charcoal. The former is a mere film that can be penetrated with a file; the latter is of sensible depth, or from $\frac{1}{32}$ in. to $\frac{1}{16}$ in. deep.

Potentiometer.—Electrical instrument similar to the Wheatstone Bridge for measuring electromotive forces and potential differences. It possesses three advantages over other methods : (1) It is a *null* method, that is, one in which no deflection has to be observed ; (2) That in testing batteries no current is taken from them, and thus their true e.m.fs unaffected by polarisation are obtained ; and (3) That the resistances of the galvanometer and battery tested do not enter into the calculation, since no current passes through them.

Pot Steel.—Cast steel (q.v. in Pt. 2, Basic Terminology), to distinguish it from steel made by Bessemer and Siemens' processes.

Pottering Down.—Poking or stoking the fuel around the crucibles used for steel melting.

Pouring Basin.—The depression, or hollow receptacle above the ingate of a mould, which receives the metal from the ladle. The basin is deep, of sufficient area, and all sharp corners of sand are scrupulously avoided.

Powder Cutting Process.—Oxy-acetylene cutting process (q.v.) in which a powder is fed into the reaction area, increasing the temperature and enabling the process to be used on stainless steels and non-ferrous materials on which the oxide produced has a higher temperature than the parent metal. In the flux-injection process the chromium oxides in stainless steels are removed by a flux which is injected into the cutting oxygen stream. In the iron-rich powder process the powder is carried to the reaction zone by a suitable gas, such as compressed air. In the oxy-kinetic process finely-divided silicate (usually quartz sand) removes the chromium oxides from stainless steel by mechanical abrasion and by the increased kinetic energy imparted to the oxygen cutting stream by the solid particles.

Powder Metallurgy.—Process of producing parts from metal powder by first compacting the powder to the required shape by the application of pressure to suitably shaped dies and then heating the briquet up to, or close to, the melting point of the metals. This stage is termed sintering and the part is often termed a sintered product. (See Sintered Bearings, Sintered Carbide Tool, Sintering.)

Power Gas.—Any gas that is used for driving gas engines, including the waste gases from blast furnaces. The thermal values are less than that of illuminating gas, but with full allowance for this, they are so much cheaper that they have created a demand for large gas engines and gas turbines, which are displacing steam engines for large power stations as well as for engineering works and factories.

Power House.—Specifically, the house in which power is generated for distribution to a works. The term has a more extended significance than the old engine house and boiler house, although it includes these. A modern power house comprises boilers, engines, dynamos, and often hydraulic pumps and air compressors; because steam, electricity, water and air all now have their applications in many large factories. The generators of power are the boilers, which consist generally of a battery of Lancashire type with economisers, or of water-tube boilers. The transmission of power directly from the engines to distant departments by shafting is lessening, being displaced by that of the electric conductor from generators to motors.

Power Press.—Any press in which sheet metals are stamped and pressed into shapes.

Pratt or N-girder.—A common type of girder, in which the web bracing consists of vertical and diagonal members alternately. The shorter vertical members are struts and the longer diagonals ties, for the sake of economy of material.

Precision Level or Engineer's Level.—An instrument based on the common spirit level, which has been developed for engineering purposes into a precision instrument for measuring straightness, flatness, change of angle or any tendency of the material to cant. It usually has a main vial, a cross vial, and at least one plumb vial. The base is accurately machined, and has a concave groove for use on cylindrical work or shafting. The main vial is usually graduated into divisions 0·1 in. long. Sensitivity is such that a bubble movement of one division can indicate an off-level of 0·0035 ins. per foot. See also Spirit Level.*

Pre-ignition.—Premature ignition of the compressed mixture in an internal combustion engine cylinder before the spark occurs. This gives rise to "pinking."

Premium Systems.—In these the workman receives, in addition to a fixed wage, a definite proportion of the value of any additional output which he may produce.

Pressure Angle.—The angle between the common normal at the point of contact of two teeth and the common tangent to the pitch circles at their point of contact.

Pressure Welding.—Term often used to describe the various methods of electrical resistance welding (q.v.). It should be strictly applied only to a modern development of forge welding, in which the edges of the parts to be joined are heated to just below the melting point of the metal and are united by forcing them together. Sheet metal components are built up by clamping the sheets between dies which are heated by gas burners. For aluminium alloys, to which the process is usually applied, the temperature of the dies is between 420° and 480° C. and the pressure is from 2 to 5 tons per sq. in. Pressure welding can also be carried out without the application of heat. (See Cold Welding.)

Prick Punch. See Dot Punch.*

Prime Number.—The ordinary change wheels used for screw cutting have composite numbers of teeth. These, in various combinations, will cut screws having an almost infinite range of composite pitches. But to cut threads having prime numbers to the inch, change wheels having prime numbers of teeth are requisite. Prime numbers can also be cut with a division plate, or with a traversing mandrel. Any number that cannot be divided without remainder by any other number is said to be prime; that is, one which has no divisor or factor but itself, and unity. The prime numbers up to 97, which alone would concern screw cutters, are 1, 2, 3, 5, 7, 11, 13, 17, 19, 23, 29, 31, 37, 41, 43, 47, 53, 59, 61, 71, 73, 79, 83, 89, 97. See also Prime.*

Principal Stresses.—The maximum and minimum intensities of stress in a material under strain. They are at right angles to one another and to the planes on which they act, and are the major and minor semi-axes of Rankine's ellipse of stress.

Principle of Least Work.—(See Least Work.)

Profile Milling Machine.—Machine in which the spindle as it rotates is pulled over against a form or copy of the shape of the article to be milled. See also Profiling.*

Profilometer.—Instrument used for measuring the quality of surface finish. Signals from a tracer unit, consisting of a stylus attached to the armature of an electro-magnet, are fed through electronic amplifiers to an oscillograph on which the surface irregularities are shown by the shape of the trace on the screen.

Progressive Gauge.—See Plug Gauge.

Projection Gauge.—Optical gauge which enables profiles of screw threads, form tools, gear teeth, cutters, etc., to be magnified by 20–100 diameters and projected on to a similarly enlarged drawing placed on a horizontal table. The work is placed in a beam of collimated light so that the shadow of its profile is projected by a lens on to an inclined mirror which reflects the image on to an overhead mirror mounted above the projection table. Some gauges (e.g., N.P.L. horizontal types, Westminster and Autotrope designs) are not fitted with mirrors.

Projection Microscope.—Instrument used for the examination of materials by reflected or transmitted light. The illumination can be converted to provide polarised light.   A range of objectives gives magnifications from $3\frac{1}{2}$ to 4000 diameters.

Projection Welding.—Method of welding in which spot welds (q.v.) are produced by forming projections on the parts at the points at which fusion is desired.   The parts are then clamped between current-carrying electrodes and light pressure is applied.   Welding commences as the projections make contact.   As the pressure is increased the projections collapse.   The current is then cut off.   Projection welding is used to attach screws, studs, nuts and similar small parts to plates and also for the assembly of components made from strips, flanges, etc.

Prony Brake.—Simple form of absorption dynamometer.   It consists of two blocks of wood clamped together with a pulley between them, the pulley being fixed to a revolving shaft ; one block has a lever attached to it which carries a weight at the outer end, the magnitude of which balances the moment of friction between the blocks and pulley. Often used for testing high-speed motors.

Proof Stress.—The proof stress for light alloy test-bars is defined by the B.E.S.A. Aircraft specifications as " that stress at which the stress-strain curves depart by 0·1% of the gauge length from the straight line of proportionality."   For phosphor bronze bars and rods the allowance is 0·5% of the gauge length.   See also Proof Strain.*

Propeller Starter.—An hydraulic machine which is clamped round the shaft of a propeller, and has rams, generally four in number, which being pressed against the propeller boss, starts it off the shaft.   Some of these starters are capable of pressing with a force of 300 tons.

Prototype.—When a new design is first conceived, facts and information drawn from the past are not necessarily applicable; and it is sometimes necessary to build and test a pattern or model suitable for evaluation of design, performance and production potential.   This " first-off " or original is generally known by designers as a prototype.   See also Pattern.*

Proving.—Proving a material differs from testing it, in that when testing we aim at learning its physical properties, but in *proving* it we wish to ascertain whether it satisfies the conditions required by the specification, for example, that it shall not contain more than a certain amount of sulphur and phosphorus ; that it shall have a tenacity of 30 tons per square inch, and a minimum elongation of 20% on a gauge-length of 8 inches.   See also Proving Machine.*

Puddle.—Well-tempered clay and sand used to render banks or dykes impervious to water, and for the interior of dams to prevent leakage.

Pulley Moulding Machine.—A pattern rim is drawn down between stripped plates.   In the more elaborate machines a number of rims are mounted permanently and a selective mechanism takes charge of the one to be moulded.   Pattern arms are changed to suit the pulley being moulded.

Pulsometer Pump.—One which is specially adapted for dealing with liquids having solid matters in combination.   It operates by the alternate condensation and pressure of steam acting on the liquid.

Pulverised Coal.—Coal in pulverised form, largely used in cement burning kilns, the coal being reduced to a high degree of fineness.   It is also used for firing water-tube boilers.

Pump Duty.—The *duty* of a pump is a practical way of expressing its overall efficiency.   For a pump driven by a steam engine the duty is the number of foot-pounds of work given out by the pump for every 1,000,000 British thermal units supplied to the engine by the boiler.

Hence it takes account both of the efficiency of the pump and of the steam engine. Formerly the term *duty* was the number of foot-pounds of work given out by the pump per bushel of coal burned in the boiler. In this case the efficiency of the boiler is also included. If the term *duty* is applied to a pump driven by an electric motor, it is based on 1,000,000 British thermal units supplied to the motor.

Punching Stress.—The stress produced when a rivet hole is punched or when a column rests upon a bed-plate of concrete or other material so as to cause a shearing stress tending to pierce the bed on which it rests, the area resisting shear being the perimeter of the hole multiplied by its thickness. The punching stress is about twice as great as the ordinary shear stress. See also Punching Strength.*

Put-logs or Put-locks.—Horizontal pieces supporting the floor of a scaffold, one being inserted into put-log holes left for the purpose in the masonry.

Pyrometer.*—Apparatus used to measure temperatures, usually those above the range of the conventional mercury thermometer. Pyrometers include: the thermo-electric type, in which use is made of a thermocouple (q.v.); the radiation type, in which the heat to be measured is concentrated by a mirror or lens on to the hot junction of a thermocouple; the electrical resistance type, which takes advantage of the fact that the electrical resistance of a metal increases with an increase in temperature; and the optical types, in which the intensity of the light from an incandescent source of heat is measured by comparison with the light from a reference filament, or in which the intensity is judged solely by the operator. See also Thermometer.*

Q

Q.—The symbol for quantity of water, etc., discharged.

Quality Control.—Term that is often applied rather loosely and in a general sense to all the aspects of manufacture that aim at ensuring an acceptable standard of quality in the finished product, including inspection, sampling schemes, laboratory experiments and statistical and mathematical analysis of control charts, while at the same time recognising the limitations imposed by mass-production methods.

Quadrant Control.—A term used in connection with a Diesel engine fuel pump to denote an assembly consisting of a toothed sector or quadrant, clamped to a sleeve. The assembly is designed for the purpose of controlling the rotation of a plunger within the barrel of the pump, and so varying the quantity of fuel delivered. See Quadrant.*

Quadrilateral.—A term used in geometry to describe any plane figure bounded by four straight lines. The principal quadrilaterals used in workshop calculations are the square, the rhombus, the rectangle, the parallelogram and the trapezium. The sum of the angles of any quadrilateral is always 360 degrees.

Quarter-Elliptic Spring.—A laminated or leaf-type of spring, half of a semi-elliptic spring (q.v.), which is so named because it resembles in shape one quarter of an ellipse. It is used for suspension purposes, or to cushion or absorb shocks. When used as a carriage spring, the thick end is secured to the carriage frame while the thin end, or main leaf, is pivoted or shackled to an axle. See Leaf Spring.

Quicksilver.—Another name for Mercury. See Mercury.*

Quill Bearing.—A term sometimes used to indicate a needle-roller bearing. Such rollers have a very small diameter. Bearings incorporating them are sometimes used in place of plain bearings at points where lubrication is difficult and where plain bearings could therefore give rise to friction and heat. See also Roller Bearing.

Quoin.—(a) The hollow into which a groin-post of a canal lock-gate fits. (b) Stones (usually dressed) placed along the vertical angles of buildings, chiefly for ornament.

Quoin-post.—The vertical post on which a lock-gate turns. The heel-post.

R

Racing Lines.—The concentric lines turned on the faces of flanges to hold red lead, yarn, or copper wire to make steam-tight joints.

Radiator.—(1) Device employed to dissipate excess heat from an internal combustion engine, consisting of a large number of thin-walled tubes or narrow passages through which cooling water, which passes around the cylinders and combustion spaces, is circulated by thermosiphon action or by a pump. Air is forced through the radiator either by a fan or by the pressure induced by the motion of the vehicle. (2) Heating unit, consisting of a series of tubes or a hollow panel heated by hot water, steam or electrical heating elements, used to heat a space by radiation and convection. In some types a fan is used to circulate air through the heating elements.

Radiography.—See Radiology.

Radiology.—The science of inspecting materials and components by means of X-rays and gamma rays, to detect hidden, internal faults and to verify the position of internal parts in plastic and other components. The leading insurance companies insist on X-ray examination of the welds of all high-pressure containers. In radioscopic inspection the X-ray beam is used to form a visible shadow-image on a fluorescent screen and the inspection is carried out visually by the operator : in radiographic inspection the X-ray shadow picture is recorded, with or without the aid of the fluorescence effect, on a photographically-sensitive film, the developed image on which is subjected to examination, and which is available for reference or comparison whenever the need arises.

Radius Gauge or **Fillet Gauge.**—A gauge used to check or determine the inside or outside radius of a circular part, e.g., the pins, journals and bearings of a crankshaft, or of curved portions of a machined workpiece. It consists of a metal case containing a number of thin leaves of steel each accurately shaped to a known radius at one end. The size of the radius, in either 64ths or decimals of an inch and sometimes in both, is stamped on each leaf. Most gauges are made with the internal and external radii of the same size on one blade.

Rain-water Pipe.—A special branch of pipe founding, done chiefly in Scotland, which is an interesting illustration of how the minimum thickness of iron can be run over a large surface, by using metal of softest quality, thoroughly hot, and employing several long, narrow runners. The thickness of the metal in these pipes ranges from $\frac{1}{8}$ in. to $\frac{3}{16}$ in.

Ram Jet.—An engine used in aircraft and rockets which achieves jet propulsion by the combustion of fuel in air which has been taken in and compressed solely by means of the ram effect of the machine's forward speed through the air. There are two main types of ram-jet—the Intermittent Jet (q.v.) and the steady-flow or Athodyd (abbrev. for " aero-thermo-dynamic-duct ").

Ramming.—The consolidation of the sand around a pattern by means of a rammer. Sand may be rammed too hard, or too soft ; in the first place causing scabs, in the second the formation of lumps on the castings. Sand requires to be rammed harder in some sections of a mould than in others, and dry sand may be rammed harder than green. The harder the ramming the more complete should be the venting. On many

4*

moulding machines pressure takes the place of ramming. See also Rammer.*

Random-courses.—Courses of stone of unequal thickness.

Random Rubble.—Rubble which is irregular in shape.

Rankine's Ellipse of Stress.—See Ellipse of Stress.

Rapson's Slide.—Mechanism used for maintaining a constant turning moment on the rudder of a ship in all positions of the tiller.

Rate Fixing.—The determination of a fair and reasonable rate of payment for a job, usually based on the results of careful time study (q.v.).

Rawhide Gears.—Pinions and wheels made of specially hard leather built up in several thicknesses, through which the teeth are cut. They are used to a considerable extent for high-speed driving, especially as a first gear from an electric motor, being durable, elastic and noiseless. See also Raw Hide Belting.*

Réaumur Scale.—A thermometric scale in which the difference between the freezing and boiling points of water is divided into 80 degrees, freezing point being 0° and boiling point 80°. It takes its name from the French physicist R. A. F. de Réaumur, 1683–1757, who introduced the scale.

Recalescence.—The stage in the shrinkage of cooling steel at which momentary expansion takes place, occurring between about 620° and 700° C. See Critical Point.

Reciprocal Deflections.—The principle of reciprocal deflections, due to Maxwell, states that " the deflection at a point B of a beam or structure in the direction of a load W_2 when a unit load acts at a point A in the direction of a load W_1, is the same as the deflection at A in the direction of W_1 when a unit load acts at B in the direction of W_2."

Reciprocal Figure.—Graphical method of determining the forces in the members of a framed structure which is statically determinate, by superposing the force polygons for the loads on its several joints.

Reciprocating Machine Tools.—The group which includes planers, shapers, slotters and allied forms. See also Reciprocating.*

Reeling.—Finishing and truing tubes and bars in a machine of that name.

Refuse Destructors.—Plants for consuming the refuse collected from factories, shops and private dwelling houses. The refuse, of which about 60% is combustible, is mixed with coal and burnt in cells at a high temperature. After complete combustion the gases may be used for steam raising.

Regenerator* (see Pt. 2, Basic Terminology).—Its function is the utilisation of the heat of the waste gases from the furnace hearth. These gases being led through the brickwork, give up a large portion of their heat thereto, which is afterwards taken up by a cool return current on its way to the hearth. There are separate regenerators for the air, and the combustible gases. (See Air Regenerator, Gas Regenerator, in Basic Terminology.)

Register.—To register is to fit one section to another, to cause parts to coincide accurately. The term is applied to the good fitting of pins and holes, rebates, rings, etc.

Register Gate.—Device which regulates the supply of water to reaction turbines. In this system the guide vanes are made in two parts. The inner ring next to the runner is stationary whilst the outer ring may be turned about it so as to diminish the flow into the inner ring, or to cut it off entirely, if required.

Reinforcement.—Steel rods or joists embedded in concrete to reinforce or strengthen it, the combination being known as reinforced concrete or Ferro-concrete.

Relay.—Device for supplying additional pressure energy from an independent source, to assist in overcoming a considerable resistance. Thus in the regulation of the speed of hydraulic turbines, the governor which controls the speed is insufficient to move the gates and requires assistance from the turbine itself or some exterior source, by means of a *relay*. This usually consists of a cylinder and piston moved by the hydraulic pressure of water taken from the penstock, or by oil under high pressure in a reservoir containing compressed air.

Repetition of Stress.—See Fatigue of Materials (Basic Terminology).

Resinoid Bond.—Phenolic resin grinding wheel bond. Resinoid-bonded wheels are made in many sizes and for many purposes. As cut-off wheels they can be operated safely at speeds as high as 16,000 s.f.p.m. for cutting off all kinds of material. Larger wheels, operated at speeds around 9,000 s.f.p.m., are used for snagging castings, and at normal speeds for finishing cams, roll grinding and saw gumming.

Re-starting Injector.—Is designed for use in locomotive and marine engines, which, being subject to shock, sometimes render the ordinary injector unreliable. In one form the steam nozzle is closed by a double-seated valve of conical form. Hence, when it is opened, an annular jet of steam issues from the nozzle, and condensing a portion of the cold feed water creates a partial vacuum, which causes the steam to flow with higher velocity, and carry the water into the boiler.

Reversibility.—Machines which restore the work supplied to them without appreciable loss are said to be *reversible*, as in the case of a pendulum in falling and rising, or a bent bow. In practice there are always losses due to friction, etc., and the recovery of energy is always less than that supplied, in which case the process is *irreversible*.

Reversing Engine.*—One used for reversing the direction of rotation of a main engine.

Reverted Train.—A reverted train of wheels is one in which the axes of the first and last wheels are coincident.

Revetment.—A retaining wall.

Ribbed Tubes.—The invention of M. Serve, which were introduced to increase the heating surface of the fire tubes of steam boilers, with a corresponding economical consumption of fuel, and an increase in steam pressure. The tubes are rolled with several deep internal radial ribs running longitudinally. They are made in iron, steel, copper and brass. Experiments have proved that 15% better results are obtained by the adoption of the Serve ribbed tubes when natural draught is employed, and of 20% when forced draught is used. The Purves flue tubes are ribbed and corrugated transversely.

Rightward Welding.—Method of welding in which the blowpipe is moved from left to right along the weld, preceding the filler rod. With rightward welding no bevelling of the edges of the plates is required up to a plate thickness of $\frac{5}{16}$ in., while a 60° bevel is sufficient for greater thicknesses.

Ring Gauge.—Gauge, in the form of an accurately ground ring, used to check the dimensions of shafts and external threads. Separate ring gauges are used for GO and NOT GO gauging, but it is recommended that gap gauges (q.v.) should be used for NOT GO gauging.

Ring Lubrication.—Some bearings are designed with a trough of oil underneath, and are provided with flat metal rings hanging from the shaft and dipping into the oil. The rotation of the journal causes the rings to revolve, and to carry oil with them.

Riser.—(1) The upright board of a step; (2) In the wiring of a building for incandescent lighting a riser is a conductor which is run vertically from floor to floor. (See also Pt. 2, Basic Terminology.) See also Air Gate* and Rising Gate.*

Rivet Furnace.—Small furnace in which rivets are heated in quantity, being preheated gradually as they descend. See also Rivet Forge.*

Riveting Cranes.—Light cranes from which portable riveting machines are suspended and traversed over the work.

Riveting Tower.—A tower built over the riveting machine in a boiler shop, in which Lancashire and Cornish boilers are slung vertically during the process of riveting up the plates.

Rivet Tail.—That portion of a rivet which is turned over in the process of riveting, or the opposite end to the rivet head, which is formed in a machine. See also Riveting Set.*

Roberval's Balance.—The common balance such as used for weighing letters, and in shops. The scales are attached to the two vertical sides of a jointed parallelogram, the other two sides turning about pins at their centres. If the balance moves, one scale descends exactly as much as the other rises. Hence it does not matter where the weights are placed on the scale pans.

Rock Drills.—Drills of chisel shape, of single or cross-bit form, which being reciprocated rapidly and slightly rotated in the intervals, produce circular holes in rock to receive blasting charges. They are operated by compressed air or electricity.

Rockwell Hardness Test.—Hardness test in which either a hardened steel ball for soft materials (B scale), or a conical diamond indentor for hardened steels (C scale), is used, the hardness number being recorded directly on a dial. The loads employed are 60 kg. for the steel ball and 150 kg. for the diamond indentor. In making a test, care should be taken to polish out deep machining marks, surface scale or other defects, as a smooth surface is necessary for a true result. The anvil on which the specimen rests should be well seated and the surface clean, as minute particles of grit tend to cause incorrect results. A load of 150 kg. should not be employed for testing parts case-hardened to less than 0·025 in., since low readings may be obtained which do not represent the true case-hardness. In such circumstances a superficial Rockwell tester or Vickers machine should be employed. Articles such as moulds for plastics should not be tested on the working face, as the depth of the impression would cause a flaw in the moulding.

Roller Bearing.—Bearing consisting of a series of rollers retained between a stationary and a rotating hardened-steel ring. The rollers may be parallel, in which case the bearing normally carries journal loads only, or may be tapered, allowing both journal and thrust loads to be carried. In some roller bearings the rollers are of small diameter in relation to their length, and are not positioned by a cage, as is normally the case with other types of roller bearing. This design is termed a *needle-roller bearing*.

Roller Conveyor.—May be of the gravity type, consisting of a plain roller-track, or may have power driven rollers rotated by chains and sprockets or by gears.

Rolling Margin.—In each case the weight per foot quoted in the manufacturers' catalogues is the minimum that can be rolled and is subject to a *rolling margin* of 2½% over. This margin is claimed by the rolling mills, and should be allowed for in all calculations of weights.

Rolling Resistance.—That part of the tractive resistance which is caused by friction between track and wheels. See also Resistance,* Rolling Friction,* Traction* and Tractive Force.*

Rolling Stock.—A general term covering all the engines, carriages, wagons, etc., which travel on rails or along a railway track.

Root Circle.—In the design of wheel teeth, the circle which passes through the roots of the teeth.

Rope Brake Dynamometer.—Simple form of absorption dynamometer, consisting of a rope which encompasses a pulley fixed on its shaft, the rope being kept in position by wood blocks laced to the rope. One end is attached to a spring balance, and the other to a dead weight of suitable value. The difference between this weight and the reading on the spring-balance multiplied by the radius at which the rope acts measures the frictional moment on the wheel.

Rotary Planing.—Commonly applied to the work of the face milling machines, hence termed rotary planers, since they do the work which would otherwise go to the common planers. See also Rotary Cutter.*

Rotative Engine.—Term used to distinguish the ordinary engine with crank and fly-wheel from one in which the reciprocating movement is not converted into circular motion. The pumps of the Worthington class are driven directly by engines which are not rotative. The term is common in specifications.

Rotor.—(1) The rotor of a dynamo refers to that part of the armature which revolves, to distinguish it from the *stator*, which is stationary. (2) A localised vector, that is, a vector which has a definite position.

Rubber Bond.—Rubber-bonded grinding wheels are used chiefly where a good finish is required. The rubber softens under the heat of grinding and acts as a cushion for the grains of abrasive, which consequently do not cut as deeply as when more rigid bonds are used. The rubber also acts as a buff to polish out the grain marks. Extremely thin wheels can be made in this bond because of its strength and toughness; for example, wheels as thin as 0·005 in. are used for slotting pen points.

Rubber Die Press.—Type of press in which a metal die is carried on a platform which is raised by hydraulic cylinders in order to bring the die into contact with a block of rubber above it. The work is thus formed by the squeezing action of the metal die against the rubber die. The pressure normally required is approximately 2 tons per sq. in. of the surface of the work. Advantages are the fact that only one accurately made metal die is required, and the ease with which the rubber die may be repaired by vulcanising-in sections as required or reconditioned by vulcanising-on a new facing layer. In some instances a separate, loose sheet of rubber may be laid over the work on the platform, thus protecting the main rubber block from wear.

Rubbing.—When it is desired to design a wheel or wheels to gear with one already existing, it is customary to rub a sheet of clean white paper on the ends of the teeth of the old wheel, in order to obtain their correct curves, to which those of the new wheel are adapted. This is called "taking a rubbing."

Rubble.—Stones, broken brick, etc., used to fill up behind the face courses of walls.

Rule of the Middle-third.—This rule requires that in the case of a rectangular cross-section of any structure, the line of pressure acting along a principal axis of the section must lie within the middle third of that axis in order that the stress, whether tensile or compressive, may not change sign.

Running-in.—Initial running of an engine or mechanism at moderate speeds and loads. in order to produce satisfactory bearing surfaces. (See Beilby Layer, Basic Terminology.) See also Running In.*

Rustless Steel.—Steel which is rendered rust-resisting by the addition of 10–15% of chromium. See also Rust* and Rusting.*

S

Safety-gap.—Two points or balls connected to an electric circuit in which abnormal voltages are apt to be set up. In case of the rest of the apparatus being endangered, a discharge takes place between the points, similar to the action of a fuse.

Salt Bath Furnace.—Type of hardening furnace in which the temperature is regulated by the employment of fused salts. See Cyaniding.

Saltpetre.—The popular name for potassium nitrate. It is a strong oxidising agent, and is also a constituent of gunpowder.

Sand Blasting.—Used extensively for cleaning castings, removing scale and dirt from forgings preparatory to tooling, brazing, galvanising, etc. Compressed air, or steam, is the agent employed for blowing the sand against the work. See also Sand Blast Sharpening.*

Sand Drying.—Is done in kilns, fixed or rotary, preparatory to grinding for foundry service.

Sand Grinding.—Necessary to crush lumps preparatory to mixing. It is done in mills of various types.

Sanding.—(1) Feeding sand on wet rails in front of locomotive driving wheels. (2) Glasspapering woodwork in a machine.

Sanding or Sandpapering Machines.—A large group used chiefly in joiners' shops. The work is laid upon or held against a flexible band charged with powdered glass. See also Glass-papering Machines.*

Saw Guard.—An adjustable covering fitted to a circular saw to afford protection to the hands of the workman.

Saw Sharpening Machines.—A large number of these are now made for dealing with band and circular saws, abrasive wheels being used and the saws being fed tooth by tooth automatically. See also Saw Sharpening.*

Saw-tooth Truss.—Form of roof truss adapted for factories. The slopes of the roof are unequal, the smaller slope being covered with slates or tiles, in the usual way, while the steeper slope is of glass. Used with spans of 20–35 feet.

Scabble.—To dress off the rougher projections of stones for rubble masonry with a stone-axe or scabbing hammer. See also Scab.*

Scaffolding.—The formation of arched accumulations of partially melted iron in a foundry cupola, due to bad unequal charging of iron and coke. The coke burns and sinks down from underneath the iron, leaving it in a more or less pasty condition, which either delays the casting or renders the iron unsuitable for its purpose, or blocks up the cupola. See also Scaffold.*

Scalar.—A numerical quantity which has *magnitude* and *sense* but no directon. Scalar quantities obey the laws of arithmetic and algebra in opposition to *vector* quantities which are also affected by direction.

Schiele Pivot.—In a footstep bearing which is at first flat, the wear at the outer circumference is greater than towards the axis of rotation and the frictional resistance is no longer uniform. The object of the Schiele pivot is to give uniform pressure and uniform wear by making the vertical section take the form of a *tractrix*, or anti-friction curve.

Sclerometer.—An instrument for testing hardness. (See Abrasion Test.)

Scleroscope.—Instrument consisting of a small hammer weighing about one-twelfth of an ounce and fitted with a diamond at one end. It is used for measuring the hardness of materials by raising the hammer and allowing it to drop from a definite height upon the material under test, the relative hardness being shown by the height to which it rebounds as measured on a vertical scale.

Score.—A term used to denote a deep scratch, groove, notch or incised line, especially of such an incision when it causes damage to the interior of a cylinder through which gas leakage may take place.

Scotch Boiler.—The marine cylindrical return-tube boiler.

Scotch Turbine.—This reaction wheel is a development of Barker's Mill. Water is admitted through a vertical supply pipe and flows outward through horizontal curved arms. The reaction of the jets due to the change of momentum produces rotation. Only made in small sizes and of no practical importance.

Scragging.—The process of testing carriage and locomotive springs by impulsive loading.

Scragging Machine.—Machine adapted for scragging springs. (See Scragging.)

Scraper Ring.—An auxiliary piston ring designed to remove surplus oil from the cylinder walls of an engine.

Screw.—A screw is a headed and externally threaded mechanical device possessing capabilities which permit it to be inserted into holes in assembled parts, of mating with a preformed internal thread or forming its own thread, and of being tightened or released by turning its head.

Screw Conveyor.—A form of conveyor in which a screw propels loose materials along a tube.

Screw Machines.—May or may not be automatic. They are used for turning and threading small screws, studs, and pins, from rod or bar fed through a hollow spindle. A turret may or may not be fitted.

Screw Thread Milling.—Cutting the threads of worm screw with a revolving milling cutter in a machine provided with the necessary adjustments. There are several designs.

Scroll Gear.—A form of the variable gears (q.v.) in which the pitch surface is in the form of a scroll, the effect being to impart a gradually increasing rate of motion to a shaft.

Seamless Copper Tubes.—The numerous accidents which have occurred in consequence of the bursting of steam pipes of brazed copper have resulted in the introduction of seamless tubes by electrolytic deposition. The copper is perfectly pure, and the pipes are homogeneous and of equal thickness and strength, and without the uncertain brazed joint. The pipes are made in all sizes, from 6 in. to 18 in. diameter, and from a little more than $\frac{1}{16}$ in. to $\frac{3}{8}$ in. in thickness. See also Seamless Tubes.*

Seam Welding.—Method of welding in which a continuous spot weld (q.v.) is formed by placing the parts to be welded between motor-driven rollers connected to a source of welding current. Tubes, ducts, casings, etc., may be welded at speeds of 6–15 ft. per min. by this method.

Secondary Battery.—In commercial secondary cells lead plates, separated by insulating material, are immersed in dilute sulphuric acid. When a current is passed through the plates, the acidulated water is decomposed, hydrogen appearing at the anode and oxygen at the cathode. This oxygen combines with the surface lead to form peroxide of lead (PbO_2). When the *charging* process has continued for some time the battery is disconnected from the dynamo, and forms an accumulator of electrical energy at about two volts per cell. On reversing the process the battery gives back the energy stored in it. These batteries have many applications, as to lighting, traction, telephone and telegraph working, motor-car ignition, X-ray working, etc.

Secondary Bracing.—Bracing in a roof truss which divides the principal rafter into panels, and supports it both as a strut and a beam.

Secondary Buckling.—Secondary buckling tends to occur in a latticed stanchion when the unsupported length anywhere between the battens or diagonal bracing has a slenderness-ratio (q.v.) greater than that of the stanchion itself.

Secondary Hardening.—High-speed steel, after quenching from 1,300° C., contains a high percentage of austenite. By secondary hardening—that is, reheating to 550–630° C. and slowly cooling—this austenite is "broken down" into martensite. The secondary hardening treatment is sometimes repeated to complete the transformation of austenite to martensite. Secondary hardening raises the cutting ability of the tool, although it does not always increase the indentation hardness.

Secondary Stresses.—Stresses which are calculated, as is usually done, on the supposition that the joints of a loaded frame are frictionless pins, can only be regarded as a first-approximation to the truth, the stresses being known as *primary* stresses. Secondary stresses are such as are produced by departure from the ideal conditions and are due to friction and the rigidity of the joints when riveted ; any eccentricity will then cause bending moment. (See Manderla's Method.)

Sector Discharger.—Method of extracting coke from vertical retorts by cutting off a section of the coke monolith below the base of the retort, at intervals of 40–50 minutes.

Sector Gears.—A term applied to a special form of toothed gearing, by which motion of an intermittent character is transmitted from one shaft to another. The pitch peripheries of the wheels, instead of being continuous, are broken up into arcs of circles of different curvatures, each pair of arcs therefore transmitting different velocity ratios. The gears are, therefore, variable (see Variable Gears), though the term intermittent would more properly be applied to them, because there is no merging of different speeds into each other, but a sudden and instant change of speeds.

Self-hardening Steel.—Also known as *Mushet steel*. It contains about 5·5% of tungsten, 2% of carbon, 1·5% of manganese, 1% silicon and 0·4% chromium. Such steel is hardened by heating to a bright red colour (about 1,500° F.) and allowing it to cool. It keeps the cutting edges of tools much better than ordinary carbon steel.

Semi-automatics.—These constitute a large class of machines which occupy a middle position between the entirely or "full" automatic machines and those which involve the constant attendance of an operator. The term is chiefly applied to turret lathes and screw machines. Usually several movements are automatic, but not all. Thus the stock may or may not be fed through the hollow spindle, and gripped in the chuck by power, the turret may be rotated and brought up to its work by hand or by power. The screwing dies may be self-opening or self-closing by hand or by power. But the point is, that somewhere in the cycle of operations the attendant has to interfere to keep the cycle progressing. There is little to distinguish them from the automatics (q.v.). They are made in many types, and in shops that deal with a miscellaneous and general class of work they are often more adaptable than the full automatics.

Sense.—The sense of any magnitude is the equivalent of positive and negative in algebra. If AB be a line having positive sense, then BA=$-$AB is the negative sense, and AB$+$BA$=0$.

Sensitive Drilling Machines.—A group designed for operating small drills only, of about $\frac{1}{4}$ in. and under, at high speeds. Over 1,000 revolutions per minute are common, and feeds at the rate of 28 in. or 30 in. per min. for $\frac{3}{4}$ in. drills. They are belt driven, with speed changes through cones or gears, and have ball bearings to revolving parts. Some are fitted with several spindles.

Sensitiveness.—The sensitiveness of a governor increases as it approaches isochronism (q.v. Pt. 2, Basic Terminology). It may be expressed as

$$\frac{\omega_1}{\omega_2-\omega_1}$$ where ω_1 and ω_2 are the angular velocities at the extreme ends of the distance through which the sleeve moves.

Separation.—See Cavitation.

Separator.—A device to prevent priming in steam pipes. There are various types, the aim being to throw down the particles of water which are mechanically mixed with the saturated steam, leaving the latter comparatively dry on its passage to the cylinder.

Separators.—Usually of cast-iron, or steel tubing, through which a bolt is passed to keep the tube in place and the steel joists in position when used for a grillage foundation (q.v.).

Series-wound Motor.—(1) In this type the whole of the current which passes in the armature circuit flows round the turns of the field coils. (2) Is of utility when the speed of the motor is required to vary with the load, as in cranes.

Servicing Schedule or Maintenance Schedule.—A document, usually in booklet form, listing the series of operations which need to be performed at prescribed intervals of use and/or of time to ensure that equipment, vehicular or static, is maintained in a condition to operate efficiently. The document is usually compiled initially by the manufacturer, but the instructions contained therein are generally modified by the user to meet particular operating circumstances.

Servo Motor.—See Relay. A servo system is one in which the muscular force of the operator is assisted either by utilising the momentum of the mechanism, or by a piston operated by air or hydraulic pressure. Electrical actuation is also used.

Shaft Governor.—A governor fastened on the shaft of an engine, in which the steam supply is controlled by the eccentric acting on the cut-off of the valve. The steam, being admitted at full pressure instead of by throttling, works by expansion in the cylinder when cut off by the valve.

Sheeting.—Covering surface with boards, sheet iron, felt, etc.

Shellac Bond.—Grinding wheels bonded with shellac are classed as elastic wheels, since shellac is somewhat elastic and softens under the heat of grinding; it is similar to rubber as a bond, but wheels so bonded cut more freely than rubber wheels and will take deeper cuts without burning. See also Shellac.*

Shielded Arc Welding.—Welding process in which the molten metal is protected by an envelope of chemically-reducing or inert gas from contact with the oxygen and nitrogen in the air. The gas used may be argon or helium, or may be liberated by a special coating on the electrode. Alternatively a chemical powder may be deposited along the line ahead of the weld as is done on certain types of automatic arc welding machine (q.v.). When argon or helium gas is used the gas is projected around a tungsten electrode which passes through the tubular nozzle of the welding torch.

Shim.—A packing piece, generally in the form of a thin flat strip of metal or plastic material having a solid cross section, frequently used to maintain a predetermined distance between two surfaces, e.g. between the two halves of bearing brasses. As wear occurs in the bearing, the shims can be removed to take up the gap to limit the slackness. See Packing Piece* also Washer.*

Shingled Bloom.—Term applied to the ball of crude malleable iron which has been subjected to the process of shingling (q.v. Pt. 2, Basic Terminology), preparatory to rolling into a slab or puddled bar. See also Shingling.*

Shipyard Cableways.—A design of hoisting machine for service in shipbuilding. Cables suspended from carriages running on cross girders at each end of a slipway receive the hoisting trolleys which are travelled along them. Electric current is employed for all motions.

Shock Absorber or **Shock Damper.**—A frictional or hydraulic damping device placed between the axles and frame of a vehicle, to check bouncing, pitching and rolling.

Shop Rivets.—Shop rivets are those which are driven in the ordinary way, in contradistinction to *field* rivets (q.v.).

Shore.—A prop or piece of timber set obliquely so as to act as a strut on the side of a building when it is in danger of falling.

Short.—Without tenacity, i.e. brittle or easily pulverised.

Short Circuit.—Connection between two parts of an electric circuit which is of low resistance compared to that of the parts connected. Also used as a verb, as " to short circuit a lamp."

Shunt —(a) To turn a train, etc., from one line of rails to another, or used as a noun. (b) In a current circuit, a connection in parallel with a portion of the circuit, thus in the case of a galvanometer, a resistance coil may be put in parallel with it to prevent too much current going through the galvanometer. This connection is a shunt. In a dynamo a special winding for the field may have its ends connected to the bushes, from which the regular external circuit also starts. The dynamo is then said to be *shunt-wound.*

Shunt-wound Motor.—Used when the motor speed must be constant irrespective of variations in load. The field magnet winding is parallel with that of the armature.

Shuttering.—Vertical " forms." (See Forms.)

Side Planing Machine.—One in which the tool is carried at the end of an arm which moves laterally in relation to the work bolted next the side of the machine framing, or on tables at the side.

Side Valves.—Poppet valves mounted side by side in the cylinder block of an engine.

Silal.—A proprietary, high-silicon-alloy cast-iron, containing about 6% silicon. It has special heat-resisting properties, and a much higher resistance to oxidation than has ordinary cast-iron. See also Silicon-aluminium Alloys, Silicious Iron,* Silicious Steel* and Silicon.*

Silencer.—A sheet-metal cylinder or box with internal baffle plates in which the exhaust gases lose pressure, expand and are cooled. It reduces the noise of the gases when discharging into the air.

Silicate Bond.—Silicate grinding wheels are those in which the abrasive is bonded with sodium silicate. Wheels produced by this method have a mild cutting action and find application in the grinding of fine-edged tools, and knives of various kinds, etc. They should not be used for rough grinding. Of the several modifications of silicate bonded wheels, one gives an extremely open structure and brittle bonding, thereby making it especially suited to surface-grinding operations using the side of a cup or ring wheel, where there is a large area of contact.

Silicon-aluminium Alloys.—Light alloys which in many cases have superseded the copper-aluminium alloys. They possess excellent casting properties, including high fluidity, which makes possible the casting of large, thin sections ; they are free from hot-shortness, and take sharper impressions in the mould than any other aluminium-base casting alloy. Because of their tendency to form a dense skin, they are particularly suitable for castings which must be free from porosity and pressure-tight. They are notably resistant to corrosion, and are consequently widely used for many marine and architectual purposes. See also Silicon.*

Simple Harmonic Motion (S.H.M.).—If a point P describes a circle with constant velocity and a perpendicular PN be drawn to the diameter, the motion of N is a Simple Harmonic Motion. A familiar example is the motion of the bob of a pendulum moving in a small arc.

Simple Pendulum.—The simple pendulum is an *ideal* pendulum supposed to consist of a mass collected at a point and supported by a string having no mass. Such a pendulum has only a theoretical existence ; its length is realised in Capt. Kater's pendulum (q.v.).

Single Curve Teeth.—Involute teeth (q.v. in Pt. 2, Basic Terminology), so called because their flanks and faces, above and below that which corresponds with the pitch line in double curve gears, are formed with a single continuous curve only. (See Double Curve Teeth.)

Single-ported Slide Valve.—The common valve, the travel of which is the same as the throw of the eccentric. The term is merely used to distinguish it from the double- and treble-ported valves.

Sintered Bearings.—Bearing bushes produced by moulding metallic powders. A typical mixture consists of 90% copper powder with 10% tin powder ; graphite is also sometimes added. The powders are moulded in self-ejecting presses under pressures of about 40,000 lb./sq. in. and the moulded pellets afterwards sintered, or furnace-heated, in a reducing atmosphere at a temperature of about 700° C., followed if required by quenching in lubricating oil. The bearings can be made slightly porous to retain oil if necessary. Self-lubricated sintered bearings are used on automobiles, aeroplanes, electric machines, agricultural machinery, in heavy plant such as conveyors and in small apparatus such as typewriters. Porous-moulded bronze bushes are also used in machinery for textile and other trades where oil splashing must be avoided.

Sintered Carbide Tool.—Intensely hard cutting tool made by sintering (q.v.) tungsten carbide, which is usually mixed with titanium carbide and an additional metal (usually cobalt) which acts as a binder. In the United States of America tantalum is often used instead of tungsten carbide. Sintered carbide tools usually take the form of tips which are brazed to a steel shank. The alloys are generally known under trade names such as Ardoloy, Carboloy, Cutanit, Elmarid, Tecometal, Wimet, etc.

Sintering.—Term used to describe the bonding together of separate particles of a powdered metal or metals under the influence of heat and pressure, the term usually being applied only when the temperature at which bonding occurs is below the melting point of any of the constituent metals. The process is used in the production of hard metal cutting tools, porous bronze oil-retaining bearings, oil pump gears, permanent magnets and other parts. (See Sintered Carbide Tool, Sintered Bearings.)

Skewback.—The bearing- or springing-joint of an arch.

Slabbing Machines.—Large milling machines of the planer type used for taking wide and deep cuts off heavy work. The cutters are generally made with inserted teeth. See also Slabbing.*

Slag Pit.—Depression made beneath the tap hole of a Siemens steel melting furnace to receive the slag after the metal has been tapped out. Also, the similar pit beneath a Bessemer converter.

Sleeve.*—A tube into which a rod, piston or another tube is inserted, as when a cylinder has worn so that steam or gas leaks past the piston, a sleeve is forced into the cylinders so that the piston fills it.

Sleeve Valves.—Hollow cylinders fitted between the piston and block and caused to reciprocate. Ports in the sleeves register at proper intervals with ports in the block, to provide for induction and exhaust. Alternative to poppet valves.

Slenderness Ratio.—The ratio of length to least radius of gyration in the case of columns or struts. It measures the relative resistance of different kinds of columns to buckling.

Slider Crank Chains.—Kinematic links, which are connected by pairs of mechanical elements in such a way that two of the links cannot turn, but must slide simply. It may be supposed to be derived from a *lever crank chain* (q.v.), two of the links of which are made of equal length and infinitely long. An illustration occurs in the rectilinear motion of an engine slide block, moving in guide bars, converted into the angular and rotary motion of the connecting rod.

Slip* (see Pt. 2, Basic Terminology).—Also, the difference between the volume swept through by the plunger of a reciprocating pump and the actual discharge. It may be positive or negative, since it sometimes happens that due to the rise of pressure in the cylinder towards the end of the stroke, the discharge valve opens and allows water to flow through it, before the plunger has completed its stroke. The slip is then negative and the discharge is greater than the cylinder volume.

Slip Gauge.—Extremely accurate gauge blocks used to verify dimensions to 0·00001 in. The gauge blocks must be kept at a constant temperature of 68° F. They may be combined into a frame to form extremely accurate snap gauges (q.v.).

Slip-rings.—Current-carrying rings made of copper or bronze and shrunk on to a sleeve, from which they are insulated with mica.

Slot Drill.—Tool used in slot drilling (q.v. in Pt. 2, Basic Terminology). It is a flat-ended double-cutting drill, having two radial cutting edges opposed to each other, and no centre point. Being traversed as it operates, it is also termed a traversing drill. See also Slot Drilling.*

Smoke Tubes or **Fire Tubes.**—The tubes of multitubular boilers, through which the flame and hot gases pass from the fire grate to the chimney.

Snap Flask.—A moulding box hinged at one corner and fastened with a snap at the corner opposite. It is used for boxless moulds made on machines.

Snap Gauge.—See Gap Gauge.

Snecked Masonry.—This is either coursed or uncoursed rubble.

Snift Valves.—These valves open for the passage of air, but close if liquids attempt to pass. They are used in hydraulic systems and for hydraulic rams to keep the air-vessels supplied with air. See also Snifting Valve.*

Snout Boring Machine.—One in which the cutters are carried on an extension of the spindle—the snout—without any other support. They are of special value in boring blind-ended cylinders, and motor cylinders generally.

Soap Film Technique.—Method of stress analysis (q.v used to evaluate stresses in parts of complex section subjected to torsion, in which a soap film is stretched over a hole of the same cross section as the part under investigation. Prantle observed in 1903 that the displacement of a thin membrane of this type, when stretched over such an opening and slightly distended by a uniform pressure, produced partial differential equations corresponding to those employed in calculating stresses in equivalent sections. The volume of the bubble formed over the experimental hole is measured, and for a given twist in a bar, the torque is proportional to the volume of the bubble. Stress at a given point is also determined by measuring the inclination of the bubble at that point by optical means.

Soda Ash.*—An alternative term for sodium carbonate, or commercial washing soda in powder form.

Soffit.*—The lower surface or intrados of an arch, architrave or projecting balcony.

Solenoid.—An electromagnet comprising a hollow coil of wire enclosing a movable iron core, the movements of which respond to the intensity of the magnetic field when current is passed. It is either shunt or series wound.

Soluble Oil.—Cutting fluid (q.v.) compounded of a mineral oil and/or a fatty oil, to which is added an emulsifier. When mixed with water a soluble oil may produce a milky fluid ; recent tendency, however, is to use transparent soluble oils, which are more stable and allow a better view of the work.

Solution Treatment.—The heating of an alloy to a temperature (below its melting point) at which certain constituents are completely soluble within the metal, followed by quenching to maintain the supersaturated solid solution. Solution treatment is applied to many magnesium and aluminium alloys, often as a prelude to age-hardening (q.v.).

Sorbite.—Structure formed when a fully hardened steel is tempered at between 550° and 650° C. It may be described as evenly distributed fine particles of carbide in a ground mass of ferrite. Its microstructure appears minutely granular.

Space-average.—When the magnitude of a force, acting on a body in the direction of its motion, is plotted on a straight base which represents the *distance* through which the force acts, the average value of the force, or the mean height of the diagram is the *space-average* of the force. (See Time-average.)

Sparking Plug.—Device used in the cylinder of an internal combustion engine, whereby an electric spark is passed between two electrodes, through the explosive mixture. The plug portion is screwed into a tapped hole in the cylinder head.

Specific Speed.—For a water turbine the specific speed is the speed at which the turbine will run to produce one horse-power under a head of one foot. Sometimes called the *unit speed* or the *type characteristic* of the turbine, because each type of turbine has its own specific speed. For a centrifugal pump, the specific speed is the speed at which the pump will deliver one gallon of water per minute under a head of one foot. Its specific speed characterises the type of pump.

Specified Length.—All sections of rolled steel are cut to a margin of one inch over or under the specified lengths. When sections are ordered to be *exact*, which means within ⅛ in. of the specified length, an extra charge is made.

Speed.—Speed, technically, differs from velocity, though the terms are constantly used as synonymous. Thus speed is the rate of motion of a point along its path, and is therefore measured by the distance moved divided by the time taken ; it is a scalar quantity, whereas velocity is a vector quantity and is the rate at which a body changes its position and is therefore the distance in the straight line joining its initial and final positions divided by the time. Speed and velocity are therefore strictly only the same for motion in a straight line. See also Speeds.*

Speedometer.—Instrument driven from the gear-box or from a wheel of a road or rail vehicle. A centrifugal or magnetic device indicates the speed and a mileage-recording device is often included.

Speed Recorder.—A tachometer incorporating mechanism by which an autographic diagram is made on paper carried on a rotating cylinder.

Speed Reduction.—Rendered necessary by the general adoption of motor drives. The practice is in favour of discarding bevel gear reductions, using spurs only. For high reductions double helical gears are used, or epicyclic trains. Raw hide pinions are common.

Spherical Engine.—The Tower engine once in vogue, which consists of a pair of quarter spheres, hinged to a disc along diameters at right angles with each other, and enclosed in a hollow sphere of the same diameter as the disc, with the curvature of which also that of the quarter spheres corresponds. The hollow sphere is steam tight, and forms the cylinder. The two spaces between the sections and disc will in the course of a revolution open and close in pairs, and this is the effect therefore of the action of steam admitted into the hollow sphere. In principle the spherical engine is a Hooke's universal joint, the sections corresponding with the bows, and the disc with the cross piece connecting the bows.

Spherical Grips.—Grips for holding specimens under test, incorporating spherical seatings that have capacity for accommodation, ensuring that the pull on the specimen shall be perfectly axial.

Spherometer.—Instrument which stands on tripod legs and which by means of a central micrometer screw measures the height of convex or concave surfaces above or below a zero mark.

Spigot.—A pin or peg used to stop a vent or to command the opening through a faucet (q.v. in Pt. 2, Basic Terminology).

Spigot Bearing.—Bearing carrying two shafts in line, and allowing them to turn independently.

Split Key.—A key (q.v.) which is split at one end similarly to a split pin, to lessen the tendency to work out of its bed.

Split Rivet.—See Bifurcated Rivet.*

Spot Weld.—Localised weld formed by electrical resistance welding. See Spot Welding.

Spot Welding.—Method of electrical resistance welding in which over-lapping edges are united at intervals by passing a heavy current through two copper electrodes which are at the same time forced together by mechanical, hydraulic or pneumatic power. Spot welding forms a cheap and satisfactory substitute for riveting and is capable of precise control of the duration and programme of heating by electronic timers (q.v.).

Sprag.—A form of pawl and ratchet fitted so as to prevent a vehicle from running backward on a hill, should the brakes fail.

Spring Base.—The distance from centre to centre of the spring cradles or hangers upon which the ends of an arched spring are pivoted.

Spring Buckle.—The forging which embraces a sheaf or bundle of arched springs, at the centre.

Spring Compass.—The amount of curvature of an arched spring.

Spring Cradle or Spring Hanger.—The forging which sustains the ends of arched springs, either with pins or with hooks.

Spring Governors.—Those designs in which over-sensitiveness and the tendency to hunt is checked by the controlling action of springs.

Spring Hammers.—Power hammers used for die forging and other purposes, in which the tup is suspended from laminated springs disposed either vertically or horizontally to lessen jar.

Spring Wheel.—One in which, by the action of springs interposed between the shaft boss and the rim, irregular driving forces and shocks are neutralised.

Sprung Weight.—A term used in automobile engineering to include all the weight of a car which is carried by the springs, including the frame, engine, radiator, clutch, transmission, body, passengers, etc.

Spur Gear Cutting.—This is now done mostly on fully automatic machines of strong design, often using cutters of high-speed steel. Single rotary disc cutters divide favour with hobs, and with the Fellows' pinion-shaped cutter, the last two operating on the generating principle. See also Spur Gearing.*

Stacking Truck.—Motor-driven truck provided with a platform or with forked arms designed to fit into pallets (q.v.). The platform or arms are elevated by a chain and sprocket, a rack and pinion or an hydraulic cylinder, enabling goods to be lifted, carried and stacked without manual effort.

Staggered.—Applied to rivets, when so placed that the rivets in one row alternate with those in the next row. Also, the spokes of a wheel are said to be staggered when they are inclined in opposite directions alternately.

Stamps.—Pieces of refined iron of about 28 lb. in weight each, obtained by nicking and breaking up the blooms shingled under the steam hammer in the refining process, for the manufacture of coke plates. See also Stamp.*

Standardisation.—An immense movement which has taken place in recent years in favour of uniformity in the dimensions of manufactured materials and goods. Many volumes of such dimensions have been published. See British Standards. In the shops also standardisation of products goes hand-in-hand with interchangeability and specialisation.

Standards.—The standards of the British yard measure and pound weight are deposited in the House of Commons. The standard yard consists of a solid bar of brass thirty-eight inches long, and one inch square, and the lines representing the length of thirty-six inches are cut upon gold plugs inserted and sunk in the bar. This is kept in a mahogany box sealed up in an oak case. The standard pound, of seven thousand grains, is a mass of platinum, in the form of a cylinder, nearly 1·35 in. high, and 1·15 in. in diameter. There are four Parliamentary copies constructed, and forty copies for distribution among foreign governments. They are made of "Baily's metal," consisting of 16 parts copper, 2½ parts tin, and 1 part zinc. See also Standard.*

Standard Corrective Gauge or Corrective Gauge.—A cylindrical step gauge made of steel, which is turned in a succession of short parallel lengths or steps, to six or eight different standard diameters, and furnished with a handle like an ordinary parallel cylindrical male gauge. Sizes advance either by sixteenths or eighths of an inch, the smaller gauges beginning at $\frac{3}{16}$ or $\frac{1}{4}$ in., and the larger ones range as high as 3 in. or 4 in. in diameter. These gauges are not made for the testing of work in progress, but for the periodical testing of the shop gauges, by which work is bored and turned.

Standing Wave or Hydraulic Jump.—When water moving through a sluice or running down an incline at a high velocity impinges on water moving slowly, the loss of momentum due to the impact causes a rise of pressure which produces a wave known as a *standing wave*. The difference of head is used to measure the discharge in the Venturi Flume (q.v.).

Standpipe.—Vertical pipe, open at the top, connected to a pipe line to ensure that the pressure head at the point of connection cannot exceed the length of the standpipe. When used in connection with water turbines, the lower end is connected to the penstock near to its connection with the turbines and the standpipe is of such a height that when exposed to the static head in the supply reservoir, the water level is within a short distance from the top of the pipe. Any sudden increase in pressure at the turbine, due to a sudden closing of the gates, produces an overflow from the standpipe. At the same time, a sudden demand for power is assisted by a fall in the standpipe, which functions similarly to an accumulator. See also Stand Pipe.*

Star Gears.—Variable gears (q.v.). Also the term star wheel signifies the knuckle type of gear, the movement of which, tooth by tooth, imparts the feed to facing heads and some tool boxes. See also Star Wheel.*

Starling.—An enclosure consisting of piles driven closely together into the bed of a river and secured by horizontal pieces at the top. The space between the rows of piling being filled with gravel or stone, the whole forms an effective protection for the foundation of a pier.

Starter Motor.—Electric motor used to start an internal combustion engine. The shaft carries a Bendix pinion (q.v.) mounted on a screw-thread which draws it into mesh with teeth on the flywheel rim.

Stator.—The stationary portion of a rotatory field induction motor. It carries the coils which receive the primary current.

Steam Blower.—A pipe and cock on a locomotive, employed to create a draught before the engine starts. The pipe leads into the chimney, and when steam is turned into it, the action is similar to that of the blastpipe in intensifying combustion when the engine is running, except that the blast is continuous instead of intermittent, as in the case of the blastpipe.

Steam Heating.—Valuable heating system when exhaust steam would otherwise run to waste. Radiators composed of coils are used for diffusion as in hot water systems.

Steam Loop.—Burnham's steam loop is a device for economising fuel, by the return of condensed water to a steam boiler without the use o₁ pump or injector, and in which it is not necessary that the water should be made to flow by gravity. A bent pipe is carried from the condenser or separator, as the case may be, consisting of two vertical portions and a horizontal portion. The vertical portion from the condenser or separator is called the "riser," the level length the "horizontal," and the vertical portion to the boiler the "drop leg." These constitute the "loop." The arrangement is such that a current of steam from the boiler causes condensation in the loop, and the water from the condenser or the separator flows in to occupy the partial vacuum thus formed, and gradually a constant flow of feed water is thus obtained.

Steam Packing.—Packing which is used in the stuffing boxes of engine cylinders, steam chests, etc., to render them steam-tight without interfering with the free movement of the rods. The term is used in opposition to water packing (q.v. in Pt. 2, Basic Terminology). The materials used are flax, or hemp, well saturated with grease ; asbestos, or divided metallic rings. The packing is placed in the stuffing box and tightened around the rod by the screwing down of a gland.

Steam Turbine.—A rotary engine, designed originally by the Hon. C. Parsons, to be coupled directly to a high-speed dynamo, now used for driving in electric light stations, and in torpedo boats and destroyers, and for passenger and warships. High-pressure steam is directed against moving turbine vanes through fixed guide-blades.

Steel Boilers.—The use of iron for steam boilers has long been abandoned for that of steel. The early difficulties which arose from hard brittle plates have been overcome, and a mild, low carbon steel plate is more reliable in every way than one of iron. There is also the advantage that plates of larger size are rolled in steel, and the number of riveted seams is diminished. Steel rivets are also generally made use of. Simultaneously with the development of the steel boiler the machinery for boiler making has grown in size and power, while hydraulic riveting and pneumatic caulking have almost wholly displaced the work of this kind formerly done by hand.

Steering Gear.*—In the case of motor vehicles the mechanism usually consists either of a worm and wheel, or of a cam and follower converting rotary movement of the steering wheel into a fore and aft swing of a lever called the drop arm ; this moves the wheels by a drag link, track rod and steering arms. (See also Pt. 2, Basic Terminology ; and Rapson's Slide.)

Steering Lock.—The maximum angular amount that the front or steered wheels of a vehicle can swivel from side to side. It is sometimes indicated by the minimum circle in which the vehicle can turn.

Stellite.—A term used in metallurgy to describe an alloy of cobalt, chrome, tungsten, and sometimes molybdenum and iron, whose good corrosion resistance and hot-hardness qualities make it especially suitable for use in the manufacture of surgical instruments and for the seating rings of exhaust valves.

" Stelvetite."—The registered trade name (John Summers & Sons Ltd.) for steel in sheet or coil to which PVC is bonded on one or both sides. The standard product is on an electro zinc plated base, but for external use and extra corrosion resistance it is also available on a hot dip galvanized base.

Stiffening Girder.—Girder used to prevent the cable of a suspension bridge from oscillating. This girder is attached to the main girder by suspension links or hangers, and is often jointed at the centre to the cable.

Stirrups (see Pt. 2, Basic Terminology).—Also, U-shaped pieces of steel used for reinforcing Tee beams of concrete against shear. See also Stirrup.*

Stitch Welding.—Method of welding in which a series of overlapping spot welds (q.v.) are produced by passing the work between the electrodes of an automatic welding machine.

Stocking Cutter.—Term applied to a milling gear cutter that roughs out the material between teeth, preceding the action of the finishing cutter. The operation is termed stocking, or gashing.

Storage Battery.—An electrical accumulator comprising positive and negative plates in a cell with dilute acid or an alkali. Electricity charged into the cell from an outside source is afterwards converted into work.

Straddle Mills.—Milling cutters that are built up to machine the outer and opposite edges of a piece of work, and generally the top face as well. They are gang mills (q.v.).

Straight-eight.—An engine with eight cylinders in line.

Straight-line Formula.—Formula which gives approximately the buckling load on a column ; so-called because it is represented graphically by a straight line, and is used as a simplification of parabolic and other formulæ of an empirical nature. It is sufficiently accurate within certain limits which occur in usual practice, but these limits should be kept in mind.

Strainer-rack.—Racks placed in the head race which supplies water to a turbine. They are usually made of flat bars held together by bolts and separators and are generally placed at an angle varying from 45° to 60° with the vertical, to prevent floating bodies from entering the turbine.

Strain Gauge.—Device used to measure the distribution and magnitude of stresses in materials. Early electrical strain gauges took the form of carbon strips which were secured by an insulating adhesive to the component under test. Extension or compression of the material along the axis of the gauge increased or decreased the resistance of the gauge to the flow of current, which was measured by a sensitive galvanometer and interpreted in units of stress. In modern stress analysis the same principle is employed but the gauge is made of a flattened coil of fine wire sandwiched between two filter papers and is cemented to the part to be tested. The gauges are orientated in accordance with the probable directions of stresses, as determined by prior calculation. As many as several thousand gauges may be used on an elaborate structure such as an aircraft. Automatic instruments are used to record the readings of the gauges. See also Strain.*

Stream-lined.—A body which is so shaped to offer as little resistance as possible to movement through a fluid at high velocity.

Stress Analysis.—Investigation of the distribution of stresses within a material, combined, where possible, with their measurement. In qualitative analysis only the distribution of stresses is ascertained; in quantitative analysis individual stresses are measured. (See Brittle-lacquer Technique, Electron Diffraction Analysis, Electronic Strain Gauge, Photoelasticity, Soap Film Technique, Strain Gauge.) See also Stress.*

Stresscoat.—A brittle lacquer coating used in stress analysis. (See Brittle Lacquer Technique.)

Stringers or **Rail-bearers.**—These carry the weight of platform and axle loads on to the cross-girders of a railway bridge to which they are attached at 5-ft. centres with the ordinary 4 ft. 8½-in. gauge.

Stripper Plate or Shipping Plate.—A metal plate used on many moulding machines, through which the pattern is withdrawn. Its function is to prevent the sand from becoming broken down by the withdrawal. Plates are filed to the pattern outlines, or, in the more intricate forms, white metal is poured around, supported by a ledge a little larger than the pattern. Also used in stamping operations to strip the work from the dies.

Stroboscope.—Device used to study rotational or reciprocating motions by presenting a series of accurately-timed images of the moving object. Modern stroboscopes are of the electronic type, in which the light flashes used to illuminate the object can be controlled with great accuracy. Among the applications of the stroboscope are the analysis of motion and detection of wear, distortion or chatter in moving parts. In addition the stroboscope can be used as a tachometer by adjusting the rate of flashing until a rotating part appears to be brought to a standstill when viewed by the light from the flashes.

Stub Axle.—Stub axles are small axles carrying the wheels and swivel pins of a vehicle for steering purposes.

Stud Welding.—Method of percussion welding (q.v.) used to weld studs to metal parts. An arc, struck between the end of the stud and the part, raises the end of the stud to melting point and produces a molten pool in the part into which the stud is driven. In the most widely used system the stud is held in a special "gun" and is pressed against the work by a spring. The contact area is cleaned by a small electric current before the trigger button is pressed to apply the welding current of 60–90 volts D.C. On pulling the trigger of the gun the full current is applied and the stud is lifted slightly to strike the arc, which is extinguished after a predetermined interval by a relay in a preset timer that controls the action of the gun.

Sub-press.—A small press or die and combination forms, for stamping delicate work, instead of making the dies, punches, etc., fitting directly to the main press.

Sub-zero Heat Treatment.—The introduction of one or more cooling periods, at a temperature well below the freezing point of water, in the normal heat treatment of steel. Briefly, the object of sub-zero cooling is to ensure complete transformation of austenite to martensite ; in many steels the transformation is not completed by a direct quench to room temperature. One cooling at −100° F. is sufficient for many plain carbon and low-alloy steels ; steels of high alloy content, however, may require several sub-zero coolings, with a tempering and air cooling cycle between each.

Suction Gas Producer.—Gas producer (q.v. Pt. 2, Basic Terminology) in which, instead of pressure, suction is employed for the air supply to the grate, in combination with a regenerative device, in which steam is mixed with the air. The " suction " is produced by the pull of the engine on the outward stroke, which causes a reduction of pressure through the plant, and so draws air in.

Suction Tube or Draught Tube.—Consists in a lengthening of the vertical discharge pipe in a water turbine until its lower end discharges below the surface level in the tail-race. This allows the turbine to be placed at any depth up to about 25 feet above the tail-race, without loss of working head. It also permits of convenient inspection of the turbine.

The end of the tube is flared at an angle not exceeding 15°, to aid the discharge. The tube is often turned at right angles so as to discharge in the direction of flow into the tail-race. This avoids deep excavation.

Sulpho-chlorinated Oil.—Cutting fluids (q.v.) which contain sulphur and chlorine in an active form. Usually based on a mineral oil, the fluid is chemically treated so that the sulphur and chlorine are contained in the same molecule; as a result, the oil film assumes a uni-molecular thickness before rupturing under load, and is capable of withstanding very much higher pressures than an oil which does not contain sulphur and chlorine. Sulpho-chlorinated oils are usually blended, in the proportions of 5–20%, with other oils of suitable viscosity.

Sulphurised Oil.—Cutting fluid (q.v.) consisting of either a mineral or a fatty oil processed to produce a free sulphur content which improves lubrication.

Sump.*—A well from which liquid is pumped. Used in mining, and in machine tools.

Supercharger.—Type of rotary pump designed to deliver petrol-air mixture to the cylinders of an internal combustion engine under pressure and so to supplement piston suction and to increase power output. The process is sometimes called " forced induction."

Superposed Frames.—Two or more frames may be superposed to form a *compound* frame, the members of which may be supposed to fulfil distinct functions in each of the component frames. Thus a lattice girder may be regarded as obtained by superposing a Warren girder upon another similar one inverted. The Fink and Bollman trusses are further examples of such compound frames, the only common members in them being those which make up the top boom. Each of the frames in such cases may be examined separately, the members which are common to both frames having to bear the stresses equal to the sum of those which are determined for the component frames separately.

Supersonic Flaw Detector.—Apparatus which can detect all types of flaws, from microscopic hair-line cracks to blowholes, porosity and segregations, and which can locate the fault accurately within the body of the material at any depth below the surface, from $\frac{1}{2}$ in. to 30 ft. Two small probes are applied to the surface of the test piece. From one, a narrow beam of supersonic energy is directed into the material. This energy is reflected back to the second probe by the opposite boundary of the material and also by any cracks or other faults. The receiving probe is connected to an amplifier which feeds a signal to a cathode-ray tube (q.v.). On the screen of the tube is traced a luminous straight line, on which are two or more sharp deflections. The first deflection marks the point of entry of the beam of energy into the material; other deflections are caused by reflections from flaws ; while the last deflection (termed the bottom mark) indicates the reflection of the beam by the far boundary of the test piece. The location of any flaw can thus be determined from the position of the probes on the surface of the test piece and from the position of the flaw mark on the screen in relation to the entry mark and the bottom mark.

Surface Meter.—Stylus-type of instrument for measuring the texture of surfaces. It magnifies irregularities from 400 to 100,000 times and provides both a graph showing a cross-section of the surface, and a number representing the centre-line average height of the texture. A pick-up unit having a diamond-pointed stylus is traversed across the surface by means of a motorised driving unit. The up-and-down movements of the stylus as it rides over the surface are converted into a correspondingly varying electric current, which is amplified and then applied to one or the other of two measuring instruments. One of these instruments is the Recorder, which provides a graph drawn on paper, representing the geometrical shape of a cross-section of the surface undulations. The other instrument is the Average Meter, which gives the centre-line average value of the undulations. (See Profilometer, Talysurf, Talyrond.)

Surge Tank.—Large tank mounted on the top of a standpipe (see above). On account of its large surface area, it reduces the range of oscillations in the standpipe, due to sudden changes of pressure.

Swabbing.—Moistening the edges of a foundry mould with the water brush (q.v. in Pt. 2, Basic Terminology) in order to effect the due consolidation of the sand. Swabbing must be only moderate in amount; if excessive it tends to produce scabs and blowholes.

Sway Bracing.—Consists of braced or plate girders connecting the main girders of a *through bridge* (q.v.) at their upper ends. It stiffens the bridge against wind pressure and from distortion due to deflection when a load crosses it.

Swell of Pulley.—The curved surface of a pulley rim, with the object of keeping the belt from working off when running.

Swing Bridge.—Bridge which turns on its centre support, in order to allow ships to pass when required.

Switch.*—Device for making or breaking an electric circuit, of which there are many designs. The contacts are of metal or carbon. In some cases a liquid, a solution of soda, serves as a resistance. Switches are used with controllers to regulate speed, as on tramcars, cranes, etc.

Switch-board.—A board or tablet to which wires are led connecting with cross bars or other switching devices to enable connections among themselves or with other circuits, to be made.

Synchronous Motor.—An alternating generator of electricity, if brought up to speed, excited and connected to a suitable alternating current supply, will run as a motor. Such motors are said to be synchronous because they will operate only at approximately synchronous speeds. They will not start under load, but have to be started by auxiliary means.

Synthetic Resins.—(1) Plastics, usually in liquid or powder form, of the thermo-setting or thermoplastic type, extensively used as adhesives in bonding a wide range of materials. The application of heat and pressure may be required, as with the Redux process of bonding aircraft metal structures. (2) Another range of synthetic resins, the cold-setting polyester and epoxy types, are self-curing when mixed with a suitable catalyst. These are now widely used, usually with glass fibre as a reinforcement, in the construction of laminated components such as boat hulls, motor car bodies and innumerable other components.

System of Limits.—See Limit System, American Standard System of Limits, British Standard System of Limits, Newall System of Limits.

T

Tachometer.—Instrument by means of which the speeds of shafts of high-speed engines are indicated in revolutions per minute. The essential mechanism is that of the high-speed spring-loaded governors.

Tacking.—A term used to refer to local joining of metal parts by welding, brazing or soldering. The object is to assist in locating and temporarily securing a number of items or pieces after alignment, prior to completing the assembly process or repair.

Tail Pin.—The back centre pin of a lathe.

Talyrond.—Instrument for measuring the roundness of parts in which an electric displacement indicator carried on an optically-worked precision spindle is rotated around the inside or outside of the part. The signal from the indicator is amplified and recorded on a polar co-ordinate graph.

Talysurf.—Laboratory instrument for measuring the quality of surface finishes. (See Surface Meter.)

Tantalum.—White lustrous metal of specific gravity 16·6 and having a melting point at 2,800° C. Used for making tantalum incandescent lamps and for alloying purposes. (See also Sintered Carbide Tool.)

Taper Gauges.—Sometimes called *Engineer's Tapers*. Used for measuring hole sizes and the inside diameter of tubing, for sizing slots, for determining the depth of key-ways, for making a quick check of clearances between machine parts, and for ensuring the correct taper of taper holes in machine work. Taper gauges are also used to ensure corresponding accuracy of dimension in the work intended to fit the hole. Several types of gauge exist, but all are essentially wedges having a specified taper, e.g., $\frac{1}{64}$ in. for every quarter-inch of length, and carrying graduations which indicate the dimensions of the wedge at the point in question. The type generally used for hole sizes, tubing, etc. is a strip or leaf of tempered steel of uniform thickness, somewhat similar to a feeler gauge but with the width tapering from $\frac{1}{16}$ in. to (say) $\frac{5}{16}$ in. depending on the length of the leaf. Another leaf will taper from $\frac{21}{64}$ in. to $\frac{9}{16}$ in. and so on. A complete gauge of this pattern generally comprises several leaves of various tapers held together in a case and with the leaves pivoted as in a pen-knife. The type of gauge used for determining key-way depths, sizing slots, etc. is a blade uniform in width but tapering in thickness from about 0·01 in. to 0·15 in. depending on the length of the gauge. This type is usually a single blade with one end shaped to provide a finger-grip. Both types of gauge are also available with graduations in thousandths of an inch for more accurate work. Another type of gauge is the plug or ring type (used for internal and external work respectively) which works on GO/NO–GO principle. See also Plug Gauge, Ring Gauge and Cylindrical Gauge.*

Taper Turning.*—Specifically, turning in lathe by means of a taper attachment, or by means of the set-over poppet. Many lathes are fitted with an attachment in the rear, in the shape of an adjustable bar to which the upper slide of the rest is connected, and by which its sliding movement is controlled. This is better than using the set-over poppet.

Tare.—The weight of a vehicle when empty.

Target Date.—A term used in engineering to indicate the date on which it is desired that an action shall be either initiated or completed.

Taylor-White Process.—Relates to the early high-speed tool steels, which are hardened in air and operated at a blue heat.

Technotherm-Rakos Welding.—Method of fusion welding in which the edges to be joined are heated by the passage of a low-voltage current from a single carbon electrode.

Tecometal.—Sintered carbide alloy used in the production of cutting tools. (See Sintered Carbide Tools.)

Telephone.—An instrument for the transmission of articulate speech by means of the electric current. A microphone (q.v.) is used as *transmitter* and the *receiver* is a telephone at the far end. The vibrations on the transmission diaphragm are reproduced on the receiving diaphragm of the telephone by the electric current, thereby reproducing the sounds originated by the voice at the transmitter.

Telpherage.—An electric transportation system used for the carrying of ore, freight, etc. Two electric conducting lines, running parallel to one another, are supported on transverse brackets fixed to a row of supporting poles. The cars are suspended from pulleys running on one or the other of the conductors. A train of cars are connected, and the current is taken in near one end and leaves at the other.

Tensometer.—Small portable testing machine, having a capacity of up to 160 tons per sq. in., on which load-extension diagrams are autographically recorded ; in addition, elongation and reduction of area are read from indicators without the necessity for calculation. Brinell hardness tests, notched-bar tests, transverse and strip tests can also be carried out.

Terminals.—The ends of any open electrical circuit of an electrical apparatus, such as the terminals of a dynamo or battery.

Theorem of Three Moments (see Clapeyron's Theorem). See also Theorem.*

Therblig.—Symbol used in a form of shorthand employed by motion study specialists. The name is derived from that of the originator of the system, Gilbreth, spelt backwards. Eighteen basic motions are covered ; for example " search " is represented by the outline sketch of an eye with the pupil turned as if searching, " inspect " by a rough sketch of a magnifying lens, and so on. Many motion study specialists, however, find therbligs too limited in scope, preferring to adopt symbols which show more clearly the exact path taken by the operator's hand.

Therm.—A unit of heat proposed by the British Association. It is the heat required to raise the temperature of one gram of water one degree Centigrade, starting at the maximum density of water (4° C.). See also Thermal Unit.*

Thermionic Valve.—Vacuum tube containing a heated cathode from which electrons are emitted, an anode which collects some or all of the electrons, and, usually, one or more additional electrodes or a grid which controls the flow of electrons to the anode. Originally developed for wireless telegraphy and telephony, the thermionic valve has numerous applications in measuring, control and timing systems used in engineering. (See Thyratron.)

Thermit.—A welding compound of powdered aluminium and iron oxide, on the ignition of which intense heat, approximately 3,000° C., is evolved.

Thermit, Cast Iron.—Plain thermit to which is added ferro-silicon and mild steel, used in the thermit welding of cast iron.

Thermit, Forging.—Plain thermit to which is added carbon, manganese, nickel or other alloying metal, and mild steel. Used for the welding of castings, steel forgings and rails.

Thermit Welding.—Method of welding in which heat is obtained by the ignition of thermit (q.v.). The thermit powder is packed into a crucible, lined with a refractory cement, which is clamped above a mould surrounding the parts to be welded. Ignition of the powdered thermit produces a small quantity of molten steel within less than a minute; this is released into the mould by tapping up an asbestos sealing disc at the base of the crucible.

Thermocouple.—Welded circuit consisting of two wires of dissimilar metals or alloys, in which an e.m.f. is generated when one junction is heated or cooled while the other junction is maintained at a constant temperature. In practice, the cold junction is formed by the attachment of the wires to the leads of a galvanometer or potentiometer by which the thermoelectric difference at the hot junction is determined.

Thermostat.—Automatic device used to maintain a constant temperature in a given space by cutting off or reducing the supply of heat if the required temperature is exceeded and restoring the supply when the temperature falls below the basic figure. The methods of actuating the regulating device vary widely with different applications. Among them are : the use of a bi-metal strip or spiral ; a bellows or capsule filled with an expansible liquid or gas, or a liquid which boils at the required temperature ; a steel bulb containing mercury which is connected by a capillary tube to a Bourdon tube ; the use of a pyrometer in which the junction between dissimilar metals, or thermo-couple (q.v.), produces an e.m.f. with rise of temperature ; and the use of the change in electrical resistance of a coil with variations in temperature. See also Thermostatics.*

Thermosiphon System.—Circulation of water by conduction, due to the difference in density of hot and cold water.

Thickness Gauge.—See Feeler Gauge.

Thoroughfare Taps.—Taps whose square heads are small enough to allow them to pass through the hole which has been tapped by them, and so save the time which would otherwise be lost in running them back.

Thread Gauge.*—Gauge used for the measurement of screw threads. Thread gauges may be of the ring or plug gauge type (q.v.) or may take the form of a set of profile gauges accurately cut to the form of the threads to be checked.

Thread Rolling.—The formation of screw threads by rolling pressure instead of by cutting tools.

Three-phase Motor.—An electric motor in which three currents are employed, each of equal amplitude and alternating with the same frequency. Used chiefly for large generating stations.

Thrilling.—Has the same meaning as knurling (q.v.).

Throat Plate.—The plate in the outer firebox of a locomotive boiler, to which the lower portion of the barrel is riveted. It is flanged forwards at the top to receive the barrel, and backwards at the sides to receive the arched sheet of the outer firebox.

Throttle Governing.—Governing an engine at the throttle valve instead of by the shaft governing of the slide valve.

Through Bridge.—A bridge in which the roadway is carried by side girders often of considerable depth, which rise above the roadway, so that the traffic passes between them. In contradistinction to a *deck* bridge, in which the main girders are beneath the level of the roadway.

Thyratron Valve.—Valve used in electronic timing controls consisting of a plate, a hot filament and a grid. When the filament is heated an applied voltage causes electrons to flow to the plate. The valve is then a conducting medium and the current flowing through the thyratron valve is used to operate a contactor which controls the main current. In order to switch off the contactor the flow of electrons to the plate is prevented by applying a controlling voltage to the grid, the current flow ceasing immediately the control voltage is reached. A timing period is obtained by means of an adjustable resistance in the charging circuit which controls the time taken to charge a condenser up to the grid control voltage.

Time and Motion Study.—Scientific analysis of the time taken and the movements made by the operator in carrying out the various elements of a complete task, or of a sub-division of an operation. The application of time and motion study in modern factories has resulted in marked savings in labour, bench space and floor space, combined with improved working conditions and lessened fatigue for the operators. The majority of time and motion study specialists prefer to treat the subject under the two separate headings: Time Study and Motion Study (q.v.).

Time-average.—When the magnitude of a force is plotted on a straight base which represents time to *scale*, the average value of the force, or the mean height of the diagram, is the *time-average* of the force. When the space-average of a force is used, the product of force by time gives *work* done, but when the time-average is used the product gives momentum generated. (See also Space-average.)

Time Recorder.—An instrument in which, on cards or tapes, the times on which work is commenced and finished are registered. A number of these recorders can be synchronised from a master clock in a factory. See also Time Sheets.*

Time Study.—Scientific analysis of the time taken to complete the various movements of a given job. For most job studies a decimal stop watch is sufficient, but for the timing of rapid motions, micro-chronometers or high-speed motion picture cameras are necessary. Time study can be used as a basis for rate fixing and as a guide for estimating times for new jobs.

Timing Gear.—The two-to-one drive connecting the camshaft to the crankshaft. The phase relationship of these rotating parts is called the *timing*, and determines the points at which the valves open and close.

Tolerance.—The difference between the high and low limits (q.v.) of the size of a dimension or part. The tolerance is the variation in size tolerated to allow for reasonable imperfections in workmanship. See also Limits of Tolerance.

Tool Grinding Machines.—A large group, some of which are designed for grinding the single-edged tools used in lathe, planer, etc., others for dealing with milling cutters, others for twist drills. Uniformity in cutting angles and in shapes is in most cases provided for by mechanical adjustments.

Toolmaker's Microscope.—Optical gauge in which a beam of parallel light is projected by a collimating lens on to a mirror which reflects the light past the work to the eyepiece. The latter incorporates a graticule on which is engraved the correct profile of the screw thread, cutter or other part to be examined. Alternatively a projection screen may be fitted over the eyepiece and transparent drawings mounted on it to act as line templates.

Tool Points.—The cutters which are clasped in a tool holder, to distinguish them from solid tools.

Tool Room.—A small department in a machine shop in which tools and gauges, etc., are made, ground, or otherwise kept in order, and whence they are passed into the stores. The practice is superseding the older one in which each man attended to his own tools, and it is a very essential element in machine shop practice, now that so many fine gauges and testing instruments are used, all of which must be kept in perfect order to be of proper service. The tool room is a microcosm of the larger shop, being fitted with its own machines.

Tool-Room Lathe.—A very fully equipped lathe of 6-in. or 7-in. centres used in the tool room in the preparation of blanks for cutters, of various tools, templets, jigs and fixtures.

Tool Steel.—Covers a wide range of carbon steels and high-speed steels, designated by tempers. Each temper, and often different consignments of the same brand, require modifications in treatment for hardening and tempering.

Torsion Bar.—Generally, a straight metal bar having serrations, flats or keys on both ends, which is designed to withstand a severe twisting action along its longitudinal axis. It is often used in the suspension system of vehicles to absorb road shock, with one end of the bar securely anchored to the frame and the other end secured to the arm carrying the road wheel. The bar is subjected to torsion by the weight of the vehicle, thus acting as a spring. See also Torsion.

Torque.—The turning effort or twist which a shaft sustains when transmitting power. See also Torsion.*

Torque Converter.—A form of fluid coupling (q.v.) in which at least one stationary ring of reaction vanes is incorporated to redirect the flow of fluid returning from the turbine to the impeller, thus assisting the action of the engine to rotate the impeller and increasing the effective torque. More generally, the term can also be applied to any device that multiplies torque, i.e., a conventional gearbox (q.v.).

Torque Tube.—A member in a motor vehicle enclosing the propeller shaft, which resists the torque reaction of the rear axle.

Torsion Meter.—An instrument developed in response to the need for ascertaining the horse-power of steam turbines. The torque of the turbine shaft in degrees is ascertained, and this and the modulus of its rigidity is brought into a formula with diameter, length and number of revolutions. The types made are mechanical, electrical, and flashlight.

Total Heat of Steam or Total Heat of Evaporation.—Comprises the latent heat, in addition to the sensible heat. It increases uniformly with temperature under constant pressure. At temperatures higher than 212° F., the sensible heat increases, and the latent heat decreases, but the total heat of steam slowly increases. See also Total Heat.*

Tough-pitch Copper.—The bulk of copper to-day is the so-called tough-pitch copper, in which a small quantity of free cuprous oxide is present, the oxygen being required to confer suitable casting properties and, in very impure coppers, to keep the impurities in the form of oxides, preventing them from passing into solid solution, in which form they would have very detrimental effects on the physical properties of the copper. See also Tough Pitch.*

Track.—The distance between the wheels of a vehicle, on the same axle. See also Tracks.*

Transformer.—An instrument by means of which electrical currents are changed in regard to voltage and amperage, from high to low, or *vice versa*.

Transition Fit.—Class of fit intermediate between clearance fit (q.v.) and interference fit (q.v.), i.e., extra light drive, heavy, medium or light keying or push fits.

Transmission Dynamometer.—An arrangement by which the work done by a machine is measured whilst transmitting it, such as the epicyclic-train dynamometer, the belt-dynamometer, the Ayrton-Perry dynamometer and others.

Transmission Lines.—The electrical conductors by which current generated at a station is carried to consumers over long distances at high voltages.

Transporter Bridge.—A bridge which does not interfere with the navigation of a stream, nor compel ascent to a high level. Passengers and vehicles are carried on a platform suspended from a carriage which runs across the lower members of a suspension bridge supported by towers on the banks at a height sufficient to clear the river traffic.

Traversing Drill.*—See Slot Drill.

Trembler.—Device used for breaking and making an electric circuit, usually in conjunction with an induction coil (q.v.). There are two kinds : (1) The " make and break," and (2) The " wipe." With the first the trembler is the contact maker, but with the second the trembler is placed on the coil.

Trepanning.—Cutting circular holes, or sinking mine shafts, using a trepan.

Trial Bar or **Proof Bar.**—A bar of blister steel, withdrawn from time to time through the tap hole of the trough or pot, in order that by inspection of its general appearance and fracture, a correct opinion may be formed of the progress of the cementation process. See also Proof Bar.*

Trimming Press.—Any press by which the waste or the fin is removed from sheet metal stampings or die forgings. See also Trimming.*

Trip Engines.—Engines of the Corliss type, and some pumping engines, the valves of which are opened by means of short levers instead of by eccentrics, as in ordinary engines.

Trolley Tracks.—Narrow-gauge railways laid down in workshops. See also Trolly.*

Troostite.—Grain structure produced in steel in two ways : (a) By tempering steel having a martensitic structure, at between 250° and 450° C. troostite is formed. This is a dark-etching constituent having traces of an acicular structure resulting from the original martensite. (b) By quenching steel at a speed insufficient to suppress fully the thermal change point, a dark-etching constituent is seen under the microscope in irregularly shaped masses. It is generally called " troostite," although of recent years this has been shown to be very fine pearlite.

Troughing or **Trough Flooring.**—Steel girders of splayed, rectangular, or arched cross sections arranged side by side to support the heavy floors of bridges or warehouses.

Truncated.—A term used in connection with screw thread forms to convey that the crest or top of the thread and/or root has been cut off, and either flattened or rounded. In the standard Whitworth thread, for example, the crest and root of the thread are truncated from the full angular form, and rounded. See also Whitworth Thread.*

Tubbing.—The cast-iron linings for pit shafts, made in segments and bolted together.

Tunnelling.—Work done under compressed air, tunnels very near to the surface excepted, the work being done in a shield with an air lock, the excavation being lined as the work proceeds with segments set in concrete or grouting.

Tunnel Shield.—A steel structure of cylindrical form and furnished with a cutting edge at the front. It is divided into numerous working chambers with air locks, and is pushed forward with hydraulic jacks as the work proceeds. Essentially it is a caisson operated in a horizontal direction instead of vertically.

Turbine.—Any mechanical device (e.g., a rotary motor) which is designed to utilize the latent energy of existing or generated fluid, steam or air pressure by directing this pressure against vanes or blades fitted to a drum or wheel, with the object of imparting motion (by re-action or impact, or both) to a rotating element such as a shaft, to enable it to perform useful work. See also Turbine, Water.*

Turbine Pump.—Rotary pump in which the work is divided into stages as in a steam turbine. It is designed for lifting against higher heads than the centrifugal pump.

Turbo-generator.—Electric generator driven by steam-turbine, for generating continuous or alternating current.

Turbo-pump.—Centrifugal pump having a ring of guide-vanes at exit from the impeller, which acts as a diffuser ring and reduces eddy losses.

Turnbuckle.—An arrangement for connecting and tightening two lengths of a tie-bar by a swivel. By turning the swivel, into which the ends of the tie-bar screw by right- and left-handed threads respectively, the two parts are drawn together.

Turning Circle.—The circle of minimum radius in which a vehicle can be turned. It depends on the maximum angle through which the front wheels can be turned and on the wheel-base. (See Steering Lock.)

Turning Cylinders or Slewing Cylinders.—The shorter cylinders of an hydraulic crane, the rams of which cause the crane to slew or turn on its centre, as distinguished from the lifting cylinders.

Turret Lathe.—This type of lathe is made in a great variety of forms. It is a labour-saving machine designed to take the place of the ordinary sliding, surfacing and screw-cutting lathe when engaged on repetition work.

Tuyère Box.—The air chamber underneath a Bessemer converter from which the air necessary for the decarburisation of the metal passes to the tuyère openings. Sometimes an air belt is termed a tuyère box. See also Tuyère.*

Twin Turbine.—A twin turbine is one consisting of two similar turbines mounted on the same horizontal shaft and discharging in opposite directions, whereby the end thrust is balanced.

Two-Throw Pump.—A double-barrelled suction pump, operated by two cranks. It is therefore not in equilibrium, as is the case with a three-throw pump.

Tyre Rolling Mills.—Are either vertical or horizontal. The tyres are expanded and shaped between outer and inner rolls in a roughing and a finishing pass. See also Tyre.*

U

Ultrasonic Cleaning.—Method of cleaning in which the part is immersed in water or a fluid solvent which is vibrated at an ultrasonic frequency, causing cavitation and the formation of minute vapour bubbles which subject the surface of the part to the scouring effect of intense micro-agitation. Extremely effective cleaning results, which penetrates tiny crevices, bores, blind holes and fine screw threads. The process is sometimes termed "cold boiling".

Ultrasonic Drilling.—Method of drilling in which a reciprocating tool, vibrating at an ultrasonic frequency, is used in conjunction with an abrasive powder to drill a hole of any required shape. The process is particularly suitable for brittle materials such as glass, ceramics, quartz and sintered carbides. The principle has also been applied to oil-well drilling units in which a nickel core weighing well over one ton vibrates the drilling bit.

Ultrasonic Inspection.—Non-destructive method of testing materials for internal flaws and other defects. The general principle is to direct an ultrasonic beam at the part and to record the "echo", picked up by a receiver, that is obtained when the beam is reflected by a flaw. The echo is usually recorded visibly as a trace on a cathode ray screen.

Ultrasonic Soldering.—Method of soldering or tinning without the use of flux in which the soldering iron or bath is vibrated at a frequency of about 19·5–21 kc/s, resulting in the oxide film on the work being continuously broken up. The process is particularly suitable for soldering and tinning aluminium and its alloys.

Ultrasonics.—The term refers to the vibration of sound at very high frequencies (20 kc/s upwards), generated by electronic apparatus and converted into mechanical vibration by magnetostriction (q.v.). Among the uses of ultrasonics in industry are ultrasonic cleaning, drilling, inspection and testing and ultrasonic soldering tinning (q.v.).

Underslung Spring.—A spring passing under the axle, instead of above it.

Unified Screw Thread.—System of screw thread forms agreed between Great Britain, Canada and the United States of America. The basic form of thread has a 60° angle ; the external (male) thread has a rounded root ; the crest is rounded for use in Great Britain and flat for use in the U.S.A. The internal (female) thread follows conventional practice in having a cleared minor diameter to facilitate tapping. With bolts and nuts made to commercial tolerances, hand assembly is practicable between Unified and Whitworth threads, up to about $\frac{1}{2}$-in. diameter. Full details of the thread are given in British Standard 1580.

Unit Construction.—The system of building the clutch housing and gear-box in unit with the engine as distinct from mounting the gear-box separately. Similarly, any other form of unified construction.

Unit of Section.—A convenient unit, or constant dimension, employed in estimating the area of a section. It is usually the square inch, sometimes the circular inch or circular mil ; or, in the metric system, the square centimetre.

Universal Grinder.—A grinding machine having a swivelling headstock which enables it to grind work at any desired angle, and which thus gives it a much wider range of application than normal grinders. In addition to flat work, plain surfaces, slots, etc., the universal grinder can be used for irregular work such as grinding the teeth of gears and sharpening milling cutters, taper reamers, taps, etc. It can also be adapted to perform the internal and external grinding of cylindrical work. See also Grinding Machines.*

Unserviceable.—A term used in engineering to indicate that an item of equipment is not fit for the purpose for which it was intended. The unserviceability may arise either from wear or from a defect in a new item. Frequently abbreviated, esp. in the Forces, to " U/S ".

Unsprung Weight.—A term used in automobile engineering to describe the weight of all those parts of a vehicle which are not carried on the springs, i.e., the wheels, axles, brakes, etc.

V

Valve Milling Machine.—One designed for milling the square or hexagonal portions of valves and cocks.

Valve Reseating.—Recutting, by means of a steel cutter or a grindstone, the seats of valves which have become worn. See also Valve Seat.*

Valve Rod or **Valve Spindle.**—The rod to which a valve of any kind is attached. See also Valve Stem.*

Vanadium Steel.—One of the alloy steels which, containing a small percentage of vanadium from 0·25% to 0·55%, possesses very great ultimate strength and high elastic limit.

var or **VAr.**—Volt-ampere reactive.

Variable Compression Engine.—A special single-cylinder test engine used to determine the highest compression ratio at which a fuel can be used in an internal combustion engine without detonation occurring. Compression ratio is varied by means of a hand-crank operating a worm-gear which raises or lowers the cylinder, thus altering the compression volume. Compression ratio is determined by means of a micrometer-head attached directly to the cylinder. When fuel tests are being made, running conditions in the engine such as ignition-advance, engine speed, temperature of cooling water, fuel-mixture strength and temperature, are closely controlled by means of special instruments. See also Knockmeter.

Variable Gears.—Toothed wheels of various outlines, which transmit varying velocity ratios during the course of a single revolution. The outlines of their pitch peripheries may be rectangular, triangular, elliptical, rudely star shaped, etc.; each depending upon the number of alterations in speed required in a single revolution. In square gears the pitch peripheries are rectangular, with convex corners; and there are eight changes in speed, four maximum and equal, four minimum and equal. The star gears are modifications of the square gears, the diagonals being longer, and the sides correspondingly deepened, so that the changes in speed are correspondingly greater. In triangular gears the pitch peripheries are of the outlines of equilateral triangles, with convex corners. There are then six variations in speed, three maximum and equal, three minimum and equal. These also may be developed into triangular star gears. (See also Elliptical Gears, Scroll Gears, Sector Gears).

Vector.—A vector quantity is one which has magnitude, sense and direction, and may be represented completely by a straight line drawn to scale. Vectors cannot be added or compounded in the same way as scalar quantities (q.v.), but can be dealt with graphically by vector addition or by a branch of higher mathematics involving imaginary quantities known as vector analysis, and formerly as *Quaternions*.

Ventilation.—Is produced by artificial means in modern workshops. The plenum and vacuum systems are each employed, in the first by a fan producing pressure, in the second by exhausting. Heating and purification of the air may or may not be included. See also Ventilator.*

Vents.—The channels, area spaces, etc., by which the air and gas are carried away from a mould. They may be cylindrical channels, pierced with vent wires, or with rods, if straight; or with ropes, if curved; or they may be bodies of ashes, between the interspaces of which the air collects as in a reservoir, to be led away quietly by the bent channels. In any case care must be taken that metal does not get into the vents and choke them. This is avoided in many ways. In a surface of large area the metal is prevented from entering by interposing a thin stratum of sand, say of $\frac{3}{8}$-in. or $\frac{1}{2}$-in. in thickness, between the face of the mould and the vent openings; and the air is forced through this stratum by the pressure in the mould. Where cores abut against each other, metal is prevented from getting between by a luting of loam, or clay water, or dry sand. This is called securing the vents. See also Vent,* Venting* and Vent Wire.*

Venturi Flume.—A channel through which water flows, so constructed that by reducing the width, a difference of pressure head is created between the channel and the constricted section. This allows the quantity of water or sewerage flowing along the channel to be calculated on the same principle as for the Venturi meter (q.v.), but an automatic instrument gives the result mechanically.

Venturi Meter.—Simple arrangement for measuring the flow of water through a pipe, which consists of introducing a short length of pipe having a reduced section, which causes a drop in pressure at this section in accordance with Bernoulli's law (q.v.). The variation in head is shown as a curve on a rotating cylinder and automatically gives the discharge.

Verdigris.—A bluish-green rust which forms on copper, brass or bronze articles after prolonged exposure to the atmosphere. See also Corrosion.*

Vernier.—Contrivance for measuring fractional portions of one of the equal spaces into which the main scale of a graduated instrument is divided. The vernier is a short sliding scale divided into a number of equal parts greater or less by 1 than the number of spaces which it covers on the main scale. In the latter case the vernier is said to be *retrograde*. The principle of the vernier scale is embodied in calipers and gauges of different shapes and in numerous machine slides, the readings usually being to $\frac{1}{1000}$ in.

Vernier Coupling.—Means of effecting fine adjustments of the ignition timing by a special mechanical coupling between its drive and the armature shaft of the magneto.

Vertical Boring Mill.—A type of machine which has appropriated much of the work formerly done on the common and the face lathes. It is used for turning as often as for boring, and in most of these machines both operations can be carried on simultaneously. The work is bolted to, or held in jaws on the revolving face-plate or table, the face of which is horizontal, and the tools are held in a box or boxes, or in a turret on a single upright head, or on a cross-rail that spans two uprights. Such machines carry two tools in the larger sizes, three in the largest machines. The capacity of the largest machines is not infrequently sixteen to thirty feet or more in diameter, and they will turn a depth of six feet in some cases.

Vertical Milling Machine.—The vertical disposition of the spindle in these is favourable to the employment of end mills. Profiling is done on vertical spindle machines. They are usually belt-driven. Many variations occur in details.

Vertical Welding.—Method of welding in which the work is carried out in a vertical direction. For plates above $\frac{3}{16}$ in. thick two operators are employed, working simultaneously on opposite sides of the plate.

5*

Vibrating Conveyor.—Consists of a flat pan-shaped or tubular trough to which is attached one or more vibrators which impart an upward and reciprocating movement to the trough. On the upward and forward movement the material is propelled forward, while on the return stroke the trough is drawn backwards from beneath it. The material is therefore largely in a state of suspension with the result that wear on the trough is reduced.

Vibrometer.—An instrument designed to indicate variations from the correct balancing of revolving machinery.

Vickers Hardness Test.—Versatile hardness test which can be used for testing very thin sections and extremely thin cases in addition to general testing. The load may be varied from 1–120 kg. The indentor is a pyramidal diamond, the indentation of which is measured by a low-power microscope. A hardened steel ball may be used for testing soft materials.

Vignoles Rail.—A flat-bottomed rail, so called from the engineer who designed it. The rail is spiked to the sleepers which support it.

Virtual Centre.—The virtual centre of a moving body is the point about which it is actually turning at the instant. Also known as the instantaneous centre of rotation. Its locus is known as the *centrode*. (See Instantaneous Centre in Basic Terminology.)

Virtual Work.—If any system of force acts in any directions at any points of a structure and the system is in equilibrium, and if we imagine the points to receive any small displacements compatible with their mechanical connections, the total work done is zero. The displacements are said to be *virtual*, because they do not actually take place. This is known as the *Principle of virtual work*, and is used for the solution of certain awkward problems in mechanics.

Viscosity.*—The resistance to the sliding motion of adjacent layers of a fluid when in motion. It is analogous to friction in solids. It decreases with rise of temperature.

Vitrified Bond.—In grinding wheel terminology vitrified bonds may be regarded as glasses resulting from fusion of ceramic materials during "firing," a process carried out in kilns at high temperatures. Vitrified wheels are therefore somewhat brittle and must be handled at all times with reasonable care.

V-notch.—A notch of V shape for measuring small discharges. It has an advantage over the rectangular notch, in that the ratio of the wetted perimeter to the area of discharge is the same for all heads. Also for a very small discharge the head over the bottom of the V is more accurately measurable. The coefficient of discharge is approximately 0·6.

Voids.—The spaces between the separate particles in a granular material : e.g. in concrete the voids are the air spaces in the *aggregate* which are filled by the *matrix* of sand and cement in correct proportion, to obtain the maximum density and strength.

Volatility.—A term used to define the ability of a liquid to change from the liquid to the vaporized state, either when heat is applied to it or when it comes into contact with a gas into which it can evaporate. See also Evaporative Cooling, also Evaporation.*

Volt.—The unit measurement both of potential difference and of electromotive force. One volt is the pressure necessary to cause one ampere of current to flow through a conductor having a resistance of one ohm. See also Ampere and Ohm.

Voltmeter.—Instrument for measuring the difference of potential between two points in an electric circuit, or the electromotive force of the current

Volumetric Efficiency.—The ratio of the volume of explosive mixture which at normal temperature and pressure would completely fill the working volume of a cylinder in an internal combustion engine to the volume actually taken into the cylinder. The ratio may be expressed : Volume of cylinder working space divided by volume of charge taken into working space. The efficiency may then be determined from an indicator diagram. See also Efficiency* and Volume.*

Volute Chamber.—The space between the periphery of the runner of a centrifugal pump and the casing into which the air or fluid discharges. This space is designed to have a section area which increases as the discharge increases towards the discharge pipe, thereby reducing the loss by impact.

Vortex.—A mass of fluid in rotatory motion. The vortex may be *free, forced,* or *compound.* In a free vortex the fluid rotates under gravity without change of energy, and the velocity varies inversely as the distance from the axis of rotation. The free surface then takes the form of a hyperboloid of revolution ; water flowing from an outlet at the bottom of a bowl, or round a river bend, are examples. A forced vortex, on the other hand, is one in which energy is added or subtracted from an external source, as in the wheel of a centrifugal pump. The energy supplied then increases the pressure head, so that if free to rise, the surface would assume the form of a paraboloid of revolution. A compound vortex is a combination of a free and a forced vortex, as in the case of water approaching the outlet orifice in a basin, when the velocity increases to such an extent that the loss by friction changes the free vortex into a forced vortex. When the fluid also moves radially as in this case, it is known as a free spiral vortex.

Vortex Chamber.—(See Whirlpool Chamber.)

Vulcanising.—Raw rubber cannot be used alone for many types of rubber goods, but is mixed with other materials, more especially sulphur and carbon black, and subjected to great heat and pressure. For repair purposes the plastic rubber and sulphur is applied and subjected to suitable heat and pressure. This treatment is known as *vulcanising.* See also Vulcanised Rubber.*

W

Wale-piece.—Horizontal timber of a quay or jetty, bolted to the vertical timbers, or secured by anchor rods to the masonry to receive the impact of vessels coming against or lying alongside.

Wales.—Long, longitudinal timbers in the sides of a ship, cofferdam, caisson, etc.

Walings.—Boards about three feet long, supported by struts, for retaining the earth when digging deep trenches.

Walking Pipes.—The jointed pipes that carry water under pressure to portable hydraulic riveting machines.

Washing Down.—A term used to denote the thinning, or tapering down of anything, to a feather edge.

Waste Gases.—The gaseous products of combustion, chiefly carbonic oxide, carbon dioxide, and marsh gas, mixed with inert nitrogen. Metallurgists now aim at utilising the combustible gases as much as possible. Hence the use of close-topped blast furnaces, and of regenerative furnaces, by which the greater portion of the heat is utilised, instead of being dissipated into the atmosphere, and of large gas engines.

Water Gas.—Produced cheaply by the decomposition of steam, which is enriched with liquid hydrocarbons from petroleum oil sprayed over hot iron bars, to be taken up by the gas.

Water Gland.—In turbine design an air-seal is produced by a paddle wheel acting as a centrifugal pump. The pressure difference over the gland is balanced by the centrifugal head produced in the water. It is used in conjunction with a labyrinth packing to prevent air leaking into the surface.

Water-hammer.—When water flowing in a cylinder or pipe separates due to cavitation (q.v.) and then the part left behind impinges on the water in front of it, the noise of the impact is known as *water-hammer.* It is liable to occur in long small pipes and in reciprocating pumps at the beginning of the forward stroke, especially with long suction pipes and big lifts from the sump.

Water Softening.—The removal of the hardness from boiler feed waters by treatment with lime and soda in commercial plants, with tanks for the removal of the precipitates.

Water Table.—A belt course of stone or slate built into the foundations of a wall to prevent the moisture in the soil from rising by capillary attraction. A damp course.

Water Tube Boilers.—These, formerly termed sectional boilers, may be classed in two groups, those in which the circulation is extremely active as in the Yarrow and Thornycroft types, and others modelled upon these, and those in which the circulation is more regular and normal, to which class belong the Belleville, the Babcock and Wilcox, the Niclausse, and others of similar types. All the boilers of the first class have small tubes, rarely exceeding from 1 in. to $1\frac{1}{2}$ in. in diameter. They are arranged over the furnace in positions which approximate more nearly to the vertical than to the horizontal—forming, in fact, with the base of the boiler, a rudely triangular outline. They are rapid steam raisers. All the boilers of the second class have large tubes ranging to 3 in. and 4 in. diameter, and they are disposed on a slope that approaches within a few degrees of the horizontal. They do not raise steam so rapidly as those of the fast or express classes, but they are less liable to need repairs, and they are better adapted for regular and steady steaming. See also Water-tube Boiler.*

Watt.—In electrical engineering, the unit of measurement of power. One watt is the measure of the rate at which work is being done in a circuit in which a current of one ampere is flowing when the e.m.f. applied is one volt. In formula, Watts = Volts × Amperes.

Wattmeter.—An instrument for reading watts (the unit of electrical energy = volt × ampere) combining therefore the functions of a voltmeter and an ammeter.

Weak Sand.—The term is applied in opposition to strong sand because it contains no horse manure or core sand.

Wear Test.—Any type of test which is used to determine the resistance of a sample to abrasion under specific conditions of loading, lubrication, speed, etc. See also Test* and Wear.*

Weep Holes.—Openings made at intervals in a retaining wall to drain off water which collects behind the wall.

Weibel Welding Process.—Method of fusion welding in which heat is generated by passing a low-voltage current through a carbon electrode to the work and returning the current through a second electrode. The edges to be welded must be flanged and must be free from burrs and irregularities.

Weld-Decay.—A form of corrosion to which austenitic nickel-chromium steels are susceptible if reheated to temperatures between 600° and 900° C. Susceptibility to weld decay can be overcome either by reheating the steel to the softening temperature of 1,100° C. and cooling in air, or by modifying the composition of the metal. See also Corrosion.*

Weld Face.—Exposed surface of a fusion weld.

Welding.*—Method of uniting metals by fusion or by a combination of heat and pressure. The earliest method of welding was forge welding, as used by the smith (see Welding, Basic Terminology). Modern welding methods may be divided into three groups : See Fusion Welding, Electrical Resistance Welding and Pressure Welding.

Welding Rod.—Filler metal, in wire or rod form, used in gas welding and arc welding processes in which the electrode does not furnish the filler metal.

Weld Iron.—Wrought iron, to distinguish it from ingot iron (q.v.).

Weld Metal.—Metal deposited by fusion in a weld.

Well-base Rim.—A wheel rim of special channel section, enabling the edge of the tyre to be pressed into a *well* ; the opposite side of the cover can then be pulled over the rim.

Wetted Perimeter.—The linear measurement of the cross-section of a channel conveying a fluid, taken along the bounding surface in contact with the fluid.

Wheel-spin.—Slipping of the tractive wheels of a vehicle or locomotive due to loss of adhesion.

Whirling of Shafts.—Owing to lack of straightness in a shaft, due to its own weight or to pulleys, etc., on it, centrifugal forces are set up by rotation, which at a high speed deflect the shaft from its statical position, and this deflection becomes dangerous at certain speeds. When rotating at one of these speeds, the condition is known as the *whirling speed.*

Whirlpool Chamber.—Sometimes called the vortex chamber of a centrifugal pump. The space surrounding the impeller into which it discharges. As the velocity of the air or fluid moving in a free spiral vortex varies inversely as the radius, the kinetic energy of rotation is partly converted into pressure energy, by which the efficiency of the pump is increased.

Whiskers.—A term applied to the hairlike deposit which is sometimes formed on electrodes and on the insulators of sparking plus. The precise cause of plug-whiskering is not known, but it has been found to result from the use of fuel to which too much tetra-ethyl lead has been added.

Wicket Gate.—A favourite system of regulation for water turbines. It consists in having guide vanes pivoted at their centres which are automatically adjustable to the speed of rotation of the turbine wheel, or can shut off the supply of water altogether, if required. The angle of the guide vanes is regulated by an ordinary pendulum governor, acting through a servo motor.

Wimet.—Sintered carbide alloy used in the production of cutting tools. (See Sintered Carbide Tools.)

Wimshurst Machine.—An influence machine for producing high potential or static electricity. It consists of two circular discs of thin glass mounted on a horizontal steel spindle one-eighth of an inch apart. On the outside of each disc are cemented 16 or 18 sectors of tinfoil or thin metal. Wire brushes attached to two curved brass rods just touch the outer surfaces of the plates. Four collecting combs are arranged horizontally on insulated supports to collect electricity from the horizontal diameters of the discs. These lie at an angle of about 45° with the other equalising rods.

Windage.—(1) The loss of energy due to air resistance. In steam turbines which run at high speeds, the loss results from the fanning action of the blades. (2) In a dynamo the real air gap between the armature windings and the pole pieces is sometimes so termed.

Wing Bar.—A fire bar which is the outermost of the series in a Cornish or Lancashire boiler. In small boilers the curvature of the fire flue is so great that the bars of ordinary depth cannot be brought sufficiently close to the sides to fill up the space. Hence in these cases shallower bars, one on each flank, are employed to fill up the space. These are the wing bars.

Wing-walls.—The retaining walls which flare out from the ends of bridges, culverts, etc.

Witness or **Witness Line.**—In machining, when faint traces of the lines marked on the work, or some of the original rough surface of the casting or forging, are intact and visible, these are said to be left as " witness," signifying that the workman has not carelessly reduced below given dimensions. Also, in lined-out work it is usual to strike a line concentric with, and outside and inside the working line, according to the character of the work, and to centre pop it. This line, remaining after the machining is done, testifies to the concentric accuracy or otherwise of the machined work.

Working Pressure.—The pressure at which a boiler is safely worked, as distinguished from the test pressure.

Workshop Microscope.—Optical inspection instrument designed for use on certain types of machine tools, enabling the operator to check the forms of threads and profiles of cutting tools and grinding wheels without taking the work to the inspection department. The workshop microscope is a less specialised version of the toolmaker's microscope (q.v.).

Worm Wheel Hobbing Machines.—Machines in which the teeth of worm wheels are cut by means of a hob which is the exact counterpart of the worm, but serrated and hardened to form cutting teeth and having an increased addendum to give a clearance to the worm teeth. Instead of the worm hob leading the wheel round, the latter is rotated at a speed which corresponds exactly with its rotation relative to the worm, so that no slip occurs, and the teeth when finished are of correct pitch and form. These machines are largely a result of the demand for good reduction gears for electric and other driving. See also Worm Wheel.*

Wrought-iron Pipe.—Common welded gas tubing is used for engineers' steam and water connections, for moderate pressures only. For pressures up to 60 or 70 lb. per inch, and in diameters up to 2½ or 3 in., it is a cheap and safe material. For higher pressures and large diameters solid-drawn tube, or copper tube is employed. Wrought-iron pipe can easily be bent at a red heat, its connections are easily made with tees, elbows, crosses, and screwed unions and sockets; and faulty tubes seldom occur. See also Wrought Iron.*

X

Xerography.—A photographic copying process in which an image is formed on a plate whose surface has been coated with selenium. The image is then transferred to a sheet of charged paper, and fixed there by heating. The process enables large numbers of copies of manuscript, typed or printed material to be obtained quickly and relatively cheaply. See also Copying Machines and Copying Paper.

X-ray.—Ultra-short-wave radiation used industrially for the examination of welds, castings and components (see Radiology). X-rays are emitted when a stream of electrons from a heated metal filament or cathode in a vacuum tube strikes a tungsten target embedded in the end of a copper anode. A low voltage of 10–12 volts is employed to heat the cathode, while a very high voltage of from 10,000 to 2,000,000 volts is applied across the anode and cathode to accelerate the electrons. The higher the voltage, the greater the penetrating power of the rays. Typical voltages used are : examination of textiles, 12,000 volts ; plastics and light alloy castings up to about 2 in. in thickness, 40,000 volts ; light alloys up to 4 in. thick and steel up to 1¼ in., 150,000 volts ; steel up to 3 in. thick, 250,000 volts ; 8 in. steel, one million volts.

X-ray Crystallography.—See Electron Diffraction Analysis.

Xylene (sometimes also called *Xylol*).—A liquid, distilled from coal tar at a temperature of about 140° C., which is used as a solvent for synthetic resins and gums.

Xylonite.—Trade name for cellulose nitrate plastic compound, more commonly known as celluloid. Celluloid plastics are tough, resistant to water and possess good light transmission properties, although the clear types tend to discolour when subjected to light for long periods. They are inflammable, but this feature has been modified during recent years by the addition of chlorine and phosphates. The plastics may be moulded by compression in a heated punch and die or blown into shape by air pressure. When mixed with solvents they are used in the manufacture of emulsions, adhesives and dopes. It is used to build up models of framed structures, toothed wheels, riveted joints, etc., for the purpose of indicating the stress distribution in the materials, by passing polarised light through the models and projecting the results on a screen. (See Photoelasticity.)

Y

" Y " Alloy—Copper-aluminium alloy containing 4% of copper, 2% of nickel and 1·5% of magnesium. It is a high-strength casting alloy used in the heat-treated form, and retains its strength and hardness at high temperatures better than the majority of other general-purpose casting alloys. It is employed for internal combustion engine components such as pistons, cylinder heads and crankcases, all of which parts are, in service, normally subjected to elevated temperatures.

Yaw, to.—To turn an aeroplane in the air about the normal axis. An aeroplane is said to yaw when the fore-and-aft axis turns to starboard or port, out of the line of flight. The angle of yaw is the angle between the fore-and-aft axis of the aeroplane, and the instantaneous line of flight.

Yellow Metal.—Also termed Muntz Metal, a brass consisting of 60% copper and 40% zinc which is very largely used for hand fabrication by native craftsmen ; many hundreds of tons are exported every year to the East. As sheets of yellow metal are chiefly produced by hot-rolling, the finished surface is usually rougher than a cold-rolled brass, and it is not a material suitable for highly polished products. See also Muntz Metal.*

Yield Point or Breaking Down Point.—That point in the stressing of a material in which the deformation increases very suddenly. It occurs immediately beyond the elastic limit, and is marked by a well-defined and sudden curve in the stress-strain diagram.

Yoked.—A term used to describe the forked type of connecting rod used in a V-type internal combustion engine, in which two connecting rods operate on the same crank-pin. In this arrangement, one rod (known as the " main " or " forked " rod) has a split or forked lower end while the other rod (the " auxiliary " or " plain " rod) has a single end which fits between the fork. Both rods are attached to one bearing. The yoking or forking of connecting rods allows the cylinders of a V-engine to be arranged in line, as opposed to their arrangement in the off-set or desaxe types of engine. See also Desaxe Engine and Forked Connecting Rod.*

Young's Modulus.—The coefficient of linear elasticity. Generally expressed as the ratio of the stretching force in dymes per unit of cross-sectional area to the degree of elongation of the material per unit of its length. See also Modulus of Elasticity.*

Z

Zinc-aluminium Alloys.—Light alloys possessing the virtues of cheapness and excellent machineability. From the foundry standpoint they possess many disadvantages, being notoriously hot-short and, in the lower zinc range particularly, subject to high shrinkage. The alloy containing 13·5% of zinc and 2·75% of copper (alloy 465), however, commonly referred to as 2-L5, or merely as L5, possesses good mechanical properties, reasonable ductility and mechanical strength. It is widely used for automobile castings such as crankcases, gearboxes, valve covers and brackets, and other components not subjected to other than low stresses. Its strength at elevated temperatures is not sufficient to allow of its use for pistons or cylinder heads. See also Zinc.*

Zirconium.—A rare metal used in the manufacture of electrical porcelains, wireless valves and electrodes. When alloyed with nickel, it is acid-resisting and non-rusting, and makes an efficient high-speed cutting tool. It is also used in the manufacture of zirconium steel, which is claimed to make a superior light-weight armour-plate, and of projectiles. Zirconium has the very high melting point of 1,900° C. It is very hard and will readily scratch glass.

Zone Refining.—A method of purifying metals in which small quantities of the metal are melted by intense local heating. During re-solidification, some of the impurities diffuse towards the portion which is last to solidify. Repetition of the operation progessively purifies the metal, the impurities eventually becoming segregated at one end of the metal bar. Some metals have been zone-refined to a point at which they contain less than one part in a million of impurity.

Zoning.—See Grid System.

Zyglo Crack Detector.—System of crack detection by which the part to be tested is immersed in a fluid which has fluorescent properties when irradiated by ultra-violet light. After removal from the fluid the parts are thoroughly washed and then examined in a dark cabinet illuminated only by ultra-violet rays. Any fluid which is trapped in a crack or flaw will seep to the surface and will reveal the fault as a glowing line, spot, pinhole or area of porosity. In case of uncertainty more positive indications may be obtained by immersing the parts in a " developing " solution and allowing them to dry before re-examining them under the ultra-violet lamp. The Zyglo system can be applied to aluminium, magnesium, brass, copper, cast iron, steel, tungsten, plastics, ceramics and glass.

PART 2

BASIC TERMINOLOGY

USED IN

MECHANICAL ENGINEERING

A

Abele (*Populus alba*), or **White Poplar.**—A species of poplar sometimes used by pattern-makers. It is of a reddish colour, light and porous, and moderately hard. Sp. gr. ·32 to ·51. A cubic foot weighs from 20 to 31·8 lbs.

About Sledge.—Signifies the ordinary swinging of a smith's sledge-hammer in a circle for the delivery of heavy blows, as distinguished from the lifting it through a small arc only, or uphand (q.v.), for light blows.

Abrasion.—The process or act of grinding, as opposed to that of cutting. This is of great economical importance in work-shop practice, since all emery-wheels and grindstones are employed for the abrasion of metals in cases where the employment of cutting tools would be inadmissible, being either too slow in their action, or too inaccurate in their results, or would fail altogether to attack the harder varieties of metal.

Absolute Pressure.—It is customary to estimate the pressure of steam in lbs. above the atmosphere, but it is also the practice in some cases to reckon it from the point of no pressure, or the point of perfect vacuum. This is called absolute or total pressure.

Absolute Strength.—The actual breaking strength (q.v.) of a bar or structure, as distinguished from the safe or working load.

Absolute Temperature.—Temperature measured from absolute zero (q.v.), and useful in dealing with gases. The temperature of steam is often given in degrees of absolute temperature. These may be translated into degrees Fahrenheit by adding 460° to the degrees on that scale, or into degrees Centigrade by adding 273½° to the degrees on that scale.

Absolute Zero.—Corresponds with the bottom of the air thermometer at which the volume of the air is imagined to have been reduced to nothing. It has of course no existence in fact, but is deduced from theoretical considerations based on the expansion of gases. It corresponds with −459·13° F. or −272·85° C.

Absorbing Power.—The capacity of timber for absorbing the preservative fluid used in its impregnation (q.v.). The absorbing power varies with the nature of the wood subjected thereto.

Abutments.—The surfaces of support of an arch, beam, or bridge, which sustain the reactions due to the load.

Acacia.—The true acacia is the wood called sabacu, which though used in shipbuilding is scarcely, if at all, employed by engineers. The timber known popularly as acacia is the wood of the locust-tree (*Robinia pseudacacia*), or false acacia. Both woods belong to the order *Leguminosæ*. The locust-tree grows in America and in Europe, and the wood is of a greenish yellow colour. It works very much like oak, though not so hard, and is used for the cogs of mortice-wheels. A cubic foot weighs when dry 51 lbs. Sp. gr. ·82.

Accident Crane.—A break-down crane (q.v.).

Accumulated Work.—Work done upon a body or system of bodies over and above that necessary to enable them to overcome external resistance. The fly-wheel of an engine affords a familiar example of accumulated work.

Accumulator.—The cylinder into which water is forced or accumulated under pressure, in order to furnish the motive power necessary for lifting the weight case (q.v.) in hydraulic machines of various kinds.

Acid Bath.—A large shallow bath used for fixing phototypes (q.v.). The liquid is composed of one part of hydrochloric acid to nine parts of water. See Blue Bath.

Acidity.—See Oils.

Acid Process.—The early process of steel making in the Bessemer converter (q.v.) as practised previous to the introduction of the Basic process (q.v.). The silicious or "acid" lining (ganister) originally employed for the converters rendered it impossible to effect the removal of the phosphorus present in the pig, because silica has a greater affinity for the oxide of iron which is expelled with the slags than phosphorus has ; the silica displaces therefore phosphoric anhydride, $P_2 O_5$, as soon as formed, and permits of its immediate reduction to phosphorus and oxygen again. Hence the necessity for a strong basic lining, by whose employment the free oxidation and elimination of phosphorus is rendered possible.

Acid Pump.—A pump whose barrel and valves are made of glass in order that they may remain unaffected by the action of the acid liquors which have to pass through them.

Acid Steel.—Steel prepared by the acid process (q.v.).

Acidulated Water.—Water in which acids have been generated, and which produces corrosion in steam boilers. The acids usually are the result of the decomposition of oils and fats carried over from the cylinder through the condenser, or from tallow introduced into the boiler in injudicious quantity with a view to the prevention of incrustation.

Action.—The action of an engine or machine denotes its mode of working. In mechanics it is an axiom that action and re-action are equal and opposite.

Actual Horse-Power, or Available H.P.—Sometimes called Dynametrical H.P. The net useful power given out by an engine. Its amount is estimated by subtracting the power absorbed by the engine itself from the indicated H.P. (q.v.). The actual H.P. may average ·7 or ·8 of the indicated H.P.

Acute Angle.—One which is less than a right angle.

Adamson's Flanged Joint.—See Flanged Seam.

Addendum.—Sometimes used to signify the point or face of the tooth of a gear-wheel, lying without the pitch circle.

Adhesive Power, or Adhesion.—The amount of bite or friction of the driving wheels of locomotives upon the rails, or of the coupled wheels in heavy engines. When the adhesive power is less than the tractive force (q.v.) the wheels slip. With clean and dry rails the adhesion is about one-fifth or one-sixth of the weight—with greasy rails one-tenth or one-twelfth.

Adiabatic Curve.—The curve which represents the expansion of a gas within a vessel, through whose substance no heat is supposed to escape, notwithstanding that the gas is doing actual work during its expansion. It is useful in estimating the economical efficiency of steam, air, and gas engines.

Adjustable Eccentric.—An eccentric sheave which is so constructed that it can be moved relatively to its shaft into a position for either forward or backward gear as required. There are numerous forms in use. Most eccentrics, however, are now fixed. See Fixed Eccentric.

Adjustable Footstep.—See Footstep Bearing.

Adjustable Level.—A spirit-level (q.v.) which is capable of adjustment for reliable indication, irrespective of any inaccuracy which may be present in the stock itself. The bubble tube is attached to a casing of metal which is hinged at one end, and provided with a screw for adjustment at the end opposite. By turning the adjusting screw which elevates or depresses the casing and simultaneously packing up the stock, until the bubble remains in the same position on the reversal of the stock, or end for end, very great accuracy may be attained.

Adjustable Stroke.—The stroke or amount of travel of the ram of a shaping or of a slotting machine, whose length can be varied according to the work in hand. It is made adjustable by causing one end of the connecting-rod of the ram to move across the face of the driving cog-wheel, or of the throw disc, nearer to, or farther from its centre, and clamping it at the particular radius desired.

Adjusting Screw.—A set-screw by means of which the position of machine parts is adjusted or regulated more minutely than would be possible of attainment by the mere setting to dimensions.

Adjustment.—The placing and setting of engine and machine parts in position. Or the more exact and precise regulation of the positions of parts already set approximately.

Adjustment Strips.—Strips of metal by means of which the exact bearing of sliding surfaces is accurately adjusted, the precise shade of contact being effected by pressure imparted to the strips from set or adjustment screws. The sliding faces and edges of machine parts are thus adjusted. Vee strips (q.v.) are the commonest form in which they occur.

Adjutage, or Ajutage.—An outlet in the side of a tank or vessel for the efflux of fluids. See Vena Contracta—Compound Adjutages.

Admiralty Rule.—What is known as the Admiralty rule for the nominal horse power of engines is (1st), Multiply seven times the area of the cylinder in inches by the mean velocity of the piston in feet per minute, and divide the product by 33,000. (2nd), Square the diameter of the cylinder in inches, and multiply by the mean velocity of the piston in feet per minute, and divide the product by 6,000.

Admission.—The period or instant at which the steam enters an engine cylinder. Or the act of entrance. Or the whole period of time during which steam is entering, from the initial point to the moment of cut off (q.v.).

Admission Corner.—That corner of an indicator diagram which corresponds with the period of the entry of steam into an engine cylinder. If this is much rounded or sloping it shows that the steam enters the cylinder too slowly, and that there is therefore too little, or no lead (q.v.).

Admission Line.— That side of an indicator diagram which corresponds with the rise of pressure due to the entering steam in an engine cylinder. The nearer the line approaches to the vertical the better; a sloping admission line indicates a too early or a too late admission, according as it slopes at an obtuse or at an acute angle, with the atmospheric line. In the first case there is too much, in the second, too little lead.

Admission Port.—The steam port or passage through which the entering steam gains access to an engine cylinder. In engines of ordinary construction, each port is alternately supply and exhaust, though it is usual to apply the latter term only to the port which exhausts directly into the atmosphere.

Advance.—See Angular Advance.

Adze.—A wood-cutting tool, whose cutting edge stands transversely or at right angles with the handle. Its bevel is ground on the inner face only. The entire outer face is slightly rounding.

Adze Block.—A solid oblong block, square in section, of iron or steel, which carries the cutters or plane irons of a wood-planing machine. It is furnished with slots to allow of endlong adjustment of the irons.

African Oak.—See Teak.

After Blow.—The blowing of air into a Bessemer converter (q.v.) after the carbon has been burnt out, in order to oxidise the phosphorus. The metal is said to be overblown when the operation is continued too long. Overblown metal possesses similar characteristics to those of burnt iron (q.v.). In the basic process the result of overblowing is not so deleterious to the iron as in the acid process, for while in the latter overblowing produces oxidation of the iron, in the former the result is mainly the more complete oxidation of the phosphorus.

Aggregate Motion.—Motion whose origin is of a compound character. In other words, motion which is produced by the concentration of two or more independent motions upon one spot at the same instant.

Agitator.—A mechanical stirrer designed for the admixture of the spiegeleisen and molten metal on its removal from the Bessemer converter to the ladle. The stirrer is attached to the bottom end of a vertical spindle, and is coated with loam to protect it from the high temperature. It is revolved at the rate of about 100 revolutions per minute. The agitator runs in fixed bearings, the ladle being brought underneath it.

Agricultural Locomotive.—See Portable Engine.

Aich Metal, or Gedge's Metal.—An alloy of copper, 60; zinc, 38·2; and iron, 1·8. It is capable of being cast, hammered, rolled or drawn.

Air.—A knowledge of the properties of atmospheric air is requisite to the engineer, since it plays an essential part in the working of pumps and pump connections, engines both condensing and non-condensing, air engines and air compressors, and gas engines. Air possesses the properties of a perfect gas, and as such comes under the law of Boyle and Marriott (q.v.). Air is a mechanical mixture essentially of 79 volumes of nitrogen, with 21 volumes of oxygen. The oxygen alone is capable of promoting combustion, so that when air is forced into a blast furnace or boiler grate 79 volumes out of every 100 are inert, and count for nothing. The pressure of the atmosphere at sea level is 14·7 lbs. per square inch, put roughly at 15 lbs. in ordinary calculations, that being the weight of a column of mercury one square inch in section, and thirty inches high, which column will just balance the atmospheric pressure. Air is a bad conductor of heat.

Air Bag.—The presser (q.v.) of a pneumatic moulding machine (q.v.). Consists of bags inflated with air, by which an elastic and equal pressure is imparted to the sand.

Air Belt.—In common cupolas the blast enters the tuyeres directly from the blast pipes, but the more modern practice is to surround the tuyere zone with an annular ring or case into which the air enters from the blast pipe or pipes, and from which it is delivered through six or eight tuyere holes into the cupola. This is termed an air belt or wind chest, and its advantage is that the tuyeres being arranged equidistantly within it, the blast is diffused more regularly than under the older method.

Air Brake.—A brake actuated either by compressed air, or by the production of a vacuum.

Air Casing.—An enclosed space enveloping a reservoir of heat in order to prevent loss therefrom by radiation. The uptakes of marine boilers are for this reason often provided with air casings.

Air Chamber.—See Air Vessel.

Air Channels.—Shallow channels or thin spaces underneath the hearths and fire bridges of reverberatory furnaces (q.v.), their function being to protect the foundations from injury by the high temperature of the furnaces.

Air Compressor.—A machine by which atmospheric air is compressed in volume in order to be used for purposes of ventilation, or as a motive power for driving machinery, rock drills, &c., in situations where the presence of exhaust steam would be objectionable. See Pneumatic Tools in Dict. of Modern Terms.

Air Crucible Furnace.—An ordinary brass furnace (q.v.) as distinguished from a reverberatory furnace, in which metal is melted without the use of crucibles.

Air Cylinder.—The air pressure cylinder, or the cylinder in which the air is compressed in an air compressor. The term is used to distinguish it from the steam or other cylinder by whose agency the compression is effected.

Air Engine.—See Hot Air Engine.

Air Furnace.—A term denoting generally any furnace in which no artificial blast is employed. Reverberatory and brass furnaces therefore come under this designation.

Air Gate.—A riser (q.v.).

Air Hole.—A small hole drilled in the closed metal moulds used for pressing cup leathers (q.v.) and similar works, as a provision for the exit of the air, which, but for this precaution would become compressed, and prevent the complete contact of the joints. Also when shafts or pins are driven into bored holes closed at one end, air holes are provided in the closed ends.

Air Jet.—A jet or blast of atmospheric air introduced into a gas producer to effect a more complete combustion of the fuel, and an increase in the volume of the combustible gases.

Air Pump.—A pump employed with condensing engines, both of the marine and land types, for the purpose of pumping out the water of condensation which accumulates in the condenser. Since the air mingled with the condensed water is exhausted by the same means, and a vacuum is thereby produced, the term air pump is employed. Air pumps are worked from an eccentric in oscillating engines, and from a rocking lever in engines of the inverted cylinder type.

Air Receiver.—An intermediate reservoir of air placed sometimes in the course of the blast main, between the blast furnace and the blowing engines, in order to render the blast pressure uniform. Air receivers are not required when the blast mains are long, and of large capacity nor when two engines are employed in alternate strokes.

Air Regenerator.—The regenerator (q.v.) through which atmospheric air passes to be heated on its way to the furnace hearth of a reheating or steel melting furnace. It is larger than the corresponding gas regenerators.

Air Spaces.—The openings between the fire bars of engine boilers are termed the air spaces. It is essential that these shou'' not be too contracted, else the bars will become unduly heated and twisted out of shape.

Air Tap.—A tap fixed in the air pipe in hot-water apparatus, to allow of the escape of air from the series, which without this means of exit would accumulate therein. The air tap is placed at the highest point in the series of pipes. Air taps in pumps and engines are called pet cocks (q.v.).

Air Thermometer.—See Thermometer.

Air Vessel.—A domed, globular, or egg-shaped vessel, attached to force pumps, long suction and delivery pipes, and hydraulic rams, for the equalisation of the flow of the liquid. The air which becomes entangled in the pipes and mingled with the water is driven into the vessel, and accumulates in its upper area under pressure, becoming an elastic cushion or buffer tending to neutralise the shocks due to the reversal of suction and delivery. Pet cocks (q.v.) are necessarily attached to all air vessels.

Ajutage.—See Adjutage.

Alder (*Alnus glutinosa*).—A genus of plants of the natural order *Betulaceæ*. The wood is of a reddish-yellow colour, and soft. It is used for friction-blocks, for patterns occasionally, and in parts of wooden pumps, being but slightly affected by moisture. Sp. gr. ·56. A cubic foot weighs 34·9 lbs.

Algebraical Signs, or Symbols.—Conventional combinations of figures, letters, signs, brackets, &c., by which certain processes in arithmetical and geometrical calculations are indicated and understood.

Alignment.—This term signifies the linear accuracy, uniformity, or coincidence of the centres of the fast and loose poppets of a lathe. It is also applied to the axial continuity of shafting and shaft bearings in general.

Alligator Shears, or Crocodile, or Cropping Shears.—Shears used for cutting off puddled bars in lengths suitable for piling, and also the crop ends of bars in general. There is a fixed lower jaw, and an upper movable jaw, whose fulcrum is set at the inner end of the cutting portion. Behind the fulcrum the lever is prolonged, and attached to a connecting-rod which receives its oscillatory movement from a crank or eccentric.

Alligator Squeezer, or Crocodile Squeezer.—A form of squeezer used for the expulsion of the cinder from, and the consolidation of the puddled ball. An upper jaw, pivoted at one end, is alternately elevated and depressed at the other by an arm worked from a crank. Between this and the lower fixed jaw the puddled ball is operated upon. The specific term is given it by reason of the serrations which are imparted to the upper jaw for the purpose of increasing its bite. Double *ll*-gator squeezers are also made, containing two pairs of jaws.

All Mine Pig.—Pig iron of the ordinary kind which is smelted entirely from ore, as distinguished from cinder pig (q.v.).

Alloy.—A homogeneous combination of two or more metals, effected by fusion, for the purpose of obtaining certain qualities more suitable for special purposes than those possessed by the original constituent metals themselves.

Alternate Cones.—When two equal cones are arranged on parallel shafts, with their bases facing in reverse directions, they are said to be alternate, and their mutual function is the production of variable motion by means of a shifting belt made to travel from end to end.

Alternate Stresses.—See Oscillating Stresses.

Alumina.—Symb. Al$_2$O$_3$. Sp. gr. 3·9. The only oxide of the metal aluminium. It is the main ingredient in emery, and is an essential constituent of all clays, determining their basic character, and suitability for furnace linings.

Aluminium.—Symb. Al. Sp. gr. 2·6. Comb. weight, 27·3. A white malleable silvery-looking metal, unaffected by the atmosphere. It is a widely diffused metal, existing never in the free state, but in combination with silicon and oxygen, in which condition it enters into the essential formation of all clays. Aluminium is now largely employed.

Aluminous Bricks.—Fire-bricks whose basis is alumina. They are employed in those portions of hearths and furnace linings where contact between the bricks and metallic oxides would occur, silica bricks being unsuitable for use in these situations.

American Rock Elm.—See Elm.

Analysis.—The determination of the percentage proportions of the elements existing in a compound material. The analysis is then said to be quantitative. Qualitative analysis simply demonstrates the presence of certain elements by their reactions on other elements or bodies, without determining their relative proportions. Analysts are employed in all large iron and steel works.

Anchor.—A chaplet (q.v.).

Anchor Bracket.—A bracket, or block, to which the fast end of a brake-strap, or that farthest away from the lever, is attached.

Angle.—An angle is formed by the inclination of two lines towards one another. The angle is plane, if the lines are straight; curvilinear, if the lines are curved. A right angle is formed where one line is perpendicular to another, an acute angle when the meeting lines form less than a right angle, an obtuse angle when they make more than a right angle. The dimensions of angles are given in degrees of arc included by the converging lines.

Angle Bearing.—A crank-shaft bearing attached to an engine bed, the centre line of its joint being placed at an angle of about 45° with the bed, the purpose of which is to effect that disposition of the metal best calculated to withstand the strains due to the motion of the crank and connecting-rod.

Angle Board.—A board upon which pattern-makers plane their angles and hollows (q.v.). It is traversed longitudinally with vee'd grooves of different depths to suit angles of different sizes, in which grooves the stuff is laid while being planed, a transverse strip near the end acting as a stop.

Angle Bracket.—A bracket, two of whose faces abut against the angular faces of a structure, while the third face or hypothenuse forms an angle therewith of 45° or otherwise. Angle brackets are of cast iron of various designs, or are built up with angle irons and plates, and are used as stays, supports, stiffeners, connections, braces.

Angle Chuck.—An angle plate (q.v.).

Angle Iron.—Angle irons are rolled malleable iron bars, whose section is that of the letter L. Their cross sectional dimensions are given as the widths of each side, and their thicknesses, thus :—3 in. by 4 in. by $\frac{1}{2}$ in., or 3 in. by 3 in. by $\frac{3}{8}$ in. as the case may be. Angle irons are equal sided when the two webs are of equal width, as 3 in. by 3 in. by $\frac{3}{8}$ in. ; unequal sided when one web is wider than the other, as 2 in. by $8\frac{1}{2}$ in. by $\frac{3}{16}$ in., or 5 in. by 4 in. by $\frac{1}{2}$ in. ; square edged when the outer edges are square in section instead of rounding, the last being most commonly the case ; obtuse angled when the two webs are not at right angles, as is also most usual, but inclined at an obtuse angle ; round backed when the angle formed by the meeting of the webs is not sharp but rounded off instead ; bulbed when the edge of one web is beaded or bulbous in section. Angle irons, or angles, as they are called for shortness, are used in almost all kinds of constructive engineering works in wrought iron and steel.

Angle Iron Furnace.—A reverberatory furnace used in boiler-makers' and platers' sheds for the purpose of heating angle irons, preparatory to bending and welding.

Angle Iron Shears.—Vee-shaped shears attached to shearing machines for the cutting off of angle irons, the outside faces of the angle iron being laid in the bottom vee.

Angle Iron Smith.—A workman in a boiler shop, whose special task consists in bending the angle irons of boiler and other plated work, and generally in preparing the angle irons of roofs, bridges, and similar built up structures in readiness for riveting.

Angle Joint.—A joint made by the meeting of plates at an angle with each other. The plates may either be distinct, or a single plate may be bent round at an angle. Joints of separate plates are commonly made with angle iron (q.v.) ; but welded joints and flanged joints, the latter being either riveted or bolted, are also usual.

Angle Motion.—See Canting Motion.

Angle of Advance.—See Angular Advance.

Angle of Flexure.—The angle measured in degrees through which torsion deflects a shaft. It varies as the length, and inversely as the diameter.

Angle of Friction.—See Angle of Repose.

Angle of Inclination.—The angle which the thread of a screw makes with the axis of its body.

Angle of Relief.—The angle formed between the back or lower part or face of a cutting tool, and the face of the material upon which it operates. An angle of relief is necessary to prevent the setting up of undue friction or grinding action between the tool and the work. It varies from about 3° in some metal-cutting tools to 25° in planes for wood working.

Angle of Repose, or Angle of Friction.—The angle of a plane surface inclined relatively to the horizon, upon which a body will, under specific conditions, just begin to slide. It varies widely with the nature of the particular materials placed in contact and with the presence or absence of lubricants.

Angle of Upset.—The angle which the longitudinal centre line of the upper works of a balance crane of the portable type makes with the longitudinal centre line of the truck when it would just begin to upset or overturn with the weight of the load. See Blocking Girders, Rail Clips.

Angle Plate, or Angle Chuck.—A special machine part or adjunct, of cast iron, being formed of two ribs of metal, standing at right angles with each other, and pierced with slot holes for the reception of bolts. Angle plates are used for the attachment of various work to the beds of planing machines and shaping machines, and to the chucks or beds of lathes, one face being bolted to the bed or chuck, the other receiving the work which is to be operated upon.

Angles.—Strengthening pieces running round the angular portions of castings. Hollows (q.v.) answer the same purpose and have a neater appearance. Angle irons (q.v.).

Angles of Cutting Tools, or Cutting Angles.—These are measured between the surfaces of the materials upon which they operate, and their cutting faces They vary from about 15° in spokeshaves to 120° in broaches.

Angular Advance.—The angle which the centre of an eccentric sheave makes with a line which is set 90° in advance of the crank-pin. The amount of the angular advance governs the extent of the lap (q.v.) and lead (q.v.) of the valve.

Angular Brace.—A drilling brace which is jointed to permit of the drilling of work at an angle in confined situations.

Angular Deflection.—Equivalent to the angle of flexure (q.v.).

Angular Displacement.—Is expressive of the distance travelled by a body which is moving in a circle, the distance being given in angular measure of the radii of the moving body.

Angular Fence.—See Fence.

Angularity of Connecting-rod.—See Obliquity of Connecting-rod.

Angular Pattern Die-Stock.—That form of die-stock in which the handles are placed at an angle with the longitudinal axial centres of the dies, in order to permit of the set-screw which tightens the dies thrusting in the line of their axial centres.

Angular Thread.—A screw thread whose cross section is that of a triangle, as distinguished from a square thread.

Angular Velocity.—A body travelling in a circle moves through a certain distance away from its starting point in one unit of time, or say in one second. If lines are drawn from the extreme points of its travel to the common centre of its radii, a definite angle is enclosed thereby. The angular velocity is measured in terms of this angle, not in degrees, but in the ratio which the arc described in one second bears to the radius.

Anhydrous.—Free from water, that is not only free from water in a state of mechanical mixture, but also as chemically combined. Unslacked lime, for example, Ca O, is anhydrous ; but mixed with water, chemical combination takes place, and Ca O HO results. Compounds of this kind are then said to be hydrated. Thus brown hæmatite is a hydrated ferric oxide $Fe_2 O_3 3H_2 O$. The term anhydride is used to designate an anhydrous substance ; as for instance, carbonic acid, or carbon dioxide, is very properly termed carbonic anhydride CO_2, since an acid, strictly speaking, must contain water.

Anhydrous Tar.—Tar which has been boiled to expel the water. It is mixed with burnt dolomite (q.v.) to form a paste for the lining of Bessemer converters (q.v.).

Animal Oils.—Lubricating oils for machinery obtained from animal sources, and being found in the adipose tissues,—the principal being sperm, ordinary whale, neats-foot, seal, and though not popularly termed oils, the solid fats, lard, tallow, &c. They are all highly valuable to the engineer for purposes of lubrication, and are noted under their several headings. Excluding those which are erroneously termed the fish oils, as sperm, whale, seal, the animal oils do not dry (see Drying Oils), and therefore do not gum (see Gumming), but they decompose and generate fatty acids which corrode the metal work with which they come in contact, and produce also residual deposits (see Glutting). Of the animal oils, sperm and tallow are those most largely used, and when obtained pure, are excellent types of lubricants of light and heavy body (see Body) respectively; but the tendency is now rather towards the employment of compound oils (q.v.).

Animal Power.—Varies with the mode of its application, and its period of duration. It is estimated as units of work (q.v.), or as a definite task prolonged through a definite time.

Annealing.—The subjection of materials which are either naturally brittle and non-elastic, or which have been rendered so by necessary processes in their manufacture, or by fatigue (see Fatigue of Materials), to the action of long-continued heat, the effect of such heating being to rearrange the ultimate molecules. The material thus becomes more homogeneous and tougher, and consequently better able to withstand strain and shock than it was previous to annealing. Steel castings, hammered works, old chains, and rods which have been a long time in use, and a multitude of other articles are subjected to the annealing process. Plates and bars newly rolled, and whose fibrous condition has therefore never depreciated or been subject to strain and fatigue, are, on the contrary, subject to a diminution of tensile strength by annealing. Hence the properties of homogeneity and ductility are in most instances obtained by a direct sacrifice of rigidity and tenacity. See also Tests.

Annealing Oven.—A section of a crucible cast-steel melting furnace, in which the crucibles are slowly baked after preliminary drying. One end of the melting house is provided with rectangular ovens or chambers of fire brick, each holding from twenty to thirty crucibles, which have been previously dried and seasoned. The crucibles are placed mouth downwards upon fire bars covered with fuel, and their interspaces filled up with small coke. They are then slowly raised to redness, and afterwards allowed to cool to a black or low red, previous to removal to the melting holes. The term is properly applied to any furnace used for annealing. The wheels of railway cars are, in America, made of chilled cast iron, and are cooled in annealing ovens or pits of various forms.

Annealing Pots.—Pots of wrought or cast iron, or steel, varying in size and shape with the work which requires annealing, their function being the preservation of plates, sheets, &c., from the action of the atmosphere during the process of annealing. They are run into an annealing furnace, and exposed to the action of the heat for several hours or days. The pots are hermetically closed and properly luted to preserve the contents from oxidation. Plates of malleable iron are annealed in pots measuring 10 ft. in length, and 3 ft. 6 in. in width, and 5 ft. 6 in. in depth; iron sheets in pots much smaller. Plates remain in the furnaces about 24 hours, and are then withdrawn and allowed to cool during a period of four days, still, however, protected from the air. Articles in malleable cast iron are annealed in cast-iron cylinders about 12 in. in diameter, by about 16 in. high, fitted with loose covers.

Annual Rings.—The circumferential layers of wood seen in a cross section of timber, and which represent the yearly additions to the woody fibres. In good timber the rings are narrow and closely packed, in inferior timber they are wide and open, estimated of course relatively to the characteristic and normal growth of the particular timber under consideration.

Annular Engine.—A steam engine, the cross section of whose cylinder is that of a ring, the piston being also annular in plan. Being of this form, therefore there are two cylinders, the outer or working one, and the inner cylinder, within which works the cross-head to which the lower end of the connecting-rod is attached. The connecting-rod passes thence upwards to the crank overhead. The communication between the piston and the cross-head is made by two rods passing to prolongations of the cross-head above. The advantage of this engine, which was designed by Maudslay, is the possession of a long connecting-rod in a restricted space.

Annular Gear.—Circular racks, curb rings, &c., with their pinions.

Annular Seating.—A seating ring-shaped in plan, upon which a circular pump valve rests.

Annular Valve.—A circular disc valve which rests upon an annular seating, and which is therefore a lift valve (q.v.).

Annular Wheel.—A cog-wheel whose teeth are fixed to its internal diameter ; called also an internal wheel. An annular wheel always revolves in the same direction as that of its pinion.

Annular Wind Engine.—A windmill (q.v.).

Anthracitic Coals.—Hard lustrous non-caking coals, rich in carbon, decrepitating and burning with difficulty, requiring a strong draught for perfect combustion, but giving out an intense heat. They consist almost entirely of free carbon, and their sp. gravities range from 1·35 to 1·92. Pure anthracite is the hardest coal known, is of a deep black colour, brittle, and clean to the fingers. The anthracitic coals so called embrace many kinds approaching to the qualities of pure anthracite, but possessing the caking property in slight degrees.

Anti-fouling Compositions.—See Incrustation.

Anti-friction Grease.—A general term applied to mixtures of tallow, palm oil, plumbago, and other lubricants.

Anti-friction Metal.—Babbitt's metal (q.v.), or white metal (q.v.), used for lining the steps and bearings of shafts and axles, and the faces of slide valves and similar moving parts. So called because of its low coefficient of friction (q.v.). The term " white brass " is sometimes applied to anti-friction metal.

Anti-friction Motion.—Motion, the amount of whose friction is reduced to a minimum by the employment of special mechanism, or of superior lubrication.

Anti-friction Rollers.—Freely revolving, or live rollers (q.v.), which sustain the pressure of a revolving spindle or shaft. The crank axles of foot lathes, and the spindles of some light running machines are often furnished with bearings consisting of such rollers.

Anti-Incrustation.—Relates to the remedies employed for preventing incrustation (q.v.) in steam boilers.

Antimony.—Symbol Sb. comb. weight 122. Sp. gr. 6·71. A bluish white,

very brittle, and readily fusible metal, whose value consists in the hardness which it imparts to certain alloys of tin and lead; forming white metals. The presence of antimony in wrought iron is injurious, proportions so small as 0·3, or 0·2 per cent. producing both cold shortness (q.v.) and red shortness (q.v.).

Anti-priming Pipe.—A pipe attached to the steam supply within a boiler. It is furnished with a perforated plate occupying its transverse section, from which the spray driven upwards by the steam is thrown back into the boiler instead of mingling with the steam in the supply pipes. See Priming.

Antiquarian.—A drawing paper of very excellent quality, measuring 53 in. by 31 in. It may be obtained either unmounted, or mounted on brown holland.

Anvil.—The block upon which the operations of smith's work are performed. The ordinary anvil is of wrought iron, faced on its upper portion with hardened steel about half an inch in thickness. Smiths' anvils weigh from one hundredweight upwards. For some special classes of work anvils are made of cast iron, either chilled or faced with cast steel. Small steel anvils of a few pounds weight only are also made for bench use.

Anvil Block.—A massive block of cast iron which is placed beneath the anvils of steam and other heavy hammers, and whose mass absorbs much of the vibration due to the blows. It is embedded in masonry or on concrete, or is laid on piles.

Anvil Chisel.—See Anvil Cutter.

Anvil Core.—The main body of a smith's anvil, to which the corners and beak are welded.

Anvil Cutter, or Anvil Chisel.—A chisel-like tool or set, provided with a shank for fitting loosely into the square hole on the face of an anvil, the cutting edge being uppermost. Iron rods and bars are laid across this, and nicked and cut off, both when hot and cold.

Anvil Head.—The upper portion of the body or core of a smith's anvil.

Anvil Stand.—The square cast-iron frame which supports a smith's anvil.

Aperture.—Specifically, denotes the extent of area of a discharging orifice (q.v.), or mouthpiece (q.v.) for liquids.

Apex.—The summit of a cone, pyramid, or conical figure generally. In engineering, it signifies the points of the triangles in a lattice, or Warren girder, or those portions of the braces which intersect with the flanges, either at top or bottom.

Apple Tree (*Pyrus malus*).—A tree of the natural order *Rosaceæ*, sub order *Pomæ*. It produces a fine, close, straight-grained, hard wood, of a reddish brown colour. Its chief use in engineering is for the cogs of mortice wheels, for which, when thoroughly seasoned, it stands unrivalled. The crab apple is mostly used. Sp. gr. ·73. A cubic foot weighs 45·5 lbs.

Appolt Oven.—See Coke.

Apprentices.—Are usually bound until the age of 21 years, commencing at 14, 15, or 16, seldom later. The apprenticeship system is not universal, many of the best firms refusing to take them. The alternative arrangement then is that a lad shall come into the shop, paying no premium, but be taught his trade so long as he continues to behave himself, and that this tacit engagement shall be terminable at pleasure on either side. Actually the lads usually stay out the ordinary term, and receive the ordinary wages; they behave better, learn their trades **more**

rapidly, and often receive substantial additions in the shape of piece work balances. This system is therefore far preferable to the hard and fast engagements of the apprenticeship indenture system.

Approximately.—A word in frequent use when describing the capabilities of machines, and their dimensions, and also specifically, their shipping weights (q.v.). In many instances, strict exactitude, if not actually impossible of attainment, is either not requisite, or could only be attained by an expenditure of trouble out of proportion to the value of the results.

Apron.—The vertical fixing in the front part of the slide rest of a screw-cutting lathe which carries the clasp nut (q.v.). The term also designates the vertical slide of a punching or shearing machine. See also Breast Plate.

Arbor.—A small shaft or spindle.

Arbor Chuck.—A lathe chuck used for turning the outside diameters and faces of cylindrical work after the hole has been just bored, the hole fitting over a mandrel or spindle which insures concentricity of outer and inner diameters. Engineers usually term arbor chucks, mandrels simply ; hence the term expanding mandrel (q.v.).

Arc.—The arc of a circle is any portion of its circumference.

Arch.—A curved structure designed to resist external pressure.

Arched Beam.—A beam whose shape is that of an arch. Used for the support of roofing.

Arch Head.—The ends of the timber beams of those pumping-engines which were constructed previously to the introduction of parallel motions (q.v.). The ends of the beams were simply struck to a radius around which the chains attached to the piston and the pump-rods accommodated themselves.

Archimedian Drill, or Persian Drill.—A drill stock whose shank is formed into a quick multiple thread, over which a nut, to which a handle is attached by means of a hinge, slides freely. By the sliding of the nut up and down the thread the shank is rotated from left to right, and from right to left alternately—the motion of the drill being therefore of a reciprocal character. Pressure is applied to a knob at the end of the shank. This drill is suitable only for light work, centering, model making, &c.

Arch Joint.—The joint plates of a rule which are hollowed into the form of an arch.

Arc of Approach.—In toothed gearing, is that portion of the arc of contact (q.v.) along which the flank of the driving-wheel is in contact with the face of the driven wheel.

Arc of Contact.—In toothed gearing, is the space included between those two points where the contact of a single pair of wheel teeth begins and ends, the measurement being taken on the pitch line to which those points are projected. The larger the arc of contact the more perfect the bearing of the teeth, and the greater the number of teeth in gear at one time. In belt gearing, it is the length of arc measured in degrees, through which a belt is in actual contact with its pulley. Thus an arc of contact of 180° signifies that the belt is in contact with the pulley around half of its circumference.

Arc of Recess.—In toothed gearing, is that portion of the arc of contact (q.v.) through which the face of the driving-wheel acts upon the flank of the driven wheel.

Arc Pitch.—The pitch of wheel teeth measured between the pitch points around the arc of the pitch line, as distinguished from chord pitch (q.v.).

Are.—A French measure of surface, being equivalent to one hundred square metres, or one square decametre. Its English equivalent is 1076·41 square feet. See Centiare.

Area.—The superficies, or superficial extent of a figure. Knowing the bounding dimensions of any figure, the area can be calculated therefrom by simple mensuration. There are certain relations subsisting between the areas of figures which it is convenient to remember.

Argillaceous Iron Ore.—See Clay-band.

Arithmetical Mean.—The number which is equally distant between two others.

Arithmetical Progression.—See Progression.

Arm.—A term signifying a rod, or a radial bar of metal. Thus we speak of the arm of a lever or of a wheel.

Arsenic.—Symbol As. comb. weight 74·9. A non-metallic substance, though closely allied to the metals, whose presence in small quantity in iron is valuable for the purpose of chilling (q.v.). It likewise increases the hardness of steel, though diminishing its toughness. Arsenic in combination with oxygen as arsenious acid $As_2 O_3$ is a bye product also in some smelting operations.

Articled Pupil.—See Pupil.

Artificial Grindstone.—See Grindstone.

Artificial Seasoning.—Timber is seasoned artificially by exposing it in a suitable chamber to a current of hot air, delivered either from a fan, or by a natural draught. Small blocks can be roughly and quickly seasoned by boiling them in water for two or three hours, and then allowing them to dry.

Asbestos.—A mineral compound composed chiefly of silica and magnesia, existing as varieties of hornblende, augite, and serpentine. The commercial asbestos comes from Italy and Canada. It is a silky, fibrous substance, owing its value as a heat-resisting body to the presence of silica. It is used in various preparations for making the steam joints of flanges, as yarn for glands, and as grease for lubrication. For the latter purpose it is made into a paste and termed asbestoline. Although but a few years have now elapsed since its introduction for these purposes, it is rapidly supplanting the hempen and other forms of packing, and the millboard hitherto in use.

Ascending Room.—See Lift.

Ash (*Fraxinus excelsior*).—Sp. gr. 0·84. A light-coloured, coarse-grained wood, employed for the spokes and felloes of wheels, for smith's hammer-shafts, for the rings upon which the cleading of boilers is fastened, and generally in work requiring great flexibility combined with moderate strength. Ash becomes more brittle with age. Its ultimate tenacity averages about 17,000 lbs. per square inch of cross-section. A cubic foot weighs 52·4 lbs.

Ash Pan.—A pan placed underneath the furnace grate or firebox in boilers of the portable type, to receive the falling ashes.

Ash Pit.—That portion of the furnace of a stationary boiler lying below the fire bars, and immediately in front of the furnace doors, which receives ashes that fall through the bars, and through which much of the draught necessary for combustion enters.

Ash Plate.—A term sometimes applied to the back plate of a boiler furnace.

A-Standard.—A vertical framing shaped as its name implies, and used as an attachment, either singly or when placed in pairs for the parts of pumps and engines.

Atlas Drawing Paper.—A drawing paper measuring 33½ in. by 26 in.

Atmosphere.—See Atmospheric Pressure.

Atmospheric Engine.—See Single-acting Engine.

Atmospheric Gas Engine.—See Gas Engine.

Atmospheric Line.—The line on an indicator diagram which divides the steam area above from the vacuum area below, and which corresponds with the height of the pointer before the indicator-cock is opened.

Atmospheric Pressure.—The pressure of the atmosphere is a vital factor in all questions relating to the pressures in engine cylinders, pumps, &c. The barometer is the instrument by which it is measured. Engineers also measure the amount of atmospheric pressure present in condensers, boilers, engine cylinders, &c., by means of the vacuum-gauge (q.v.). Its pressure is equal to 14·7 lbs. at sea level, equivalent to that of a column of mercury at 32° F.,29·022 in. high, or a column of water at 62° F. 33·9 ft. high.

Atmospheric Pump.—See Lift Pump.

A-truss.—A roof truss whose outline is that of the letter A spread out. It is either a simple truss, consisting of two rafters and a tie beam only, or it is a braced truss.

Attachment Screw.—A screw employed for fastening portions of mechanism together, in contradistinction to adjustment screws (q.v.) and guide screws (q.v.).

Auger.—A wood-boring instrument of large size for bolt holes. Augers are either "shell" or "twisted," the former being the stronger, the latter boring more accurately.

Australian Copper.—Copper pyrites (q.v.) and malachite (q.v.) imported from Australia.

Autogenous Soldering.—The union of metals effected without the intervention of solder, by the mere fusion of the surfaces in contact. See Burning on, Welding.

Automatic.—An action or movement is said to be automatic when it is effected without the direct intervention of the hand, being due to some special portion of the machine designed to effect that action or movement.

Automatic Expansion.—Expansion (q.v.) of steam effected by the governors of an engine, the grade of expansion being diminished when increased work is thrown upon the engine, and vice versâ. The advantages of its employment are economy of working, regularity of motion, and the maintenance of the full boiler pressure in the cylinders.

Automatic Expansion Gear.—The arrangements of gear by which automatic expansion (q.v.) is effected.

Auxiliary.—Small auxiliary engines are usually provided for the supply of boilers with feed water while the larger engine with its pumps is standing idle. Auxiliary parts are carried to sea in case of breakdowns. An auxiliary screw or screw-blade is often taken in the event of the fracture of a blade.

Available Horse-Power.—See Actual Horse-Power.

Avoirdupois.—The name given to the English system of weights and measures of which the grain and the pound are units. The pound is the tenth part of the weight of a gallon of distilled water taken at 62° F. and 14 7 barometric pressure. The pound is also equal to 7,000 troy

grains   The various weights and measures are arbitrary quantities having no mutual inter-dependence as in the metric system (q.v.)

Axe.—A wood-cutting tool whose cutting edge is parallel with the axis of the handle, and whose bevels or cutting angles are symmetrical on each side.

Axial Pitch.—The pitch of a screw or helix measured in a direction parallel with the axis. The term is specially applied to many threaded screws, to distinguish the pitch of a single helix only, from that termed divided axial pitch (q.v.). and from normal pitch.

Axis.—Signifies a central line considered in relation to certain geometrical or mechanical relations. The axis of a cylinder or sphere is the right line which passes through the centre of all the corresponding parallel sections of the same. The axis of a parabola, ellipse, or hyperbola likewise divides the curves symmetrically. See also Axis of Symmetry, Neutral Axis.

Axis of Symmetry.—An imaginary central line around which a symmetrically developed body is formed, and in which the centre of gravity is found.

Axle.—A shaft which carries the driving, travelling, or truck wheels of a locomotive, railway waggon, or trolly.

Axle Box.—The complete bearing arrangements for an axle. It comprises an outer casing of cast iron with internal adjustable brass bearings or bushes, and grease or oil chambers. The whole structure is sustained by springs, and guided by horn plates. Axle boxes are called oil axle boxes, or grease axle boxes, as they are constructed for using one or the other lubricant. Also denotes the outer casting only which contains the bearings or bushes for an axle.

Axle Grease.—A special preparation used for the lubrication of axles. Numerous mixtures are employed into which tallow, palm oil, spermaceti, plumbago, in various proportions enter.

Axle Guard.—See Horn Plate.

Axle Keep.—See Keeper.

Axle Wad.—A washer of papier-maché, or of wood, placed in the back part of an axle box, to protect the axle from the entry of dust and dirt. It is divided across the middle with bevelled edges

B

B.—See Best.

Babbitting.—The process of lining bearings with Babbitt's metal (q.v.), or with white metal.

Babbitt's Metal.—An alloy used for lining the bearing parts of engines and machines which are subject to much friction. There are many inferior kinds of such metal sold, but the proper composition is as follows : 4 lbs. of copper, 12 lbs. of Banca tin, 8 lbs. regulus of antimony, and 12 lbs. more of tin while the mixture is in a molten condition. Pour the antimony into the tin, and mix with the copper in a separate pot off the fire. This is termed the hardening, and the actual lining metal is composed of 1 lb. of this mixture to 2 lbs. of Banca tin, so that the final composition is 4 lbs. copper, 8 lbs. regulus of antimony, and 96 lbs. tin.

Back Balance.—A circular disc or plate, formerly much used, and still employed to a limited extent in the engines of river and other slow-running steamers, for assisting the reversal of the eccentric for forward or backward gear. The eccentric is loose upon its shaft, and the back balance disc maintains it in equilibrium against the catches or stops.

Back Centre.—The pivot or dead centre upon which the back or tail end of the mandrel of a lathe headstock runs.

Back Cut-off Valve.—A sliding plate, moving on the back face of the main slide-valve of an engine for the purpose of regulating the point of cut-off for the steam, and by which its grade of expansion is governed. It is worked independently of the main slide from a separate eccentric, and is usually made capable of adjustment relatively to the steam passages in the main slide.

Back Firing.—The escape of a portion of the gaseous charge of a gas engine at the ports. This is prevented by the rapid closing of the slide immediately after ignition.

Back Gear.—An arrangement of toothed wheels by which the power of a driving-belt is proportionally increased. The term is understood as applying to lathes for metal-turning, and to drilling and other machines of similar character. Taking it as it occurs in the lathe, the main or driving mandrel is hollow, and while its outer portion or quill A carries the driving-cones B and a pinion-wheel C, an inner spindle D carries the mandrel nose and centres and a cog-wheel E. This cog-wheel gears into a pinion F, similar to the one upon the cone spindle, but which runs upon another or back spindle G, set parallel with the first. This second spindle also carries a wheel H similar to the one E upon the mandrel, and working into the mandrel pinion C. The wheel and pinion on the back spindle are rigidly connected together, and this hinder spindle is capable of receiving a backward movement to throw it out of gear. When out of gear the larger wheel E on the main mandrel is coupled to the cone pulley B on the quill, and the lathe is driven only from the belt, in a direct manner. But when the cone pulley B and the mandrel wheel E are uncoupled, and the wheels of the back spindle are thrown into gear, the lathe is driven through the cones B to the pinion C on the quill, thence to the wheel H on the back spindle, and through its pinion F back to the wheel E upon the main mandrel. The power gained is calculated by taking the product of the number of teeth in the pinions and of those in the wheels, and dividing the latter by the former. Treble back gear is made use of for some lathes of special construction, where very slow speeds and great powers are required.

Back Gear Eccentric.—See Backward Eccentric.

Back Guys.—The timbers which connect the tops of the masts of derrick cranes (q.v.) with their sleepers. Also the hinder ropes or chains of shear legs.

Backing Off.—The cutting or bevelling away of the hinder or "leaving" portion of a screw tap, in order to allow the "entering" or cutting part to work more freely. The effect of backing off is to reduce the amount of friction and consequent grinding between the sides of the tapped hole, and of those portions of the tap which do not actually cut.

Backing Out.—The running back of a tap or die after the thread has been cut.

Backlash.—The backward surge of a pair or of a train of toothed wheels when the driving pressure is variable in amount. Its amount is equal to the "clearance" allowance between the flanks of the teeth which are in gear. Excessive backlash should be avoided as causing shock and producing much noise. Its remedy is minimum of tooth clearance and regular running.

Back Link.—The second link which is attached to the beam of a beam engine, or between the main link (q.v.) and the centre of the beam, in the arrangement of levers known as parallel motion.

Back Plate.—A plate of cast iron bolted against the back or outer edges of the bars of moulding boxes which have to be put in the foundry pit for casting vertically. The purpose of the employment of back plates is to prevent the pressure of the head of liquid metal from forcing the sand outwards, and so allowing the metal to run away from the mould.

Back Pressure.—The pressure on the piston due to the steam left in the end of an engine cylinder and in the exhaust passages on the commencement of the return stroke, and which offers a certain amount of resistance to the piston. The amount of back pressure can never be less in non-condensing engines than the atmospheric pressure, nor in condensing engines than that of the imperfect vacuum present in the condenser. It is always more than these because of the friction which the passages oppose to the steam ; hence passages should be short and direct, if long and tortuous they injuriously increase the back pressure. The amount of back pressure should not exceed from 2 to 3 pounds per square inch, reckoned above the atmosphere in non-condensing engines, and above the vacuum-gauge in condensing engines. Also termed counterpressure.

Back Pressure Valve.—A valve which is used for preventing the reflux of fluid or liquid in a pipe. Hence it is a retaining, or foot valve (q.v.). When applied to a steam boiler it is termed a check-valve (q.v.).

Back Saw.—Any saw whose plate or blade is stiffened with a metallic back. Tenon and dovetail saws are back saws.

Back Shaft.—The shaft which runs at the rear of a self-acting lathe along its whole length, and through which motion is transmitted from the headstock to the slide-rest. It is used for sliding and surfacing only, and is capable of reversal for traversing the saddle up or down. See Reversing Motion.

Back Stay.—See Back Steady Rest.

Back Steady Rest.—A lathe traversing rest, distinct from the slide rest, used for the support of long shafts, and of slight cylindrical pieces of work generally which are being turned up. A common form consists of an internal bearing or angle block attached to a vertical carrier or bracket, the bearing being adjusted both above and behind the shaft so that the latter lies in its angle. The bracket is bolted to the saddle of the slide rest, and travelling therefore with it, follows the cutting tool and provides a constant support to the shaft immediately behind the tool where the tendency to spring is greatest during the turning process. Also called a back stay ; or back following stay ; or following steady.

Backward Eccentric.—The eccentric which opens the slide-valve to steam supply when its engine is required to run backward. Also called back gear eccentric.

Backward Gear.—An engine is said to be in backward gear when the relative arrangement of eccentrics, slot links, levers, and rods is such, that the engine will on the admission of steam, run backward.

Baffle Plates, or Bafflers.—Plates of metal so disposed in the fire-boxes and flues of steam boilers as to throw the flame and hot gases against the surfaces most suitable for the economic and complete generation of steam—that is against the best heating surfaces. Also a plate of thin sheet iron attached at a little distance away from the inner face of a furnace door to admit air above the fire, the air entering through a revolving or sliding grid in the door. The use of the plate is also to prevent burning and buckling of the door, the efficiency of the protection being due to the interposition of the film of air between the door and the baffle plate.

Bafflers.—See Baffle Plates.

Balance.—In order that bodies should balance each other there must be equilibrium or equality of moments; that is, the sum of the forces tending to move any body in one direction must be equal to the sum of the forces tending to turn the body in the opposite direction.

Balance Ball.—A spherical weight which is attached to a crane chain just above the hook for the purpose of overhauling it. It fulfils the same purpose as a pear weight (q.v.), but is employed for the smaller cranes chiefly.

Balance Box.—The box which counterbalances the load lifted by a balance crane. It is usually a cast-iron box loaded with small weights or "kentledge."

Balance Crane.—A crane in which the load is counterbalanced by a weight attached to the tail or hinder end, the amount of weight depending on the radius at which the load is lifted, the length of the tail, and the disposition of the mass of the crane itself. Balance cranes are made both fixed and portable, and are worked by hand or steam. See also Blocking Girders.

Balance Cylinder, or Balancing Cylinder.—A small steam cylinder attached to the top of the slide-chest in some large marine engines, which performs the double function of acting as a guide to the valve-rod, and of relieving the valve gear of the weight of the valve and valve-rod. A piston moves in the cylinder, and is open to steam below derived from the valve chest and to the condenser vacuum above. The valve therefore moves in equilibrium. Also a small steam cylinder or cylinders attached to the bottom cover of large marine engines for the purpose of counteracting the dead weight of the piston. A rod passes from the piston of the main cylinder to the piston of the balance cylinder, and the pressure of steam introduced underneath the latter is the counterbalancing agency.

Balanced Crank.—See Disc Crank.

Balanced Valve.—An equilibrium valve (q.v.).

Balanced Wheel.—A rapidly rotating wheel which is balanced by the turning of its rim true, and by the drilling of holes in or near the rim if found necessary, so that the wheel turns equally free or comes to rest in any position indifferently. All high-speed wheels and pulleys are, or should be, thus balanced. The term is also applied to a wheel which is counterbalanced by the addition of a weight to a portion of the rim which lies opposite to another part with which it is connected, such as a crank or connecting-rod, its function being to balance or counterpoise the weight of the crank or connecting-rod.

Balance Spring.—See Spring Balance.

Balance Weight.—(1) The weights placed in the driving-wheels of locomotives are termed balance weights. (2) A weight slid over the end of a lever and attached thereto by a set screw, its use being to counterbalance a moving part. The function of the weight is not always that of counterbalancing merely, but also in addition the lifting of the working part—the lever, drill, brake, chisel, or whatever it may happen to be—out of or away from its place without the intervention of the workman or attendant. (3) Applied generally to any weight used as a counterpoise, such as on the driving-wheel of a foot lathe, or on a crank, or on reversing gear. (4) Also the small weights put into a balance box (q.v.).

Balancing.—The removal of irregularities in weight from those portions of a revolving body where it is in excess, in order to equalise the strains upon it due to momentum and centrifugal force. Hence rapidly revolving pulleys and wheels are, where possible, turned all over. When that is not practicable they are hung on a free spindle, and the weight of the heaviest sides reduced in succession by drilling out holes, or in other ways removing superfluous metal. Such a pulley or wheel is then said to be balanced. Another mode of balancing is when a balance weight (q.v.) is attached to a driving or a fly-wheel, to counterbalance the mass of a crank, or an eccentric sheave, or similar projecting portion on the axle. All good mechanism and rapidly revolving parts are thus balanced.

Balancing Cylinder.—A balance cylinder (q.v.).

Balk.—A squared log of timber.

Ball.—The mass of spongy puddled iron which is just ready to be withdrawn from a puddling furnace to be passed to the squeezers or hammer, as the case may be. A puddled ball weighs from 60 to 80 pounds.

Ballast.—The foundation for permanent way laid on the formation level (q.v.). It is about 1 ft. 7 in. in depth, and is variously composed of gravel, shingle, broken brick, burnt clay, or slag. It requires to be hard, to allow of the passage of rolling stock without the crushing of its materials, and well drained and porous to prevent the accumulation of water. The material which is placed in the balance box of a balance crane is also called ballast. It may consist of iron balance weights, or blocks of stone, or masses of rubble or scrap iron.

Baller.—The workman at the balling or reheating furnace (q.v.).

Balling.—See Balling Up.

Balling Furnace.—The puddling furnace in which the process of balling (q.v.) is performed.

Balling Up, or Balling.—The last stage of the puddling process in which the pasty mass of iron is gathered together in lumps or balls on the furnace bed, being sufficiently self-coherent to be rolled towards the bridge of the furnace, to be withdrawn through the working door for subsequent shingling.

Ball Joint.—A universal joint, sometimes employed for piping. The globular end is retained in its hollow seating with a gland screwed over it. This form of joint is used with the connections of centrifugal and other pumps.

Balloon Boiler, or Haystack Boiler.—A squat circular vertical boiler of large dimensions, with a hemispherical crown. It is fallen nearly into disuse.

Ballooning.—The lifting up of the fine impalpable mud and scale in steam boilers to the surface of the water by the ebullition of the bubbles of steam. In large boilers, scum troughs (q.v.) are provided for the collection of this sediment.

Ball Pane.—See Pane.

Ball Valve.—A lift-valve of globular form which closes an annular seating. It is employed in quick-running force pumps, and its advantage consists in the fact that as it continually shifts its position on its axis its wear is nearly uniform. It is enclosed in a cage open on four sides, the cage being sufficiently high to permit the requisite amount of lift, the internal height of the cage limiting the lift of the ball. The cage consists of an upper and lower portion, which are screwed together after the insertion of the ball.

Banca Tin.—Straits tin (q.v.). It is sold in blocks weighing from 40 lbs to 120 lbs. each.

Band.—A term which taken in its strictest sense designates a flat belt, though both terms are loosely applied in a general way to the cords and belts used for driving purposes. See Belt.

Banding, or Bonding.—The embracing and securing of the lagging (q.v.) around steam cylinders and boilers with broad bands of sheet brass or of hoop iron.

Band Pulley.—See Pulley.

Band Saw.—An endless saw running on revolving pulleys, and used for cutting curved work both in wood and iron. A tension apparatus (either a spring or a weighted lever) maintains the saw at its proper degree of strain, and a canting table allows of the work being sawn either square or upon the bevel. Band saws vary from $\frac{1}{8}$ in. to 4 in. and 5 in. in width, and travel on an average at the rate of 4,000 ft. per minute.

Banjo Frame.—A bow connecting-rod (q.v.).

Banking Up.—(1) The beating down of green coal (q.v.) around the central portions of a smith's fire, or around a piece of forging laid therein. (2) The covering up of the fire of an engine boiler in order to check or moderate the formation of steam for a time, either because the engines have to be slowed down or stopped for a while. Banking up is effected by pushing the fuel back towards the bridge (q.v.) and covering it over with small coal, wetted and beaten down.

Bar.—(1) The bridge of metal which separates a valve port or steam passage from its adjacent port in the valve faces of steam cylinders. Also the similar bridges in double and treble ported slide valves. The thickness of the port bars is of little importance, if the slide valve is accommodated thereto, but they commonly measure a trifle more than the thickness of metal between the passages. (2) The flat or round or other sections in which wrought iron and steel are sold. They are described under their specific names. The bars are brought to their required shape by rolling. (3) The stays or bridges in foundry moulding boxes which afford support to the sand enclosing the pattern. Vertical or cope bars as a rule have their edges kept $\frac{1}{2}$ in. away from the pattern, and $\frac{3}{4}$ in. from the joint of the box. Flat or drag bars do not follow the outline of the pattern, being used in the bottom half of the mould or in the drag-box. Bars are purposely cast as rough as possible to ensure the adherence of the sand, which adherence is further assisted by washing them with clay water, and also frequently by hanging lifters (q.v.) from their top edges.

Bar Clamp.—A tool used for the purpose of clamping or squeezing closely together long timber joints, which are glued or dowelled, the jaws of the clamp operating on the outer edges of the wood. It consists essentially of a long bar of iron of rectangular section, having a fixed head or jaw at one end, carrying a square threaded pinching screw and a movable head or jaw sliding along the bar, and capable of being set in any position at distances of from 2 in. to 4 in. apart by means of a pin, or of ratchet teeth. The movable head being set in the approximate position required for various widths of board, the pinching up of the clamping screw in the fixed head affords the necessary degree of pressure.

Bare.—A workshop term which signifies that a dimension is very slightly under a definite size. In most cases it will mean a difference so slight as to be unappreciable by a rule measurement, and only to be detected by calipers or gauge. A $\frac{3}{8}$ in. or a $\frac{3}{4}$ in. is a definite dimension, but "bare" would signify something less than this, or say the difference between a slack and a driving fit.

Barffed.—Ironwork which has undergone Barff's process (q.v.) is sometimes said to be barffed. Thus the iron plugs of plug cocks so treated are termed "Barffed plugs."

Barff's Process.—A patented process employed for the protection of the surfaces of iron from rust, effected by artificially coating them with a film of magnetic oxide. The iron is first heated to redness, and steam is then passed over it. The iron decomposes the steam, liberating oxygen, which latter immediately attacks the iron, forming magnetic or black oxide, Fe_3O_4.

Barker's Mill.—A piece of mechanism which illustrates the not e of the reaction of water. It consists of an upright tube to whose l wer end two horizontal tubes, set opposite to one another, are attached, openings being provided therein near their ends and in the reverse direction to one another. When water is poured into the vertical tube it issues from the side openings in the horizontal tubes, and drives the mill round in a direction contrary to that of the effluent water. See Turbine.

Bar Lathe.—A lathe whose bed is made in a single piece, usually of a triangular section, one side of the triangle being beneath and horizontal. Used only for lathes of light construction.

Bar Mill.—A rolling mill where rectangular iron bars are manufactured.

Barometer.—An instrument used for measuring the pressure of the atmosphere. It consists of a glass tube containing mercury, and standing in a vessel containing mercury, so that while the atmosphere presses on the surface of the mercury in the vessel there is nothing above that in the tube save its own vapour, the amount of whose pressure is disregarded. The height to which the mercury rises in the tube is called the height of the barometer. It is used for comparison with the vacuum-gauge (q.v.) in order to ascertain the amount of perfection existing in the vacuum present in the condenser of a steam-engine.

Barometric Pressure.—The amount of pressure of the atmosphere as measured by the barometer. It is measured in atmospheres, and fractions of atmospheres. The pressure of a single atmosphere is equal to 29·905 in. of mercury at 32° F. at London, or 14¾ lbs. on the square inch, or by the metric system, 760 millimetres of mercury at 0° C. at Paris, equivalent to 1·033 kilos per square centimetre.

Barrel.—(1) The body of a pump within which the piston moves. (2) The cylindrical shell of a locomotive or portable engine boiler. (3) The drum or cylinder around which the chain is wound in hoisting machinery. Barrels are either plain or grooved spirally, and the grooves may either be made to take the flat of alternate links or the links continuously, but lying at an angle.

Barring Engine.—An engine used for the initial turning of larger engines.

Barrow Ladle.—A foundry ladle mounted on a low carriage provided with two wheels in front, and handles and legs at the hinder part. Used for wheeling melted metal about the foundry.

Barrow Pump.—A force pump mounted on a hand barrow for portability, and furnished with hose pipe for suction and delivery. Frequently termed California pump, because of its value in mining and agricultural districts.

Bar Stays.—Boiler stays (q.v.) which are solid rods screwed at their ends, as distinguished from tube stays or stay tubes (q.v.).

Base Circle.—See Fundamental Circle.

Base Plate, or Bed Plate.—The foundation plate or support for a piece of machinery. It consists usually of a comparatively thin plate, well stayed and stiffened with deep outer and inner ribs or flanges, and furnished with all necessary facings for the attachment of accessory portions, and also with lugs for holding or bolting it down.

Basicity.—Refers to the proportion of metallic oxide present in, and the absence of silicon from a furnace lining, a cinder, flux, or slag. The fettling (q.v.) of a puddling furnace depends for its efficacy on its basic character, as does also the magnesian limestone of a Bessemer converter.

Basic Process.—The process of Messrs. Thomas & Gilchrist, by means of which the phosphorus and sulphur are eliminated from the pig iron in the Bessemer converter. It consists in the substitution of magnesian limestone or dolomite, which is composed almost entirely of metallic oxides, and is therefore highly " basic," for the silicious ganister which is used as a lining in the acid process (q.v.).

Basic Steel.—Steel produced by the basic process (q.v.).

Basil.—A term sometimes applied to denote the bevelled edge of a drill or chisel.

Basket.—A small rose or strainer for suction hose pipe.

Bass Scrubs.—Brushes made of ordinary bass, used by fettlers for brushing the sand off castings when they leave the foundry. See Fettling.

Bastard Cut.—A very coarsely-cut file, a trifle finer than the rough or first cut. In a twelve-inch bastard-cut file the lines of teeth would number eight to the linear inch.

Bastard Pitch.—The pitch of a toothed wheel which is not of a definite dimension, but is either over or under an exact pitch.

Bastard Wheel.—Occasionally used to denote a bevel wheel whose amount of bevel is so slight that it approaches very nearly to a right angle or to a spur wheel. Also one accommodated to an existing wheel, though not absolutely correct.

Bath.—The mass of boiling metal in a steel melting furnace of the open hearth type.

Battens.—Sawn pine timber measuring not more than 9 in. by 3 in. in cross section, or not less than 6 in. in width ; when in sizes less than that being termed scantling or quartering ; 7 in. by 2 in. is the common section of battens. Cleets are also termed battens. See Cleet.

Battering Off.—The finishing of the surface of forged work, effected by the hammer while the iron is becoming of a low red or black heat.

Battery of Boilers.—Steam boilers arranged in series, for the supplying of power to a large factory, one or two of the boilers being spare or auxiliary, for use when either of the others are undergoing repairs.

Bauxite.—A hydrated aluminous ferric oxide. It contains about 60 per cent. of alumina, 20 per cent. of ferric oxide, 15 to 20 per cent. of water, and from 1 to 3 per cent. of silica. It is a highly refractory body, and is used in the manufacture of Bauxite bricks (q.v.).

Bauxite Bricks.—Refractory fire-bricks used for the lining of furnaces of various kinds. They are made by mixing calcined bauxite (q.v.) with clay or plumbago.

6*

Bay.—In a lattice, or a Warren girder, the space included between two adjacent apices is called a bay.

Bayonet Engine.—A horizontal engine is said to be of the bayonet type when the bed-plate is curved round to one side of the crank. The curved portion carries the bearing for the crank-shaft. This outline is by a stretch of the imagination supposed to be similar to that of a bayonet, hence the term.

Baywood.—See Mahogany.

B.B.—See Best, Best.

B.B.B.—See Best, Best, Best.

Beach Chuck.—A common form of die chuck containing three dies sliding in a hollow cone which is tightened or relaxed upon them. Used for holding small drills and pieces of wire or rod.

Bead, or Beading.—(1) The thickened edge of a sheet of metal, thickened or turned over to stiffen and strengthen it. (2) An ornamental fillet or strip curved around the edge of a casting for the same purpose. (3) The semicircular termination of the spigot end of a pipe. Its function is to form a stop for the material used in caulking (q.v.). (4) Angles (q.v.) and hollows are frequently called beads.

Beaded Tube.—The ends of boiler tubes, after being riveted over the tube plates, are beaded or rounded, for the sake of appearance, with a beading tool (q.v.), just as rivet heads are finished with a die or snap. The process is termed beading.

Beading.—See Bead, Beaded Tube.

Beading Tool.—A tool used by boiler-makers for finishing off the riveted ends of boiler tubes (q.v.). Its shape is similar to that of a drift (q.v.), except that at a short distance behind its narrow end it is shouldered out into a hollow curve which is the counterpart of the bead it is designed to mould on the ends of the tubes. The beading tool is struck by a hand hammer.

Bead Sleekers.—Moulders' sleeking tools, rounded in the direction of their length and also rounded in their cross section similarly to a plain semicircular bead bent around a cylinder. Used for smoothing the impressions of beaded work in the sand moulds. Also called bead tools.

Bead Tools.—(1) Bead sleekers (q.v.). (2) Metal turners' scraping or finishing tools made hollow, or the exact reverse of the shape of a round-nose tool, and used for finishing the circular edges of beadings.

Beak Iron.—The conical-shaped piece which projects from the end of a smith's anvil. Also termed beck.

Beam.—A piece of timber or other material, or a built up structure, whose length is greater than its width or depth, and whose strains are those due to leverage, or tension, or compression. Beams are made in a vast variety of cross sections, and are solid beams or built-up beams, symmetrical or unsymmetrical, semi beams, compound beams, Warren and other girders, &c. The term beam does not necessarily imply that the material is subjected to transverse strain only, since a tie-beam (q.v.) is only in tension. The term girder (q.v.) is understood to refer to beams subject to transverse stress, and either resting on a single support, as a cantilever, or on both extremities. The term beam is applied to so many different kinds of structures, both elementary and compound, that it is always necessary to denote by some specific term the structure under consideration, and under their specific names the typical beams in use with engineers will be found described.

Beam Compass.—See Trammels.

Beam Engine.—A steam-engine in which the connection between the piston and connecting-rods is made through a beam whose point of oscillation is set midway between the centres of the two rods. Beam engines, except for pumping and some special work, are becoming obsolete.

Beams,—relative strength of.—The strength of beams depends—their cross-sections remaining the same—upon the manner in which their loads are applied. Taking the strength of a beam supported at one end and loaded at the other as unity, a similar beam having the same load equally distributed would suffer only half the stress. If the same beam be supported at both ends and loaded in the centre, it will be subject to one-fourth only of the stress. If supported at both ends and the load distributed, it will receive but one-eighth of the stress it sustains in the first instance.

Bean Shot Copper.—Globules of copper obtained by pouring the melted metal through a perforated ladle into a vessel of hot water.

Bear.—See Punching Bear.

Bearer.—(1) The cheek of a lathe-bed. (2) The support or bearing-bar which carries the firebars of an engine boiler.

Bearing.—The support or carrier of a rotating shaft.

Bearing Metals.—Anti-friction and white metals, brass, and gun-metals, and the various alloys used for making or lining the bearings of journals.

Bearing Neck.—The portion of a rotating shaft which is in contact with its bearing.

Bearing Ring.—A ring welded up from iron bar of square section which forms the bearer or support for the firebars in a vertical boiler.

Bearing Springs.—The springs which support the weight of an engine or truck, as distinguished from buffer and draw-bar springs.

Bearing Surface.—(1) The area of the surface upon which a shaft rotates. (2) The surfaces of bearing parts which are in mutual contact. The larger the total bearing surface the less the amount of friction per unit of surface, since pressure is then distributed over a larger area.

Beating.—A term applied to the regular thudding sound produced by the engines of a locomotive or a steam vessel.

Beaumontague.—This term is applied in the shops to any compound employed for the filling up of holes for purposes of concealment. It is applied to the chalk and varnish compound of the pattern-maker, to the white metal of the founder, and to the salammoniac and borings of the fitter. As used for filling up blow-holes (q.v.) in castings, it consists of beeswax, resin, lamp-black and iron borings.

Beck.—A beak iron (q.v.).

Bed.—A term of general application, as engine-bed (q.v.), coke-bed (q.v.), a base-plate (q.v).

Bed Charge.—The quantity of coke which forms the coke-bed (q.v.) of a cupola.

Bedding.—(1) The bringing of a piece of mechanism to a close and proper fit upon its base or foundation. (2) Or the material or structure on which a piece of mechanism rests. (3) A seating. (4) The laying of a piece of machinery on its foundation.

Bedding In.—The simplest method of iron moulding in which the pattern, instead of being rammed on each side in a jointed box and rolled over, is laid or embedded in the sand of the foundry floor. The sand is then tucked in and rammed around and over it.

Bed Plate.—See Base Plate.

Beech (*Fagus sylvatica*).—A genus of the order *Cupuliferæ*. A light-brown coloured wood, hard and close-grained, used chiefly by engineers for the cogs of mortice-wheels. Its sp. gr. ranges from ·75 to ·85. Its ultimate tenacity ranges from 11,500 to 22,200 lbs. per square inch of section. A cubic foot weighs from 46·8 to 53 lbs.

Bee-hive Oven.—See Coke.

Beeswaxing.—The coating over of iron patterns while warm with a thin film of beeswax, for the purpose of importing a glossy skin to their surfaces, and thus facilitating their withdrawal from the sand.

Beetle Head.—The monkey of a pile-driver.

Bell Centre Punch.—A centre punch (q.v.) which is self-centering. The actual punch is enclosed within a tube whose lower end or mouth is expanded or tapered downwards, resembling a bell. This being dropped over and embracing the end of a round bar ensures that the enclosed punch shall pop on the exact centre of the bar. The mouth adapts itself to bars of differing diameters.

Bell Chuck.—A cup chuck (q.v.).

Bell Crank Lever.—A lever having two arms which meet at an angle of 90°, the fulcrum being situated in the meeting point of the angle. Employed much in machine work when motion has to be conveyed to rods which lie 90° apart.

Bellied.—Curved underneath, the depth increasing towards the centre. The term is therefore applied to plated and ribbed portions of machinery and structures which have this outline.

Bellied Core.—See Chambered Core.

Bellied File.—The ordinary taper file (q.v.), so called because it is rounded very much on its edge, and slightly in the thickness, so that the greatest thickness is about in the middle. Taper files are used more for the opening out of small holes than for ordinary flat surface filing. See Hand File.

Bellied Girder.—A fish-bellied girder (q.v.).

Bell Metal.—A hard, brittle, sonorous alloy of copper 16, and tin 5.

Bell Mouthed.—When the open end of a vessel or pipe expands or spreads out with an increasing diameter, resembling a bell, this term is applied to it. Also called trumpet-mouthed.

Bellows.—A smith's bellows consists of two chambers separated by a diaphragm, so that while the motion of the bottom board is intermittent, the blast is nearly continuous. Ordinary kitchen bellows are used in the foundry for blowing away the parting sand from the faces of patterns, and also for disposing of superfluous blacking or loose particles of sand from the moulds.

Belly.—The belly of a blast furnace is a narrow zone at the junction of the two truncated cones which form the body and the boshes.

Belly Helve.—See Helve Hammer.

Belt.—(1) A band or strap, sufficiently flexible to act as a transmitter of power over smooth pulleys. A belt acts by friction only. The materials for belting are leather, india-rubber, gutta-percha, and cotton, described under their specific headings. (2) Head metal on a cylinder or pipe.

Belt Coupling.—The union of the ends of a belt. It is effected by the aid of laces or of belt screws.

Belt Dressing.—Leather belts are apt to become hard and crack in dry situations, and various dressings are used as preservatives to maintain them supple. Pure tallow dried into the belt by the heat of the sun or in a warm room is a good preservative. Some apply castor-oil. One recipe gives castor-oil 2 quarts, tallow 1 lb., powdered resin, 1 oz., hard soap 2 oz., melted together. For belts which have already become hard, neatsfoot oil and resin are recommended. Resin mixed with tallow is also recommended for use with belts which have to run in damp places.

Belt Fork, or Strap Fork.—A pair of prongs standing out from a strap bar and enclosing a space within which the belt or strap of a machine fitted with fast and loose pulleys runs. The edges of the belt either run against the sides, if plain iron rods, or against freely revolving rollers slid over the rods as spindles, the latter method diminishing friction.

Belting.—See Belt.

Belt Lacing.—See Lacing.

Belt Perch.—A bar of wood or a metal rod placed alongside a belt-pulley on which to rest the belt when unshipped for repairs, in order to prevent the latter from becoming hitched in the revolving shaft, and being itself damaged or inflicting damage on the machine or attendant.

Belt Pipe.—Applied to a steam or exhaust pipe when it surrounds its cylinder.

Belt Pulley.—See Pulley.

Belt Punch.—A plier-shaped punch used for piercing holes in belting for the reception of the laces or screws used for the coupling together of the ends.

Belt Rivets.—Copper rivets used for fastening the joints in leather belting.

Belt Screws.—Male and female screws used for uniting the overlapped joints of belts. Used in preference to laced joints.

Belt Shifter.—A bar of iron carrying a pair of forks or prongs, used for shifting a belt backwards and forwards between loose and fast pulleys.

Belt Shipper.—Usually a length of round bar iron or of gas tubing held from below, by which an overhead belt is slipped over its pulley. Patent belt shippers are also used.

Belt Shipping.—The placing of belts on their pulleys, or the transference of a belt from one pulley to another, usually performed while the pulleys are in motion.

Belt Stretcher.—A piece of mechanism employed for stretching new leather belts in readiness for use. But belts are very commonly suspended in a loop from the roof, and loaded only with ordinary weights. All new belts must be stretched, otherwise they would soon have to be taken up and relaced.

Belt Tension.—The ultimate strength of leather belts varies from 3,000 to 5,000 lbs. per square inch. But at laced joints the strength is only ·3 of those values, or from 900 to 1,500 lbs. per square inch. The working tension should not exceed about 300 lbs. per square inch. A good rule is 20 lbs. per inch of width for each $\frac{1}{16}$ inch in thickness of the belt. The tensile strain is found by multiplying the number of the H.P. to be transmitted by 33,000 and dividing the product by the velocity of the pulley in feet per minute. This represents the strain in pounds upon the driving side, independently of the initial tension producing adhesion between the pulley and the belt. This initial tension should be sufficient in amount to prevent slipping on either of the pulleys at the moment of starting.

Belt Tightener.—A contrivance used for pulling the ends of belts together for coupling up. Clamps are attached to the opposing free ends of the belt and then drawn together by means of screws and nuts.

Bench.—A strong support or table, more or less massive, upon which certain kinds of work are performed. See Pattern Bench, Core Bench, Brass Bench, Fitters' Bench, Vice Bench.

Bench Drilling Machine.—A small drilling machine made to bolt to a work bench, and actuated by hand or by power. Numerous types and sizes are made, either single, double, or treble geared. The hand wheel or winch handle is sometimes placed at the back, more commonly at the side of the machine, and drives the drill-spindle through bevel-wheels. The left hand only is free to manipulate the work.

Bench Hook.—A stop for sawing light work on the bench without damaging the bench itself. It is a block of wood about 12 in. long, furnished with a projecting stop at each end placed on opposite faces of a central web. One face of the web being laid on the bench, the lower stop is pressed against the bench edge, while the upper one takes the thrust of the wood which is being cut by the saw.

Bench Lathe.—A light lathe fitted with short standards to permit of its being bolted upon an ordinary work bench.

Bench Planes.—The planes which are always kept on the bench as being in constant use. They are jack, trying, and smoothing planes.

Bench Rammer.—A small round-ended rammer (q.v.) provided with a short handle (12 in. to 15 in.), used for making cores of moderate size, or those which can be made on the core bench, or the smaller moulds whose flasks are laid upon a work bench.

Bench Work.—Used to distinguish work carried on at the bench or vice as distinguished from lathe work, and also to distinguish the work of the fitter (q.v.) from that of the erector (q.v.).

Bend.—The curved part of a pipe by which its direction of motion is changed. "Quarter bends" include four in the circle; eighth bends, eight to the circle; and sixteenth bends, sixteen to the circle. Their angles are 90°, 45°, and 22½° respectively. When practicable, bends should be of large radius to diminish the friction of the liquids which they convey.

Bending, or Flexure.—The curvature of a beam about its axis or central plane. (See Neutral Axis.) The amount of bending action is measured by the bending moment (q.v.). The safe or working load must be estimated in reference to the intensity of the stresses due to bending, so that the moment of resistance of any section must be equal to the bending moment at that section. Diagrams and formulæ are given in most engineer's reference books for obtaining the bending moments and working loads of beams of various sections, variously supported or fixed.

Bending Machine.—See Straightening Machine.

Bending Moment.—The resultant or sum of the straining forces acting upon each side of a given section. The bending moment of the external forces at either side of any cross section equals the moment of resistance of that section.

Bending of Pipes.—See Pipe Bending.

Bending Rolls.—Heavy rollers of cast iron or steel set in strong standards, and used either for the straightening of crooked plates or for bending them into arcs of circles or into complete cylinders. The centres of the rolls are made adjustable relatively to each other with this object, their bearings running in strong housings, the upper pair of which are elevated or depressed at pleasure by means of a screw. The ordinary bending rolls contain three rollers, but are typical of numerous machines of modified characters used for bending, straightening, and rolling iron of various sections, both hot and cold.

Bending Strain.—See Transverse Strain.

Bent Gouge.—An outside firmer gouge curved in the direction of its length, and used by pattern-makers for hollowing out the concave portions of core-boxes. A spoon gouge is a short, more quickly curved form of bent gouge.

Bent Lever.—A form of the lever in which the arms are inclined at an angle towards each other. In estimating the mechanical efficiency, instead of taking the lengths of the actual arms, the lengths of the perpendiculars from the fulcrum to the lines of directions of the forces are taken.

Bessemer Converter.—The vessel employed for the production of mild steel directly from cast iron, by the patented process of its inventor. The converter is a pear-shaped vessel lined with a refractory material —ganister (q.v.) or magnesian limestone (q.v.)—and provided with tuyeres for the blowing in of air under pressure. It swivels upon trunnions, by means of which it is tilted up and the molten steel poured out when the process is complete. The blowing in of air oxidises the carbon to carbonic oxide. Speigeleisen is then added in sufficient amount to impart the quantity of carbon requisite to make steel of a definite degree of carburisation.

Bessemer Pig.—Pig-iron of a special quality prepared for conversion into Bessemer steel made by the acid process. It is made from hæmatite ores, and contains not less than 2 per cent. of silicon, from 3 to 5 per cent. of graphitic carbon, not more than $0 \cdot 1$ per cent. of phosphorus, about 1 per cent. of manganese, and as little as possible of sulphur and copper. White iron is unsuitable. The use of silicon consists in the high calorific power generated by its oxidation. Bessemer pig is classed in Nos. 1, 2, and 3. For the basic process pig rich in phosphorus and manganese and poor in silicon is preferred.

Bessemer Plant.—The plant used in the production of Bessemer steel (q.v.). It usually consists of a pair of converters, a casting pit, hydraulic cranes, ladle, slag pit, cupola for melting the speigeleisen, and the cupolas for melting the pig in readiness for running into the converter.

Bessemer Process.—The process of the decarburisation of cast iron effected by blowing atmospheric air through it while in a fluid condition, followed by a measured amount of recarburisation. (See Bessemer Converter.) The acid process (q.v.) originally introduced has been superseded by the basic process (q.v.). The Bessemer process is sometimes termed the pneumatic process, in allusion to the blowing through of atmospheric air.

Bessemer Skulls.—Skulls (q.v.) formed around the lining and the mouth of a Bessemer converter. Their formation is due to what is known as a cold blow, that is, there is not sufficient silicon in the pig to maintain by its oxidation the heat necessary for perfect fluidity. See Bessemer Pig

Bessemer Steel.—The mild steel produced directly from the pig in the Bessemer converter (q.v.). Used for boilers, rails, tyres, girder work, general smith's work, &c.

Best, or B.—A brand of wrought-iron plate or bar equivalent to No. 3

quality, or that grade which is only just superior to the commonest. It is obtained by piling, reheating, and rerolling either No. 1 and No. 2 iron, or all No. 3 brands. The "best" plates of the first-class houses are, however, equal to the "best, best" and "treble best" of other firms.

Best, Best, or B.B.—A brand of wrought-iron plate or bar indicating a superior quality obtained by piling, reheating, and rerolling "best" or No. 3 bars.

Best, Best, Best, or Treble Best, B.B.B.—A brand of wrought-iron plate or bar indicating the best quality of plate made. It is obtained by re-piling, reheating, and rerolling the "best, best" quality.

Beton.—Concrete (q.v.).

Betty.—A crowbar (q.v.).

Between Centres.—Signifies the chucking of lathe work between the centres of the headstock and poppet, as distinguished from the attach-ment of work to face and other chucks. The centres are either the dead centres or the prong chuck and dead centre.

Bevel, or Bevel Square.—A tool used for testing the accuracy of work which is cut to an angle or bevel. It consists of a stock and blade. The blade is movable and is fixed or set by a screw. In the best bevels the blade is also capable of movement along the screw by means of a slot.

Bevel Gearing.—Gearing comprising combinations of bevel-wheels (q.v.).

Bevelled Washers.—See Timber Washers.

Bevel Ring.—A ring or washer used for insertion between the flanges of abutting pipes which do not stand parallel with each other. When thin it is made of lead, if thick of cast iron. See also India-rubber Washer.

Bevel Square.—See Bevel.

Bevel Wheels.—Toothed wheels the pitch planes of whose teeth meet in the common centre of their axes. In most bevel-wheels these axes are at right angles with each other, but in some instances they are set at other than right angles. When the wheels are of equal diameter and the bevel of the pitch plane is 45°, they are termed mitre-wheels.

Bib Cock.—A cock whose nozzle is bent downwards after the fashion of an ordinary cistern tap.

Bicarbonate of Lime.—See Carbonate of Lime.

Bight.—The hanging loop of a chain or rope which falls below the pulleys in lifting tackle.

Bilge Pipes.—The pipes through which bilge water on board ship is removed by means of the bilge pump (q.v.). They are made of lead.

Bilge Pump.—A pump attached to marine engines for pumping out the bilge water in the event of excessive leakage.

Billets.—Short rectangular lengths of piled bar or welded scrap iron used for producing the smaller sections of finished iron in the rolling-mill, or for smith's use. Ordinary billets may measure about 18 in. long by 3 in. square.

Billeting Rolls.—See Roughing Rolls.

Binding Straps.—The loops which secure the cutting tool against the face of the tool-box of a planing or shaping machine. The straps are attached to the tool-plate and encircle the shank of the tool, which is then pinched by a clamping screw.

Binding Wire.—Iron or brass wire of small gauge used for binding joints which have to be brazed, or for fastening india-rubber hose around brass unions. For brazing purposes the term binding wire may signify either the iron wire used to confine the jointed pieces or the brass wire which acts as the solder.

Bisecting Scale.—A flat scale fully and symmetrically divided on each side of its centre line. It is usually graduated for $\frac{1}{4}$ in., $\frac{1}{2}$ in., and 1 in. scales.

Bismuth.—Symbol Bi. Comb. weight 210, sp. gr. 9·8. A pinkish white metal which does not oxidise at the ordinary temperature. It is chiefly useful as an alloy with other metals to form antifriction compounds for bearings, and to produce alloys which expand in cooling.

Bit.—(1) The jaw of a smith's tongs. (2) A copper bit (q.v.). See also Bits.

Bite.—(1) The grasping power of the jaws of lathe-chucks, vices, nippers, wrenches, tongs, and similar tools. (2) The friction of a rope around a pulley or capstan. (3) Signifies gripping action in general.

Bits.—Boring tools for wood, comprising centre, shell, nose, gouge, countersink, and other forms. They are actuated by the brace or stock.

Bit Stock Drill.—A drill whose shank is square tapered to fit into the socket of a common brace.

Bituminous Coal.—Smoky burning, non-caking, or caking and coking coals, containing more hydrogen and oxygen than the anthracites, but less than the lignites. The bituminous caking coals are only suitable for gas and coke making, but the semi-bituminous, non-caking, free burning coals are largely used for steam boilers, either alone or mixed with anthracitic coals (q.v.). The semi-bituminous coals burn readily, and give out much smoke.

Bituminous Paint, or Tar Varnish.—A protective coating used for pipes and structures which are laid under water. It is variously made, but is composed essentially of the bituminous products of coal mixed with mineral oils. One recipe gives 30 gallons of coal tar, fresh with all its naphtha retained, 6 lbs. of tallow, $1\frac{1}{2}$ lbs. of resin, 3 lbs of lampblack, 30 lbs. of fresh slaked lime finely sifted. Mix immediately and apply hot.

Black Band Ore.—This ore is of the same essential composition as the clay band (q.v.), but it contains in addition a quantity of bituminous matter intermixed therewith. It is mined in Scotland and in Prussia.

Black Copper.—Fine metal (q.v.), which, not being of sufficient purity, or containing only about 60 parts of copper to the 100, is subjected to the processes of recalcining and resmelting. Black copper, as it is then called, contains from 70 to 80 per cent. of copper.

Black Diamond.—See Carbonado.

Blackening.—See Blacking.

Blackening of Screws.—Small set screws in model and other work are often blackened to protect them from rust. They are heated to a black heat and dipped in a bath of oil.

Blacking, or Blackening.—Ground charcoal, coal or coke dust, or plumbago, used by moulders for the purpose of preventing sand burning (q.v.). It is sleeked over the surface of the mould with a brush and trowel, or simply dusted over. It prevents the immediate contact of the heated metal with the sand through the generation of a film of carbonic oxide, or of carbonic acid gas produced by the union of the oxygen with the carbon constituting the bulk of the blacking. Blacking is also mixed in certain proportions with the sands used in green sand moulds.

Blacking Bag.—A porous muslin or calico bag used for containing foundry blacking (q.v.), which is then dusted finely over the faces of the moulds.

Blacking Mill.—A large revolving closed cylinder, containing heavy rollers rotating freely upon its internal diameter, by whose crushing action small coal or coke is ground into blacking (q.v.) for foundry use. Instead of cylindrical rollers some blacking mills are furnished with ordinary spherical balls revolving in a corresponding annular groove. Sometimes called coal or coke mill.

Blacklead.—This is used for imparting a smooth skin or coating to some foundry patterns to facilitate delivery. Also for coating the faces of cast-iron chilling moulds, with or without the addition of oil. Mixed with soft soap it is used as a lubricant for wood which is revolving in a lathe steady during turning.

Black Nuts.—Nuts (q.v.) which are not polished but left rough as forged. They are used for common and out-door work, and for those portions of machine work which are concealed. See Bright Nuts.

Black Oils.—Crude mineral oils of good body which have been subjected to one series of purifications only to remove their mechanical impurities, and volatile oils, but which have not been filtered to improve the colour. They are used for cylinder lubrication.

Black Oxide of Copper.—An ore of copper containing about 80 per cent. of pure copper. It is found in Chili.

Black Pin.—Signifies a stud or pin which is not turned but left black from the forge. Black pins are often used in the rougher kinds of work, and particularly in those joints which are simply connections, and not pivots.

Black Plates.—Plates of thin sheet iron rolled and cut to size, ready to be brightened for the tinning process in the manufacture of tin plates (q.v.).

Black Print.—See Phototype.

Black Red Heat.—Denotes that temperature of wrought iron or steel in which the red colour is just visible by daylight. It may be roughly taken as corresponding with 1,000° F.

Black Sand.—Old sand (q.v.).

Black Smoke.—See Smoke.

Black Tin.—Tin ore which has been calcined and washed, and is thus prepared ready for smelting. It contains about 60 per cent. of tin.

Black Varnish.—Ordinary shellac varnish stained by the addition of lamp-black. It is used by pattern-makers.

Black Wash, or Wet Blacking.—A solution which is used in loam moulds (q.v.) and dry sand moulds for the same purpose as blacking (q.v.) is used for green sand moulds. It consists of a mixture of clay water, and ground charcoal or coal, applied with a swab or brush, and dried in the stove.

Black Work.—Work which has not been machined or polished. In some instances the term would apply to metal work which had been machined on a bearing section, but not elsewhere. And in other cases where no portion, working or otherwise, had been machined.

Blade.—A thin lamina or plate, as the blade of a square or of a screw.

Blank.—A piece of metal prepared specially to be shaped or ground to some particular shape. Thus, gear-wheels whose teeth are cut out of the solid are cut from prepared " blanks," and the term is generally applied to block-like castings or forgings which are to be definitely shaped by means of machinery.

Blank Bolts.—The rough forgings of the bolts previous to screwing.

Blank Cap.—A covered cap used for screwing into the open end of a hose union when not in use.

Blank End.—The end of a pipe or cylindrical casting which is closed up so that no through passage exists.

Blank Flange.—A solid flange used for bolting to the end of a pipe or cylinder for the purpose of closing it up permanently, or for a temporary purpose only.

Blank Holes.—Rivet holes in boiler, ship, bridge, and other plated works which are so very inaccurate that when the plates are placed in position the holes do not correspond within a distance equal to their own diameters. In such cases the holes have to be redrilled, or the work condemned.

Blank Socket.—A stop end (q.v.).

Blast.—The volume of air forced into those furnaces whose combustion is quickened by artificial means. They embrace blast furnaces, cupolas, locomotives, and some marine boilers. See Blast Furnace, Cupola, Blast Pipe, Forced Draught, Tuyere. Also Bessemer Converter.

Blast Fan.—See Fan.

Blast Furnace.—The furnace employed for the smelting of iron from its ores. Its longitudinal section is that of two truncated cones of unequal height placed base to base. (See Boshes, Throat.) Blast furnaces vary from forty to one hundred feet in height, and have a capacity of from 10,000 to 50,000 cubic feet. They are built of brick or of sheet iron lined with a refractory material. Once lit they are kept in blast for several years. Their details of structure vary with the nature and quality of the ore and fuel available. The blast, from which the furnace derives its name is either cold blast (q.v.) or hot blast (q.v.). It varies in amount and pressure with the character of the ore and fuel, and quality of pig to be produced. Roughly, from five to six tons weight of air may be required per ton of iron produced, at a pressure of from two to six lbs. per square inch. The blast is urged by blowing engines (q.v.), through a blast main into the tuyeres.

Blast Main.—The pipe which conveys the blast (q.v.) from the blowing cylinders to the tuyeres of blast furnaces. Its capacity is large in order to prevent the current from being spasmodic in character.

Blast Nozzle.—The extreme end of a blast pipe or tuyere which is pierced with the blast orifice.

Blast Orifice.—The mouth or opening of the blast pipe (q.v.) of a locomotive, through which the exhaust steam finds exit into the chimney. The diameter of the orifice is often made capable of variation. See Variable Blast.

Blast Pipe.—The pipe which conveys the exhaust steam from the cylinders of a locomotive or similar high-pressure engine into the chimney. The introduction of the blast pipe marked a most important era in the history of steam locomotion, its agency in producing a rapid forced draught being of vital importance. The escape of exhaust steam through a contracted orifice, by creating a partial vacuum in the chimney, causes the heated gases from the furnace to traverse the tubes with an acceleration of speed, with the immediate effect of producing more rapid combustion. See Variable Blast.

Blast Pressure Gauge.—A gauge used to indicate the pressure of the blast in the blast mains of smelting furnaces and cupolas. It is placed in communication with the current by means of a stop cock, and the height to which a column of mercury is raised thereby indicates the blast pressure. The graduations are those of ounces of pressure on the square inch, or of inches of water, or of both in combination in the same gauge.

Blast Stove.—See Hot Blast Stove.

Blazed Pig, or Glazed Pig.—An inferior class of pig iron, highly siliceous, which is frequently produced after the first blowing in of a blast furnace, as well as under certain conditions of regular working. Its inferior quality is due to the presence of silicon in large quantity derived from an excess of fuel.

Blazing Off.—A term used in tempering operations, signifying the determination of a definite tempering heat, by the flashing point (q.v.) of a grease or fat in which the article to be tempered is quenched.

Bleeding.—The red streaks of rust which weep through the scale adherent to the insides of boilers, and which reveal the presence of corrosion in the plates underneath.

Blende, or Zinc Blende.—An ore of zinc, occurring as a sulphide in black or brown lustrous crystals—hence its name. It contains 67 parts of zinc to the hundred. It is found in veins in many English districts, and in Europe and North America.

Blind Holes.—Rivet holes which are punched so inaccurately that when the plates are brought together the holes do not coincide within the extent of an entire diameter. See also Half Blind Holes.

Blister Copper.—Metallic copper obtained by the roasting of fine metal (q.v.) in order to expel the sulphur, the metal being then run into sand moulds. Its blistered appearance is due to the ebullition and escape of sulphurous acid gas, SO_2

Blisters.—Are defects present in boiler plates of inferior quality due to the non-expulsion of cinder or sand in the original rolling process.

Blister Steel.—That quality of cast steel prepared by the cementing or converting process. So called because its surface is covered with minute blisters or bubbles caused by the expansion of carbonic oxide gas, the gas having been generated by the action of oxide of iron upon carbon.

Block.—The bearing cheeks or pieces, either wood or metal, which carry the pulleys in lifting tackle. The term also commonly includes the pulleys.

Block Carriage.—The travelling frame which carries the chain sheaves upon the horizontal jib of a crane. It is travelled or racked along the jib by racking gear (q.v.).

Blocking, or Packing.—Short lengths and odd ends of timber balks, planks, deals, and battens which are used in the works, when erecting machinery and various other structures, the blocks being employed for packing up portions of the same.

Blocking Girders.—The girders which are attached to the under side of the truck frames, both back and front, of travelling cranes. They are made considerably longer than the width of the frames, and are placed in a transverse direction to prevent overturning of the crane when lifting cross-ways with a full load, an accident which would happen without the broader base afforded by the girders. The girders are supported upon blocking placed underneath their projecting ends. See Angle of Upset.

Blocking Up.—The elevating and supporting of masses of machinery by means of cranes, jacks, and blocking.

Block Setting Crane.—A Hercules (q.v.), or a Titan crane (q.v.), because designed specially for the setting of concrete blocks in harbour works.

Block Tin.—See Doubles.

Bloom.—(1) Literally a lump, from the Saxon "bloma." Signifying therefore, any mass of iron prepared and purified in readiness for the formation of a special forging. (2) A mass of puddled ball which has undergone the process of shingling or squeezing. (3) A mass of piled iron brought into an outline of a shape and size roughly corresponding with that of the article which has to be rolled therefrom.

Bloomery.—See Finery.

Blooming Down.—The rolling down of steel ingots into blooms.

Blooming Mill.—A Shingling Mill (q.v.).

Blooming Rolls.—See Puddling Rolls.

Bloom Steel.—Steel made by the open hearth processes, and rolled into blooms.

Blow.—The forcing of the blast into the tuyeres of blast furnaces and of cupolas, for the purpose of reducing ore in the former, and of melting the iron in the latter, is called the blow, or blowing, or the blast. The forcing of air through the molten metal in a Bessemer converter is also called the blow, but has reference to a definite period of the process, that period during which the carbon is being oxidised or burnt out. A blow is, in foundry language, caused by the imprisonment of air and gas in the metal of a mould. See Blow Hole.

Blowers.—Machines employed for the production of artificial currents of air, and utilised for the production of blast, ventilation, &c. They are for the most part either blowing cylinders (q.v.) and pistons, or rotary blowers (q.v.). The advantage of blowers over fans is that they can be driven at a much slower speed, with a corresponding economy of motive power, and increased safety, and that the amount of blast which they will deliver is of a positive character, and therefore measurable and under control.

Blow Hole.—Blow holes are hollow cavities in castings, caused by the presence of air or gas in the moulds into which the castings were run, and which become entangled among the molten metal, in consequence of imperfect venting. Blow holes of a bluish colour imply the presence of sulphur in the metal.

Blowing.—The application of the blast to a foundry cupola.

Blowing Cylinders.—Cylinders employed for pumping the air under pressure into the blast main (q.v.). They are double acting, having valves in both top and bottom. A large blowing cylinder will discharge as much as 50,000 cubic feet per minute under a pressure of from two to ten lbs. per square inch.

Blowing Engine.—An engine used for driving blowers. For blowing cylinders the engine will be of a reciprocating type, for rotary blowers a rotary engine (q.v.) will be used, or a three-cylinder engine (q.v.).

Blowing Fan.—An ordinary fan used for blowing.

Blowing Off.—The periodical driving out of the lower portions of the water of a steam boiler, in order to prevent accumulation of solid matters, and consequent incrustation (q.v.).

Blowing Through.—The sending of a jet of steam through the cylinders, valves, and condenser of a condensing steam engine in order to create a vacuum before starting. In high-pressure engines, blowing through is accomplished through the ports and pet cocks to clear the cylinders of the water of condensation.

Blown.—A casting is said to be blown when it is honeycombed with blow holes (q.v.).

Blown In.—Blast furnaces are said to be "blown in," or to be "in blast" when they are in full working order. They remain so except when repairs are needed, or trade is too bad to make them pay. Blowing in is an operation which takes three or four weeks to fully effect.

Blown Out.—Blast furnaces are blown out, or "out of blast," when they are doing no work, being extinguished either through badness of trade or want of repairs.

Blow Off.—The pipe and cock used for emptying an engine boiler of its contents. It is situated at the lowest part of the boiler. The practice of blowing off is one preventive of incrustation (q.v.).

Blow-off Bend.—A bend pipe which connects the blow-off cock (q.v.) with the blow-off seating (q.v.) in a Lancashire or Cornish boiler, and through which the blowing off takes place into the ash pit.

Blow-off Cock.—A plug cock, by whose opening the blowing off of a boiler is effected.

Blow-off Seating.—The blow-off cock, or its bend, is attached to a separate casting called a seating, in Lancashire and Cornish boilers. The seating is fitted to the curve of the boiler, and furnished with a flange for the attachment of the bend or cock.

Blows.—Signifies the number of charges (q.v.) which are drawn from a blast furnace, or a cupola in a given time.

Blow-through Valve.—A valve in use with condensing marine engines. Its purpose is the clearing of the cylinder, condenser, and air-pump of the air which they contain, and the supplying its place with steam, which being immediately condensed, creates a vacuum. This is necessary before the engines can be started. The valve when opened makes a communication between the steam in the valve casing and the condenser.

Blue.—See Prussian Blue.

Blue Bath.—A bath of dilute hydrochloric acid, containing 8 or 9 per cent. of acid, in which phototypes (q.v.) are bleached after having been developed in the prussiate bath, and washed in the water tray. It becomes blue through contact with the prussiate of potash. Also called acid bath. This bath is lined with gutta-percha.

Blue Billy.—A fettling (q.v.) used for puddling furnaces in the Cleveland district. It is a residual product derived from the roasting of copper pyrites.

Blueing.—The heating up or letting down of the temper of steel until it assumes a blue colour. A dark blue corresponds with a temperature of about 570° F., and is the colour for springs.

Blue Line Phototype.—See Phototype.

Blue Lines.—See Dimension Lines.

Blue Metal.—See Fine Metal.

Blue Print.—A phototype (q.v.) in which white lines appear on the blue ground.

Blue Process.—See Phototype.

Blunt File.—A file which is nearly but not quite parallel throughout its length. The term blunt relates to the point, sometimes called "blunt pointed," to distinguish it from a taper file (q.v.).

Board.—Thin timber as distinguished from planks. Anything below about two inches in thickness would be called a board, above that a plank. Strictly speaking, however, a plank should not measure less than three inches in thickness.

Bob.—The pear-shaped or globular weight depending from a plumb-line. Sometimes applied to the beam of an engine.

Bob Lever.—See Rocking Handle.

Body.—(1) The main portion of a pattern, casting, or forging, as distinguished from its auxiliary or subsidiary portions. (2) As applied to oils, signifies the degree of consistence or viscosity of the oil. A heavy oil is one with much body, a light oil signifying, on the other hand, a thin fluid oil. The mineral oils when purified have little body, so also has sperm ; while castor oil and some of the vegetable and animal oils and fats, as lard and tallow, have a good body. Body must not, however, be confounded with resinous properties which some of the vegetable oils possess and which cause gumming (q.v.). The oils having most body are used for heavy shafts, for engine cylinders, and for high temperatures ; those having less body for light machinery. The fluidity of an oil increases with increase in temperature, and with light oils firing and evaporation frequently take place at comparatively low temperatures.

Body Core.—The main or principal core in a mould, as distinguished from branch or smaller cores. The body core of a cylinder is that which forms the bore, in opposition to the passage cores. The body core of a pipe is that in the main pipe itself, as distinguished from branch or Tee cores.

Bogie.—A swivelling framework which carries the axle of a pair of locomotive or carriage wheels, and by means of which the main framing is enabled to accommodate itself to curves of short radius. A bogie truck is a short truck resting on four wheels, and pivoted at its centre to the frames of an engine or carriage, a truck being pivoted at one, or at each end, of the engine or carriage. See Bogie Engine. Its use is to enable the engine or carriage to run round sharp curves, which it effects by substituting its own short wheel base (q.v.) for the ordinary long wheel base.

Bogie Engine.—A locomotive provided with a bogie. In the single bogie engines the leading wheels are the bogies, in the double bogie or Fairlie engines the leading and trailing wheels are bogies.

Boil.—The period during which the carbon is being burnt out of the iron in a Bessemer converter (q.v.). The carbon during this period is being oxidized to carbonic oxide CO, and burns at the mouth of the converter.

Boiled Oil.—See Linseed Oil.

Boiler.—The vessel in which the steam used in the driving of an engine is generated. Forms and details are numerous and varied. Cornish boilers have one fire flue, Lancashire boilers have two flues. Locomotive boilers and those of portable engines are multitubular. Vertical boilers are those in which the circular section is in a horizontal plane. Boilers require setting when of the Cornish or Lancashire type, the setting consisting of brickwork and an arrangement of flues built therein. Vertical boilers require no setting. Where there are no smoke tubes, as in Lancashire and vertical boilers, there are cross tubes passing through the fire-box for the purpose of exposing a larger surface to the action of the fire. Marine boilers are a type by themselves, consisting chiefly of the return multitubular class. Sectional boilers are those in which the steam is generated in a multitude of distinct tubes communicating with one steam chest or receiver. Details of these and of other types of boilers will be found under their specific headings. See also Water Tube Boilers in App

Boiler Bear.—See Punching Bear.

Boiler Capacity.—See Heating Surface.

Boiler Circulation.—See Circulation.

Boiler Coating.—Non-conducting compositions are used to prevent the radiation of heat from steam boilers. Such compositions are smeared over while in a plastic state and become hardened by the heat. Various materials are employed, comprising felt, silicate cotton, cements, silicate compositions, asbestos, &c. They are mostly patented preparations.

Boiler Corrosion.—Is either internal or external. Internal corrosion will result either from the presence of acidulated water (q.v.) attacking the plates below the water level, or from superheated steam in the steam chamber. External corrosion is the result of leakage, and of contact with damp foundations and seatings.

Boiler Crown.—The uppermost plate in the shell of a vertical boiler. It is of a hollow discoid shape, flanged around the edges, and riveted to the outer shell plates. Sometimes it is of a flat form only and stayed. The boiler crown proper is that belonging to the outer shell. The fire-box crown is that over the top of the furnace or inner shell.

Boiler Dome.—The superheating chamber of a multitubular boiler. It is usually of one piece, welded up and flanged at top and bottom, and encased in a brass mounting.

Boiler Explosions.—These are almost invariably due to overpressure (q.v.) of steam, the overpressure taking place by reason of the boiler being taxed beyond its capacity ; the chief sources of overpressure being weakness due to faulty original design, to weakness caused by wear and tear, by bad workmanship, by sudden strains put upon a structure, which, though strong enough to withstand ordinary pressures, is not able to sustain a sudden and excessive stress applied thereto. Unstayed or imperfectly stayed surfaces are examples of bad design, as are also the use of plates, stays and rings too weak for the pressures which they are designed to carry. Corrosion and grooving are examples of wear and tear. Among the errors of bad materials and workmanship may be classed the use of laminated and pitted, or of brittle plates, excessive drifting, hard caulking, &c. ; while overheating, with the consequent weakening due to increase in temperature, the accumulation of deposit over the fire, lowness of water, are illustrations of sudden straining ; and also the too rigid attachment of boilers to their seatings or foundations cause overstraining by preventing freedom of expansion and contraction.

Boiler Feed.—The water supply of a boiler.

Boiler Feeder.—(1) The agency by which the water supply of a boiler is maintained. Usually it takes the form of a force pump (q.v.) or of an injector (q.v.). (2) The reservoir from which the supply of water for a boiler is drawn. (3) A feed water heater (q.v.). (4) An old-fashioned arrangement for regulating the water supply consisted of a float at the water level connected by a chain with a lever attached to a valve, which last, being lifted by the lowering of the float, allowed a definite amount of water to flow into the boiler.

Boiler Fittings.—Those additional portions of boiler work which, though not considered as parts of the shell, are yet essential to its completion. They are usually charged for extra, and comprise man-hole and mud-hole doors, fire bars and their bearers or rings, furnace doors, dampers and frames, the various fittings being subject to modification by the special type of boiler to which they are attached.

Boiler Float.—In old egg-end and waggon boilers feed was regulated by means of a float and chain, which rising and falling with fluctuations in level indicated the height of the water.

Boiler Incrustation.—See Incrustation, Soda, Carbonate of Lime, &c.

Boiler Leakage.—See Leakage.

Boiler Maker.—The work of a boiler maker consists in marking out the plates of boilers to size and shape, in marking and punching the rivet holes, in riveting together and caulking up the seams, in fitting tubes, gussets, stays, strengthening rings and steam chests. Here his work ends, and the boiler is handed over to the engine fitter to receive its fittings and mountings.

Boiler Makers' Dolly.—See Dolly.

Boiler Mountings.—Those additional portions of boiler work, extraneous to the shell, but which are essential to the proper working of the boiler. They are usually charged for extra, and comprise safety and stop valves, blow-off cock, steam and water gauges, try cocks, back pressure valve, and fusible plugs.

Boiler Plate.—The plate from which steam boilers are manufactured. It is not a special brand of plate, but ordinary best best, or treble best, or Lowmoor plates. Usually treble best is used for the shell, and Lowmoor or other similar quality for the fire-box. Sometimes the crown fire-box plates are Lowmoor, and the body treble best. The specification always fixes the quality of plates used in contract boilers, though usually leaving the manufacturer to make his choice from several houses. The tensile strength of ordinary wrought-iron plates should be 21 tons per square inch with the grain, that is in the direction in which the plates are rolled. Lowmoor plates will bear a strain of 24 tons with the grain. The tensile strength of mild steel plates should range from 26 to 32 tons.

Boiler Pressure.—The pressures in steam boilers vary with the type, the age, and the condition of the boiler. In externally fired, waggon, and egg-ended boilers they will vary from 15 to 20 lbs. per square inch above the atmosphere. In new Cornish and Lancashire boilers, well stayed, from 45 to 60 lbs., reduced when worn to 25 or 35. In locomotives and portables, from 120 to 150 lbs., gradually reduced with age to 100, or less. In marine return tube boilers, from 100 to 180, according to size and construction. The pressure is adjusted with the safety valve, and indicated by the pressure gauge.

Boiler Prover.—A force pump and pressure gauge combined, employed for testing boilers by hydraulic pressure.

Boiler Scaling Hammer.—See Scaling Hammer.

Boiler Section.—See Effective Section, and Gross Section.

Boiler Shell.—The outer body, or casing, pertaining to a boiler; which encloses the water and fire spaces in internally fired boilers; and the water space only in externally fired boilers.

Boiler Shop.—A lofty building in which the work of boiler making and the construction of works in angle iron and plate is carried on. A few forges are ranged down one side, the whole of the central portion being left free for the building up of work, which is moved about by a powerful overhead traveller. The machines are placed in an open part of the shop to allow freedom of manipulation for the bars, angles, and plates.

Boiler Smithy.—A department of the boiler shop containing smiths' forges, where angle iron, tubes, and the lighter work of that class are bent and welded.

Boilers, Setting of.—The laying of stationary boilers upon suitably prepared foundations of brickwork. The flues are built in the brickwork to form either a split draught (q.v.) or a wheel draught (q.v.), and a certain amount of inclination (q.v.) is given to the boilers.

Boiler Stays.—Screwed rods or smooth tubes which connect and stay the flat ends of steam boilers. Bar stays (q.v.) are screwed into the ends of shells or fire-boxes, and are either secured with nuts or riveted over. Tube stays (q.v.) are riveted and beaded over. Gusset stays (q.v.) are riveted. Copper, steel, and wrought iron are the materials used for stays.

Boiler Testing.—The testing of the strength of steam boilers, or their capacity to withstand the internal stresses due to steam pressure, is mostly performed by the pressure of water produced by a test pump, the amount of pressure being recorded by one or more dial gauges placed in communication with the interior of the boiler. The test pressure is usually from one and a half to twice that of the actual pressure to which it is intended to work the boiler. The test should continue for about half an hour, during which time the pressure gauge should not go back at all. If it goes back, that indicates leakage. During the time of the test, the boiler should be carefully examined to detect leakage at the seams and rivet heads ; any alterations in diameter of the shell, or in the shape of the end plates indicating bulging stresses should also be recorded, and noted, and compared with the condition of the boiler when the test pressure is removed, to ascertain the amount of permanent set, and also to detect sections of inherent weakness of structure. Boilers are sometimes also tested with steam, a practice which was formerly universal, but is now resorted to chiefly in the case of locomotive and portable boilers, it being considered that testing by steam causes the test to approximate more nearly to the conditions of actual practice.

Boiler Tubes.—Tubes used for increasing the efficiency of a steam boiler by the enlargement of its heating surface (q.v.) without increasing its bulk. See Field Tubes, Galloway Tubes, Multitubular Boiler, Tube Stays.

Boiling.—The process of purifying tin by holding stakes of wet wood under the molten metal while it is in the refining basin. The steam evolved causes an ebullition of the metal to the surface where the impurities present therein become oxidised. After being allowed to settle the metal is ladled out and cast into ingots. The upper stratum is called refined tin, the lower common tin.

Boiling Point.—The temperature at which a liquid throws off bubbles of vapour or steam, producing an ebullition at the surface. Pure water in an open vessel, at a pressure of 30 inches of mercury, boils at 212° F. or 100° C. The presence of solids in solution or an increase of pressure raises the boiling point. The boiling point is lowered by a reduction in pressure.

Bollard.—A cast-iron pillar used for the mooring of vessels on quay walls. Also the pillars of cast iron bolted to the bulwarks of vessels for the coiling of a rope when paying out or heaving to.

Bolster.—(1) A plate or disc of metal which sustains articles of wrought iron or steel through which holes are being punched. A hole in the bolster a trifle larger than the punch permits the button to pass through, while its adjacent edges by sustaining the edges of the plate around the punched hole prevent buckling or bending of the same. (2) The bearings that fit within the housings (q.v.) in forge and mill rolls, and which sustain the rolls.

Bolt.—A fastening employed by engineers. In the ordinary form of bolt the head is forged with the shank—the end opposite or the tail being screwed to receive the nut by which the bolt is secured. Both heads and nuts may be either rectangular or hexagonal in form. To prevent the tendency to turn round, the shoulder or neck underneath the head is often made rectangular to fit into a square hole. The various forms of bolts will be found described under their respective headings.

Bolt Cutter.—A machine for cutting the heads of bolts, worked by hand or by power.

Bolt Head.—The flattened expansion at one end of a bolt. It is usually rectangular or hexagonal in form, its shape generally corresponding with that of the nut.

Bolt Machine.—A machine used in the smithy for the forging of the tails or shanks of bolts. It consists essentially of a series of stamps or dies, the lower pieces being fixed, the upper ones movable in a vertical direction. The movable dies are actuated by eccentrics turned on the main driving spindle above, and the spindle is provided with a fly wheel for the equalisation of the motion. The diameters of the stamps vary, in descending order, so that the bolts are drawn down by being passed along a succession of dies of diminishing diameters.

Bolt Oliver.—See Oliver.

Bomontague.—See Beaumontague.

Bonding.—See Banding.

Bonnet.—(1) The movable cap or cover which is made removable to permit of the introduction of a pump valve into, or its withdrawal from, its seating. (2) The hood of a smith's forge.

Bonneted Safety Valve.—A safety valve covered with a casing or bonnet enclosing the valves and communicating with a waste pipe, through which the waste steam is conveyed away without the building. The bonnet and waste pipe prevent the steam from filling the shed or building when the boiler is closely roofed in.

Boom.—(1) A term applied to the jib of a derrick crane (q.v.). (2) The upper or lower flange of a built-up girder, lattice or otherwise.

Borax, or Tincal.—Symbol $Na_2 B_4 O_7 + 10 H_2 O$ A sodium compound of the element boron. Used for welding mixtures in steel and iron, and for brazing gun metal and copper. It is also added in small quantity to the zinc thrown last of all into the brass melting-pot or crucible to prevent its oxidation and loss in vapour. It is prepared for use by previous driving off of the water of crystallisation at a high temperature.

Bore.—(1) The internal diameter of a pipe or cylinder whether it be a rough or a machined casting. (2) Also signifies to turn the internal diameter of hollow cylindrical work. (3) The bore of a wheel is the size of the hole or eye which receives the shaft, and designates the diameter after it has been bored out. (4) The bore of a pipe is the internal diameter as a rough casting. The dimensions of pipes are never given otherwise than as the bore and length.

Boring.—The operation of making or finishing circular holes in wood or metal.

Boring Bar.—A stiff cylindrical bar provided with a screw feed to the cutter head. Used for boring cylinders. The bar is fixed in standards or between lathe centres, and revolves when cutting, the work being bolted down to the bed.

Boring Collar.—See Cone Plate.

Boring Flange.—See Drill Plate.

Boring Head, or Cutter Head.—The ring which carries the cutters of a boring bar. It is actuated by a screw sunk in a recess in the bar itself, and driven by gearing from one end. A boring head may also be fixed on a mandrel, and revolved without endlong movement, the work being made to travel by means of a slide rest against the cutters.

Boring Lathe.—See Boring Machine.

Boring Machine.—(1) A machine specially constructed for boring holes in cylinders, bosses, bearings, and the like. It may be horizontal or vertical in type. The cutter in the former revolves in a circle whose axis is horizontal; in the latter, in a circle whose axis is vertical. (2) Boring machines are used for wood with brace bits; for metal, with drills and cutters. Lathes are also boring machines, the work being either fixed or in motion. When the work is fixed, a boring bar set between centres holds the tool and carries it round. When the work revolves the tool is held in the slide rest. (3) In boring mills the tool is attached to a bar, being passed through a slot therein and secured with wedges, and both revolves and travels, the bar passing through a hollow mandrel in th headstock. (4) For boring large cylinders the axis of the cutter is vertical, the cylinders standing on end.

Boring Mill.—See Boring Machine above, and Vertical Boring Mill.

Boring Rest.—A lathe rest whose outline is that of the letter L, the horizontal arm, which is uppermost, being slotted to take the flat of a boring tool or drill, and so prevent it from turning round. The hinder end of the boring tool is centred on the poppet mandrel, and by it fed along to its work. Such rests are used with ordinary lathes, and also with lathes of special construction called boring lathes, in which the poppet mandrel is fed forward automatically, through the medium of a back shaft and gearing.

Borings.—See Turnings.

Boring Tools.—Brace bits and similar tools for wood, and turning tools and cutters for metal, embracing a large number of types. The ordinary boring tool consists of an ordinary roughing or finishing tool, cranked or turned round at the end, right or left handed; the shank being rounded for a few inches behind the cutting point. Boring and drilling differ in this respect—that boring is performed by a cutter having a single cutting edge, while drilling is performed by two cutting edges placed on opposite sides of the axis of the tool. Boring is also performed by cutters set in boring heads and boring bars.

Bosh.—See Water Bosh.

Boshes.—That area or section of a blast furnace (q.v.) which extends from the section where the diameter is widest down to the crucible. The section of the boshes taken vertically is that of a truncated cone, and the angle is dependent upon the character of the ore and of the fuel, the condition being that the charge shall not sink down too rapidly.

Boss.—(1) The centre or hub of a wheel. (2) In a more general sense the circular bearing parts of castings which carry shafts, pins, or studs. (3) A circular disc pierced with a central hole and cast on framed work for increasing the length of bearing parts without adding unnecessarily to the thickness of the remaining or plated portion of the structure.

Bossed Up.—A forging is bossed up when a circular disc of a sensible thickness is formed upon its face, being either made from the solid iron by upsetting and finishing to outline, or by welding on. Turning the edges of such bosses in the lathe, for the production of bright work, is also called bossing up.

Bossing Machine.—A steam hammer used for welding the bosses on the wheels for rolling stock.

Botting.—Closing the tap hole of a cupola with clay after a ladle of molten metal has been tapped out.

Bottle Jack.—A screw jack of light construction, wh :h is slightly conical in elevation, and provided with a handle at the side, by which it is carried. Its shape is, therefore, not unlike that of a jug or bottle, hence the name applied to it.

Bottle Tight.—Signifies that the seams, rivets, fittings and mountings of a steam boiler make such close and perfect joints that there is not the slightest leakage when tested under water or steam.

Bottling Up.—The temporary confinement of the steam in the tubes of a sectional boiler, due to its too rapid generation. The steam accumulates until it acquires sufficient energy to force its way to the steam dome. When this occurs, cool water rushes in to supply its place, and undue strain is put upon the tubes due to the sudden change in temperature.

Bottom Board.—The board upon which the joint of a pattern is laid while being rammed up. Its purpose is either to stay and sustain a weak pattern, or to make the moulder's sand-joint, hence called joint board.

Bottom Card.—An indicator card, taken from the bottom of a vertical or oscillating cylinder. In a bottom card the lead line (q.v.) is on the right-hand side.

Bottom Face.—That face of a mould or casting which is downwards when pouring. The bottom face is always sounder than the top face (q.v.), since the sullage floats to the top, and the liquid pressure consolidates the lower metal.

Bottom Flue.—A flash flue (q.v.).

Bottom Fuller.—A fullering tool (q.v.) which is laid on the anvil, or set in the square hole near its end.

Bottoming.—The rubbing or grinding of the points of wheel teeth in the roots of the teeth of the wheels into which they gear. It is due to want of bottom clearance or to badly-fitting shafts.

Bottoming Tap.—A plug tap which is not rounded off at all at the end, but finishes with square edges in order that it may cut its thread perfectly to the bottom of a drilled hole.

Bottom Part.—The lower part or section of a foundry moulding box. With bedded-in work the bottom part would signify the portion of the mould in the foundry floor.

Bottom Rake.—The angle of relief (q.v.) in cutting tools.

Bottoms.—An impure alloy of copper with antimony, tin, lead, iron and sulphur, which results when fine metal (q.v.) is roasted for the purpose of obtaining a pure regulus from which to manufacture best selected copper.

Bottom Tool.—The lower half of a fullering tool (q v.).

Bott Stick.—A light iron rod of about five or six feet in length, having a small disc-like expansion at one end which receives the stopper of clay used for botting (q.v.). The other or pointed end of the rod is used for tapping the hole for the egress of the metal.

Bouche.—A bush (q.v.).

Bourdon's Gauge.—See Dial Gauge.

Bow Compass.—See Spring Bows.

Bow Connecting-Rod, or **Kite Connecting-Rod,** or **Banjo Frame.**—A form of connecting-rod employed in steam pumps, where compactness of arrangement is sought after. It is triangular in outline, and the crank which drives the fly-wheel is enclosed by the bow.

Bow Drill.—A fiddle drill. Made by stretching a cord between the ends of a rod curved like a bow. The drill is inserted in a reel or small pulley around which the cord is twisted. The drawing of the bow backwards and forwards revolves the reel and with it the drill in opposite directions alternately. Bow drills are useful only for light work, and are employed in engineers' workshops chiefly for drilling up the ends of shafts and other cylindrical work which has to be centred in the lathe.

Bower's Process.—Has for its object the same result as Barff's process (q.v.), but air instead of steam is the oxidising agent employed.

Bow File.—A riffler (q.v.).

Bowling Hoop.—A ring whose single section is that of an arch with side expansions or flanges for the reception of rivets. Employed for uniting the sections of furnace shells in horizontal boilers.

Bowling Iron.—Iron plate produced by the Bowling Iron Company, in Yorkshire. Its quality is considered about equal to that of Lowmoor (q.v.) and is therefore often substituted for the latter for the crowns of the fire-boxes of boilers, and other work where first-class material is in request.

Bowl Sleepers.—Pot sleepers (q.v.).

Bows.—Short drawing compasses, similar in shape to those of larger dimensions excepting that the head is made circular in order that it may be rolled with facility between the thumb and forefinger. See Spring Bows.

Bow Saw.—A narrow-frame saw, held in tension by the leverage afforded by the twisting of a cord. These saws are used for cutting curves, but the band saws have mostly superseded them in workshops.

Bowstring Girder.—A girder in which the outward or horizontal thrust of a curved beam is sustained by a horizontal tie beam forming the chord of the arc.

Box.—(1) A bearing for a shaft. (2) A box coupling (q.v.).

Box Coupling.—A shaft coupling made as a hollow cylinder to fit over the abutting ends of the shafts to be coupled together, and held in place with keys. Sometimes called muff coupling.

Box End.—A connecting-rod end having no loose strap end (q.v.), but in which, instead, the brasses are thrust into a slot from one side and slid along to their seatings, the side flanges on one face being removed to permit of their sliding in. The brasses are tightened with a cottar (q.v.).

Box Filling.—The filling up of a moulding box with its body of sand enclosing a pattern. This is usually labourer's work, and is distinct from the ramming. The moulder rams while the labourer is engaged in filling in.

Box Girder.—A wrought-iron girder built up with two parallel joists or girders of I-iron, of equal depth, united at top and bottom with flat plates, riveted to the flanges of the joists. Also termed a tubular girder.

Boxing Up.—The construction of heavy patterns by a process of rectangular framing together of thin boards in preference to cutting them from solid planking.

Box Link.—A slot link (q.v.) whose internal faces are recessed, so that the shifting-block is partly embraced by the edges as well as by the internal faces.

Box Metal.—Sometimes applied to the metal used for bearings; it may be gun metal, or a white metal. One recipe gives copper 32, tin 5; another gives zinc 75, tin 18, lead 4·5, antimony 2·5.

Box Nut.—A nut made for the covering and protection of the end of a bolt. It is similar to an ordinary nut, with the addition thereto of a dome-shaped closed end. The screwed part is also terminated internally in a circular recess larger in diameter than the deepest portion of the Vee of the thread, in order that the cutting tap shall clear itself. Box nuts are used on the covers of locomotive cylinders and in similarly exposed situations. The screwed bolt ends and their nuts are thus both alike protected from rust or accident, hence there is no difficulty in slackening back when necessary. Also termed cup nut.

Box Safety-Valve.—A bonneted safety-valve (q.v.).

Box Spanner.—A spanner used for those bolts whose heads are sunk below the surface of the material into which they are inserted. The box-spanner end is circular without and square or hexagonal within to receive the nut. The circular portion drops into a round hole somewhat larger than the head of the nut and bored concentrically around it.

Box Standard, or Boxed Standard.—The standard, or main framework, of a machine or engine, which is hollowed internally to obtain the maximum of strength with the minimum of material. See Hollow Structures.

Box Wood (*Buxus sempervirens*).—Sp. gr. 1·04. A hard, tough, close-grained wood of a pale yellow colour, belonging to the order *Euphorbiaceæ*. Its valuable properties of hardness, toughness, and freedom from susceptibility to warping and shrinking, have led to its employment in the manufacture of rules and scales for workshop and office use. Previous to being worked up it undergoes several years of seasoning. It is also employed for some small tools as planes and spokeshaves, and for tool handles. A cubic foot weighs 64·8 lbs.

Boyle and Marriott, Law of.—The law of Boyle and Marriott, so named after its discoverers, is that the volume of a gas varies inversely as the pressure to which it is subjected. In engineering this law has its application in the expansive working of steam in a closed cylinder, and in calculations pertaining to gas and air engines.

Brace.—A tool used for actuating bits and drills for boring wood or metal, and varying in character with the class of work which it is designed to do. The brace for boring wood is made of wood and iron in combination, or entirely of iron. It is turned directly by the hand alone. For light drilling in metal, a brace having a couple of bevel wheels to actuate the drill is used. For heavier work a smith's brace is employed, turned, as in the first instance, by hand, but the requisite pressure is imparted by a screw at the top. For the heaviest drilling a ratchet brace is used, in which the drill is moved by a long lever, and the pressure is derived from an arm which receives the reaction of the drill, the feed being imparted by a screw and nut.

Brace Bits.—The ordinary bits used for wood boring, and having square tapered shanks to fit in the socket of a common brace.

Braced Girder.—A built-up lattice girder.

Braced Truss.—A truss, either braced with single braces or counter-braced with diagonal ties and struts, to sustain the stress of a moving load without deformation. See Bracing, Counter Bracing.

Braced Work.—See Bracing.

Bracing.—The staying or supporting of inherently weak structures with rods and ties. The object of bracing is the conversion of transverse stresses into those of a longitudinal character. Tensile or compressive braces take the form of triangles, since that is the only figure which maintains its form unaltered while the lengths of its sides remain constant. Bracing is practised in all built-up structures, as bridges, roofs, cranes, girders of various kinds, and the calculation of the strains on the different members comprising the structure is obtained readily by graphic methods, or by the method of moments.

Bracket.—A rib occupying an angle in a casting, or in built-up wrought-iron work and placed there for the strengthening of a plate or flange. The term, however, is one of very wide application in engineering, being used to denote castings themselves which have a resemblance, however remote, to the typical bracket. It is often employed to denote bearings of many different kinds, as, for instance, those which are bolted to walls to carry plummer blocks for shafting, or those which depend from timber, also to carry shafting, and castings bolted upon bed plates, and so on.

Bracket Pedestal.—A wall bracket (q.v.).

Brad.—A cut nail, rectangular in section and having a head or lip on one side only. Used for pattern work and by moulders for mending and staying broken and weak sand. See Sprigging.

Brad Awl.—A boring tool used previous to the insertion of brads and nails. It first divides the fibres by means of its cutting edge, and then thrusts them aside, but does not extract them.

Brad Punch.—A blunt pointed steel rod used for hammering the heads of nails slightly below the surface of the wood into which they have been driven, when, for the sake of appearance, it is undesirable that the heads of the nails should be seen, the head of the punch being struck by the hammer, similarly to the centre punch.

Brake.—(1) The mechanism by which a moving part is brought to a standstill. (2) The frictional arrangement by which the descent of heavy loads is regulated. The ordinary brake consists of a flexible strap of wrought iron, lined with friction material, embracing the periphery of a smoothly turned iron wheel, and is capable of being tightened by the intervention of a lever, or of a hand wheel and screw. Friction-disc and internal-expanding brakes are also widely used to-day.

Brake Blocks.—Blocks of friction material used in a brake.

Brake Cylinder.—The cylinder of a steam or of a water brake.

Brake Drum.—A large wooden drum used for winding the rope or chain which lifts and lowers the cages or the trucks on colliery or quarry works. Called also a winding drum. "Brake drum" refers to the method by which its action is controlled, being that of a brake.

Brake Handle.—The handle of a brake lever.

Brake Horse-Power.—The horse-power of an engine or machine taken off a brake attached thereto. The instrument employed for the purpose is the friction dynamometer (q.v.), and the advantage of its employment is that the net useful work given out is then directly obtained, whereas when the horse-power is taken by the indicator, a doubtful allowance has to be made and deducted for the power absorbed by the engine itself, due to friction and other losses.

Brake Lever.—The lever by means of which the action of a brake is controlled. It may be a hand or a foot lever.

Brake Power.—The frictional resistance developed by a brake. It is expressed in any convenient units, or as an equivalent to a definite amount of mechanical force against which it is set off.

Brake Shoe.—Shoe lined with friction material forced against a wheel or drum by the brake levers to arrest its motion.

Brake Strap.—The encircling band of hoop iron to which the brake blocks (q.v.) are screwed in the friction brake of a crane or hoisting machine.

Brake Wheel.—A wheel whose periphery is turned to receive the pressure of the brake strap and blocks, the whole with the necessary levers constituting a friction brake. The brake wheel may be either a distinct casting or a ring simply cast on the face of one of the toothed wheels.

Braking.—The application of brake power to a machine.

Bramah's Press.—The hydraulic press (q.v.) as first invented, embodying the practical application of the principle that hydraulic pressure is directly proportional to the head and to the area of surface.

Bran.—Is used for rubbing over the plates of sheet iron used in tin-plate manufacture, in order to dry them after the pickling process.

Branch.—An offset from a pipe which diverts the contents of the pipe into another channel, or which conducts into the pipe.

Branch Pipe.—A pipe having offsets or branches, that is, short outlet pipes attached to the main axis.

Brands.—(1) Certain marks used to distinguish the qualities of wrought iron and steel plates and bars. They are not uniform tests, since the similar brands of different makers will often indicate different qualities of iron. By the distinguishing marks cast or stamped by manufacturers upon the materials of construction, their own special qualities of iron, and the blast furnaces or rolling mills at which they are produced, are at once apparent. Sometimes the initials of the firm or company are given, often the initials are quite arbitrary, often far ciful, like telegraphic code words, frequently there is a heraldic device in addition—a lion, a tortoise, a gun, &c. These then indicate some definite qualities of iron, even though the B B or B B B themselves are omitted. Crown quality is often represented by the drawing of a crown to which the letter B, or B B are added. Keys to all the brands used by all the houses in the kingdom are supplied in Ryland's well-known Directory of the iron trades. (2) Brands, or branding-letters, are block projecting letters cast upon the ends of light rods or handles, and used for burning manufacturers' initials and names into timber, packing cases, &c.

Brasque.—A mixture of coke or coal dust, with or without powdered gas carbon and coal-tar. Used for the lining of various furnaces to prevent the corrosion of certain slags.

Brass.—(1) An alloy of copper and zinc, or copper, zinc and lead. It has a wide range of uses in all branches of engineering. The term is, however, used rather loosely, being also applied to gun metal (q.v.). (2) The term is commonly applied to the bearings for the journals of shafting, the half bearing being termed a brass, the two bearings a pair of brasses, and the bearings with their seatings a plummer block or pillow block. These bearings are not actually made in brass, but in gun metal, which is harder and more durable.

Brass Bench.—The bench of the brass moulder consists of a plain table, either of wood or iron, upon which the moulding flasks are manipulated, and of a bin or trough containing the sand.

Brass Borings.—The borings and turnings of brass castings collected in the shops are separated from those of iron by magnetting (q.v.) to be remelted.

Brass Contraction.—See Contraction, Contraction-rule.

Brasses.—See Brass.

Brass Finishing.—The later stages of the manufactor of brass cocks, valves, lubricators, and similar engine and pump fittings. It is a special branch carried on in a department or in a factory by itself. It combines turning, milling, grinding and burnishing.

Brass Foundry.—The department in which brass moulding is performed is usually either separate from, or a section divided off from the iron foundry, special materials and appliances being employed therein.

Brass Furnace.—The furnace in which brass and gun metal are melted. It is an air furnace (q.v.), or an "air crucible furnace" (q.v.), and is built in brickwork partly below and partly above the level of the foundry floor, the draught passing through it from a grating in front to the chimney behind. It is usually made to hold but one crucible at a time, an increase in melting power being obtained by increasing the number of furnaces, which are then ranged side by side. The crucibles are lifted out of the mouth or top of the furnace with the crucible tongs. When large castings are required a reverberatory furnace is often employed.

Brass Moulder.—The work of moulding and casting brass is a special department of foundry practice, almost invariably performed by a workman specially trained to it, who seldom touches iron moulding. Good brass moulders receive higher wages than ordinary green sand moulders, and do much of their work by the piece.

Brass Sieve.—See Sieve.

Brass Tubing.—Is used to a slight extent in engineering for cutting off into hand railing, sheathing, distance pieces, &c. Its thickness is given by the wire gauge. The common tube is soldered or brazed; but the best tubes used for condensers are solid drawn, and usually made of an alloy of brass. See Muntz Metal.

Brass Wire.—Is put to a variety of uses in engineer's work. It is hard when unannealed, soft when annealed. Its dimensions are those of the wire gauge.

Brasswork.—The small mountings of steam boilers, gauges, hose unions, cocks, valves, whistles, and similar works are made in common brass, various mixtures being employed. When mixing brass, care must be taken to use the same mixtures for work which is in immediate proximity, since the colours vary with the mixtures. Brass is sometimes variously named according to its colour, as red brass, yellow brass, white brass.

Brazed Joint.—A joint united by brazing (q.v.) as distinguished from a soldered joint, a weld, or a riveted or screwed joint.

Brazing.—The union of metallic surfaces by means of a film of an alloy interposed. The joints to be brazed are cleaned, bound with wire, put into a clear fire, sprinkled with borax, and heated until the alloy melts. Also called hard soldering.

Brazing Metal.—An alloy composed of 98 parts of copper and 2 of tin.

used for casting the flanges of copper steam pipes, and the facings of sluice valves which have to be cast to their iron casings. It is necessary that the proportion of copper should be high, else the flanges would melt at the temperature required for brazing.

Brazing Wire.—Soft brass wire of small gauge used for binding around joints which have to be brazed. The joint being heated and sprinkled with borax, the wire melts and runs in.

Break.—A brake (q.v.).

Breakdown.—A breakdown is said to happen when some portion of an engine or machine fails or gives out, so that the mechanism depending thereon is either partially or wholly brought to a standstill. Breakdowns frequently call into play the highest ingenuity and ready resource of the workman or attendant. Breakdowns at sea are, from the circumstances of the case, more troublesome than those on land, and they are more especially so in the case of crank shafts, and similarly massive portions, for which auxiliary (q.v.) and spare parts (q.v.) cannot be taken.

Breakdown Crane.—A form of crane, especially constructed for railway use, to be employed in clearing the line after accidents, hence called, "accident crane." It is made compact and strong, and cast iron is employed as little as possible in its construction, in order to lessen the risk of fracture by rapid transit and the shocks of shunting, &c. It is necessarily a balance crane (q.v.), but the balance-box is made to slide inwards, or to haul backwards as required, and the whole is mounted upon tyred wheels, with springs, axle-boxes, buffers and draw-bars, as in ordinary rolling stock.

Breaking.—When molten iron is poured into a ladle, its surface shows a multitude of continually varying curves, due to the rising up of the metal from beneath. Different qualities of metal have different aspects of striation, so that a founder can distinguish between hard and soft iron while yet in the ladle, and can also roughly judge of its temperature. This striation is termed the breaking of the metal.

Breaking Down.—The sawing of logs of timber into planking.

Breaking-down Rolls.—Roughing-down rolls (q.v.).

Breaking Joint.—(1) Used in reference to the joints in metallic piston rings. The joints of the rings alternate one over the other to prevent escape of the steam. Sometimes called cross joint. (2) The separation of the joints of steam and water pipes, sockets, flanges, cylinders, &c., for purposes of repair. (3) The placing of the longitudinal seams of cylindrical riveted structures alternately in relation to each other, in order that the plates shall afford each other mutual support. This practice is common in the barrels of steam boilers. (4) Where a fagot or pile is so built up that the layers of which it is composed are not continuous over the whole width, but are made in different widths, so that the joints overlap, they are said to break joint.

Breaking Pieces, or Spindles.—Short lengths of shafting used for coupling up the engine with the bottom rolls of a forge train, or the rolls with each other. The breaking pieces are made weaker than the necks of the rolls in order that, in the event of overstrain, they will break and so prevent damage to the machinery, hence their name. They are coupled to the journal ends by means of wabblers (q.v.).

Breaking Strength.—Corresponds with that amount or limit of stress at which a structure or beam gives way, or is ruptured.

Breaking Weight.—The load necessary to break a beam or a structure

The actual breaking weight is a variable quantity even for the same particular class of materials, depending partly upon the quality selected, and also essentially upon the manner in which it is applied, whether suddenly by impact (q.v.), or gradually, or as the last in a long series of variable loads. See Dead Load, Fatigue of Materials, Live Load, Variable Load.

Break Lathe.—A lathe of large size, whose bed is deepened in front of the headstock to receive wheels and other work of large diameter, and the length of whose gap or break is capable of variation by sliding the bed along a base plate.

Breast Brace.—Any brace (q.v.) which is furnished with a knob in order that the pressure on the drill or bit may be communicated thereto by the breast of the workman.

Breast Hole.—The arched hole in front of and just above the base of a foundry cupola, where the fire is lit, and through which the cinder and slag are extracted after the casting work is over. It is closed by a sheet-iron door or breast plate during the time that the blow is on.

Breast Plate.—(1) The plate of sheet iron which covers the breast hole of a cupola. (2) A sheet of metal laid against the breast to receive the thrust of the drill spindle used with the fiddle drill.

Breast Wheel.—A water-wheel in which the water meets the buckets near the horizontal line which passes through the axis of the wheel. When the water flows in at a point above the horizontal line, the wheel is termed high breast, and when at a point below, low breast.

Breeches Pipe.—A bend pipe having two legs or branches running either parallel with each other, or divergent. The branches unite into one at their point of bifurcation. The casting which connects a single inlet suction pipe with the two barrels of a lifting pump furnishes an illustration of a breeches pipe.

Breeze.—Small or dust coke. Used sometimes for grinding into blacking and facing for foundry use.

Brick.—Common clay bricks are used in foundries to form the framework or backing upon which the loam is plastered in loam moulding. Curved bricks specially made are used in the smaller cylindrical work. See Bauxite Bricks, Dinas Brick, Fire-Brick, for those kinds which are used for the linings of furnaces. Iron moulders use bricks made of loam (see Loam Brick), in certain sections of their work.

Brick Arch.—An arch built up of fire-brick placed transversely across the fire-box of a locomotive boiler, in front of the tubes, and sloping downwards and backwards towards the fire-bars, its purpose being to deflect the flame and hot gases backwards and so prevent them from passing into the tubes too rapidly.

Bricking Up.—The building up of the outline of a loam mould with courses or layers of bricks. See Brick, Loam Brick, Loam Mould.

Bridge.—(1) A structure by which a road or railway is carried over a river. Bridges are sometimes made of cast iron, seldom of timber, though both materials were largely employed formerly. Wrought iron is the material now chiefly used. (2) An arched guide casting attached to the cover of a lift pump or a force pump, and through whose central boss the free end of the piston or plunger-rod travels. (3) In a reverberatory furnace, the wall which divides the fuel chamber from the hearth. (4) The barrier which stretches across the fire-box of an engine boiler, at the farther end of the fire-grate. It is usually built of brick laid upon a girder-like casting, or upon a bar of wrought iron. Its purpose is to throw the flame upwards to the heating surface and to prevent too rapid escape of the heated gases.

Bridge Cylinder.—See Foundation Cylinder.

Bridge Plates.—A cheap quality of iron similar to ship plates. Its tensile strength is low, about 18 tons, or even less, to the inch, and it is brittle, having but 2 or 3 per cent. of ultimate set (q.v.).

Bridge Rail.—See Flat Bottomed Rail.

Bridle.—(1) The loop which is forged on a slide-valve rod to embrace the back of the valve. It may be square or circular in plan. (2) Sometimes applied to the strips in a steam chest between which the valve travels. (3) A loop or clip used for holding test pieces in a testing machine.

Bridle Rod.—See Radius Rod.

Bright Nuts.—Nuts whose ends and faces are polished for the sake of appearance. All nuts exposed to view on machines, and on good indoor work are bright. Their ends are turned in a centering machine, and their facets are planed while strung on a bar in a planing machine, or they are done with a milling cutter, or on a special machine.

Bright Red Heat.—A stage of temperature in smith's work when the black scales on the surface of iron are thrown into relief against the red background, and which corresponds roughly with a temperature of 1800° F.

Bright Work.—Denotes those portions of metallic surfaces which are polished, either to diminish the friction of bearing parts, or for the sake of appearance. The brightening is effected by machine cutters of various kinds, or by files, emery-wheels, emery-papers, or powders.

Brine Pump.—A pump used for drawing off a certain portion of the water periodically from a marine boiler to prevent excessive saturation.

Brine Valve.—Sometimes applied to the blow-off valve of a marine boiler.

Brittleness.—In the materials of engineering this is usually a concomitant of hardness. The hardest and most highly tempered steel is the most brittle; white iron is more brittle than grey, and chilled iron than any other. But brittleness is also a sign of inferiority, as in wrought iron and mild steel. The brittleness of castings and malleable work is reduced by annealing (q.v.).

Broach.—A boring tool, employed for the purpose both of enlarging and imparting accuracy to tapered and parallel holes. Broaches are made of various sectional forms, and their cutting angles range from 90° to 130'. Sometimes called rose reamers. See also Rose Bit.

Broaching.—The enlarging and smoothing, and truing of drilled holes by means of a broach or reamer. Broaching is chiefly done in a drilling machine or a lathe.

Broad Brush.—A draughtsman's colour brush, broad and thin, used for tinting over large areas.

Broad Gauge.—Seven feet between rails, introduced by Brunel on the Great Western Railway.

Brob.—A spike driven alongside of a butt-jointed timber standing at right angles with another timber. Brobs are driven into the longitudinal piece to prevent the abutting timber from slipping sideways.

Broche.—A broach (q.v.).

Broken Glass.—Frequently thrown on the surface of molten brass while yet in the crucible, in order to prevent oxidation from taking place.

Bronze.—See Gun Metal.

Broom.—A large oblong wire brush (q.v.) used for brushing the sand from iron and steel castings.

Brown Hæmatite.—A hydrated oxide of iron, embracing ores widely differing both in appearance and in quality. They receive different specific names in different districts, but all agree in being composed essentially of peroxide of iron. Fe$_2$ O$_3$, and water.

Bubble Tube, or Spirit Glass.—The tube of a spirit-level (q.v.) which contains the enclosed spirit.

Bucket.—(1) A scoop-shaped vessel or grab (q.v.) used in dredging operations, and for the hauling of grain. (2) A mud bucket (q.v.). (3) The piston of a lift pump. (See Pump Bucket.) (4) The outlets for water in turbines. (5) The receptacles for the water in over-shot and breast wheels. These were formerly made of wood, but iron is now employed. Their section also is curved instead of being angular, as was the case when wood was the material used. (6) Ordinary galvanized iron buckets are used by moulders for damping and re-mixing sand which has become dried and friable by being cast in. The intimate mixture of the sand and water is effected by turning it about with the shovel.

Bucket Air Pump.—The ordinary form of marine engine air pump, provided with piston, foot and head valves.

Bucket and Plunger Pump.—A form of double-action pump in which a bucket and plunger are combined on a single rod, the plunger being uppermost. By its combined action half the contents of the barrel are discharged during the up, and half during the down stroke. The volume of liquid ejected at each stroke is equal to the displacement of the plunger.

Bucket Valve.—The flap valve (q.v.) of the bucket of a lift pump (q.v.).

Buckle.—The localised inequalities of a plate of wood or metal caused by unequal and localised strains and tensions. Moisture or dryness, or want of homogeneity of grain, will produce buckling in timber. Unequal internal stress due to different densities, or qualities, or rates of cooling, will cause buckle in metals.

Buckled.—See Crippling.

Buckled Plates.—Plates of cast or wrought iron having a curved or slightly angular section if taken transversely in any direction, the rise of the angle or curve being uppermost. The object of buckling is to strengthen the plate by placing it in the condition of an arch, and so to enable it to withstand the injurious effects due to the vibration of moving loads. Buckled plates are used in bridge construction for supporting the foot ways.

Buckling.—See Crippling.

Budding Spanner.—A shifting spanner in which the movable jaw is adjusted by means of a small skew rack and wheel on the back of the lever handle.

Buff.—An appliance for polishing and finishing metallic surfaces. Usually a cylindrical body, having a covering of emery. The emery is cemented with glue or other composition, and the buff runs on a spindle.

Buffer.—An elastic spring by which the shock of contact between railway waggons is deadened.

Buffer Bar.—See Buffer Beam.

Buffer Beam, or Buffer Bar.—A beam of timber or wrought iron which carries the buffers in locomotives and rolling stock.

Buffer Box.—The casing which encloses the buffer spring and buffer rod.

Buffer Disc.—The spheroidal disc against which the contact of buffers takes place.

Buffer Plunger.—A buffer rod (q.v.).

Buffer Rod.—The rod which carries the buffer disc.

Buffer Spring.—A strong coiled steel spring enclosed in the buffer box, which resists and deadens the force due to the collision of buffers.

Buffing.—The process of polishing by means of leather and emery, or other powder.

Buff Wheel.—See Buff.

Building Up.—(1) Patterns are said to be built up when they are constructed by laying several courses of segments one above the other. (2) Works in wrought iron and steel are built up by the piling (q.v.) of small bars, or when numerous pieces are welded together, as distinguished from drawing down, upsetting, or otherwise manipulating in a single piece.

Built-Up Pulleys.—See Wrought-Iron Pulleys.

Bulb Angles.—See Angle Iron.

Bulb Bars.—Rolled iron bars having a bulb, or bead on one edge. They range from 6 in. by $\frac{7}{8}$ in., with $7\frac{1}{2}$ lbs. weight per foot, to 12 in. by $\frac{7}{8}$ in., with 22 lbs. per foot.

Bulb Tees.—Rolled tee-iron having a bulb or bead on the centre web. They range from 6 in. by 4 in., by $\frac{7}{8}$ in., to 12 in. by $6\frac{11}{16}$ in. by $\frac{1}{2}$ in.

Bulger Ram.—A round-ended ram employed in forcing metal plates into apertures, in experiments on bulging stress (q.v.).

Bulging Stress.—The stress to which flat plates are subjected when supported around their edges, and strained at the centre. Or the stress of cylinders when subjected to internal pressure.

Bull Dog.—The material used for the lining or fettling (q.v.) of puddling furnaces. It is a mixture of oxide of iron, chiefly ferric oxide, though ferrous is present in small quantity, with silicon and titanic acid.

Bullet Compasses, or Cone Compasses.—Compasses having a bullet at the end of one leg for setting in a central hole from which a circle has to be scribed.

Bullock Gear.—Mechanism for the utilisation of animal power in the driving of well and other pumps, and for much light work besides of a general character. It consists of the long poles to which the cattle are attached, and which by their leverage drive a pair of bevel-wheels, and through these the machinery connected thereto. Also termed, indiscriminately, cattle gear, and horse gear.

Bunker.—The enclosed space which contains the furnace coal or coke for the use of an engine boiler. Sometimes called coal bunk.

Bunker Plate.—A plate of sheet iron which encloses the area utilised for the deposit of boiler fuel.

Burden.—The charge of a furnace. See Charge. Light Burden, Heavy Burden denotes the excess of fuel or of ore respectively.

Burnett's Fluid.—A fluid used for the impregnation (q.v.) of timber. It is composed of chloride of zinc diluted with water, and applied under pressure.

Burning.—Wrought iron becomes burned when it is allowed to remain in the fire too long at a temperature equal to or greater than that of the welding heat (q.v.). It can then barely be utilised by repeated reheatings and hammerings, but is best discarded.

Burning On.—The process by which a broken-off or incomplete portion of a casting is made good, or replaced. A sand mould of the portion to be burned on is made and placed in proper juxtaposition to the old casting in the bed of the foundry floor. The molten metal is then poured in, but at the same time permitted to run away through a gate or outlet at the side. By this device the molten metal is allowed to flow for a certain time over the broken face of the old casting until that portion is in a state of local fusion, when the outlet is stopped with sand, and the flow of metal arrested. When cold, perfect amalgamation will be found to have taken place.

Burnish.—(1) To polish a metallic surface by the friction of another metallic surface thereon. (2) The envelopment of a body by the bending over of a metallic ring around its edges. (3) Sometimes applied to the process of buffing (q.v.).

Burnisher.—(1) A half-round file ground smooth, or a piece of steel of similar section used to burnish metallic surfaces with. (2) A blunt tool by which pressure is imparted against sheet metals during the operation of spinning.

Burnishing.—See Spinning.

Burnt.—See Sand Burnt.

Burnt Casting.—See Sand Burnt.

Burnt Iron.—Cast iron which has been long subjected to the action of heat is said to be burnt, and becomes rotten, and red in colour, owing to the absorption of oxygen.

Burnt Steel.—Cast steel which has been overheated so that its nature as steel has been abstracted, due probably to the absorption of carbon and the formation of oxides. Burnt steel is considered valueless, but there are methods sometimes adopted for improving it by bringing it into contact with certain elements, or by reheating and hammering. But its lost tenacity can never be fully restored, and it is scarcely worth the while to attempt it, because the burnt end can be broken off and thrown away, leaving the remaining portion good.

Burr.—(1) The turned-over undetached line or edge of metal which results when a metallic edge or surface is subjected to abrasion. (2) The turned-up edge of metal which commonly results from the operation of punching or drilling, and cutting operations in general.

Burra Burra Copper.—Malachite (q.v.) obtained in Australia.

Bursting.—(1) The breaking asunder of a rapidly revolving part when the stresses due to centrifugal force overcome the cohesive strength of the material. (2) The bursting of a boiler or hydraulic cylinder is due to the excess of internal pressure, as distinguished from that of collapsing pressure (q.v.).

Bursting Pressure.—The pressure necessary to produce actual rupture in a closed vessel, as a boiler or pipe.

Bush.—The internal cylindrical lining of a bearing. Usually of brass, or gun metal, or some kind of antifriction metal. Bushes are used either for purposes of economy, as when lining an inferior metal with one more expensive, or for convenience, so that they may be replaced when worn, without replacing the main casting. A bush is undivided, while a brass is divided.

Bushing.—The fitting and driving in of bushes into their seatings.

Butt Coupling.—The ordinary form of box coupling used for connecting shafting, the abutting ends of the shafts being enclosed by the box keyed around them. See Box Coupling.

Butterfly Valve.—A pair of flap valves (q.v.) arranged back to back.

Butt Joint.—A smith's welded joint, made by the abutting together of two pieces of metal. Butt joints are sometimes strengthened with dowels. See also Welding in Dict. of Modern Terms.

Butt Measurement.—The measurement of a dimension by thrusting one end of a rule against the portion whence the measurement is taken.

The term is employed to distinguish it from measurements taken with trammels, compasses, or dividers, or measurements set off by marking from the rule divisions themselves.

Button.—The circular disc of metal pushed through a plate by punching

Button-Headed Screws.—Small set or attachment screws whose heads are hemispherical in outline. They are driven in with a screw driver.

Button Sleekers.—Egg sleekers (q.v.).

Buttress Thread.—A screw thread whose section is triangular, but which has one face at right angles with the axis of the screw, the second face alone being sloped. These threads are only employed in cases where they have to resist a force always acting in one direction.

Butt Riveted Joint.—See Butt Riveting.

Butt Riveting.—When the edges of two plates are brought one against the other, and are riveted into a covering plate or strip embracing the two; the term butt riveting is used to distinguish it from lap riveting (q.v.).

Butt Strip.—The covering strip used for the connection of the plates in butt riveting (q.v.).

Butt-Welded Tubes.—Malleable iron tubes whose joints are simply abutted in welding, as distinguished from lap-welded tubes.

Buzz Saw.—A term sometimes applied to a circular saw, derived from the buzzing sound which it produces.

B. W. G.—Birmingham wire gauge. See Wire Gauges.

C

Cab.—The hood or covering which arches over the footplate of a locomotive engine.

Cabinet Nest.—A palette (q.v.) containing several independent saucers, circular in plan, and fitting closely one above another, to prevent drying up of the ink or colour when not required for immediate use.

Cable-laid Rope.—A rope composed of three common ropes instead of strands, laid or twisted from left to right, or in the reverse direction to the strands in the several ropes composing it.

Cage.—See Lift.

Cage Valve.—A ball valve (q.v.).

Caisson.—A water-tight casing sunk into the bed of a stream and used in the construction of the foundations of bridges and structures laid under water, the work of excavation being carried on within it.

Caking Coal.—Coal which has the property of giving off abundance of gas, and hardening subsequently. The caking coals are largely used in smiths' fires, and for gas making.

Calamine.—The principal ore of zinc. It consists of the oxide in combination with carbonic acid contaminated with oxide of iron. It is found in Flintshire, the Mendip Hills, Alston Moor, and in many parts of Europe.

Calcination, or Calcining, or Roasting.—The heating of metallic ores in order to expel some of the foreign and injurious ingredients mingled therewith, preparatory to their reduction to the metallic condition. The ores of iron are roasted in kilns, to drive off the carbonic acid and water present. In the extraction of copper the roasting of the ores is

7*

effected in a reverberatory furnace to expel arsenic and sulphur. The arsenic combines with oxygen to form arsenious acid, the sulphur yields sulphurous acid gas, or remains in small quantity in the ore as sulphate of copper. The calcination of tin ore is also effected in a similar furnace, and for the same object.

Calcining.—See Calcination.

Calcining Furnace.—The reverberatory furnace in which the calcination (q.v.) of ores is carried on.

Calcium.—Symbol Ca. Comb. weight 39·9. Sp. gr. 1·58. One of the metals of the alkaline earths, the metallic base of lime, CaO, and carbonate and sulphate of lime, of the limestones, mortars, and concretes. The metal calcium does not occur in the free state in nature, being too readily oxidizable, but its various compounds are of immense utility and interest to the engineer.

Calender Rollers.—The heavy grooved rollers which feed the timber along to the saws, or to the cutters, in frame saws, and planing machines for wood.

California Pump.—See Barrow Pump.

Caliper.—An instrument used for measuring the diameter of circular work, at sections intermediate from the ends, where rule measurement is impracticable. Calipers are termed "outside" when they are bellied for measuring outer diameters—"inside" when the legs are straight and only bent round at the points to measure internal diameters. See Combination Caliper.

Caliper Rule.—A short steel rule, a portion of which is attached to a well fitting slide, both rule and slide being graduated into minute portions of an inch. The slide is drawn out until an object such as a bar or plate can be embraced between the opposing portions of the divided rule, and the thickness read off directly from the dimensions on the slide.

Calking.—Caulking (q.v.).

Caloric.—A term used by the older writers on mechanics, which expressed a false and misleading theory. Caloric was applied to signify the material principle of heat—heat being long believed to be a subtle elastic fluid which permeated the molecules of all bodies. See Heat.

Caloric Engine.—See Hot Air Engine.

Calorie.—The French practical unit of heat. It is the quantity necessary to raise one kilogramme of water 1° C., the temperature being taken between 0° and 4° C.

Calorific Intensity.—The actual value of a fuel as a thermal agent. It signifies the net available amount of heat given out by the combustion of a definite quantity of fuel in a given time after the modifying influences of the presence of moisture, the nature and action of the chemical constituents of the fuel relatively to each other, the nature and specific heats of the ultimate products of combustion, and the way in which the fuel is burnt, are allowed for. These arithmetical calculations are so much affected by these modifying influences that the calorific intensities of various fuels are ascertained instead by means of direct experiment.

Calorific Power.—Signifies the amount of heat given out by a substance during combustion, independently of its rate of combustion.

Calorimeter.—The sectional area of a boiler flue, given in square inches, is sometimes termed the calorimeter of the boiler.

Cam.—A plate or cylinder which transmits variable reciprocal motion to a piece of mechanism. Cams which are plates are provided either with an edge of definite and special outline, volute, or heart-shaped for example, or with an irregular definitely shaped groove on one face. Cams which are cylinders have a groove cut on the periphery, which, instead of being circular in one plane moves to the right or left as it passes around.

Camber.—The amount of curve given to an arched bar or structure. It is applied specifically to the leading and trailing springs of locomotives and trucks. It is given in terms of the versed sine of the arc, and usually with the load on.

Camel Hair.—Largely employed for manufacturing driving belts.

Campaign.—The life of a blast furnace lining, estimated by the number of months or years during which it remains in blast.

Cam Shaft.—The shaft in a cam-worked punching and shearing machine, which actuates the slides. Usually, however, in machines of modern manufacture, an eccentric is employed instead of cams or wipers. Similarly applied to any shaft which carries cams or wipers.

Candles.—See Puddlers' Candles.

Cannon.—A hollow spindle or shaft which has a motion of its own, and which also permits of the independent motion of a shaft through its internal diameter. Its applications to machinery are numerous. Sometimes termed a quill.

Cantilever—A beam fixed at one end and loaded at the other, or loaded uniformly. Sometimes called a semi-beam.

Canting Fence.—See Fence.

Canting Motion, or Angle Motion.—The provision for placing the table of a band, or the fence of a circular saw at an angle with the saw for the purpose of cutting bevels.

Canvas Hose.—Hose piping (q.v.) made of woven canvas. It is not so durable as the rubber hose (q.v.).

Canvas Packing.—Steam packing formed of canvas, with a core of india rubber.

Cap.—The upper or loose portion of the bearing of a shaft. Made either in iron or brass, and screwed in place with bolts or with studs. The bore of the cap may be in actual contact with the shaft, or its function may be simply that of confining the top half of a divided brass bearing in place.

Capacity.—Another term for yield. Thus the capacity of a manufacturing firm will be, say, 200 or 300 tons per week, or perhaps given as so many tons per furnace.

Capillarity.—The capillarity of fluids is taken advantage of for the lubrication of bearings, the oil being fed into the bearings along a filament of cotton wick. The capillarity of any oil is therefore a recommendation, if not obtained at a sacrifice of some other more important property.

Capping.—The shrouding (q.v.) of gear wheels.

Capstan.—A cone-shaped drum used for hauling in ropes or chains. A warping cone (q.v.).

Capstan Lathe.—A lathe provided with a capstan tool rest (q.v.).

Capstan Tool Rest.—A thick perforated disc or circular tool holder, attached to the upper part of a lathe rest and used for carrying drills and tools of various kinds. The drills are set radially in the holes around the periphery of the disc and fixed by set-screws put in from the top. Any number of duplicate pieces of work can thus be operated upon with different sized tools, with precisely similar results in each case, by slewing round the capstan and bringing the desired tools into alignment with the axis of the work centred in the lathe. Also called from its shape, a turret or turret-rest.

Carbide of Iron.—White iron (q.v.).

Carbon.—Symbol C. Comb. weight, 11·97. A non-metallic element, whose various combinations with iron chiefly produce the essential differences between the qualities of that metal. It is also the solid element of combustion. Carbon is allotropic. See Graphite, Combined Carbon, Charcoal, Hydro-carbons, &c.

Carbonado, or Black Diamond.—An uncrystallized variety of carbon found in Brazil. It is as hard as the diamond, but free from its liability to split. It is used for turning down and truing emery-wheels.

Carbonate of Copper.—See Malachite.

Carbonate of Lead.—Symbol Pb CO₃. A compound of carbonic acid with lead oxide, thus Pb O + CO₂. White lead is essentially a combination of the carbonate with the hydroxide, thus 2Pb CO₃ + Pb H₂ O₂. The white lead of commerce is invaluable as the basis of oil paints.

Carbonate of Lime.—Symbol Ca CO₃. The principal compound to whose deposition the incrustation of steam boilers and water pipes is due. It is in the first place held in solution in the water as a bicarbonate, the excess of carbonic acid holding it in solution. This excess being driven off by heat, the carbonate of lime remains as a floury or muddy deposit, its precise condition varying with the nature of the salts with which it is usually accompanied. In the presence of heat it hardens and forms an injurious scale. See Clark's Process, Soda, Washing-out, &c.

Carbonate of Magnesia.—This carbonate is held in solution in some feed-waters, and on becoming deposited produces incrustation (q.v.). It is usually present in smaller quantity than is the case with the salts of lime. See also Dolomite.

Carbonates.—Salts formed by the union of carbonic acid with bases. The carbonates always effervesce when brought into contact with acids, with the evolution of CO₂. The carbonates of iron, lead, copper, magnesia, and lime are those which chiefly concern the engineer.

Carbon Cores.—Cores of solid carbon prepared for foundry use, and employed where heavy liquid pressures have to be resisted.

Carbon Dioxide.—See Carbonic Acid.

Carbonic Acid, Carbon Dioxide, or Carbonic Anhydride.—Symbol CO₂. Produced by the combustion of carbon in excess of air; hence it plays an important part in combustion and in smelting operations. In the presence of reducing agents it gives up one atom of its oxygen, being thus reduced to the lower oxide CO. See Carbonic Oxide.

Carbonic Oxide.—Symbol CO. A gas produced by the combustion of carbon in a limited supply of air. It owes its value to the metallurgist to its power as a reducing agent, appropriating an atom of oxygen from metallic ores, thus reducing them to the metallic condition, itself being converted into carbonic acid (q.v.).

Carbonization.—See Carburization.

Carburization.—This refers to the quantity of carbon present in iron. The limit of carburization for cast iron is about 4·7 per cent.; in the best wrought iron not more than ·01; while the many varieties of cast iron and steel occupy positions intermediate with these.

Cards.—See Indicator Cards.

Card Wire.—A brush made of very fine steel wire, used in combing cotton wool. A strip of this when nailed to a piece of wood is used by fitters for cleaning the teeth of files which have become choked up with particles of abraded metal.

Carnot's Principle.—This principle is, that the amount of work done by a heat-engine is independent of the nature of the intermediary agent employed, being dependent upon its temperature alone.

Carpentry.—Carpentry occupies a secondary position in most engineering works, being confined chiefly to the framing together of heavy structures in wood ; to the making of packing-cases and the cutting out of templets for boiler makers and erectors ; the making of blocks for the iron turners and the general repairs about the factory. In railway works, however, the carpentry and joinery department is one of the most important.

Carriage Hoist, or Locomotive Hoist.—This consists of a set of shear legs (q.v.), erected in repairing sheds for the purpose of lifting the trucks or engines off their axles when repairs are necessary. The ends of the shear legs are sunk in cast-iron shoes buried in the ground.

Carriage Ladle.—A foundry ladle mounted on a low carriage for convenience of portability.

Carriage Traverser.—See Traverser.

Carrier.—A looped, or heart-shaped, iron or steel forging, or a steel casting, made to clip a piece of metal rod which has to be turned between the dead centres of a lathe. It is tightened upon the rod by an adjusting screw, and is itself driven by means of a short projecting pin affixed either to a face or catch plate, or to the mandrel itself. See Driver.

Carrier Wheel.—An idle wheel (q.v.).

Cartridge Paper.—The commonest drawing-paper used in offices for rough work. It is made in sheets measuring 30 in. by 22 in., and 40 in. by 27 in., and in continuous lengths of 54 in. wide.

Case Hardening.—A process by which a thin film of hardened steel is formed on the surface of wrought iron. It is of especial value in cases where a hard bearing surface is wanted without the expense of employing steel, or without the risk of the warping and curving which steel almost invariably suffers during hardening. Case hardening is effected by heating the iron to a bright red in contact with substances rich in carbon, or carbonaceous matters, and then quenching in cold water. Small articles are heated while enclosed in iron boxes in contact with leather clippings, horns, hoofs, bones, and other nitrogenous matters, the depth to which the hardening penetrates being dependent on the duration of the period of their exposure to a high temperature. A surface film is often imparted by heating the iron, rubbing it over with the yellow prussiate (ferrocyanide) of potash, placing it in the fire for a few moments and then quenching in water.

Case Hardening Box.—A box of cast iron, circular or square in shape, provided with a cover and suitable handles for lifting. Articles to be case hardened are placed in this, together with nitrogenous substances, and the box put into a furnace for a definite period, lasting from a few hours to several days, according to the depth of hardening required ; the hardened skin extending from the thickness of a mere film to $\frac{1}{32}$ in. or $\frac{1}{16}$ in.

Cast.—(1) The twisting, curving, and warping of timber. (2) The running of molten metal into moulds.

Cast Holes.—Holes are cast for several reasons. It is the custom to cast all holes which have to receive black bolts, the holes being more cheaply cast than drilled, and a close fit being non-essential. Cast holes are made larger to the extent of from $\frac{1}{16}$ in. to $\frac{1}{8}$ in. than the bolts which they are to receive. Holes are also cast to lessen the labour of drilling, being cast small, and broached or rymered out afterwards. Large lightening and clearing holes are always cast. Core prints usually indicate the positions of cast holes on patterns. Cast holes are necessarily rough, and their accuracy not quite reliable.

Casting.—(1) The art of making metal work by the pouring it, while in the liquid condition, into moulds of sand, or metal, or plaster of Paris. (2) The metal work which is produced by the pouring of the same into its mould, in contradistinction to that which is forged, or drawn out under the hammer.

Casting Ladle.—A foundry ladle (q.v.) More specifically the ladle which receives the metal from a Bessemer converter, whence it is poured into the ingot moulds.

Casting On.—The process of uniting cast to wrought-iron work, by pouring the former around the latter, while arranged in due position. It is chiefly resorted to in light ornamental work. See also Burning On.

Casting on End.—See Casting Upright.

Casting Pit.—A pit in which are arranged the moulds for the casting of steel ingots. See also Pig Mould and Foundry Pit.

Casting Upright, or Casting on End.—This signifies the pouring of metal into a foundry mould, the latter being set in a vertical position the while. It is practised in the case of long cylindrical work which is wanted of special soundness, either for the purpose of withstanding heavy strains or having to be turned or bored with a clean surface free from blowholes. Work cast on end is sunk into the foundry pit (q.v.), a head (q.v.) is cast on, and the metal, though poured of necessity from the top, is often led down a vertical ingate or runner to the bottom of the mould, whence it rises upwards. This is preferable in certain cases to pouring from the top, as less damaging to the mould, and also producing a sounder casting, since the falling metal would carry down and enclose air-bubbles with it.

Cast-iron.—Strictly speaking, a compound of iron and carbon ; in pure iron, soft iron, or malleable iron the amount of carbon being nil. It is on the percentage of carbon present that the different qualities of cast iron mainly consist. The two extremes of quality in cast iron are the grey and the white. The former contains sometimes as much as 4·7 per cent. of carbon, mainly in the graphitic condition, the latter containing 2 or more per cent., chiefly in chemical combination. Between these extremes occur the various shades of grey and mottled, each suitable for its specific purpose. The presence of graphitic carbon imparts fluidity to the iron and renders it soft, so that it is readily turned and filed. Cast-iron scrap is mixed with foundry pig to impart toughness or other qualities as desired. Cast iron is affected by the presence of foreign elements, as silicon, sulphur, phosphorus, manganese, as noted under those terms

Cast-iron Pipe.—Used for steam and water, and for fluids generally. Its thickness is always excessive in amount, to allow of a sufficiently large margin of safety after it becomes weakened by corrosion. It is cast around a core (q.v.), and flanges, sockets and spigots, as required, are cast on.

Castor Oil.—An excellent lubricant for heavy bearings, having much body or viscosity, but becoming dry or gummy after long exposure to the atmosphere. Its sp. gr. is about 960, taking water at 1,000, and its flashing point 550° F. Cold-drawn oil is that expressed without the aid of heat, and is the best ; in the preparation of the second quality the heat of steam is employed. Castor oil is adulterated with black poppy or other inferior oils, but the adulteration may be detected by mixing with absolute alcohol, in which the castor oil is completely soluble, leaving the inferior admixture as a deposit. The castor oil plant (*Ricinus communis*) belongs to the natural order *Euphorbiaceæ*, and grows in the south of Europe and other warm regions.

Cast Steel.—Blister steel (q.v.) which has been broken into small pieces, melted in a crucible, and poured into ingots.

Cataract.—A vessel containing water, a plunger, and valves by which the valves in the steam passages in pumping engines are opened and closed automatically.

Cataract Rod.—A vertical rod attached to the lever of a cataract and which directly actuates the levers for the opening and closing of the steam passages of pumping engines.

Catch.—See Hitch.

Catch Pawl.—See Dog.

Catch Plate.—A miniature face-plate which carries the driver (q.v.) of a turning lathe.

Catch Water.—See Interceptor.

Catenary.—The curve assumed by a cord suspended freely from two points.

Cattle Gear.—See Bullock Gear.

Caulking.—(1) Making a joint water or steam tight by filling it in with rust cement (q.v.). (2) Burring or driving up the edges of boiler plates along the riveted seams to make them steam and water tight.

Caulking Ring.—A caulking strip (q.v.) of circular form.

Caulking Strip.—A strip of sheet metal interposed between the body of a wrought or cast-iron structure and a cast-iron piece attached thereto. Its employment is rendered necessary by the impossibility of caulking the cast iron itself. Caulking strips are used between boiler shells, and the flanges of cast-iron man-hole, safety valve, and stop valve seatings.

Caulking Tool.—A blunt-ended rectangular strip of steel used for closing up the edges of boiler plates. It is driven against them by sharp blows from a hammer.

Caustic Soda.—Symbol Na. HO. A hydrate of soda. It is used in solution for cleansing the outsides of condenser tubes from the accumulations of grease and oil deposited thereon by the steam exhausted from the cylinder. See Soda.

Cement.—Few cements are used in mechanical engineering. Under this head may be classed the glue (q.v.) of the patternmaker, the black wash (q.v.) of the moulder, the iron cement (q.v.) for hydraulic work, and the concrete (q.v.) of the hydraulic engineer, and cements for leather and other materials.

Cementation.—(1) The process of making steel from malleable iron by heating the latter in contact with carbon in a furnace. Bars of malleable iron are broken off and placed in layers in pots or troughs having intermediate layers of charcoal powder sifted over and between them. The surface is lastly covered over with wheel swarf (q.v.), and the whole mass is heated together during several days, a portion of the carbon being transferred to the iron. The charcoal powder is called cement, hence the derived name. The process is also termed the converting process. This is the first stage in the manufacture of cast steel (q.v.). (2) Also applied to the manufacture of malleable cast-iron goods.

Cement Copper.—Copper extracted from the water which is pumped out of copper mines. The water being pumped into tanks containing scrap iron, the sulphate of copper in the water exchanges its sulphur with the scrap iron, forming sulphate of iron, and copper deposited. It is therefore almost in a state of chemical purity.

Cement Tester.—A form of testing machine (q.v.) used especially for the testing of the tensile and crushing strengths of cements, the specimen, specially moulded for the purpose, being clipped by shackles for tensile stress, and a hydraulic machine being employed for crushing tests.

Center.—See Centre.

Centiare.—A French measure of surface, comprising one square metre (q.v.).

Centigrade.—See Thermometer.

Centigramme.—A French measure of weight, being the one-hundredth part of a gramme (q.v.), and equivalent to the ·1543 part of an English grain.

Contilitre.—A French measure of capacity equal to the one-hundredth of a litre (q.v.) and corresponding with ·0176 of an English pint.

Centimetre.—A French measure of length, being the one-hundredth part of a metre (q.v.), and the ·3937 part of an English inch.

Central Forces.—A body revolving around its axis has a tendency to fly off from the centre—this is its centrifugal force. The force which prevents it from thus flying off is called the centripetal force. These are central forces, and are, of course, equal and opposite.

Centre, sometimes spelt Center.—(1) That fixed point about which the radius of a circle or of an arc of a circle moves. (2) A wheel centre (q.v.). (3) See also specific heads following.

Centre-bit.—A boring bit for wood, comprising centre point, nicker and cutter attached to a shank, the nicker and cutter revolving about the centre point and being actuated by a brace.

Centre Dab.—A centre pop. See Centre Punch.

Centre Gauge, or Screw Cutter's Gauge.—A thin metal gauge having angular recesses cut in its edges, and used for a variety of purposes. Its angles are 55°. Hence it becomes a templet for turning the cone points of lathes, for grinding the angles of screw tools, for setting those tools in the slide rest, and for determining the number of threads per inch of a screw.

Centre Line.—A line scribed, or otherwise marked off, upon a piece of work as a basis from which to obtain other dimensions equally divided or symmetrical on both sides. A centre line may, of course, be either rectilinear or curved. Dimensions are almost invariably taken from centre lines, seldom from edges or outer faces.

Centre of Compression.—The line where is located the resultant of the compressive forces in the lower section of a beam.

Centre of Gravity.—The centre of gravity of a body is that point about which it will be balanced though placed in any position, hence it is the centre of parallel pressures. It may be determined experimentally by suspending the body in different successive positions and hanging a plumb line against the face. The common point of the numerous intersections of the plumb line will correspond with the centre of gravity.

In regular figures or solids the centre of gravity corresponds with their geometrical centre. The common centre of gravity of two bodies is in a point which divides the distance between their individual centres of gravity in the inverse ratio of their weights. The common centre of gravity of more than two bodies combined in one system is found by first obtaining the common centre of any two of them, and then obtaining the common centre of those two with a third, and so on till all are included.

Centre of Gyration.—That point in a revolving body in which its momentum is concentrated.

Centre of Moments.—That point in a rigid body about which the moments or forces act, or the fulcrum.

Centre of Oscillation.—That point in the axis of a vibrating body in which, if the whole matter were concentrated, the body would continue to vibrate in the same time. It lies in the same axis as the centre of gravity, but is necessarily situate farther from the point of suspension.

Centre of Percussion.—That point in a body revolving about an axis at which, if it struck an immovable obstacle, all its motion would be destroyed.

Centre of Tension.—The line where is located the resultant of the tensile forces in the upper section of a beam.

Centre Plates.—Plates of wrought iron or of brass, which are used for the purpose of temporarily securing jointed and dowelled patterns while they are being turned between lathe centres. The plates are usually screwed to both halves of the pattern at its ends, but some small plates are held with clamp-like teeth only.

Centre Punch.—A steel punch about three or four inches long having one end ground conical, and terminating at its apex in a sharp point. It is used for indenting small circular " pops " or depressions upon the ends and surfaces of metal work, as a guide for centring in the lathe, as well as for drillers, machine hands, or fitters in cutting to outline. See Bell Centre Punch.

Centres.—(1) A term of wide application in the workshop. Dimensions are almost invariably given and taken from centres, and not from edges. Centres are obtained by the intersections of lines, by compass measurement, by scribing blocks, by trammels, and rules. (2) The centres of lathes are the common axis of the headstock and poppet. The point chucks or dead points are also called centres. (3) We constantly speak of valve centres, trunnion centres, centres of motion, centres of shafting, of connecting-rods, &c. The term has therefore an infinity of applications.

Centres of Motion.—The centres around which wheels or levers turn. Centres of motion may be fixed centres (q.v.) or movable centres (q.v.)

Centre Square.—A tool used for finding the centre of a circle, or of an arc of a circle without compasses or geometry. It comprises a straight-edge and two pins, or the pins may be absent, and their place be taken by edges, the essential being that the straight-edge shall be normal to the lines joining the pins, that is radial from the centre of a circle, whose periphery would touch the pins or edges as the case may be. When the pins are placed against the circumference of a circle, the radial straight-edge points towards its centre. By sliding the pins to different points around the circumference a sufficient number of intersections are obtained by means of the straight-edge to determine the position of the centre.

Centre-weighted Governor.—A high-speed governor, having a heavy

weight sliding on the central spindle, whose gravity has to be overcome by the centrifugal force of the balls. It is sensitive, and a small governor has the efficiency of a much larger governor of the common type.

Centrifugal Force.—See Central Forces.

Centrifugal Pump.—A type of pump by which water or liquid is drawn in, and delivered in a continuous stream by the rotation of a fan, or a combination of radial vanes enclosed in a well-fitting casing.

Centring.—(1) A temporary structure of woodwork, upon which the operations of bridge or arch building are carried on, and which is removed on the completion of the work. (2) Marking off the centres upon work which has to go into the lathe to be turned.

Centring Chuck.—A piece of mechanism for obtaining and punching the exact centres on the ends of bars of circular and square section, which have to be put between centres in the lathe. It consists of a self-centring scroll chuck, to be held in a vice, fastened to a bench or dropped over the end of the shaft; when the shaft is centred it is struck with a punch set in the centre of the chuck.

Centring Machine.—A machine in which iron bars are clipped by a self-centring chuck or jaws, while cutters set on a revolving cone or spindle face the ends and mark the centres.

Centring Rest.—A rest, whose outline is that of the letter L, used in conjunction with a square centre (q.v.) for marking countersunk centres in the ends of cylindrical rods. Being screwed on the slide rest it is made to thrust the revolving bar into a position truly concentric, while the square centre cuts or rymers its way into the end of the bar, central with the outside.

Centripetal Force.—See Central Forces.

Chain.—Chain is made by the welding together of oval or circular loops. Chains are either open link (q.v.) or stud link (q.v.). The sizes of chains are given in terms of the diameter of the iron from which they are made; thus inch-chain signifies that the diameter of the bar from which the link is made is one inch in diameter. The strength of a chain is estimated by the area of the cross section, that is of two bars of the material in the link. The length and width of a link always bears a definite proportion to the diameter of its iron. Thus the length of an ordinary short-link chain is $5\frac{1}{2}$ times its diameter, and its width is $3\frac{1}{2}$ times, the proportions of links varying with their types.

Chain Barrel.—See Barrel.

Chain Feed.—A mode of feeding balks of timber along to frame saws by means of an endless chain led around sheave wheels.

Chain Joint.—A riveted joint where the ends of two outer plates overlap and embrace the ends of a central plate.

Chain Pump.—A lift pump consisting of a series of rectangular or circular discs, connected at short intervals to a chain, and travelling within a closed pipe in the direction in which the water is to be lifted. The plates travel at a rapid rate, and fit so closely to the pipe that little waste takes place at the sides.

Chain Riveting.—Rows of rivets placed in parallel lines, both in the longitudinal and transverse directions.

Chair.—A cast-iron seating or support for metal rails. It is spiked down to the sleeper, and the rail is wedged into it by wedges driven in sideways.

Chalk.—Essentially a carbonate of lime, $Ca\ CO_3$, though usually contaminated with particles of silicon, alumina, or magnesia. It is used much by engineers for whitening rough or dark metallic surfaces to render centre and dimension lines clearly visible ; for transferring the mark from a chalk line (q.v.), for whitening one of two opposite surfaces when jointing, for marking the edges of straight-edges, and for numerous kindred purposes.

Chalk Line.—A piece of fine twine wound upon a reel. It is used for marking straight chalked lines upon timber. Being unwound and whitened with chalk, and strained from each end of the board, the central portion of the string is lifted, and allowed to rebound sharply upon the timber, some portion of the chalk being thereby transferred from the taut line to the board.

Chambered Core.—A core (q.v.) whose diameter is enlarged in its central portion, either for the sake of lessening the weight of metal around it or to save time in boring the hole, which is then bored only near the ends. Or if the core is thin, and especially if it is both thin and square, to obviate the disadvantage of a probable bending of the core, and consequent difficulty of fitting a bar through it, a chambered core is very commonly employed. Also termed bellied or belly core, or roach belly core.

Chamfer.—A bevel imparted to edges otherwise rectangular.

Chamfering Machine.—A machine by which the bevels of nuts and the rounding ends of bolts are formed. It is usually a centring machine (q.v.) in which the various cutters are inserted as required.

Change Wheels.—Cog-wheels of fine pitch, whose function is the transmission of motion from the mandrel of a lathe head to the guide screw, for the cutting of screws of various pitches. The wheels are interchangeable, both on the mandrel, the guide screw, and the intermediate stud ; hence their name. A combination of change wheels may consist either of a single or of a compound train (q.v.). The pitch of change wheels is not usually estimated in parts of an inch, but by numbers ranging from 14 to 6, and called 14 pitch, 6 pitch. The former is equal to $\frac{1}{14}$ in. full, the latter to $\frac{1}{2}$ in. full. (See Diametral Pitch.) A set of change wheels consists of 22, commencing at 20 teeth, and rising by fives to 120, and one 60 or 90 wheel extra, for cutting a screw of equal pitch with the guide screw. But wheels go as low as 15 and rise to 150 or 200, and some of these are often added to lathes, over and above the set so called.

Changing Hook.—A ram's horn hook used in foundries for changing or transferring ladles of molten metal from one crane to another contiguous thereto.

Channel Iron.—Rolled wrought iron bar, whose section is that of three sides of a parallelogram ⌊_⌋, used in bridge and girder work, and for structural purposes generally.

Chaplet, or Stud.—A metallic support, or stay, given to a core within a mould, as an aid to, or altogether independently of prints. Chaplets are used underneath cores to support them, or above, or at their sides, to sustain upward or lateral pressure due to the head and flow of metal. Chaplets are of various forms, noted under their specific headings.

Chaplet Block.—A pyramidal block of wood driven into the sand of a mould to sustain the thrust of a chaplet nail (q.v.). It affords the breadth of base necessary in the yielding sand. See Tinned Nail.

Chaplet Nail, or Moulders' Nail.—A nail having a broad flat head, used as a chaplet (q.v.) for the lighter kinds of work.

Charcoal.—Wood which has been desiccated and carbonised, while heated to a temperature of about 650° F. in the absence of air, by which process of dry distillation the volatile ingredients, water, wood spirit, pyroligneous acid, tar, and various gases are driven off. It is employed sparingly in some smelting operations. Foundry moulds, when small, are skin-dried (q.v.) by means of charcoal. It is used also for charcoal blacking (q.v.).

Charcoal Blacking.—Blacking (q.v.) for moulders' use, made from oak charcoal, as distinguished from coke blacking, coal dust, prepared blackings, and plumbago.

Charcoal Furnace.—A blast furnace in which charcoal is the fuel employed. Its use is confined to Sweden.

Charcoal Iron.—Iron which has been smelted with wood charcoal as fuel. It is the best iron, its superiority consisting in its freedom from sulphur, which is never wholly absent from that smelted with coal or coke.

Charcoal Plate.—Tin plate for which the sheet iron has been refined with charcoal, as distinguished from coke plate (q.v.). It is superior in quality to the coke plate.

Charge.—(1) The gaseous or combustible mixture of air with gas, petrol, paraffin or Diesel oil fuel, brought together in the cylinder or explosion chamber of an internal combustion engine. (2) The mass of material, ore, fuel, and flux, or metal, fuel, and flux, which is introduced into a blast furnace, or into a cupola at one time. Also called the burden.

Chargeman.—A workman who has charge of a contract job, and who allots to the men under him their separate portions of work, and is responsible for the correctness of the work when done. All large jobs are let to chargemen, who, being competent men, are little interfered with by the foreman.

Charging.—(1) The supplying of furnaces with their ores and fuels. (2) The priming or fetching of a pump with water.

Charging Door.—(1) The door through which the charge is introduced into a reverberatory furnace. The charging door is situated to one side of the furnace. (2) The door of a blast furnace, or a cupola, situated near the top, and level with the charging platform.

Charging Platform.—The platform from which the charge (q.v.) is delivered into a blast furnace or a cupola.

Charles, Law of.—May be expressed thus: That the volumes of gases under constant pressure increase by equal fractional amounts, no matter what the nature of the gases. Sometimes termed the law of Guy Lussac, or the law of Dalton.

Chaser.—(1) Ordinary screwing dies are sometimes called chasers. (2) The actual cutting dies or screwing tools, which are gripped and set by the collets (q.v.) in a screwing machine. (3) A comb-like tool of steel, whose edge is a counterpart of a screw section. It is employed for striking or cutting screws by hand in a lathe, or for smoothing and finishing them when already cut by other means. Chasers are internal or external, according as they are used for cutting inside or outside threads.

Chasing.—The act of cutting screw threads with a chaser, or of smoothing and finishing threads which have already been cut by other means. It is effected by pressing the chaser against the revolving metal, and giving it a longitudinal travel at the same time, corresponding in amount with the pitch of the screw.

Chatter.—Caused in machine work by want of rigidity in the cutting tools. It occurs chiefly in cutters set in a revolving cutter bar, where the bar is not sufficiently stiff, or where the cutters stand out to too great a distance, or in wide scraping tools.

Check.—(1) A check is a joint composed of two portions, male and female, fitting the one into the other, and so forming a guide or steady. A dowel fitting loosely into a hole, a shallow stud or pin fitting into a recess, are illustrations of checks. The mitre-wheels of the tool boxes of planing machines are checked together ; so also are couplings, and frequently flanges. Numerous illustrations of checking occur in machine work. Loam boards (q.v.) furnish an illustration of indirect checking, the halves of the loam mould being checked, the boards therefore being cut in the reverse direction. (2) A round brass disc of about the size of a half-crown, stamped with a definite number, and provided with a hole for hanging upon a nail on a board called a check-board. The number corresponds with the name of a workman. See Check System.

Checked In.—This term signifies the employment of a check in parts fitted together.

Checking In.—The fitting together of corresponding parts or checks (q.v.).

Check Nut.—A lock nut (q.v).

Check Rail.—On railway curves of very quick radius it is sometimes the practice to lay an extra rail on the inner side of the inner rail of the curve, leaving just sufficient space for the flange of the wheel to pass freely between. This relieves the pressure of the wheel flange against the outer rail and prevents it from mounting the rail. Hence called a check rail.

Check System.—A system of keeping workmen's time, which was once used in the more important factories. Details differ, but the method is essentially this. A board is hung just without the entrance gates, upon which are hung a number of checks (q.v.) corresponding with the number of workmen in the place. Each workman as he passes in takes off his particular check and drops it into a box called the check-box, which is a proof that he has gone in to work. The gate-keeper collects the checks from the box, and by comparing their numbers with the numbers in the book giving the names and numbers of the workmen, is able to enter each man as in or out, as the case may be. He returns the checks to the board in time for the next bell-ringing.

Check Valve.—A wing valve, inserted in a feed pipe between the boiler and feed pump of an engine in order to prevent the return of the water from the boiler to the pump. It is free to open towards the boiler for the passage of water, but is closed by the boiler pressure.

Cheeks.—A term having a wide and varied application in engineering. In general it signifies the flanks or side supports for pieces of mechanism. Thus it is applied to the bearers of a lathe, the standards of pumps and engines, side frames (q.v.), the jaws of a vice, &c.

Cheese Head Rivet.—A rivet, the shape of whose head is that of a short solid cylinder.

Cheese Head Screw.—A screw whose head is cylindrical in form, and nicked or slotted across to receive a screw driver. It is used in the lighter parts of machine work.

Chemical Change.—This denotes an interchange between the ultimate atoms of the elementary bodies from which substances of wholly different characters are formed ; differing in this respect from mechanical mixtures, in which the arrangement of atoms remains the same. Heat is the agent which usually produces or quickens chemical action. The reactions which go on in the various melting furnaces are essentially due to chemical changes, new bodies being formed from the elements reduced from their original compounds.

Chequering.—(1) The roughing over of footplates to prevent the foot from slipping upon them. Diamond-shaped chequers are formed upon cast-iron footplates, while on those in wrought-iron crossing ridges are stamped. (2) The brickwork in regenerative furnaces (q.v.) and stoves, through which the products of combustion pass.

Chequer Plates.—Footplates of cast or wrought iron whose surfaces are covered with chequering (q.v.).

Cherry Red.—Bright Red Heat (q.v.).

Cherry-wood (*Cerasus*).—A hard, reddish or brownish coloured wood, close grained, and of small size, the logs averaging 10 or 12 inches in diameter. If well seasoned it is suitable for cutting and turning small patterns, and as such is used in country districts. Sp. gr. ·715. A cubic foot weighs 45 lbs.

Cheval Vapeur, or Force de Cheval, or French Horse-Power.—The French unit of work done by engines. It is equal to 75 kilogrammetres (q.v.) per second, 4,500 kilogrammetres per minute, or 542·4825 English foot pounds per second, or 32,549 foot pounds per minute. Expressed decimally one cheval vapeur equals ·9863 English horse-power.

Chili Copper.—Black oxide of copper obtained in Chili.

Chill.—The metallic mould into which specially mixed molten iron is run to produce a chilled casting. See Chilling.

Chilled Box.—A wheel boss whose central hole has been chilled against an iron core.

Chilled Roller.—A roller chilled around its circumference. Employed where durability and smooth hard surfaces are required, as in bending and metal rolls of various kinds. See Chilling.

Chilled Wheels.—A wheel chilled around its periphery or "tread." Employed for contractors' waggons running on rails, for the trucks of portable cranes, and similar work. In America they are used for railway cars.

Chilling.—The hardening of the surfaces of iron castings by pouring certain mixtures of metal (whose qualities and proportions are learned by experience only) into cold metallic moulds. The graphitic carbon in the metal is believed to enter into chemical combination with the iron at the surface, producing a steely skin. Chilling will penetrate from ⅛ in. to 1 in. inwards from the surface, contingent entirely upon the character of the mixtures employed. Chilling insures great durability of wearing surface.

Chimney.—The tube or funnel for the exit of waste steam or smoke from an engine or boiler. It should bear a definite proportion to the fire-grate area, the proportion varying with the type of engine.

Chinese Ink.—The finest quality of the ink which generally goes under the name of Indian Ink. It is used for mechanical drawings. Chinese ink dries glossy ; Indian ink dries dead black.

Chinese Windlass.—A contrivance by means of which a large weight may be raised very slowly by a slight expenditure of power. Two drums, or cylinders, differing but slightly in diameter, are contained on one axis, and a single coil of rope is wound in opposite directions on each, so that while it is winding on the larger cylinder it is being unwound from the smaller, and *vice versâ*. The nearer the diameters of the barrels approximate to one another, the greater the mechanical gain.

with of course the necessary loss in time corresponding thereto. See Differential Principle.

Chipping.—The removal of minute particles or chips of metal from surface faces or from edges, both in cast and wrought iron, gun metal, copper, &c. It is effected by means of a chipping chisel (q.v.), and is resorted to when the metal to be removed is too large in quantity to be attacked with the file alone.

Chipping Chisel, or Cold Chisel.—A tool used by fitters, boiler-makers, smiths, and engineers generally. It is made from steel rod, ranges from 4 in. to 10 in. or 12 in. in length, for different purposes ; its cutting faces are ground to a double bevel to an angle of about 45°, and highly tempered. It is used for cutting metal by the force of the impact derived from the blows of a chipping hammer.

Chipping Face.—The face of a chipping strip (q.v.).

Chipping Hammer.—A fitter's hammer weighing about a pound, commonly used to deliver the blows upon the head of a chipping chisel.

Chipping Piece.—A chipping strip (q.v.).

Chipping Strips.—Narrow and thin metal strips cast around the edges or across the face of those portions of castings where a good face bearing is necessary, the casting having its bearing upon the faces of the strips only. They are used to lessen the labour of fitting large metallic surfaces. Called chipping strips because they are usually chipped over with a chisel instead of being shaped in a machine.

Chisel.—The type of the cutting tools, whose essential principle is that of the wedge. Specifically,—chipping, cross-cut, firmer, paring, mortice chisels, described under their respective headings.

Chisel Rod.—Steel rod of a flat section with rounded edges, made expressly for the forging of cold chisels. It is cut off by the tool smith and drawn down to the proper wedge shape.

Chloride of Mercury.—See Kyanising.

Chloride of Zinc.—See Salammoniac, Burnett's Fluid.

Chops, or Chaps.—The jaws of a vice.

Chord.—The straight line which unites the ends of an arc of a circle.

Chord Pitch.—The pitch (q.v.) of a wheel measured along the chord of the arc between the pitch points. Arc pitch (q.v.) is that usually employed.

Chords, Line of.—One of the sectorial scales (see Sector) by means of which angles are obtained and measured. It consists of two diagonal lines, each divided into sixty equal parts and marked C. To obtain an angle therefrom the sector is opened out until the distance between the brass centres at the terminations of the lines corresponds with the radius of the arc of the circle to be taken. Then, if under 60°, the direct measurement for the chord is taken across or transversely to the legs to those division numbers which correspond with the number of degrees in the angle required If over 60° two or more successive measurements are necessary.

Chords, Scale of.—A scale of chords is obtained by striking an arc of 90°, dividing it out into degrees, drawing chords from one extremity of the arc to each division in succession, and transferring them to a straight line. The 60° division in the scale will always be the radius by which an arc is to be struck when it is desired to take off the length of any particular chord from which to obtain an angle.

Chuck.—Any attachment through the medium of which work is secured to the mandrel of the headstock of a lathe for the purpose of turning

The commonest forms are the fork, the bell or cup, the drill, self-centring, and face chucks in their various modifications. They are made to fit the mandrel by a female screw which fits accurately to that on the mandrel nose. There are, besides, the various complicated geometric and ornamental chucks which do not come within the range of the engineer's appliances.

Chucking.—The process or act of attaching lathe-work to the various chucks. Facile and correct chucking is an art which, though apparently simple, is only to be attained by long practice, and the economy of a metal turner's time is largely dependent upon his skill in this respect.

Chucking Reamer, or Straight Shank Reamer.—A rear whose shank is circular and parallel, to be used in a self-centring chuck.

Chute.—An enclosed trough which conducts the water to a water-wheel.

Cinder Bed.—A coke bed (q.v.).

Cinder Frame.—Wire work sometimes placed in front of the tubes of locomotives for foreign use, to arrest the passage of fragments of ignited fuel.

Cinder Pig.—A very inferior class of pig iron, obtained by smelting slag and cinder of puddling or reheating furnaces, together with proportions of inferior ore. The slag is rich in oxide of iron, but contains also a large proportion of phosphorus and silicon, which yields an inferior product.

Cinematics.—See Kinematics.

Circle.—A plane figure described by a right line moving around a fixed point called the centre.

Circular Inch.—The area of a circle of one inch in diameter, as distinguished from a square inch. The pressures in steam cylinders are frequently calculated in circular inches. To obtain the number of circular inches in a given diameter it is only necessary to square the diameter.

Circular Motion.—The circular motion of a shaping machine is that arrangement of tool box parts and feed gear by which the outline of an arc of a circle is imparted to the ends of lever rods and similar articles, which cannot be shaped in a lathe. The circular motion is given by means of an endless screw actuating a quadrant rack on the tool box.

Circular Nut.—A nut whose outline is circular instead of hexagonal. Circular nuts are provided with a round hole or holes for the insertion of a tommy (q.v.), by which they are tightened up.

Circular Pillar Drilling Machine.—A drilling machine in which the table embraces and swings around a circular pillar which carries the gear. A toothed rack is hollowed at the back to fit the pillar, and is itself embraced by the table. Into whatever position the table is slewed the rack is also carried round, so that the pinion which gears with it, and which has its bearings in the table, is always in gear, and always therefore in a position for lifting the table. A worm and worm-wheel actuate the pinion spindle.

Circular Pitch, or Circumferential Pitch.—The ordinary method of estimating the pitch of toothed wheels by dividing the circumference at the pitch line by the number of teeth in the wheel. The term is used in opposition to diametral pitch (q.v.).

Circular Plane, or Compass Plane.—A plane used for working out hollow sweeps whose curves run in the direction of the length of the plane. It is either made of wood, with its face rounded longitudinally and provided with a front adjustable stop to alter the curve within narrow limits, or it is an iron plane whose face is an elastic strip of steel, the amount of whose curve is regulated by a screw and interlocking levers. In this last form the curve can be made external or internal at pleasure.

Circular Saw, sometimes called **Buzz Saw**.—A saw whose teeth are divided around the edge of a circular disc running upon a central spindle. These saws vary from a few inches in diameter to 7 feet. They are used chiefly for the rougher kinds of sawing. Circular saws are also used for cutting off iron, both cold and red hot. Then they are as much as $\frac{3}{8}$ or $\frac{1}{2}$ in. in thickness, their teeth are short and square across, and their lower portions run in water for cold sawing.

Circular Table.—A circular cast-iron plate, which sustains the work which is being operated upon in drilling and slotting machines.

Circular Valve.—A valve circular in plan, as distinguished from a slide valve.

Circulating Pump.—The pump which circulates water through the tubes in the surface condensers of marine and other condensing engines. The cold water is drawn through the tubes by the suction of the pump. The pump is worked from the engine itself, or from a separate engine. Sometimes a centrifugal pump is used for circulating.

Circulating Tubes.—The inside smaller tube in Field's tubes (q.v.), or the cross tubes (q.v.) of vertical boilers, or Galloway tubes (q.v.), or the ordinary forms used in multitubular boilers or in surface condensers.

Circulation.—The circulation in a steam boiler is due to the bubbling up of the lighter boiling water from the heating surfaces through the cooler water at the upper portions, which latter then descends to take its place. The efficient circulation of the water is necessary to the rapid generation of steam and the prevention of deposit and incrustation (q.v.). It is promoted by the introduction of water tubes, so arranged that convection shall readily take place from their surfaces, and by keeping as small a quantity of water as possible below the heating surfaces. The circulation of water in a condenser is necessary for the rapid condensation of the exhaust steam, and is effected by the circulating pump (q.v.). See also Hot-water Apparatus, Surface Condensation, Jacketing.

Circumference.—The circumference of a circle is the line described by the radius (q.v.) moving about the centre.

Circumferential Pitch.—(1) This pitch is, in a screw wheel, the distance between the points in which a plane at right angles with the axis cuts two contiguous threads. (2) Circular Pitch (q.v.).

Clack.—That portion of a pump-valve or bucket which is lifted by the action of the water or air. It is made of leather stiffened with plates of brass or iron. Applied more specially to a flap-valve (q. v.).

Clack Box.—The box or chamber in which the clack (q.v.) of a pump works, or that portion which contains the valves that open and close to suction and delivery.

Clam.—See Vice Clamps.

Clam Nut.—A clasp nut (q.v.).

Clamp.—Clamps are tools used for holding portions of work together, both in wood and metal.

Clamping Screw.—Any screw by which a piece of work or a tool is pinched or held in place. The screws which confine the cutting tool in the binding straps of planing and shaping machine tool boxes are illustrations of clamping screws.

Clark's Process.—A process which has been adopted to a limited extent for the purification of the hard feed-waters of steam boilers from the lime held in solution. It depends for its efficacy on the action of lime on carbonic acid, with which it readily combines. In water which contains bicarbonate of lime in solution, the lime is held in solution by the carbonic acid present, which being driven off by boiling deposits the lime on the boiler plates. But if lime be added to the water in sufficient quantity it will combine readily with the carbonic acid, neutralizing it, and will be precipitated as carbonate, together with the carbonate disengaged by its action. In the carrying out of this process tanks have to be provided for the filtration of the purified water.

Clasp Nail.—A malléable iron nail, rectangular in section, whose head is formed by two opposite projections pointing downwards. Used by pattern-makers for the temporary holding down of segments when building up work, the toughness of the wrought iron permitting of their extraction after the glued joints have set.

Clasp Nut.—The movable nut by which the slide-rest of a screw-cutting lathe is put into connection with, or released from the leading screw. It is worked by a cam plate when double; by an eccentric pin when single. Sometimes termed clam nut and clip nut.

Claw Coupling.—A loose coupling (q.v.) used in cases where shafts require instant connection or disconnection. It is somewhat similar in outline to a flange coupling (q.v.), but instead of being plain, projections or claws are cast upon each face, which engage in corresponding recesses in the faces opposite. The claws usually number two or three. The coupling is thrust in and out of gear through a pin or a fork taking easily into a groove turned in the circumference of a bossed up hinder portion. It is often termed a claw clutch.

Claw·Nut.—The clasp nut (q.v.) of a lathe.

Claw Wrench.—A wrench having a loose jaw, pivoted in such a manner that its bite increases with the pressure put upon it. Its principle is that of the pipe wrench (q.v.), and it is used both as a spanner and also for pulling the core rods and wires out of castings in the task of fettling (q.v.).

Clay.—Essentially a silicate of aluminium resulting from the decomposition of felspar. Used for making bricks, clay wash, and clay water, for the linings of furnaces and ladles, for stopping tap holes, and making crucibles, its value consisting in its plasticity, its quality of hardening, and its refractory nature.

Clay Band.—A carbonate of iron mixed with clay. It is mined in Staffordshire, Yorkshire, Derbyshire, and South Wales. Its composition is variable. It is the chief ore of iron worked in England. Called also argillaceous iron ore, and clay iron-stone.

Clay Iron-stone.—See Clay Band.

Clay Pit.—A pit from which the clay used in foundry operations is dug.

Clay Plug.—See Plug.

Clay Wash, or Clay Water.—A solution of clay in water. Used in foundry work for painting over the bars and lifters of moulding boxes, to cause the adhesion of the sand thereto ; and for cores.

Clay Water.—See Clay Wash.

Cleading.—The covering put around a boiler or engine cylinder for the purpose of preventing the radiation of its heat into the surrounding atmosphere. The cleading consists of wood strips tongued together with hoop iron, and bonded with hoop iron or brass. Often thin sheet iron is substituted for the wood cleading. There is a space between the boiler or cylinder and the cleading, filled with felt, or similar good non-conducting material. See Lagging.

Clean Boiler.—A steam boiler free from incrustation, scale, or muddy deposits.

Clean Casting.—A casting having a clean skin (q.v.). To produce a clean casting, the mould must be properly vented, sand suitable for the nature of the casting must be used, and the surface sleeked over with plumbago.

Clean Cut.—A cutting tool is said to produce a clean cut when the cut surface is not grooved, or wavy, or ridged, but continuous and smooth. The cleanness of a cut depends on the proper cutting angle being maintained, on the degree of force applied to the tool, too much pressure causing rough cutting ; and partly on the nature and homogeneity of the material itself.

Cleaner.—A moulder's tool of steel or brass, used for smoothing and dressing the various irregularly shaped portions of sand moulds. The typical cleaner is a thin long flat tool, its blade running in a longitudinal direction, while a short end is turned up at right angles thereto. The latter is used as a lifter. The term cleaner is applied to tools of various shapes.

Clean Fire.—A boiler fire free from clinkers and ashes.

Clean Hole.—A hole which is drilled or bored without showing a wavy or ridged outline. Clean drilling is effected by having a drill properly ground, and by cutting at a definite speed, and at a moderate rate of feed only. Cleanness of surface is increased by broaching or rymering, or by lapping.

Cleaning.—Engines and machines are cleaned by rubbing over their bright portions with sponge cloths (q.v.) or waste (q.v.), and oil.

Cleaning Up.—Smoothing over the surface of a foundry mould with trowel and cleaners, and blackening in readiness for closing and casting.

Clean Lift.—When a pattern is withdrawn or lifted from a foundry mould, the lift is said to be clean when the edges and sides are not torn away, or are torn away to a very slight extent only. The cleaner the lift, the less the amount of mending up necessary.

Clean Metal.—Applied to metals in general to indicate the absence of scoriæ, scurf, scale, oxide and similar foreign matters which detract from their value as metals.

Clean Mould.—A foundry mould which is properly sleeked and blackened, and from which all loose particles of sand are removed, in readiness for closing and casting.

Clean Scrap.—Forged scrap, from which all traces of cinder have been removed by hammering.

Clean Skin.—A casting or forging is said to have a clean skin when its exterior surface is free from scabs, pits, blisters, or other excrescences or depressions.

Clean Thread.—The thread of a screw is said to be clean when it is smooth and sharp, instead of being ragged, imperfect, and coarsely cut.

Clean Water.—The feed water of boilers is said to be clean when it is free from mud and other visible sedimentary matters. Clean water may nevertheless be hard, and produce scale, though not muddy deposit.

Clearance.—A term in frequent use, meaning generally the amount of space, open or free, between contiguous parts. Thus clearance of wheels means the space between contiguous teeth, clearance of brasses the space between their shoulders or flanges and the collars upon their shafts. Clearance of steam means freedom to exhaust. The difference between the distance across the flanges or throats of railway wheels and the gauge of the rails, is also termed clearance. This is equal to ¾ in. or 1 in., and is necessary to allow of end play in running round curves, and also for inaccuracy in the rails.

Clearance Angle.—The angle of relief (q.v.).

Clearing Hole.—The term is used in opposition to tapping hole (q.v.). It signifies a hole full to the specified size, so that a turned stud or bolt of the same nominal diameter will pass freely yet closely through it.

Clear Oils, or Pale Oils.—Lubricating oils which have been subjected to filtration and purification to free them from their dark natural colour. See Black Oils.

Cleet.—A block of wood which furnishes a steady point of attachment for a part of a structure, as a batten.

Cleft.—A shake in a balk of timber which is radial or running along the course of the medullary rays.

Clenching, or Clinching.—The turning and hammering over of the points of nails against a wood face to secure their adhesion under rough usage. The clenching transforms the nails into rude clamps.

Click.—The detent or catch of a small ratchet wheel. When of large size it is termed a paul (q.v.).

Click Wheel.—A small ratchet wheel (q.v.).

Clinching.—See Clenching.

Clinker.—The slag, or vitrified material which accumulates from the fuel in a smith's or a boiler fire.

Clinkering.—The periodical removal of the clinkers from a smith's or from a boiler fire.

Clips.—(1) Slight castings provided with side flanges or ears placed on opposite sides to each other. The clips are curved between the ears to embrace the outside of a hose pipe, and are used to effect a rapid union between the hose and the metal nozzles. The flanges are cast in brass, and clasped together with screw bolts. (2) Thin, flat, rectangular, washer-like castings, or more generally castings having one end rounded, and recessed slightly on their front, or straight edges, to clip, or cover the edge of the flat bottom flange of a rail. A fang bolt or a spike passes through a hole in the clip, enabling the latter to retain the rail in place.

Clip Drum.—See Clip Pulley.

Clip Nut.—See Clasp Nut.

Clip Pulley, or Clip Drum.—A rope pulley whose rim, of a vee-section, is constructed of movable clips instead of rigid sides. The clips are three or four inches long, and are hinged on pins whose axes are ranged in the direction of the periphery of the pulley. The positions of the pins are such that when a rope bites between the clips the effect is to pull them towards each other with the result of increasing the bite of the rope, the amount of which bite is, therefore, in direct proportion to the pull of the rope. Clip pulleys are used for hauling at steam ploughing implements, for mines and inclined planes, and for the transmission of power through long distances, especially in those cases where winding drums are not convenient.

Clogging.—The thickening of lubricating oils due to the absorption of oxygen, and to the presence of dust. Machinery is said to clog when its lubricating oil becomes thick and dry.

Closed Stokehold.—The stokehold of a steamer which is closed to all admission of air, save that which is supplied through blowing fans. The advantages claimed for closed stokeholds are increased efficiency of boilers, due to the more rapid combustion obtainable, permitting of a diminution in their bulk, and the exact adjustment and regulation of the air supply under calm weather and in any direction of the wind.

Close Grained.—Iron is close grained when its crystals are of moderate size and densely packed. The term is, however, relative, since all heavy castings which are not cast under pressure are somewhat open and porous in their central portions. It is customary in specifications to stipulate for close-grained iron for special classes of work, as, for example, engine cylinders and bright working parts. Timber is close grained when of slow growth, as evidenced by the small size of the annual rings.

Close Link Chain.—Ordinary open link chain, the length of whose links does not exceed five times the diameter of the iron from which it is made. Its width is three and a-half diameters. Called close link to distinguish it from circular link chain.

Close Mouth.—Applied to punching bears and punching machines which are open back and front for the passage of bars, but closed at the sides. Used for rails and bars. The term is used in opposition to open mouth (q v.).

Close Topped Furnace.—The modern form of blast furnace in which the mouth is closed by a cup or cone or some other suitable arrangement, the waste gases being led down to heat the blast.

Closing Hammer.—A hammer used for closing the seams of boiler plates.

Closing Up.—(1) The riveting or burring over of a rivet head, either by hand or by hydraulic pressure. The length of rivet required for closing up in hand riveting is $1\frac{1}{4}$ times the diameter, for snap head and conical rivets ; once the diameter for countersunk rivets, and a trifle more than these, an $\frac{1}{8}$ in. or $\frac{1}{4}$ in. in machine riveting. (2) Covering up or placing on of the top box or cope of a foundry mould in readiness for casting.

Clothing.—The felting and wood coverings placed around steam pipes and boilers to prevent radiation and loss of heat therefrom. See Cleading.

Clout Nail.—A strong malleable iron nail, having a large flat head. Used for fastening leather on wood. These nails are also often employed, after cutting off their heads with a cold chisel, as pins for securing the cogs of mortice wheels in place.

Clutch.—The medium by which a temporary connection is made between separate spindles or portions of shafting. It may be a claw clutch, having claws, two or more in number, which take into corresponding recesses in the opposite half, or it may be some form of smooth friction clutch (q.v.). Also termed disengaging coupling.

Clyburn Spanner.—An adjustable spanner, one of whose jaws receives movement by means of a screw and milled head.

Coach Screw.—A wood screw with a coarse vee-thread and a square head, used generally for bolting timber work together.

Coal.—Used in smelting operations, for engine fires, in the manufacture of coke, for mixing with foundry sands, and for foundry blacking. Coals are anthracitic or bituminous, caking or non-caking.

Coal Box.—(1) The box or bunker which carries the coal in a locomotive which is not provided with a tender. (2) The trough or receptacle for the coal used in the smith's fire.

Coal Bunk.—See Bunker.

Coal Dust.—(1) Ordinary coal pulverized in a blacking mill. It is used for admixture with foundry sand to the extent of 8 or 10 per cent. When the coal becomes oxidized the sand is said to be burnt. (2) Used also for facing moulds, both as a dust and in the form of black wash (q.v.).

Coal Mill.—See Blacking Mill.

Coarse Feed.—The feed of a machine tool set for heavy cutting. The term is relative, but in ordinary light work anything over $\frac{1}{16}$ would be coarse feed. In massive work $\frac{3}{4}$ in. or 1 in. feed is common.

Coarse Fibre.—The fibre of wrought metal which is large and rough, indicating a low tensile strength.

Coarse Grain.—The crystallization of metal when the crystals are of large size.

Coarse Grit.—The texture of an emery-wheel or grindstone which is suitable for rough grinding only.

Coarse Hard.—A class of emery-wheel used for edge-grinding, for trimming castings, and in general for rough work.

Coarse Metal.—See Copper Matt.

Coarse Pitch.—The pitch or size of wheel teeth considered in relation to other teeth.

Coble Ores.—Red copper ores (q.v.) imported from Cuba.

Cock.—A cylindrical valve, consisting essentially of a shell, plug, and cover. A passage-way is formed through the plug, which being brought opposite to the entrance and discharge openings, allows the fluid to pass through. When the plug-openings are brought opposite the blank walls of the shell, the cock is closed. Cocks are of various kinds, as bib, flanged, gland, socket, two-way, three-way, four-way, described in their sections. Sometimes the term cock is extended to include any valve which is opened or closed by hand.

Cock Metal.—Sometimes used to designate the metal used in the very commonest brasswork, as the cheaper class of cocks and valves.

Cock-wheel.—See Idle-wheel.

Coefficients.—(1) Numerical values deduced from experiments, and used in engineering formulæ and calculations. Hence there are coefficients for friction, for elasticity, for tension, for the flow of water, &c. Thus, for example, the ratio of the stress set up in a body by a deforming force, to the change in dimension, is called the coefficient of elasticity for that material. (2) Any factor or factors of an algebraical product regarded in relation to the other factors. If a figure, it is called the numerical coefficient. See Modulus.

Cog.—The tooth of a gear-wheel. More properly used to distinguish the wooden teeth of mortice-wheels from those of iron.

Cogged Bloom.—A bloom or crude mass of steel which has been passed through the cogging mill in readiness for rolling into rails or other sections.

Cogging.—(1) The fitting in and working of the cogs of mortice-wheels. (2) The rolling of steel blooms from ingots.

Cogging Engine.—An ordinary rail-mill engine used for driving the cogging mill.

Cogging Mill.—A rolling mill in which steel blooms are rolled out. It is similar to a blooming mill.

Cogwheel.—A toothed wheel. Sometimes used to distinguish a mortice-wheel (q.v.) from a wheel whose teeth are of iron.

Cohesive Strength.—The strength of a material resisting forces tending to rupture it by tension.

Coiled Spring.—A spiral spring (q.v.).

Coils.—Coils of iron piping conveying steam are used for the heating of workshops. The pipes are often provided with rings or collars, large and thin, for the better radiation of the heat.

Coke.—The solid residuum produced by the destructive distillation of coal. It is either a residual product of gas-making, then called gas coke, or the chief product, and termed oven coke or hard coke. The making of oven coke is sometimes performed in kilns of rectangular shape, but usually in ovens. The beehive ovens consist of a row of chambers which are either circular or square, and as much as ten feet in diameter in some cases. The coal is charged into these, and the heat necessary to effect its decomposition is obtained by passing air over the upper surface of the incandescent pile, so that the process of coking is effected downwards and inwards, from the top of the mass. Essentially the objects to be obtained are, the introduction of a proper amount of air to burn the gases, but not the carbonaceous matters, uniform and rapid coking, and the prevention of loss of heat by radiation. The beehive ovens are those chiefly used, but Appolt, Carves, Coppée and Pernolet ovens are also employed. Sulphur is the most deleterious body present in coke, and one which cannot be removed by any process at present known. A good coke should be hard, brilliant, crystalline, free from sulphur and from dark and dirty patches.

Coke Basket.—See Moulder's Basket.

Coke Bed.—(1) The first layer of coke which is introduced into a cupola, previous to the throwing in of the iron. Its weight bears a definite relation to that of the iron to be melted, but varies with the condition of the furnace. (2) Before moulding and casting pieces of work, plates, &c., having large superficies, and which are bedded in, provision is made for carrying off the gas generated in the mould by means of a coke bed. This is a porous stratum of coke and clinker in pieces of various sizes, laid to a depth of several inches below the mould; the porosity of which allows the gas to escape as it to a reservoir, to be drawn off by vent pipes (q.v.).

Coke Dust.—Used by moulders for the same purpose as coal dust (q.v.).

Coke Furnace.—A blast furnace in which coke is used as fuel.

Coke Mill.—See Blacking Mill.

Coke Oven.—See Coke.

Coke Plate.—Tin plate for which the sheet iron has been refined with coke, as distinguished from charcoal plate (q.v.). "Coke plate" now commonly refers to plate made from puddled iron. This is inferior to charcoal plate.

Cold Air Machine.—See Refrigerator.

Cold Bend.—See Forge Test.

Cold Blast.—Iron smelting furnaces were originally universally fed with a current of air at the ordinary atmospheric temperature. But in the year 1828, at the Clyde works, Neilson introduced the practice of heating the blast before feeding it into the furnace. Hence the distinguishing term "Cold blast" as opposed to hot blast. Iron produced under the old system is called "cold-blast iron," and is of superior quality, and much sought after by iron founders for scrap. Cold-blast iron is now made to a very limited extent only, and is expensive, its manufacture being confined to about a dozen firms in the Kingdom.

Cold Blast Iron.—See Cold Blast.

Cold Chisel.—A chipping chisel (q.v.). The term is probably used to distinguish it from a smith's chisel used for hot iron.

Cold Iron Saw.—A circular saw, thick in proportion to its diameter, having short teeth, sharpened square across the edge of the disc. It runs at a slow speed, and is usually driven by power, though saws worked by hand by the intervention of bevel gearing are in use for portable purposes. A small circular saw is employed for cutting slits in the heads of screws, and other articles where shallow and narrow grooves are required.

Cold Riveting.—Small rivets in thin plates are hammered up without being heated in the fire, hence the term.

Cold Rolled.—Bars and plates rolled without being previously heated.

Cold Rolling.—The practice of rolling iron plates cold produces a material having a high tensile strength, but with a corresponding sacrifice of ductility and toughness. Its effect is, therefore, the reverse of annealing. The surface of iron when cold rolled acquires a greater smoothness and polish than when rolled hot.

Cold Sand.—See Hot Sand.

Cold Sawing.—The sawing of iron while cold with a cold iron saw (q.v.).

Cold Set.—A smith's set, or chisel-like tool made thicker than the hot set (q.v.), and used for nicking cold metal.

Cold Shortness.—That condition of wrought iron and steel in which it is impossible to work it below a dull red heat without fracture or cracking at the edges, by reason of its brittleness. This condition is due to the presence of phosphorus, silicon, arsenic, and antimony, the first-named being the most common cause of the quality described.

Cold Shorts.—See Cold Short.

Cold Short.—(1) Wrought iron or steel having the quality of cold shortness (q.v.). (2) When cast metal pours dead or thick it sometimes happens that in those portions of the mould where the mass is thin, or comes into contact with the sand, that section of the metal will, if poured too slowly, soon began to thicken and partly solidify, so that the metal which comes after does not amalgamate properly therewith, but forms instead an imperfectly united contiguous or superimposed layer. The surfaces in contact form a " cold shut," and complete fracture is always liable to take place there. Running short of metal in the ladle, and slight delay in emptying the contents of a second ladle into the mould, is also a fruitful source of cold shuts. Also termed cold shorts and cold shots.

Cold Tests.—The testing of the tensile strength of iron and steel bars and plates, by bending, while cold, to a certain angle, both with and across the grain without fracture.

Cold Water Pump.—An ordinary lift or force pump, as distinguished from a hot water pump (q.v.).

Cold Water Test.—The ordinary hydraulic test, for pressure only, to which steam boilers are subjected, as distinguished from a hot water test (q.v.).

Collapse.—The failure of a tube or cylinder due to the stress of pressure applied externally.

Collapsible Core Bar.—See Core Bar.

Collapsing Pressure.—The pressure which, applied to the outside of a tube, causes it to fail by bending or crumpling inwards. In its usual application, it has reference to the tubes and fire-boxes of steam boilers.

Collar.—A ring formed on a shaft, either by forging in the solid, or by being made as a separate casting or forging, bored or turned, and held in place with a set screw or a split pin. The use of a collar is to prevent endlong play in the shaft by providing a face that shall work against the bearing flanges of brasses, or to retain loosely running gear, or gear not otherwise confined endways, in place relatively to the longitudinal direction of the shaft.

Collar Bearing.—A bearing provided with several rings or collars, to take the thrust of a shaft (see Thrust Bearing), or in the case of a vertical shaft to provide adequate surfaces for lubrication.

Collar Gauge.—See Cylindrical Gauge.

Collaring.—The clinging to and wrapping of a rolled bar around the bottom roll of a rolling mill. See Stripping Plate.

Collar Tools.—See Swage Tools.

Collecting Vessel.—A cylindrical vessel enclosed in a steam boiler for the purpose of collecting the muddy ingredients contained in the water, and which would otherwise produce scale and cause incrustation. The vessel is perforated to admit the water which bubbles over into it, carrying over also the sediment, the last settling down into the smooth water within the vessel, to be subsequently blown out at intervals.

Collet.—The disc or ring, as the case may be, by which screwing dies are held fast. Also the clip by which nuts are held in a screwing machine.

Colour Brushes.—The brushes used by draughtsmen are camel hair, Siberian, or sable hair, mounted in crow, duck, goose, swan, or eagle quills, their value increasing in the same order. See Broad Brush, Softener, Wash Brush.

Columbier Drawing Paper.—A paper measuring $34\frac{1}{2}$ in. by 24 in.

Column.—A column may be considered as a beam set on end, and receiving pressure in the direction of its longitudinal axis. But the conditions of these vary with the length. (See Long Column, Short Column.) The resistance of a column to flexure is diminished by rounding the ends, increased by flattening them, still further increased by extending their area, and by fixing them both.

Combination Caliper.—A caliper pivoted near the centre of the two legs, making thus four movable ends. Of these the two to the one side of the pivot are used for inside, and the two on the other for outside measurements.

Combination Chuck.—A universal chuck (q.v.) which is also endowed with the property of independent action of the jaws. It is the geared form of chuck principally which is made for combination by a cam movement through which the pinions are thrown out of gear with the rack.

Combination Gauge, or **Compound Gauge.**—A dial gauge which has two sets of registers, as pressure and vacuum, pressure and head of water, pressure and heat of steam.

Combination Machines, or **Compound Machines.**—Machines designed and adapted for performing several different processes, either at different periods of time or simultaneously.

Combined Carbon.—Carbon which has entered into true chemical combination with iron to form white iron, chilled iron, or steel. The compounds so formed are sometimes termed carbides of Iron. See Graphitic Carbon.

Combined Steam.—Superheated steam (q.v.) and wet steam (q.v.) allowed to mingle together before use. It is advantageous to employ steam of this character in order to diminish the evils of boiler corrosion on the one hand, and of priming on the other. Its temperature should not exceed 310° F.

Comb Tools.—Chasers (q.v.).

Combustible Mixture.—See Charge.

Combustion.—The chemical union of bodies with oxygen. It is accompanied by the evolution of heat, sometimes, though not necessarily with light, since the slow rusting or oxidation of iron is combustion as truly as the vaporisation of a metal in the electric arc. Combustion results in various well-defined products. (See Products of Combustion.) There may be partial or complete combustion (q.v.). The laws of combustion receive practical application in calculations bearing on the air supplies of furnaces, fuel, fire-grate areas, calorific power and intensity, specific heat, and kindred subjects.

Combustion Chamber.—That portion of a boiler flue in which the hot gases are burnt. It is situated between the fire-grate and the smoke-flue proper.

Combustion, Products of.—See Products of Combustion.

Coming to Nature.—A term used by puddlers to signify the stage of the accession to the pasty condition of the ball of iron undergoing the operation of puddling. Also called drying.

Common Jaw Chuck.—A lathe face chuck whose jaws are stepped in three sections, that being the ordinary form.

Common Rafters.—The lighter rafters which lie between the principals or principal rafters and cross the purlins in a roof structure.

Common Slide Valve.—The ordinary plain D-slide valve in which the amount of openings of the ports, both for steam and exhaust, are equal, as distinguished from the exhaust relief valve (q.v.).

Common Thread.—An ordinary Whitworth screw thread as distinguished from a gas thread.

Common Tin.—See Boiling.

Compass.—A drawing instrument used for measuring and transferring distances, and for describing arcs and circles. It consists of two legs movable about a sector joint. The legs may be single or double jointed, and may each be an integral portion of the compass, or one may be removable to permit of the substitution of separate legs for pen, pencil, or point, and for a lengthening bar (q.v.). Workmen's compasses are either made plain without any means of tightening or setting minutely, or they are wing compasses. See also Bows, Dividers, Hair Compass, Spring Bows, Proportional Compasses, Triangular Compasses, Wholes and Halves.

Compass Caliper.—A scribing tool having one leg caliper shaped, the other straight like a compass and pointed. It is used for scribing lines from the end of a piece of work, the curved leg being slid along in contact with the end, while the point of the straight leg scribes a line parallel therewith.

Compass Saw.—A short narrow saw, tapering towards the point, used for cutting sweeps and curves by hand. Sometimes termed a table saw.

Compensating Collars.—Annular rings or collars inserted on the spindles of drilling machines between the feed screw and the grooved spindle, to form hard wearing surfaces, and by whose adjustment the wear of the spindle and collars can be taken up.

Complement.—The complement of an angle is its difference from 90°.

Complete Combustion.—Combustion, where all the elements contained in fuel, or in gaseous charges, enter fully into chemical combination with atmospheric air. In chemical language their atoms are fully satisfied. The regulation of the fuel and air supply of furnaces should, for economical reasons, approximate as nearly as possible to these conditions.

Component.—The components of a force are the various forces which combine to make the resultant (q.v.).

Composition of Forces.—The process or method of ascertaining the resultant (q.v.) of a series of single forces from a consideration of the magnitudes and directions of those forces. It is effected either by calculation or by a graphic method, lines being drawn whose relative lengths represent the magnitudes, and whose directions represent the directions, of the various single forces.

Composition of Levers.—See Compound Lever.

Compound Adjutages.—Adjutages, where the shape of the pipe is modified to increase the flow of liquid, the head remaining the same. See Vena Contracta, Converging Mouthpiece.

Compound Engine.—An engine comprising two or more cylinders with their parts. The steam, after doing work in the smaller or high-pressure cylinder, is allowed to exhaust into a larger or low-pressure cylinder, or cylinders, and to do useful work under expansion. The advantage obtained is that the irregularity of motion and the liquefaction of steam which occur when the variations of initial and terminal pressures are confined to one cylinder, are minimised by dividing them among two or more cylinders, so that both pressure and temperature become more equalised. High ratios of expansion are also obtained without the employment of complicated cut-off gear.

Compound Expression.—An algebraical expression consisting of more than one term. Compound expressions consisting of two, three, or more terms are called binomial, trinomial, and multinomial respectively.

Compound Gauge.—See Combination Gauge.

Compounding.—The addition of a low-pressure cylinder to an engine which has hitherto possessed a high-pressure cylinder only.

Compound Lever.—A system of levers by whose combination great power is developed within a restricted space, the mechanical gain equalling the product of all the long arms divided by that of all the short arms. A testing-machine (q.v.), a wheel train (q.v.), and a weigh-bridge (q.v.) afford examples of compound levers.

Compound Machine.—See Combination Machines.

Compound Oils.—The advantage of a compound oil is believed to consist in this : that certain advantages of single oils are gained and their disadvantages neutralized, thus—an animal oil alone is liable to develop acid, to the consequent corrosion of the bearings with which it is in contact ; a vegetable oil alone is apt to dry and gum, or become sticky and clog the bearing ; a mineral oil alone is usually thin, has a low firing point, and is liable to become squeezed out and evaporated or volatilized. But by a combination of the three, the body of the animal oil is retained while its tendency to decomposition is lessened, the good lubricating property of the vegetable is utilised without much gumming taking place, and the flashing point of the mineral is raised or counteracted, while its fluidity and its cleansing action are utilised.

Compound Screw.—A differential screw (q.v.).

Compound Sliding Table, or Compound Table.—The table belonging to a machine tool which is used for bolting the work upon, and provided with at least two movements, one longitudinal, the other transverse. Commonly a circular movement around its vertical axis is also included. The table of a slotting machine furnishes an illustration of such a combination.

Compound Stress.—A stress of two kinds acting at a single time upon a structure, as torsion and bending, or bending and tension.

Compound Surface-condensing Engine.—A compound engine (q.v.) provided with the means of surface-condensation (q.v.). Most marine engines are of this type.

Compound Table.—See Compound Sliding Table.

Compound Tallow Cup.—A grease cup furnished with two cocks, one leading from the cup into a globular intermediate chamber, the other from the chamber into the cylinder or steam-chest, thus keeping the tallow clean and preventing waste.

Compound Train.—A train of change-wheels (q.v.) used in screw-cutting, in which there are two or more intermediary wheels on the stud. All screws of fine pitch, or say over about ten to the inch, are cut with compound trains.

Compressed-air Lift, or Pneumatic Lift.—A form of lift or hoist in which air takes the place of water as a motive power, the cage being supported by an air cylinder. It is used for the charging of blast furnaces.

Compressed Steel.—Cast steel subjected, while in a metallic ingot mould, to hydraulic or steam pressure, being compressed while still in a molten condition. The pressure applied varies from 6 to 20 tons per square inch, depending on circumstances. It is compressed to the extent of about $1\frac{1}{2}$ inch per foot of length, its specific gravity being increased thereby. The object of compression is the production of a sounder ingot than can be obtained in an open mould.

Compressibility.—That property of gases—as air, steam, carbonic acid, &c.—according to which their volumes are reduced by pressure. Pressures and volumes vary in inverse proportions in all true gases. Liquids are practically incompressible, though not absolutely so. Solids are compressible in different degrees.

Compressing Cylinder.—(1) The cylinder of an air-compressor within which the compression of the air takes place. (2) A cylinder used in some gas engines for compressing the air used in the charge, and distinct from the working cylinder (q.v.) in which the charge is exploded.

Compression.—(1) A body is in compression when it is subject to forces tending to crush it in the direction of that axis which is continuous with the line of direction of the pressure. (2) The resistance of the steam left in the end of an engine cylinder and the passage leading thereto on the return stroke of the piston, and which is due to the early closing of the port. This is also termed cushioning. The compression of steam in an engine cylinder should never be allowed to rise above the initial pressure of the steam, otherwise the compressed steam will force its way against the entering steam into the steam-chest, reducing the pressure. (3) The pressure exercised during its instroke by the piston of a gas engine, on the combustible mixture admitted behind it.

Compression Bar.—A bar which is being subjected to compression stress.

Compression Coupling.—See Cone Vice Coupling.

Compression Engines.—Those gas engines in which the mixed charge is subject to compression previous to ignition. The advantage of employing compression is that a smaller percentage of gas is consumed thereby than is the case when the charge is ignited at atmospheric pressure.

Compression Line.—The line on an indicator diagram which shows the rise of pressure near the termination of the piston-stroke due to compression. The presence of a loop at the termination of the line indicates that the steam has been compressed above its initial pressure.

Compression Strength.—The strength necessary to enable a bar or structure to resist compression or crushing.

Compression Stress.—The stress due to compression.

Concentrated Load.—A load localised upon a beam, or girder, or structure. Under the same conditions of fixing, a concentrated load will produce twice as much stress as a distributed load (q.v.) will produce.

Concentric Chuck.—A concentric jaw chuck (q.v.).

Concentric Jaw Chuck.—A chuck in which all the jaws are moved uniformly towards, or from the centre by a common mechanism, thus avoiding the trouble of testing the accuracy of the work by the trial and error method. Concentric jaw chucks are usually of small size, and are made both for lathes and for screwing machines.

Concrete, or Beton.—Hydraulic lime, sand, and gravel, whose proportions vary with circumstances.

Concrete Block.—A rectangular block of concrete or artificial stone, employed for the foundations of breakwater and harbour works. Such blocks are made by mixing the concrete, depositing it in moulds, and allowing it to set. They are often made to 40 tons weight.

Concrete Mixer.—A revolving vessel of cast or wrought iron, used for mixing the cement, sand, gravel, and water, which form the ingredients in concrete blocks. There are several patented mixers constructed with the view of throwing the materials over from one side to the other, on an average four times during each revolution.

Condensation.—The reduction of gaseous bodies to the condition of liquids by the influence of cold and of pressure, acting either singly or in combination.

Condenser.—A vessel in which the condensation of gases is effected. Specifically, the vessel in which the condensation of the exhaust steam from an engine cylinder is effected. It is either a jet condenser (q.v.), or a surface condenser (q.v.).

Condenser Door.—The rectangular or round-ended cast-iron plate which closes the end of a surface condenser near the ends of the tubes. When the door is unscrewed and removed, the ends of the tubes are open to inspection.

Condenser Tubes.—Tubes which traverse the condensers of marine and other engines of the surface condensation type, having steam without, and cold water circulating through their interiors. They are usually drawn tubes of brass or Muntz metal, about $\frac{1}{8}$ in. thick and from one to two inches in diameter.

Condensing Engine.—An engine whose exhaust steam is sent into a condenser, instead of being exhausted into the atmosphere.

Condensing Surface.—See Condensing Tube Surface.

Condensing Tube Surface, or Condensing Surface.—The total area of the internal diameters of the tubes in surface condensers. This area bears a definite relation to the engine power, but varies with circumstances.

Conduction.—The transfer of heat from the hotter to the colder parts of a body. Hence conduction depends upon the fact of inequality in temperature existing in the several portions of the body. The transfer of heat through solids, as through boiler plates, is due to conduction. See also Convection, Radiation, Transmission of Heat.

Cone.—(1) A round pyramid having a circular base. Its sections are : from vertex to base, a triangle ; transversely and parallel with the base, a circle ; if obliquely through both sides, an ellipse ; on a plane parallel with the side, a parabola ; when the plane makes a greater angle with the base than the cone, a hyperbola. (2) A driving pulley used for belting, and formed into steps or sections of various diameters, for the governing of different speeds. (3) The top of a blast furnace is often furnished with a cone for the purpose of preventing the escape of the waste gases into the air, and also for regulating the quantity of the charge.

Cone Bearing.—Cone bearings are adopted for the mandrels of lathes partly from motives of economy, partly for convenience of taking up the wear. The cones are made distinct from the headstock, and driven into holes bored in the main castings. They are made of hardened steel or of gun metal, occasionally in white metal. In small lathes, there is a single cone only, and a back centre ; in larger ones there are two cones, usually placed in a direction reverse to one another. Their wear is taken up with back nuts, pressing against the movable cone at the tail of the headstock.

Cone Clutch.—A common form of friction clutch in which the power necessary for driving is effected by the bite of smooth turned conical surfaces, male and female respectively, tightened against each other, either by the application of lever or of screw pressure.

Coned Neck.—Applied to that portion of a lathe mandrel which runs in the front cone bearing.

Cone Gear.—Cone clutches (q.v.) employed in hoisting machinery to drive the lifting drum, gears, &c., by the simple friction of their surfaces.

Cone Key.—A form of key used for retaining a wheel or pulley in place when the hole in the wheel is larger than that portion of the shaft upon which it is keyed. The wheel is bored slightly conical and a conical ring turned to fit in the bored hole and to embrace the shaft. It is then slotted into three parts, forming three separate keys. The wheel is thus maintained concentric with the shaft, and will pass over a larger to a smaller section without the necessity for splitting it.

Cone Plate, or Boring Collar.—A lathe appendage used for boring holes in the ends of work, such as spindles, shafts, mandrels, &c., which are too long to be centred and held firmly by a grip or face chuck, and which cannot be placed between centres because one end is wanted free for boring. It consists of a small poppet-like bearing, bolted to the lathe bed, and carrying a circular plate or disc perforated with a series of holes conical in section. The plate is centred on the poppet so that by the slacking of a central screw any one of the holes can be brought into coincidence with the lathe centre, and there tightened. The holes being of different diameters, and tapered in section, will adapt themselves to spindles of different diameters. The boring tool is held in the slide rest, and is fed forward into the spindle when boring.

Cones.—(1) Sometimes used for producing variable motion in shafting. (See Alternate Cones.) And (2) as friction clutches. (See Cone Clutch.) (3) The nozzles of injectors are called after their functions—steam cones and water cones.

Cone Tubes.—Galloway tubes (q.v.)

Cone Vice Coupling.—A form of box coupling (q.v.) consisting of an outer barrel whose interior diameter is bored doubly conical, and two sleeves turned without to fit the conical portions of the barrel, and within to fit the shafts. These sleeves are slotted through on one side to provide for their compression inwards, which compression is effected by three screw bolts passing through both, pulling them together so that they tighten on the barrel and on the shaft simultaneously. Sometimes called Compression Coupling.

Congelation.—The thickening of lubricating oils in cold weather ; hence the necessity for selecting oils with reference to the climate in which they are to be used.

Conical Rivet.—A rivet whose head is conical in section. It is easier to hammer than the ordinary rivet in a restricted space where a small hammer only can be brought into requisition.

Conical Spring.—A helical spring (q.v.).

Conical Turning.—The turning of tapered work in the lathe. It is accomplished between centres by setting over the poppet, and upon the face chuck either by setting over the headstock, or more usually by swivelling the top slide of the slide rest to the angle required.

Conical Valve.—A form of lift valve for a pump. Its sides are cone-shaped in section and it fits on an annular seating, and has no wings. It sometimes takes the place of the common lift valve with parallel wings.

Conic Frustra.—The development of tubular boiler plates is that of the frustra of cones.

Coning.—The turning of the taper on the diameters of railway wheels and crane and turntable rollers. The object in coning railway wheels is, that as the train runs round a curve, and the wheels are thrown outwards by centrifugal force, the outer wheel may run on the rail where its diameter is greater, and the inner where the diameter is less, so that the difference in length of inner and outer curves shall be compensated for by the differences in the diameters of the wheels.

Conjugate Axis.—The right line which crosses the transverse axis of a curve at right angles. It is the shortest diameter in the case of an oval or an ellipse, and is parallel with the base in that of a parabola.

Connecting-Rod.—The rod which converts the rectilinear motion of a piston-rod into the rotating motion of a crank.

Connecting-Rod.—Marine pattern. See Marine Pattern Connecting-Rod.

Connecting-Rod.—Obliquity of. See Obliquity of Connecting-Rod.

Conservation of Energy.—The principle of the conservation of energy is, that force is never lost or wasted—that potential and kinetic energy mutually replace each other. Motion is convertible into heat, heat again produces motion ; the diminution of friction renders more power available for useful work, heat dissipated in engines and furnaces is useful work lost. The engineer, of all men, should never forget that energy is indestructible, and his aim should be to utilise and to translate it into the forms, and by the methods most economical for his purposes.

Consolidated Emery-Wheels.—Solid emery-wheels (q.v.).

Constant.—A number deduced from actual experiments made upon the strength of a particular material, and used as a basis in calculation affecting the strength of structures made in that material, but differing in dimensions. Knowing, for example, the weights required to break off, or to induce permanent set in a test bar of cast iron measuring 3 in. by 2 in. by 1 in., we can use those weights as constants, by

which to estimate the stresses in structures of the same material, but differing in length, breadth, and depth.

Constant Load.—A dead load (q.v.).

Constant Travel.—The travel of a slide valve which is not rendered capable of variation for purposes of variable cut off. See Varying Travel.

Continuous Beam.—A beam which rests on more than two supports.

Continuous Brake.—A brake which acts upon a series of wagons simultaneously, as distinguished from the separate brakes formerly applied to each truck.

Continuous Feed.—(1) Usually means a delivery of feed water which is not intermittent, as from an injector, or from a pump provided with an air vessel. (2) The feed given to a machine tool, or the work, by means of cog or worm-wheels, and a screw, as opposed to intermittent feed (q.v.).

Continuous Paper.—Drawing paper and tracing paper, prepared and sold in long rolls instead of in sheets. Any length can then be cut for specially long drawings.

Contracted Vein.—See Vena Contracta.

Contraction.—The diminution in length which all metals (bismuth and some alloys of bismuth alone excepted) undergo in cooling down from their fusing points. The amount of contraction for different qualities of the same metal varies, as also does the contraction of the same quality under different conditions of mass and of cooling. The contractions of cast and wrought iron, steel, brass, gun-metal, lead, tin, &c., are very variable, and a knowledge of the amounts to be anticipated under varying conditions is indispensable.

Contraction of Area.—The reduction which a bar undergoes in diameter or area previous to fracture, when elongated by tension. The greater the amount of contraction the better the quality of the bar. It varies from about ten per cent. in plates, to twenty per cent. in round and square bars.

Contraction Rule.—A rule used for the construction and measurement of foundry patterns, whose length exceeds that of the ordinary or standard rule by the amount due to the contraction (q.v.) either of iron or brass. A contraction rule for iron is longer than the standard in the ratio of $\frac{1}{8}$ in. in 15 in.; one for brass in the ratio of $\frac{1}{4}$ in. in 10 in.

Contract Note.—A piece-work note (q.v.).

Convection.—The setting up of currents by the heating of a liquid in a vessel. The portions in direct contact with the source of heat, being first warmed, become lighter than the others, and rise from the heating surface, allowing colder currents to take their place, which in turn become warmed and give place to others. Hence convection is entirely due to differences in Sp. gravity. The heating of the water in steam boilers is the cause of its circulation, and the greater the heating surface, or the surface suitable for convection, the quicker the circulation, and the more rapid the generation of steam. The circulation also in hot water apparatus is due to convection, and the greater the expansion the greater the upward force of the ascending currents.

Converging Mouthpiece.—A mouthpiece placed against the side or the bottom of a vessel for the passage of liquid. Its longitudinal section is that of a truncated cone, and its function is the prevention of the loss of effect due to Vena Contracta (q.v.). The larger aperture is next the vessel, and the smaller aperture is equal in diameter to that of the actual jet of water which has to be delivered.

Converter.—See Bessemer Converter.

Converting.—The term applied generally to the manufacture of steel made by the cementation process. Also to the Bessemer process for the direct production of mild steel.

Converting Furnace.—See Cementation.

Converting Pots.—The troughs in which bar iron undergoes the process of cementation (q.v.).

Converting Process.—The process of cementation.

Cooling.—By the judicious cooling of the heavier portions of castings the setting up of internal stresses is avoided or minimised. Thus the bosses of wheels are often cooled by raking the sand away from them while the casting is yet white hot. Thick flanges and ribs are cooled in like manner. Without such precautions, numbers of castings would be lost, owing to the heavier portions, which remain hot longest, continuing to contract after the lighter portions had set fast, and so tearing them asunder. A massive casting when well proportioned should always be allowed to cool down while enveloped in its sand, rather than be stripped and exposed to the air.

Cooling Surface.—The superficial area in a condenser which is exposed to steam on the one side and to water upon the other. The area of cooling surface may be, roughly, one-half or three-quarters that of the heating surface of the engine.

Cope.—Used to designate the upper portion of a loam mould, or the top flask used in green sand moulding.

Cope Off.—Signifies the lifting of a loose pattern piece in the top box, or cope (q.v.).

Cope Ring.—The ring which carries the bricks and loam forming the cope (q.v.) of a loam mould. It is of cast iron furnished with lugs around its circumference for the attachment of chains or links for lifting.

Copper.—Symbol Cu. Comb. weight, 63. Sp gr. 8·93. A metal employed for a variety of purposes by engineers. It is not strong, but is tough. It is used for pump-rods, boiler tubes, rivets, steam pipes, wire, fire-boxes, &c. It is of especial value in the formation of alloys.

Copper Bit, or Soldering Iron.—A pointed piece of copper riveted to a cleft rod of iron to the opposite end of which a wooden handle is attached.

Copper Drift.—A short cylindrical piece of copper held with a twisted hazel-rod, and struck with a hammer, its use being to prevent the ends of shafts or similar portions of bright finished work from being bruised and burred over, which would happen if they were struck directly with a steel-faced hammer. The copper, therefore, acts as an elastic medium or cushion.

Copper Glance.—A valuable ore containing copper and sulphur. It is mined in Cornwall.

Copper Hammer.—A flat paned, double-ended hammer, made of copper, and employed for the same purpose as a copper drift (q.v.).

Copper Lining.—A copper bush or liner, driven into the working barrels (q.v.) of iron pumps, to prevent the formation of rust.

Copper Matt.—See Matt.

Copper Pipe.—Copper is used for the steam and various other pipes of large engines, its utility consisting in the readiness with which it can be curved to any form, and in the ease with which it accommodates itself by expansion and contraction to variations of temperature, without risk of tearing off the flanges. The flanges of copper pipes are

8*

brazed on (see Brazing Metal), a hole being bored through the flange to receive the pipe. Copper pipes are filled with molten lead before being bent, in order to prevent wrinkling up, and the bending is performed in the case of large pipes in a screw press.

Copper Pyrites.—The most important ore of copper, consisting of copper, sulphur, and iron. It is mined in Cornwall and Devon, Sweden, Siberia, and other parts.

Copper Wire.—Fine copper wire of about 20 S. W. G. is used for making flanged joints steam or water tight. It is coiled spirally a few times over the face of the flange, and the ends lapped. The tightening up of the bolts flattens it out, and makes a perfect joint, which has also the advantage of being more readily broken than a red-lead joint.

Copy.—See Former.

Copying Frame.—See Printing Frame.

Copying Machines.—A large class of machines whose purpose is the production of many similar articles from a single templet, or copy. Of this class are the various tracing, and housing, and spoke and handle turning machines, in which the cutters are affixed to blocks which move simultaneously with a guide, following the outline of the copy, which last is generally made in metal where a very large number of articles are required.

Copying of Drawings.—See Phototype, Tracing.

Copying Paper, or Ferro-prussiate Paper.—A sensitised paper used for the reception of phototypes (q.v.). It is prepared as follows : ammonia citrate of iron, 1 part; water, 5 parts. Ferrocyanide of potassium, 1 part; water, 4 parts. Mix together in equal quantities. The paper saturated with this has a yellowish hue, and owing to the various washings which it has to undergo is made very stout.

Copying Principle.—The principle embodied in the construction of copying machines (q.v.).

Corbels.—Short cantilevers built into, and projecting from the walls of workshops to support the rails for the travelling wheels of overhead travelling cranes.

Cordage.—See Ropes.

Core.—(1) A central portion. (2) In foundry work a body of sand, either green or dried, which takes out the central or inner portions of a casting. Cores are either made in boxes called core-boxes, or if circular and of large size, are struck to shape upon a revolving bar, a mixture of loam being used for the body, and its due outline being imparted to the mass by revolving it against the bevelled or chamfered edge of a templet board, called a core board, or a loam board. When the core is finished and dried, it is put into the mould and the metal poured between it and the outer mould. After the casting has cooled, the core is withdrawn from its interior. (3) The central portion of a core packing (q v.). (4) The body of a smith's anvil. (5) The wood removed by a boring tool. (6) A wooden or iron model of the interior of the lining of a Bessemer converter, between which and the outer casing the slurry used for lining the converter is rammed.

Core Bar.—A stout iron bar upon which a circular core is struck up. The bar is made to revolve on the bearings of core trestles (q.v.) and foundry loam is laid upon it and struck to shape against the edge of a board cut to the required outline. When large in diameter it is called a core barrel. Core bars will vary in diameter with the sizes of the cores which they are to receive, ranging from half an inch to a foot or more.

For cores of largest size they are encircled with core plates (q.v.). All bars are hollow tubes, and are pierced with numerous holes for the escape of the gases generated in casting. Light core bars are made from gas pipe, heavier ones are made of cast iron. Good bars should have bearing necks turned at their ends to run in the trestles. Bars are revolved with a winch handle turned by a core boy (q.v.). In the case of pipes of large diameter whose production is a speciality, the core bars are made collapsible, in order to their more ready withdrawal from the casting, and also to save the cost and labour of putting on hay bands, the loam being daubed directly on the bars. An ordinary collapsible bar consists of three longitudinal segments, maintained circular in section during the striking of the core, by means of internal cones wedging them outwards. After the casting is cooled, the cones are knocked back, and two of the segments collapse or fall inwards, so that the whole bar is loose in its hole.

Core Barrel.—See Core Bar.

Core Bench.—A bench consisting of a flat plate of iron, upon which small foundry cores are rammed up in their boxes. A small heap of core sand occupies one portion of the bench, leaving the remainder free for manipulation of the boxes.

Core Board.—The board against whose edge a core is struck up on a revolving bar. It is commonly termed a loam board, loam being the material employed.

Core Box.—A box commonly made in wood, though often of iron or brass, in which the core for a hollow casting is rammed up. The box is in its interior a counterpart of the shape of the core, and is so constructed that when the latter is finished the box can be separated and withdrawn piecemeal, leaving the core intact.

Core Boy.—A boy who rams up the smaller cores in foundries and turns the core bars for the core-makers. In all foundries of moderate size there are several such boys, and as little responsibility attaches to the work it is the first task the moulder's apprentice is set to perform.

Core Carriage.—A low carriage upon which are laid the cores, and the loam patterns of moulds which are to be run into the drying stove.

Cored Hole.—(1) A cast hole which is cored with a dry sand core instead of delivering as a hole directly from the pattern. (2) Generally any hole in a casting which is not bored in the machine shop. It is customary in drawings to distinguish rough cast holes from those which are to be bored, by this term.

Core Irons, or Core Rods.—Rods of wrought iron from $\frac{1}{8}$ inch or $\frac{1}{4}$ to $\frac{1}{2}$ inch in diameter, according to the size of core, which form a skeleton or framework upon which a foundry core is made and dried. Slender cores would fall to pieces, apart from the assistance derived from core irons. After the castings are made the irons are withdrawn by the hand or by pincers.

Core-maker.—A workman employed in foundries, whose duty is the striking up or ramming up of cores. The core-maker is usually a moulder who has taken to this special department of the work, and having acquired the distinctive knowledge relative thereto does not quit it.

Core Oven.—A small oven built over an ordinary stove fire or over a brass furnace, and used for drying small cores.

Core-packing.—Packing both for steam and water, which is stiffened and rendered more durable by the introduction into it of a core, usually of india-rubber.

Core Plates.—Circular plates, cast in open sand, for fitting upon a core bar when the core is of large diameter. The plates form a skeleton or framework, supporting a large mass of hayband and loam, without unduly increasing the weight of the bar. They are secured to the central bar with wedges. To get them out of the casting they are usually broken to pieces by blows delivered from a bar or a hammer. For this reason they are cast thin. They are also pierced with holes to allow a free passage to the gases, and to render their fracture easy.

Core Ring.—A ring of cast iron which carries the core for a loam mould.

Core Rods.—See Core Irons.

Core Sand.—Sand used for making cores. It is a mixture of various sands whose names and qualities vary with the localities in which they are found. It has the property of hardening when dried. Horsedung is mixed with it to render it sufficiently porous for the escape of the gases generated during casting.

Core Stove.—See Foundry Stove.

Core Trestles.—Strong cast-iron trestles having vee'd-bearings upon their upper edges for the reception of revolving core bars. The core bar is turned by a boy set at a handle inserted into its end, the vees forming a sufficiently good bearing for its journals.

Coring out.—The taking out or forming of the interior portions of castings with cores, as distinguished from self-delivery (q. v.).

Coring up.—The placing of the cores in their positions in a foundry mould, in readiness for casting.

Cork.—Used for glass-paper rubbers. See Rubber.

Corliss Engine.—A very economical type of engine, the working of whose valves is controlled in an automatic manner from the governor. The inlet and exhaust valves are distinct from each other, and while the latter are always opened to their full extent, the former are only opened so much as is permitted by the position of the governors. The valves are moved by rods from a disc or wrist-plate, but are disconnected at every stroke of the engine, and are then closed instantaneously by a dash-piston and spring. The advantages of Corliss engines are that they are strictly automatic in action, the steam supply being proportioned to the requirements of the engine at the moment, and that the steam passages are extremely short, preventing loss due to back pressure (q.v.) and the waste inseparable from long passages, and that the sudden or positive cut-off of the valves is substituted for the gradual cut-off of the ordinary slide-valve.

Corliss Valve.—The valve of a Corliss engine. The valve forms a segment of a circle and revolves through an arc of a circle in an annular seating, alternately covering and uncovering the steam port. The supply valve is closed by means of a dash-piston (q.v.).

Corner Sleekers.—Square corner sleekers (q.v.).

Cornish Boiler.—An internally fired, horizontal, cylindrical stationary boiler, having one furnace flue only.

Cornish Engine.—A beam engine of massive type, used among the mining districts of Cornwall, and which, variously improved and modified was accepted as a standard form for pumping engines in general. Formerly all Cornish engines were single acting, but later they were double acting, and worked expansively.

Cornish Valve.—See Double Beat Valve.

Corollary.—A theorem or problem deduced from a previous related proposition.

Corrective Gauge.—See Standard Corrective Gauge.

Corrosion.—The rusting or oxidation of metals by contact and chemical union with oxygen in a damp atmosphere. Iron-work is protected from corrosion in many ways. A common method is to heat it to a temperature of about 300° F., and apply to it a coating of pitch or tar. Another is to apply linseed oil cold. Oil paint, when not objectionable, is largely used. Coating with magnetic oxide by Barff's process (q.v.), or Bower's process (q.v.), is employed for bright work. Tallow is also rubbed over bright work when leaving the shops. A coat of pure white lead paint is often given. Galvanizing is resorted to for work which has to be exposed to the weather. See also Boiler Corrosion.

Corrugated Furnace Tubes, or **Fox's Corrugated Tubes.**—Furnace tubes for boilers, both land and marine, corrugated in their longitudinal sections. They have come into very general use, their superiority consisting in the fact that the strains tending to bulge the end boiler plates are prevented or minimised by the yielding character of the corrugations which absorb the linear expansion due to heat.

Corrugated Iron.—Thin sheet iron, of from 18 to 26 S.W.G. It is rolled out in alternate equal curves of elevation and depression, or ridges and valleys, by which corrugation the rigidity of the sheets is increased. The pitch, or distance between the centres of alternate curves, varies from 3 in. to 5 in. Corrugated iron is used principally for light roofing, being united with small bolts passing through the summits of the ridges.

Corundum.—Crystalline alumina, of which emery (q.v.) is a variety.

Cos.—The cosine of an angle.

Cosec.—The cosecant of an angle.

Cot.—The cotangent of an angle.

Cottar, or Cotter.—A tapered rod or pin, either flat or round in section, used for the purpose of wedging the ends of rods, or of pins into bosses, or of strap ends (q.v.) over their rods. Cottars pass through slotted holes prepared for their reception through the rods and bosses, or the strap ends, as the case may be, and in consequence of the clearance given to the holes on their opposite sides afford a means both of tightening rods, and of taking up the wear of bearings. They are held by their own friction simply when the amount of taper is small, or are kept with set-screws, or have their ends or "tails" screwed and tightened with a nut when the taper is considerable in amount. The gib and cottar arrangement is that in which the cottar is formed in two parts, whose outer faces are parallel with each other, the requisite taper being given to their inner or sliding faces. The fixed half is then termed the gib.

Cottar Files.—Narrow files, used for the cleaning out of grooves, keyways, cottar-ways, and similar narrow passages. They are made in three degrees of coarseness. Cottar files are either parallel or tapered in their length ; in the latter case they are called entering files, because they enter into and enlarge or open out small narrow holes.

Cottar Way.—The oblong slot which receives a cottar (q.v.). Its sides are parallel with each other, but its ends are slightly tapered, to allow of the tightening action of the cottar, the total taper being from about 1 in 30 to 1 in 15.

Cotton.—Cotton is used for steam packing, for ropes for the transmission of power at high speeds, and to a limited extent for belting.

Cotton Card.—See Card Wire.

Cotton Cords.—Driving cords used for working overhead travellers. They are about ¼ in. or ⅜ in. in diameter, and are remarkably soft and pliable. They run in grooved pulleys, and under similar conditions to ropes (q.v.).

Cotton Tapes.—Strands of cotton used for purposes of lubrication, the oil feeding along the tapes by capillarity alone.

Cotton Waste.—See Waste.

Counterbalancing.—The addition of a definite amount of weight to a moving mass, usually a wheel or crank, in order to equalise the forces or moments around a revolving shaft. Counterbalancing is an important factor in rapidly revolving mechanism. It is frequently, though not quite in strictness, termed balancing (q.v.).

Counterbore.—A kind of flat countersink, used for boring the seatings for cheese-headed screws (q.v.). It is attached to a pin, shank, or mandrel which fits the hole drilled for the body of the screw.

Counterbraced Brace.—A brace is counterbraced when it is capable of acting both as a strut and a tie. See Counterbracing, Rolling Load.

Counterbraced Girder.—A girder which is provided with counterbracing (q.v.), in order to enable it to withstand the action of a rolling load.

Counterbracing, or Cross Bracing.—Diagonal bracing introduced into a truss or a girder, for the purpose of giving additional support to the beam and relieving it of transverse stress.

Counterpressure.—Back pressure (q.v.).

Counter Shaft.—A short shaft, intermediate, or between the line or main shafting in a workshop, and the particular machine or machines which it has to drive, its use being to transmit the rotation of the line shafting to the machine, and to modify the speed. Countershafts are necessary, because of the different speeds at which different machines require to be driven.

Countersink.—(1) A hole recessed conically for the head of a screw or rivet, in order that the head shall lie level with the surface of the material into which it is fitted. (2) The bit with which a hole is countersunk. It is often called a rose bit.

Countersinking.—The recessing of the orifice of a hole by drills or countersink bits, to receive the heads of screws and rivets.

Countersunk.—A hole recessed with a countersink bit is said to be countersunk.

Counterweight.—A balance weight (q.v.).

Country Cut.—Timber which has been sawn into planks and boards in the districts where it was felled. It is usually unequal in thickness, and the saw teeth being coarse leave the surfaces of the boards rough, so that working it up is a wasteful process.

Couple.—A mechanical couple consists of two forces which are equal and opposite, and therefore in equilibrium.

Coupled Wheels.—Locomotive running wheels are coupled or connected with coupling rods, in order to ensure uniform running; that is, the weight necessary for adhesion is distributed equally between all the coupled wheels. They are called four, six, or eight coupled, according as the wheels on two, three, or four axles are thus united.

Coupler.—The ring or loop which is slid along and tightened over and around the handles of smiths' tongs, in order to make them take a firm grip on the work. See also Hooks and Eyes.

Coupling.—(1) A draw bar (q.v.). (2) The uniting together of lengths of shafting or pieces of mechanism, wheels or engines. (3) The medium through which shafting, &c., is united. Shafting is united by flange couplings (q.v.) or box couplings (q.v.). (4) Engines are said to be coupled, when in a compound arrangement, high and low-pressure cylinders are not made to disconnect and work independently of one another. (5) The coupling for railway trucks consists of two links of wrought iron united by a right and left-handed screw. The screw being fed by a pendent lever, the coupling is either tightened or extended.

Coupling Boxes.—Loose, freely fitting boxes used for coupling together the breaking pieces (q.v.) of puddling rolls with the roll necks. They fit, one-half over the roll neck, and one-half over the breaking piece, and are held with stops.

Coupling Rod.—A connecting-rod which couples or unites the motions of two cranks, rendering their action simultaneous.

Cover.—Lap (q.v.).

Covering of Pipes.—Steam pipes are clothed with numerous non-conducting compositions to prevent cooling and condensation by the atmosphere. Felt, fossil meal, slagwool, sawdust, asbestos, and numerous patent compositions are employed, and the leaving of air spaces in combination with wool, cotton, or felt is also practised.

Covering Strip.—The strip of metal plate which covers a butt riveted joint. Called also welt, and fish plate.

Cover Plate.—A term of general application, signifying any plate which covers in an open space, or a portion of a machine.

Coversin.—The coversed sine of an angle.

Cowburn Valve, or Dead-Weight Safety-Valve.—A form of safety-valve in which no lever or spring is employed, but the valve is instead held down by what are termed pendent weights, that is, annular weights dropped over a casing attached to the upper portion of the valve. These valves have not so neat an appearance as the ordinary types, but possess the advantage of having an amount of weight directly corresponding with the pressure in the boiler. Grouped valves are those in which, instead of having a single opening as is usual in ordinary valves, the total area is divided between several smaller areas of one square inch each, each separate valve opening being therefore weighted to as many pounds of dead weight as the required pressure, so that for 50 lbs. pressure there would be 50 lbs. on each separate valve of 1 in. square.

Cow Hair.—This is used for loam moulds for the same purpose as horse dung (q.v.).

Cow-mouthed Chisel.—A round-nosed chisel ground to a thick angle and used for chipping hollows in metal work.

Cowper Stove.—A stove, constructed on the regenerative principle, and employed for increasing the temperature of the blast (see Hot Blast) for smelting purposes. The hot waste gases which come off from the top of the blast furnace are brought over and made to combine with air in a vertical tube, whence the heated products of combustion pass down through a chequer work of brick by which the temperature is regulated, and thence into the blast main.

Cowrie Pine.—See Kauri Pine.

Crab.—(1) A low type of hoisting crane which has no jib, but is provided with a snatch block or running pulley, dependent from the barrel. Or the lifting chain passes from the barrel to a snatch block dependent from shear legs (q.v.). Crabs are made to work by hand or by steam, and are fixed to the ground, or travel on gantry beams. They are of all sizes, and single, double, and treble geared. (2) A claw clutch is called a crab.

Crab Winch.—A crab (q.v.).

Cradle.—(1) The end girders of overhead travelling cranes which carry the running wheels are termed cradles. (2) The term is also applied generally to carriages and movable bearings running beneath a beam.

Cramp.—A tool for squeezing, made both in wood and metal.

Crane.—A machine used for the hoisting of heavy weights by the intervention of gear-wheels, chain barrel, chain, and jib. The jib (q.v.) is a beam or strut standing at an angle with the ground line, and the lifting chain runs over a pulley at its top end, and depending therefrom sustains, and lifts, and lowers the load. Cranes are worked either by hand, steam, or hydraulic pressure. See details under specific headings.

Crane Chain.—Ordinary close link chain (q.v.).

Crane Hook.—The hook which depends from the lifting chain, and to which the load or the sling chain (q.v.) is attached. In a properly shaped hook the centre of gravity of the load should be suspended immediately beneath the centre of the swivel. See also Ram's Horn.

Crane Jib.—See Jib.

Crane Ladle.—A foundry ladle (q.v.) of the largest type, and which is therefore slung in the crane. It may hold from four or five to twelve or fifteen tons of metal, according to the requirements of the foundry.

Crane Post.—The pillar upon which the stresses due to the load on a crane are mainly concentrated. At its head it sustains the tension of the ties and at the ground line (q.v.) the thrust of the jib. Its strains are therefore those of a cantilever tending to break off across the ground line. Hence it is the custom to make the longitudinal outline of a post conical or parabolic in form, tapering from the ground line upwards. All types of cranes are not alike provided with posts, crabs and overhead travelling cranes of various types being the exceptions.

Crank.—A lever which rotates about the axis of a shaft. The usual form of engine crank is that which consists of a web of metal stiffened with a rib or ribs, a boss to receive the crank shaft (q.v.), and a smaller boss to take the crank pin (q.v.). Often the crank is a simple bending in the crank shaft, or axle. Sometimes a crank is circular, in which case it is termed a disc crank (q.v.).

Crank Axle.—The axle of a locomotive upon which the cranks are formed.

Crank Circle.—The circle described by the crank pin. Specially the circle in a valve diagram (q.v.) which represents the path of the crank. See Crank Path.

Crank Disc.—See Disc Crank.

Crank Handle.—A small lever having a handle at one end and a boss at the other, the latter being made to fit over the squared ends of spindles, so that they may be rotated by the turning of the handle.

Cranking.—In cutting tools, cranking signifies the hollowing or curving of the tool immediately behind the cutting edge. It is performed in order that the tool shall have a tendency to spring back and out, rather than hitch into the work under the stress of a heavy cut, and also to give a body of metal for grinding, to prevent the need of frequent reforging. In principle the cutting edge should not stand higher than the face of the tool shank.

Crank Path.—The circle described by the crank pin (q.v.).

Crank Pin.—The pin or movable journal which unites the connecting-rod of an engine or pump with the crank.

Crank Plate.—See Disc Crank.

Crank Shaft.—The shaft upon which the main boss of a crank is keyed, and which, therefore, receives and transmits the circular motion of the crank.

Crank Web.—The central plated portion of an ordinary cast-iron crank, which contains the bosses and is stiffened with a rib or ribs.

Crank Wrist.—A crank pin (q.v.).

Creaking.—The peculiar sound emitted by a bar of tin when bent alternately backwards and forwards.

Creaser.—A fuller (q.v.).

Creases.—A term applied to the moulding tools of the coppersmith which fulfil the same purpose as the top and bottom tools (q.v.) of the smith.

Creasing Tool.—A tool used by coppersmiths for making beads and tubes. It is of a T-shape, having the horizontal portion furnished with grooves of various sizes.

Creeping.—The very slight loss of speed which results when drums are driven by rope gearing, due to the slipping of the rope. This, though almost infinitesimal in amount, is sufficient to cause a slight difference in the number of revolutions performed by the driving and driven drums when a considerable space of time—an hour, for example, is taken into account. In some instances, as in the case of belting, creeping is probably also due to the elasticity of the belt itself, which stretches on the tension side and contracts on the slack side, so causing a slight amount of loss.

Creosoting.—The injection of timber which has to be exposed to atmospheric influences, with creosote, in order to increase its durability. The timber is first deprived of its moisture, which is then replaced with creosote. The durability of the wood is enhanced thereby fourfold, or more.

Creosoting Cylinder.—A strong wrought-iron cylinder, in which railway timbers, sleepers, &c., are exhausted of their moisture *in vacuo*, and saturated with creosote pumped in under pressure.

Creosoting Plant.—The appliances used in the saturation of timber with creosote, consisting of cylindrical receivers, vacuum and pressure pipes and connections.

Crimson Lake.—The colour used to distinguish steel in sectional drawings.

Crippling.—When the elasticity of a beam or structure is destroyed so that it becomes permanently and excessively bent and wrinkled, without, however, undergoing actual fracture, it is said to be crippled, or buckled.

Crocodile Shears.—See Alligator Shears.

Crocodile Squeezer.—See Alligator Squeezer.

Crook Bit Tongs.—Smith's tongs whose jaws are bent round at right angles with the handles, one jaw being furnished with a bit or nib to prevent the slipping of the work. These tongs allow rods to be held and to be passed alongside of and parallel with the handles.

Crop Ends.—The ends cut off from rails after they leave the rolling mill.

Cropping.—(1) The cutting or shearing off of puddled bar into lengths suitable for piling (q.v.). (2) The sawing off of the rough ends of bars and rails.

Cropping Machine.—A shearing machine used for the cutting off of ends and short lengths of iron bars.

Cropping Shears.—Alligator Shears (q.v.).

Cross.—(1) A four-armed beam used in foundries for lifting heavy moulding boxes and loam moulds about, the weights being suspended in links slung from the ends of the arms. (2) A wrought-iron pipe connection having four ways or openings arranged in the form of a cross, for connecting four pipes meeting at right angles.

Cross-bending, or Cross-bending Strain.—As usually understood, signifies the strain acting in a transverse direction on horizontal-engine beds, pump beds, and base plates, due to the thrusts of the rods. To reduce this to a minimum the centres of the cylinders and rods are kept as low down as possible, lessening the leverage thereby.

Cross-bracing.—See Counterbracing.

Cross-breaking.—The actual fracture of a beam or structure due to tensile forces operating thereon.

Cross-cut Chisel.—A chipping or cold chisel, which is very narrow, or from $\frac{1}{8}$ inch to $\frac{1}{4}$ inch wide. It is used for cutting grooves in various directions across a plane surface, in order to facilitate the work of subsequent chipping.

Cross-cut Saw.—A saw provided with two handles, one at each end, and used for cutting heavy timber—as deals, balks, &c.—across the grain. The teeth are of the shape of an equilateral triangle, so that the thrust is given from each end of the saw alternately, the cutting being equally efficient in both directions.

Cross-cutting.—The removal of material in a transverse direction. Commonly applied to the action of a chisel or a saw.

Crossed-arm Governor.—A governor in which the points of suspension of the rods from which the balls depend are on the opposite sides of the central axis around which they revolve. See Parabolic Governor.

Crossed Belt.—A driving-belt which undergoes a twist between the driving and driven pulleys, passing from the upper side of one to the lower side of the other. With a crossed belt the pulleys revolve in opposite directions.

Crossed Rods.—When the eccentric rods of reversing-engines cross each other on their way to join the ends of the slot-link, the centres of the sheaves being between the axle and the link, this term is applied to them; when they do not cross, the rods are said to be open.

Cross Girder.—Any transverse girder or beam which unites longitudinal girders, side frames, or standards together, and becomes at the same time a distance-piece, stretcher, or strengthening beam, or the support also, or base, for machinery. Examples occur in the cross or central girders of steam cranes.

Cross Grain.—Signifies a section of timber taken at a low angle with the direction of the longitudinal growth of the fibres. Or when cutting action takes place in such a direction as to produce a tendency to tear up the grain; as, for instance, when the front of a plane moves in the same direction as the downward slope of the grain fibres, the tool is said to work against the grain, or cross-grain. The term cross-grain or cross-grained also alludes to a curly or interlacing condition of fibres, more noticeable in hard than in soft woods, so much so in many specimens that an ordinary plane will tear up the fibres when applied in any direction.

Cross Head.—(1) That portion of an engine or pump which unites the piston and connecting rods and slide-blocks together. (2) The upper or transverse beam of an hydraulic press, which transmits the pressure to the weight to be lifted. (3) A cast-iron plate attached to the top of the plunger of an accumulator, and to which the weight case is fastened. (4) The end of a wheel press (q.v.).

Crossings.—Gaps in the rails of permanent way, through which the flanges of wheels can pass where one line of rails crosses another.

Cross Joint.—See Breaking Joint.

Cross-over Road.—A short diagonal line of rails on permanent way, provided with a pair of points or switches at each end, and connecting two parallel lines of rails together.

Cross Pane.—See Pane.

Cross Section.—A transverse section, always understood, unless otherwise noted, to be at right angles with the longitudinal axis of the piece of work or drawing through which the section is taken.

Cross Slide.—The horizontal slide or bridge which carries the tool box of a metal planing machine. The tool box is traversed across it by means of a horizontal screw, and the slide is elevated and depressed with vertical screws actuated by bevel-wheels.

Cross Tail.—In a side-lever engine (q.v.) the rod which unites the side levers with the connecting-rod.

Cross Tubes.—The heating tubes in a steam boiler—usually applied to boilers of the vertical type. The tubes pass through the fire box, the fire therefore surrounding them, and assist in maintaining a rapid circulation of the water. A mud door is placed opposite the end of each tube, so that it may be cleaned readily.

Cross Tube Boiler.—An ordinary vertical boiler (q.v.) provided with cross tubes (q.v.).

Crow.—See Crow Bar.

Crow Bar, Crow, or Pinch Bar.—A round iron bar flattened to a chisel-like expansion at one end, and used for raising a heavy weight through a short distance, for pinching trollies and wagons along a line of metals through a limited distance, and generally for the application of a large leverage for a temporary purpose.

Crown of Boiler.—See Boiler Crown.

Crown Plate.—The commonest quality of wrought-iron plate manufactured.

Crown Wheel.—A bevel-wheel upon a vertical shaft, having its teeth facing uppermost.

Crow's Feet, or Ticks.—The angular lines < > which are used in drawing, to indicate the points between which a dimension is given.

Crucible.—(1) Crucibles are deep cup-shaped vessels used for melting brass, steel, and other metals. They are manufactured of fire clay, or of plumbago. The clay is ground, sifted, and mixed with water, and with a certain proportion of burnt clay obtained from old crucibles, and kneaded and wrought into shape. They are either wrought into shape by hand, or pressed into moulds made of plaster of Paris, which absorbs the water from the surface of the crucible. Afterwards they are dried and heated up to redness. Crucibles should stand the strongest heat to which they may be subjected without fusion, or cracking, or much wasting away; they should also withstand the corrosive action of the materials melted in them, and resist the pressure of the tongs and the weight of the contained metal. The nature of the constituents employed in their construction varies therefore with the materials which they are to melt. (2) The lower portion of a blast furnace, into which the reduced metal sinks, is called the crucible, or the throat.

Crucible Cast Steel.—Ordinary cast steel (q.v.), as distinguished from Bessemer and other mild steels.

Crucible Tongs.—Tongs used by brass and steel founders for the lifting of crucibles out of the melting furnace. They embrace the sides of the crucible, and the handles are clamped with a coupler (q.v.).

Crude Oil.—Oil newly expressed, and unpurified.

Crushing Strain.—The strain necessary to cause the failure of a material by compression only. With rigid metals (q.v.), as cast iron and steel, and short specimens, the crushing strain produces shearing at an angle. With flowing metals (q.v.) the specimens give way by bulging.

Crushing Strength.—The crushing strength of a material is assumed to be equal to the weight which will just crush a prism of one inch square, and from one, or one and a half, to four or five diameters, in height only.

Crutch.—The cross, or transverse arm of a foundry ladle.

Crutch-handle Shovel.—An ordinary foundry shovel, the head of whose handle is in the form of a letter T, or like the head of a crutch.

Crystallization of Iron.—This is affected by the conditions under which it is cooled. If cooled rapidly against a cold metallic surface it becomes chilled, and the crystals are long and needle-like. If cooled slowly the crystals are large and the grain is coarse. Crystals which are near the surface always arrange themselves at right angles to the surface. Crystals near the surface are always smaller than those nearer the central portions. The crystals of graphite mingled among those of the iron are also affected by the conditions of cooling, remaining uncombined in metal cooled slowly, but entering into chemical combination when cooled rapidly.

Cube.—(1) A square prism, bounded by six equal faces, perpendicular to each other. (2) The cube of a number is the product of the number multiplied twice into itself.

Cube Root.—The cube root of a number is the number which multiplied twice into itself would produce its cube.

Cup Chuck, or **Bell Chuck.**—A hollow cylindrical chuck, made in various modifications and sizes for turning work. Cup chucks for metal turning are provided with set-screws for pinching rods and bars during turning. For wood turning the timber is driven into the cylindrical hole, and holds by its own friction. Small chucks are made of brass, large ones of iron.

Cup-head Bolt.—A bolt having a hemispherical or cup-shaped head. It is used chiefly on timber work, and is prevented from turning either by a lip underneath the head, or a square-shouldered shank.

Cup-head Rivet.—See Snap-head Rivet.

Cup Leather.—A leather packing used for the pistons of hydraulic machines. Its cross section between the sides of the plunger and the bore of the cylinder is U-shaped, so that the sides of the leather are pressed both against the sides of the plunger and of the cylinder with increasing force as the pressure of the water increases. Cup leathers and hat leathers (q.v.) are made by being pressed into moulds of cast iron. See Cup Leather Press.

Cup Leather Press.—The mould in which cup leathers (q.v.) and hat leathers (q.v.) are moulded into shape. It consists of two blocks, an upper and a lower, of cast iron turned truly, the space between the two blocks corresponding with the intended sectional shape of the leather. The latter is softened in warm water, and squeezed between the two halves of the mould, by means of a central bolt, and allowed to harden in place.

Cup Nut.—See Box Nut.

Cupola.—The furnace in which cast iron is melted for foundry use. It is a tall cylindrical hollow structure, circular in plan, made of wrought-iron plates riveted together, and lined with fire-brick. It is provided with charging hole, breast hole, tapping hole, tuyeres, and hearth. Cupolas vary in dimensions with foundry requirements, or from two to four and five feet in diameter, the height being from five to six diameters.

Cupola Blast Furnace.—A term applied to the lighter blast furnaces constructed in recent years, as distinguished from the older and more solid forms. Cupola blast furnaces are built of wrought iron cased with brickwork.

Cupola Fan.—An ordinary fan provided with radial vanes, used for supplying the tuyeres of a cupola with air.

Cup Shake.—A shake in a balk of timber which is circumferential, separating the annual rings.

Curb, or Curb Ring.—An internal ring of teeth used for turning or slewing cranes around. A pinion revolving in a fixed bearing on the upper portion of the crane, is, by suitable gearing, made to travel round the curb ring, carrying the crane along with it.

Curb Ring.—See Curb.

Curved Roof.—A roof arched in end view. Curved roofs are usually covered with corrugated iron.

Curve of Expansion.—The curve which represents the expansion of a gas or vapour under diminished pressure.

Curvilinear.—A line or a path of motion which is curved.

Curving of Patterns.—Foundry patterns which are long, narrow, and ribbed are curved in the direction of their length to compensate for the curving of their castings, due to the unequal cooling produced by unequal distribution of metal. The direction and amount given to the curving is always the reverse of that which is expected to take place in the casting. Its direction and amount is solely a matter of experience.

Cushioning.—A piston is said to be cushioned when the valve is opened to lead. Cushioning takes place just before the end of the stroke and the lead may average $\frac{1}{16}$ in. in engines running at moderate speeds.

Cut.—The removal of a shaving from a piece of work in the lathe, or in planing, boring, and similar machines. Thus, it is common to speak of the breadth, length, and depth of a cut. Iron turners and machinists speak of the progress of an automatically cutting tool as the cut; thus the cut is said to be "on," or "off," according as the tool is cutting or not cutting.

Cut-off.—The termination of the period of admission of steam into an engine cylinder. Cut-off took place in the older engines only at the termination of the stroke, but is now effected at fractional parts of it, as $\frac{1}{3}$, $\frac{2}{3}$, $\frac{3}{4}$, &c. The point of cut-off is regulated by the amount of lap on the slide valve.

Cut-off Plate.—A cut-off valve (q.v.).

Cut-off Valve, or Cut-off Plate.—The second valve in the arrangement for working steam expansively by means of a separate slide. The cut-off valve is usually a plate provided with cross bars and ports, or is solid.

Cut Sprigs, or Cut Brads.—Used by moulders for mending up broken sand, the sprigs binding the sand together. They are also used for thrusting into the thinner and weaker portions of moulds to support the sand.

Cutter.—(1) Any cutting tool fixed in a machine, for the automatic cutting of wood or metal. The angles of cutting tools and their shapes vary with the nature of the materials to be operated upon. Cutters are held fast in cutter blocks, tool boxes, and cutter bars. (2) Sometimes applied to a cottar (q.v.).

Cutter Bar.—See Tool Holder.

Cutter Block.—That portion of a wood-working machine which carries the cutter or tool. The cutter block may be rectangular, or circular, or discoid, in shape, according to the class of machine to which it belongs. Cutters are almost invariably held in place with screws in preference to wedges.

Cutter Head.—A boring head (q.v.).

Cutter Holder.—A tool holder (q.v.).

Cutter Pin Drill.—A modified pin drill (q.v.) in which the pin becomes the guide for a cutter, the extent of whose radius is independent of the size of the body of the drill itself. That part of the body immediately above the pin is swelled into a socket slotted through transversely for the reception of the cutter, whose shank is cottared into the slot, and whose cutting edge turns down at right angles therewith. A hole being drilled for the pin in the first place, the cutter describes a circle around this, and removes a ring of metal equal in radius to the radius of the cutter. Cutter pin drills are used for holes of two or three inches and upwards in diameter, and are particularly useful for making the holes in iron and steel boiler-plates for the reception of the tubes.

Cutting Angle.—See Angles of Cutting Tools.

Cutting Blast.—A cupola blast of too high velocity for the nature of the fuel and charge. A cutting blast will blow away much of the fuel. The pressure should seldom exceed 2 inches of mercury.

Cutting Edge.—The edge of a cutting tool against which the division of the shaving takes place. On the proper formation of the edge depends the efficiency of the instrument, and the angles which produce the edge vary with the material to be cut. See Angles of Cutting Tools.

Cutting Face.—That face of a cutting tool against which the material is removed, and along which it curls or is thrust aside. Commonly the cutting face is the top face.

Cutting Gauge.—A gauge furnished with a narrow knife-like cutter in place of the ordinary marking point, and used for cutting off narrow parallel strips of thin stuff.

Cutting Nippers.—A pincer-like tool, having sharp edges of hardened steel, and used for cutting off wire of small gauge.

Cutting-off Machine.—A machine used for cutting off lengths of bar iron in the smithy, by which the labour of marking off, nicking, breaking, and squaring up is saved. The iron is passed through a hollow mandrel, is stopped by a gauge plate, and is parted by cutters.

Cutting Pliers.—Pliers (q.v.) which, in addition to the flat and roughened jaws, are furnished with a pair of sharp nippers placed to one side for cutting off wire.

Cutting Tools.—Tools which shear, or scrape, or cut shavings from materials, in opposition to those which abrade or grind away. Thus chisels and gouges, planes, turning tools in all their forms and modifications may be taken as types of cutting tools.

Cutting Tools (Speed of).—The speeds of cutting tools vary within extremely wide limits, being quickest for wood-working machinery and slowest for the hardest metals; thus bearing out the mechanical axiom, that speed and power are in inverse ratios to each other.

Cutting-up.—The gashing of the broken edges or faces of a sand mould preparatory to adding fresh sand for mending-up.

Cyanogen.—A compound of carbon with nitrogen, which is believed to play an important part in the transference of carbon to malleable iron in the process of steel making and in the case-hardening of iron.

Cycle.—A cycle of operations signifies the complete circle of work gone through by an engine in one complete revolution, embracing the admission of steam, its expansion and condensation.

Cycloid.—A curve formed by a point in the circumference of a generating circle, rolling upon another circle as a base line.

Cycloidal Teeth.—Wheel teeth whose curves are cycloidal in form.

Cylinder.—(1) A solid whose ends are circles and whose cross sections at any intermediate points are also circles. It may be conceived to be formed by the rotation of a line around the circular ends. (2) In engine work the closed receiver in which the power of the steam, or gas, or water, is made to actuate a disc or piston.

Cylinder Bit, or Half-round Bit.—A boring tool used for metal or wood, whose section at the cutting face is that of a semicircle. Its cutting face is sloped at an angle of about 4°. It is employed for the boring of very long holes where much accuracy is essential. It is entered first into a shallow hole prepared for its reception with a gouge or a drill. Also termed a D-bit.

Cylinder Bore.—The internal diameter of an engine cylinder when finished in readiness to receive the piston. The bore and the length of a piston stroke are among the necessary factors for estimating the horse-power of an engine.

Cylinder Cock.—See Pet Cock.

Cylinder Covers.—The covers or ends of an engine cylinder made separately from the body of the cylinder itself, for convenience of casting, boring, repairs, and examination of the latter. Sometimes the bottom end is cast solid with the cylinder, and the top or front cover only bolted on. The top cover is always furnished with the necessary stuffing box and gland for the piston rod. In large engines the covers are dished or buckled, the better to withstand the enormous steam pressure. Covers are attached to the cylinder flanges by bolts or studs.

Cylinder Escape Valve.—A valve supported by a spiral spring, attached to the ends of marine cylinders to permit of the escape of the water of condensation, and of priming. It is enclosed in a cap of metal, to protect the engineer and attendants from becoming scalded in case of a sudden blow-off of the hot water. Also termed Relief Valve.

Cylinder Flanges.—The flanges cast upon an engine cylinder to which the covers are bolted. They are always external, except in some few cylinders of special construction, where a bottom flange is internal and a top flange external.

Cylinder Lubrication.—Engine cylinders are too often lubricated with tallow of inferior quality, by whose employment destructive corrosion is set up. Animal fats and oils are not suitable for cylinders, but specially prepared hydro-carbon oils should be used. See Oils.

Cylinder Lubricator.—It is necessary that the pistons of engine cylinders should be freely lubricated, and various lubricators are employed for the purpose, both single and compound tallow-cups, and self-acting lubricators and impermeators (q.v.).

Cylinder Mandrel.—The mandrels of all lathe poppets of good workman-ship are made hollow, so that the traversing screw is contained therein and travels within the body. Hence termed Cylinder Mandrels.

Cylinder Metal.—Strong or slippery iron (q.v.) used specially for steam cylinders and liners. Blænavon, Madeley Wood, and hæmatite iron, mixed with selected scrap in varying proportions, make good cylinder metal.

Cylinder Oil.—Oil prepared specially for the lubrication of the pistons of engine cylinders, in preference to tallow, which decomposes, forming fatty acids that corrode the iron.

Cylinder Passages.—See Port.

Cylindrical Gauge.—A gauge composed of two pieces, a plug gauge or solid cylinder furnished with a handle, and a collar gauge or hollow cylinder into which the plug gauge fits. These gauges are used as templets for boring and turning parts of machines which are required to correspond in dimensions, the plug gauge being inserted into a bored hole and the collar gauge being slid over the spindle or shaft which is required to fit the bored hole. The use of gauges is not subject to the errors incidental to the handling of calipers. See Difference Gauge.

Cylindrical Mouthpiece.—The passage of effluent liquid through short parallel tubes produces less contraction than in a thin plate (see Vena Contracta), the coefficient of discharge for different tubes vary-ing with the length. Thus the coefficient for a tube whose bore is $\frac{1}{4}$ or $\frac{1}{3}$ the length is ·81, where the bore is $\frac{1}{4}$ to $\frac{1}{12}$ the length ·77, and from $\frac{1}{16}$ to $\frac{1}{18}$, ·68. If the tubes are not parallel the coefficients are different: for a converging mouthpiece, the angle of whose sides is 13$\frac{1}{2}$, the coefficient for the narrow end is ·94. With a diverging mouth-piece, angle 5°, the coefficient for the narrow end is ·92, for the broad end ·55.

D.

Dabbing On.—Specially applied to the making of some forms of butt joints (q.v). Thus the spoke ends of railway wheels are dabbed on.

Damper.—A plate, valve, cover, or other suitable contrivance for regu-lating the amount of draught in a boiler or furnace flue.

Damper Weight.—A weight used to counterbalance that of the damper of a steam boiler in order to render it easy of adjustment.

Damping.—See Watering.

Damping Down.—Slacking down (q.v.).

Damp Steam.—Wet steam (q.v.).

Dam Plate.—A cast-iron plate which forms the backing or support for the dam stone (q.v.).

Dam Stone.—A stone occupying one side of the hearth of a blast furnace over which the slag flows, and through which the molten metal is tapped.

Dangerous Section.—That section of a bar at which the bending moment (q.v.) is greatest.

Dash Piston.—The piston sliding in the dash pot (q.v.) of a Corliss engine, whose release by a spring closes the steam or exhaust valves, and whose concussion is deadened by the springs or buffers enclosed in the dash pot.

Dash Plates, or Division Plates.—Plates sometimes fixed in a marine boiler in line with, and over, the tubes, in order to prevent the crown plates of the fire-box from being exposed to the action of the flame when the ship heels over heavily. But for these plates the water in the boiler would run to the lee side, and leave the plates temporarily bare.

Dash Pot.—A small cylinder forming part of Corliss-engines and similar types, which functions as a buffer case to the sharp closing of the steam and exhaust valves. It is furnished with springs of steel or india rubber, which deaden the blow of the dash piston (q.v.).

Datum Line.—Any base or fundamental line from which dimensions are taken, or graphic calculations made.

Day Work.—Work done without any definite price being given for quantity, the workmen being occupied only so many hours at it. Day work is chiefly done in pattern shops and in foundries, and in the other departments on jobbing work, and work which will not admit of accurate estimate of cost, or which is required of first-class excellence. But in foundries and other shops where machine work is done, day work is not the rule. See Piece Work.

D-Bit.—A cylinder bit (q.v.).

Dead.—(1) A common term of emphasis, as dead level, dead square, dead true, &c. (2) Molten metal is said to be dead when it pours thick, sluggish, and viscid, due to insufficient melting, or to having remained too long in the ladle. Dead metal is liable to produce cold shuts (q.v.), though for heavy classes of work it must not be poured too hot. (3) Timber is dead when the woody walls of the vessels become partially decomposed, turning brown and friable. Alternate exposure to dryness and moisture is the most fruitful source of dry-rot.

Dead Axle.—An axle which is not a driving axle (q.v.).

Dead Blow.—(1) A blow which is not of an elastic character. To strike a dead blow the handle of the hammer must be grasped firmly. (2) Indicates the nature of the blow given by those steam hammers in which no steam is introduced between the piston and the cylinder end for cushioning. See Elastic Blow.

Dead Centre Lathe.—A lathe whose front point centre does not revolve with a mandrel, as is the case with ordinary lathes. The work alone is revolved between dead centres. Dead centre lathes are only used for work requiring such accuracy that the revolution of the mandrel would be objectionable.

Dead Centres, or Dead Points.—(1) In a reciprocating engine or pump, when the axis of the piston-rod, the crank shaft, and the crank pin are all in line, the engine or pump is said to be on dead centres. Hence an engine on dead centres cannot be started by the steam pressure alone, but requires extraneous help to turn it partly round, and start it on its rotation. In a double engine having cranks at right angles, if one is on dead centres the other is in the best position for starting, so that, practically, there can be no dead centres in the case of such an engine. (2) The plain conical centres of the headstock and poppet mandrels of a lathe. These centres are used when long cylindrical work is being turned, the ends being centre punched and slightly drilled to receive the centre points. The work is then revolved by a carrier and driver.

Dead Eyes.—Bearings without any line of division or jointure, such as exists in a plummer block ; the hole being bored through solid metal. A dead eye answers the same purpose as a plummer block, but is capable of no adjustment for wear, except by rebushing, and is consequently used for rougher and more temporary kinds of work.

Dead Head.—(1) Head metal (q.v.). (2) The metal in the runner (q.v.) of a mould. (3) The feeding head (q.v.). (4) The poppet (q.v.) of a lathe.

Dead Level.—Simply used in the sense of being perfectly or quite level. A term of emphasis.

Dead Load.—A load whose pressure is steady and invariable, neither being removed, diminished, nor increased at any time. Structures subject to a dead load undergo less stress than those on which the load is variable (see Live Load) and therefore a lower factor of safety (q.v.) suffices.

Dead Load Safety-valve, or Cowburn Valve.—A type of valve which is loaded without the interposition of a lever or spring, the load consisting of annular weights which are dropped over a cylindrical seating. The weight on the valve is increased or diminished according to the number of rings employed.

Dead Melting.—The perfect fusion of metal, obtained by allowing it to remain in the furnace in the fluid state for some little time previous to withdrawal, and so increasing its temperature some 200° or 300° above the melting point.

Dead Parallel File.—See Parallel File.

Dead Plate.—The cast-iron plate immediately within the furnace door of an engine boiler, which receives the fuel before it is passed onwards to the grate surface. The function of the dead plate is the caking and partial coking of the coal.

Dead Points.—See Dead Centres.

Dead Pouring.—The pouring of metal which has been allowed to thicken or become dead (q.v.). It is adopted with the heavier castings.

Dead Size.—The exact or precise size or length, that is, exclusive of any allowances for contingencies or for machining. A finished dimension, or the opposite of a rough dimension.

Dead-smooth File.—The finest-cut file made. A 12-in. file of this description would contain 88 lines of teeth to the linear inch.

Dead Spindle.—The mandrel of the poppet (q.v.) of a lathe, as distinguished from the live spindle (q.v.).

Dead Water.—The water which lies below the heating surface in a steam boiler, and where circulation is extremely sluggish. In Cornish and Lancashire boilers the flues are brought forwards under the bottom to heat the dead water.

Dead-weight Safety Valve.—See Cowburn Valve.

Dead-weight Test.—A test to which rails, bars, and machinery are subjected, the load being a dead load (q.v.).

Dead Wheel.—In an epicyclic train (q.v.) the wheel around whose centre the remainder of the train revolves is called the dead wheel.

Deal Frame.—A reciprocating or frame sawing-machine used for the cutting of deals and boards. It is smaller than the similar frame saw for cutting logs. The driving or feed rollers also revolve in a vertical instead of in a horizontal axis.

Deals.—Sawn pine timber measuring usually not less than 9 × 3 and not more than 12 × 3 in cross section. Timber exceeding 11 in. in width is often conventionally termed deals, but properly comes under the head of planking.

Decagramme.—A French measure of weight containing 10 grammes (q.v.), and equivalent to 154·34 English grains.

Decalitre.—A French measure of capacity containing 10 litres (q.v.), or 2·201 English gallons.

Decametre.—Ten French metres (q.v.), corresponding with 393·7079 English inches.

Decarburization.—The removal of carbon from combination with metals in the processes of the manufacture of malleable iron in the puddling furnace, and of steel in the Bessemer converter. (See Carburization.) The decarburization of malleable iron castings is effected by the presence of a cementing material, as red hæmatite iron ore.

Decigramme.—A French measure of weight, being the tenth part of a gramme (q.v.), and containing 1·5434 English grains.

Decilitre.—A French measure of capacity, being one-tenth of a litre (q.v.), and corresponding with ·1761 of an English pint.

Decimal.—See Fraction.

Decimal Equivalent.—A decimal number,—integer and fraction combined, or fraction alone, which corresponds in value with an integer and vulgar fraction, or vulgar fraction alone. It is useful for facilitating calculations involving the assistance of fractions.

Decimal Gauge.—See Wire Gauges.

Decimal Pitches.—Millimetre Pitches (q.v.).

Decimetre—A French measure of length, being the tenth part of a metre (q.v.), and equivalent to 3·93707 English inches.

Deck Crane.—A fixed crane used on board ship. The central part of the post is fixed tightly in a plate which is bolted to the deck, and its lower end is dropped into a step or toe plate, fastened to a lower deck. The gearing, side frames, and jib of the crane are made to slew around the post.

Deck Winch.—A steam winch (q.v.) bolted to the deck of a ship for hoisting or lowering goods from, or into the hold, and generally for doing work whereby manual power can be saved.

Deep-well Pump.—A pump specially constructed for deep-well operations, the term having reference to the nature of the pump connections —as strainers, retaining valves, &c.—rather than to any particular type of pump.

Deflection.—(1) The bending of a beam or structure under an applied load, and which may be increased until the breaking strain is reached. The deflection may be small in amount, and the beam or structure return to its original form, or it may increase and accumulate, becoming permanently set (q.v.), or it may go farther and lead to fracture of the beam. (2) The movement of the needle or hand of a gauge or recording instrument.

Deflector.—This in a general sense denotes a bridge, or a bell, or trumpet-mouthed opening, used for deflecting hot gases or liquids out of a straight course, and having its applications in boiler construction.

Deformation.—The alteration in form which a structure undergoes when subjected to the action of a load, the load producing either an elastic strain (q.v.), or a permanent set (q.v.).

Deliver.—A term signifying the manner in which a pattern leaves or lifts from the sand. **If the mould remains unbroken it is said to be a good delivery, and the contrary when it tears the sand in the act of withdrawal. See Taper.**

Delivery.—See Deliver.

Delivery Box.—The upper or delivery chamber of a series of two or three throw pumps, into which the liquid is lifted by the pistons and from which it is delivered.

Delivery Pipes.—The series of pipes through which the liquids drawn up by pumping machinery are ejected.

Delivery Valve, or Discharge Valve.—The pump valve through which the pump contents are ejected into the delivery pipes.

Demy Drawing Paper.—The smallest size made in sheets, measuring 20 in. by $15\frac{1}{2}$ in.

Density.—Density has reference to the amount of matter contained in a body. Density and specific gravity are proportional. Specific density of a body is its density estimated in relation to that of another body. Platinum is the densest, hydrogen the rarest, element.

Deoxidation.—The removal of oxygen from metallic oxides, effected during the processes of calcination, smelting, puddling, and converting, the gaseous products being carbonic oxide and carbonic acid.

Dephosphorization.—The removal of phosphorus from combination with iron and steel. In the blast furnace it is scarcely eliminated at all ; in the puddling furnace it is removed by the basic materials used for fettling, combined with exposure to the action of the air ; in the Bessemer converter by the basic lining.

Dephosphorization Process.—The basic process (q.v.).

Deposit.—See Incrustation.

Depreciation.—The loss in value which machinery sustains with the lapse of time, which amount has to be written off the prime cost annually. The loss due to wear and tear is added to this. The total amount is usually taken at 6 per cent. for engines, 10 for boilers, 8 for machines, 5 or 6 for millwork and gearing, 45 for belting.

Derrick.—A form of crane in which the radius of the jib is rendered capable of alteration by means of chains or guys passing over the top of the mast. Derrick cranes proper are fixed cranes (q.v.), but portable cranes, both hand and steam, are provided with movable jibs, whose mechanism for derricking is made in various types.

Derrick Chains.—The chains by which the jib of a derrick crane is raised or lowered. They are wound and unwound off a barrel called the derrick barrel, the engine power being the motive power used.

Derrick Gear.—The arrangement of chains, barrel, and worm gearing by which the jib of a derrick crane is raised and lowered.

Derricking.—The act of raising or lowering the jib of a derrick crane.

Design.—(1) The drawing out of a machine or structure. (2) The working out of mechanical ideas. The art of designing is not so original as some would suppose, since few engineers dare to strike out new and untried paths. The study of new designs and the records of the Patent Office show that a design is in most cases the utilisation of previous experiences, and the application of old principles to new modifications in practice.

Desulphurization.—The removal of sulphur from combination with iron and steel. It is largely removed during the calcination of the ore ; its elimination in the blast furnace depends on the nature of the flux. With an excess of lime a portion of the sulphur unites with the slag ;

if silica be largely present, the sulphur, instead of uniting with the lime, will remain in the pig. A high temperature is also favourable to the elimination of sulphur, so that the grey irons which are produced at a higher temperature than the white irons contain less sulphur than the latter. In the puddling furnace sulphur is removed by the oxygen in the fettling, in the Bessemer converter by the oxygen in the air which is blown through.

Detail Drawing.—One which is not a general drawing (q.v.), but embraces some portion or section only of a machine.

Determinable Quantity.—A quantity whose value or amount can be determined or fixed from other quantities already known. Thus, the temperature and pressure of any given vapour being known, its volume can be determined, and is therefore a determinate quantity.

Detruding Action.—When a hole is punched in a plate the action is that of detrusion, and the plate is strained more than if the hole were drilled.

Development.—(1) A drawing is said to be developed when certain working details are drawn in full. Thus a propeller blade is developed when the various transverse sections are shown ; the section of a turbine is developed when the curves of the vanes are fully drawn ; a plate or templet is developed when it is so marked out that if cut to the developed lines and then bent, it will form the envelope of some definite geometrical figure, as a frustrum of a cone, a segment of a sphere, the knee of a pipe, &c. (2) Iron or steel nicked and bent suddenly shows a crystalline fracture, but when bent slowly the fracture is of a fibrous character. This latter mode of fracture is termed its development, and the fibres are said to be developed, or drawn out.

Devil.—A light iron lattice-work frame, either circular or rectangular in outline, filled with burning charcoal, and used in foundries for drying the surfaces of green sand moulds. It is either suspended in the mould, or laid upon iron rods placed over it.

Diagonal.—A straight line joining opposite angles in a triangular, quadrangular, or polygonal figure.

Diagonals.—Bracings placed at an angle across rectangular framings, to prevent alteration taking place in the shape of the frames. Hence the braces, or struts and ties of a lattice girder, are its diagonals.

Diagonal Scale.—A scale used for drawing purposes, which has, in addition to the primary divisions, one or both end divisions corresponding in length with the primary ones, divided into ten equal parts. Eleven equidistant and parallel lines are also drawn in a longitudinal direction, and diagonals drawn from the tenth division on one outer longitudinal to the ninth on the other, and so on in succession. With this scale one hundredth-parts of the primary divisions can be taken off.

Diagonal Stays.—Stays used for stiffening the end plates of steam boilers, in preference to, or to assist, the longitudinal stays. They reach from the inner face of the plate to that of the shell, in a diagonal direction.

Diagonal Winch.—A steam winch whose cylinders are placed diagonally on the side frames.

Diagram.—An outline drawing, or graphic construction, made for the purpose of illustrating or elucidating some problem in mechanical or arithmetical study.

Diagram of Work.—It is customary to estimate the work done on a body by a graphic method. A parallelogram is constructed and a curve enclosed therein represents the ratio between pressures and volumes.

The parallelogram is divided by crossed lines perpendicular to each other, termed respectively lines of pressure and lines of volume.

Dial Gauge.—(1) A dial having an index hand actuated by a curved or Bourdon tube, or a capsule through which pressure, or vacuum is indicated. (2) A sensitive measuring instrument or comparator. See Dict. of Modern Terms.

Diameter.—The diameter of a circle is a line passing through its centre, and uniting the circumference or sides.

Diametral Pitch.—A mode of expressing the pitch of toothed wheels in terms of their diameter. It is obtained by dividing the number of teeth by the pitch diameter. Thus the diametral pitch of a wheel 6 in. in diameter, and containing 48 teeth, would be 8. Conversely, the diametral pitch being given, the number of teeth divided by the pitch gives the diameter at the pitch line. It is the rule to give the sizes of change wheels for lathes in diametral pitch.

Diamond Point.—(1) A pointed tool used by pattern makers for turning the inner faces of work which, owing to their obliquity, or to their being bounded by cylindrical walls, could not be reached by ordinary chisels. The cutting edges of a diamond-pointed tool will stand at an angle of from 20° to 40° with the body, both edges being ground symmetrically, so that in plan they form the shape of a letter V. (2) A narrower tool but of similar shape is used by metal turners. (3) A tool in which the cutting angles are more obtuse is used by fitters and boiler makers for the purpose of cutting holes in boilers and other thin plates, and for nicking round pipes and tubes which have to be broken off. See also Diamond Tool, Dict. of Modern Terms.

Diaphragm.—A built-up distance piece of plate and angle-iron which unites two parallel girders together.

Die.—(1) A female or internal screw used for cutting outside screw threads. It is made of hardened steel, and fluted, to present two, three, or four separate cutting edges. The larger dies or those above ¼ in. in diameter, are made in halves and set in a die stock. The smaller dies are cut in a screw plate. Also termed screw dies, or screwing dies. (2) The movable block embraced by a slot link (q.v.). (3) An iron or steel mould used for the forging of eyes, bosses, tie-rod ends, truss-rod ends, forming sheet metal, etc. The die or matrix may be in one piece only, or it may comprise an upper and a lower portion, pinned or checked together. The work is generally formed in the dies under a steam or drop hammer or power press.

Die Box.—The box of a screwing machine by which the dies are embraced.

Die Chuck.—A lathe chuck furnished with steel jaws or dies sliding in grooves, and pinched or slackened with screws. The pinching jaws are Veed or bevelled, and one or all the jaws may be moved in some chucks at pleasure, so that work can be chucked either centrally or eccentrically. Die chucks are also self-centring, and usually contain two or three dies.

Die Forging.—The practice of forging repetition work in dies.

Die Stock, or Screw Stock.—The frame which encloses the dies used for screw cutting. It is furnished with lever handles for turning it in the circular direction, and with a set-screw for regulating the distance between the halves of the dies. Die stocks are made in various and well-known forms.

Difference Gauge.—A cylindrical gauge (q.v.) is sometimes furnished with two plug gauges to one collar gauge, one of the plugs being slightly larger in diameter than the other ; the difference between the two being just that which is sufficient to permit of easy movement, or of tight fitting of a mandrel in a cylinder.

Differential Motion.—The motion of one part to or from another, the two parts forming a mechanical combination. It is a motion whereby mechanical gain is obtained. The Chinese windlass, where two axles of unequal size move together, embodies the differential principle.

Differential Pulley Block.—A pulley block which owes its efficiency to the slight difference in the sizes of two sheaves fixed together on one axis. An endless chain embraces the two sheaves and a movable pulley below. The mechanical gain is proportional to the difference in the diameters of the two sheaves. A most valuable property of this pulley block is that, owing to the excessive amount of friction set up, the load cannot run down, but remains stationary in any position in which it happens to be left suspended.

Differential Screw.—Two screws of unequal pitch cut on the same spindle. On turning the spindle a nut on one of the screws moves by an amount equal to the difference in the pitches.

Differential Strains.—The strains due to a variable load.

Digger.—A term sometimes applied to a grab (q.v.) used for excavations.

Digging Shovel.—A round-mouthed or pointed-end shovel used in foundries for digging the sand in the floor. The square-ended and crutch-handled shovels are used for box filling and sand wetting.

Dimension.—A definite measurement shown on a mechanical drawing, as length, width, thickness. Dimensions are given on all working drawings, whether drawn to scale or not, but dimensions are not inserted in shaded and general drawings.

Dimension Lines.—Those lines upon an engineer's drawing which indicate to what parts or lines the dimensions figured have reference. They may not be drawn in black, but in blue, or sometimes in red ink.

Dimension Saws.—Applied to a couple of circular saws mounted radially from a common spindle, one for cross cutting, the other for ripping. The saws are moved upwards or downwards as required with a worm and worm-wheel.

Diminishing Socket.—A wrought-iron socket inserted between, and connecting two pipes of different bores.

Dinas Brick.—A refractory brick made from a highly siliceous clay found in the Vale of Neath, and used for the roofs of reverberatory furnaces.

Dip.—The amount by which the upper edge of a paddle-wheel blade is immersed beneath the surface of the water when the blade is in the vertical position.

Dip Crank Shaft.—A crank shaft in which the crank is formed by the simple bending of the bar, instead of being slotted out of the solid. It is stronger than the slotted form because of the continuity of the fibres.

Direct Acting.—An engine is said to be direct acting when the action of the piston is transmitted directly to the crank shaft. Nearly all engines are therefore direct acting.

Direct-acting Slide Valve.—A slide valve whose length of travel is the same in amount as that of the throw of the eccentric, as opposed to the indirect-acting slide valve (q.v.).

Direction of Force.—The line along which a force acts, whether it be productive of actual motion or of pressures or tensions only. Forces are assumed to act only in straight lines.

Direct Process.—(1) The method of producing malleable iron direct from the ores without preliminary smelting into pigs. (2) The process of making steel in the Bessemer converter by bringing the iron direct from the smelting furnace.

Disc Area.—See Screw Area.

Disc Crank, or Crank Disc, or Crank Plate.—A crank of circular outline in which the metal is so disposed that the varying motion of the connecting-rod is suitably balanced. The crank shaft is keyed into a boss in the centre of the disc, and the crank pin is riveted into a hole near the periphery. The half of the disc upon the side opposite to the crank pin is thickened up to act as a counterbalance to the rod. Its action is more regular therefore than that of a common crank.

Discharge.—The liquid ejected from a pump or from the vanes of a turbine (q.v.). See Vena Contracta, Cylindrical Mouthpiece, Nozzle.

Discharge Valve.—(1) A delivery valve (q.v.) (2) A self-acting valve attached to the side of a steamer, through which the water is discharged from the circulating pump, hence called side discharge valve.

Discharging Tube.—An adjutage (q.v.).

Disconnecting Engine.—A double engine, usually of the compound type, in which the cylinders can either be used in combination, or each separately from the other.

Disc Valve.—A modification of the flap valve (q.v.) used extensively in large pumps, and of which the air pump of a condensing engine furnishes a familiar type. It consists of an india-rubber disc hinged at the centre, resting on a perforated grid, through which the water obtains ingress, and lifts the valve. The disc is prevented from lifting too high by a superimposed perforated guard-plate dished or hollowed in the upward direction.

Disc Wheel.—A wheel having its central portion solid or plated instead of being provided with arms. The disc may be flat as in change wheels, and the lighter classes of toothed wheels generally, or it may be dished as in the wheels of trucks and trollies. The latter form is strong, because it will accommodate itself to the expansion and contraction due to changes in temperature, or to internal and to external stresses.

Disengaging Clutch.—A clutch whose function is the throwing of a line of shafting, or a train of wheels, into and out of gear.

Disengaging Coupling.—A clutch (q.v.).

Disengaging Gear.—The mechanism used for throwing clutches into and out of gear. It is usually either a lever or a hand wheel and screw.

Disengaging Nut.—The clasp nut of a lathe, being disengaged when not required for use or for screw cutting.

Dishing.—The hollowing and rounding of the surface of a disc, to render it more elastic or stronger.

Displacement.—The quantity of liquid discharged from a plunger pump at each double stroke is equal to the amount displaced by the plunger.

Displacement Cylinder.—An auxiliary cylinder belonging to some gas engines, by which the constituents of the charge are forced into the working or power cylinder.

Displacement Lubricator.—A lubricator which acts by the difference in the sp. gr. of oil and of water. An impermeator (q.v.) is one form of displacement lubricator.

Displacer Piston.—An auxiliary piston in some gas engines whose function is the expelling of the residual gases or products of combustion

from the cylinder. In other cases a flushing charge (q.v.) is used for the same object.

Distance Piece.—Any casting whose chief or only function is to maintain engine or machine parts at a fixed distance asunder.

Distress.—A beam or structure is in distress when it is subjected to undue or excessive stress, or to an amount exceeding the working stress.

Distributed Load.—When a load is spread evenly over the whole extent of surface of a beam or structure it is termed a distributed load. A girder will sustain a distributed load of twice the total weight which it would sustain if concentrated. See Concentrated Load.

Diverging Mouthpiece.—A mouthpiece for the discharge of liquid, whose longitudinal section is that of the frustrum of a cone, the smaller diameter being placed next its tank. See Cylindrical Mouthpiece.

Divided Axial Pitch.—In multiple threaded screws the axial pitch (q.v.) is divided into as many portions as there are threads, and the distance between any of these divisions is called the divided axial pitch, that is, where the centre of each successive helix cuts the axial line.

Divided Bearing.—Any bearing which is not solid, and which contains, therefore, provision for adjustment for wear. Bearings are commonly divided into two portions, but in some special cases into three.

Divided Normal Pitch.—The distances between the centres of the successive helices of many-threaded screws, measured perpendicularly to the threads. It is, therefore, less than the divided axial pitch (q.v.).

Divided Steam Chest.—A steam chest divided, either through the centre of the stuffing-box parallel with the cylinder bore, or obliquely, that is, at an angle with the bore. The object of dividing is for convenience of removing the cover for examination and repairs, without the necessity of removing the valve and valve rod—the rod in the first case passing through the joint of the divided box ; in the second, through the box, which is cast wholly with the cylinder.

Dividers.—A form of compass used for taking off or setting out minute and exact dimensions, being suited for more delicate adjustment than common compasses. These are of various kinds, and employed both in the drawing-office and in the workshops. See Spring Dividers.

Dividing Machine.—A machine, constructed in various forms, for the exact marking off of degrees, points of equal division, wheel teeth, &c.

Division Peg.—A peg of hardened steel, attached to an elastic rod called the index spring, and used in conjunction with a division plate (q.v.). The peg fits into any of the holes drilled in the plate, and holds it in steadily while operations of drilling or shaping are being performed.

Division Plate.—(1) A plate of brass fastened to the front of the driving pulley of a lathe headstock. It is pierced with numerous concentric circles of holes ; those in the best and most complete lathes numbering 360, 192, 144, 120, 112, 96, 60, 12. By means of these almost any number of divisions that are required can be obtained. A division peg (q.v.) sets the pulley in any position, while the work in the lathe is being operated upon by drills or cutters set in the cutter frame and driven from overhead gear. (2) The plate of a wheel-moulding, or of a wheel-cutting machine, through which the handle shaft passes, and which is notched for setting the handle at quarter, half, three-quarter, or full turns. See also Dash Plate.

Division Plate Wheel.—The change wheel on the handle shaft of a wheel-moulding or of a wheel-cutting machine.

Dog.—(1) One of the jaws or clips of a jaw chuck (q.v.). (2) The carrier (q.v.) of a lathe. (3) The catch which prevents the winch shaft of a crane from sliding along when the gears are changed. Called also Catch Paul. (4) A spike (q.v.).

Dog Chuck.—See Jaw Chuck.

Dogtail Trowel.—A small heart-shaped moulders' trowel, provided with a curved metal handle, the handle being imagined to resemble a dog's tail.

Dog Wheel.—See Ratchet Wheel.

Dog Wrench.—A form of spanner whose handle is turned around similarly to a winch handle.

Dolly.—A tool used by boiler-makers for holding under the heads of rivets during the act of riveting. It consists of a round iron bar, furnished with a head, flat or cup-shaped; the size, shape, and angle of inclination of the bar varying with the size and shape of the rivets, and their position relatively to the work. It is used for the same purpose as the holding-up hammer (q.v.).

Dolomite.—Magnesian limestone (q.v.).

Dome.—See Steam Dome.

Dome Cover.—The brass covering which encases the steam dome of a locomotive.

Donkey.—A donkey-engine (q.v.).

Donkey-Boiler.—A small boiler, usually of the vertical type, used for driving small engines.

Donkey-Engine.—An auxiliary engine used on board ship for pumping, or other light work, independently of the main engines.

Donkey-Pump.—The independent feed pump of a steam boiler driven directly by a small engine, the whole being attached to a single base-plate which is bolted to the boiler.

Double-Acting Engine.—An engine in which the steam acts upon both sides of the piston alternately, either against atmospheric pressure, or against a vacuum produced by condensation. All engines are now made double-acting, just as formerly all were made single-acting.

Double-Acting Piston.—A piston acted upon by gaseous or fluid pressure on both sides alternately. The pistons of engine cylinders and of double-acting pumps are of this type.

Double-Acting Steam Hammer.—A steam hammer which admits steam above the piston as well as below.

Double-Action.—Double action is in principle that of an alternate reciprocal pressure or movement. A series of actions equal and alternate in opposite directions.

Double-Action Pump.—One which throws water at each stroke, as distinguished from a lift pump.

Double-Armed Pulley.—A pulley which has two sets of arms. Pulleys over ten or twelve inches in width have commonly a double set of arms. They are set parallel with each other at a little distance within the outer edges of the rim which they support, and the boss may either be continuous between the arms, or a clear space may exist between the bosses of each set.

Double-beat Valve.—A ring-shaped lift valve having two seating faces, by which the steam is admitted on both sides at once, hence called an equilibrium valve. In our ordinary slide steam valve the steam pressure is unequal, because the area on the steam chest side is greater than the area open to the port side, and this pressure has to be overcome before the valve can be moved. The double beat being in equilibrium is more

easily moved. It is also termed a Cornish valve, because used extensively in the large pumping engines of that district.

Double Belting.—Belting formed of two thicknesses of leather. It varies from $\frac{1}{4}$ to $\frac{1}{2}$ an inch in thickness, and is employed for heavy driving, as for main driving pulleys, and the more powerful class of machines. It is made by cementing two thicknesses of leather together, and then sewing or riveting.

Double Bogie Engine.—See Bogie Engine.

Double Butt Strips.—Two covering strips employed in making a butt riveted joint, one strip being placed on each side of the abutting plates. These are much stronger than the joints with single butt strips (q.v.).

Double Cards.—Indicator cards taken from both ends of an engine cylinder.

Double Contraction, or Double Shrinkage.—Where a wood pattern is made for the casting of a metal pattern, from which the actual mould is to be taken, two amounts of contraction are allowed on the wood pattern. Thus if the contraction of the casting is $\frac{1}{8}$ in. per foot, $\frac{1}{4}$ in. per foot is allowed in the first or wood pattern.

Double Cut.—This refers to the crossing of the lines of teeth of files at an angle, as distinguished from the single or float-cut files.

Double-cutting Drill.—A drill which is ground to cut with equal facility in a right or left-hand direction. The cutting edge lies therefore in the longitudinal axis and the angles are symmetrical on both sides. These drills form but a small class.

Double-cylinder Engine.—A steam engine having two cylinders. Its action is steadier than that of a single-acting engine owing to the relative arrangement of the pistons and cranks. One piston is at half, while the other is at full stroke, and the cranks being therefore at right angles with each other, the dead points are passed with little of the jerky motion which accompanies the revolution of a single-cylinder engine.

Double Driver Chuck.—A driver chuck which has two stems or pins for driving, so that the pressure of one is counterbalanced by that of the other. It is used for work requiring special accuracy.

Double Disc Valve.—A tapered plug valve having two faces fitting against two opposite seatings.

Double Elephant.—A drawing paper of good quality measuring 40 in. by 27 in. It may be had rough or smooth, and either unmounted, or mounted on brown holland.

Double-ended Boiler.—A marine boiler having furnaces and flue doors at each end, and fired therefore from each end.

Double-ended Bolt.—A bolt having no solid head, being screwed at each end alike for the reception of nuts.

Double-ended Machine.—Applied to those punching and shearing machines in which two sets of operations can be carried on at one time.

Double-faced Hammer.—A hammer is double faced when both sides of the head are alike flatted, so that there is no pane.

Double Fagoted.—Iron which has been subjected twice ever to the process of fagoting (q.v.) for the purpose of increasing its density, homogeneity, and tensile strength.

Double-fished Joint.—A fished joint having two fish plates, or covering strips laid against opposite faces, as distinguished from a single-fished joint (q.v.).

Double-flued Boiler.—See Lancashire Boiler.

Double Gear.—The employment of two pinions and two wheels in com bination for the purpose of gaining mechanical efficiency, as in the lifting gear of cranes. Back gear (q.v.) is often termed double gear. See Double Geared.

Double Geared.—A lathe or drilling machine is said to be double geared when it is provided with the ordinary back gear; that is having a single back gear spindle and set of wheels only, as opposed to an ungeared or plain hand-turning lathe on the one hand, and to a treble-geared lathe on the other. Double geared does not mean that double power is gained thereby, since the power gained is usually about nine to one, but refers only to the employment of the second spindle and wheels. The term double geared is also applied to ~illing machines having back gear not of the ordinary form, but wheels which can be slid up and down on a vertical spindle into, or out of gear with the wheels on the main spindle.

Double-handed Tool.—A tool furnished with two handles, as ordinary die stocks and cross cut saws.

Double-headed Rail.—A rail for permanent way, the shapes of whose top and bottom sections are alike, so that when one face becomes worn the rail is turned over in the chair and the opposite edge utilised.

Double-helical Gear.—See Helical Gear.

Double-ported Slide Valve.—An exhaust relief, or equilibrium valve. It is designed for nearly equal steam pressure on both sides and to give a greater area of opening to exhaust than to supply, and this is effected by allowing steam to enter the cylinder through passages in the pierced body of the valve as well as at the ends, and to exhaust into the hollow, or D, of the valve which contains the ports through which the steam gains entry. Having separate ports for supply and exhaust, it is obvious that by giving to the latter an increase of area over that of the former, and by adjusting the travel, the escaping steam will be allowed greater freedom of exit. Having double openings for steam, the cylinder is necessarily double ported, that is, there are two ports on each side of the exhaust instead of one, and the travel of the valve is only half that which it would be with single ports.

Double Puddling Furnace.—See Puddling Furnace.

Double Purchase.—A term used to designate the mechanical efficiency obtained by the employment of double gear (q.v.). It is equal to the product of the radii of the wheels divided by that of the pinions.

Double Reading Scale.—A draughtsman's scale on which there are two primary divisions, one at each end, subdivided.

Double Riveting.—Where there are two lines of rivets in a lap joint, or four in a butt joint, the term double riveting is employed, as distinguished from a single line in the former and two lines in the latter.

Doubles.—Sheet iron plates prepared for tinning. Their thickness ranges between No. 20, B.W.G. (·035 in.) to No. 25, B.W.G. (·020 in.). The term is also applied to the plates already tinned, to which an extra thick coating has been given, and which, besides, have been hammered between a polished hammer and a polished anvil to render the union of the metals more intimate. Also termed Block Plates.

Double Shaper.—A shaping machine having two rams and two tool boxes, for a double set of operations. Sometimes one head is made for straight cutting only, the other for circular cutting.

Double Shear.—A rivet which unites a joint made with double fish-plates will have two sections subject to shearing strain; hence the term. See Single Shear.

Double Shear Steel.—Shear steel (q.v.), in which the process of welding is repeated by reheating the bars, doubling them up, and welding them together once more.

Double Shrinkage.—See Double Contraction.

Double Stroke.—Two piston strokes, or one forward and one backward. A term used in calculations relating to the power of steam engines.

Double-threaded Screw.—A screw comprising two distinct helices, winding parallel with each other around the body. Employed to give an increase in the rate of travel.

Double Valve Box.—A valve, or clack box, provided with two valves only. Having no intermediate valve (see Treble Clack Box) the stream delivered is intermittent.

Double-webbed Girder.—A box girder (q.v.), having two webs or plates connecting the top and bottom flanges.

Double-welt Joint.—A double-fished joint (q.v.).

Dovetail.—An interlocking joint used for wood-work, so called from the fancied resemblance which the pin bears to the tail of a dove.

Dovetailing Machines.—Those used in the cutting of dovetails. They are of two types, according as the cuts are made by spindles or by saws. In the former a revolving bit cuts away the wood, in the latter it is removed by a circular saw set at an angle while the work is being passed before it through a measured distance.

Dowels.—Pins either of wood or metal used by pattern-makers to retain portions of patterns temporarily in position during the process of moulding, which portions, if firmly nailed or screwed, would prevent good delivery from the sand. Dowels are now mostly made of metal, either malleable iron or brass, the better sort being cast attached to flat plates, which are recessed into the pattern joints flush with their surfaces, and there screwed.

Down.—See Running Down.

Down-comer, or **Down-take.**—The vertical pipe which conducts the waste gases from the top of a close-mouthed blast furnace into the blast main.

Down-take.—(1) The short passage leading from the back end of the fire flue of a Cornish boiler to the brickwork or smoke flues which pass to the front of the boiler. (2) A down-comer (q.v.).

Downward-flow Turbine.—See Turbine.

Draft.—A drawing. Drawing is often termed drafting.

Drag.—The bottom part or section of a founder's moulding box.

Drag Bar.—The rod by which a locomotive and its tender, or railway wagons, are coupled together.

Drag Bolt.—The bolt which couples up drag bars.

Drag Hook.—The hook attached to back and front of locomotives and railway wagons for coupling up.

Drag Link.—(1) A link employed for connecting or disconnecting the cranks of coupled engines. (2) The rods by which the slot links of valve gears are moved over for forward or backward gear are called drag links.

Drag Plate.—The casting which forms the footplate or platform of a locomotive. It is bolted between the frames, and carries the drag bar and other attachments. It is made heavy to equalise the weight on the wheels, averaging about three tons.

Drag Rope.—A rope used for the purpose of pulling logs of timber up to a circular saw. It is coiled around a drum which is driven through intermediate gearing from a pulley on the saw spindle, by which means the speed of the rope is reduced with a corresponding increase of power.

Drag Spring.—A spring attached to a drag bar (q.v.) to lessen the shock of concussion.

Drag Surface.—The forward face of a screw propeller, or that from which the water glides off.

Drain Cock.—A pet cock (q.v.).

Drain Pipe.—A small copper or brass pipe leading from the pet cock of an engine cylinder. The drain pipes are led away to any convenient spot for discharging the water of condensation. On board ship they are led into the bilge or the hot well.

Draught.—(1) The tapering or gradual thinning down of the vertical sides of foundry patterns, to permit of their delivery from the sand. (2) The quantity of air which passes through a furnace in a given time is termed its draught. Natural draught is that produced by the exhaustion of air from the furnace and the creation of a partial vacuum therein. This is attained by the rapid ascent of the highly heated gases. Forced draught is that produced by the injection of a jet of steam into the chimney, which hastens the escape of the gaseous products of combustion. In draught by compressed air the pressure is produced by various forms of fans and blowers.

Draught Bar.—A drag bar (q.v.).

Draughtsman.—The engineer upon whom the task of designing machinery, and of making both general and working drawings, devolves. A draughtsman, to be practical, must have an intimate knowledge of the conditions of work in the several departments of the factory, and should be able to combine with this the best theories and fair mathematical acquirements. A designer, and a copyist or tracer, though both alike called draughtsmen, stand on very different footings, the first being a true engineer, the second being rated below an ordinary mechanic.

Drawback.—A plate or a skeleton framework of cast and wrought iron, used for carrying some portion of a founder's mould, which has to be temporarily removed, either for the proper withdrawal of the pattern or for the convenience of the cleaning up of the mould faces. The plate with its contained sand is lifted or drawn back, and afterwards replaced in its original position, in readiness for casting.

Draw Bar.—See Drag Bar.

Draw Bench.—The bench used by wire drawers.

Draw Cock.—A pet cock (q.v.).

Draw Filing.—The polishing of a metallic surface by drawing a smooth file along it; the file moving to and fro in a direction transversely to its own longitudinal axis, so that the cutting action of the teeth scarcely comes into play. Also called poker filing.

Draw Hook.—The central hook of a railway wagon attached to the draw bar, and flanked by side hooks.

Drawing.—(1) The delineation of machinery and machine parts in plans, elevations, and sections to proportional scales. In these there can be no perspective, hence they are conventional drawings only, and do not represent truly correct views to a spectator standing in any fixed position. But for convenience of measurement, such scaled drawings are the only ones which are of use in the workshops. (See Phototype, Hand Sketch, Tracing.) (2) The manufacture of wire, piping, and tubing, by pulling or drawing the material of which they are composed through perforated plates.

Drawing Board.—(1) A board used in the drawing office for pinning down or stretching drawing paper upon. The cheaper boards are clamped at the two ends only ; the better ones are panelled, or enclosed in an entire frame. Good boards should be tongued in addition, and those of the largest size should have adjustable joints to permit of the expansion and contraction of the wood. Pine and mahogany are the materials used for drawing boards. Their dimensions are given similarly to those of the various drawing papers, but a little extension in size is made. (2) The workshop boards, on which portions of mechanism are drawn to full size, are mostly of large dimensions, several feet square, and the pencil drawings are made directly upon the boards themselves, whose surfaces are whitened with chalk to show up the lines.

Drawing Down.—The reducing or thinning down of forged work under the hammer ; hammers, fullers, and flatters, being employed for the purpose. When drawing down, the dimensions of the cross section may be diminished all around alike, or be diminished in one direction and increased in the other.

Drawing Fires.—The raking out of the furnace fires of an engine boiler.

Drawing In.—A cutting tool is said to draw in when it cuts deeper than it should do. This is due to malformation of the same kind as that which produces a hitch or catch. See Hitch.

Drawing of Castings.—The setting up of internal molecular stresses in castings, due to bad proportioning, or to unequal cooling, or to both in combination, the tension being variable in adjacent sections. See Cooling, Feeding, &c.

Drawing Office.—The office or department of an engineer's factory in which the work of the draughtsman is carried on. It is furnished with sloping desks, stools, boards, paper, &c., and with nests of drawers for the storage of the drawings.

Drawing of Patterns.—The lifting (q.v.) of foundry patterns from the sand. See also Rapping, Delivery, &c.

Drawing of Temper.—The heating of a steel article to redness, and allowing it to cool slowly in air. This is the reverse of hardening or tempering.

Drawing of Tubes.—See Tube Drawing.

Drawing Paper.—This comprises several varieties, their qualities increasing in the following order : Cartridge, Imperial, Double Elephant, Antiquarian, and Emperor; and their dimensions as follows: Demy, Medium, Royal, Super-royal, Elephant, Imperial, Columbier, Atlas, Double Elephant, Antiquarian. See sizes under those headings. Paper is either hot pressed, or not pressed at all, or else is rough ; the latter being for general or ordinary drawing, the first for pencil drawing, and the medium quality for common work.

Drawing Pen.—A pen formed of two blades of steel, approaching at the points, and adjustable with a milled-headed screw, the ink being enclosed between the blades. Used for marking ink lines of uniform width upon drawings. Two such pens mounted side by side for drawing distinct but parallel lines form what is called a road pen.

Drawing Pins.—Short, broad-headed pins, used for holding down both drawing and tracing paper upon drawing boards.

Draw Knife or **Drawing Knife.**—A wood-cutting tool, having a long and narrow blade, attached at the ends to two handles, which stand at right angles therewith. The handles are of wood, and the tangs, into which the ends of the blade are prolonged, are passed right through these, and riveted over brass plates at their ends. The tool being drawn by the handles towards the workman cannot come loose,—the handles being held firmly by the rivets. The tool is used for cutting thick and heavy chips off the rough edges of boards in order to lessen the labour of the plane.

Drawn On or Pulled On.—This refers to the method commonly practised of attaching wheels, pulleys, &c., to their axles or shafts when their mass of metal is such that the blows of a hammer would have no useful effect. Long bolts, whose heads are held in a massive cross attached to one end of the axle, pass on the outside of or else through the wheels between the arms, and the tightening of their nuts against a bar or washer plate stretched across the wheel face slowly but effectively pulls the latter along the shaft into its required position. By the same means wheels are drawn off their shafts. Hydraulic presses are also used.

Draw Plate.—A plate of hardened steel, or of ruby, used in wire and tube drawing.

Dredger.—A machine used for removing mud and sand from the bed of a stream or harbour, either by means of scoops or by suction. See Grab, Ladder, Mud Bucket.

Dresser.—A smith's tool with a round top face and flat sides, over which the forked ends of connecting and other rods are finished to shape. It fits by a square shank into a hole in the anvil.

Dressing.—See Belt Dressing.

Dressing off.—Fettling (q.v.).

Drift.—(1) A steel tool, commonly of rectangular but often also of some special section, used for the purpose of enlarging and of finishing parallel or tapered holes. Drifts are either smooth, or serrated on their flanks, acting by compression in the first case, and by shearing and abrasion in the second. (2) A rectangular strip of steel slightly curved in its length, used for driving keys out of their beds. (3) A cylindrical tapered punch used for pulling rivet holes whose boundaries overlap into coincidence with each other. Excessive drifting is to be condemned because of the strain which it puts upon the plates.

Drifting.—(1) The shaping and enlarging of holes by the use of a drift. (2) The pulling of overlapping rivet holes into line with each other.

Drill.—A tool for boring holes in metal or in wood, and revolved either by some form of hand brace or by a special machine. Drills may be flat towards the point and simply bevelled, or else of the twist type. The flat drills may have either single or double cutting edges.

Drill Bow.—The bow of a bow drill (q.v.).

Drill Box.—The reel of a bow drill (q.v.).

Drill Chuck.—A lathe chuck made for the special function of holding drills. The hole for the drill is sometimes square tapered only, but the better chucks have circular holes, while the best are also self-centring.

Drilled Holes.—Holes are drilled in all the best work in preference to being cast. Holes are always drilled for bright bolts notwithstanding that in cast iron the holes may have been first cored small. The best boiler plates are drilled in preference to being punched, but drilling leaves a burr which has to be removed to insure close frictional adherence of the plates. Holes are drilled both by hand and by machine.

Drilled Plates—This has reference to the mode in which the holes are pierced through boiler and other plates. In the best boilers, and in some special classes of work, the rivet holes are specified to be drilled instead of being punched. By drilling, the tenacity of the plates is slightly increased, whereas by punching they are weakened.

Driller.— A machinist who attends to a drilling machine, maintaining it in working order, setting and fixing the work on its table, and in some shops grinding the drills and regulating their speed.

Drill Feed.—The cutting feeds for drilling machines are of several kinds. They usually consist of a screw and hand wheel, or of a screw and cog-wheel whose boss forms the nut of the screw, or of a rack and pinion, actuated by hand gear, or of a ratchet and paul.

Drilling Cramp.—A pair of rods suspended from a cross bar of which they form a rigid connection. The rods are curved at their lower ends to receive pipe lengths, which rest in the hollows of the curves, as in cradles. They are drilled from above while thus held, the thrust of the drill being taken by the cross bar. Drilling cramps are used specially for outdoor work where a machine is not available.

Drilling Machine.—A machine which holds and actuates a drill. The forms of these are exceedingly numerous, and are described under their distinctive headings. Essentially a drilling machine consists of a table, a drilling spindle, driving gear, and suitable feed motion. Machines vary in size from those of massive proportions driven by power, to light structures driven by the foot only.

Drilling Oil.—A cheap oil used for the lubrication of the edges of drills and drill-like cutters.

Drilling Pillar.—A vertical pillar clamped to a work-bench, and furnished with a sliding adjustable arm from which pressure is exerted through the medium of a screw on a drill revolving between the screw and the work lying on the bench.

Drilling Spindle.—See Drill Feed, Drill Spindle.

Drilling Table.—The table of a drilling machine upon which work is either laid or clamped while being drilled. It is ordinarily made adjustable for height, frequently also for lateral movement, and is furnished with slots for the reception of the bolts used for holding down the work, and the angle-plates to whose sides small work is frequently bolted.

Drilling Templet.—A templet (q.v.) of cast or wrought iron, which becomes a guide for the drilling of holes. The templet is attached to its work, and the drill is guided in its movements through holes drilled in the templet itself.

Drill Plate.—A circular plate which is fitted over the mandrel nose of a lathe poppet to receive the pressure of a drill revolving in the drill-chuck, the work which is being drilled resting against the plate.

Drill Socket.—The socket which receives the shank of a drill, either on the lower end of the drilling spindle of a machine or in the drill chuck of a lathe.

Drill Spindle.—The vertical spindle of a drilling machine, which carries the drill, and revolves ; and through whose vertical movement the feed is operated. It is a compound structure consisting of the actual spindle which revolves, and the mechanism, usually a screw, or a rack, or lever motion, which imparts its feed. The upper and lower portions are connected with a swivel joint, having hardened compensating collars, or the lower part is bored to receive the actual spindle which slides through it

9*

Drill Stock.—A drill holder used in hand-drilling; that is, it includes the socket which receives the drill shank, together with the plain body to which it is attached.

Drip Can.—See Soap Water Can.

Drip Cup, or Drip Pan.—A light shallow tray of tin or iron, placed underneath a bearing or machine to catch the waste oil and prevent it from dirtying subjacent parts.

Drip Oil Can.—See Soap Water Can.

Drip Pan.—See Drip Cup.

Driven.—In gearing signifies the wheel, or wheels, actuated by a driver (q.v.). In calculations involving trains of gearing the product of the series is in geometrical, not arithmetical ratio. Hence the rule is: Multiply the radii of all the drivers together and the radii of all the driven, and divide the latter by the former for the mechanical efficiency.

Drive Pipe.—The feed or inlet pipe of a hydraulic ram (q.v.).

Driver.—(1) In gearing signifies the wheel which drives another. (See Driven. (2) A projecting piece of metal used when turning between dead centres, to impart rotation to the work. It is either attached to a face plate, or is carried in a slot in the point centre, and bearing against a projection on the carrier which embraces the end of the work, so drives it round.

Driver Chuck, or Running Centre Chuck.—A point chuck in the head-stock mandrel which carries work being turned between centres.

Driving Axle.—An axle which communicates motion; hence called a live axle, in opposition to a dead axle (q.v.). The axle of a locomotive to which the motion of the connecting-rods is directly communicated is an illustration.

Driving Band.—See Belt.

Driving Chain.—A pitch chain (q.v.); or an ordinary chain whose links fit into suitable recesses cast around the periphery of an ordinary sheave.

Driving Chuck.—(1) A lathe chuck fitted with a driver (q.v.) for actuating the carrier (q.v.). The chuck is either a circular-face plate, having a pin projecting from its face; or it is a dead centre enlarged at the neck, and pierced with a slot for the reception of an ∟-shaped driving piece. (2) Sometimes applied to the cup chuck, because the wood is driven into it by blows from a hammer.

Driving Fit.—A bush, a shaft, or spindle is said to be a driving fit, when the blows of a hammer are required to send it down into the hole bored for its reception. The amount of force required will vary with circumstances. It must not be so much as to burst the metal around the hole, nor so slight that the fitting will wear or work loose in time.

Driving Gear.—A term of general application, signifying the immediate arrangement of wheels, pulleys, belts, levers, &c., by which motion is communicated.

Driving Home.—A shop term which signifies the driving of a wheel or a shaft, or any portion of mechanism, to its permanent position and final resting place.

Driving Side.—That side of a belt which drives its pulley. The side most suitable for driving will depend upon the relative positions of the driving and driven pulleys. In a horizontal arrangement it is the lower or slack side; in pulleys arranged vertically either side may be the driver. When belts pass over guide pulleys placed at different

angles from the main pulley, they can only **drive on** the sides which are in the same plane.

Driving Springs.—The springs which carry the axle boxes of the driving axles of locomotives. They are frequently volute springs instead of being built up, and a pair are put under each axle box.

Driving Surface.—The after face of a propeller blade, or that which thrusts against the water.

Driving Wheels.—The wheels of a locomotive which are attached to the driving axle.

Drop.—(1) That stage of the pig-boiling (q.v.) process in which the ebullition or boil ceases. (2) Also that stage of the process of the manufacture of steel in the Bessemer converter at which the flame suddenly shortens, indicating the complete decarburisation of the metal, and which is the signal for the shutting off of the blast. See Boil.

Drop Bottom Boxes.—Boxes used in the laying or deposition of blocks of artificial stone or concrete by gravity alone. The bottoms of the boxes are made to unfold and open downwards, allowing the enclosed block to drop into its place.

Drop Bottom Cupola.—The bottoms of many cupolas are now furnished with hinged doors made to fall downwards by the knocking away of a pin. The advantage of their use is, that the labour of raking out the fire after the blowing is avoided, the whole mass falling bodily on the dropping of the doors. Such cupolas are necessarily supported on short columns, to give sufficient space for the embers underneath.

Drop Bottom Skip.—See Skip.

Dross.—The sullage, scurf, oxide, and other impurities which are skimmed off the top of molten metals, or which accumulate in the head or in the riser.

Drum.—(1) A large barrel which coils the chain in lifting-tackle. (2) A wide plain belt pulley. (3) A stepped belt pulley, for driving at different speeds.

Drum Head.—The upper portion of a capstan.

Drummer.—A term applied to a smith's hammerman in many parts of the country.

Drunken Saw.—A circular saw tightened on its spindle, at an angle with the axis of the spindle, by means of bevelled collars. The turning round of the collars through an arc of the circle allows of variation of the angle within certain limits. These saws are used for the cutting of grooves in timber, and the angle at which the saw is placed governs the width of the groove. Sometimes called the wabble saw.

Drunken Screw.—A screw whose outline is irregular or wavy.

Dry Bottom.—A puddling furnace is said to have a dry bottom when the slag is not allowed to accumulate, but is drawn off as frequently as possible.

Dry Brush.—Iron-moulders term the brush with which they sweep the sand away from the joints of moulds and of patterns a dry brush, to distinguish it from the wet or water brush (q.v.).

Dry Copper.—See Poling.

Drying.—(1) Drying has the same signification as coming to nature (q.v.). (2) The expelling of the moisture from a mould or a core, in a drying-stove, previous to casting.

Drying Oils.—Certain vegetable oils which become dry and solid on exposure to the oxygen of the air. Linseed, hemp, poppy, almond, and colza are drying oils. The presence of litharge or lead oxide hastens the drying process. Drying oils are therefore used for paints. (See Linseed Oil.) Owing to this rapid absorption of oxygen, spontaneous combustion (q.v.) has frequently taken place.

Drying Stove.—A stove used for the drying of foundry moulds and of large cores. It is a brick-built chamber of rectangular form adjoining the foundry, of which it forms a part. Into it run lines of rails for carrying the core carriages. It is heated by a coke fire made in a pit in the floor, the doors being hermetically sealed, and the smoke passing away through a flue at the back. The temperature of the stove should not exceed about 400° Fahr., else the cores will become burnt. The time during which a mould or core should remain in the stove will vary from two or three hours to as many days, being dependent entirely on its bulk.

Dry Moulding.—Moulding in dry sand (q.v.).

Dry Puddling.—Puddling carried on in the reverberatory furnace, where decarburisation and oxidation are chiefly effected by the oxygen of the air. It is so called because the metal is never allowed to pass beyond the pasty condition. It is now almost superseded by the process of pig boiling (q.v.).

Dry Sand.—Foundry sand consisting of a mixture of strong sands, coal dust, and dung. Called dry because moulds made in it are dried in the stove, in contradistinction to green sand which is not dried. (See Green Sand, Skin Drying.) Drying is always resorted to when castings of superior quality are desired.

Dry Sand Moulding.—The moulding of work in dry sand (q.v.). The distinction between dry sand moulding and loam moulding (q.v.) is that the latter is always effected without the use of a full pattern, while in the former a pattern is employed, and also that the sand in the former case is mixed and manipulated like ordinary green sand, while in loam work it is mixed with water to render it plastic.

Dry Steam.—Steam which has neither been superheated (q.v.) on the one hand, nor mixed with the water of priming on the other, but remains in the normal condition, which experience has proved to be the most suitable for use in engine cylinders. Called also saturated steam.

Dry Uptake.—When the uptake of a marine boiler is placed without the shell away from contact with water and steam it is termed dry, to distinguish it from a wet uptake (q.v.).

Dry Wood.—Timber from which the sap has been removed by seasoning. It should be allowed to remain in the log or balk for two or three years after felling. When sawn into boards it should be piled in a rack on an open building exposed to currents of air, thin strips of wood placed crosswise alternating with the boards in the piles, to allow of the access of air to both sides of the timber. The boards thus " stripped " should remain a twelvemonth or two years, depending on thickness, before being used.

Ductility.—That property of metals in virtue of which they can be drawn out into wires. This property depends partly upon malleability, but chiefly upon the tenacity of the particles composing the metal.

Dull.—A tool is said to be dull when it cuts or abrades with difficulty.

Dull Metal.—Cast metal is said to be dull when it is allowed to part with some considerable portion of its heat in the ladle before being poured, so as to enter the mould in a somewhat thick condition. Dull metal is used by preference for heavy work, since it shrinks less, crystallises larger, and cools sounder than hot metal. The limit to the use of dull metal is that at which cold shuts form, and the metal does not possess sufficient fluidity to fill the minor ramifications of the mould. Metal is dulled by being allowed to stand in the ladle, or by the throwing of cold runner heads into the molten mass.

Dull Red Heat.—This corresponds with a temperature of about 1290° **F.**

Duodecimals.—A system of multiplication to obtain the superficial and solid areas of rectangular figures, by the use of units of square feet, square inches, superficial primes, solid primes, solid feet, solid inches, solid seconds ; the multiplication taking place crosswise ; the product of feet into inches giving primes. A prime is 12 superficial inches, that is, a piece 12 in. long and 1 in. broad. It is used in timber measurement, and in rectangular measurement of tanks, &c.

Duplex Bear.—A punching bear (q.v.) which is actuated by a screw moving double levers.

Duplex Lathe.—A self-acting lathe furnished with two rests, one on each side of the bed, for the taking of two cuts at one time off a piece of work. The hinder tool is therefore inverted.

Duplex Machines.—Machines in which two sets of similar operations are performed at one time.

Duplex Planing Machine.—A double planing machine having two beds and two tables.

Dutch Wheel Crane.—A whip crane (q.v.).

Duty.—The duty of a steam engine is the number of pounds raised one foot high by the burning of a bushel of Welsh or Cornish coal. The average duty is 60,000,000 of pounds raised one foot high, or 60,000,000 of foot pounds.

D-Valve.—Applied to the common slide valve, because the face of the valve is hollowed in the form of a letter " D," and the outside rounded to the same outline.

Dwarf Pillar.—A very short pillar. Dwarf pillars are used for the support of the beds of reverberatory furnaces and the sides of blast furnaces.

Dying In.—The merging of a hollow radius into a plane surface is so termed by workmen to distinguish it from the meeting of abrupt angles.

Dynametrical Horse-Power.—See Actual Horse-Power.

Dynamical Values.—Theoretical values converted into terms of work units. Thus thermal values (q.v.) multiplied by foot pounds (q.v.) give dynamical or practical values in foot pounds. Yet since in estimating these dynamical values no allowance is made for losses due to imperfection of mechanism, and loss of heat and work consequent thereon, a percentage only of the calculated dynamical values can be taken in practice.

Dynamic Fatigue.—The fatigue of materials (q.v.) which results from a dynamic load.

Dynamic Load.—A rolling or a moving load, as distinguished from a dead load.

Dynamics.—That branch of mechanics which treats of the laws of forces that produce motion in the bodies and structures upon which they act. See Statics.

Dynamometer.—An instrument for measuring the intensity of a force ; usually consisting of some form of bent or spiral spring, whose strength has been tested and registered. A special form of dynamometer is that contrived by Prony, for determining the friction of a shaft. A pair of friction blocks, capable of being tightened around a revolving shaft by means of bolts, are attached to a lever. The amount of friction of the shaft for any given pressure is determined by the product of a weight suspended from the lever, and just sufficient in amount to counteract the tendency to revolution, into the length of the lever itself. See Hydraulic Dynamometer, Dict. of Modern Terms.

E.

E.—The modulus of elasticity (q.v.).

Ear.—A lug (q.v.).

Early Cut Off.—This term relates to the ratio of the expansion of steam in an engine cylinder. Any cut off under one-half the stroke may be properly termed early.

Ebonite.—A compound of india-rubber with sulphur and other ingredients. It is used for draughtsmen's set squares and scales, for which its qualities of hardness, moderate elasticity, and smoothness of surface render it suitable. Ebonite differs essentially from vulcanised rubber in being charged with an excess of sulphur, and subjected to a greater heat for a longer period of time.

Eccentric.—A piece of mechanism employed in engines for converting the rotatory motion of the crank shaft into a reciprocating rectilinear motion upon the slide valve or the feed pump. It is effected by giving to a ring upon the shaft a definite " throw " or eccentricity, equal in amount to the half of the travel of the valve, or of the plunger stroke. Eccentrics are also much used in general machine construction. Sometimes spelt excentric.

Eccentric Hoop.—See Eccentric Strap.

Eccentricity.—The deviation of the centres of two circles from one another.

Eccentric Lug.—The projecting portion of an eccentric strap, to which the eccentric rod is attached.

Eccentric Rod.—The rod which receives the motion of the eccentric strap and transmits it to the valve, or pump, or other rod to which movement is to be imparted.

Eccentric Sheave.—The body of an eccentric, or the actual eccentric itself, which is forged or keyed directly upon the crank shaft, and whose throw is communicated to the eccentric strap.

Eccentric Strap or Eccentric Hoop.—The belt which encircles the eccentric sheave, and transmits its motion to the eccentric rod. It is grooved on its inner face to fit into a corresponding groove on the rim of the eccentric, without which it would slip off sideways. Being grooved, therefore, it is made in two pieces, which embrace the sheave, and are held with bolts passing through projecting lugs. Straps are made in iron or in gun metal, sometimes iron castings are employed, having gun-metal liners.

Eccentric Throw Out.—The motion by which the wheels on the hinder mandrel of a back-geared lathe are thrown out of gear, a small eccentric made fast to the mandrel being used for the purpose. The term distinguishes this form of throw-out gear from a backward or linear throw-out motion, in which the mandrel bearings slide in linear guides, and also from an endlong linear motion, by which the teeth are slid out of contact in their longitudinal directions.

Economiser.—An apparatus, or arrangement of pipes or reservoirs, in which the feed water for steam boilers is heated up to, or higher than, the boiling point. Feed water heaters, therefore, are economisers.

Edge-Planing Machine.—See Plate Edge Planing Machine.

Edge Runners.—Loam and mortar mills are so called, because the rolls run on edge.

Edge Tools.—Cutting tools.

Eduction.—The escape of the exhaust steam from a cylinder.

Eduction Overlap.—Exhaust lap (q.v.).

Eduction Port.—An exhaust port (q.v.).

Eduction Trunnion.—The trunnion of an oscillating cylinder through which the exhaust steam passes on its way to the condenser.

Effective Area.—The effective area of a screw blade is that obtained by the projection of the blade on a plane at right angles with its axis. It is, therefore, less than the superficial area.

Effective Heating Surface.—See Heating Surface.

Effective Pressure.—The net amount of pressure upon the piston of an engine, over and above that which is necessary to balance the pressure of the atmosphere.

Effective Section.—The effective section of the shell of a steam boiler is a fractional portion of the gross section (q.v.) obtained by deducting therefrom the loss of strength due to the riveted seams.

Efficiency.—The efficiency of a joint, structure, machine, or part, is the ratio which such bears to some understood standard of reference. The efficiency of a riveted joint is its percentage strength, estimated relatively to that of the solid plate. The efficiency of a machine is its modulus (q.v.).

Egg-end Boiler.—A horizontal, cylindrical, externally fired steam boiler, having hemispherical ends. It is now little used.

Egg Sleekers.—Moulder's tools, whose faces are a segment of a sphere. Used for cleaning the hollow faces of work of hemispherical sections. Also termed button sleekers.

Eighth Bend.—A bend pipe having socketed and spigoted ends, whose length equals one-eighth of the circumference of the circle to whose radius the curve of the bend is struck.

Ejector.—See Injector.

Ejector Condenser.—A condenser in which the water is drawn from the feed tank into the condensing chamber by the action of the steam itself.

Elastic.—Strictly speaking all bodies are elastic, since all when subject to stress experience strain. But it is customary to divide bodies into elastic, and non-elastic or rigid, according to their amount of sensible elasticity ; so that a body when subjected to a given stress at a given temperature, which does not increase when the stress is prolonged, and disappears when the stress is removed, is called an elastic body. See Elastic Limit.

Elastic Blow.—(1) A blow from a hammer which is not firm and dead. (2) Signifies the nature of the blow given by that class of steam hammers where steam is introduced between the piston and the cylinder ends to produce cushioning.

Elastic Core Packing.—A packing sometimes used for stuffing boxes, consisting of india-rubber enclosed in canvas, the india-rubber constituting the elastic core. Sometimes called elastic steam packing.

Elastic Force.—That property which steam possesses in common with gaseous bodies generally of undergoing expansion and compression, with corresponding diminution and increase of pressure.

Elasticity.—This has reference specifically to the extent to which a bar or structure may be elongated by tensile stresses, without remaining permanently extended on the removal of the tensile force.

Elasticity, Modulus of.—See Modulus of Elasticity.

Elastic Limit.—That point at which the molecular actions within a bar or structure tending to resist deformation cease to balance the strains imposed thereon ; in other words, when the straining action is so great as to produce permanent change of form. Materials should never be strained to this limit. See Permanent Set.

Elastic Nut.—A form of nut by which compensation is made for the wear of the nut and its screw. An elastic nut **may** either be sprung, by sawing a longitudinal groove through one side of it, and hammering it upwards, or it may consist of two separate nuts, one above the other, kept pressing in opposite directions on the screw by an interposed spring ring.

Elastic Steam Packing.—See Elastic Core Packing.

Elastic Strain.—The amount of strain imposed upon a structure by a load not sufficiently great to cause the material of the structure to exceed the elastic limit (q.v.).

Elastic Strength.—The greatest stress which a bar or structure is capable of sustaining within the elastic limit.

Elastic Washer.—(1) A form of washer employed in place of a locknut. It consists of a split steel ring whose faces near the split ends are turned slightly upwards and downwards, thereby diverging from a true plane. These are sprung or compressed level by the pressure of the nut, and their elasticity prevents the latter from slacking back. (2) Washers of vulcanised india-rubber, employed chiefly in pump work.

Elbow.—A bend pipe whose direction of outline is abruptly angular, without any radius or rounding of one portion into the other. The two arms of the elbow meet at 90° of angle. Employed chiefly for connecting wrought-iron piping.

Elephant.—A drawing paper, measuring 28 in. by 23 in. and weighing 72 lbs. to the ream.

Elevation.—In drawing signifies that view of a structure which it bears to an observer standing beside it and looking against its vertical face. A sectional elevation supposes the structure cut vertically through the centre, and the observer standing as before.

Ellipse.—One of the conic sections, the curves of whose ends are equal, and whose centres, or foci, lie in the conjugate axis at equal distances from the centre.

Ellipsoidal Rivet.—A rivet the section of whose head is elliptical in form.

Elm (*Ulmus campestris*).—A tree of the natural order *Ulmaceæ*. It is a coarse, open-grained wood, much given to warping, and is very sparingly used in engineers' works. It is employed for foundry strickles, being used to strike the wet loam to shape. Its tenacity is about 14,000 lbs. per square inch. A cubic foot weighs 34·3 lbs. Sp. gr. ·55. The American elm (*Ulmus Americana*) is whiter and harder than the English elm, and also more straight-grained.

Elongation.—See Reduction of Area.

Emery.—A species of corundum composed of oxide of iron, alumina, silica, and a small proportion of lime. It is crushed and pounded to different degrees of fineness, made into a variety of preparations, and used for different purposes, its value depending upon its hardness.

Emery Buff.—A wood roller, having emery powder glued upon its periphery and side faces for the polishing of metallic work. It is revolved at a high speed.

Emery Cloth.—Powdered emery glued on thin cloth, and used for removing file marks, and for polishing metallic surfaces.

Emery Grinder.—An emery wheel (q.v.).

Emery Grinding.—The abrasion of metallic surfaces upon an emery wheel, buff, or lap.

Emery Lap.—See Lap.

Emery Paper.—Powdered emery glued on paper and used for polishing metallic surfaces. It is not so flexible as the emery cloth.

Emery Planer.—An emery surfacer (q.v.).

Emery Powder.—Crushed emery passed through sieves of different mesh, superfine flour emery being the finest, corn emery the coarsest used by engineers. Used for abrading and polishing metallic surfaces.

Emery Stick.—See Lap.

Emery Surfacer.—A form of emery grinder (q.v.) in which a broad solid emery wheel is employed as a substitute for the cutting tools of shaping machines. The work to be surfaced is affixed to a sliding table, and passed underneath the revolving wheel.

Emery Wheel.—A wheel made of powdered emery cemented together, or of emery cemented to a wood centre. It is revolved at a high speed, and is used for grinding. An emery buff (q.v.) is also a wheel, but is of fine texture and used for polishing only. Emery wheels are cemented together with a silicate insoluble in water. They are made in about ten grades of coarseness.

Emperor.—A drawing paper of finest quality. It is made in sheets measuring 72 in. by 48 in.

Empirical Rule.—Any rule or equation which is not deduced from purely mathematical or physical considerations, but which is based upon experience, convenience, or custom. It is therefore not scientifically, but only practically and approximately accurate. Most of the formulæ employed by engineers are empirical in that they are based chiefly on the results of previous experiment and practice.

Enamelling.—The coating of water pipes internally with a smooth paint in order to diminish the friction of the fluid, and loss of head consequent thereon.

Encastre.—Encastre signifies the firm fixing of the end of a cantilever, or the ends of a beam in a wall or other support. A beam is stronger when encastre than when simply supported.

Enclosed Wheels.—Reaction wheels (q.v.).

End Elevation, or End View.—A view on a drawing showing the end of a structure, as distinguished from its side view.

End Grain.—That face of a piece of timber exposed by the cutting of its fibres transversely to their course.

Endless Chain.—A chain united at the ends and used as a carrier of power, such as a pitch chain (q.v.), or an ordinary chain which is made to fulfil the purpose of a pitch chain by fitting into recessed chain wheels.

Endless Rope.—A rope united at the ends, and used as a carrier of power, as when the traversing motion is given to an overhead travelling crane by a man hauling at the rope below. Endless ropes run in smooth sheaves, or in wave wheels, the effect of the last-named form being to increase the bite of the rope.

Endless Saw.—A band saw (q.v.).

Endless Screw.—A worm (q.v.), so called because it drives continually in one direction, without reversal of its motion.

End Links.—The links which terminate the standard lengths, usually 15 fathoms long, of chain as ordinarily manufactured. They are therefore used as shutting links (q.v.), being about two-tenths thicker than the ordinary links.

End Measurement.—Measurement taken with a caliper or micrometer gauge, as opposed to line measurement (q.v.). This is far more accurate than line measurement.

End Play.—The play at the ends of a revolving part, that is, play in a longitudinal direction. It is essential in some cases, in others it is an evil. End play is necessary in the crank pins of oscillating cylinders, while in the back centres of lathes it is an evil.

End View.—See End Elevation.

Energy.—The capacity for performing work, or, using an older term, *vis viva*. It may be potential energy, or that which remains in abeyance until called into exercise, such as the energy present in fuel, or in a machine which is wound up, or in a lifted weight, or in the steam in a boiler. It may be kinetic energy, or energy in motion, or actual energy, doing tangible work. The energy of a moving body is proportional to the square of its velocity.

Engine.—A motor or prime mover (q.v.), automatic in its action, as distinguished from a machine which can only receive motion from a motor external to itself.

Engine Beam.—The beam of a beam engine.

Engine Boiler.—See Boiler.

Engine Counter.—See Counter.

Engine Cylinder.—See Cylinder.

Engineering.—See Mechanical Engineering.

Engine Fire.—The fire underneath an engine boiler.

Engine Friction.—The total friction developed by an engine independently of that of any machinery which it has to drive. Wide bearing surfaces, balanced valves, and proper lubrication are the means whereby its amount is reduced. It ranges from about 10 to 50 per cent., the last being an excessive amount, and only to be met with in engines of bad construction.

Engine Lathe.—A somewhat vague term, but as generally used signifies a lathe of moderate or large size, designed to perform the operations of engineer's work, and driven by the motive power derived from the engine.

Engine Pit.—A pit from 2 ft. 6 in. to 3 ft. 6 in. deep, dug between the rails in locomotive repairing sheds, and over which the engines are brought in order that the workmen may be able readily to get at the under work.

Engine Power.—The power of an engine is obtained by multiplying the piston speed into the total average pressure upon the piston. From the product the amount of engine friction has to be deducted, and the result is a remainder which divided by 33,000 gives the net horse-power.

Engine Register.—A counter (q.v.).

Engine Shaft.—The driving or crank shaft of an engine.

Engine Work.—The mechanical details connected with the construction of engines, as distinguished from other classes of machine work.

Entablature.—An overhead table or frame sustained by vertical columns arising from a base plate, and between which and the base plate certain machine parts are carried, as the working parts of forging presses, pumps, &c.

Entering File.—See Cottar Files.

Entering Tap.—A taper tap (q.v.).

Envelope.—(1) The covering of a solid. (2) The setting out or development of plates, and boiler-makers' work may be regarded as the development of the envelopes of solid bodies.

Epicyclic Train.—A train of gearing in which the centres of motion (q.v.) are not fixed, but movable in space.

Epicycloid.—A cycloidal curve which is formed by a generating circle rolling upon and without a fundamental circle. The faces of external or spur-wheel teeth are so formed.

Equal Forces.—Forces which act in opposite directions and balance each other.

Equality of Moments.—The condition of equilibrium in a system of forces, by which when power is gained time is lost, and *vice versâ*. When opposing forces balance one another, so that equilibrium results, there is said to be equality of moments.

Equalling File.—A very thin, flat, and generally parallel file, sometimes also furnished with one safe edge.

Equal-sided Angles.—See Angle Irons.

Equation.—An assemblage of quantities and signs, which placed to right and left of each other, and separated by the sign =, are numerically equivalent to each other.

Equation of Moments.—The representation of the various moments of force about a structure, put into the form of an equation.

Equilibration.—The application of balance weights to the driving wheels of locomotives and other rapidly revolving bodies, in order to produce steadiness of motion.

Equilibrium.—(1) Forces are in equilibrium when they neutralise each other's influence. (2) A body resting upon its base is in equilibrium when its centre of gravity lies vertically over the base on which it is supported. See Stable, and Unstable Equilibrium.

Equilibrium Governor.—A governor whose balls and arms are counterbalanced.

Equilibrium Ring.—A metallic ring attached to the back of the slide valve in large engines generally, for the purpose of lessening the amount of its friction, which friction becomes excessive in large valves exposed on the one side to the full steam pressure, and on the other to atmospheric pressure, or to the vacuum of the condenser. The ring encloses a space on the back of the valve equal in area to that of the exhaust port, into which space, therefore, no steam can gain admittance. The ring is commonly and properly put into a recess turned on the inside face of the valve casing, occasionally, however, on the back of the valve itself. It is kept pressed against the face by means of set-screws thrusting against india-rubber rings and metallic springs.

Equilibrium Slide Valve.—A slide valve which is provided with an equilibrium ring (q.v.) upon the back. Or a valve which is of the gridiron, double or treble ported type, the steam having free access to the outer and inside hollow portions See Double-ported Slide Valve.

Equilibrium Valve.—A valve in which the pressures on each face are equalised, or nearly equalised, so that it moves with the least expenditure of power and the minimum of friction. See Double-beat Valve, Equilibrium Ring, Equilibrium Slide Valve.

Equivalent.—A number or quantity numerically equal to another number or quantity, but expressed in different terms.

Erecting.—The final building up of machines or engines in their entirety in readiness for working. All the work of the turners, planers, slotters, drillers, and fitters is brought to the erector, presumably ready to go together, little or no adjustment being necessary at his hands. He therefore builds or erects the work which has been prepared in the other departments.

Erecting Shop.—A large, lofty, roomy building used for the erecting (q.v.) of engine and machine work. It may be a building separate from the fitting and machine shops, or a section adjoining these. It is provided with vices, a powerful overhead traveller, and rails, pits, foundation and base plates, or other conveniences, according to the nature of the engines and machines which are manufactured.

Erector.—A fitter or working engineer whose special task is that of erecting (q.v.).

Escape Valve.—See Cylinder Escape Valve.

Estimate.—See Tender.

Evaporation.—The conversion of a fluid into vapour.

Evaporative Value.—Denotes the relative capacities for vaporising water possessed by different types of steam boilers, by various arrangements of heating surfaces, or by fuels of different characters. These values are expressed either in horse-powers, or in units of work, or in thermal units.

Even Pitch.—A screw is said to be of even pitch when the number of threads per inch in it either corresponds with, or is some aliquot part of, the pitch of the threads of the leading screw of the lathe in which it is being cut. Thus with a leading screw of 4 threads to the inch, 4, 8, 12, &c., threads per inch would be even pitches, and the tool is in the right position for cutting in any position of the clasp nut.

Excavator.—A term applied to a class of machines whose function is the digging out and removal of earthy material in the preparation of foundations and harbour works. See Grab.

Excentric.—See Eccentric.

Exhaust.—The passage through which the spent steam of an engine cylinder obtains egress to the outer air, or to the condenser.

Exhaust Edges.—The inner edges, or the edges of the hollow or D portion of a slide valve, by which the exhaust steam is cut off.

Exhausting Fan.—Essentially an ordinary fan (q.v.), but having the sides enclosed to form an exhaust pipe.

Exhaustion.—The escape of spent steam through the exhaust pipe (q.v.). Sometimes termed eduction.

Exhaust Lap.—The reduction or narrowing of the inner faces of a slide valve to less than that distance which would correspond with a length measured between the inner edges of the steam ports, by which difference the ports are closed earlier than they would be if their edges coincided exactly with those of the arch of the valve. The result is that the steam is cushioned or compressed. The term exhaust lap, or its equivalent, eduction over-lap, is used in contradistinction to steam lap (q.v.).

Exhaust Line.—The bottom line of an indicator diagram, which represents the manner of the exhaustion of steam in an engine cylinder.

Exhaust Pipe.—The pipe which conveys away the exhaust steam from the exhaust port from an engine cylinder.

Exhaust Port.—The opening or means of exit provided for the escape of the exhaust steam from an engine cylinder. Its area exceeds by about one-half that of the steam ports in order to diminish the evil of back pressure. Sometimes called the eduction port.

Exhaust Relief Valve.—A slide valve whose construction is such that a greater width is opened to exhaust than to steam ; in other words, the port is fully opened to exhaust, but only partly opened to steam.

Exhaust Steam.—The steam which, having spent its force and accomplished its work in a cylinder, is allowed to pass away into the atmosphere or into the condenser.

Expanding.—Boiler tubes are expanded when their ends are opened out, and tightened in their tube plate. See Tube Expander.

Expanding Bit.—A boring bit, having a cutter or cutters capable of radial adjustment to enable the one tool to bore holes of different diameters.

Expanding Mandrel.—A form of lathe chuck, which is made to expand within the work, instead of, as is usually the case, to clip it upon the outside. The work is therefore bored first and turned afterwards. With this chuck, work of various diameters, within the range of the mandrel, can be centred truly, and when turned can be released without the risk attendant upon driving off with a hammer. The commonest form is that of two cones, or rather, frustra of cones, whose apices point towards each other, and between which the work is grasped by the tightening up of a bolt. There are several patent mandrels.

Expanding Metal.—Bismuth, and alloys of bismuth, which are therefore employed for the close filling up of holes.

Expanding Ring Clutch.—A form of clutch, or coupling, in which the friction of a split ring against a bored hole, or recess, is the effective agent. The female portion of the clutch contains a parallel bored hole. Within this fits a metallic ring divided transversely, and capable of being forced outwards against the sides of the hole by means of a wedge, the wedge being actuated by a lever or a screw. On the release of the wedge frictional contact ceases.

Expansion.—(1) The period during which the steam in an engine cylinder is increasing in volume, extending from the instant of admission or cut off to that of release. (2) The increase of steam in volume, corresponding with diminishing pressure. (3) The extension of bodies corresponding with increase of temperature, the extension being linear, superficial, or cubical, being in the ratios one, two, and three respectively.

Expansion Coupling.—A form of coupling in which two thin sheets of steel corrugated circumferentially are interposed between two ordinary flanged coupling faces which they connect, so that a certain amount of end play within the limits of the elasticity of the steel discs takes place. Other expedients are also made use of.

Expansion Curve.—The falling curve in an indicator diagram which corresponds with the falling pressure due to the expansive working of the steam in the cylinder, and which reaches therefore from the point of cut off to the commencement of exhaust.

Expansion Gear.—The whole of the arrangements by which the expansion of steam in an engine is accomplished and regulated.

Expansion Engine.—A steam engine worked expansively. See Expansive Working.

Expansion Hoop.—A ring which forms an expansion joint in a boiler flue, usually a Bowling hoop (q.v.).

Expansion Joint.—(1) A sliding joint inserted in a series of steam pipes in order to prevent the fracture which would result from the expansion and contraction of rigidly fixed pipes under changes of temperature. A turned spigot end of pipe passes through a socket-like joint (really a stuffing box) into the pipe beyond. A gland is fitted around the spigot end and within the stuffing box, packed with tow or gasket, to render it steam or water tight. The spigot pipe and gland are, therefore, free to slide one over the other, and so accommodate themselves to the expansions and contractions of the entire series of pipes. In a very long series, or where there are several bendings in the series, there will be several expansion joints. (2) The expansion of long boiler tubes is allowed for by a Bowling hoop, or an Adamson's flanged seam.

Expansion Plates.—See Cut-off Plate.

Expansion Ring.—See Expansion Hoop.

Expansion Rollers.—Rollers placed underneath the ends of long iron girders and roof principals, to allow of freedom of movement under the expansion and contraction due to heat. See also Pipe Roll.

Expansion Tank.—In apparatus for heating by hot water, the supply cistern is increased in capacity beyond that necessary for actual supply, in order to allow of the expansion of water due to heat without the risk of overflowing.

Expansion Valve.—The valve or valves used in the steam chest of an engine cylinder worked expansively. When a single valve only is used and the cut-off effected by the direct adjustment of the eccentric, it is, strictly speaking, an expansion valve, but the term is more properly used to designate an arrangement of two valves, that is a main valve having a cut-off valve (q.v.) behind it.

Expansive Working.—When steam in an engine cylinder, instead of being allowed to enter the cylinder at full pressure until the termination of the stroke, is cut off at some fractional portion of the stroke, and so caused to do work by its own expansion simply, the term expansive working is applied to it. The steam may be expanded in one only, or in two or more cylinders. See Compound Engine.

Explosion.—(1) In internal combustion engines, the burning of the charge of liquid or gaseous fuel, mixed with air. The term "explosion," while common, is misleading, since progressive burning of the charge should take place. If the mixture explodes, loss of power is experienced; the term detonation is then generally used. (2) In steam boilers the result of internal steam pressure exceeding in amount the tensile strength of the boiler plates, or that of their seams.

Explosion Chamber.—The hinder extension of the cylinder of a gas engine in which the charge is exploded.

Exposure.—The placing of a sheet of ferro-prussiate, or copying paper (q.v.), in a printing frame, with a negative print in the printing frame (q.v.), employed in the development of phototypes. The time of exposure varies with the weather and the nature of the print, the blue line phototype requiring less time than the white line. In sunshine an exposure of less than a minute suffices for blue lines: in white line, under similar conditions, from five minutes to half an hour. In dull

weather, for blue lines, from half an hour to three-quarters ; for white lines, from one to two hours. Fine cloth tracings copy best, tracings next best, paper drawings worst of all.

Expression.—A combination or assemblage of algebraical symbols.

External Forces.—Forces which act upon bodies, or systems of bodies, from without. These are, therefore, the forces which produce strains.

Externally Fired.—Applied to those steam boilers which have no internal fire box or furnace flue. They embrace the egg-end, balloon, and wagon boilers.

External Screw, or **External Thread.**—A screw cut upon the outside of a cylinder, therefore a male screw.

External Screw Tool.—A screw tool or a chaser of the form adapted for cutting external screws.

External Thread.—An external screw (q.v.).

Extrados.—The back or upper surface of an arch.

Eye.—(1) The looped or ringed end of a rod or lever, used as a medium of connection with another rod or lever, by means of a joint pin (q.v.). (2) The eye of a wheel or pulley is the central hole or bore through which its shaft or axle passes. (3) The eye of a furnace is that spot or area embraced or commanded by the sight holes (q.v.).

Eye Bolt.—A bolt provided with a hole or eye at one end, instead of the usual head. The eye is slipped over a round pin or stud, which receives the pull of the bolt.

F.

Face.—The face of a cog-wheel signifies the breadth of the teeth ; the face of a belt pulley, the breadth of the rim. The face of an anvil is its upper surface. The face of a casting is that surface which is turned or polished.

Face Chuck.—A face plate (q.v.)

Face Lathe.—A lathe chiefly or exclusively used for surfacing (q.v.). These lathes are generally a distinct class, having short beds, or only so much bed as is sufficient to sustain the saddle of the rest ; deep gaps reaching as low as the foot of the bed, or frequently lower still, being extended into a deep pit sunk into the floor of the workshop, so that wheels, and rings, turntables, &c., of twenty or thirty feet in diameter, can be turned.

Face Plate, or **Face Chuck.**—An intermediate metal plate, which is fitted between the mandrel nose of a lathe and chucks of the best class. It is attached to the chuck by screws, or by a taper plug (see Taper Plug), and to the mandrel nose by a screw, or by a taper plug. Often called a flange chuck. But the term face plate is more commonly applied in the shops to the ordinary face chucks, which are provided either with slots for bolts, or with movable dogs or jaws. See Jaw Chuck.

Face Plate Coupling.—See Flanged Coupling.

Facing.—A portion of a machine elevated above the general surface or area, for the attachment of another piece which has to be bolted thereto accurately. The amount of planing, filing, or chipping necessary to produce a fit between the two parts is thereby confined to the exact extent of surface where it is required, without reference to the general level of the casting

Facing Machine.—A centring machine (q.v.), which by the insertion of suitable cutters is employed for the facing of bolt heads, axles, studs, &c.

Facing Sand.—Sand used in iron foundries to line the faces of the moulds with a protective coating, in order that the molten metal shall not, by combining with the ordinary sand, form a skin of hard rough silicate. Facing sand is composed of coal dust, mixed with proportions of new or unused sand, and black or old sand. It is rammed against the face of the pattern, to the depth of about an inch.

Factor.—A single number which is multiplied into another number or numbers to form a product.

Factor of Safety.—When a calculation of the ultimate strength (q.v.) of a structure, or a portion of a structure, has been made, it is necessary to provide for contingencies dependent upon inferior material, wear and tear, sudden and unexpectedly applied loads, &c., and this provision takes the form of a multiplier, and is called the factor of safety. It will range from four to eight or ten in different classes of work. Sometimes termed margin of safety.

Fagot.—A mass of piled iron prepared in readiness for reheating and welding. It is either puddled bar of uniform, or of various sections, or scrap; and arranged in definite grouping according to the purpose for which the fagot is required.

Fagoted Scrap.—The ends, cuttings, or remnants of wrought-iron bar and plate of the smithy and boiler shop, piled into fagots and welded under the steam hammer. It is tougher than ordinary bar and is used for the best work. Sometimes called hammered scrap.

Fagoting.—(1) The laying of lengths of puddled bar in bundles or fagots for the purpose of reheating and rolling. (2) Likewise the making up of a fagot (q.v.) of any kind.

Fahrenheit.—See Thermometer.

Failure.—Materials and structures are commonly said to fail when they are unable to endure or accomplish the work for which they were selected and designed; whether they become crippled merely or undergo complete fracture.

Fair.—This has the same meaning as flush (q.v.).

Fairbairn Crane.—A fixed well crane whose jib is arched in order to obtain sufficient clearance beneath for the lifting of bulky loads. The section of the jib is that of a box girder (q.v.).

Fairlead.—A casting which takes the friction of the rope during the mooring of vessels.

Fairlie Engine.—See Bogie Engine.

Fall.—The amount of the descent of water to a mill wheel.

Falling Weight.—The dead weight of a steam hammer tup, piston rod, and head, irrespective of momentum or of steam pressure.

Falling Weight Test.—A test to which rails and bars are subjected, the loads being impulsive loads (q.v.), produced by the falling of weights. Rails and some test bars are subjected to this test.

Fall Pipe.—A drive pipe (q.v.).

False Bottom.—The bottom of a loam mill (q.v.) or mortar mill, formed of loose plates in order to permit of their removal, when worn out, without the necessity of replacing the entire pan.

False Core.—A term applied by brass moulders to a drawback (q.v.).

False Key.—A turned pin driven into a hole drilled, one half in the end of a spindle, and the other half into a boss, by which it is embraced, the centre of the pin being on the circumference of the spindle or shaft. Sometimes it is only put in for a temporary purpose, to be removed for the insertion of a permanent key. Also termed a glut.

False Water.—When steam is generated very rapidly in a boiler, the immediate effect is a rapid rising of the water level in the gauge cocks, due to the increase in volume caused by admixture of steam. This sudden increase is termed false water.

Fan.—A blowing machine, either of rotary or centrifugal type, and used for the production of furnace blast or for ventilation. (See Exhaust Fan.) It is considered better that the fan should be placed eccentrically in its case to allow for the accumulation of an increasing volume of air in the direction of the exit. The disadvantage of a fan in comparison with a blower is that it has to be driven at an excessively high, almost at a dangerous speed, with a great expenditure of motive power ; that its quantity of blast is not so measurable or invariable as that of a blower ; that the friction of its bearings is excessive, and that its working is not so much under control as that of the blower.

Fang Bolt.—A bolt in which the nut is a triangular plate with teeth for biting into timber, the bolt being tightened by revolving the head and shank. Fang bolts are used for attaching ironwork to wood.

F. A. S.—Free alongside ship. Seller engages to deliver goods, at his risk, alongside ship, but excluding loading.

Fast Coupling.—A shaft coupling united in a permanent manner with bolts or keys, as distinguished from loose sliding couplings.

Fastening Down.—The securing of the various parts of a foundry flask or moulding box together with cottar or screw bolts, to prevent their separation by the liquid pressure of the molten metal.

Fast Feed.—A quick feed (q.v.).

Fast Head.—A term sometimes applied to the headstock of a lathe, as opposed to the movable head or poppet.

Fast Pulley.—A pulley which is employed to transmit motion through the medium of a belt to a machine. It is keyed upon its shaft, and is called fast to distinguish it from the loose pulley which is not keyed on, and which simply carries the belt when the machine is not in use.

Fathom.—6 ft. or ·001336 of a mile, and equivalent to 1·8287 metres. Chains and ropes are sold by the fathom.

Fats.—See Oils.

Fatigue of Materials.—Materials which have been long subjected to severe straining actions, or to moderate straining actions often repeated, deteriorate in strength, and will then break under loads which they had previously sustained with safety. This appears to be due to molecular change and to the accumulation of permanent set (q.v.) to such an amount as precludes the possibility of further elongation. In cases where this fatigue has supervened it is customary to subject the material to a process of annealing by heating to a red heat in a furnace. Crane chains are frequently treated thus, with the result of restoring their lost tenacity.

Faucet.—The socket (q.v.) of a pipe. A term frequently applied to a water cock of small or moderate size.

Feather, or Sunk Key.—A parallel key which is partly sunk into a recess in its shaft, so as to form an integral part of the shaft, and over which its wheel or clutch slides. Feathers are used when machine parts have to be thrown into and out of gear, the wheel or clutch being driven by the feather, but having freedom of motion longitudinally.

Feathering Board.—See Feathering Float Paddles.

Feathering Float Paddles.—Paddle-wheels (q.v.) provided with float boards free to move eccentrically in relation to the main wheel, so that the boards shall dip into and leave the water as near as possible in a vertical direction, thus economising power.

Feathering Screw.—A propeller screw whose blades are rendered movable, so that they can be placed parallel with the line of the keel when it is desirable that the vessel shall sail under canvas alone.

Feed.—The amount or distance of the transverse of a lathe or other machine cutter taken transversely to the depth of the cut. The term is also applied to the feed gear itself.

Feeder.—(1) A head of metal cast over a disproportionately heavy part of a casting, in order to supply the necessary metal to compensate for the sinking in of the heavy section, due to shrinkage. Sometimes called a shrinking head. (2) The opening made in a foundry mould for the introduction of the feed rod.

Feeder Head.—Head metal. The mass of metal which has been utilised for feeding a mould. See Feeder, Feeding.

Feeding.—As a mass of molten metal solidifies the outside will set first, the inner continuing fluid for some time after. As the inner portion cools it will contract, and as a consequence shrink upon itself, leaving a depression on the top outside face. To avoid this evil it is customary to feed the heavier portions of castings by inserting a small metal rod through the runner or riser, as the case may be, and by its motion keep a passage open for the inflow of fresh metal to compensate for the contraction. Feeding is continued until the mass becomes too pasty to allow of the rod being moved any longer. The feeding metal is either supplied in small quantities from time to time by a hand ladle or by having a sufficient mass of molten metal in the head in the first place.

Feeding Head.—See Feeder Head.

Feeding Rod.—The rod used for feeding (q.v.), usually $\frac{1}{4}$ in. or $\frac{3}{8}$ in. in diameter, and about 3 or 4 ft. long.

Feed Pipe.—The pipe which conveys the feed water from the feed pump to the boiler of an engine.

Feed Pump.—(1) A force pump which supplies steam boilers with feed water. (2) A force pump used for supplying gas engines with the necessary supply of air.

Feed Rod.—See Feeding Rod.

Feed Screw.—The screw by means of whose rotation through an arc of a circle an exact and measurable amount of feed is imparted to the cutting tool of a machine.

Feed Tank.—The tank which contains the supply of feed water for a steam boiler.

Feed Water.—The water used for the supply of a steam boiler. Sometimes it is used quite cold, sometimes warmed in a feed water-heater (q.v.). In condensing engines it is warmed in the condenser. As the durability of the boiler depends largely upon the character of the feed water, it should be chosen, if possible, with reference to its softness; hard water promoting incrustation (q.v.).

Feed Water-Heater.—A kind of boiler in which the supply water for a steam boiler is heated, preliminary to being taken up by the pump or injector. It consists usually of an arrangement of tubes enclosed in a cylindrical shell, the exhaust steam being employed as the heating agent.

Felt.—Used as a non-conducting sheathing for steam boilers and engine cylinders. It is of a dark brown colour, and consists of the woollen refuse from paper mills. Felt owes its property of matting, or "felting," to the well-known jagged character of animal hairs, as seen through the microscope.

Felting.—The covering of steam pipes with felt to prevent loss of heat by radiation. The felt is either retained in place by coils of wire, or by wood lagging.

Female.—The recessed portion of any piece of work into which a dowel or a stud fits is called the female portion.

Female Screw.—A screw which is cut in a hole or on a hollow surface.

Female Gauge.—An internal or bored gauge.

Fence.—A metal plate attached to a saw-bench for the guidance of the stuff edgeways, and the insuring of parallelism in the cut. Fences are made to stand at right angles, but are capable of an angular adjustment, or side cant, for bevel sawing. Hence called canting-pieces.

Fencing In.—The enclosure of machinery and revolving parts, in order to diminish the risk of accidents. Fencing in is compulsory. Belting, gearing, shafts, &c., when they are in proximity to workmen are fenced in.

Ferric Oxide.—The peroxide of iron, $Fe_2 O_3$, is the condition in which iron occurs in the red hæmatites, and, combined with water, in the brown hæmatites. It is also the condition in which the black band and clay band (q.v.) ores are fed into the blast furnace, the carbonate and an atom of oxygen having been driven off by the process of calcination.

Ferro.—See Ferro-Manganese.

Ferro-Manganese.—A variety of pig-iron which contains a large proportion of manganese. It differs from spiegeleisen in the larger percentage of manganese present, the term ferro-manganese being applied to those pigs containing upwards of 20 per cent. of manganese. Usually termed ferro, simply.

Ferro-Prussiate Paper.—See Copying Paper.

Ferrous Carbonate.—The spathic iron ores, the clay and black band ores, which are by far the most important ores of iron reduced in smelting furnaces, are essentially ferrous carbonates—$Fe CO_3$, that is ferrous oxide, FeO, combined with carbonic acid, CO_2, thus, $FeO \times CO_2 = Fe CO_3$. During calcination the carbonic acid is driven off, and partial oxidation taking place also leaves the ore in a ferric condition.

Ferrous Oxide.—See Ferrous Carbonate.

Ferrule.—A ring of hoop iron enclosing a round core print, and left in the mould to receive the core. The reason of using a ferrule is to prevent the sand from being pushed away by the entrance of the core in cases where the latter is thrust in after the final closing of the mould; and the founder is therefore not able to see the interior of the mould afterwards in order to ascertain if the core has carried in any sand with its passage downwards. The ferrule prevents the core from coming in contact with the sand and so removes the risk referred to. (2) Rings of hard wood used for holding condenser tubes in their plates. The ferrule fits between the outside of the tube and the hole in the plate, and being swelled by the action of the water renders the tubes water-tight. (3) Rings of metal enclosing and confining the wood around the tangs of edge tools to prevent splitting.

Fettlers.—The men whose business it is to clean off the foundry castings

They are paid by the piece, usually so many shillings per ton depending on whether the castings are heavy or light. There are special rates for brass fettlers.

Fettling.—(1) The dressing-off of castings after they leave the foundry sand, including the chipping off of runners, joint marks, fins and scabs, the cleaning out of cores, and the brushing of all sand off their surfaces. The tools used are chisels and hammers, files, rods and wire brushes. Also called trimming. (2) The infusible lining or coating of oxides of iron employed for the protection of puddling furnaces and purification of the iron. See Bull Dog, Wheel Swarf, &c.

Fibre.—Malleable iron which has been subjected to the process of rolling develops a fibrous structure, and is stronger in the longitudinal direction than in the transverse, in the proportion of about twenty-three to seventeen. Yet, though a bar of wrought iron will show a distinct fibrous structure if drawn out or bent slowly, it will if nicked and broken off sharply show a crystalline fracture, as though the fibre were only developed in the process of bending.

Fibrous Iron.—Bar iron, whose fibrous texture results from the intermixture of layers of cinder, and is therefore a source of weakness and a mark of inferiority.

Fiddle Drill.—See Bow Drill.

Field's Tubes.—Vertical boiler tubes employed for circulation and heating purposes. They are open at the top or water end, and closed at the lower or fire end, are suspended from the furnace crown, and descend into the fire which plays all around them, excepting at the end whence the steam and water have egress. Within the outer, or water tube, a second smaller tube is dropped, being suspended from a conical casting resting in the upper end of the outer tube. The efficiency of Field's tubes depends on the fact that a heated column of water is lighter than a colder column. The heated water rises in the outer tube, and the cooler water descends through the inner one to supply its place. The two waters are prevented from mingling at the top by the conical or bell-mouthed deflector attached to the top of the inner tube.

File.—A tool for removing particles of metal or other material by abrasion, or by the cutting of a multitude of fine points. Files are made in a vast number of sizes, and shapes, and degrees of coarseness, and are cut by hand or by sand blast. See details under special headings.

File Handle.—A short stout handle, into which the tang of a file is driven. A file handle should be large and globular at the end or heel which fits into the hollow of the hand, as any other shape soon tires the muscles.

Filing.—The art of filing is such that it can be acquired by practice alone, the difficulty consisting mainly in producing a flat surface, the tendency invariably being to produce one of rounding contour. The economic importance of filing in the shops has been much diminished by the universal employment of planing, shaping, and milling machines, and of emery wheels.

Filing Block.—A piece of board laid upon a work bench, or held in a vice, upon which thin portions of metal are laid flat while being filed.

Filings.—The minute particles of metal abraded by the teeth of a file.

Filled Rail.—A point rail, or a stock rail, which has one or both sides filled up flush, to give excess of strength to compensate for the cutting away or tapering of the point rail, or the weakness of the stock rail due to the cutting out of the gap for the crossing.

Fillet.—A term loosely applied to many diverse meanings. Hollows (q.v.) are often called fillets; mouldings, beadings, flanges, chipping strips, and similar parts are designated thus. In a general way it applies to any thin strips used either for strength or ornament.

Filling In.—See Box Filling.

Fin.—Any thin wafer-like expansion of metal occurring on the side or edge of a larger portion. Fins occur on the joints of castings, and of forgings pressed in dies.

Final Pressure.—The pressure at which the steam is exhausted, or discharged from an engine cylinder. Also called terminal pressure.

Fine Feed.—The feed of a machine cutter is fine when minute in quantity.

Fine Grain.—The crystallisation of fractured metal is fine when the crystals are of small size.

Fine Grit.—A grade of emery wheel, or grindstone, in which the particles are of small size, and suitable for the later and more delicate processes of abrasion.

Fine Hard.—A quality of emery wheel useful for grinding tools, and for light work in general.

Fine Metal.—Copper matt which results from the fusion of calcined coarse metal with slags and ores. It is called blue metal when it contains sixty to seventy per cent. of copper, white metal from seventy-five to seventy-eight per cent., and pimple metal above seventy-eight per cent. See Matt, Refined Iron.

Fine Pitch.—A toothed wheel is said to be of fine pitch when the teeth and teeth spaces are of small dimensions. The term is relative.

Finery.—A puddling furnace (q.v.).

Fine Soft.—The finest quality of emery wheels (q.v.) used for polishing brass work, and for sharpening tools.

Fine Teeth.—The teeth of saws, wheels, files, &c., are fine when of small dimensions, estimated relatively to other teeth.

Finger.—A narrow projection used as a guide or index in various kinds of work, foundry or otherwise.

Finished Dimensions.—The dimensions of a piece of work after turning, boring, planing, &c., as distinguished from rough dimensions (q.v.).

Finishing Cuts.—After the major thickness of metal has been removed from the surfaces of castings and forgings, it is usual in all but the roughest classes of work to take a last fine or finishing cut with a scraper or spring tool.

Finishing Rolls.—See Puddling Rolls, Mill Rolls, Plate Mill.

Finishing Tools.—Commonly applied to those cutting tools used by metal turners, whose cutting edges are broad and straight, for the removal of the ridges left upon work by the roughing tools (q.v.). They include square and right and left hand side tools, and spring tools.

Fir.—See Spruce, and Pine Wood.

Fire Bars.—The grate bars of the furnace of an engine or other boiler. Single gratings would be too large and cumbrous, hence separate bars are used, and replaced as they burn out.

Fire Box.—That part of a boiler which receives the fuel and the water tubes. It is commonly applied to the furnaces of boilers of the locomotive type, and to those of vertical form.

Fire Box Crown.—The upper plate of the furnace in a vertical or locomotive boiler. It is flanged, and riveted to the outer plate of the fire box.

Fire Box Shell.—The plates composing a fire box (q.v.). The shell is double in locomotive and portable boilers, comprising inner and outer portions, the outer being of iron, the inner, which is exposed to the fire, of copper, and the two are maintained apart by means of short stays passing through the water space. The sides of the inner shell are tapered upwards to facilitate the disengaging of the steam from the surface, the water space therefore being an inch or more in width at the upper than at the lower portion.

Fire Brick.—Fire bricks of various shapes are employed for the permanent lining of cupolas, blast, reverberatory, and other furnaces. Over these is a lining of clay, ganister, or other highly refractory material.

Fire Brick Arch.—A curved arch set at an angle to the tube plate of a locomotive fire box in order to deflect the flame before it passes into the tubes. See Brick Arch.

Fire Bridge.—See Bridge.

Fire Clay.—Fire clays are very numerous, but they consist essentially of silica, alumina, and water in various proportions. They are found in the coal measures chiefly; Stourbridge, in Worcestershire, being especially noted for its clays, and from them are manufactured bricks and crucibles for foundry use.

Fire Door.—(1) A furnace door. (2) The door of a steam boiler through which the fuel and firing irons are introduced.

Fire Flues.—The flues of Lancashire and Cornish boilers. Strictly speaking, the front portions of those flues which contain the grate bars.

Fire Hardening.—See Hardening.

Fire Hole.—The door of a furnace, or that portion of the grate immediately within the door through which the stoking is effected.

Fire Tube Boiler.—A multitubular boiler (q.v.) as distinguished from a water tube, or sectional boiler.

Fire Hole Ring.—A ring of wrought iron encircling the fire hole of a locomotive or portable boiler, and serving to unite the inner and outer fire boxes.

Firing.—(1) The application to, or the quickening of the source of heat in a steam boiler is termed firing, or "firing up." (2) The ignition of a charge in a gas engine. (3) A wooden brake is said to fire when the heat of friction produces sparks. (4) A bearing is said to fire when it becomes red-hot. (5) Oil is said to fire when it spontaneously ignites.

Firing Chamber, or Lighting Chamber.—The small cavity or chamber through which the charge of a gas engine is ignited. At the moment of firing it is shut off from communication with the outer air.

Firing Hole.—The door in the side of a reverberatory furnace through which the fuel is introduced to the grate area.

Firing Tools.—The ordinary tools used in stoking. See Hook, Rake, Slice.

Firmer Tools.—The ordinary short chisels and gouges of wood workers, so termed in order to distinguish them from paring tools (q.v.).

First Motion.—A term of general application, as first motion shafts, first motion belts, first motion wheel, &c., meaning the one which first receives, and then communicates, power to its successors.

Fish-bellied.—Girders and ribs are said to be bellied (q.v.), or fish-bellied, when they are curved underneath, the depth of curve increasing towards the centre.

Fish-bellied Girder.—A form of girder, whose under side forms a drooping curve, the girder being deeper in the centre than at the ends.

Fish-bellied girders are usually built up of wrought-iron plates and angles, and require no trussing.

Fish Bolt.—A bolt employed for fastening fish plates and rails together.

Fish Joint, or Fished Joint.—A joint made between rails through the medium of fishplates (q.v.).

Fish Oils.—Oils used for lubrication, embracing sperm, whale, seal, cod, &c. See Animal Oils, Oils, &c.

Fish Plate.—A plate of metal covering the butt joints of boilers, rails, and other work. Any plate covering a riveted joint is called a fish plate.

F. I. T.—Free in truck. Engages to load goods in railway trucks without extra charge.

Fitter, or Engine Fitter.—A working engineer whose duties consist in the fitting together of machine or engine parts, after the preliminary stages of turning, planing, drilling, &c., have been accomplished.

Fitter's Bench.—A stout wooden bench, to which is attached the vice or vices, and a drawer for the reception of tools. The bench is low enough to give a proper working height to the vice jaws.

Fitting.—That section of mechanical engineering devoted to the bringing together and adjusting of the different portions of engines, machines, &c., after they have left the hands of the turners and machinists. In a limited sense, however, the working up and finishing of the smaller portions of metal work which cannot be done in the machines.

Fittings.—Commonly applied to the essential parts or adjuncts of an engine, or boiler, or machine. Specially applied to boiler fittings (q.v.).

Fitting Shop.—The shop in which the operations of fitting are carried on. It may occupy a ground floor, or an upper storey, and may be separate from the machine and erecting shops, or combined with these in one building. Its chief feature consists in the parallel vices bolted to long benches, ranged around the shop against the windows, each fitter having a distinct vice, and separate drawer in which are kept his files and necessary small tools. Each fitte has usually a small surface plate against his vice, and in addition there are large plates in various parts of the shop. These, and a few light wall cranes, and one or more light overhead travellers, nearly complete the equipment of a fitting shop.

Fix.—The fettling (q.v.) of a furnace.

Fixed Centre.—A centre whose position is localised in space, in opposition to a movable centre (q.v.).

Fixed Crane.—A crane set down for use only in one place, in opposition to portable cranes, or those which are constructed to travel.

Fixed Cutters.—Plane iron cutters of great width set in a box, or shoe, and used for facing, planing, and match boarding. The cutters are fixed in the bed of a machine, and the stuff is pushed over them at a rate of about one hundred feet per minute.

Fixed Eccentric.—An eccentric sheave fixed upon its shaft, the movement for forward or backward gear being given through the link work. Fixed eccentrics have mostly superseded the old adjustable eccentric (q.v.).

Fixed Expansion.—That expansion which is constant, being due only to lap on the valve.

Fixed Head.—The head or tool box of a shaping machine, which is attached to the ram, and has motion only in the direction of its stroke.

Fixed Oils.—See Oils.

Fixed Pulley.—A pulley (q.v.) whose axis is fixed in space, the pulley revolving thereon.

Flame Box.—Sometimes applied to that portion of the shell of a steam boiler which contains the smoke or flame tubes.

Flame Furnace.—A reverberatory furnace (q.v.).

Flame Plates.—The top or crown plates of a boiler flue, or fire box. These being exposed to the action of the flame are usually of Low Moor (q.v.) or Bowling (q.v.) iron.

Flanch.—See flange.

Flange.—A rib, or offset on a casting, used either for the purpose of imparting steadiness of base, or of providing for bolt attachments through holes cast or drilled therein. Cast-iron flanges are also bolted, riveted, or screwed to structures of wrought iron. Occasionally spelt flanch.

Flanged Bend.—A bend pipe furnished with a flange at each end.

Flanged Chuck.—A face plate (q.v.).

Flanged Coupling.—The ordinary shaft coupling in which the abutting shafts are keyed into the bossed portion of each half coupling, and bolts passing through the flanges maintain the two halves in position. Also called face-plate coupling.

Flanged Beam.—A rolled beam. See Joists.

Flanged Girder.—A flanged beam (q.v.).

Flanged Joint.—See Flanged Seam.

Flanged Nut.—A nut having a broad flange turned solid with its bottom face. It is used instead of employing a separate washer, its purpose being the covering over of holes of large diameter.

Flanged Pipe.—A steam or water pipe, which is provided with flanges at the ends as a means of attachment to other pipes or connections. In cast-iron pipes the flanges are a part of the casting. In wrought-iron pipes they are separate castings screwed over the pipe ends upon their outside diameters.

Flanged Seam.—A seam used in furnace tubes, by which the tubes are materially strengthened, and compensation made for their endlong expansion. It may consist of a ring of $\top$-iron riveted to the tube plates on each side. Or it may be formed by the riveting of a ring, arched or corrugated, with a single corrugation in the section; the flue tubes being attached to narrow borders or flanges spreading out from the base of the arch. Or it may be an Adamson's flanged seam, in which the ends of the flue tubes are curved or dished outwards, and riveted together through an intermediate ring. The first form allows of no endlong expansion. The last two permit of free expansion longitudinally.

Flanged Socket.—A very short pipe, having a flange at one end and a socket at the other.

Flanged Spigot.—A very short pipe, having a flange at one end and a spigot at the other.

Flanged Wheel.—A truck or trolly wheel, having a flange or flanges at the edges to keep it on the rails. It may either have one flange, being then a single-flanged wheel, or two flanges, or a double-flanged wheel.

Flange Rail.—A flat-bottomed or flat rail, as distinguished from a double-headed rail.

Flange Washer.—See India-rubber Washer.

Flanging.—The bending over of the edges of wrought-iron and steel plates to form narrow flanges for the purpose of attachment to other plates, either by riveting or by welding. Flanging is mostly done hot, and the edge is beaten over by wooden mallets or iron hammers, or where a quantity of similar work has to be accomplished, in dies, screwed down or squeezed together under a forging press. Flanging is boiler-makers' work.

Flanging Machine. —A machine used for bending over the flanged portions of boilers and other built-up structures in wrought iron and steel. These machines are actuated by steam power, or by hydraulic pressure, and the laborious operation of flanging by hand is saved by their employment.

Flanks. —(1) The sides of wheel teeth lying below the pitch line. (2) The sides of an arch.

Flap Valve. —A hinged valve, circular or rectangular in outline, and usually made of leather, stiffened with iron or brass plates riveted in each face. Flap valves are used in ordinary lifting pumps, both in the piston and at the suction end of the barrel. The disc valve (q.v.) of an air pump may be considered a flap valve hinged at the centre and lifting around the circumference.

Flash Flue. —The flue underneath an egg-end, or similar externally fired boiler.

Flash Point. —The constant temperature at which an oil or fat flashes momentarily into flame. A good oil should not flash below 500° or 600° F. Some oils flash below 200° F.

Flask, or Moulding Box. —The frame enclosing the founder's work. The box receives the sand in which the pattern is rammed up, and from which it is withdrawn, leaving the mould ready for the inflow of metal. Flasks are parted into two or more horizontal sections to enable the moulder not only to withdraw the pattern, but to clean and black the faces of the mould. They are held together temporarily while the metal is being poured, by pins passing through lugs cast upon their sides; and the various parts are secured by heavy weights or by cottared bolts. See Fastening Down, and Weighting Down.

Flat Bar Iron. —Wrought iron of rectangular cross section. Its sections range from 1 in. $\times$ $\frac{1}{4}$ in. upwards.

Flat Bit Tongs. —Smith's tongs whose jaws are flat and parallel.

Flat-bottomed Rail. —A railway rail whose base is spread out into a broad flat flange of from 5 in. to $5\frac{1}{2}$ in. wide, to be spiked directly to the sleepers, or to be wedged into chairs. When the rail is solid throughout it is termed a Vignoles rail; when it is arched like a bridge and hollow underneath it is termed a bridge rail.

Flat File. —A file of rectangular section whose width is greater than its thickness. It is either a tapered or a parallel file.

Flat Gouge. —One which has the least amount of curvature given to gouges, the amount of curvature varying, however, with the width. See Middle Flat Gouge, Quick Gouge.

Flat Iron. —See Flat Bar Iron.

Flat Rammer. —A flat-ended iron tool used in foundries for ramming over large surfaces of sand. It is furnished with a stout handle of three or four feet in length, and the workman, holding this in front of him, with both hands gives repeated blows downwards upon the sand bed, finishing the surface therewith.

Flat Rope. —See Wire Rope.

Flats. —Flat Bar Iron (q.v.).

Flat Sweep. —A flat sweep or curve signifies one which is relatively of less curvature than others with which it may be compared.

Flatter. —A species of hammer used by smiths. It is like a set hammer (q.v.) but broader on the face. It is held by a handle upon the work and struck with the sledge. Its use is to finish over broad surfaces which have been brought to size by the sledge and set hammer.

Flaw.—A crack or incipient fracture in a casting or forging. Flaws are often difficult to detect by inspection alone, but their presence is usually found by ringing with a hammer, as sound metal always has a sonorous ring. See also Crack Detection, Dict. of Modern Terms.

Flax.—Used for steam packing (q.v.).

Flesh Side.—The rough side of a leather belt.

Flexible Ball Coupling.—See Flexible Coupling.

Flexible Coupling.—The flexible ball coupling consists of two discs attached to the shaft ends, and hollowed on their faces to embrace a ball placed beween the two. Projecting jaws on the faces of each coupling also fit loosely into corresponding recesses on the opposite disc face and drive the shafts. The coupling faces are retained in contact by means of screw bolts passing loosely through them, and rendered flexible endways by elastic washers under their nuts. The coupling with its shafts is therefore perfectly free to turn through moderate angles around the ball as an axis.

Flexible Crank Shaft.—A crank shaft in which the strains due to the rigidity of an unyielding mass compelled to revolve under conditions of strain due to the want of alignment of its bearings, are reduced and minimised by the introduction of flexible joints in its length. The joints consist of a modification of the flexible ball coupling, but a bush and ring plate, convex and concave, are introduced instead into the crank web, and so form a flexible pivot or centre around which the straight shafts may rotate through a small angle.

Flexible Hose.—Piping made either of india-rubber or leather, and used for the conveyance of liquids where metal pipes would be unsuitable by reason of their rigidity.

Flexure.—See Bending.

Flitch.—A plate of metal or of wood bolted to an otherwise weak and unstayed beam or structure in order to strengthen and support it.

Flitch Beam.—A combination beam in which outer timbers enclose a central joist of rolled iron, screw bolts passing through the whole to bolt the members together.

Flitch Plate.—A broad thin plate or rolled bar used in building up flitch beams and plated work. The width may range from 6 to 24 inches.

Float.—A buoy which is used to indicate the height of the water, either in tanks or boilers. In ordinary cold-water tanks it consists of a disc, or block of light wood, attached by a piece of jack chain, running over a pulley, to an indicator set in a convenient position for observation. In steam boilers it is made of some heavier substance, as stone or iron, which is unaffected by the liquid, and which is rendered buoyant by the attachment of a counterpoise on the outside, the counterpoise being so proportioned in reference to the specific gravity of the float as to render it as susceptible to the variations in water level as a float of wood.

Float Boards.—(1) The rectangular boards attached to the arms of a paddle wheel (q.v.). (2) The boards which receive the impulse of the water in an undershot wheel (q.v.).

Float Cut.—This signifies a file having single lines of cutting teeth only, as distinguished from double-cut files, or those with crossing rows of teeth.

Float Gauge.—A water gauge, where the height of water in a steam boiler is registered by means of a float (q.v.).

Floating.—When the lever of a weigh-bridge or of a testing machine is in equal balance it is said to be floating.

Floating Derrick.—A derrick crane placed on board a floating hull for transferring goods to and from vessels, independently of proper wharf and dock accommodation.

Floating Mill Wheel.—A water wheel, having its bearings in a boat moored in the stream of a rapidly flowing river, which turns the wheel and performs work for which it is suitable.

Flogging Chisel.—A chipping chisel of large size, which is struck with a light sledge, one man holding the chisel while another strikes. It is a tool which is now seldom used.

Floor.—The sand bed of a foundry is termed the floor, and the work is said to lie on the floor, the men kneel on the floor, patterns are bedded in the floor, &c.

Floor Plates.—Foot plates.

Floor Rest.—In large pattern-makers' lathes, used for wheels and similar work of large diameter, a rest is provided, which is carried on a heavy standard, resting upon the floor. The pattern-maker then has more control over the work than with the ordinary bed rest.

Flour of Sulphur.—Used frequently mixed with sal ammoniac and iron borings, for the making of rust joints. Its presence in large quantities makes the cement more slowly setting.

Flow Gate.—A term sometimes applied to a riser (q.v.).

Flowing Metals.—Metals of the ductile class which yield to, and change their form, under impact, or tensile or compressive strain. Wrought iron, mild steel, lead, copper are the flowing metals regarded from the point of view of an engineer.

Flue.—The flues of boilers are those parts which carry off the smoke and waste gases, and produce the draught necessary for combustion. Cornish boilers have one, Lancashire boilers two internal flues; besides which they are embedded in brickwork, through which external flues are constructed to produce what is called a split draught (q.v.), the draught being capable of regulation by means of dampers. (See also Wheel Draught.) The copper tubes of a locomotive, or portable boiler, and the short stack answer to the flues of ordinary boilers, while the action of the blast pipe compensates for the shortness of the tubes.

Flue Bridge.—A bridge of brick enclosing a hollow frame of cast iron, which passes transversely across a reverberatory furnace between the hearth and the stack.

Flue Plates.—The ends of a horizontal boiler to which the flues are fastened, or the fire-box crowns (q.v.) of vertical boilers.

Flue Surface.—The area of the flues of a boiler, as distinguished from the grate surface; being exposed to the action of hot gases, though not to actual flame.

Flue Tube.—A furnace tube (q.v.).

Fluffy.—Timber is said to be fluffy when the sawdust is stringy, and moist or greasy instead of granular and sharp.

Fluid.—A body or substance which yields to any pressure imposed upon it from without, and therefore cannot sustain a longitudinal pressure in any direction without receiving corresponding lateral support. In fluids at rest the pressure is therefore equal in all directions. Fluids are either liquids or gases.

Fluid Compressed Steel.—See Compressed Steel.

Fluidity.—The various brands of cast iron possesses different degrees of fluidity when melted. No. 1 Scotch is the most fluid of all, and is therefore used for thin, finely marked and ornamental work. It is also mixed with stronger brands to increase their flowing properties. White iron runs thick and pasty. Metal can be rendered less fluid, as is sometimes desirable, by allowing it to cool in the ladle previous to casting. As a general rule the metal is run hot and fluid for thin castings, cool and thick for heavy ones.

Fluid Pressure.—Pressure is transmitted by fluids in all directions alike; the transmitted pressure is equal in all directions, and it is directly proportional to the area of the surfaces.

Flume.—A channel conducting the water to a water wheel.

Fluor Spar.—A compound of fluorine and calcium, used in some metallurgical operations as a flux. Its fluorine combines with silicon, forming fluoride of silicon, while the calcium enters into combination with oxygen to form lime, which again combines with silica to form a silicate of lime, and is then able to form a flux with the clay and oxide of iron present in the furnace.

Flush.—Parts are said to be flush or fair when their surfaces are on the same level.

Flush Bolt.—A bolt whose head is let into a countersunk hole so that its top face stands level with the face of the plate into which it is sunk.

Flushing Charge.—A charge of air or water swept through the combustion chambers of gas engines to remove the residual gases or products of combustion. Sometimes a displacer piston is employed instead.

Flush Topped.—A term applied to the firebox casing of a locomotive boiler when it forms a continuous line with the boiler casing instead of being shouldered upward.

Fluted Reamer.—A reamer fluted longitudinally, to cut at its sides, to distinguish it from a rose reamer or broach (q.v.).

Flutes.—Grooves, semicircular in section, which are cast in ornamental columns.

Flux.—The substance thrown into a smelting or a melting furnace for the purpose of combining with those infusible and other matters which require to be separated from the metal, and which alone could not be rendered fluid at the temperature of furnaces. Also, substance used to protect surface of molten metal from contact with the atmosphere in welding, brazing and soldering, and to improve the " flow " of the metal.

Fly.—(1) The loaded lever which actuates the screw of a fly press. (2) Cast iron is said to fly when it breaks through unequal contraction due to bad proportioning, producing internal stresses, or to the effects of cold or of heat acting upon it, or to both in combination.

Fly Cutters.—Cutters set in a cutter block or chuck held in a lathe or in a facing and chamfering machine, and used for shaping the ends of metal rods presented thereto. The cutting edges are set in a diagonal direction, sloping towards the centre. Also applied in general to any revolving cutters which are used for shaping definite sections, as those of wheel teeth, beads, and other irregular forms whose sections are uniform.

Fly Nut.—See Wing Nut.

Fly Press.—A punching machine employed generally for stamping out any thin works in metal, and for punching and gulleting out saw teeth. The punch is actuated by a vertical screw and lever, the lever being weighted with heavy balls to impart due momentum.

Flywheel.—A heavy wheel used in engines and machinery where reciprocal motion is converted into motion of a circular kind. The flywheel, by absorbing some of the energy of the machine, carries the disc or crank over dead centres, and also maintains uniformity of motion by alternately absorbing and giving out work. Flywheels are used with engines, pumps, reciprocating or frame saws, jigger saws, &c.

Foam Cock.—A scum cock (q.v.).

Foam Collector.—A scum trough (q.v.).

Foaming.—Priming (q.v.).

F.O.B.—Free on board. An engagement to deliver goods on board ship without extra charge.

Foci.—The centres from which the end curves of elliptical figures are struck. They are set in the transverse axis (q.v.).

Follower.—A wheel which is driven by another wheel. See **Driver, Driven.**

Following Edge.—That edge of the blade of a screw propeller which leaves the water behind it, as distinguished from the leading edge (q.v.).

Following Joints.—The rings of which cylindrical boilers are built, being lap jointed, fit one within the other. Instead, therefore, of being parallel cylinders, they are necessarily frustra of cones, and as the joints all lap in one direction, they are termed following joints.

Following Steady.—A steady (q.v.) which is attached to the back of the side rest of a lathe, and which embraces the work behind or after the tool and follows it along with the rest. See **Back Steady Rest.**

Foot.—(1) A base or flange which sustains a casting or structure. (2) A common standard of measurement.

Foot Board.—See Treadle.

Foot Brake.—A brake which is applied or released by means of a lever actuated by the pressure of the foot.

Foot Bridge.—An arched bridge which carries a footstep bearing.

Foot Drill.—A light drilling machine driven by a treadle (q.v.).

Foot Lathe.—A light lathe of from three to six inch centres, driven from a treadle and crank actuated by the foot. Foot lathes are seldom employed in engineers' shops, and then only for special classes of work, models, &c.

Foot Lever.—A lever worked by the pressure of the foot alone. Foot levers, as a rule, are provided with a balance or counterweight to throw back the lever automatically when the pressure of the foot is withdrawn, and the part on which the foot presses is flattened out and roughed up to prevent slipping.

Foot Pound.—A unit of work used in calculations, signifying one pound lifted one foot high. See Units.

Foot Step, or Footstep Bearing.—A bearing closed at its bottom end, to sustain the end thrust of a vertical shaft or spindle. It is, therefore, a bearing socket, called also a step, and toe step. Footstep bearings are sometimes made adjustable by placing the bearing in the centre of a ring casting provided with a bridge carrying a set-screw below and four set-screws around the ring. By slackening and tightening the set-screws, the bearing is adjusted both sideways and vertically.

Footstep Bearing.—See Footstep.

Foot Ton.—A unit used in calculation, signifying one ton lifted one foot high. See Units.

Foot Valve.—The lowermost valve in a pump.

F.O.R.—Free on rail. Signifies the placing of goods on the railway without extra charge.

F.O.T.—Free on truck. The same as F.I.T. (q.v.).

Force.—The power or cause which impresses or tends to impress motion on matter.

Forced Draught.—Forced air supply to a furnace, with or without the employment of the closed stokehold (q.v.). The latter, however, permits of a better and equable regulation of the forced draught, which is then compelled to pass into the furnaces. The draught is fed by a fan or fans, and the pressure is equal to about 1½ in. of water. The advantage of the adoption of forced draught on steamers is that ventilating pipes can be reduced greatly in size, and the supply of air be rendered nearly uniform under all conditions of weather and wind.

Force de Cheval.—See Cheval Vapeur.

Force, Moment of.—The moment of a force is the product of the force into the perpendicular distance of its direction from a given point. The moment represents its effect or leverage on a body moving about the given point.

Force Pump.—A pump in which the water is lifted by the force due to atmospheric pressure acting against a vacuum, but in which it is expelled again by a plunger ram instead of by a bucket. The backward motion or withdrawal of the plunger in its cylinder produces a vacuum into which the water rushes ; then, by the forward motion of the plunger, the aperture through which the water entered is closed by a foot valve (q.v.), and the water is forced out through an upper or delivery valve. Also termed a plunger pump.

Fore Carriage.—The framing or bogie which carries the two front wheels of a portable engine.

Fore Gear Eccentric.—See Forward Eccentric.

Forehearth.—That portion of the hearth of a blast furnace which lies to the front of the hearth proper, beneath the tymp arch and behind the dam plate.

Foreman.—The individual who has charge of a department of work. In engineering factories there are foremen of pattern-makers, moulders, fitters, turners, erectors, smiths, boiler-makers, carpenters, and labourers. A working foreman is one who divides his time between supervision and manual labour, and is therefore little more than a leading hand. Foremen are always recruited from the ranks.

Fore Plane.—A plane intermediate in size between a jack and a smoothing plane.

Fore Plate.—A plate extending in front of the bottom roughing and finishing rolls used in puddling, to receive the shingled bloom from the bogie on which it is brought to the rolls.

Forge.—(1) The structure upon which a smith's fire is built, consisting of hearth, tuyere, chimney, bonnet, water and coal troughs. Forges are either fixed or portable, and are either made wholly of iron, or of iron and brick combined. (2) Also applied to the mill where the rolling of puddling bar is carried on.

Forgeability.—The capacity of a metal for being drawn down and worked into a definite shape on the anvil. See Red Short, Cold Short.

Forge Crane.—Similar in type to a foundry crane (q.v.).

Forged Scrap.—The odds and ends of wrought-iron scrap in a smithy are commonly reheated and balled up under the steam hammer into slabs for use. This is economical not only because of the saving in material and the superior quality of the scrap over new bars, but also because the men are able to make it a stock job during times of temporary slackness of orders.

Forged Work.—Work in wrought or malleable iron, as distinguished from founder's or cast work.

Forge Fire.—(1) A smith's fire. (2) A puddling furnace (q.v.).

Forge Pigs.—Pigs of iron most suitable for puddling purposes, consisting of the white irons in which the carbon is entirely in the combined condition.

Forge Rolls.—The train of rolls by which slabs and blooms are converted into puddled bars.

Forge Scale.—The flakes of black oxide of iron which form rapidly on malleable iron exposed at a red heat to the action of the air. It is called forged scale because it accumulates at the smith's forge, being struck off on the anvil.

Forge Test.—A bending test applied to plates of wrought iron and steel. The plates are bent both hot and cold, and also with and across the grain ; and the angles to which they are to be bent in each case without fracture are mentioned in the specification relating thereto. Plates of Lowmoor and Bowling iron are often excepted from forge tests, their brands being sufficient guarantee of their quality. A good boiler plate of any thickness should bend hot with the grain to an angle of 125° without fracture, and across the grain to an angle of 100°. If bent cold, a $\frac{1}{2}$ in. plate should bend to 35° with the grain, and 15° across.

Forge Train.—The series of puddle rolls employed for rolling out shingled bloom after it leaves the steam hammer. The rolls are grooved in diminishing series, and their action is reversed when two rolls only are employed, or the driving power is in one direction, when three high rolls are used. Termed puddling-rolls, as distinguished from mill-rolls (q.v.).

Forging.—(1) The art of making forged, or smith's work. (2) The work produced by the smith.

Forging Machine.—A machine in which an upper row of stamps, hammers, or dies is actuated by eccentrics upon a single power-driven shaft. Fixed dies are set below, corresponding with those on the upper row, and between these work is swaged into shape. The dies and stamps can be changed at pleasure. Sometimes called a bolt machine.

Forging Press.—(See Hydraulic Forging Press.) The advantages of a forging-press over a steam hammer are, that being lower it permits of the use of overhead travelling cranes, is less likely to break down, its action being quiet, and is therefore less expensively kept in working order.

Fork.—See Strap Fork.

Fork Chuck.—See Prong Chuck.

Forked Connecting-Rod.—A connecting-rod having a forked end (q.v.), so that the cross head is enclosed by it, instead of being itself enclosed, as in the solid-ended form.

Forked End.—The end of a lever, or of a connecting-rod, is forked when it bifurcates or divides into arms, in order to receive the end of another lever or rod within its bifurcation, the two being connected with a joint pin (q.v.).

Forked Strap.—The cleft end of a pump-rod, which embraces the end of a wooden rod or spear.

Formation Level.—The level of the tops of the embankments and bottoms of the cuttings of a railway upon which the ballast is laid.

Former, or Copy.—The templet used for the cutting of wheel teeth, and other works in copying machines.

Formula.—An arithmetical or algebraical assemblage of figures, letters, signs, and symbols, arranged in a conventional manner, by means of which, processes of calculation are obvious by simple inspection without verbal explanation.

Forward Eccentric.—The eccentric which opens the slide valve to steam when the engine is to run forward. Sometimes called fore-gear eccentric.

Forward Gear.—An engine is said to be in forward gear when the eccentrics, slot links, and levers are arranged for the running forward of the engine.

Fouling.—(1) The incrustation of steam boilers, and steam and water-pipes with scale. (2) The interference of machine or structural parts generally with each other, hindering their action; being due to errors in design or in construction.

Foundation Bolts.—Bolts employed for holding or screwing down base plates, beds, steps, machines, cranes, engines, pumps, and similar work to the foundations in which they are to be held. Also termed hold-down bolts.

Foundation Cylinder.—(1) The foundations of fixed cranes (q.v.) are often laid in concrete, where the natural soil is not sufficiently rocky to sustain the upward pull. The concrete is sometimes enclosed in a cylinder of cast iron sunk in the ground, and termed a foundation cylinder. (2) Foundation cylinders are used for sinking into the beds of rivers for the support of bridge piers. They are made in cast iron, having internal flanges for bolting together. Their length is about equal to their diameters, which may measure four, six, or eight feet.

Foundation Ring.—A foundation cylinder (q.v.).

Foundations.—See Machine Foundations.

Founding.—See Moulding.

Foundry.—The shed or workshop in which the operations of moulding and casting are carried on. It is usually confined to a ground floor covered to a depth of about two feet with sand, and furnished with cranes, boxes, sand bins, the necessary tools, drying stoves, and cupolas; the latter being without the building, but having their tap holes opening within the wall. Brass foundries are smaller, and their appliances are lighter in character, and the furnaces are within the building.

Foundry Crane.—A crane employed in foundries for lifting about moulding boxes, ladles of metal, cores, &c. The framework is of timber or wrought iron. The jib stands horizontally, and carries a block carriage, the post makes a right angle with the jib, and takes the gearing whose shafts revolve in cast-iron cheeks bolted thereto, and the thrust beam makes an angle of about 45° with the post and jib. Large foundry cranes are centred at top and bottom in shoes or steps in which they also swivel, and stand in the centre of the shop. Small cranes are attached to the foundry walls. For the lightest class of work small cranes are made without footsteps, being simply hinged to the wall in order to leave the floor space clear for work. In large foundries, overhead travelling cranes are superseding the old fixed cranes.

Foundry Ladle.—See Ladle.

Foundry Mould.—See Mould, Moulding, and details under their headings.

Foundry Pig.—Foundry pigs are those suitable only for making castings, being grey or mottled, as distinguished from forge pig (q.v.).

Foundry Pit.—A large and deep pit sunk in a foundry floor, and lined with cast-iron plates, or bricks. Used for the reception of moulds which have to be placed on end for the purpose of sound casting. Foundry pits are either sand pits, or open pits. Open pits are simply circular flat-bottomed excavations in the foundry floor for the reception of moulds which require no support, beyond that afforded by the flasks; and which, having no sand rammed around them, are called open. They are used merely for the convenience of keeping the pouring gate low enough down to receive the molten metal poured from the crane. The walls of pits are built of stone, brick, or iron. Sand pits are so termed because the mould is of so weak a character as to require the support of sand rammed around in the space between it and the walls of the pit. All pits must be quite dry.

Foundry Sand.—The material into which molten metal is run in order to impart the definite outlines required in castings. Sand is used in preference to other substances because its refractory nature resists the destructive influence of metal at high temperatures, because its porosity permits of the free escape of the gases generated in casting, and because of its compact and adhesive nature by virtue of which it can not only be moulded into any shape, but is also able to resist great liquid pressure. The more silica present in the sand, the greater its suitability for foundry purposes; alumina and magnesia should also be present in small quantity. The sands employed in moulding are those from the coal measures and the new red sandstone, also from the greensand and chalk. The sands are named according to the localities in which they are found, as London, Belfast, Falkirk, Worcester, Cheshire; or red yellow, green, according to their colours. Foundry sand is always rammed damp, though sometimes dried, before the pouring of the metal. See also Green Sand, Dry Sand, Core Sand, Coal Dust, Facing Sand, Parting Sand, Silica, &c.

Foundry Stove.—See Drying Stove.

Four-Cutter Machine.—Machines for planing wood, having four sets of cutters, one top, one bottom, and one on each side, for planing all four faces of the stuff at once.

Four-Cylinder Engines.—Of steam engines, compound engines in which two cylinders are high and two low pressure.

Four-way Cock.—A cock having four branches.

Foxey.—Timber is said to be foxey when there is an excessive proportion of green sapwood present in it.

Fraction.—A part of anything. A vulgar fraction is represented by two numbers, one placed over the other, the one above the line being the numerator, the one below the denominator. A proper fraction is one whose numerator is less than its denominator, an improper fraction one whose numerator is greater than its denominator. A mixed number contains integer and fraction combined. A decimal fraction is one whose denominator is done away with, a dot, the decimal point, being placed before the numerator, and placing a figure to the right of the dot makes the fraction a tenth of what it was before. The denominator is always therefore, 10, 100, 1,000, or some other power of ten.

Fractional Pitches.—The pitches of screw threads which are not aliquot

10*

parts of the inch, and therefore not directly divisible by, or into, the pitch of the leading screw in the length of one inch. To obtain the change wheels for cutting these screws it is necessary to multiply the integer and the number which represents the pitch of the leading screw by the denominator of the fraction, thus: Supposing it were required to cut a screw of $6\frac{1}{4}$ threads per inch with a leading screw of 4 threads to the inch, we have $\frac{4}{6\frac{1}{4}} \times 4 = \frac{16}{25}$ and deduce the change wheels (q.v.) then in the ordinary way; as, for instance, by factors $\frac{16}{25} = \frac{4 \times 4}{5 \times 5} = \frac{40 \times 40}{50 \times 50}$, and thence get wheels in an ordinary train which are not equals $\frac{40 \times 20}{50 \times 25}$, and these will cut the fractional pitch.

Frame Saw.—See Bow Saw, Deal Frame.

Framing.—The skeleton of a locomotive which sustains the boiler, machinery, and axles. It is a plate framing, when plates of sheet iron are used in its construction, a bar framing when iron bars are employed. The latter type is seldom used in England, but it is the American practice.

Free Burning Coal.—Non-caking or but slightly caking coals. Anthracite is difficult to burn, but the anthracitic and the semi-bituminous coals burn easily.

Free End.—In a cantilever the end which is not fixed is always designated the free end.

Free Piston Gas Engine.—A form of gas engine in which the piston is driven in an upward direction by an explosion of gas and air in a closed cylinder below. As the pressure of the gas diminishes with its increase of volume, the piston descends by the weight of the atmosphere pressing upon it. This is known as Otto and Langen's patent, but has mostly fallen into disuse.

Free Sand.—Sand used in the making of foundry cores or moulds. Being free or open, it permits of the escape of the gases, and requires little venting.

Freezing Point.—The temperature at which water passes into the solid state. It is determined by placing a thermometer in a vessel containing pounded ice or snow well moistened with water. The latent heat (q.v.) of the mixture prevents alteration of the temperature until the ice is all melted, or the water all frozen. It corresponds with 32° F. or 0° C.

French Curves.—Curves or sweeps cut in wood or plastics. They are radial, but of varying sweeps, blending together, and are used for laying on drawings in order to obtain curves which will present a graceful appearance without the trouble of drawing them by hand. They are made in useful sets.

French Horse-Power.—See Cheval Vapeur.

Fretting.—The abrasion of the edge of a cutting tool in the process of sharpening on an oil-stone.

Friction.—The resistance to motion which is set up when two rough surfaces are moved one over the other. The amount of friction is independent both of the rate of motion of a body and of the extent of surfaces in contact.

Frictional Gearing.—See Friction Gearing.

Frictional Resistance.—In a general sense, the resistance of surfaces due to friction. Specially applied to the resistance to slipping of riveted joints by the contraction of the rivets.

Friction Block.—A brake block (q.v.) or any block of wood, the frictional resistance of whose surface is utilised in the arresting of motion.

Friction Brake.—See Dynamometer, Brake.

Friction Clutch.—A friction coupling (q.v.).

Friction, Co-efficient of.—This signifies the ratio between the perpendicular pressure, or the load, and the friction existing between surfaces in contact. Thus the friction divided by the load gives the co-efficient, and from the co-efficient multiplied by the load the friction can be deduced. The co-efficient may be expressed in lbs. per ton necessary to move a load, or as a vulgar fraction. But it is customary to employ decimal numbers. The co-efficient of friction is denoted by the Greek letter μ.

Friction Coupling.—Consists essentially of discs or rings of wood, or metal, or both combined, whose surfaces are brought into close contact by screw pressure. The commonest existing forms are bevelled; that is, they form frustra of cones fitting one to the other; sometimes also an internal split metallic ring is pressed outwards by a lever against a recess bored in an encircling plated ring.

Friction Disc.—A contrivance similar in principle to cone gear, but in which frictional surfaces are substituted for toothed gearing. A pair of revolving discs, and furnished with a limited movement also longitudinally with their axes, enclose a smooth turned wheel whose axis is at right angles therewith. This last wheel is keyed to a spindle but receives no motion save through the disc wheels, and then only when contact is made by moving the inner face of one or the other disc wheels into contact with the periphery of the spindle wheel. Evidently the shaft of the disc wheels moving in one direction, the effect of bringing first one and then the other disc into contact is to cause the driven wheel and spindle to rotate in opposite directions. This form of gearing is conveniently used for reversing the motion of traversing cranes. The working faces of the discs are blocks formed of alder or other suitable wood held in place with a bevelled retaining ring.

Friction Gearing.—(1) Gearing, whose driving force is produced by the friction only of the peripheries of the wheels. The driving faces are either of wood, or millboard, or leather. This gearing is seldom used, being unsuitable for the transmission of much power. (2) In some forms of this gearing the wheels are provided with turned rings and grooves on their circumferences which are the counterparts of each other, and in which, therefore, the amount of bite is increased.

Friction Hoist.—A light hoist driven by the friction of the smooth turned surfaces of pulleys, the proportion between the diameters of the pulleys being the same as though toothed gearing were employed, so that their velocity ratios still depend upon their relative diameters.

Friction of Motion.—The power required to keep a moving body in motion. This is less in quantity than the friction of repose (q.v.).

Friction of Repose.—The power necessary to set a body moving from a state of quiescence. This is greater in quantity than the friction of motion (q.v.).

Friction Ring.—A loose elastic metallic ring used in some forms of friction clutches. It is divided in one portion of its circumference and pressed outwards against a female portion by means of a lever.

Friction Rollers, or Friction Wheels.—Small rollers which revolve in bearings, and sustain an axle in the depression formed by the contiguity of the upper portion of their peripheries. They are used for light running machinery, as the crank axles of lathes. In these there is no sliding friction (q.v.).

Friction Strap Brake.—See Strap Brake.

Friction Wheel.—Any wheel which drives or is driven by friction, as when contact only takes place between smooth or grooved surfaces without the intervention of teeth.

Front Top Rake.—See Top Rake.

Frosting.—Ornamental scraping performed for the sake of good appearance only, by a series of light finishing cuts performed at different crossing angles to reflect the light at those angles. It is done either with a scrape, or with a bit of oil-stone, and is performed on most light machines and engine parts exposed to view. It is so slight in amount that it does not affect the truth of the surface in the least degree after the scraping proper is completed.

Frustum.—The solid included between two transverse sections of a cone or wedge, or parabolic figure; the sections being taken at different heights.

Fuels.—Fuels play an important part in engineering processes, both in the first smelting of the metal and in subsequent remeltings, puddling, the heating of steam boilers, &c. Fuels are solid, as coal, coke, and charcoal, or liquid, or gaseous, or patent fuels.

Fulcrum.—The point on which a lever turns.

Fulcrum Plate.—In an ordinary lift pump worked by hand and attached to a back board, the metal plate which receives the stud about which the handle moves is called the fulcrum plate.

Full.—(1) A workshop term which signifies that a dimension is slightly larger than it should be, but to so slight an extent that it is to be detected by calipers rather than by rule measurement. (2) The full of an eccentric is the portion of the sheave which is situated at the greatest distance from the crank shaft upon which it is fixed.

Full Bore.—See Full Waterway.

Fullering.—The process of forming grooves in smith's work, both as simple grooves merely, and as a series of grooves for flattening out a surface. Also a mode of caulking adopted for boiler plates, which, however, differs from caulking proper in that the whole of the edge of the plate is hammered or burred over instead of a portion of the edge only. Fullering is less likely to groove the face of the adjoining plate than the ordinary caulking.

Fullering Tool.—A round-faced smith's tool used for fullering. See Top Fuller, Bottom Fuller.

Full Gate.—When the governor or regulator of a turbine is opened to its full extent so that the whole width of the vanes is utilised for the reception of the water, and the turbine works to its limit of efficiency, it is said to be at full gate.

Full Gear.—In an engine furnished with reversing gear when the die blocks are at either of the extreme ends of the slot links, so that the valves are set in their extreme positions for forward or backward running, the engine is said to be in full gear. If worked in any other position of the die blocks excepting mid-gear (q.v.), the position is denoted by the fractional portion of the cut off, as $\frac{1}{2}$, $\frac{5}{8}$, $\frac{3}{4}$, $\frac{7}{8}$, &c.

Full Shrouding.—See Shrouding.

Full Size.—A term used to distinguish drawings made to the actual size of the work which they represent, from those made to a smaller scale. The term full size is usually written on the sheet.

Full Thread.—A screw thread which is cut clean and sharply to its proper depth, instead of being merely scratched into a fractional depth only

Full Waterway, or Full Bore, or Full Way.—A waterway is said to be full when the area of the opening given in the valve or plug, as the case may be, is equal in extent to the area of the suction bore.

Full Way.—See Full Waterway.

Fully Divided Scale.—A scale (q.v.) used for drawing purposes on which all the primary divisions are subdivided. See Open Divided Scale.

Fundamental Circle, or Base Circle.—A curve which is rolled over by a generating circle in the production of cycloidal curves, used in the striking of wheel teeth.

Funnel.—The furnace chimney of a marine engine, built of sheet iron plate riveted together. Its area bears a definite proportion to the engine power.

Furnace.—A general term which may signify any structure in which metal is melted, refined, or remelted. (See Air, Blast, Reverberatory, Brass, Steel, Puddling Furnaces, &c., and Cupola.) Also an engine or other fire.

Furnace Bars.—See Fire Bars.

Furnace Grate.—See Grate.

Furnace Hoist.—A light hoist used for elevating the metal, fuel, and limestone to the platform of a cupola or melting furnace.

Furnace Lining.—A coating of refractory material used as a protection between the casing of a furnace and the intense heat of the fuel. The linings vary with the classes of furnaces. Cupolas are relined with fire clay or sand at every lighting of the fire. Blast furnaces are lined with a refractory fire clay; converter bottoms are lined with ganister, or with magnesian limestone, the lining being removed at every few heats, or from about five to ten. The lining burns away most rapidly round the tuyeres (q.v.). See Fettling, Tuyeres.

Furnace Shell.—The shell (q.v.) of the furnace, or fire box, of a steam boiler.

Furnace Tube.—The tube within which the fuel is enclosed in an internally fired boiler (q.v.).

Furring Up.—The partial closing of steam or water pipes by the deposition of solid matters which were previously held in suspension in the steam or water.

Furrowing.—See Grooving.

Fusibility.—The readiness with which metals pass from the solid into the liquid condition.

Fusible Alloy.—An alloy so compounded as to melt at a definite and low temperature. See Fusible Plug.

Fusible Plug.—A plug composed of soft and easily melted metal or alloy inserted in a brass casing which is screwed into the furnace crown of a steam boiler, and which melts when the water level falls as low as that of the crown, and so allows of the escaping water and steam to extinguish the fire and to give timely notice of the shortness of water. It melts at a low red heat.

Fusing Point.—The temperature at which a metal or an alloy melts potassium comes lowest, platinum highest in the scale.

G.

G.—In mechanical and mathematical calculations G stands for the force of gravity (q.v.) at any particular latitude.   It may be defined as a number expressing the velocity produced in a falling body in unit of time, or a number expressing twice the distance through which a body falls in unit of time ; or a number expressing the weight of the unit of mass in absolute measure.

Gab.—A hook, or open notch, in a rod or lever, which drops over a spindle, and forms a temporary connection between valve or other motions.

Gable.—The outer ends of the cranked portion of a crank shaft.   Dressing this down square is termed cutting the gable.

Gab Lever.—Generally any lever which is connected up by means of a gab.   Specifically the lever which forms the connection between the slide valve spindle and the eccentric rod in some forms of marine engine valve.

Gage.—See Gauge.

Gager.—See Gaggers.

Gaggers, or Gagers.—Short conical or pyramidal projections, cast upon core plates and the plates for loam moulds, to assist the adhesion of the loam.   The term is sometimes applied also to lifters (q.v.).

Gain.—Although no machine, actual or ideal, can give out more work than is actually put into it, yet it is customary when speaking of mechanical exchanges to use the term gain, or mechanical gain, when one form of energy is exchanged for another.

Galena.—The most important ore of lead.   It is a sulphide, that is, a compound of lead with sulphur—Pb S, containing eighty-six and a half per cent. of lead.   It occurs in veins, has a metallic lustre, and a dark colour, and is very heavy.   Sp. gr. 7·5.   It is found abundantly in various parts of England, Scotland, Spain, Saxony, and the United States.   Argentiferous galena is an ore from which silver can be profitably extracted, though it may contain but two parts of silver in the thousand.

Galloway Boiler.—A horizontal, internally fired boiler, which is a modification of the Lancashire boiler.   The two flues pass at their hinder end into a broad single flue containing a number of Galloway tubes (q.v.) arranged in a vertical manner, which absorb the heat from the flues, and so increase the efficiency of the boiler.   The hinder flue in boilers of more recent construction has its upper and lower curved faces struck from a common centre.   The Galloway tubes may number thirty or more, according to the size of boiler.

Galloway Tubes.—Water tubes of a conical form longitudinally, circular in cross section, and flanged outwards at the ends ; placed across the tubes of Lancashire and Cornish boilers to increase the circulation and the heating surface, and also as a secondary consideration, to assist in staying the tubes.

Galvanic Action.—The action upon one another of electro-positive and electro-negative metals, causing wasting away of the positive metal.   It is particularly noticeable where iron and copper, or zinc and copper, or brass and iron, are in contact in the presence of acidulated water.

Galvanized Iron.—Iron, which after having undergone preliminary cleansing, has been dipped in a bath of molten zinc. **The process has** nothing to do with galvanic action.

Galvanising.—The coating of iron with a film of molten zinc.

Gamboge.—The yellow colour used to indicate brass and gun metal on drawings.

Gang Saws.—A number of saws arranged in parallel positions in a swing frame (q.v.). Used for sawing logs, or balks of timber, into planks and boards.

Gangue.—Is a term applied to the earthy matter of iron and other ores. It consists chiefly of clay.

Ganister.—A refractory material used for the lining of Bessemer converters (q.v.), and of steel moulds. It consists of a highly siliceous material cemented with fire clay ; the silica equalling about ninety per cent.

Gantry.—The trussed beams, or the girders, as the case may be, which carry the crab, or crane, in overhead travelling cranes.

Gantry Crane.—An overhead travelling crane carried on a gantry (q.v.).

Gap Bed.—A lathe bed having a portion recessed out in front of and below the headstock, to receive work larger in diameter than the height of the lathe centres, under ordinary conditions, will allow of. The bed is strengthened correspondingly below the gap. The gap, when not in use, is filled up with a gap bridge (q.v.).

Gap Bridge.—The piece of casting or the bridge which closes up the gap of the bed of a lathe, when not required for use.

Gap Lathe.—A lathe provided with a gap bed (q.v.).

Gas.—A fluid (q.v.) which possesses the property of indefinite expansion in all directions, so that it immediately fills any vessel into which it is introduced (see Liquid, Boyle and Marriott—law of). A perfect gas is one which expands at a uniform rate with uniform increase in temperature. Saturated vapours do not absolutely fulfil the conditions of perfect gases.

Gas Coke.—The soft coke which results from the dry distillation of coal in gas works. It is too friable for use in cupola furnaces, but owing to its cheapness it is employed in the drying stoves in foundries.

Gas Engine.—A general term denoting any motor which is actuated by the explosive force generated by the compression and subsequent ignition of gaseous compounds. Commonly, coal gas and air are employed, but various hydrocarbons have also been made use of. In gas engines the extremes of temperature are greater than in steam engines, hence under suitable working conditions the greater should be the duty obtainable from them.

Gaseous Fuel.—See Producer Gas.

Gaseous Mixture.—The charge (q.v.) of a gas engine.

Gaseous Steam.—Superheated steam (q.v.).

Gas Furnace.—See App.

Gasket.—Plaited rope well greased with tallow, and used for the packing of stuffing boxes and pistons in engine and pump work.

Gas Pliers.—Two-jawed tongs, usually having two sizes of serrated or grooved circular holes, for the holding or gripping of gas pipe which is being manipulated.

Gas Ports.—The inlet passages by which the gas obtains access to the cylinder of a gas engine.

Gas Producer.—A gas-making furnace in which coal gas is obtained by the distillation of fuel for the special purpose of heating ingot and

other furnaces. The gas is mixed with air before being burnt. See Producer Gas.

Gas Pump.—A small force pump used in some gas engines for forcing gas into the combustion chamber.

Gas Regenerator.—The regenerator (q.v.) through which producer gas passes on its way to the furnace hearth, in a reheating or a steel-melting furnace. It is smaller than the corresponding air regenerator.

Gas Slide.—The slide valve of a gas engine, by means of which the access of gas to the combustion chambers is regulated.

Gas Stocks and Dies.—Taps and dies made specially for cutting gas threads, being of a finer pitch than engineers' stocks and dies.

Gas Tap.—See Gas Stocks and Dies.

Gas Thread.—A screw thread of fine pitch, employed for wrought iron and brass tubes.

Gas Tongs.—See Pipe Tongs.

Gas Tubing.—See Wrought-iron Pipe.

Gate.—(1) The orifice by which molten metal is poured into a mould. (2) The annular opening through which the water passes into the vanes of a turbine (q.v.).

Gauge.—(1) The distance between the inner edge of rails or tramways, or permanent way. It ranges from 18 inches to 7 feet. (2) A term applied to a multitude of measuring tools. (See Cutting, Cylindrical, Marking, Mortice, Pressure, Screw, Vacuum, and Wire Gauges.) (3) Mixed gauge refers to the laying down of two lines of rails—a broad and a narrow gauge—side by side. The instrument by which the distance between rails is measured, is termed a gauge. It is a bar of iron having two offsets at right angles to the bar, the distance between the offsets, which lies between the rails, giving the gauge.

Gauge Cocks.—The cocks attached to the upper and lower portions of the water gauge upon a boiler, and which are opened periodically to test the freedom or otherwise of the water passage from obstruction.

Gauge Glass.—A stout, well-annealed glass tube, attached to the front of an engine boiler to indicate the height or level of the water within. Unions above and below, with cock attachments, connect the glass tube with the steam and water spaces respectively, and allow of either connection being shut off in case of a glass becoming broken.

Gauge Plate.—An adjustable plate fixed to shearing, cropping, and cutting-off machines, for insuring the uniform length of short pieces of bar or plate which are to be cut off in large numbers.

Gauge Rod.—A rod of iron from ¼ in. to ½ in. in diameter, and used for measuring the internal diameters of portions of work in cases where great accuracy is essential. Thus, when internally fired cylindrical steam boilers are being subjected to the hydraulic test, it is usual to gauge the internal diameters of the flue in various portions of the circumference, to ascertain if there be any material distortion, or tendency to assume an oval shape under pressure. Also, when shafts have to be turned to fit the eyes of wheels or pulleys, or pistons have to be turned to fit engine cylinders, the pulleys, wheels, or cylinders being away from the workshop, gauge-rods are filed to fit the bores exactly, and the shafts or pistons turned to the lengths of the rods. The ends of the rods are filed off tapered, or nearly to a point, to afford a very narrow point of contact only.   See also Plug Gauge, Dict. of Modern Terms.

Gear.—A term of very general application, signifying arrangements of toothed wheels, valve motions, pump work, ropes, lifting tackle, &c.

Gear Cutters.—Cutters of hardened steel formed on the edge of a circular disc whose section is that of the tooth spaces which they are intended to cut. Sometimes a single cutter is set in a revolving block. See Milling Wheels.

Gear Cutting.—The formation of the teeth of wheels by means of revolving cutters, or by planing.

Geared Brace.—See Geared Hand Drill.

Geared Chuck.—A form of universal chuck (q.v.).

Geared Engine.—An engine in which gearing is interposed between the crank shaft and the first-motion shafting, either for the purpose of increasing power and diminishing speed, or the reverse.

Geared Fly Wheel.—A fly wheel furnished with teeth on its periphery. Used with Corliss and other engines.

Geared Hand-Drill, or Geared Drill, or Geared Brace.—A form of drill-stock held against the breast, but in which the drill is revolved by a pair of bevel-wheels worked by hand.

Geared Headstock.—A headstock provided with back gear (q.v.).

Geared Ladle.—A safety ladle (q.v.).

Geared Pump.—A pump which is driven by an engine through the medium of gearing, usually with a reduction of speed.

Gearing.—This has the same general meaning as gear (q.v.), but is more properly applied to gear-wheels (q.v.). Gearing, or to gear a wheel, also signifies the driving in and working of the wooden cogs of mortice wheels.

Gearing Chain.—See Pitch Chain.

Gearing Down.—Signifies the reduction of speed from a prime mover to the mechanism which it actuates, by the intervention of gear wheels, with, of course, an accompanying gain in power.

Gear Wheel.—Any form of toothed wheel, as distinguished from running wheels, pulleys, fly wheels, friction wheels, &c.

Geat.—A gate (q.v.).

Gedge's Metal.—Aich metal (q.v.).

General Drawing.—A complete drawing of an engine or machine, in which many matters of detail may be omitted but in which the main outlines of its construction are apparent.

General Joiner.—A combination machine used in wood-working. It is furnished with saws and cutters for sawing, planing, tenoning, moulding, boring, grooving, and all the ordinary operations of the workshop.

Generating Circle.—The circle, which rolling upon a fundamental circle (q.v.), generates, by means of a point in its circumference, a cycloidal curve. Used in the striking out of wheel teeth.

Generating Surface.—Heating surface (q.v.).

Geometrical Mean.—The geometrical mean of two numbers is obtained by multiplying the two numbers together and extracting the square root of the product.

Geometrical Progression.—See Progression.

Gib.—The fixed portion of the gib and cottar arrangement, in which the bearing is kept in place by a loose strap end. The gib is the portion provided with hook ends, whose function is to prevent the straps from spreading or opening out. See Cottar.

Gib and Cottar.—See Cottar, Gib.

Gibbet.—The triangular framework of a crane, consisting of post, jib, and strut.

Gib-Headed Key.—A key having a set-off standing at right angles with the thicker end, for convenience of drawing it back in situations where the use of a drift is not practicable.

Giffard's Injector.—See Injector.

Gimbal Joint.—A Hooke's joint. See Universal Joint.

Gimlet.—A boring tool which prepares the way for the entrance of wood screws. Its point is that of a taper screw and its body that of a cylinder having a spiral groove, hence called a twist gimlet ; or a hollowed, nearly semi-cylindrical shell, called therefore a shell gimlet.

Gimp Nail.—A gimp pin. A small, round-headed upholsterers' nail, useful to pattern-makers for fastening pattern letters to name-plates (q.v.).

Gin.—A rude portable, yet effective, hoisting machine, the well-known tripod form, from whose apex the pulleys and gin-block (q.v.) for lifting are suspended for working heavy tackle. A barrel and winch are fixed between the two legs which form the shears.

Gin Block.—A single-sheave pulley of a hollow-rim section, having its bearings in a skeleton frame suspended from a hook. A rope is passed over the pulley, one   d of which is held by a man, the load being attached to the end op. site. There is thus no gain in power but only an alteration in the direction of motion. Called also whip-gins, rubbish pulleys, monkey wheels.

Gin Pulley.—The pulley of a gin block (q.v.).

Girder.—A beam, usually of timber or of cast or wrought iron, used for supporting a superstructure. Girders are of various sections, the commoner form, when in cast iron, having a section like the letter I, in which, however, the bottom flange where the metal is in tension is much wider than the top. In wrought iron, the Warren, or trussed girder, is that usually employed. Girders act as levers in which, the cross sections being the same, their strength is in inverse proportion to their length, but the conditions of strength vary with the mode in which the load is distributed, and with the way in which the ends of the beams are fixed. (See Beams, Relative Strength of.) Girders are understood to be subject to transverse or bending strains only.

Girder Stays, and Roofing Stays.—Used to stiffen the fire-box roofs of locomotive and portable boilers. Two plates of wrought iron are riveted together at a little distance apart, distance pieces being interposed to allow of the passage of bolts between, and then placed edgeways across the fire-box roof. From these, bolts provided with clip-washers depend, and are fastened inside the fire box. There will be several such stays on the roof, at a pitch, or distance apart, of about 4 in. Formerly a thicker solid plate, pierced with holes for the bolts, was employed, but is superseded by the lighter built-up girder. Girder stays rest on the fire box only near the ends, so that there is a hollow water space of about 1 in. in depth underneath them, which permits of circulation and prevents the accumulation of deposit and burning of the fire-box crown.

Girt.—The circumference of round timber.

Girth.—The circumferential measurement of a rope is termed the girth.

Git.—A gate (q.v.).

Give.—A joint, riveted or otherwise, is said to give, when it slips, or slides, or breaks away.

Gland.—The small casting which closes the mouth of the stuffing box in engines and pumps, and receives the wear of the piston rod. When the rod is in place the screwing down of the gland presses the packing close around the rod, and prevents leakage of steam or of water.

Gland Bolt.—A stud bolt (see Stud) which has a bearing collar or flange

turned upon it. It is screwed down until the lower shoulder of the collar bears against the flange or face of the stuffing box. Sometimes called a pillar bolt.

Gland Cock.—A plug cock, in which the plug is packed with a gland over it. These are superior to the common plug cocks.

Glass.—Broken glass is sometimes used as a flux for hard gun-metal.

Glass Barrel Pump.—See Acid Pump.

Glass Gauge.—See Gauge Glass.

Glass Paper.—The abrading and smoothing material used by wood workers. It is made from crushed bottle glass, sifted to assort it into sizes ; the powder is then dusted through a sieve on to stout paper, one of whose surfaces is moistened with glue. Formerly sand was the material employed, hence the term sand-paper.

Glass-papering Machines.—Machines for glass-papering the surfaces of wood-work. They consist, in one form, of a table underneath which revolves a cylinder around which the glass-paper is rolled, and which projects just above the surface of the table through a slit. The work in being slid over the table is brought into contact with the periphery of the cylinder. In the other form the glass-paper is attached to a disc revolving at the end of a radial movable arm, and the work being laid upon a table underneath, the disc is brought to bear upon any portion of the surface where it may be required.

Glass-paper Rubber.—See Rubber.

Glazed Pig.—See Blazed Pig.

Glazing.—The filling up of the interstices of the surface of a grindstone or emery-wheel with the minute abraded particles detached in grinding. The surface has then a smooth and polished appearance, and ceases to cut efficiently. Glazing arises from several causes, as an insufficient supply of water, or the grinding of substances for which the nature of the stone or emery-wheel is unsuitable, or from bringing so broad a surface to be ground that the abraded particles are imprisoned thereby, and forced or squeezed between the rigid particles composing the stone or wheel.

Glenboig Brick.—A fire-brick used for regenerative furnaces.

Globe Valve.—A valve whose casing is of a spherical shape. Sometimes called diaphragm valve, because an annular valve seating contained within parts the chamber into two portions. The term globe valve relates only to its outline ; its applications are numerous, but they are chiefly those of a stop or regulating valve.

Glue.—A cement for wood made by boiling down the intestines of animals. Russian glue is the best. It is prepared for use by boiling with water, the glue contained in an inner vessel, being surrounded by the water contained in an outer vessel. See Marine Glue, Glue Cement, Hydraulic Glue.

Glue Cement.—A waterproof cement used for attaching paper labels to castings which are to be shipped abroad. It is composed of glue, four parts, boiled oil one part, oxide of iron one part. The whole well boiled and mixed, and applied while hot.

Glue Heater.—A tank, or water bath, containing steam and hot water, by means of which glue is heated and kept warm in the workshop.

Gluing.—(1) The art of gluing consists in making a good joint of the surfaces in apposition, then after the application of the glue to rub out all that can be rubbed out by moderate pressure, combined with sliding motion, before setting the joint. (2) The sticking of oil in and about bearings, due to its oxidation or drying. Also termed gumming.

Glue Pot.—The double vessel used in the preparation of glue. It consists of an outer pot containing the boiling water, and an inner one containing the glue. The glue is thus never allowed to become burnt, which burning would impair its adhesive property.

Glut.—A gun-metal block having a face hollowed out to fit against the bossed-up end of the valve-rod in a knuckle joint, and by which the wear is continually being taken up, the glut being tightened by a wedge and screw, or by a cottar. See also False Key.

Glutting.—The choking, or partial stoppage, of engine cylinders and passages, and of condenser tubes, with a carbonaceous deposit from inferior oils used in lubrication. Animal oils, including tallow, suet, and lard, are found to produce both glutting and corrosion, the decomposition of the fats causing the formation of fatty acids, and the deposition of carbon. These objections do not apply to the use of mineral oils.

Glycerine.—(1) Glycerine is used for mixing with pipeclay (q.v.) in making models of ornamental castings, when it is desirable that the clay shall not set too quickly. (2) Glycerine is also the basis in lubricating oils.

Goliath Crane.—A crab mounted on a gantry, which gantry is attached to end standards provided with flanged driving wheels, so that the whole structure travels bodily upon rails laid upon sleepers ; the difference between a goliath and a traveller therefore is, that the former travels in the mass while the latter travels only to the limited extent permitted by the end beams. Goliaths, like travellers, are worked both by hand and by steam power.

Goose Neck.—The bent rod by which the tap hole in a casting ladle is opened and closed.

Gorge.—A term sometimes applied to signify the groove of a sheave pulley in which the rope or chain runs.

Gouge.—Gouges are either long paring, or short firmer tools, the former being ground on the inner or hollow curve, the latter on the outer or convex face. They range from $\frac{1}{8}$ in. to 2 in. in width. Smith's gouges or "hollow sets" are curved tools embraced by the ordinary withy handles, and struck by the sledge hammer.

Gouge Bit.—A bit used for boring holes in wood. Its cross section is that of a gouge, and it is simply rounded at the cutting point, without having a nose or lip.

Gouge Slip.—A thin slip of oil-stone, whose edges are rounding in cross section for the purpose of abrading the internal or hollow faces of gouges when sharpening.

Governor.—A piece of mechanism by which the speed of a steam engine, or turbine, or water-wheel, or other motor is regulated. The forms of governors vary much, but the essential principle is that of heavy balls caused to rotate by the motion of the engine and which by the operation of central forces open or close a throttle valve. See Pendulum Governor, Crossed-arm Governor, Parabolic Governor, Weighted Pendulum Governor.

Governor Arms.—The rods which are attached to the governor sleeve on the one end and to the governor balls on the other.

Governor Balls.—Suspended, freely revolving balls of cast iron, through whose centrifugal force the opening of the throttle valve of an engine or other motor is regulated.

Governor Cut-off.—See Automatic Expansion.

Governor Sleeve.—The hollow cylinder which slides vertically on a governor spindle and carries the governor arms.

Governor Spindle.—The vertical pillar or spindle which carries the governor sleeve, and around which it revolves.

Governor Valve Gear.—The arrangement of valve gear through which governors are made to regulate the opening and closing of the induction valve in automatic expansion (q.v.).

Grab.—A form of dredger bucket used in the minor and more local operations of excavation. It consists of two half buckets meeting in the centre, and actuated by an arrangement of chains, and hooks, and other special tackle, in such combinations that the grab closes on the earth, and the jaws bite the more firmly the harder the upward pull of the chains. Also the grab being lifted and coming in contact with a ring or other arrangement suspended overhead, it is automatically opened and discharges its contents. Grabs are half tine or whole tine according [to the length of the steel jaws with which they are furnished; they are termed buckets simply when there are no serrations, but only smooth meeting edges; the former are used for rocky and gravelly and stiff mud bottoms, the latter for soft mud and grain.

Grade.—(1) A term often applied to the steps of belt pulleys. (2) The various commercial qualities of irons and steels.

Gradual Load.—A load gradually applied to a structure, and which therefore furnishes the most favourable condition of stress.

Grain.—(1) The arrangement of the woody fibres in timber. With the grain, signifies longitudinally in the course of the fibres as the tree grows. Across the grain, signifies a transverse section. Crooked grain, or cross-grain, denotes a wavy and interlacing condition of the fibres; open-grain, wood of a soft and rapid growth; close grain, the reverse; harsh grain, coarse and gritty in nature. (2) The grain of iron signifies the disposition of its crystals and laminæ, and we may say in some cases its cinder or scale. The grain of timber runs in the direction of its growth; that of iron is irregular when cast, fibrous when rolled, the fibre being developed in the direction of the rolling. The grain of wood and of rolled iron are alike strongest in the direction of their length. Boiler plates are rolled in the direction of their length for this reason. The forge test (q.v.) also recognises this difference of grain.

Grain Rolls.—Rolls made of a tough quality of cast iron not chilled, to distinguish them from chilled rolls.

Grain Side.—The hair or smooth side of a leather belt. With a smooth turned pulley this should be used as the driving side, since it makes a closer fit with the rim, to the consequent exclusion of air.

Grain Tin.—Refined tin of purest quality, which is obtained by heating the ingots and allowing them to fall from a height, when they break up into irregular prismatic fragments. Grain tin is used in tin-plating.

Gramme, or Gram.—The standard French measure of weight. It is the weight of a cubic centimetre (q.v.) of pure water of the temperature of 4° C., taken at Paris. It is equivalent to 15·4323 English grains.

Granular Iron.—Bar iron which shows a granular fracture due to the absence of cinder, and is therefore a guarantee of its strength and purity.

Granulated Copper.—See Matt.

Graphic Methods.—The methods employed for ascertaining strains upon structures, velocity ratios, and such like, by the laying down of lines to a uniform scale. Used to distinguish it from the mathematical and geometrical methods of obtaining similar results.

Graphic Statics.—See Graphic Methods.

Graphite.—Carbon in an uncombined state. It is found in most abundance in the grey variety of cast iron, and is readily distinguished on the face of a fracture, as flakes, or crystals, darker in colour than the metal, and distributed over its surface.

Graphitic Carbon.—See Carbon, Graphite.

Grasshopper Engine.—A term applied to an old and scarce type of beam engine, in which the working beam, instead of being pivoted at the centre, is pivoted at one end. The connecting-rod is thus attached at a point midway between the pivoted end and the end which is actuated by the piston-rod ; and the friction on the pivot is correspondingly reduced.

Grate.—The area which contains the burning fuel in the furnace of a steam-engine boiler, or of a reverberatory furnace, or air oven.

Grate Area, or Grate Surface.—The area in square feet covered by the fire-bars of an engine boiler. That is, equivalent to the area over which full combustion can take place in consequence of the due access of air to the fuel. As a very broad rule, one foot of grate area per nominal horse-power is allowed in Cornish and Lancashire boilers. It is usual to estimate the grate area in relation to the weight of coal burnt ; hence, it becomes the reciprocal of the rate of combustion. Cornish and Lancashire boilers, with good draught, will require ·07 square feet per lb. of coal, while locomotives will only require ·0: square feet per lb.

Grate Bars.—Fire-bars (q.v.).

Grate Surface.—See Grate Area.

Grating.—The perforated plate in a foot valve, or air pump, whose function is to prevent the entry of solid matters along with the water.

Graver.—A hand tool for metal-turning, formed of a square bar, and ground from one of the edges to an angle of about 45°. This gives two cutting faces of 60° each, right and left respectively, in addition to the point formed by the angle of the bar and the face.

Gravity.—The force which draws all bodies towards the centre of the earth or to its surface, in a direction perpendicular to that surface.

Gravity Wheel.—A water-wheel in which the weight of the water alone is utilised ; the water resting in the buckets until discharged. Overshot wheels (q.v.) are gravity wheels.

Grease.—See Tallow.

Grease Axle Box.—See Axle Box.

Grease Box.—The upper portion of an axle box (q.v.), which is made the receptacle for the grease used in lubrication.

Grease Cock.—A cup, with pipe, and stop cock, screwed into the cover or body of an engine cylinder, and which receives and regulates the supply of grease used for the purpose of lubricating the piston.

Grease Cup.—An oil cup (q.v.).

Greasy Steam.—Steam which becomes its own lubricant by a mechanical admixture of grease therewith. The grease is mingled with the steam before its entry into the cylinder, by means of an impermeator (q.v.) fixed on the steam chest.

Green Coal.—The small coal newly laid on a smith's fire.

Green Heart. (*Nectandra Rodiæi.*)—Sp. gr. 1·05 to 1·09. A hard, tough, and rigid timber, of the natural order *Lauraceæ*, used largely in engineering works. It is imported from British Guiana, and is of a yellowish, or greenish-yellow colour, and very durable. A cubic foot weighs from 65 to 68 lbs.

Green Sand.—Ordinary foundry sand moistened with water and not dried, as distinguished from dry sand (q.v.) and loam (q.v.).

Green Sand Moulding.—The moulding of work from patterns, in green sand (q.v.).

Green Wood.—Timber from which the sap has not been removed by seasoning and drying.

Grey Copper Ore.—An abundant ore of copper, containing copper, iron, sulphur, antimony, and arsenic, in variable quantities. It is mined in Cornwall and Frieburg.

Grey Iron.—Pig, or cast iron, which, on being fractured, has a grey coarsely crystalline structure. It is a moderately strong and very fluid iron, suitable for fine castings, and for admixture with scrap, to produce a good, tough, and very strong quality for ordinary foundry use.

Grid.—(1) A skeleton-like frame of cast iron, which is used to carry a core, or drawback, or some other portion of a mould. (2) The grating of an air-pump valve.

Gridiron Valve.—A type of slide-valve in which the ports are subdivided transversely by narrow bars or bridges, the ports in the cylinder face being similarly subdivided also, the object of the divisions being to obtain the necessary steam-way with a diminished amount of travel of the valve. Gridiron valves may be double ported (q.v.), or treble ported (q.v.).

Grinder.—(1) Any appliance by which work is shaped or finished by grinding. Usually it is applied more particularly to lead-laps and emery-wheels rather than to actual grindstones. (2) The workman who has the emery-wheels of a workshop, and their special work, under his charge.

Grinding.—The abrasion of metallic surfaces on a grindstone, or emery-wheel, or lead-laps, or grinders. Grinding is resorted to not only for the 'roughest but for the finest works, not only to remove metal in quantity but to render more accurate work which has already been shaped carefully in lathes and other machines. All cutting tools are ground, so also are lathe mandrels and cylindrical gauges. Grinding, especially since the introduction of emery-wheels, occupies a higher economical importance in the workshop than ever.

Grinding Clamps.—A divided lap (q.v.) used for grinding mandrels and cylindrical holes. The halves of the clamps are capable of adjustment by means of set-screws. See Grinder, Grinding, &c.

Grinding-in.—The bringing of the conical seating surfaces of poppet valves or plugs and cocks of a circular shape to an exact fit by means of emery-powder. The powder, mixed with oil, is smeared sparingly and evenly over the surfaces which are then ground by turning the valve or plug through portions of a revolution, each two or three successive turns commencing at different starting-points in the circle in order to equalise the wear.

Grinding Line.—A line marked along the twist of a twist-drill, coincident with its axis, as a guide to the grinder in maintaining the cutting point in exact position.

Grinding Machines.—Machines of various types employed for the abrasion and shaping, and truing-up of metallic surfaces, emery-wheels of various qualities being the grinding agents. Spindles and mandrels are ground while revolving between centres by means of a revolving and traversing wheel, or a lap; surfaces are ground while being traversed underneath a revolving emery-wheel. Irregular works are ground whilst held against ordinary wheels mounted on spindles. Taps and cutters of various kinds are also treated in these machines.

Grinding Rest.—A support, or a guide, for a cutting tool during the process of grinding on a stone. In the turning or fitting shop it generally consists of a bar, or a bridge, of square iron, laid across the top of the water-trough in front of the stone, and upon which the tools are laid. For grinding wood-cutting tools various forms of rests have been devised, but except for the knives of rotary cutters workmen do not hold them in much value. See also Twist Drill Grinder.

Grindstone.—A revolving stone against which tools and materials are abraded by grinding. Grindstones are natural sandstones, Newcastle being the best; they are also artificial compositions. Grindstones revolve in troughs containing water, but should not run actually in the water, as that softens them and causes them to wear unequally. The water should be fed on to them from a drip-can. For ordinary tools they should make about 90 or 100 revolutions per minute.

Grindstone Truer.—Grindstones are trued, or turned up, by hand by means of a pointed bar of steel, which is constantly rotated to present a new cutting point to the stone. Grindstones are trued mechanically by means of a threaded roller of steel, clamped in a frame and allowed to rotate against the surface of the stone.

Grip Chuck.—A jaw chuck (q.v.).

Grit.—Grindstones are specified as of hard grit or soft grit, according to their density or texture.

Grit Stone.—A natural grindstone as distinguished from an artificial or composition stone.

Gromet Washer.—A grummet washer (q.v.).

Grooving, or **Furrowing.**—(1) An action which injuriously affects certain portions of boiler plates. It is the cutting round, or corroding away, of those parts which are subject to continual leverage, and the grooving action is partly due to mechanical strains constantly repeated, and partly to the attacks of acids in the lines of strain. It is particularly observable in the seams of those plates which are not properly stayed. (2) The recessing of the edges of boards to receive the tongues of corresponding boards. Grooving is performed by revolving cutters, or by a small thick circular saw having widely-spaced teeth.

Grooving Saw.—A drunken saw (q.v.).

Gross Indicated Horse-power.—See Indicated Horse-power.

Gross Pressure.—Absolute pressure (q.v.).

Gross Section.—The gross section of a steam boiler is the total number of inches contained in its circumference. If it were a solid plate this would also be its effective section (q.v.).

Ground Line.—The horizontal line along which the strains in a crane post (q.v.) tending to break it off are concentrated in greatest force. This is not necessarily on a level with the ground, but coincides with the upper face of the foundation-plate or ring in which the post is fixed, in a fixed crane, or in which the rollers revolve in a portable crane.

Ground Wheels.—The running, or travelling wheels of a portable crane.

Grouped Safety-valve.—See Cowburn Valve.

Grout.—A coarse mortar used for cementing iron work into stone.

Grouting.—The fastening of iron work, as bolts and eyes, into masonry by the aid of grout (q.v.).

Grub Screw.—A screw which has neither nut nor head of any kind, but which is simply slotted across the top end for the insertion of a screwdriver, by means of which it is driven home.

Grummet Washer, or **Gromet Washer.**—A washer made of a strand of rope, or spun yarn, or tar twine, twisted into the form of a ring. It is employed for insertion under the heads of square-shouldered bolts, in built up cast-iron water tanks; its purpose being to prevent leakage round the heads where the rust caulking does not completely fill up.

Guard.—See Shield.

Guard Plate.—The curved plate in an india-rubber disc valve, which prevents the india-rubber from opening or lifting beyond the distance limited by the plate.

Guard Rail.—In permanent way, a short rail laid down at a crossing near the outer rail to guide the wheel flanges. It fulfils a similar purpose as a check rail (q.v.) on a curve.

Guard Strap.—The strips of sheet iron which arch over the tops of the wheels of locomotives, as a protection against injury to the drivers.

Gudgeon.—The cross shaft which stands at right angles to a piston-rod, or pump-rod, and which forms the medium of communication between the rod and the crosshead, or slipper blocks.

Guide.—(1) Any attachment or contrivance by which a crosshead, or slipper block, or pin of an engine, pump, or piece of mechanism, is coerced to travel in one path only, whether linear or curvilinear, is termed a guide or guide-rod, guide-bar, or slide-bar. (2) A chisel-shaped attachment to rolling mills, which lifts the rail or bar off the rolls on the leaving side. Without the guide, the bar would   around the roll.

Guide Bars.—See Slide Bars.

Guide Blocks.—See Slide Blocks.

Guide Iron.—A piece of iron rod, about ½ in. square, which being bent to the contour of a curved pattern pipe, becomes a guide by which the core maker strickles up its core without requiring a core-box. It is equally available for striking up a pipe pattern in loam.

Guide Plate.—See Ramps.

Guide Pulley.—A pulley which neither acts as a driver or a driven, but simply as a guide for changing the direction of motion, or to take up the slack or strain on an extensible belt or cord.

Guide Rod.—See Guide.

Guide Screw.—The leading screw of a self-acting lathe. Used for traversing the slide rest, and for initiating the threads of screws.

Guillotine Shears.—A special type of shearing machine, used for the cutting up of puddled bars and slabs ready for piling (q.v.). The shears are similar to those of ordinary shearing machines, but instead of being set transversely to, they are parallel with, the plane of the machine framework.

Gulleting.—The deepening of the roots of the teeth of circular and gullet saws.

Gullet Saw.—A saw whose teeth are deepened and hollowed in their roots. Most circular and pit saws are gulleted, and the dust runs away with greater freedom from such saws, the gullets being sloped outwards on alternate sides.

Gullet Teeth.—The teeth of a gullet saw (q.v.).

Gumming.—Lubricating oils are gumming or non-gumming, as they become thick and sticky, or otherwise. See Saw Gumming.

Gun Metal.—An alloy of copper and tin in various proportions, employed largely in engineering construction. It is tougher than cast iron, does not heat rapidly with friction, and is easily turned and planed. Its tensile strength is less than that of wrought iron, but greater than that of cast iron, and equals about 11 tons per square inch of section. It is employed for bearings, and in pump work where the use of iron would be objectionable, by reason of its liability to rust. Gun-metal is hard or soft as it contains greater or less proportions of tin. Called also bronze. A tough mixture contains copper 90, tin 10 parts.

Gusset.—An angle-iron bracket used to stiffen an angular portion of a structure. See Gusset Stay.

Gusset Plates.—The webs or plated portions of gusset stays (q.v.), as distinguished from their angle-irons.

Gusset Stay, or Gusset.—A triangular plate of wrought iron or steel, whose function is to support the inherently weak flat ends of Cornish and Lancashire boilers. In large boilers about five gusset stays are attached to each end, being secured to the shell and to the end-plates by angle-irons.

Gut Bands.—Gut bands are used for driving some of the smaller foot-lathes. They are united by hooks and eyes, sometimes by splicing. Gut is prepared from the intestines of the sheep, but sometimes from those of the horse, ass, or mule ; never from the cat.

Gutta Percha.—A vegetable exudation which is largely employed by hydraulic engineers for the valves of hydrants and other water valves. It is used for taking impressions of ornamental work for which foundry models have to be made. The gutta-percha is placed in cold water to cool, to prevent its contraction.

Guys.—The tension ropes or chains which act as braces or supports to shear legs, or which sustain the mast or boom of a derrick crane.

Gyration, Centre of.—See Centre of Gyration.

H.

Hack Hammer.—A chisel-shaped cross-paned hammer (see Pane), attached to a stout handle, and used for the purpose of hacking (q.v.) grindstones.

Hacking.—The dressing off of the projections upon the face of a grindstone with a hack hammer (q.v.). When a stone becomes very uneven it is difficult to use the ordinary pointed grinding tool with effect. The hack hammer renders the task easier, by first reducing the chief projections by a series of cross hatchings, or chequered cuts.

Hack Saw.—A light frame saw used for cutting metal. The saw has a thin narrow blade, and equilateral triangular-shaped teeth, and is stretched between a frame of wrought iron, the requisite tension being imparted to the saw by means of a thumbscrew. A handle is affixed to one end of the frame, and downward pressure is imparted by laying the left hand upon the back or top of the frame. This saw is in constant use by fitters and machinists, for cutting off odds and ends of metal where the concussion of a chipping chisel would be objectionable, and for cutting fine slits in work of the exact width of the saw thickness.

Hæmatite.—See Brown Hæmatite, Red Hæmatite.

Hair Compass.—A drawing compass in which provision is made for minute adjustment for distance. The upper portion of one of the legs is formed into a spring to which a screw, having a milled head, is attached, and by means of which the exact amount of the required movement is regulated.

Hair Felt.—See Felt.

Hair Side.—The grain side (q.v.) of a leather belt.

Half-blind Holes.—See Half-lap Holes.

Half-centre.—Half-centre is sometimes used to denote the position of the crank-pin of an engine when midway between the two dead centres or dead points. The rotational effect is greatest at half-centre.

Half-crossed Belting.—Belting is said to be half-crossed when it drives between two pulleys whose axes are at right angles with each other. Such belts can only run in one direction.

Half-lap Coupling.—A box coupling in which the liability of the shafts to slip is prevented by making a short length of one shaft overlap a corresponding length on the other, one-half the sectional area being removed in each case. The diameter at the joint is increased to maintain equal strength, and a little taper is imparted to the lapped parts.

Half-lap Holes, or Half-Blind Holes.—Rivet holes in boiler and other plates which are punched so inaccurately that when the plates are brought together they do not correspond within the extent of one-half their diameters, the edge of one coming to the centre of its fellow when the plates are in position. In this case the holes should be broached or reamered out, and larger rivets inserted, though it is too often the custom to pull them together with a drift. See Drift, Blind Holes.

Half-lap Joint.—See Halving.

Half-lattice Girder.—A Warren girder (q.v.).

Half Pattern.—One of the halves of a pattern which is jointed through the centre for convenience of moulding.

Half-rip Saw.—A hand-saw, from 28 in. to 30 in. long, containing about three and a half teeth to the inch. It is used for cutting with the grain chiefly.

Half-round Bit.—See Cylinder Bit.

Half-round Chisel.—See Round-nose Tool.

Half-round File.—A file which is flat on one side and convex on the other, the amount of convexity never, however, equalling the half of a circle, being always much less, though variable in amount. The amount of convexity is indicated by the terms, "round-edged flat," "extra thin flat," "flat," and "high-backed half-round."

Half-round Iron.—Rolled wrought-iron bar used in various engineering constructions. It is either solid half round, the section being that of a solid hemisphere, or hollow half round, the section being that of a semi-cylinder. Solid half rounds are made from about $1\frac{1}{2}$ in. × $\frac{3}{4}$ in. to $3\frac{1}{2}$ in. × $1\frac{3}{4}$ in., and hollow half rounds from $1\frac{1}{2}$ in. × $\frac{7}{8}$ in. thickness to $3\frac{1}{2}$ in. × 1 in. thickness.

Half Section.—A sectional view which terminates at the centre line dividing the object represented into two symmetrical portions. Where a half section is given, the complementary half is usually an outside view or elevation.

Half Shrouding.—See Shrouding.

Halving.—The making of joints in pattern work, effected by cutting half the thickness from the face of one piece, and the other half from the back of its fellow, so that when brought half to half their outer surfaces are flush.

Hammer.—There are not many different types of hammers used in factories, and those being ordinary fitter's hammers, the smith's hammers, hand and sledge. Hammers are sold by weight, and range from half a pound to 14 lbs.

Hammer Block.—The steel face of a steam hammer which is attached to the tup (q.v.) by means of a dovetailed joint.

Hammer Hardening.—When cold sheet metal is subjected to long-continued hammering, its fibres are thereby rendered dense, hard, and brittle. This is termed hammer hardening in order to distinguish it from fire hardening. It is therefore the reverse of the process of annealing (q.v.).

Hammer Head.—The hammer itself as distinguished from its handle.

Hammerman.—A smith's striker or mate, the man who hammers the heavier work upon the anvil with a sledge.

Hammer Marks.—The marks of the hammer left on smith's work. Freedom from hammer marks is one of the tests of good work. The use of flatters tends to obliterate these marks.

Hammer Scale.—The scale squeezed out from a bloom of puddled iron under the steam hammer. It is employed for coating the bottom or hearth of the puddling furnace, because of its richness in oxygen, and consequent value as a decarburising agent.

Hammer Shaft.—The handle of a sledge hammer.

Hammer Tongs.—Smith's tongs bent round at right angles with the main axis, for the clipping of pieces of punched work. The opening or loop enclosed by the jaws is of large area. Called hammer tongs because used in the manipulation of hammers in course of manufacture.

Hand.—A word commonly used as affix and prefix, as hand screw, hand feed, hand shears, hand hole, right hand, left hand, &c.

Hand Brace.—See Brace.

Hand Brake.—A brake which is applied or released by means of a lever moved by the hand, as opposed to a foot, steam, or water brake.

Hand Crane.—A crane whose motive power is supplied by the energy of a man or men turning the handle or winch.

Hand Drill.—A drilling machine worked by hand power. Or a drill actuated by a bow, or a hand brace, or a ratchet brace.

Hand Drilling.—The drilling of holes by means of a ratchet brace, or other suitable hand drill, as distinguished from machine or power drilling. The practice of hand drilling is of necessity resorted to in all outdoor work, but also to a considerable extent in the workshop where the use of machines is available, as in cases where work is in course of erection and cannot be brought to the machines, or where an individual piece, as a long column, or a girder, or large awkward casting, is too ungainly or bulky to be placed under the machine at all, or without an expenditure of trouble and time out of proportion to the cost of drilling the holes by hand. Where the proportion of cost of transit and setting of work under the machine is heavy, hand-drilling specially performed by a man accustomed to it is cheaper.

Hand Expansion Gear.—Variable expansion gear, adjustable by hand, usually consisting of two screws, right and left-handed respectively, by which the positions of a pair of cut-off valves are adjusted relatively to the slide-valve.

Hand Feed.—(1) The feed (q.v.) of the cutting tools of machines of various kinds, which is effected by the hand of the workman alone, in opposition to power feed (q.v.). (2) The passing of a board over the cutters of a wood-planing machine by hand, as distinguished from roller or automatic feed. (3) The thrusting of timber through a circular saw, in opposition to the mode of feeding by drag ropes or rollers.

Hand File.—Or flat file. The ordinary engineer's file, which, though slightly rounding or bellied lengthways, is flatter than the bellied file (q.v.).

Hand Fitted.—Metal work and machine work fitted by hand labour, as opposed to work performed by machines.

Hand Gear.—Gear, usually toothed, worked by a winch handle, in contradistinction to power gear.

Hand Hammer.—An ordinary fitter's or smith's light hammer, weighing from one to one and a half pounds.

Hand Holes.—Holes cut in crankcases, tanks, boilers, machinery casings, and similar parts, through which the hand is passed for purposes of cleaning and repairs. Cf. Man Hole.

Hand Hook.—A hook wrench (q.v.).

Handing.—Making symmetrical work right and left hand respectively, and altering patterns from left to right hand and *vice versâ*.

Handing Round.—The turning round of the handles of a die stock (q.v.) which in the larger kinds is done by two men who pass, or hand round the handles from one to the other.

Hand Ladle.—Or hand shank ladle. A small ladle used in foundries for carrying molten metal to the moulds. It is called a hand ladle because it is carried by one man without assistance. It holds about a half-hundredweight, and is provided with a handle about three feet in length.

Handles.—See Tool Handles.

Hand Lever.—A light lever worked by hand power only. Hand levers are employed in hundreds of machines and for an infinity of purposes. The part grasped by the hand is usually turned to a shape convenient for the grasp.

Hand Lift.—A lift (q.v.) worked by the hand hauling at an endless rope; the rope passes over a sheave and thus actuates a worm, worm-wheel, and suitable spur-gearing. The whole of the gearing is placed at the head of the lift.

Hand Rammer.—A rounded hammer-like tool furnished with a short iron handle, and used in foundries for ramming the sand round patterns, and into core-boxes.

Hand Rest.—A tee-rest (q.v.), so called because used for unassisted hand-turning, as distinguished from an automatic or slide rest.

Hand Rivetting.—The turning over of the heads of rivets by men working hand hammers only, with or without the intervention of a die or snap, as opposed to machine rivetting.

Hand Saw.—A saw about 26 in. in length, the ordinary one-handled saw of the largest size used in the workshop. It is used for the common work of sawing stuff of moderate thickness both longitudinally and across. Rip, half rip, and panel saws may be considered modified hand saws, from which they differ in the size of the teeth, and to a slight amount in length. A hand saw proper contains from five to six teeth to the inch.

Hand Screw.—A pair of small wooden vice chops, tightened by two wooden hand screws, and used for the temporary clamping together of jointed pieces of wood.

Hand Shank Ladle.— A hand ladle (q.v.).

Hand-shearing Machine.—A small machine used for cutting tin plate and thin sheet iron by shearing action. These machines are all worked by means of levers, one blade being fixed, the other movable ; but they differ in details of construction.

Hand Sketch, or Sketch simply.—A rough kind of free-hand drawing made without rule or compasses, and therefore, though not to scale, it is a working sketch because furnished with necessary dimensions. Hand sketches are used for rough and temporary purposes, or for those jobs which will not pay for the making of regular drawings.

Hand Surfacing.—See Surfacing.

Hand Tap.—A tap worked with a tap-wrench by hand, as distinguished from a machine tap. The shank of a hand tap always terminates in a rectangle or in a flat expansion, to enter the hole in the tap-wrench.

Hand Tools.—All the tools, cutting and otherwise, actuated by the hand alone, as distinguished from machine tools.

Hand Traverse.—A traverse or cross movement of a machine part which is effected through mechanism actuated by the hand alone. It is therefore the opposite of automatic.

Hand Traversing.—See Traversing.

Hand Turning.—Turning accomplished by the unassisted aid of hand tools, both for wood and metal, as distinguished from turning done by the aid of the slide-rest in self-acting lathes.

Hand Vice.—A small spring vice held in the hand and used for the manipulation of wire and of small and light work in general.

Hand Wheel.—A wheel turned by the hand for the purpose of actuating screws, worm-gearing, clutches, levers, &c., by which motion is imparted to some portion or portions of a machine beyond the immediate reach of the attendant.

Hand-wheel Feed.—Any feed motion imparted to the feed-screw by means of a hand wheel. It is chiefly in drilling machines that the spindle is fed by means of a hand wheel. Sometimes the hand wheel moves vertically, carrying the screw along with it ; in other cases the screw alone moves through the boss of the hand wheel, the latter being stationary. See Stationary Hand-wheel Feed.

Hang.—The particular style, or angle, or fashion, at which a smith or a fitter holds his hammer.

Hang Down, or Hanger.—A bearing suspended from a roof or beam for the journal of a shaft, hence its name. It is held in place with bolts, and the bearing parts consist of brasses and a cap in the usual forms.

Hanger.—See Hang Down.

Hangers.—The looped suspension rods which depend from the transverse beam attached to a foundry crane, and which receive the swivels of a moulding-box slung therefrom.

Hanging.—Hanging means the fixing of a pulley, fly-wheel, grindstone, &c., upon its shaft or axle. When the central hole is bored to the same diameter as its shaft there is practically no hanging required, so that the term is only strictly applicable in cases where the hole is larger than the axle, as in those of rectangular and polygonal sections, and where therefore the pulley, or stone, or wheel has to be adjusted and set by means of wedges or keys.

Hanging Tube Boiler.—A type of vertical boiler in which a number of small water-tubes closed at the lower end hang from the firebox crown and are exposed directly to the play of the fire and hot gases. See Field's Tubes.

Hang To—A term having several applications. A file hangs to its work when it cuts without slip. A saw hangs to, when it feels as though being drawn into the timber. A pattern hangs to the sand when it delivers with difficulty.

Hard Brass.—(1) Brass which has not been annealed after drawing or rolling. It is highly elastic, and is therefore used for springs and work of a similar character. (2) Hammered brass, and brass which contains a large proportion of tin. Gun-metal is often called hard brass, and is therefore suitable for bearings.

Hard Coke.—Oven coke (q.v.).

Hardening.—Hardening is the result of an increase in the density of a metal. The hardening of steel is effected by heating it to a temperature higher than that which it is to receive permanently, and then suddenly cooling it in oil, or water, or other suitable solution. The temperature for hardening is controlled by the quality of the steel and the purposes to which it is to be applied. See also Case Hardening, Hammer Hardening, Babbitting.

Hardening Mixtures.—Fluids or semi-fluids employed for hardening and tempering works in steel. Pure spring water is the best for general purposes, but for special work a multitude of media are used, and most workmen have some special mixture of their own which they deem endowed with peculiar virtues. The water is frequently medicated with common salt, or with other ingredients, and is used either cold or slightly lukewarm. For small delicate works and tools, oils and fats are used in preference to water. See Blazing Off. For special work fusible metallic alloys are employed.

Hard Firing.—Rapid stoking of engine fires and quick succession of charges.

Hard Grit.—Hard grit signifies a grindstone or emery wheel of hard texture.

Hard Iron.—Cast iron, which is dense and close grained. It is obtained by making suitable mixings of various brands with scrap, and is used for wearing parts, the liners of engine cylinders, and cog-wheels.

Hard Lead.—Lead which has been smelted but not " improved." The hardness is due to the presence of antimony, copper, and iron in minute proportions, which united, render the lead unsuitable for the purposes for which it is commonly employed.

Hard Metal.—See Soft Metal.

Hardness.—See Toughness.

Hard Ramming.—Hard Ramming relates to the amount of pressure employed when ramming foundry sand around patterns embedded therein. Excessively hard ramming is productive of scabs. Too soft or too light ramming results in lumpy and swollen castings, because the pressure of metal causes the sand to yield before it. Ramming should be harder at a distance away from the pattern than in its immediate vicinity. The proper amount of pressure varies therefore, and is a matter for the exercise of experience and judgment.

Hard Solders.—The solders used in brazing (q.v.) and generally for uniting the harder and more infusible metals.

Hard Tap.—When the clay which closes the tap hole of a cupola is pierced with difficulty it is said to be a hard tap.

Hard Water.—Water containing a large quantity of carbonate and sulphate of lime in solution. Hard waters are objectionable for use in water piping and in steam boilers, since they produce calcareous deposition.

Hard Wood.—Of the woods employed in engineering construction, oak, ash, beech, elm, mahogany may be considered hard.

Hatch.—A sluice; usually applied to a sluice of large size set in the course of a stream for regulating the supply of water to a mill wheel.

Hatchet.—A double bevelled cutting tool whose edge is parallel with the axis of its handle, and is actuated by percussive force only.

Hatchet Stake.—A tool used by coppersmiths for bending their thin sheets. Its edge is sharp, somewhat like that of a hatchet, hence its name, and the sheet metal is turned over the edge.

Hat Leather.—The leather ring packing used for hydraulic pistons. It is of the same section which would be produced by bending a strip of angle iron into a ring with one edge facing inwards. Sometimes it is called a cup leather (q.v.), though strictly speaking this term has reference to a leather of quite a different form. Hat leathers are commonly used in pairs, or back to back.

Hawse Pipe or **Hawser Pipe.**—(1) A stout cast-iron pipe or tube with well-rounding edges which lines the hawse or chain hole in a ship's side. (2) A pipe which guides the barrel chain in some forms of grabs.

Hawser Rope.—A cable laid rope (q.v.).

Hay.—See Hay Band.

Hay Band.—Common hay spun into bands of $1\frac{1}{4}$ in. or $1\frac{1}{2}$ in. diameter, and of several yards in length is used in the making of foundry cores. The band is wrapped closely round a core bar, and the thickness of loam struck upon it. The hay then forms a porous mass through which the gases generated in casting readily find egress from the mould.

Hay Band Spinner.—A long lantern wheel made of wood, and set spinning on an axle fixed in a wall, and on which the core maker or core boy spins his bands. It is rotated by one hand, while the hay is fed on to it and twisted with the other.

Hay Rope.—Hay Band (q.v.).

Haystack Boiler.—A balloon boiler (q.v.).

Hazel. (*Corylus avellana*).—An English wood of the order *Cupuliferæ*, whose qualities of toughness and elasticity render it invaluable to smiths for the handles of flatters, swages, chisels, &c., where metallic handles would produce a painful jar in the hands due to the hammer blows. See Hazel Rods.

Hazel Rods.—The rods or handles of smiths' small tools are made by binding hazel rods of about $\frac{1}{2}$ in. in diameter around the head and twisting them tightly. The rods are supplied in bundles, and are soaked in water and warmed previous to being bent.

Head.—(1) A term of wide application, signifying the broad end or expansion of screws, nails, bolts, &c., hammer heads, head metal, &c. (2) The height of a column of liquid, or the pressure equivalent to that height. This enters into all calculations concerning the pressure on mains and valves, the power of turbines and water wheels. (3) The head of a rail is the upper surface upon which the wheels travel.

Headers.—Headers refer to the laying of courses of bricks for the linings of blast and cupola furnaces; being set with their heads or ends towards the body of the furnace, thus giving the maximum of thickness to the lining. See also Water Tube Boilers in App.

Heading Tool.—A smith's tool used for shaping the heads of screw bolts. The tail of the bolt passes through a hole in a plate, and the bar while white hot is flattened out over the surface to form the head.

Head Metal, or Dead Head.—A mass of metal, usually in the form of a ring cast on the upper end of cylindrical work to form a receptacle for the dross, sullage, scoriæ, &c., which collects upon the top of molten metal. When the casting is removed from the mould the head metal is turned off, leaving the actual casting smooth and free from these foreign impurities.

Headstock.—(1) The fixed head or poppet of a lathe which carries the revolving mandrel and rigger. The poppet proper or the back poppet is sometimes called the movable headstock. (2) The end timbers in the under frame of a railway truck.

Head Valve.—The uppermost or delivery valve of an air pump, so termed in opposition to the foot or suction valve.

Heart and Square Trowel.—A double-ended moulder's trowel, whose two ends are united by a metallic handle, one trowel being heart-shaped, the other rectangular. They are made in small, medium, medium large, and large sizes.

Heart Cam.—A cam heart-shaped in form, used for the conversion of rotary into rectilinear motion.

Hearth.—(1) That portion of a reverberatory furnace (q.v.) upon which the ore or metal, as the case may be, is exposed to the action of the flame which reaches it from the fireplace after passing over the bridge. (2) The conical portion of a blast furnace which lies below the boshes, and into which the molten metal descends. (3) The bottom of a cupola upon which the coke bed is laid. (4) The open area of a smith's forge in front of the tuyere upon which the fuel is placed.

Hearth Back.—A plate of cast iron which forms the back of a smith's hearth. It is supplied with a nozzle for the entry of the nozzle of the water tuyere, and with holes for bolting up into place.

Hearth Plate.—A cast-iron plate which sometimes covers a furnace hearth.

Heart Trowel.—A moulder's trowel, whose outline is cordate or heart-shaped. It is named long or broad, according to its proportions.

Heart Wood.—The wood in, and immediately around the centre of a tree. The heart wood is best in young timber; in old trees it is of little value.

Heat.—(1) Heat is produced by the motion of the ultimate molecules of matter, which motion results either from mechanical action, the friction of bearing surfaces for example, or from chemical affinity, as the union of the carbon or hydrogen in fuel with the oxygen of the air, or from electricity. Heat is convertible into mechanical work, and work, conversely, into heat. The more rapid the vibration of the material particles, the greater their separation and expansion and the greater the quantity of heat evolved. Hence heat is simply a mode of motion, and not a material substance, as was formerly supposed. (2) A term commonly employed in smithies and boiler shops to signify the getting hot of any piece of metal which is to be operated on. Thus a man will be said to get so many heats in a definite time, which represents therefore the amount of work which he is able to do. Or he will shape a definite section of a forging at a single heat, that is, without reheating the work in the fire. The same term is applied by puddlers to the working through of a single charge of iron.

Heater.—See Feed Water Heater.

Heating Apparatus.—The apparatus by means of which buildings are warmed up to a definite temperature. The heating is effected by the circulation of hot water in pipes due to convection, or of steam in steam pipes provided with radiating plates.

Heating of Bearings.—Bearings are said to heat when their temperature rises so much above the normal that their axles either stick fast or are subject to great extra frictional strain, or are themselves rapidly abraded. The temperature may be such that the hand can barely tolerate, or so high as a low red heat. Heating is due to want of lubrication, and the remedy is to slacken the bolts, and pour on water first and oil afterwards. If the bearing is very hot, the first application of water should be hot, since cold water might produce fracture. The presence of sulphur also assists the cooling action of the oil.

Heating Surface.—(1) The entire superficies of a steam boiler exposed to the flame and incandescent gases on the one side, and to the water upon the other. (See Transfer of Heat.) (2) The total area of the brick chequering or partitioning in regenerative furnaces. (3) The pipe surface in a pipe stove (q.v.).

Heat, Unit of.—The British practical unit of heat is 772 foot pounds, or the Joule. See also Thermal Units.

Heavy Cut.—The work of a cutting tool on metal is said to be heavy when it is removing thick and broad shavings. The term is relative, because what would be a heavy cut in small work would be a light cut in work of larger dimensions.

Heavy Firing.—Close or thick firing.

Heavy Oil.—Thick oil (q.v.).

Heavy Running—The reverse of light running (q.v.).

Hectare.—A French measure of surface, being equivalent to one hundred square ares (see Are), or ten thousand square metres (q.v.).

Hectogramme.—A French measure of weight, containing one hundred grammes (q.v.), and equivalent to 3·5274 English ounces.

Hectolitre.—A French measure of capacity, containing one hundred litres (q.v.), and being equivalent to 22·009 English gallons.

Hectometre.—One hundred metres (q.v.), or 3937·079 English inches.

Heel.—The thick or broad end of a wedge-shaped piece, the broad end of a railway switch for example.

Heel Tool.—A turning tool for metal which was commonly employed before the use of the slide-rest became general. The cutting end resembled in shape that of a boot, the heel being supported upon the rest, the toe being the actual cutting edge. The handle rested in an oblique direction against the shoulder of the workman.

Helical Gear.—Toothed gear in which the wheel-teeth instead of being at right angles with their faces are set at some other angle therewith. Such gear may be single or double ; in the former the teeth being in one diagonal plane only across the wheel, in the latter they are in two diagonal planes placed symmetrically on each side of a plane cutting transversely through the axis of the wheel, and midway between its faces, the last form being mostly employed. Helical gears are a modification of the old stepped gears, the steps being blended into one continuous face, hence the teeth slide down one another with more regularity than is the case with ordinary wheels. In principle they are screw gears.

Helical Teeth.—See Helical Gear.

Helical Spring.—See Spiral Spring.

Helve.—(1) The handle of an axe, hatchet, or adze. (2) A helve hammer (q.v.).

Helve Hammer.—An antiquated form of smiths' hammer, sometimes called tilt or trip hammer, in which the hammer is attached to a lever lifted by cams or wipers, and allowed to fall with the force due to its own

momentum. Helve hammers are of two kinds ; one, the nose or frontal helve, in which the cams act upon the lever which carries the hammer, at the extreme end, or opposite to the fulcrum ; the other, or belly helve, where the cams act upon the lever between the hammer head and the fulcrum. The latter gives more room for the manipulation of the forging or of the bloom than the former.

Hemp, or Tow.—Hemp or tow is used in foundries by core makers for winding on small core bars previous to the daubing and sticking on of the loam. It fulfils the same purpose as the hay-bands in larger cores, becoming a rough basis for the attachment of the loam, and also a porous mass through which the gases generated by casting escape into the hollow of the bar. Hemp is also largely used by moulders for filling up the interstices of cores which are laid finally in their places, the mould being yet unfinished, the object being the prevention of the accidental falling in of dust and sand into the lower, or already finished portions. The hemp is removed just previously to the final closing of the mould.

Hempen Ropes.—Ordinary ropes for the transmission of power, and for haulage, &c., are made of coarse hemp, once chiefly imported from Russia, hence the ropes are frequently termed Russian ropes. The ropes are usually tarred, and are thus distinguished from white or un-tarred ropes. (See Manilla Ropes.) The prepared hemp is first spun into yarns, these into strands, the twist of the strand being in an opposite direction to that of the yarns, and the strands laid or twisted into the rope. Tarring, when practised, is effected when the yarns are spun, and before they are twisted into the strands. The hemp genus belongs to the natural order *Cannabinaceæ*.

Hercules Crane.—A crane of a large and powerful type, used in harbour works for the setting of concrete blocks. It is mounted on a travelling framework, and has a horizontal jib which slews either through a portion of, or the whole of a revolution. Steam is the motive power used.

Hexagon Nut.—The ordinary six-sided form of nut which has superseded the primitive square form, owing to its superior neatness, and to the fact of its being more readily turned in a narrow space, the handle of the spanner being moved through a smaller arc to obtain a fresh bite.

High Breast Wheel.—See Breast Wheel.

High Flashing Point.—An oil is said to have a high flashing point when it will take fire at a high temperature only.

High Press.—A workshop abbreviation employed to signify the high-pressure cylinder in a compound engine.

High-pressure Engine.—(1) An engine which exhausts its steam directly into the atmosphere. (2) Any engine, condensing or non-condensing, which is driven by high-pressure steam.

High-pressure Cylinder.—The smaller cylinder of a compound engine which receives the steam direct from the boiler, and in which it is first expanded, and from which it is exhausted into the adjacent low-pressure cylinder.

High-pressure Steam.—Steam whose pressure is sufficiently above that of the atmosphere to enable it if required to drive an engine which is unprovided with a condenser.

High-speed.—A term having various applications in practice, being of general rather than of special signification. High-speed engines may be considered to embrace any steam engines running at over 200 to 300 revolutions per minute ; or internal combustion engines running at over

1,000 revolutions per minute. Rotary and three-cylinder steam engines also are termed high-speed without special reference to any definite rate of revolution. High-speed belting applies to belts for fans, wood-working machinery, centrifugal pumps, &c., in opposition to those for line and counter, and other slowly driving shafts.

High Speed Bearings.—Bearings whose length exceeds their diameter by from four to six times. Their value consists in the distribution of the excessive friction of rapidly revolving shafts over a large extent of surface, with a consequent diminution of heating. They are made in various forms ; in one of the commonest the actual bearing is made to swivel in its plummer block or hanger, and is thus rendered capable of accommodating itself to slight inequalities in the line of the shafting, or to inaccuracy in the setting of the main block or hanger.

High Speed Engine.—See High Speed.

Hippopotamus Hide.—Sometimes used for the covering of buff wheels, being more elastic and durable than leather.

H Iron.—Rolled wrought-iron bar whose section is that of the letter **I**. Used extensively for building up engineering structures.

Hitch.—A cutting tool moving automatically in a tool holder is said to hitch or catch when it is pulled into the work to a depth greater than it is intended to cut. Hitching is due either to malformation of the tool, to an unsuitable cutting angle, or to an improper position of the cutting point relatively to the shank.

Hob, or Hub.—A master tap. Also, a cutter used in gear-cutting.

Hogger Pump.—The upper portion of a deep mine pump.

Hogging.—The curving or distortion of the furnace tubes of boilers, caused by the local expansion of the plates due to heat. Hogging is chiefly observable immediately after the lighting of the fire, and before the temperature has become approximately uniform.

Hoist.—See Lift.

Hoisting Crab.—See Crab.

Hoisting Engine.—An engine specially fitted up for ordinary hoisting or lifting purposes. It is usually attached to its vertical boiler, and to-gether with a hoisting drum and suitable gearing is secured to a low truck having unflanged wheels for convenience of portability. The truck is drawn by horses to the locality where the services of the crane are required. The lifting chain passes from the drum to a gin or pulley block suspended from temporary shear legs or other convenient support.

Hold-down Bolts.—See Foundation Bolts.

Holding up.—The maintaining of a firm pressure against the heads of rivets while their closing up is being effected, a holding-up hammer being used for the purpose.

Holding-up Hammer.—An ordinary hammer used for holding up (q.v.). The handle is of ash, and is pressed against a convenient fulcrum, the hammer being held against the rivet head, not directly, but with the force due to the leverage exerted on the handle.

Hold up.—A term commonly used to express the capacity of a piece of wood or metal to finish to a particular size. If it is under size, or its surface is too winding to permit of finishing to thickness ; or if circular, if its diameter is too small ; or if, though originally large enough, it has not been centred truly, it is said not to hold up.

Holliper.—See Oliver.

Hollows.—The inside curves imparted to the otherwise angular parts of castings. They are inserted in order that the crystals of the metal may arrange themselves in the strongest position, that is, radially to the face, which in a casting having purely right angles would leave a line of break, and consequent source of weakness. Pattern hollows are made in wood, metal, or leather.

Hollow Blows.—Blows delivered by a hammer upon a substance which is either unsupported, or insufficiently supported by an opposing block. The blows are consequently elastic in character.

Hollow Column.—(1) A hollow column which contains the same quantity of metal as a given solid column, is stronger than the solid column, and the strength nearly equals the difference between that of two solid columns whose diameters are equal to the external and internal diameters of the hollow column. (2) The transverse strength of a thin hollow cylinder is to that of a solid cylinder of equal weight as the diameter of the former is to the radius of the latter.

Hollow Half Rounds.—See Half-round Iron.

Hollow Mandrel Lathe.—A lathe whose headstock mandrel is made hollow throughout to allow of the turning off of short pins, screws, or other cylindrical work from a long piece of rod. The rod is slid through the hollow mandrel as each successive pin or screw is turned and cut off.

Hollow Moulding.—A term sometimes employed to designate the cylindrical and hollow forms of foundry work, as distinguished from flat moulding.

Hollow Plane.—A plane whose face and cutting iron are hollowed out, and which therefore planes rounding in section. Used for working the rounding portions of beads and edges.

Hollow Set.—A smith's gouge, used for dressing off the circular portions of forged work.

Hollow Spindle.—Applied to those lathes, screw-cutting, and cutting-off machines in which the bar that furnishes the materials for use, is made to traverse to an exact and measurable distance at a time through the mandrel of the headstock, which is made hollow for that purpose. See Hollow Mandrel Lathe.

Hollow Structures.—The larger machine frames, standards, parts, &c., are made hollow from motives of economy. A given amount of material properly disposed in a hollow form will possess more strength than the same quantity solidly arranged. It is for this reason that machine framings, base plates, columns, box girders, &c., are made hollow.

Hollow Tools.—Bottom swages (see Swage) are termed hollow tools.

Homogeneous.—Of the same quality throughout. Cast metal is homogeneous when the crystals are as nearly as possible of the same size, wrought metal when the fibre is clean throughout, and not an intermixture of various qualities together with layers of scale and rubbish intervening. Again, piled works, as iron rails for example, usually lack the quality of perfect homogeneity by reason of the different qualities of puddled bar and slab employed in their construction, the presence of cinder, scale, or oxide, and imperfect welding; while steel works rolled from the ingot have the quality of homogeneity in perfection.

Hone, or Oil-stone.—A slaty stone used for whetting edged tools after they have left the grindstone, to impart the keen and fine edge required for clean cutting. The qualities of hones vary greatly, some being soft and therefore fretting quickly, but producing a coarse edge; others being hard and slow of action, but giving a keener edge. For ordinary shop use, a Turkey or a Charnley Forest hone are the best, but Washita and Grecian hones are in frequent demand, though not so reliable as the first named.

Honeycombing.—That kind of boiler corrosion which takes the form of numerous minute recesses, or small blank holes or pits, due to the action of acids, or to want of uniformity in the quality of the plates, or to galvanic action, or to all combined.

Hood, or Bonnet.—The upper covering or canopy over a smith's forge which conducts the smoke into the chimney.

Hook.—A firing tool having a narrow prolongation of the end of a long bar, turned down at right angles therewith, and used for clearing the interspaces of the fire bars of boilers, of clinkers and ashes.

Hook Bolt.—A bolt which, instead of a square head, has an end hooked or turned round at right angles. These bolts are often placed against the outside faces of the work which they hold together.

Hooked Tommy.—A tommy used for turning round-headed screws and circular nuts, when, owing to the shallowness of their holes, an ordinary tommy would not hold sufficiently well. A hooked tommy is therefore curved to follow the circular shape of the heads.

Hooke's Joint.—See Universal Joint.

Hooks and Eyes.—Cylindrical belt fasteners used for leather or gut belts of circular section.

Hook Tool.—A form of tool used for vertical slotting, which is placed transversely to the axis of the ram of the machine.

Hook Wrench, or Set, or Hand Hook.—A smith's tool used for taking work out of winding or out of twist. It is formed of a piece of iron bent round to form three sides at right angles with each other successively, two being short and one of the two parallel sides being prolonged to form a handle. It is slipped around the work like a spanner, and the handle serves as a lever to pull out the twist.

Hooped Spindle.—A spindle provided with an eye at one end, either circular or rectangular in outline, which eye embraces a boss cast upon a valve. Often applied to slide valves in place of the ordinary valve rod with lock nut arrangements.

Hoop Iron.—Thin iron used for securing the corners of packing cases. It is made in widths ranging from ⅝ in. to 2½ in., its thicknesses being given in the numbers of B.W.G., ranging from No. 21 to 12.

Hoop L.—The trade mark of the Swedish malleable iron from the Dannemora mines, so called from the letter L being placed in the centre of a circle.

Hoop Tongs.—Smith's tongs whose jaws are bent at right angles with the handles. The jaws lay parallel with each other for the embracing of rings and hoops of flat bar iron.

Hooter.—A steam whistle used instead of a factory bell to summon the men to work. Hooters can be heard a dozen miles away in a still morning.

Hopper.—A box or receiver used for the purpose of feeding supplies of materials to machines or furnaces of various kinds. It is often furnished with sloping sides, and a valve or cover provides the means of regulating the fall of the contents. Concrete mixers, blast furnaces, gas producers, are furnished with hoppers.

Horizontal Boiler.—A boiler, the longitudinal axis of whose barrel is horizontal. Horizontal boilers embrace Cornish, Lancashire, Galloway, Wagon, and marine types.

Horizontal Boring Machine.—See Boring Machine.

Horizontal Crane.—A portable steam balance crane whose cylinders are set in a horizontal direction, and whose gearing is kept as low down as possible. See Vertical Crane.

Horizontal Engine.—An engine the longitudinal axis of whose cylinder and piston rod is horizontal.

Horizontal Lathe.—A boring machine whose axis is set in a vertical direction for the purpose of boring large engine and other cylinders, and rings which would sink and become distorted by their own weight if laid upon their sides in an ordinary lathe or boring machine.

Horizontal Water Wheel.—A water motor composed of two wheels placed horizontally one above the other, the upper one remaining fixed, the lower one being made to revolve by the pressure due to the head of water. The water passing through curved buckets or vanes in the upper wheel is directed by them against a series of buckets in the lower wheel, the latter being free to revolve on a vertical spindle. Horizontal wheels are mostly superseded by turbines.

Horizontal Winch.—A steam winch (q.v.) in which the cylinders are placed horizontally on the side frames.

Hornbeam. (*Carpinus.*)—A wood of the natural order *Cupuliferæ*. It is of a light colour, very stringy, tough, and of moderate hardness. Used by engineers for the cogs of mortice wheels. A cubic foot weighs 47 lbs. Sp. gr. 0·76.

Hornblock.—A casting of steel or iron riveted to a locomotive framing to receive the axle box, and by means of which the latter is constrained to move in a vertical plane only.

Horn Centres.—Small transparent discs of horn used by draughtsmen to place the points of their compasses upon when describing circles, to avoid piercing the paper itself.

Horn Plates.—The wrought-iron or steel frames whose internal edges act as guides to the axle boxes of rolling stock. They are bolted to the truck frames. Also termed axle guards, pedestals, and housings.

Horns.—The curved levers which are pivoted at the side of a planing machine, and which being knocked over by the tappets give the necessary feeds to the tool, and the reversing movement to the table.

Horse Dung.—This is used in foundries for mixing with the sand from which cores are struck up. It owes its value to the quantity of undigested hay which it contains, for this, being desiccated by the drying in the stove, leaves a network of porous channels in the core or loam through which the gases generated by casting find exit.

Horse Gear.—See Bullock Gear.

Horse Hair.—Sometimes mixed with loam for the same purpose as horse dung (q.v.).

Horse Hair Cushion.—A cushion of horse hair placed in the bottom of some types of axle boxes, and which being saturated with oil, forms a reservoir for a worsted pad (q.v.) placed above.

Horse-Power.—A conventional term whose significations require to be defined to have any practical use. See Actual, Indicated, Nett Indicated, and Nominal Horse-Powers.

Horse-Power transmitted.—The powers which belts, wheels, shafting, &c., are capable of carrying or transmitting from the prime movers to the mechanism which they have to drive, this power being expressed in the unit of 33,000 foot pounds.

Horseshoe Gauge.—A fixed gauge used for calipering metal work. It is really a fixed caliper, being cut from a solid piece of metal, and its outline approximates to that of a horseshoe, its internal narrowed portion being the part where the measurement is taken. These gauges are kept as standards for repetition work.

Hose.—Flexible piping of leather or of india-rubber used for the conveyance of water from force and other pumps where the employment of a rigid channel of transmission would be inadmissible or inconvenient. Leather hose is generally fastened down the seam with copper rivets. India-rubber hose is also cemented entire and seamless at a high temperature, in combination with pressure.

Hose Coupling.—The mode of jointing together of the ends of hose pipes. Screwed flanges of brass, male and female, are attached to the abutting ends by means of copper wire bound around them. Or flanges with bolts are similarly attached. Or brass clips only are made to grasp them. Or a third screwed collar slips over and unites the independent ends. See Union.

Hose Clip.—See Hose Coupling.

Hose Pipe.—India-rubber hose pipe is formed by the alternate winding of friction sheet india-rubber around a mandrel, and subjecting the whole to the vulcanising process, or by extrusion through a die.

Hose Reel.—A drum mounted upon wheels, around which flexible hose is wound for convenience of transit.

Hot Air Engine.—A motor in which heated air is the agent employed, and of which the essential principle is the alternate heating and cooling of the air used. It contains provision for the rapid heating of the air, which being expanded does work, giving motion to a piston called the power piston ; the air is then allowed to escape and passes through a regenerator (q.v.) to the compression cylinder, where it is rapidly cooled ; thence back through the regenerator to the power cylinder to undergo the same cycle. The methods by which the heating and cooling are effected and the types of regenerators constructed vary in different types of engines, but the principles are the same in all cases.

Hot Blast.—Until the year 1828 it had been the universal practice among ironmasters to smelt the ore by the aid of a current of cold air poured into the tuyeres. But in that year Neilson, of Carron Foundry, in Scotland, introduced the practice of raising the temperature of the blast, with a resulting economy of fuel, hence the distinguishing term "hot blast." The temperature as now used ranges from 700° F. to 1,300° F. The iron is inferior and weaker than cold blast iron smelted from the same ores, owing to the presence of silicious compounds in larger quantity, and cold blast scrap is always eagerly sought after by the dealers.

Hot Iron Saw.—A circular saw made of thick plate and provided with short dumpy teeth, which is used for cutting off iron bars. The bars are previously heated to a low red heat, and the lower portion of the saw runs in cold water.

Hot Liquor Pump.—A lifting pump (q.v.) which is placed below the hot liquid being pumped, in order that the presence of steam may not interfere with the production of a vacuum, which is the case with pumps working under ordinary conditions. Feed pumps for boilers never pump so great a quantity when the feed water is warm as when it is cold.

Hot Metal.—Molten iron and brass are said to be hot when their temperature is higher than that required for the class of work for which they are intended. If too hot the sand becomes oxidised and produces a rough casting. If too cold, the metal does not fill up the mould properly and probably becomes cold shut (q.v.). **Thin works require** hotter metal than thicker ones to counteract the chilling influence of the mould. Hot metal scintillates more than cooled or dead metal, and the founder judges of the temperature by its appearance. See Breaking.

Hot Sand.—Foundry sand which has been watered and mixed immediately after being knocked out of a flask in which a recent casting has been made. The use of hot sand is injurious both to the pattern and to the mould; the sand sticking to the pattern and producing a bad delivery, and the steam in the mould causing risk of a spongy casting. Hence sand should be allowed to become cold before being used over again.

Hot Set.—A smith's set, or chisel-like tool, made thinner than the cold set (q.v.), and used for the nicking and cutting of hot metal.

Hot Short.—Brittleness in a metal when hot. Wrought alloys show lack of ductility when hot-worked. Cast alloys tend to crack under the stresses which may be set up while cooling in the mould.

Hot Tests.—The testing of the tensile strength of iron and steel bars and plates, by bending them hot to a certain angle, both with and across the grain, without fracture.

Hot Water Apparatus.—The principle is the circulation of hot water in pipes due to the difference in the sp. gr. of columns of water at different temperatures. The cold water being the heavier drives the heated water before it.

Hot Water Test.—A test to which steam boilers are subjected, the use of hot water following that of the cold water test. The hot water by expanding the plates tests the soundness of the caulked joints.

Hot Well.—A reservoir for the hot water pumped out of an engine condenser by the air pump.

Housings.—(1) The framings which support the rolls in rolling mills (q.v.). (2) Horn plates (q.v.).

Housing Screws.—The screws which pass through the caps of rail mill housings for the adjustment of the rolls.

H.P.—See Horse-Power.

Hub.—(1) Variant of Hob (q.v.). (2) A term often applied to the boss of a wheel.

Hunting Cog.—A device of the old millwrights to prevent the unequal wearing of teeth which is supposed to result when the same teeth on wheel and pinion come successively into gear without variation. In other words, when the required number of teeth to produce the necessary velocity ratio is determined on, an extra tooth is thrown into wheel or pinion so that they may be primes to each other, or have no common divisor. Thus 72 and 24 should be 73 and 24, or 72 and 25. The extra tooth is termed a hunting cog.

Hydrated.—See Anhydrous.

Hydraulic Bear.—A punching bear (q.v.) in which the punch is actuated by the power of water under pressure. It consists of a small cistern of water, a force pump, and ram, with levers. The cistern is uppermost and encloses the force pump, the ram is below it, and immediately over the punch which it actuates. One lever works the pump, another lifts the ram and punch after the hole has been pierced.

Hydraulic Belt.—An endless belt, made of woollen material, which absorbs and holds water readily. The lower bight or bend runs in water, the upper bend passes over a pulley. The belt travels at a high speed, and so lifts the water to the pulley, whence it is discharged.

11*

Hydraulic Crane.—A crane in which the pressure due to a head of fluid, or to pressure obtained by pumping fluid into an accumulator, is utilised for the lifting of loads. The fluid is allowed to flow from the accumulator into a hydraulic cylinder, or cylinders, to whose ram one end of the lifting-chain is attached; the chain passes thence over pulleys to the jib and the load, the purpose of the pulleys being to increase the speed of lift beyond that of the rate of travel of the ram. The increase in the rate may be 5 or 6 to 1. Hence there is a corresponding loss of power, such loss being allowed for in the original designing of the crane. Neither is there any gearing, but the chain passes directly from the sheave pulleys to the jib over the top of the post. It is usual to work several cranes from a single accumulator.

Hydraulic Cylinder.—The lifting cylinder of a hydraulic press into which fluid is pumped under pressure, lifting the enclosed piston.

Hydraulic Engine.—See Water Pressure Engine.

Hydraulic Force Pump, or Hydraulic Pump.—A force pump, used to pump fluid under great pressure for testing purposes, or for hydraulic presses. Hydraulic pumps are usually of the piston type, a number of pistons often being arranged radially around a crank, or operated by a swash plate (q.v.).

Hydraulic Forging Press.—A form of hammer used for stamping iron or steel articles to definite shapes in dies. The moving head, which corresponds with the tup (q.v.) of a steam hammer, is actuated by water stored in an accumulator, hence the force exerted is of a compressive nature rather than that due to impact. The working faces may be either flat surfaces, or dies, according to the class of work in hand.

Hydraulic Gauge.—A dial pressure gauge, constructed to sustain and to indicate very high pressures. Usually gauges of this description are made for a maximum pressure of five or of ten tons to the inch, or, reckoned in pounds, of from 0 to 6,000, or 10,000, or 20,000 lbs. per square inch. Also, and more correctly, termed a hydrostatic gauge.

Hydraulic Glue.—Glue which has the property of partially resisting the action of moisture. It is made by dissolving glue in skimmed milk, or by adding a little linseed oil and oxide of iron to ordinary glue.

Hydraulic Hammer.—A hydraulic forging press (q.v.).

Hydraulic Jack.—A lifting jack actuated by a small force pump inclosed within it, and worked by a lever from the outside. Such jacks lift either from the foot or from the top.

Hydraulic Leather.—A cup leather (q.v.)

Hydraulic Lift.—A lift worked by water power, the necessary head being obtained from a tank situated in the top of the building where it is erected. The water pressure actuates a ram or plunger moving watertight in a cylinder. The cage is attached to the head of the ram, the accumulator (q.v.) being sunk in the ground. A lift is direct-acting when the stroke of the accumulator piston is equal to the extreme travel of the lift; or indirect when the stroke of a short plunger is multiplied by sheave wheels and ropes. The weight of the lift itself is counterbalanced by weights.

Hydraulic Motor.—A motor whose source of energy is derived from the power of water.

Hydraulic Pipe Prover.—A piece of apparatus employed for testing pipes by water pressure. The pipe to be tested is made fast between standards.

the flange faces being rendered water-tight by means of india-rubber washers, and the water pumped to the required pressure by means of a force pump having an escape valve, and a weighted lever for the regulation of the exact amount of pressure required.

Hydraulic Piston.—A plunger piston used in force pumps and hydraulic cylinders generally. It is a solid piston, no liquid passing through its body.

Hydraulic Press.—A machine in which the incompressibility and pressure of a fluid are utilised to produce an immense effect with little expenditure of power. It comprises the pressing area, actuated by a plunger or ram moving in a cylinder, and a force pump by which the pressure is imparted. See Hydraulic Force Pump. Hydraulic presses are used for compressing a wide range of materials and for squeezing oils, &c., out of their seeds.

Hydraulic Pump.—See Hydraulic Force Pump.

Hydraulic Ram or **Water Ram.**—A machine in which water is raised from a lower to a higher level by its own momentum. Water is allowed to flow with the velocity due to its head into a horizontal drive pipe supplied with two valves opening in opposite directions. The momentum of the water soon closes the farthest valve which is made to open inwards, and the arrested fluid then opens the other valve outwards and passes into a supply pipe provided with an air vessel. Being thus relieved of pressure, the inward opening valve falls and the water passes through it until it has again acquired sufficient momentum to open the supply valve once more. An efficiency of about 70 per cent. is obtained from hydraulic rams.

Hydraulic Riveter.—A riveting machine consisting of one movable and one lever jaw at whose extreme ends the dies are placed. The movable jaw is actuated by cylinders placed in communication—the one with an accumulator and the other with an exhaust container.

Hydraulics, or **Hydrodynamics.**—Treats of the motion of liquids, as opposed to hydrostatics. All questions and problems relating to the flow and efflux of water or other liquids in pipes, from orifices, rams, friction of liquids, &c., are embraced by hydraulics.

Hydraulic Shearing Machine.—A shearing machine in which fluid pressure is the motive power.

Hydraulic Test.—The testing of the strength of structures which have to sustain high pressures by the forcing in of water from a force pump. Pipes, air vessels, steam boilers, &c., are thus tested to about double their anticipated working pressure.

Hydraulic Tube.—The stoutest solid drawn wrought-iron tube made. It is about twice as thick as common tubing, and is used for hydraulic pump connections and for high steam pressures.

Hydraulic Wheel Press.—A wheel press in which hydraulic pressure is employed instead of that of screw bolts for pulling wheels on or off their axles. It embraces cylinder and ram, pump, and gear for drawing back.

Hydrocarbons.—Compounds of hydrogen and oxygen, either solid, liquid, or gaseous. They are organic compounds, highly inflammable, and are important constituents in fuels and in lubricating oils.

Hydrocarbon Oils.—Mineral oils (q.v.).

Hydrochloric Acid.—Symbol, HCl. A compound of chlorine and hydrogen, with water. It is used for a flux for soldering. Dilute HCl. is used sometimes for rusting metal patterns and the faces of chilling moulds.

and for dissolving the scale from boiler plates. Called also **muriatic acid**, and **spirits of salts**.

Hydrodynamics.—See Hydraulics.

Hydrogen.—Symbol, H. Comb. weight, 1. A gas possessing high calorific power when in union with oxygen. It is the lightest body known, all other elements being referred to it as unity. It is formed in foundry moulds by the decomposition of the moisture contained therein, and burns at the vent holes with its characteristic blue flame.

Hydrostatic Gauge.—A hydraulic gauge (q.v.).

Hydrostatic Press.—A hydraulic press (q.v.).

Hydrostatics.—Treats of the equilibrium and pressures of fluids at rest.

Hygroscopic Water.—Water which is not in the condition of chemical combination. Applied to fire-clays and other absorbent bodies. It is capable of being driven off by the application of a moderate amount of heat.

Hyperbola.—One of the conic sections cutting the side of the cone at an angle less than a parabola. The curve of expansion of a gas under the conditions of Boyle and Marriott's law is that of a hyperbola.

Hyperbolic Logarithm.—The common logarithm of a number multiplied by 2·30258505. Conversely the hyperbolic logarithm multiplied by ·43429448 gives the common logarithm. Hyperbolic logarithms are employed to facilitate calculations relative to the expansive working of steam.

Hypocycloid.—A cycloidal curve formed on the interior of a base or fundamental circle. The curves of the flanks of wheel teeth are commonly hypocycloids.

Hypotheneuse.—The diagonal which joins the sides of a right-angled triangle.

I.

I.—The usual algebraic expression employed in mechanical calculations to designate the moment of inertia (q.v.).

Ideal Engine.—See Perfect Engine.

Idler.—See Idle Wheel.

Idle Wheel, or Idler.—A wheel introduced into a train of gearing for the purpose of filling up a hiatus, or of changing the direction of motion without influencing the velocity ratio. Called also a cock wheel.

Igniter.—The agent by which the gaseous charge of a gas engine is ignited. It is either a jet of gas or a spiral of platinum wire rendered incandescent by electricity or by a gas jet, or some refractory material kept constantly heated to a sufficiently high temperature.

Ignition.—The lighting of the charge in gas engines, effected directly either by a jet of gas or by an electric spark, or indirectly through the heating of platinum wire. See also Ignition System, Dict. of Modern Terms.

Ignition Slide.—An ignition valve (q.v.).

Ignition, Speed of.—See Speed of Ignition.

Ignition Valve.—The valve of a gas engine which opens to permit of the ignition of the charge, but closes as soon as this is effected.

i.h.p.—Indicated horse-power (q.v.).

Impact.—The sudden fall of a load upon a beam or structure. The deflection of beams varies nearly as the velocity of impact.

Impact Wheel.—A water wheel which is driven by the percussive force of water acting at right angles to the floats, and at a tangent to the circumference of the wheel. Turbines (q.v.) are impact wheels.

Imperial.—A drawing paper of ordinary quality, sold in sheets measuring 30 inches by 22 inches. It can be had either rough or smooth.

Impermeator.—A form of self-acting lubricator used for engine cylinders. It charges the steam with greasy matter, and depends for its efficiency on the difference in the specific gravities of water and oil. It is screwed into the steam pipe or the valve chest, and consists of a brass cylinder containing in the top a chamber and cock for filling in the oil, and in the bottom two valves, one communicating with the steam chest downwards and with a small tube of brass reaching upwards nearly to the top of the interior of the brass cylinder, and the other leading direct from the steam chest into the bottom of the brass cylinder. The cylinder having been nearly filled with oil, the first valve is opened; steam rushing up through the tube, condenses, and, forming water, falls to the bottom, floating the oil up to the top, which overflows into the brass pipe, and so affords a constant supply through its valve as long as any oil remains in the vessel. In some forms a glass index gauge affixed to the outside of the impermeator indicates the height of the fluid.

Imposed Load.—The load which is extraneous to a structure, as distinguished from that due to the structure itself.

Impregnation.—Timber for outdoor use is impregnated with various fluids or salts, to enable it the better to resist the decomposing influences of the atmosphere. See Burnett's Fluid, Creosoting, Kyanising.

Improving Furnace.—A reverberatory furnace in which the foreign components found in hard lead (q.v.) are oxidised out.

Impulse.—The term is applied to the motion of the piston of a gas engine which is due directly to the explosion of the charge; thus an impulse is said to occur at each revolution, or an impulse at every two revolutions, as the case may be.

Impulsive Load.—A load applied suddenly to a structure. The structure is thus subject to the accumulation of energy due to the impact of the load, or that gathered in its motion, in addition to its actual dead weight.

Inch.—The English standard of length, usually subdivided into eighths and subdivisions of the eighth, as sixteenths, thirty-seconds, &c. The importance of the inch in engineering cannot be overestimated. Areas, cubical contents of volumes, are estimated in terms of the square and cubic inch. Heads, pressures, are estimated in inches. The weight of the solid contents of materials is deduced from the number of cubic inches which they contain. The sections of materials are given in inches when calculating their strains and stresses.

Inch Pound.—A unit of calculation signifying one pound lifted one inch high. See Units.

Inch Ton.—A unit of calculation denoting one ton lifted one inch high.

Inclination of Boilers.—Lancashire and Cornish boilers are inclined forwards about half an inch in ten feet to ensure proper drainage through the blow-off cock. Fire bars are inclined backwards about one in ten to allow of the fuel being moved rapidly away from the dead plate.

Inclined Cylinder Engine.—A type of marine engine in which the cylinders are inclined towards each other at an angle of about 102°, making a triangle with the base or ground line. They are connected by cranks to a common crank shaft.

Inclined Plane.—A plane which is inclined to the plane of the horizon, the angle which it makes therewith being termed its inclination. In the inclined plane the power bears the same relation to the weight which it will sustain that the height of the plane bears to its length. Hence the greater the height in relation to the length the greater the power which will be required to sustain a given weight. The wedge and screw are each examples of common applications of the principle of the inclined plane.

Incompressibility.—A property of liquids which is utilised for the transmission of power in hydraulic cranes, lifts, pistons, &c. The compressibility of water under the pressure of one atmosphere is ·000048, an amount altogether inappreciable in practice.

Increase Twist Drill.—A twist drill in which the angle of the twisted groove increases as it recedes from the point in order to afford greater freedom of delivery to the abraded fragments.

Increasing Pitch.—A screw is said to be of increasing pitch when the distance between each blade, or rather between each successive turn of the helix, supposing the blades were thus prolonged, increases in amount ; or the pitch may increase in the direction of the length of the blade from the centre to the circumference.

Incrustation.—The coating of the internal portions of an engine boiler with carbonate and sulphate of lime, and other solids deposited from the feed water. Various remedies are employed to prevent and to remove this as far as possible.

Independent Jaw Chuck.—A dog or jaw chuck whose jaws move independently of each other, as distinguished from a concentric jaw chuck (q.v.).

Independent Machines.—Alludes to the practice now becoming common of driving boilermakers' and the larger kinds of engineers' machines by a small independent engine, so that each machine can be stopped and started independently of a main line of shafting.

Independent Whip Crane.—See Platform Crane.

Index.—(1) The figure above and to the left of an algebraic symbol, which expresses the number of times by which it has to be multiplied into itself. Also called the exponent. (2) The finger or pointer of a float gauge, which indicates the height of the water level in a steam boiler or tank.

Index Peg.—A division peg (q.v.).

Indian Ink.—The lines of mechanical drawings are done with indian ink in preference to common ink, as being less liable to run, and not soon fading. It is sold in cakes, and also fluid in bottles ready for use.

India-rubber, or Caoutchouc.—The natural juice of a tree called the syringe tree (*Siphonia elastica*), found in Cayenne and about the lower part of the Amazon River, in South America. It dries on exposure to the air, and its property of elasticity renders it highly useful for an infinity of purposes. It is chiefly used in the vulcanised form, which is a preparation of the gum with sulphur, and also variously with sulphate of zinc, plaster of Paris, pitch, whiting. In this condition it is used for belting, hose pipe, washers, air pump, and other valves, &c. See also Ebonite.

India-rubber Belting.—Belting made by subjecting several thicknesses or plys of india-rubber cloth to heat and pressure combined, the required temperature being about 280° F.

India-rubber Core Packing.—Packing which is provided with an internal

core of india-rubber; it is made by painting strips of friction cloth with rubber solution, and rolling them around an india-rubber or other core.

India-rubber Insertion Sheet.—See Insertion Sheet.

India-rubber Pipe.—See Hose Pipe.

India-rubber Ring.—Rings of solid india-rubber of circular cross section are used as steam and water tight packings for gauge glasses, and sometimes also for condenser tubes and pipe flanges.

India-rubber Ring Joint.—A pipe joint in which a ring of india-rubber supplies the place of ordinary caulking. Accumulator and hot water pipes are jointed with india-rubber rings.

India-rubber Spring.—Buffer springs (q.v.) are often made of india-rubber.

India-rubber Tube.—Solid india-rubber tube is made by pressing rubber, rendered plastic by means of heat, through a funnel-shaped receptacle, at the termination of which is a die corresponding in diameter with that of the pipe required, the bore being formed by a centre core in a similar manner to lead pipe. It is afterwards vulcanised, a metal mandrel being inserted to preserve the shape.

India-rubber Washer.—Used for hydraulic work. A flange washer is a flat ring; a socket washer a ring of circular cross section; a bevel washer a flat ring washer, whose faces are bevelled in relation to each other; a joint washer, one made to any outline required.

Indicated Horse-Power.—(1) The mean effective pressure of the steam in an engine cylinder multiplied by the area of the piston in square inches and by the piston speed in feet per minute, and the product divided by 33,000, gives the Gross i.h.p., or the power exerted by the steam. To obtain the net horse-power, the friction of the engine has to be deducted therefrom. (2) The power as denoted by the indicator (q.v.).

Indicating.—Indicating an engine signifies the taking of an indicator diagram (q.v.) and deducing the horse-power therefrom.

Indicator.—An instrument used for showing the pressure and behaviour of the steam in an engine cylinder. It consists of a piston working freely yet closely in a cylinder, the area of the piston being exactly half a square inch. A spiral spring of definite strength is placed above the piston, and the piston rod is connected by levers to a pencil of brass wire. A drum of 2-in. diameter forms a part of the instrument, and this drum receives a semi-rotatory motion from the engine during its forward stroke, being pulled back by a spring during the backward stroke. The instrument is screwed into the cylinder in any convenient spot, usually at the covers, and by means of a stop-cock the inner portion of the indicator cylinder can be opened to the steam in the cylinder or to the atmosphere; the upper side of the piston is always open to the air. When the under side is open to the air the line which the pencil traces on the paper on the drum is horizontal, giving the atmospheric line; when the steam is admitted below the piston the latter is forced up, and the pencil traces a line around the moving paper, which indicates the action of the steam, and from which the pressure at any point of the stroke and the mean pressure can be deduced.

Indicator Card.—The slip of prepared paper which is wound upon the drum or cylinder of an indicator (q.v.), and upon which the indicator diagram (q.v.) is traced.

Indicator Cock.—The cock by which a communication is made or broken between the piston of the indicator and the engine cylinder into which it is screwed.

Indicator Diagram.—A diagram of work traced by the pencil of the indicator upon an indicator card. From it are deduced both the behaviour and the mean effective pressure of the steam in the cylinder, and thence the indicated horse-power. To obtain the mean effective pressure the area of the card is divided transversely by equidistant ordinates, ten or more as most convenient, perpendicular to the atmospheric line, whose average lengths are measured, and the whole added together and divided by the number of divisions.

Indifferent Equilibrium.—The equilibrium of a body which is neither stable nor unstable. If when a body is moved longitudinally the centre of gravity moves in a horizontal line, as in a sphere, the equilibrium is said to be indifferent.

Indirect Acting Slide Valve.—A slide valve whose motion is not derived directly from the eccentric throw and rod, but from an intermediate rocking shaft or double-ended lever, attached, one end to the valve rod, and the other end to the die block of the slot link. This form of valve gear is used on some locomotives.

Indirect Action.—The motion of parts whose linear stroke or travel is not the same as that of the portion of mechanism from which they derive their motion. The motion is therefore derived indirectly through the medium of levers, as distinguished from direct action.

Indirect Process.—The mode of production of malleable iron from pig iron, as distinguished from the direct process. The indirect processes include the open hearth or wet puddling treatment, and the ordinary dry puddling in the reverberatory furnace. Inferior ores can be reduced better in the indirect, than in the direct processes.

Induction.—The admission of steam into a cylinder.

Induction Port.—The port which is open to the admission of steam into a cylinder for the time being. Each of the steam ports are termed induction ports, in opposition to the exhaust port, or the one through which the exhaust steam escapes into the atmosphere.

Inertia.—That property of matter in virtue of which it resists the tendency of forces impressed upon it from without to cause it to change its position.

Inertia, Moment of.—See Moment of Inertia.

Ingate.—The opening or orifice through which the molten metal is poured into a mould. Frequently termed a gate simply. Sometimes used to include the whole of the passage or runner (q.v.).

In Gear.—Wheels are said to be in gear when their teeth are mutually engaged with each other, pitch line to pitch line.

Ingot.—(1) Metal after being purified by melting is run into metallic moulds usually containing the founder's brand. These moulds are oblong in shape, and the rectangular block of metal ready for remelting is called an ingot. (2) Cast steel before being hammered or rolled is melted and poured into an ingot mould (q.v.) to render it homogeneous.

Ingot Crane.—A crane used for lifting steel ingots into the soaking pits (q.v.).

Ingoting.—The melting of brass or gun-metal scrap and pouring it into ingots for the purpose of future remelting. The object of ingoting is the preparatory purification of the metal by the removal of the dross from its surface when molten.

Ingot Mould.—The mould in which ingots are cast, varying from the small branded moulds for copper and brass to the large and massive moulds made for casting ordinary steel ingots. These last are conical in form, used for the casting of the steel from the ladle or converter before it |is passed on to the rolling mills. Such moulds are made in cast iron, averaging from 4 to 6 in., or more in thickness, according to the size of the ingot. They are accurately bored. The ingots may weigh from five to six tons each. Sir Joseph Whitworth cast some ingots under a hydraulic pressure of several tons to the square inch, for which the moulds were made of special construction—an outer cylinder of steel banded with steel hoops, enclosing an inner lining of cast-iron lagging, which contained within it again a lining of refractory material. Some moulds also are oblong in section.

Initial Condensation.—The condensation which steam undergoes when issuing from the boiler into a cylinder colder than itself. Jacketing (q.v.) prevents this from taking place. The term is employed to distinguish it from condensation which takes place in an unjacketed cylinder, both on the first entry and during subsequent expansion and doing of work. See also Re-evaporation.

Initial Pressure.—The pressure at which steam is admitted into a cylinder.

Injection.—(1) The drawing or induction of water into a steam boiler by means of an injector (q.v.). (2) The pumping of water into the condenser of a steam engine.

Injection Cock.—The cock through which the supplies of water to the condensers of marine engines pass.

Injection Condenser.—See Jet Condenser.

Injection Orifice.—The orifice of the injection pipe in a jet condensing engine. Its diameter is estimated in reference to the quantity of water requisite to effect the condensation of the steam. Its area may be roughly taken at $\frac{1}{16}$ of a square inch per cubic foot of water evaporated from the boiler.

Injection Pipe.—The pipe which conveys the water to a jet condenser (q.v.).

Injection Valve.—See Injection Cock.

Injection Water.—The water pumped into a jet condenser. It is drawn either from the sea or from the bilge.

Injector.—An instrument used for affording a continuous supply of feed water to a steam boiler. It depends for its efficiency on the principle that a jet of high pressure steam, issuing from a boiler at a high velocity, will induce a current of cold water to flow into the same boiler against the pressure of the water in the boiler ; all injectors, however their details may vary, act by producing an induced current of feed water in this way.

Inlet Valve.—A foot valve (q.v.).

In Line.—Work is said to be in line when it is in the same centre, or in the same plane. To get several distant points in line, measurement is taken from a known level surface, spirit levels are tried upon straight edges, or a fine chalk line is strained tightly, or a theodolite is employed.

Insertable Teeth.—A practice introduced of making teeth of large circular saws distinct from the saw plate itself, attaching them thereto with screws. Its advantage is supposed to consist in the readiness with which broken teeth are replaced.

Insertion Joint.—A joint rendered steam or water tight by the insertion of an india-rubber disc or ring.

Insertion Rubber.—Vulcanised india-rubber, in which layers of rubber alternate with layers of canvas. It is named 2, 3, 4, 5, 6, &c., ply, according to the number of layers of which it is composed.

Insertion Sheet.—Thin sheeting composed of india-rubber and brass wire. Used for making steam joints.

Inside Calipers.—See Caliper.

Inside Crank.—An engine crank whose position lies between the crank-shaft bearings.

Inside Cylinders.—Locomotive cylinders placed within the framing and smoke box and connected up to the cranks.

Inside Fire Box.—The actual fire box of a locomotive boiler, which encloses the fuel and flame. It is usually of copper, and is stayed to the outside fire box (q.v.) with short stays.

Inside Framing.—A form of locomotive framing where the wheels are inside the main frames.

Inside Gouge.—A paring gouge, so called because it is ground on the inside or hollow face. See Paring Gouge.

Inside Jaw.—The jaw of a lathe chuck which is outside of the work which it clips.

Inside Lap.—Exhaust lap (q.v.).

Inside Lead, or Internal Lead.—The lead to exhaust in an engine cylinder, produced by making the length of the hollow of the slide valve measure in the direction of its travel more than the distance between the exhaust edges of the ports. Its purpose is to allow of the free escape of the exhaust steam from the passage.

Inside Screw Tool.—See Internal Screw Tool.

Inside Tool.—A turning tool bent at a right angle for getting at the internal portions of turned work.

Inspection.—All the larger and more important works erected by engineers have to pass inspection by a properly appointed person before the contract is declared completed. The inspection in the larger contracts proceeds continuously with the progress of the work, but in many cases is done only when the order is declared by the contractors to be completed. Inspection to be thorough, embraces the testing of the quality of the materials used, the suitability of the minor details of construction to the purposes for which they are designed, the excellence of workmanship, both general and minute, and the capability of the structure or machine. Inspectors are usually men who have had a practical as well as a theoretical training, and large power of altering and amending minor details, and of condemning bad material and workmanship, are vested in them. In almost all specifications it is stipulated unconditionally that all work is to be executed to the satisfaction of the inspector, and that he shall have the power of ordering such alterations as he may deem necessary.

Inspection of Boilers.—See Boiler Testing.

Instantaneous Centre.—The momentary centre around which a system of bodies or links in a piece of mechanism may be supposed to rotate for an instant. Notwithstanding that the relative positions of such bodies or links may be constantly changing, yet at any instant they will be turning round a common centre, which centre, however, shifts in space with each new relative position of the links. The determination of the virtual centre, as it is sometimes called, is of use in estimating the relative velocity, ratios, and forces acting upon bodies. The *instantaneous axis* is perpendicular to this centre; the *centrode* is the curve described in space by the instantaneous centre as it travels from one position to another. The term instantaneous is used to distinguish it from a *permanent* centre, or one which is fixed in space.

Instantaneous Grip Vice, or Sudden Grip Vice.—A vice which is provided with levers, a toggle-joint, and rack, by means of which it is enabled to clasp work without the loss of time involved in turning a screw.

Instroke.—In a gas engine, is the reverse of an outstroke (q.v.).

Instrumental Arithmetic.—Mathematical results obtained either by simple inspection, or by simple calculations aided by inspection, instead of by elaborate calculations. The slide rule, the sector, and the various computing scales are the aids to instrumental arithmetic.

Interceptor.—A T-shaped cylindrical vessel employed in connection with marine engines to prevent particles of water from being carried over with the steam into the cylinders. The steam in its passage through the interceptor meets with a diaphragm plate by which the water is thrown down, to be subsequently let off by a drain cock. Called also catch water.

Interchangeable.—In the best work where several machines or engines of the same type are being ordered, it is the custom to stipulate that all parts shall be interchangeable, so that the delay of marking each section and selecting the special parts for each machine or engine is saved. Also the manufacturer is better able to supply spare parts at a future time for the repair of any one of the machines or engines. When any kinds of machines are made in large quantities by a firm, their sections are also interchangeable, so that a piece can be taken from one machine and attached to any other of similar type and size.

Intermediate.—A term applied to a shaft, pipe, lever, flange, or any portion of mechanism placed between other portions of the same character with which it is immediately connected.

Intermediate Receiver.—A vessel or casing employed on some compound engines as a steam chamber or reservoir between the high and low pressure cylinders. It is rendered necessary when the cranks of the two cylinders are set at right angles with each other, so that when one piston is at full the other is at mid-stroke. Its effect is also to equalise the back pressure in the high pressure cylinder, and to diminish the variations in its temperature.

Intermediate Shaft.—A shaft placed between first and third motion gearing, acting as a carrier of motion between the two, with or without change of power.

Intermittent Feed.—The feed given to a machine tool, or to the work, by means of a pawl and ratchet movement, as distinguished from continuous feed (q.v.).

Intermittent Load.—A live load (q.v.).

Internal Calipers.—See Caliper.

Internal Corrosion.—See Boiler Corrosion.

Internal Flange.—A flange running round the inner diameter of a pipe or cylinder. Used on foundation, and other large cylinders and pipes.

Internal Flue.—The flue of an internally fired boiler, which is therefore enclosed within the shell. Examples occur in Lancashire, and Cornish, and marine boilers.

Internal Forces.—Forces which act between the different parts of a body or systems of bodies taken as a whole. These, therefore, are distinguished from external forces, and produce stresses.

Internal Gear.—When spur wheels or pinions engage with teeth set on the internal diameter of a ring, the gear is called internal. This is the reverse of spur gearing (q.v.).

Internal Lead.—See Inside Lead.

Internally Fired Boiler.—A boiler whose fuel is burnt in a tube or tubes within the boiler itself. Of this type are the Cornish, with one flue, and the Lancashire, with two flues, the smaller double flues giving the same amount of heating surface without the weakness of the larger single one. The marine return flue boiler, and the locomotive and portable with internal tubes, and the vertical with uptake and cross tubes, or Field tubes, are also internally fired.

Internal Pressure.—The pressure on the interior of a cylinder is obtained by multiplying the pressure in pounds per square inch into the area of a plane, which divides it longitudinally through its largest diameter, the forces acting on each side of that plane. In a cylindrical shell the intensity of longitudinal stress is only half as great as that in the circumferential direction. In a thin spherical shell the stress is only half as great as in a cylindrical shell taken under similar conditions of diameter, thickness, and pressure.

Internal Screw, or Internal Thread.—A screw-thread cut upon an inside cylinder.

Internal Screw Tool, or Inside Screw Tool.—A chaser or a screw tool (q.v.) cranked or bent at right angles, and used for cutting or chasing the threads of internal screws (q.v.).

Internal Stresses.—Stresses set up in the internal portions of castings, and due to unequal contraction caused by differences in their mass. The stresses are tensile in character. To prevent internal stress, regard must be had to suitable proportioning of parts and to proper cooling (q.v.).

Internal Wheel.—See Annular Wheel.

Intrados, or Soffit.—The interior or under surface of an arch.

Inverse Squares.—The principle or law of inverse squares is of very general application in engineers' calculations relating to beams, bars, and structures.

Inverted Cylinder.—An engine cylinder in which the piston rod is vertical, and passes downwards through the bottom cover.

Inverted Cylinder Engine.—A vertical engine in which the cylinder is inverted, and therefore above the piston rod, connecting rod, and crank. Sometimes called the steam-hammer pattern. Most marine engines are now built on this type.

Involute Teeth.—Teeth whose flanks are formed by the unwinding of a cord from a base line. They differ from cycloidal teeth in this, that the centres of a pair of such wheels may be varied within certain limits without affecting their true contact and regularity of motion. They are used to-day in a wide range of machines and mechanisms. Owing to the small angle of their tangent lines, however, it is impossible to use small pinions with this gear.

Inward Flow Turbine.—See Turbine.

Iron.—Symbol, Fe. Comb. weight, 56. The most valuable metal in existence. It is never found absolutely pure, owing to its avidity for oxygen, with which it combines in three proportions, forming Ferrous oxide Fe O, Ferric oxide $Fe_2 O_3$, and the Magnetic oxide $Fe_3 O_4$. It combines with chlorine, forming Ferrous chloride $Fe Cl_2$, Ferric chloride $Fe_2 Cl_6$; with sulphur, forming Ferrous sulphate $Fe SO_4 + 7 H_2 O$, Ferrous sulphide Fe S, and Ferric sulphate $Fe_2 (SO_4)_3$; with carbonic acid to form Ferrous carbonate $Fe CO_3$. See Cast Iron, Wrought Iron, Malleable Cast Iron, Steel, &c.

Iron Borings.—The borings from the machine shops are preserved and used for the caulking (q.v.) of water-tight joints.

Iron Cement.—The material used for making rust joints (q.v.) It consists of iron borings (q.v.), passed through an ⅛ or ¼ sieve (depending on the thickness of the joint space), and mixed with sal-ammoniac (about an oz. to the hundredweight of borings) and damped. Sulphur is sometimes added.

Iron Contraction.—See Contraction, Contraction Rule.

Iron Founding.—See Moulding.

Ironing.—The sleeking or smoothing over of the face of a founder's mould.

Iron Man.—A term applied by boiler-makers and platers to a riveting machine.

Iron Moulder.—See Moulder.

Iron Moulding.—See Moulding.

Iron Ore.—The ores of iron are numerous, comprising the magnetic, the red and brown hæmatites, and the various ironstones. The hæmatites yield the best irons, but the clay and black band ironstones afford the largest commercial products. For details see the various headings.

Iron Oxide.—See Rust, Ferric Oxide, &c.

Iron Oxide Paint.—Paint used as a protective coating for castings, prepared from an iron oxide earth originally found at Brixham, in Devonshire. It is prepared for use by mixing with linseed oil, and is cheaper than lead paints, while its affinity for iron renders it more suitable as a protection for that metal than the ordinary lead paints.

Iron Pattern.—See Metal Pattern.

Iron Plates.—See Plate.

Iron Ropes.—See Wire Ropes.

Iron Sheets.—See Sheets.

Iron Stone.—A generic term applied to several ores of iron, having different compositions, but possessing the common characteristics of being largely admixed with earthy matters. These are the principal sources of commercial iron of English manufacture. See Black Band Ore, Clay Band.

Iron Tubing.—Wrought-iron tubing which is either welded or solid drawn (q.v). Welded tube is either butt, or lap welded, is thin and unsuitable for withstanding high pressures, and is, therefore, employed for gas and water, steam at comparatively moderate pressures, also for plain handrailing and a multitude of miscellaneous purposes besides. Solid drawn (q.v.) tubes vary very much in thickness, and being homogeneous are more reliable.

Iron Turner.—Iron turning is quite a distinct section of engineering, the turner seldom or never leaving the lathe to go to the vice, either for fitting or erecting. His skill consists in proper grinding and setting of tools, quick and secure chucking, and a knowledge of the most suitable speeds at which to drive work in different materials and of different dimensions. A good iron turner commands as high wages as a fitter, but boys at low wage now do a vast quantity of the repetition work of the shops.

Iron Turning.—Iron turning is a most important branch of engineering work, since all circular bearing parts and revolving and close fitting work have to be turned in the lathe. Most iron turning as now done is effected in slide rest lathes, and the number of men skilled in hand-turning is diminishing proportionately. Special turning is done in special lathes, from the light 4 or 5 in. screw cutting lathes, to the large break lathes employed in wheel turning.

Iron Wire.—See Wire.

Isochronous.—Literally, equal timed. Signifies the passing of a pendulum governor through arcs of different lengths in equal time. The nearer the governor of a steam engine or other motor approaches to isochronism the greater its value as a regulator of motion.

Isosceles Triangle.—A triangle having two sides only of equal length. This is the form of triangle formed by the bracing of lattice girders.

Isothermal Curve.—The curve which represents the expansion of a perfect gas in a closed vessel. A perfect gas, it may be noted, is one whose pressure at a constant temperature varies inversely as the space it occupies. An isothermal curve follows the law of Boyle and Marriott (q.v.).

Isothermal Lines.—Lines of equal temperature, as opposed to adiabatic curves (q.v.); produced on diagrams of work under varying pressure with constant temperature.

J.

J.—See Joule.

Jack.—A piece of mechanism employed for lifting heavy loads through short distances with a minimum of expenditure of manual power. See Screw Jack, Hydraulic Jack, Bottle Jack, Tripod Jack.

Jacket.—See Steam Jacket.

Jacket Casing.—The casing which encloses the steam or water used in jacketing; or more simply the Jacket. See Steam Jacket.

Jacketing.—The fitting of a steam jacket (q.v.) to a cylinder for the greater economy of heat. In some large cylinders the covers also are jacketed by being made hollow, to receive steam, and hollow pistons are made with the same object.

Jacking Up.—(1) The elevation of masses of machinery and heavy structures by means of jacks. Blocking pieces and wedges are inserted as the work advances. (2) The planing off the rough outsides of boards with the jack plane is also called jacking up.

Jack in the Box.—The tool box in jim crow machines (q.v.).

Jack Plane.—The first plane used by wood-workers for roughing down timber, the cutting iron being more curved than in the other finishing planes. It takes coarse and narrow shavings off.

Jag.—A roughed-up, or barbed, or projecting portion of metal produced by nicking underneath, or in front of it, with a cold chisel, or with a smith's chisel, or by casting.

Jag Bolt.—A tail bolt whose shank or tail is roughed up by jagging.

Jaggers.—Rough projections or studs of a pyramidal or conical form, cast on loam plates for foundry use. They are simply stamped into the mould of the plate, their efficiency depending partly upon their roughness. Sometimes called prods.

Jagging.—When a wrought-iron bar, a shaft, or an eye, is cast into a piece of work, the portion which is embraced by the casting is serrated or notched in order to prevent it from being pulled out subsequently. This serration is termed jagging.

Jammer.—A spring chaplet.

Jamming.—The sticking of cocks and safety valves in their seatings, due to a wedging action induced by various circumstances. Safety valves jam by reason of corrosion or distortion; cocks through too slight an angle being given to their plugs, and also to corrosion.

Jam Nut.—A lock nut (q.v.).

Jaw.—An opposing piece in a machine or tool, used for clamping, or crushing, or cutting. Vice jaws, the jaws of shearing machines, of pliers and pincers, of tongs, and of spanners, are cases in point.

Jaw Chuck.—A lathe chuck, consisting of a face plate furnished with movable clips or jaws, the jaws being moved to or from the centre with screws, and held fast with clamping screws. Called also dog chucks, and the jaws, dogs.

Jaw Nut.—A lock nut (q.v.).

Jemmy.—A short crowbar (q.v.).

Jenny.—See Block Carriage.

Jet.—The current of water which issues from an adjutage (q.v.).

Jet Condenser.—The older form of condenser in which the steam is condensed in a closed box, by means of a spray of injection water scattered from a jet.

Jib.— The strut or thrust member of the framing of a crane. It is always in compression. It is either made in wood or in wrought iron ; either rolled joist iron, or braced angle iron framework, or a combination of both being employed.

Jib Crane.—A crane provided with a jib or compression beam, as distinguished from an overhead crane, or shear legs, or a crab, or a winch.

Jib Legs.—Timber legs pivoted to the jib pin of an accident crane, and reaching to the ground ; they afford a firm and broad base to the jib when lifting, and so prevent the crane from overturning. When not in use they are shipped over the side frames.

Jigger Saw, or Jig Saw.—A thin narrow-bladed saw to which a reciprocating motion in the vertical direction is imparted by a crank and levers, used for the same purposes as band saws. In workshops the latter have mostly superseded the former, and small jig saws are now almost entirely relegated to the use of the fret worker.

Jil Barrow.—A barrow without sides, being simply a flat carrying surface provided with handles and a wheel, or wheels. Used for carrying castings from the foundry, and about the shops.

Jim Crow.—A portable rail straightener or rail bender (q.v.).

Jim Crow Machines.—Metal planing machines, driven by a screw and by pulleys and wheels of equal sizes, and in which, the table having no quick return, the tool box is revolved round by cords, to cut in both directions. Jim crow alludes to the rapid whisking round of the tool box at the termination of the stroke. Sometimes called jack-in-the-box machines.

Jimmy.—A short crowbar (q.v.).

Jinny.—The travelling or block carriage of an overhead traveller from which the bight of the chain, and the snatch block depend.

Joggle.—A projecting pin on a casting for the purpose of affording steadiness to it when set in place. Used chiefly in castings which are bolted upon timber, the timber having a shallow mortice cut to receive the joggle. The latter, as a rule, is of a rectangular form. A raised ridge which receives the thrust of a plummer block or other bearing is also called a joggle.

John Bull.—A strong bar of wrought iron having a flat square foot at one end, and a short movable bar sliding at right angles along the body and carrying a set screw for adjustment. It is used to take the thrust of a ratchet brace during hand drilling, the foot being bolted to the work or to a bench, and the movable bar receiving the thrust of the end of the brace.

Joints.—Very many kinds of joints are made in engineers' work, for wood, and metal, and moulds, including glued, dovetailed, red lead, rust, planed, riveted, welded, sand joints, &c., described in their places.

Joint Board.—A turn-over board, used not so much for the purpose of keeping a pattern true during the process of ramming up, as to save the moulder's time in making sand joints, though the term is applied loosely to turn-over boards and bottom boards. The pattern is laid upon the board, and the face of the board without the pattern is made to the contour of the intended joint, and the sand in the flask being rammed thereon receives the joint contour.

Jointing.—(1) In moulders' work, jointing signifies the dividing off of the various portions of which the mould is composed. This is commonly done by the intervention of a thin layer of sand between the mould parts, of a different quality from that used in the body of the mould itself. Joints of drawbacks are sometimes made with sheets of interposed brown paper. (2) The jointing of iron pipes is done with red lead and boiled oil when flange faces are brought together; jointing of sockets is effected with melted lead, or with gasket, or with india-rubber rings. Hydraulic joints are made with sal-ammoniac and iron borings. piston joints for hydraulic work are made with leather pressed into the form of a cup, piston joints for steam with a metallic expanding ring.

Joint Pin.—A pin connecting the two parts of a knuckle joint (q.v.). It is prevented from moving out of place by a split pin (q.v.).

Joint Ring.—A ring employed for the making of water-tight or gas-tight pipe joints. It is usually a ring of lead whose faces are slightly hollowed out, so that when the ring is squeezed between opposing flange faces by screw bolts it is slightly flattened and fills up all unevenness on the surface. Copper-asbestos rings are used for high-temperature joints.

Joint Washer.—See India-rubber Washer.

Joists.—Rolled wrought-iron girders of an I section, employed for a variety of purposes in constructive engineering.

Joule.—The unit of heat, so called after the experimentalist who demonstrated the relations between heat and work. A joule is equal to 772 foot pounds and is designated by a J in mathematical calculations.

Journals.—The turned portions of shafts, or those parts which revolve in the bearings.

Journal Box.—Sometimes applied to a bearing or an axle box.

Joy's Gear.—A form of valve gear in which no eccentrics are employed, the valve rod being worked directly through a coupling rod or link from the connecting rod. The necessary travel is imparted by causing the up and down or oval movement of the coupling link to move a die block to which one end of the valve rod is attached, that end sliding up and down in a slot link moving in a fixed centre, and placed at an angle with the rod. The reversal of the engine is effected by altering the direction of the angle of the slot link about its fixed centre.

Jumper, or Monkey.—(1) A smith's hammer, cylindrical and solid in form, suspended from a chain and swung therefrom, battering-ram fashion. It is used for dealing blows against the ends of rods laid horizontally for the purpose of jumping them up or upsetting them. (2) The sparks or scintillations which fly off from molten iron in the ladle are also termed jun.pers.

Jumping In.—The forcing or springing in of the packing, or spring ring of an engine piston into its cylinder.

Jumping Up.—The same as upsetting (q.v.), in allusion to the repeated knocking of the end of a bar of iron down upon the anvil in order to thicken or increase the diameter of the heated portion.

Jump Joint.—A belt joint made by bringing the ends together end to end, and then lacing ; it is therefore a butt joint.

Junction Valve.—An ordinary two-way valve which unites two pieces of piping.

Junk Ring.—A ring of cast or wrought iron screwed down on one face of the piston of a steam engine or pump, and by means of which the gasket or rope packing which surrounds the piston is pressed against the cylinder bore rendering the contact water or steam tight.

K.

Kauri Pine, often spelt Kowrie, Cowrie, Kaurie, Cowdie. (*Dammara Australis*).—A New Zealand wood of the natural order *Coniferæ*. It is an excellent wood for pattern work, being large, straight grained, silky and smooth in fibre, and standing well. A cubic foot weighs 32 lbs. Sp. gr. ·512.

Keelsons.—A name given to the wrought-iron saddles or standards upon which marine boilers rest.

Keeper.—The lower movable step or bearing in an axle box. There is no bearing of the axle on the keeper excepting when the engine or truck is lifted up for repairs. Hence it is hollowed out internally.

Kentledge.—The loose balance weights supplied with a balance crane.

Kerf.—The width of the cut produced by the teeth of a saw, the cut itself being termed a saw kerf. The width of the kerf depends on the amount of set (q.v.) given to the saw teeth.

Key.—(1) The wedge-shaped strip of iron or steel used for preventing wheels from slipping round upon their axles. A flat is filed upon the shaft and a groove cut through the eye of the wheel, and these are brought opposite to each other so that the key prevents the one from slipping over the other. There is no wedging action sideways but only radially. A sunk key is one which is sunk into a recess in the shaft instead of simply lying upon a flat surface, and is used where great tension is present. A key which allows of a wheel sliding along it for the purpose of being thrown in and out of gear is called a feather. (2) Keys are the wooden wedges which are driven between a rail and its chair. The timber of which keys are made is compressed before being cut, for the same reason that trenails are compressed so that they may swell by the subsequent absorption of moisture from the atmosphere, and so hold tightly. (3) A square or hexagonal spanner or wrench used for tightening the screws of lathe chucks and other parts. (4) A two-pointed prong by which the joints of drawing compasses are tightened.

Key Bed.—See Key Way.

Key Boss.—A small swell or projection cast on the outside of the boss of a wheel or pulley, opposite and outside the key way, in order to maintain the same thickness of metal around that as around the central hole. Without this addition the cutting of the key way would become a source of weakness.

Key Chuck.—A jaw chuck whose screws are actuated by a key or square spanner turning the square heads of the screws.

Key Hole Saw.—See Pad Saw.

Key Screw.—A spanner or screw wrench.

Key Seating.—A key way (q.v.).

Key Stone.—The centre voussoir or crown of an arch.

Key Way, or **Key Bed,** or **Key Seating.**—A shallow recess cut through the eye or bore of a wheel for the reception of the key (q.v.).

Key Way Tool.—A slotting machine tool used for the vertical cutting of key ways, the wheel lying horizontally upon the table of the machine. The tool is cranked so that the cutting edge stands out in front of the shank, and the width of the cutting edge is the same as that of the key way to be cut.

Key Wrench.—A screw wrench or spanner (q.v.).

Killing.—The addition of zinc to spirits of salts or hydrochloric acid (q.v.) to form a chloride of zinc, as a flux to remove the oxide from surfaces which have to be soldered.

Kiln.—A furnace in which the ores of iron are calcined in order to drive off the carbonic acid previous to smelting.

Kilo.—Abbreviation for kilogramme.

Kilogramme.—A standard French measure of weight containing a thousand grammes (q.v.). It is equivalent to 2·2046 English pounds.

Kilogrammetre.—A French unit of work, being equivalent to a kilogramme lifted one metre high.

Kilolitre.—A French measure of capacity, containing a thousand litres (q.v.), and being equivalent to 220·09 English gallons.

Kilometre.—One thousand French metres (q.v.), corresponding with 39,370·79 English inches, or 3280·9 feet.

Kinematics.—The study of the laws which regulate the actions of bodies in motion.

Kinetic Energy.—See Energy.

King Post.—The central upright in a roof truss against which the rafters abut, and which supports the tie beam.

King Rod.—An iron tension rod which in an iron roof truss takes the place of the king post in a wooden roof truss, depending from the ridge and uniting with the tie rod.

Kingstone Valve.—A wing valve used on board ship for condenser water suction, and for boiler blow-off. It opens towards the sea, but is self-closing in case of fracture of the spindle.

King Truss.—A truss formed with a king post.

Kink.—A sharp bend or angle produced in a piece of metal by a blow or a strain. Also the getting out of place of chain links, so that they form a knot or bunch.

Kish.—The name given to the black scales of graphite, which separate and float on the surface of a slowly cooling mass of molten iron. The whole of the scum is also called kish.

Kite Connecting Rod.—See Bow Connecting Rod.

Knee.—An elbow pipe. The difference between a knee and a bend is that in a knee the branches meet in an angle, in a bend they merge into each other with a curve. (1) A curved or angular piece of timber or iron connecting two portions of work together. (2) A toggle joint (q.v.).

Knife Edges.—The bearing edges of weighbridge levers. They are made of hardened steel, and the edges are formed by the meeting of two faces sloping towards one another at an angle of about 90°.

Knife File.—A file of triangular or thin wedge-shaped section similar to that of a knife.

Knife Key.—A key with prongs for tightening the joints of compasses, but containing, in addition, a rigid knife blade at the opposite end, and a file placed intermediately.

Knocker Out.—A term often applied to the horns (q.v.) of a planing machine, against which the tappets strike.

Knocking.—The noise caused in a pump on the reversal of the motion for suction and delivery when no air vessel, or an air vessel of insufficient area, is provided.

Knotting.—A compound used for filling or covering knots to prevent absorption of oil paint. It is composed either of shellac and methylated spirit, or naphtha, or of red lead and glue.

Knuckle Joint.—One in which an eye at the end of a rod is embraced by the forked end of a second rod, the two being connected with a joint pin (q.v.).

Knurling Tool.—See Milling Wheel.

Kyanizing.—The preservation of timber by impregnation with a solution of corrosive sublimate or chloride of mercury, $Hg\,Cl_2$. The proportions are one pound of sublimate to ten gallons of water for maximum strength, or one pound to fifteen gallons as a minimum. About twenty-four hours per inch of thickness are required for saturation.

L.

Labourers.—Unskilled labour is largely employed in most of the departments of engineering. In the foundry the labourers turn out the castings, water and mix the sand, haul about the flasks or moulding boxes, attend to the cranes, do the box filling, &c. In the turning, fitting, and erecting shops, they wait on the mechanics, attend to the cranes and haulage generally. In the boiler shops they hold up the dollys, and lift and hold up the plates at the machines. In the smiths' shop the hammer-man or striker is the sole labourer. In the yard the work connected with the package and haulage of finished work is done by them.

Lac.—See Shellac.

Lace.—A strip or thong of white leather used for uniting the free ends of leather belting. See Leather.

Lacing.—The union by means of laces of the ends of leather belting used in driving machines. The distance around the pulleys being taken, the belt is made to overlap in a loop somewhat smaller than that given by the actual measurement in order to allow for after stretching, and the overlapping ends are pierced with holes through which the lace is threaded. Belt screws have, however, largely superseded laces for this purpose.

Lacquer.—A varnish applied to metal work to protect it from the tarnishing influences of the atmosphere. Its basis is shellac (q.v.), which is dissolved in spirits of wine; colouring matters are added. The article to be lacquered is warmed and the solution applied with a camel-hair brush. A recipe for a brass lacquer is as follows : shellac, 8 ozs. ; sandarac, 2 ozs. ; annatto, 2 ozs. ; dragon's-blood resin, ¼ oz. ; spirits of wine, 1 gallon. Cellulose lacquers, prepared by dissolving nitro-cellulose or acetyl-cellulose in suitable solvents, with the addition of resins, plasticisers and pigments, are extensively used. They may be brushed or sprayed.

Ladle.—An open iron vessel lined with some heat-resisting substance, as fire-clay. Used in foundries for pouring the molten metal into the moulds. Ladles vary in size from those holding half-a-hundredweight to those carrying twelve tons or more. The smaller sizes are hand ladles, the larger ones are moved by the crane. The difference between a ladle and a crucible is that in the latter the metal is not only melted but poured therefrom, while the former is only a temporary receptacle for metal which has been melted elsewhere.

Lagging.—This has several significations. When boilers and engine cylinders are cleaded (See Cleading) with wood strips, the strips are called lagging or lagging strips. The strips of oak which extend from end to end, and form the peripheries of large winding drums, are termed lagging. The similar pieces with which the patterns of engine cylinders and pipes are built up, are so designated. The process of fitting of such strips is also termed lagging, or lagging-up.

Lamellar, Lamellate.—Made up of thin plates. Term chiefly used in describing the micro-structure of metals.

Laminated.—Denotes generally the condition of thin sheets, fibres, plates, strips. Frequently applied to the character of the fibres of iron, and of metallic bodies, and structures.

Laminated Plates.—Plates of wrought iron are said to be laminated when there is a lack of homogeneity and perfect union of the various layers of which they are composed. Laminated plates are apt to blister when used for steam boilers. The presence of imperfect union can be tested by tapping with a hammer.

Laminated Spring.—A curved spring composed of thin plates superimposed one over the other, as distinguished from helical and coiled springs.

Lamp Bend.—A wrought-iron bend pipe, one of whose arms departs from a curve, being straight for some distance. Its form is such that the pipes which it connects stand at an angle greater than a right angle.

Lancashire Boiler.—A horizontal, cylindrical, internally fired boiler, having two flues.

Lancashire Files.—The smaller classes of files are made in Lancashire, but most of the ordinary files come from Sheffield.

Land Boiler.—Used in contradistinction to marine boiler, but more commonly understood of boilers built into brickwork, as of the Cornish and Lancashire type, rather than of locomotives and portables, which are a type in themselves.

Land Engines.—A term frequently employed to distinguish ordinary stationary engines from those of marine and locomotive types. It designates no type of engine in particular, whether horizontal, vertical, beam, or otherwise, or high or low pressure; but the term, as originally employed, had more special reference to the various forms of condensing, beam, and pumping engines.

Lantern Brass.—A hollow brass sometimes introduced into the stuffing boxes of engines, and supplied with steam, so that in the event of leakage occurring it shall be that of steam instead of air.

Lantern Frame Pattern.—A type of inverted cylinder engine in which the cylinder standard is columnar or cylindrical in outline, hollow within, and pierced at the sides with openings reaching nearly to the guide facings. The crosshead works in the bore of the column, and the crank bearings are cast in the base. It is a rigid and cheap form of standard.

Lantern Wheel.—An old-fashioned type of cog-wheel used in mills. It was formed by round pins or staves, arranged equidistantly between two discs of hard wood, which discs were kept at a distance apart, exceeding by a small amount the width of the teeth of another, or mortice wheel, which engaged with the staves. The staves are driven fast into holes bored in the disc. Sometimes called a trundle or trundle wheel.

Lap.—(1) A body of soft metal, such as lead, tin, or brass, which forms the matrix or support for the emery powder or pumice powder used in grinding surfaces of hardened steel, chilled iron, or other substances too hard to be attacked with ordinary cutting tools. It is usual to make the lap of the same outline as the work to be ground. For the commonest work, laps made of lead are employed, for the better class tin is used, while the finest laps are made of soft brass. (2) The amount by which a slide valve covers the steam port at the termination of the piston stroke, consequently the lap of a valve indicates the amount of cut-off of the steam in the cylinder. (See Exhaust Lap and Outside Lap.) (3) The extent to which the joints of riveted plates pass one over the other. The amount of lap should not be more than one and a half times the diameter of the rivet, measured from the centres of rivets to the edge of the plate. (4) A single turn of a rope or chain around a barrel.

Lap Circle.—The circle on a lap diagram (q.v.), whose radius represents the lap (q.v.) of the valve. If the laps are unequal there will be two lap circles.

Lap Joint.—A joint produced by the overlapping of contiguous faces of metal.

Lapped Valve.—A slide valve provided with lap (q.v.). All valves as now made have lap, but in the early days of the steam engine there was no expansive working and consequently no lap. Hence the term.

Lapping.—The polishing and truing up of spindles, bearings, and circular bearing parts generally, with laps of lead or other material.

Lap Riveting.—When the edges of two plates are placed one over the other and riveted, this is termed lap riveting, to distinguish it from butt riveting (q.v.).

Lap Weld.—A joint which is welded by the overlapping of plates, the overlapping edges being thinned down for the purpose of maintaining an even thickness.

Lap-welded Tube.—Malleable iron tubes for steam and water pipes are lap welded, that is, the plates overlap one another at their joints, instead of merely abutting at the weld.

Lard.—Used as a lubricant of good body, either alone or in combination with a lighter oil.

Last Coat.—The smooth or finishing coat of loam applied to a loam pattern, mould, or core. See Rough Coat.

Latent Heat.—The quantity of heat which can be communicated from one body to another without changing its temperature ; or that amount of heat which does not raise the temperature of a body but effects its fusion simply. Hence called the latent heat of fusion.

Lateral Strain.—A strain which bears against the side of a structure, being essentially a transverse strain.

Lateral Traverse.—The amount of end play (q.v.) allowed to the trailing axles (q.v.) of locomotives in order to facilitate their running round sharp curves. It may average 1½".

Lath.—A strip of wood used for the precise marking off of distances, centres, &c., directly from, or upon work, in preference to rule measurement. Measurement is not taken from the ends, but from one line to another scribed across the face of the strip. Thus engineers take laths of valve faces and cylinder ports for comparison, and as a permanent record of dimensions.

Lathe.—A machine employed for the production of circular work. The work is either suspended between the centres of fast and movable heads, or attached to the fast head or headstock alone, through the intervention of chucks of various kinds. The cutting tools are either actuated by hand, or are held in a slide rest, which is either moved by hand, or is automatic. The heads and the rest are supported on a bed or bearers. Lathes are actuated by the foot alone, or by power, and they are constructed for hand turning, or are self-acting. The size of a lathe is given in the height of the centre above the face of the bed, and in the length of the bed. Almost all mechanical operations can be performed in the lathe : turning, boring, drilling, shaping, screw cutting, dividing, besides those outside the range of engineer's work and which come under the head of ornamental turning.

Lathe Bearers.—The sides or cheeks of a lathe bed.

Lathe Bed.—The longitudinal support for the heads and the rest of a lathe. Lathe beds are made in wood or metal, and usually consist of two bearers, cheeks, or shears. When made in wood they are of rectangular section, but when in iron are of I or girder section, united into one casting by cross pieces or ribs. The top faces and edges of the cheeks are planed, and in lathes fitted with self-acting slide rests, the outer edges are of a Vee'd section, so forming a guide for the saddle of the rest. Occasionally small lathe beds are formed of a single shear of a triangular bar section. Beds rest at their ends upon standards.

Lathe Can.—A soap suds can (q.v.).

Lathe Centre Grinder.—An emery-wheel driven from overhead, and used for grinding the hardened conical points of lathe centres in place.

Lathe Cheeks.—The sides of a lathe bed.

Lathe Dog.—A carrier (q.v.).

Lathe Heads.—The headstock (q.v.) and poppet (q.v.) of a lathe.

Lathe Planer.—A piece of mechanism sometimes attached to a lathe for the surfacing of metal by rectilinear cutting. There are several forms of lathe planers, mostly cumbersome, not very accurate, adopted almost exclusively by amateurs, and mostly unnecessary, since plain surfacing can be done on the ordinary face chucks.

Lathe Shears.—The sides of a lathe bed.

Lathe Standards.—Stout **A** frames of cast iron provided with ribs, flanges, and feet, to form the supports of lathe beds. They rest on the ground and are bolted to the under side of the bed at the ends. In the larger lathes there may be one or more intermediate standards besides.

Lathe Tools.—The tools used especially in connection with lathe work, embracing not only those special to the lathe, but many others besides. For wood work they embrace the gouge, chisel, round nose, side tools, diamond points, and parting tools. For metal work, roughing down, and finishing tools of various shapes, and ground to different angles for the turning of cast and wrought iron, steel, and brass. Besides these there are various gauges, calipers, punches, drills, reamers, &c.

Lathe Work.—Has reference to the work commonly accomplished in the lathe, which embraces practically almost all branches of mechanical operations. See Lathe.

Lattens.—See Trebles.

Lattice.—Diagonal bracing, forming struts and ties.

Lattice Bars.—The bars forming the struts and ties of a lattice or open frame girder.

Lattice Girders.—Girders of wrought-iron made by uniting top and bottom angle irons with lattice work or diagonal bracing, the other flanges of the angle irons being attached by riveting to top and bottom plates.

Lattice Jib.—A crane jib. built up with wrought-iron lattice bars (q.v.).

Lattice Web.—A girder web made by latticing, as opposed to a plated web, or a web built up by a single system of triangulation, as in a Warren girder.

Latticing.—The combination of two or more systems of triangulation in an open webbed girder. The number of apical points being proportionately increased, the weight is distributed over a greater number of sections, and the stress in each diagonal reduced in proportion to the increase in the number of triangulations.

Laws of Motion.—The laws of motion as formulated by Newton are three : —First law, every body continues in its state of rest or of uniform motion in a straight line, except in so far as it may be compelled by impressed forces to change that state ; second law, change of motion is proportional to the impressed force, and takes place in the direction of the straight line in which the force is impressed ; third law, to every action there is always an equal and contrary reaction, or the mutual actions of any two bodies are always equal and oppositely directed.

Laying-down Board.—A bottom board (q.v.).

Laying Out.—The setting out or marking out of work to full size. Usually applied more especially to the development of boiler-makers' and platers' work.

Lay Shaft.—A small secondary shaft, which is placed beside, or at the end of a horizontal engine, for the purpose of actuating the valves. It is driven from the crank shaft by means of bevel or spur-wheels.

Lead.—Symb. Pb. ; comb. weight, 206·4 ; Sp. gr., 11·3. A bluish grey metal used in engineering work as an ingredient in brass and gun metal, in solders for making the joints of cast-iron pipes, and in the form of tubing for a variety of purposes.

Lead Glance.—Galena (q.v.).

Lead Hammer.—A hammer made of lead and occasionally used for the same purpose as a copper hammer (q.v.).

Lead Lap.—See Lap.

Lead Letters.—See Pattern Letters.

Leading Axle.—The front axle of a locomotive.

Leading Edge.—That edge of the blade of a screw propeller which cuts the water, as distinguished from the following edge (q.v.).

Leading Hand.—A workman who, by reason of his superior skill, occupies a leading position in his shop, to whom the best work is given, and who is looked up to by the remaining hands. His wages average a few shillings weekly over those of the other men.

Leading Screw, or Guide Screw.—The screw which runs longitudinally in front of the bed of a self-acting lathe, and whose pitch is the first factor in the cutting of screws; occasionally called a main screw.

Leading Springs.—The springs which carry the axle boxes of the leading wheels of locomotives and rolling stock, and which sustain and minimise the shocks due to concussion. They usually consist of about sixteen steel plates of about 15 × ⅜ in. section, the largest being thicker, or say ½ in.

Leading Wheels.—The front wheels of a locomotive.

Lead Joint.—Lead joints are used in pipe connections. Sometimes a sheet of lead is screwed up between flanges, but the employment of the metal is more common in socket and spigot joints. The lead is poured between the socket and spigot and hammered or stemmed down ; that is, driven closely home with a drift and hammer.

Lead Line.—The left-hand vertical line in an indicator diagram which represents the amount of opening to lead.

Lead of a Valve.—The amount by which a slide valve is opened at the termination of the piston stroke for the admission of steam for cushioning. It varies from $\frac{1}{32}$ in. in slow running, to $\frac{1}{4}$ in. in quick running engines.

Lead Paints.—Ordinary paints, so called because their basis is white lead.

Leaf Wood.—A term used to designate all timber not included in the pine or coniferous family.

Leakage.—The loss of feed water from steam boilers. It is due chiefly to the expansion and contraction of the plates caused by alternate and sudden heating and cooling, which produces opening of the seams and rivets. In some cases leakage is due to joints being originally difficult to make, or to bad workmanship.

Leaning Threads.—The threads of screws where one side leans over more, or makes a greater angle with the longitudinal axis than the side opposite. Such screws are seldom used except in special cases where the pressure is always in one direction.

Lean-to Roof.—A roof which has only one slope.

Leather.—A material widely used for belts (q.v.) for driving machinery It is used single or double, single belting ranging from 3-16 in. to 5-16 in. in thickness, and double from ⅜ in. to ¾ in. Its ultimate strength (q.v.) is from 3,000 lbs. to 5,000 lbs. per square inch of section. The leather is spliced, cemented, or united by sewing to make up the necessary length, and is finally laced (see Lace) or screwed in the workshop to unite the free ends.

Leather Belting.—See Leather.

Leather Lace.—See Lace.

Leathers.—The packings, and certain portions of the valves of pumps. See Cup Leather, Hat Leather, Flap Valve, U Leather, &c.

Leather Hollows.—Strips of leather largely used by pattern-makers to form the hollows in wood patterns. The reason why they are preferred to hollows made of wood is that the time occupied in laying them around curved portions of work is much less than that occupied in cutting wooden hollows. They are cemented in with a special composition supplied with the hollows.

Leather Hose.—Hose pipe made of leather, and either cemented or sewn down the joint. It is the best and most durable. See Canvas Hose, Rubber Hose.

Left-hand Engine.—An engine which stands to the left of its fly wheel when viewed from the end of the cylinder.

Left-hand Screw.—A screw which turns from left to right, or in the reverse direction to a common wood screw.

Left-hand Tools.—Side tools ground to an angle on the right hand side, and which, therefore, cut from left to right.

Left-hand Twist Drill.—A twist drill in which the twist runs from left to right up the shank.

Lengthening Bar.—An appendage to compasses used for drawing purposes. It is a brass leg or shank which is made to fit into the socket of the compass leg at the one end, and over the stem of the ink or pencil point at the other ; its purpose being the adaptation of the compass to the measurement of long distances, or the striking of arcs of large radius.

Lengthening Bearers.—Castings or wooden cheeks made to attach to the poppet end of a lathe bed to extend it for a temporary purpose, as the turning of a long rod or shaft. The poppet being slid over the lengthening bearers the longitudinal capacity of the lathe is increased in proportion to the length of the bearers.

Let In.—A shop term, which signifies the sinking in of one portion of wood or metal into another. Thus, rapping plates are let in to patterns, brass rings are let in to sluice cock faces, &c.

Letters.—See Pattern Letters.

Letting Down.—The lowering of the temperature of hardened works down to the colour which indicates the temperature at which they are to be quenched for tempering.

Level.—(1) Truly horizontal. In a horizontal plane. (2) A spirit-level (q.v.).

Levelling.—The bringing of wrought-iron and steel plates superficially true preparatory to working and cutting them out. They are levelled by the striking of blows delivered from a hammer or wooden mallet. See Levelling Block.

Levelling Block.—A large flat cast-iron plate stiffened on its under side with flanges or ribs upon which plates of sheet iron and steel are laid while being levelled. It is used by boiler-makers and platers.

Lever.—A rigid bar turning upon an axis or fulcrum. The arms of a lever are those portions of the bar standing away from and pivoting round the fulcrum. The arms are straight or bent ; in the former case it is said to be a straight, in the latter a bent lever. The principle of the lever is that the product of the power multiplied into the length of its arm is equal to the product of the weight multiplied into the length of its arm. A lever of the first kind has the fulcrum between the power and the weight. In a lever of the second kind the power and the weight are on one side of the fulcrum, the weight occupying the intermediate position. In a lever of the third kind the power and weight are on one side of the fulcrum, but the power occupies the intermediate position.

Lever Box, or Lever Bracket.—A hollow casting which carries the various levers connected to the motions of a crane for slewing, lifting, travelling, &c.

Lever Bracket.—A lever box (q.v.).

Lever Chuck.—A concentric chuck actuated by a lever instead of screws.

Lever Drill.—A form of drilling apparatus similar to a ratchet brace (q.v.) but without the ratchet, the lever in this case passing through a hole in the piece which carries the drill socket.

Lever Jack.—A form of jack comprising a simple lever for lifting, and a standard for support.

Lever Press.—A form of press used for the compression of substances for export and other purposes. It is acted upon by a long lever.

Lever Pumps.—The air and circulating pumps of marine engines which are commonly worked by levers driven from the cross heads of the piston rods.

Lever Safety Valve.—A safety valve weighted through the intervention of a lever, as distinguished from spring, and Cowburn valves. See Safety Valve Lever.

Lewis.—A contrivance for lifting stone and concrete blocks by means of dove-tailed irons attached to a shackle piece. These fit into an under-cut dove-tail in the stone, and are tightened by means of a wedge.

Lewis Bolt.—A bolt having a jagged and tapered tail. Used for insertion into masonry, where it is held with lead.

Life.—Expressive of the total period during which a structure remains efficient.

Lift, or Hoist, or Ascending-room.—(1) An apparatus used in workshops and warehouses for elevating heavy weights directly to the upper parts of the building without the intervention of stairs. A lift consists of a platform or a rectangular chamber carried up by means of hand power or steam, or hydraulic pressure. (2) The upper part of a mould which is lifted off or taken away from the pattern is called the lift.

Lifter.—(1) A moulder's tool, usually called a cleaner. It derives its name of lifter from the use to which it is put in lifting loose fallen down sand from the bottoms of flange and rib moulds. (2) Lifters, or gaggers, are hooks of cast iron or wrought iron which are hung from the bars of a moulding box into the mould, in order to afford support to the sand.

Lift Hammer.—An old-fashioned smith's hammer, which is lifted by a spring pole overhead, and depressed by the placing of the workman's foot on a treadle. The hammer, the overhead pole, and the treadle are alike connected to a pivoted horizontal bar of wood. Sometimes called an oliver or bolt oliver.

Lifting.—The raising up or springing of the top part of a mould, caused by the liquid pressure due to the head of the molten metal. The tendency to lift is counteracted either by screw bolts connecting the top and bottom boxes, or by dead weights, or by both combined. The term lifting is also applied to the drawing of a pattern out of the mould.

Lifting Blocks.—(1) Packing pieces put under the headstock, poppet, or rest of a lathe to increase its capacity for a temporary purpose. They may be either wood or metal packings. (2) Pulley blocks (q.v.).

Lifting Irons, or Lifting Straps.—Pieces of sheet or hoop iron attached by means of screws to the sides of deep wooden patterns, to enable the moulder to withdraw them from the sand.

Lifting Jack.—See Jack.

Lifting Plates.—Plates of wrought or malleable cast iron furnished with holes both for rapping and screwing, and let into or screwed on the faces of patterns; and by which they are lifted from the sand, a lifting screw being inserted into the tapped hole in the plate.

Lifting Pump.—See Suction Pump.

Lifting Ram.—The smaller ram of the two in a hydraulic forging press, and which lifts the crosshead and tup after each stroke.

Lifting Screw.—Used for lifting patterns from the sand. It is a round metal rod, having an ordinary screw-thread cut at its lower end, and bent into a loop at the end opposite. The screw is turned into a tapped hole in a lifting plate, and the finger, or a crane hook, inserted into the loop for the purpose of lifting or pulling it upwards.

Lifting Straps.—See Lifting Irons.

Lifting Tackle.—See Tackle.

Lift Valve.—A circular disc valve fitting on an annular seating, and guided in its lift by three or four feathers projecting into the body of the seating. Also called puppet, and mushroom valve. A lift valve has a full waterway (q.v.) when the amount of its lift is equal to one-fourth its diameter.

Light Cut.—In metal work a cut is said to be light when the shavings removed are thin and narrow. But a cut which would be a light one when taken off a large piece of metal would be a heavy one if taken off smaller work.

Light Firing, or Open Firing.—The making up of a thin fire, caking it on the dead plate, and frequent renewal of the charges.

Lighting Chamber.—See Firing Chamber.

Lighting Cock.—The jet by which the charge in the cylinder of a gas engine is fired.

Light Oil.—A thin oil (q.v.).

Light Running.—A wheel and axle, a lathe crank, an engine shaft, or journal of a shaft, are said to run light when there is the minimum of friction due to good bearing surface, good fitting, and proper lubrication.

Lignites.—Brown non-caking coals containing a large proportion of water.

Lignum Vitæ. (*Guaiacum officinale.*)—A wood of the natural order *Zygophyllaceæ*, found in Jamaica, Cuba, and St. Domingo. It is of a dull brownish-green colour, hard, and cross-grained, the fibres interlacing at various angles. The heart wood is chiefly used, and its uses in engineering are for the linings of the shaft bearings of propeller screws, and turbines which work in and are lubricated by water alone. A cubic foot weighs from 40 to 80 lbs. Sp. gr. from ·65 to 1·33.

Lime.—Symbol, Ca O. Sp. gr. 3·18. The oxide of the metal calcium, obtained by burning limestone in a kiln. It is an ingredient in concretes, in limestones which are used as fluxes, and is present as carbonate and sulphate in the incrustations of steam boilers.

Lime Bag.—A linen or muslin bag containing lime powder, used for dusting on the lower joint faces of foundry moulds and on the tops of cores before lowering the top flask part on. The lime touches and leaves an impression on the upper mould face if the joints are close, but non-transfer of the lime indicates openness of the joint, which the moulder sets himself to see the cause of, and to rectify.

Limestone.—Carbonate of lime. Symbol, Ca CO_3. There are numerous limestones, their value to the ironfounder consisting in the property which they have of combining with the infusible siliceous and earthy matters in the ore and the pig, and thus producing readily fusible compounds. Their value to the engineer consists in their property of cementing together to form concrete blocks.

Limiting Angle of Resistance.—The limiting angle of resistance signifies the angle of repose (q.v.).

Limit of Elasticity.—See Elastic Limit.

Limit of Weight.—In the manufacture of bar iron, beams, and plates, there are certain definite weights ordinarily manufactured. Beyond these weights it is understood that special quotations are necessary. But the limit of weight applies to particular houses only, depending upon the nature of the plant which they possess.

Linear Advance.—The direct measurement which represents the amount by which a slide valve is set forward for lap and lead beyond a line 90°

ahead of the crank. The linear advance is measured on the sheave parallel with its centre.

Linear Velocity.—The velocity with which a body moves along a path, either straight or curved, the body meanwhile not changing its position relatively to its axis. The body simply undergoes a motion of translation. Linear velocity, when uniform in character, is usually measured by the number of feet moved over in one second.

Line.—(1) A mathematical abstraction which has neither breadth nor thickness, but only length. (2) In mechanics, an affix or prefix, as centre line, chalk line, line of centres, in line, pitch line, &c., &c.

Line Colours.—Draughtsmen's colours used for drawing lines. Black is the colour for actual definition lines. Red is used for section lines, and blue for dimension lines (q.v.).

Line Measurement.—Measurement taken by means of the divisions of a rule, as opposed to end measurement (q.v.). This is not an accurate mode of measurement.

Line of Action.—The line of action of a force is the direction in which it acts upon a material point, which line must be straight.

Line of Centres.—When a crank and its connecting rod are in one plane, or when a coupling or other rod is in one plane with the pins which it connects, they are said to be in the line of centres. See Dead Centres.

Line of Chords.—See Chords, Line of.

Line of Lines.—See Lines, Line of.

Line of Pressures.—See Diagram of Work.

Line of Volumes.—See Diagram of Work.

Liner.—(1) The bush of a pump barrel or of a cylinder. (2) A thin strip of metal, leather, or wood, placed between parts of machinery in order to permit of the taking up (q.v.) of their wear, and generally to permit of exact adjustment or of better bearing. Liners are put underneath brass bearings to raise them up, and thus compensate for the lowering down due to wear. (3) The wearing surfaces of friction clutches are sometimes lined with copper. Liners are inserted between the feet or flanges of eccentric rods in the operation of valve setting. Marine engine propeller shafts are provided with brass liners where they run through the stern bush.

Lines, Line of.—A scale of equal parts on the sectorial scales, marked ∟, by means of which proportionate distances can be obtained by direct measurement. By combined lateral and transverse measurement, both equal and unequal proportionals can be obtained, hence scales of equal parts, and reduced or enlarged drawings can be made therefrom.

Line Shafting.—The main shafting in a factory, to distinguish it from the shorter intermediate countershafts (q.v.). Line shafting is commonly run at from 90 to 100 revolutions per minute.

Linings.—Cements prepared from fire clay or from ganister; used for covering the interiors of foundry ladles, cupolas, blast furnaces, and converters, to protect their outer iron casings from the action of the intense heat.

Lining Out.—The marking of the working lines on castings and forgings with scribing blocks and compasses; called also marking out, and setting out.

Lining Up.—The introduction of packing pieces under bearings to compensate for their wear.

Link.—See Slot Link.

Link Arrangement.—An arrangement of a link and sliding block common in shaping machines by means of which the ram is furnished with a quick return (q.v.) motion.

Link Block.—The movable sliding or die block in a slot link, to which the valve rod is connected.

Linking Up.—Altering the position of the slot links in an engine fitted with expansion gear, in order to produce an earlier cut-off. The link is raised and lowered by means of a hand wheel and screw, and by its means the amount of lap of the slide valve is altered.

Link Motion.—The arrangement of levers, and slot and drag links, by which the position of a slide valve in relation to the cylinder ports is regulated for backward or forward gear. Called also reversing gear. Very many kinds of link motion are in use.

Link Reversing Motion.—The ordinary type of reversing motion for engines, effected by one of the forms of slot links operating on the two eccentrics, as distinguished from reversal provided for by the direct reversal of a single eccentric on its shaft.

Link Work.—A term sometimes applied to the motion work (q.v.) of engines, sometimes to the coupling or connecting rods of wheels and cranks.

Linseed Oil.—The oil expressed from the linseed or seed of flax. When boiled, either alone or with litharge, or white lead, it acquires the property of drying rapidly, and is therefore used in the composition of oil paints. The clear boiled oil is brushed over castings, when they come from the foundry, to prevent them from rusting previous to inspection. This is done because the oil being transparent will conceal no defects, while paint would cover them up.

Lip.—(1) The mouth of a foundry ladle, from which the metal is poured. (2) The nicker or lancet-like edge on a centre bit or similar tool, which cuts the circumscribing circle during the process of boring.

List Pot.—A vessel which contains a layer of molten tin about ¼ in. deep and in which the list or thickened rim of a tinned plate is placed to effect its removal. When melted the list is detached by a blow on the plate.

Litre.—The Standard French measure of capacity. It contains 61·02705 cubic inches. It is the cube of a decimetre (q.v.), each of whose sides, therefore, measures 3·93708 English inches, and it contains 1·7607 English pints, or ·2201 gallons.

Live Axle.—A driving axle (q.v.).

Live Head.—A term sometimes applied to the headstock (q.v.) of a lathe in opposition to the poppet (q.v.) or dead head.

Live Load.—When a structure is subjected to the alternate and often repeated imposition and relief of weight, the term live load is used to distinguish it from a dead load (q.v.). Also the greater the range of variation of stress, and the greater the number of repetitions of the load, the more severe will the straining actions become, and hence larger factors of safety are required in the case of structures subject to live, than in those subject only to dead loads. Called also variable load.

Live Ring.—A live ring consists of a number of conical rollers arranged in circle between an outer and inner ring of flat wrought iron, into which the ends of the roller spindles are fastened. In some cases the rollers almost fill up the ring, and then they are not provided with spindles. These rollers are employed for turntable centres chiefly.

Live Roller.—A roller which does not revolve on a spindle, but is free to move around or along its path. Live rollers are used for turntable centres, and the slewing motion of large cranes.

Live Spindle.—A spindle which communicates motion, as the mandrel of a headstock or live head (q.v.).

Live Steam.—The entering steam in a cylinder, as distinguished from that which is exhausting.

Load.—The total amount of weight borne by a structure tending to deform or break it, its own weight included. In the case of a moving body it includes the reaction due to friction and inertia. Loads are classed as live loads (q.v.), dead loads (q.v.), impulsive loads (q.v.).

Loam.—Strictly speaking a stiff impure clay, but as the term is employed in foundries it signifies a mixture of clay, sand, and horsedung, ground up with water in a loam mill (q.v.), and used in loam work. Though of sufficient consistence when cold to be struck up to any desired outline by means of the chamfered edge of a loam board, it dries hard when exposed to the heat of the foundry stove.

Loam Board.—A board having an edge cut to the outline of the sectional shape of the work which it is intended to strike up ; the edge being chamfered also in thin knife-like fashion to prevent the loam from being dragged out after the board. The term loam board is applied to boards for striking both patterns and moulds.

Loam Brick.—A cake of loam made roughly to the size and shape of an ordinary brick. These are used for building up those portions of loam moulds where the unyielding nature of the common bricks would cause the casting to break by preventing its due contraction when cooling, and also in cases where it is probable or desirable that some portion of the brick framework may have to be cut away during the progress of the moulds.

Loam Cake.—A flat slab made of loam (q.v.) and dried. Used to form the face of a flange or similar expansion, or some portion of a draw-back, or as a convenient piece from which a portion of a mould or of a pattern can be cut with facility with saws and rasps and glasspaper.

Loam Mill.—A mill essentially like a mortar mill (q.v.), but employed for the mixing of loam (q.v.) for foundry use.

Loam Mould.—A mould made in loam.

Loam Moulder.—An iron founder whose work lies chiefly in the making of moulds in loam or dry sand, in opposition to moulders who work in green sand. Loam moulders do not use complete patterns, but strike up their work with the edges of loam boards while the loam is in a plastic state. Loam moulds, unlike green sand moulds, have to be dried before the metal is poured in.

Loam Moulding.—The process of moulding in loam (q.v.) as distinguished from green sand moulding (q.v.)

Loam Pattern.—Large patterns of a circular or segmental outline which can be produced by the aid of strickles or of loam boards are often made in loam, and moulds are taken from them as from ordinary wood patterns.

Loam Plate.—A flat plate of cast iron used to form the base or foundation for striking up a loam mould (q.v.). It is usually provided with projecting lugs around its circumference by means of which it is lifted, and with gaggers or prods on its surface to hold the loam.

Loam Work.—Work moulded in loam. It is usually considered more difficult than work in green sand (q.v.), and loam moulders command the higher wages. It is specially adopted in cases where perfect soundness of castings is required, and also where it is desired to save the cost of a pattern.

Lock Nut.—A thin nut screwed down upon another. Its purpose is to prevent the slacking back of the main nut under excessive vibration.

Lock Saw.—A pad saw (q.v.).

Lock-up Safety Valve.—A safety valve whose spring is enclosed in a padlocked casing.

Locomotive.—A steam engine complete with its boiler, its fittings, and mountings, fixed in a carriage or framing, and provided with suitable wheels, axles, buffers, drawbars, &c., to enable it to draw loaded wagons upon the permanent way.

Locomotive Boiler.—See multitubular boiler.

Locomotive Crane.—A travelling crane propelled by steam, either driven by a pair of cylinders independently of those which actuate the lifting gear, or by means of clutch work connected with the engine shaft. Properly speaking the former are locomotive cranes and the latter travelling cranes merely.

Locomotive Hoist.—See Carriage Hoist.

Log.—(1) A balk of timber, either squared or rough. (2) The logarithm (q.v.) of a number.

Logarithms.—A series of numbers which have a certain relation to the natural numbers, and by which arithmetical calculations are facilitated. Logarithms are the indices of numbers in geometrical progression.

Long Columns.—When the length of a column exceeds its diameter by from 25 to 30 times it comes under the category of long columns, which yield under pressure by bending alone, in the same manner that a beam supported at both ends will yield.

Long D-Valve, or Long Valve.—A long slide valve used is some of the older beam engines, so called because it extended over the ports at each end of the cylinder, the ports being situated near the ends of the cylinder to avoid the waste of steam consequent upon long passages.

Longitudinal Elevation, or Side Elevation.—A view showing the side of a structure, as distinguished from its end view.

Longitudinal Seams.—The seams or joints running lengthways in a boiler. They should always be made to break joint; never be put "in line."

Longitudinal Section.—A sectional drawing taken through a structure in the direction of its length.

Long-toothed Gauge.—A gauge with a movable tooth or marker used for gauging lines which are not in the same plane as the stem. The head is made long to permit of variation in vertical movements corresponding with those of the marker.

Long Valve.—See long D-valve.

Loose Centres.—Heads very similar to lathe poppets, provided with screw mandrels, and centre points. They are used as supports for some classes of work, both when being lined out and when being shaped on a planing machine. Hence called machine or planer centres. The centre lines, and keyways of shafts when suspended between these centres are marked with a scribing block; lines and distances are also marked off and squared up from a surface plate or marking-off table. When clamped or bolted to a planing machine table the work between the centres can be partially or completely rotated to bring different sections under the action of the cutting tool, so that both circular and irregular shaping can be done in a planing machine by the use of loose centres.

Loose Coupling.—A shaft coupling which is made capable of instant disconnection, either by claw engagements or by frictional surfaces, as distinguished from permanent or fast couplings (q.v.).

Loose Eccentric.—A single and adjustable eccentric sheave used for reversing engines which are not provided with the ordinary link reversing gear.

Loose Gland.—A form of gland used in making the joints of hot water piping. It is a loose ring furnished with two lugs and bolt holes, and is slipped over the spigot end of a pipe. An india-rubber ring is then placed in front, the spigot is slid into its socket, which has corresponding lugs and bolt holes, and the two are bolted together. The iron ring fitting loosely around the spigot allows of expansion, while the india-rubber ring makes the joint watertight.

Loosening.—Rapping (q.v.).

Loosening Bar.—A round and pointed bar of iron used for rapping patterns.

Loose Pieces.—Certain portions of pattern work, which standing out beyond those faces which have to be lifted in a vertical direction, cannot be drawn along with, and at the same time as the pattern, without dragging up the superincumbent sand, and are therefore attached loosely or temporarily during ramming, being afterwards withdrawn sideways into the space left by the main pattern. Sometimes pieces are made loose in the top to lift with the top box and be afterwards withdrawn. Loose pieces are held during ramming with dowels, skewers, or dovetails.

Loose Pulley.—The idle or carrier pulley of a pair, on which the belt runs when the machine which the belt has to drive is not in use. When the machine has to be driven the strap is shifted from the loose to the fast pulley (q.v.).

Lorry, or **Lurry.**—A low truck used for running loads about on yard and other tramways.

Loss.—In mechanical exchanges there is no actual loss (see Conservation of Energy). But it is customary to use the term loss to designate the difference between the calculated or theoretical and the actual work obtained from a mechanical arrangement, the loss being due to the imperfections of the arrangement.

Loss of Head.—The diminution in the weight and pressure of a liquid column. Loss of head is due to the friction of long pipes and to the presence of quick bends.

Loss of Heat.—This is due to transmission by conduction, or by radiation. In engine boilers, cylinders, and steam pipes this is reduced by the practice of cleading (q.v.) or felting (q.v.).

Lost Motion.—The difference in the rate of motion of driver and driven parts, due to bad or loose fitting, slips, &c. The term is frequently used in reference to drill spindles, belting, &c.

Lost Pass.—The backward pass of a bar or rail over the top of a two high mill (q.v.) which is not provided with reversing motion.

Low Breast Wheel.—See Breast Wheel.

Low Flashing Point—An oil is said to have a low flashing point when it will take fire at a low temperature.

Low Freezing Point.—A low freezing point is a valuable property in a lubricating oil in England and in the colder climates, since bearings are less liable to become gummed than when the oil freezes readily.

Low Moor.—The best quality of wrought-iron plate manufactured in the United Kingdom is made at Low Moor, in Yorkshire.

Low Press.—A workshop abbreviation, signifying the low-pressure cylinder in a compound engine.

Low-pressure Cylinder.—The larger cylinder in a compound engine into which the expanded steam exhausts from the high-pressure cylinder.

Low-pressure Engine.—An engine which exhausts its steam into a condenser.

Low-pressure Steam.—Steam which is either below or but a few pounds in excess of the atmospheric pressure.

Low Red Heat. -In forged work is a colour corresponding with a degree of temperature midway between a black red heat (q.v.) and a bright red heat (q.v.). It may be roughly taken at 1,290° F.

Low Water Alarm.—An apparatus which is used to announce when the water in a steam boiler falls to a dangerously low level. In modern practice the water gauge (q.v.), try cocks (q.v.), and fusible plugs (q.v.) are the only indicators of water level used. In the old boilers, arrangements of floats and weights were employed, by which, at low water, a jet of steam was directed against the steam whistle, or sometimes a float and index board alone were used.

Low Water Safety Valve.—A low water alarm (q.v.), which is constructed to discharge the steam when the water falls below a safe level.

L-Rest.—A lathe rest for hand turning made in the shape of the letter L, the head on which the tool rests being flat and short, and the leg which fits in the socket being of the usual length. This rest is useful for short work where the ordinary T-rest would be too long, and for attaching to a slide rest for temporary hand turning.

Lubricant.—An unguent interposed between two bearing surfaces to prevent them from coming into actual contact, and becoming abraded by their mutual friction. The lubricants are commonly oils and tallows. When bearings are submerged in water the latter furnishes the means of lubrication. There are certain qualities which a good lubricant should possess, the first of which is, that as compared with others, a minimum quantity of heat only should be generated during a maximum number of revolutions. In addition to this a lubricant should have a low freezing point, so as to remain fluid all the year round; a high flashing point (q.v.), in order to diminish the risk of spontaneous combustion (q.v.); freedom from excessive tendency to gumming (q.v.), and from acid, which is generated in some bad oils to the injurious corrosion of the bearing and other parts with which it comes in contact; suitable body or viscosity for heavy or light machinery, as required; good power of capillarity, to insinuate its way between bearings; freedom from rancidity; and durability and uniformity of action. All these qualities are capable of being put to actual test, and the highest-priced oils often prove to be the cheapest in the end.

Lubrication.—The distribution of an unguent over bearing surfaces to preserve them from the heating which results from the friction of surfaces in actual contact. The nature of the lubricant selected will depend altogether upon its suitability for the particular purpose for which it is designed, the lubricants having most body being chosen for heavy machinery, the light thin oils for light machinery. A thin oil would be squeezed out of a heavy bearing at once, and a thick oil would demand an increase of driving power to overcome its viscosity. Lubrication is effected by various oil cans or oil feeders, self-acting or otherwise, described under their various heads.

Lubricator.—A contrivance for supplying a regular amount of oil or

12*

grease to a journal or bearing. It is sometimes of the needle form, that is, a length of wire coming to the journal acts as a carrier of oil from an inverted glass receiver, whose mouth is closed with a bit of tube. The feeding takes place partly by capillary attraction, partly by the vibration of the shafting. If the bearings become heated the oil is rendered more fluid and flows easier. See Axle Box, Impermeator, &c.

Lug.—An ear or projection upon a casting for the reception of bolt or other attachments.

Luminosity.—Luminosity is due to the radiation of heat, the intensity of the heat rendering visible the radiations of the different coloured rays; at a red heat the red rays are visible, but at a white heat all the rays of every colour are blended and visible. The shades of luminosity vary from a black red (about 1,000° F.) to a white heat (about 2,700° F.).

Lumps.—Special fire bricks, made expressly for the linings of blast furnaces.

Lurry.—A lorry (q.v.).

Lustre.—Metallic lustre is due to the reflection of the light rays from a smooth surface. A rough surface, being made up of a number of minute surfaces, reflects and scatters the rays of light; a smooth surface reflects them almost wholly back. Hence the lustre depends upon the density of a material, and upon the degree of smoothness or polish imparted to it.

Lutes.—The materials used for rendering air-tight those vessels which have to be exposed to the heat of a melting furnace, as crucibles, annealing pots, &c. Stourbridge clay in powder, made into paste with water, and loam, are used.

Luting.—Luting is the mode of connecting pipes, or tubes, or vessels to prevent the entrance or escape of gases. In brass foundries crucibles are often luted by placing an empty one over the one containing the metal, or by placing a cover on it.

M.

M.—The Greek letter μ, the symbol of the coefficient of friction.

Machine.—A machine may be defined as an assemblage of parts, some fixed, others movable, by which motion and force are transmitted.

Machine Centres.—See Loose Centres.

Machine-cut Pattern.—The teeth of the change wheels of lathes are either cut out by a wheel-cutting machine or are cast to shape from patterns. In the latter case the patterns themselves are cut in a machine, wood patterns not being sufficiently reliable. The wheels made from machine-cut patterns are the better of the two, the hard skin which results from contact with the sand rendering them more, durable than the others.

Machine Cutters.—Usually understood not of cutting tools, but of the various milling wheels, reamers, &c.

Machine Drilling.—The drilling of work under a power-driven drilling machine. All light work, all repetition work, and much of the heaviest work is done under a machine. In exceptional cases hand drilling (q.v.) is resorted to.

Machine Foundations.—Machines doing heavy work, drills, planing, slotting, shaping, punching, shearing, and other machines of this class, are bolted to massive foundations of stone or concrete, tail bolts being sunk into the foundation to which the machine bases are bolted.

Machine-moulded Wheels.—Cog wheels, both spur and bevel, moulded by the aid of a machine, and in the making of which sectional portions only of a pattern are necessary. (See Wheel-moulding Machine.) The essential pattern parts required are a tooth block containing two teeth and a core box for spur wheels, and striking boards also, in addition to these, for bevel wheels. The wheels have the advantage of great accuracy of pitch and cleanliness of surface. They take longer in moulding than those made from a pattern, but their superiority to these enhances their value, notwithstanding that the cost of a pattern is saved.

Machine Moulding.—This embraces the moulding of wheels and ordinary work by the aid of special machines. In the wheel-moulding machines the pattern teeth alone are carried round an exact and measured distance after the ramming up of each successive tooth, and the ramming is done by hand. But in most moulding machines for common work almost all the work is done automatically, even to the ramming of the sand in some cases, so that all the moulder does is to actuate the levers and finally clean and close the mould.

Machine Riveting.—Riveting performed by a single application of steady pressure at the same instant upon the tail and the head of a rivet.

Machinery.—Machinery comprises the various machines in a factory. These are assessed at a definite sum, and a certain amount written off their value each successive year for depreciation (q.v.). It is usual in estimating the charges on work to enter a definite sum per day for the use of each machine, to cover the cost of depreciation. This may range from a few shillings to a pound or more.

Machine Shop.—The shop in which the operations of engineering requiring the use of machines, as distinguished from fitting and erecting, are carried on. The turning shop or turnery is usually included in the same department. The smaller machines are ranged in rows, leaving room for the workmen to pass around them freely. The larger ones are placed as most convenient for the moving of heavy masses of work. All are driven through the medium of countershafts, deriving their motion from the row or rows of line shafting overhead, and are served by overhead travellers, of light or heavy construction according to the class of work done in the shops.

Machine Tap.—A tap for use specially in screwing machines, as distinguished from the hand taps, which are actuated by a tap wrench.

Machine Tools.—These embrace the various machines used in the departments of engineers' works, by turners, drillers, slotters, planers, boilermakers, &c., together with the various cutting, punching, and shearing tools attached to, and forming a portion of the same.

Machine Vice, or Vice Chuck.—A small long parallel jawed vice, used on the tables of planing, shaping, and drilling machines, for holding small and irregularly shaped pieces of work which cannot be conveniently bolted to the ordinary tables.

Machining.—Denotes generally the operations performed by machines on metal work ;—turning, planing, shaping, boring, &c., are comprehended under the general term of machining. It is used both in specifications and in the shops.

Machinist.—Machinists are a class of men distinct from fitters, erectors, and turners, and signifies those who have charge of planing, drilling, slotting, shaping, and similar machines. They are not apprenticed, but are simply recruited from the ranks of unskilled labourers and handy men, and receive wages ranging midway between those of labourers and mechanics.

Magnesian Limestone.—A compound of carbonate of lime and magnesia which is employed for the lining of converters in the basic process.

Magnesite.—An impure magnesia found in Euboea, and other places, used for magnesite bricks.

Magnesite Bricks.—Fire bricks used as linings in converters and rotary puddling furnaces. They are formed of impure magnesia calcined and mixed with from 15 to 30 per cent. of raw and partially calcined magnesia, and from 10 to 15 per cent. of water, the whole being dried and burnt.

Magnetism.—The adherence of particles of metal to the points of drills, and metal turning and other tools, is due to the development of the residual magnetism therein by friction.

Magnetite.—A magnetic iron ore abounding in the north of Europe and America, largely quarried for commercial purposes. The Dannemora iron from Sweden is a variety of this ore, and so also is the loadstone.

Magneting.—The separation by the action of magnets of particles of iron from brass and copper turnings, previous to remelting.

Magneting Machine.—A machine used for magneting (q.v.) to save the time occupied in the use of hand magnets.

Magnitude of Force.—The magnitude of a force is the aggregate of the units of force which compose it, estimated in lbs. or tons, or other unit terms.

Mahogany. (*Swietenia mahogoni*.)—Sp. gr. ·56 to ·85. A close-grained red-coloured wood, belonging to the natural order *Cedrelaceæ*. The principal varieties of mahogany are :—African ; Spanish, which comes from Cuba and St. Domingo ; Honduras, or bay wood, which comes from Honduras and around the Bay of Campeachy in Central America. The latter is that chiefly used for common work, being cheaper and more readily worked than the Spanish. Mahogany is used for making the best class of patterns.

Main Bearings.—The bearings for the crank shaft of an engine.

Main Centre.—In side-lever engines is the shaft upon which the side levers vibrate.

Main Chuck.—A lathe chuck intermediate between the mandrel and the actual driving chucks.

Main Cylinder.—The principal or working cylinder of an engine, as distinguished from balance cylinders (q.v.), oil cylinders (q.v.), &c.

Main Driving Belt.—The first motion belt, which comes direct from the motor of a workshop to the main driving pulley.

Main Driving Pulley.—The first or principal pulley on the line shafting (q.v.) of a workshop or factory, or a pulley which has to do specially heavy work in comparison with others.

Main Frames.—The frames of a locomotive which carry the boiler, axle boxes, cylinders, &c.

Main Link.—The link which is attached to the end of the beam in the arrangement of levers known as parallel motion (q.v.).

Main Screw.—A guide screw (q.v.).

Main Valve.—Denotes the slide valve proper, when the expansive working of the steam is provided for by a separate expansion or cut-off valve (q.v.).

Major Axis.—See Transverse Axis.

Make.—See Yield.

Making Joints.—The bringing together and securing with proper cements, or other steam or water tight agents, the joints of steam and water pipes.

Making Up.—Mending up (q.v.).

Making-up Piece.—A mending-up piece (q.v.).

Malacca Tin.—Called also Banca tin, Straits tin. It is sold in pyramids weighing about 1 lb. each.

Malachite.—Carbonate of copper. Green malachite is found in the Ural mountains and in Australia. Blue malachite is found at Burra Burra, in Australia. The blue variety contains more carbonic acid than the green.

Male.—When a stud or a dowel fits into a recess it is said to be the male portion of that particular piece of work.

Malleability.—The quality possessed by metals of becoming extended under rolling, pressure, or hammering.

Malleable Cast Iron.—Cast iron in which the combined carbon has been converted into graphite. In the Whiteheart process the castings are annealed in contact with substances rich in oxygen at a high temperature for several hours, or even days, according to circumstances. The substances employed are oxides of iron, usually hæmatites, which part with their oxygen in order to combine with the carbon in the castings, thus reducing them to a condition approaching that of wrought iron. In the Blackheart process the iron is annealed in an inert packing, resulting in little oxidation.

Malleable Iron.—See Wrought Iron.

Malleable Nails.—Pipe nails (q.v.) made of malleable iron to prevent the risk of blow-holes in the casting.

Mallet.—(1) A heavy wooden hammer used for the delivery of blows on the handles of chisels and gouges without incurring the risk of splitting them down, as is the case when iron hammers are employed for that purpose. (2) Round-faced wooden mallets are used by moulders for rapping patterns during their withdrawal from the sand, to facilitate the detachment of the sand. Similar mallets are used by boiler-makers for levelling or bending their plates of wrought iron and steel.

Manager.—The individual who has a general supervision of an engineering works. There are two kinds of management, one general, the other more limited. The general manager has charge of the entire factory, including the offices, and undertakes or supervises the estimates, tenders, expenses, &c., and is usually an educated engineer who has been duly articled and gone through the shops and offices, and who receives the salary of a professional man. The works manager has no control over the offices, but only superintends the shops. He is usually a practical man who has risen from the ranks.

Manchester Principle.—The system of diametral pitches (q.v.).

Mandrel.—Sometimes erroneously spelt mandril. (1) A cylindrical rod. (2) A revolving spindle of wrought iron or steel used for chucking lathe work upon. (3) The spindle of a circular saw is a mandrel. A lathe mandrel is either the spindle of the headstock, or that of the poppet. A smith's mandrel, or a "nut" mandrel, is a round rod upon which nuts are finished to shape.

Mandril.—See mandrel.

Manganese.—An element which is invaluable to the steel manufacturer owing to its entering more readily than iron into combination with oxygen and sulphur. Iron containing from 5 to 20 per cent. of manganese is called spiegeleisen; if it contain more than 20 per cent. it is called ferromanganese.

Manganese Bronze.—An alloy of manganese, aluminium, iron and tin. It is not a true bronze, since it contains little tin; more correctly termed a high-tensile brass. It is useful when toughness is an essential, as in propeller blades.

Man Hole.—An oval opening in the shell of a boiler through which the attendant gains access to the interior for the purpose of examination and cleaning. It is always stayed either with a ring of wrought iron or with a casting.

Manilla Ropes.—Ropes used sometimes in preference to hemp for the transmission of power. They are white or untarred, and are stronger and more durable than those of tarred hemp, so that to do the same amount of work a rope of manilla need not be so large as a hempen one. The ropes are made in Manilla, in the Philippine Islands, from the fibre of leaf-stalks of the *Musa hoglodyarum*, a species of plantain or banana.

Mare's Grease.—A yellowish brown grease imported from Monte Video and Buenos Ayres, being the fat of the mares slaughtered for their hides, bones, and grease. Used for purposes of lubrication.

Margery's Fluid.—Sulphate of copper diluted with water, and used for the impregnation (q.v.) of timber. It is applied under pressure.

Margin of Safety.—The factor of safety (q.v.).

Marine Boiler.—Marine boilers are usually of the return tubular (q.v.) type, but many variations in the arrangements are necessitated by different conditions.

Marine Engine.—A generic term which denotes an engine used for propelling a vessel. Those commonly used in the present day are of the inverted cylinder (q.v.) type and of the compound surface condensing class. See also Trunk Engine, Oscillating Engine, and Steeple Engine.

Marine Glue.—India-rubber, shellac, and mineral oil combined to form a hard black cement. It is melted by the application of heat and is used for making joints in packing cases water tight, by saturating strips of canvas with it and sticking them along the faces of the joints.

Marine Pattern Connecting Rod.—That form of rod in which the bearing end has two brasses, with or without an iron cap, and secured by bolts to flat expansions of the wrought-iron end of the rod; as distinguished from box end, or strap end forms.

Marker Out, or Liner Out.—A workman whose special duties consist in marking out the centres, working lines, &c., of metal work in readiness for the machinists and fitters.

Marking Gauge.—A gauge used for marking the thickness of timber which has to be planed to an equal dimension throughout.

Marking-off Table.—A planed cast-iron plate or table, strengthened and stiffened with flanges upon the under side, and employed as a basis for marking off the centrés and working lines on rough castings and forgings before they go to the machines to be planed, turned, and bored. Being planed true on face and edges and blocked up and levelled, the square and scribing block can be employed with accurate results. The marking-out table is a necessary adjunct in all fitting shops of moderate and large size, so that the sole responsibility of lining-out rests with the two or three men who are employed thereon perpetually.

Marking Out.—See Lining Out.

Marks.—See Brands.

Marlborough Wheel.—A cog wheel whose teeth are elongated sufficiently to gear with two other narrow wheels on different spindles.

Mass.—The quantity of matter in a body, not to be confounded with weight, since the mass remains constant in any part of the earth's surface, while the weight depends on the force of gravity, and therefore on the latitude.

Mast.—The vertical timber in a derrick crane.

Master Jet, or **Master Light.**—The outer gas jet of the two in those gas engines where the inner jet is extinguished by the force of the explosion. Its function is to relight the inner one in readiness for the next explosion.

Master Light.—A master jet (q.v.).

Master Tap.—A hob or hub (q.v.) used for cutting steel dies.

Master Wheel.—A dividing wheel used for cutting the teeth of gears. It is properly of large diameter, in order to the minimising of errors due to its own pitching out, such errors being reduced in the case of all wheels smaller than itself. If the master wheel is of small diameter its errors are magnified when cutting wheels of larger diameter.

Masting Shears.—Shear legs erected over the edge of a dock, and used for lifting and lowering the masts of ships into their hulls. The gearing is usually placed at some distance behind the shears and the power is transmitted thereto through chains.

Mast Winch.—A winch (q.v.) which is worked either by hand or steam, and fastened to a ship's mast or to a pillar.

Match Plate.—A board or plate of wood or metal upon whose opposite faces two different portions of a pattern are put for moulding. The boxes containing the impressions, when brought together, constitute a complete mould. See Plate Moulding.

Mate.—It is customary to term any two men who work together, mates, but the term is more particularly applied to smiths and their hammermen, and to boiler-makers, platers, and angle-iron smiths and their strikers, because they always keep together, and only change in the case of illness or dismissal.

Matrix.—(1) A mould (q.v.) is a matrix, though the term is seldom applied thus. (2) The earthy matter which contains metallic ores is called the matrix.

Matt, or **Regulus.**—Copper matt is the resulting product after the oxide of iron present in the ore has been caused to combine with silica. It is granulated by being poured while molten into a vessel with a perforated bottom through which a stream of water is set running. Regulus may be coarse metal or fine metal, the former being produced in an early (called the second) process, the latter in a subsequent (called the fifth) process.

Maul.—A wooden hammer or mallet. Sometimes also applied to a smiths' sledge-hammer.

Maximum Dimensions.—In the manufacture of iron bars and plates there are certain dimensions given by most houses as the maximum. For sizes exceeding these special quotations are necessary.

Maximum Pressure.—The utmost pressure which is brought to bear upon a body or structure. Commonly has reference to the pressure of elastic fluids and liquids. From the maximum pressure and the rates of expansion the mean pressure is deduced.

Maximum Strength.—This refers to the disposition of a definite quantity of material in such a manner that a structure of the strongest section containing that quantity of material shall be designed. Illustrations are to be obtained in the sectional and longitudinal forms of timber beams, iron flanged girders, and columns.

Maximum Weight.—The utmost weight or load which a body or structure has to sustain. It has reference to the pressure of dead and variable loads. On the maximum load the factor of safety is based.

Mean.—The term is of wide application, signifying the average, as mean head, mean height, mean quantity, mean pressure, &c.

Mean Effective Pressure.—The average pressure upon a piston, minus the resistance due to back pressure.

Mean Pressure.—The average pressure upon the piston in a steam cylinder taken through the entire stroke.

Mean Strength.—The average of the strengths of similar bars or beams, as deduced from experiments.

Measurement.—Correct measurement is of the utmost importance in engineering. For common work direct measurement with the rule is sufficient, for better work, calipers, and gauges, both cylindrical and the micrometrical are employed. For sheet metals, wire gauges are used. Measurements are given in inches, fractions and multiples of the inch, and of the millimetre, and in the numbers of the wire gauges.

Measuring Machine.—The measuring machine of Sir Joseph Whitworth consists of two heads mounted on a bed, one fixed, the other movable. Each headstock is provided with a mandrel and graduated wheel. The mandrels are actuated by screws. Knowing the number of threads in the screw and the number of graduations in the hand wheels and the teeth in a worm wheel driving one of the screws, the exact amount of forward movement can be deduced.

Measuring Rod.—See Standard Rod.

Measuring Tape.—A narrow tape or riband enclosed in a circular casing from which one end is drawn out as required for measurement. Tapes vary from 33 to 100 feet, and from 20 to 25 inches in length. They are divided into feet and inches, and parts of inches, and into chains. They are made of linen strengthened with wire, and of ribbon steel. Tapes alter in length with the moisture in the atmosphere, but should not be more than ¼ out in their total length. They are used for out-door work chiefly, and for taking circumferences or arc measurements of large pieces of work.

Mechanical Centres.—Usually denote the centres of gravity, gyration, oscillation, and percussion. See appropriate headings.

Mechanical Engineering.—The art of construction of mechanism generally, comprising both prime movers and machines. It embraces designing, drawing, pattern-making, moulding, smiths' work, turning, boring, drilling, shaping, planing, fitting, millwright's work, boiler and platers' work, and erecting.

Mechanical Equivalent.—When one piece of mechanism is substituted for another, the mechanical effect remaining the same, it is termed a mechanical equivalent.

Mechanical Puddler.—A puddling (q.v.) furnace in which the operations of hand puddling are performed automatically.

Mechanical Rabble.—A rabble (q.v.) to which reciprocal motion is given by means of gearing and levers.

Mechanical Stoker.—A term applied to various pieces of automatic apparatus for supplying uniform amounts of fuel to steam boilers. **In**

one form an endless chain takes the place of fire bars, and slowly travels the fuel from the front of the furnace to the back, by which time it is consumed. In another form the fuel is fed from hoppers placed over the fire doors, the feeders being worked by gear direct from the engine.

Mechanics.—That branch of science which treats of the effect of force upon matter. It is conveniently divided into the two sections of statics (q.v.) and dynamics (q.v.), and embraces in its widest sense the actions of solids, fluids, and liquids.

Mechanism.—Commonly understood of an assemblage of parts which, without necessarily constituting a complete machine, embraces the essential principles on which the machine is constructed. Thus the mechanism of a drilling machine would include the gear and spindles apart from the framework, the mechanism of an eccentric would be regarded independently of the details of its attachment, and that of an engine distinct from the mode of its fixing.

Medium Drawing Paper.—Measures 22 × 17½.

Medium Hard.—A quality of emery wheel useful for shaping edges of tools, saws, and trimming castings.

Medium Soft.—A quality of emery wheel useful for general surface work. For grinding narrow edges it wears away too rapidly.

Mellow.—Timber is mellow when it is thoroughly dry and slightly aged.

Melting.—The fusion of solids effected by the application of heat. The melting of metals is carried on in cupolas, in reverberatory furnaces, and in crucibles. The distinction between melting and smelting is, that the former applies to the fusion of metals which have been previously reduced from their ores, the latter signifies the reduction and melting down from their ores.

Melting Down Refinery.—The usual form of refinery (q.v.) in which the charge consists of selected pig and scrap, as opposed to a running in refinery (q.v.).

Melting Furnace.—See Blast Furnace, Reverberatory Furnace, Cupola, &c.

Melting Holes.—The chambers which receive the crucibles containing crucible cast steel (q.v.).

Melting House.—That department of a steel works devoted to the melting of cast steel.

Melting Point.—See Fusing Point..

Member.—Any separate piece or unit portion of a structure is termed a member thereof.

Mending-up, or Making-up.—The repairing of the broken edges of a founder's mould which have become damaged by the withdrawal of the pattern. It is done by laying a strip of wood of the proper outline against the broken edge, and ramming sand against it.

Mending-up Piece, or Making-up Piece.—The strip of wood, plain or curved, as the case may be, which is used for mending-up (q.v.). Sometimes a strip of lead is bent round a curved edge instead of a sweep cut out in wood.

Mensuration.—The measurement of surfaces, or of solid contents.

Merchant Iron.—Finished bar iron of various sections.

Merchant Mill.—The entire plant of rolls and accessories used for the making of tee, angle, and bar iron of various sections.

Merchant Rolls.—See Mill Rolls.

Merchant Train.—A train of rolls which reduces puddled bars to their finished sections and sizes ready for the market.

Mercurial Gauge.—A form of pressure gauge much used on steam boilers before the introduction of Bourdon's spring gauge. It consisted of a suction tube partly filled with mercury, open to the air in the longer limb and in communication with the boiler in the shorter limb. The mercury in the longer limb sustained a float from which a wire passed over a pulley and downwards to an index finger which pointed to the divisions upon a scale. The height of the float varying with the pressure caused the position of the finger to indicate the amount of pressure. In high-pressure boilers the construction was modified to avoid the inconvenience of a long tube. A hole closed at the top and opening into the mercury at the lower end was used, and the amount of compression of the air in the tube according to Boyle and Marriot's law (q.v.) furnished the means of deducing the pressures.

Mercury.—Symbol Hg. Comb. weight 199.8. Sp. gr. at 0° 13·596. The only metal liquid at ordinary temperatures freezing at−40° F. The pressure of steam is often estimated in inches of mercury. Used in barometers and thermometers and in mercurial gauges.

Mesh.—The size of the openings in the grating of an air-pump, or of a Kingston valve, or of a moulder's sieve.

Meshed.—Used sometimes with the same signification as in gear (q.v.).

Metal.—Metals are elementary bodies. Those used in engineering are iron, copper, tin, lead, zinc, manganese, each described under its proper heading.

Metal Hollows.—Hollows made of metal and used by pattern-makers for the same purpose as those made of leather (see Leather Hollows). To enable the metal to bend with facility, the flat or back portions of the hollow are grooved out longitudinally.

Metallic Packing.—A kind of packing used for stuffing boxes (q.v.), formed of woven wire.

Metal Patterns.—Foundry patterns are made of metal when too weak to stand foundry usage if made in wood, or when curves have to be imparted thereto which could not well be given to wood. Hence all ornamental works are made in metal, and all works where a large number, say several scores or hundreds, have to be cast from the same pattern. Iron, brass, tin, and lead, are the metals chiefly used ; iron and brass being employed for permanent work, while lead and tin are used chiefly to be bent into various outlines, from which the actual permanent patterns in the harder metals are finally moulded. Patterns in iron are rusted and varnished, or protected with a coating of beeswax.

Metallurgy.—The processes by which metals are separated from their ores, and combined to form alloys.

Metal Saw.—See Hack Saw, Hot Iron Saw.

Metal Spinning.—The process by which light articles in the malleable metals are made to assume circular and moulded shapes by means of pressure applied to them while in rapid rotation in the lathe.

Meteoric Iron.—This is of interest as being the only state in which iron is found in a nearly pure condition in nature. It is of extra mundane origin but is of no commercial value.

Method of Moments.—Signifies the calculation of the bending strains in a structure, by estimating the moments of the forces acting thereon, as distinguished from the graphic methods (q.v.).

Methylated Spirit.—See Spirits.

Metre.—The French unit of length, containing 39·37079 English inches. It is divided into a thousand equal parts termed millimetres. A square metre is equal to 1·196 square yards or 10·764 square feet. A cubic metre is equal to 1·308 cubic yards or 35·3156 cubic feet.

Metre Rule.—A rule divided into fractions derived from the metre (q.v.). These are sometimes made a metre long, but sometimes the millimetre divisions are put on an ordinary two feet or one foot English rule. Both wood and steel are used in their construction.

Metre Scale.—A scale (q.v.) used for drawing purposes, in which the metre (q.v.) is the unit and millimetres (q.v.) the subdivisions. It may be twelve inches, or it may be one metre long. Usually the subdivisions of the inch and the metre are placed side by side for comparison.

Metric System.—The French system of weights and measures, of which the unit is the metre, originally supposed to be the 1-10,000,000th part of a meridian of the earth. The metric system is purely decimal, hence its value.

Mica.—Plates of mica are inserted in the sight holes of cupolas in order that the furnace man may see the progress of the melting of the charge.

Micrometer Caliper.—A small caliper used in workshops for the best classes of work, and constructed on the same principle as Whitworth's measuring machine. It consists of a horseshoe-shaped piece, into one end of which a traversing mandrel is screwed. The mandrel is provided with a fine threaded screw having a definite number of threads to the inch, and its head is graduated into an equal number of divisions. By advancing the screw through a portion of a revolution as indicated by the divisions on the head, the fractional portion of an inch can be gauged with accuracy.

Middle Cut File.—A Lancashire File having a degree of coarseness midway between a rough and bastard cut (q.v.). It is seldom used.

Middle Flat Gouge.—Any gouge whose amount of curvature is neither very flat on the one hand nor quick on the other.

Middle Part.—The central portion of a three-parted moulding box. Sometimes the central portion of the mould.

Mid Feather Wall.—The brick wall which divides the two flues in a wheel draught (q.v.), so called because it is in the centre of the boiler, and is very thin where the boiler rests upon it, or about three or four inches in width.

Mid Gear.—The link motion of an engine is said to be in mid gear when the arrangement is such that neither backward nor forward motion is possible. Engines at rest should always be placed in mid gear.

Mil.—The thousandth part of an inch. A term used to denote the sizes of wire gauges, the diameters being given as so many mils. The term is believed to have been first employed by Mr. Cocker of Liverpool, who attempted to introduce a new wire gauge.

Mild Centred Steel, or Soft Centred Steel.—Steel whose central portions are softer than the exterior. It is first brought into the form of rods, &c., only partially carburised in the cementation process carburisation being arrested at a definite stage. Used sometimes for engineers' taps where hardness without excessive brittleness is desirable.

Mild Steel.—Steel which contains a very low percentage of carbon, approximating therefore to the condition of wrought iron. It may contain from ·05 to ·20 per cent. of carbon. It welds but does not temper, and is suitable for boilers, ships' plates, rivets, and wire.

Mill.—A term of general application. A building with its machinery; a boring machine; a blacking mill, or a revolving blacking grinder; a rolling mill, or a forge mill for bar iron manufacture, &c.

Mill Bar.—Bar iron rough from the puddler's rolls, as distinguished from merchant bar (q.v.).

Milled Head.—The circular head of a pinching, or set, or adjustment screw, whose edge is cut into a succession of ridges to enable the fingers to grasp it without slipping.

Mill Furnace.—A reheating furnace (q.v.).

Mill Gearing.—Deriving its name from the old corn-mills, and comprising the work of the millwright, embraces cog wheels, pulleys, shaft bearings, and belting.

Millier.—The French metric ton, containing one million grammes (q.v.), or a thousand kilogrammes, and equivalent to 2204·6212 English pounds, or 19·6841 hundredweights, or ·9842 of an English ton.

Milligramme.—A French measure of weight, being the one-thousandth part of a gramme (q.v.). It is equivalent to ·0154 part of an English grain.

Millimetre.—A French measure of length, being the thousandth part (·001) of a metre (q.v.), and being the ·03937 of an English inch.

Millimetre Drills.—Drills whose diameters are made to French millimetre dimensions instead of to English eighths of an inch.

Millimetre Pitches.—Screw pitches whose unit is the millimetre. These, when required, are commonly cut on lathes having English leading screws, by a proper arrangement of change wheels. Thus, the metre equals 39·375 or $39\frac{3}{8}$ inches. A pitch of one millimetre equals therefore 1000 threads in $39\frac{3}{8}$ inches. In the same length of leading screw having four threads to the inch, we have $39\frac{3}{8} \times 4 = 157\frac{1}{2}$, or 157·5 threads of $\frac{1}{4}$ "pitch;" hence the ratio subsisting between the two is $\frac{157\cdot5}{1000}$, which reduced to its lowest denomination equals

$$\frac{157\cdot5}{1000} \div 2\cdot5 = \frac{63}{400}.$$ For a lathe having a leading screw of $\frac{3}{8}$ pitch the ratio would be $\frac{63}{600}$, for one of $\frac{1}{2}$ pitch $\frac{63}{800}$. To deduce the wheels we multiply the numerator by the pitch of the thread, and then proceed as with ordinary change wheels. Thus, to cut a thread of 10 mm. pitch, and having a leading screw of four to the inch, the equation would stand $\frac{63 \times 10}{400} = \frac{630}{400}$, breaking up into multiples $\frac{63 \times 10}{20 \times 20}$; but to obtain wheels in the set we substitute $\frac{63 \times 30}{20 \times 60}$, and 63×30 will be drivers, and 20×60 the driven wheels.

Millimetre Rule.—See Metre Rule.

Milling.—The shaping of metals by means of slowly revolving tools or milling cutters (q.v.). Surface contour can be more rapidly and accurately effected by milling than by the ordinary operations of planing and shaping with the single-edged cutting tools, and the practice of milling is constantly increasing in economical importance in the workshops.

Milling Cutters.—Discs or circular cutters made of steel and serrated around their edges to an exact counterpart of the sectional shape which they are intended to produce. Work of nearly any form and temper is capable of being figured with milling cutters. They can be used in the lathe, but commonly they are fixed in a milling machine (q.v.).

Milling Machine.—A machine in which metal work is reduced to shape when attached to a table and passed under a rotating serrated cutter. Both plain and irregular surfaces can be thus shaped, and the value of the machine consists in the uniformity of outline which can be imparted to numerous similar parts. Nut faces, the flutes of taps and reamers, and similar repetition works in iron, steel, and brass, are done in milling machines.

Milling Tool.—See Milling Wheel.

Milling Wheel, or Knurling Tool.—A small wheel running loosely on a pin set in the cleft end of a bar of iron, and used for roughening or milling heads of screws. The edge of the wheel is made into a counterpart of the pattern which it is intended to produce, and the pattern is formed by pressing the wheel with sufficient force against the revolving work to cause the wheel also to revolve, and so impress its pattern upon the work. The wheel itself is cut by pressing it against a hob set slowly revolving in the lathe. This of course precedes hardening.

Mill Pick, or Mill Bill.—A tool like a double wedge mounted in a short handle and used to dress the faces of millstones.

Mill Race.—The narrow space between the floats of an undershot water wheel (q.v.) and the masonry within which the water is confined.

Mill Rolls, or Merchant Rolls, or Mill Train.—The merchant rolls of a rolling mill. They are employed for the production of finished iron from the puddled bar after it has been cropped, piled, and balled or reheated. They are similar in the main to the puddle rolls, but differ in some matters of detail. They consist of sets of roughing or billeting, and finishing rolls, and are either two high, or three high.

Mill Scale.—See Hammer Scale.

Mill Tail.—See Tail Water.

Mill Train.—See Mill Rolls.

Millwright.—A workman of a class whose numbers are diminishing owing to the increasing specialisation of engineers' work. The occupation of the millwright originated in the development of corn and other mills driven by water power, at a period when modern engines and machinery were unknown, and when the factory system had not arisen. Owing to these conditions and the exigencies of circumstances, the millwright was compelled to perform all the various tasks which are now included under the head of engineering, and which are divided into several distinct departments, embracing the working both in wood and metal.

Mine.—Ore (q.v.).

Mine Tin.—Tin ore obtained by mining as distinguished from stream tin (q.v.). It is found in veins in quartz, granite, and clay slate, associated with arsenical and copper pyrites, specular iron ore, and wolfram.

Mineral Oils.—These are used for lubrication and in the furnaces of steam boilers. Mineral oils when used for lubrication do not generate injurious acids like the animal oils. They are manufactured from bituminous shale, or are imported from America or from Russia, from petroleum springs. They are subjected to filtration or to distillation, to free them from grit, tar, and volatile oil, and are used both pure and as compound oils (q.v.).

Minium.—Red Lead (q.v.).

Minor Axis.—See Conjugate Axis.

Minus Lap.—A term applied sometimes to the internal or exhaust lead on a steam valve. It is the lead to exhaust diminishing the amount of cushioning, and is used with very quick running engines where smoothness of working is essential. See Inside Lead.

Miter.—A frequent way of spelling mitre (q.v.), as miter wheel, miter board, &c.

Mitis Castings.—See Wrought-iron Castings.

Mitre.—Lines meeting at an angle of 45° with each other form a mitre.

Mitre Board.—A board used by wood workers for cutting mitred joints. Blocks are screwed on a plain board at angles of 45° with its edge, and form a guide for the plane, which is laid upon its side as in shooting, or sometimes for the saw, though the latter is more commonly used with the mitre box (q.v.).

Mitre Box, or Mitre Block.—A templet or guide for sawing mitre joints. It has a bottom and two sides, and two saw kerfs opposed to each other at an angle of 90°, cutting the sides therefore at angles of 45°.

Mitre Iron.—Bar iron of angular section.

Mitre Joint.—A butt joint whose ends are cut at an angle of 45°, the abutting sides therefore forming an angle of 90°.

Mitre Valve.—The mitre of a safety valve signifies the annular seating, turned to an angle of 45° in section, upon which the valve itself rests.

Mitre Wheel.—A bevel wheel whose pitch cone is placed at an exact angle of 45° with its axis. Hence pairs of mitre wheels working together are always of equal diameter, pitch, and number of teeth, and connect shafts which stand at right angles with each other.

Mixed Gauge.—See Gauge.

Mixed Oils.—See Compound Oils.

Mixing Chamber.—A chamber at the end of the cylinder of a gas engine, in which the gas and air become mingled previous to ignition.

Model.—A pattern is sometimes, though incorrectly, termed a model. A model is really a counterpart or copy upon a small scale of a piece of mechanism. The object of making a model is either to give a good idea in due proportion of a large and expensive piece of work, or as a work of art and beauty, or, as in a working model, to test the practicability or utility of an untried design.

Modulus.—A constant multiplier or coefficient employed in mechanical calculation. It expresses the ratio of the effective value of a machine as compared with its theoretical value, or the difference between the work expended and that given out, the loss being due to friction chiefly.

Modulus of Elasticity.—A ratio of stress and strain. The strain may be a change of length (Young's Modulus), a twist or shear (Shear Modulus, Modulus of Rigidity, Modulus of Torsion) or a change of volume (Bulk Modulus).

Modulus of Resistance.—Equivalent to the modulus of rupture, or to a lesser quantity, as the modulus of elastic resistance.

Modulus of Rupture.—A constant number which represents the weight necessary to break a bar of any given material of definite length, breadth, and depth, and used in calculating the strength of similar bars differing therefrom in dimensions only.

Moment.—The measure of the importance of a physical agency or of a force. Thus the moment of a lever is equivalent to the power turning its arm around the fulcrum, in other words expresses its effect or result. See Moment of Inertia ; Force, Moment of.

Moment of Inertia.—The sum of the products of each particle of a moving body multiplied into the squares of the distances of the particles from their neutral axis. It is represented by the letter I. If the mass of every particle of a body be multiplied by the square of its distance from a straight line, the sum of the products so obtained is called the moment of inertia of the system about that line, which is also called the axis. The moment of inertia for any cross section is found by dividing the total area into separate small layers or areas, and multiplying the area of each element so obtained by the square of its distance from the horizontal axis taken through the centre of gravity, and then adding the products together.

Moment of Resistance.—When a bar is subject to bending, the internal stresses set up therein by the bending action constitute a mechanical couple. The amount or moment of those stresses is equal and opposite to the bending moment. This is termed the moment of resistance for that particular section.

Moment of Rupture.—The moment of resistance at fracture.

Moments, Equality of.—See Equality of Moments.

Momentum.—The mass of a body multiplied by the velocity imparted thereto, equals the momentum. The velocity may remain constant and the mass vary, or the mass may remain constant and the velocity vary.

Monkey, or Ram.—(1) The longitudinal weight which is made to fall on the heads of piles when being driven into the soil. See Pile Driver. (2) A jumper (q.v.).

Monkey Wheel.—See Gin Block.

Monkey Wrench.—A screw wrench (q.v.).

Monkbridge Iron.—Plates manufactured by the Monkbridge Company, and held in high estimation for boiler work.

Mortar Mill, or Mortar Mixer.—A machine used for mixing mortar. It consists essentially of a shallow cast-iron pan with two revolving chilled wheels driven by bevel gearing from an engine. A false or movable bottom of loose plates is provided in order that they may be renewed when they are worn out.

Mortice.—A recess cut in a piece of timber which, with its corresponding tenon, forms a joint used in timber work.

Mortice Chisel.—A stout wood-worker's chisel driven with the mallet, and used for cutting mortices where percussion and leverage are rendered necessary.

Mortice Gauge.—A gauge used for marking the thicknesses of mortices and tenons. It consists of a stem and head similar in outline to those of an ordinary gauge, but the stem is furnished with two marking cutters, one fixed, the other movable, which can be set to equal the thickness of any mortice. Hence both edges of the mortice or tenon are marked at one time, and the head being movable permits of the marking off of a tenon from any distance inwards from the edge of the stuff.

Mortice Wheel.—A wheel in which wooden cogs are used instead of iron teeth. The rim is pierced with as many mortices as there are cogs, and the cogs are fitted into these mortices with corresponding shanks, the long grain running radially towards the centre of the wheel. The cogs are thicker and shorter than those of ordinary wheels, and the teeth of the iron wheel into which they work are correspondingly thinner. There is no flank clearance allowed as in ordinary gearing.

Morticing Machine.—A machine for cutting mortices in wood, either by means of a chisel, or by a circular cutting bit. When the latter is used the end is either left rounding or made square afterwards by a chisel. The earlier morticing machines had a positive stroke, that is, a dead thrust from the beginning, which strained the chisels. Now they are made with an adjustable stroke whose force is increased gradually by means of a system of levers.

Motion.—See Laws of Motion.

Motion Bars.—The guide bars of an engine cross head. **See Slide Bars.**

Motion Block.—A cross head (q.v.). See Slide Blocks.

Motion Disc.—See Wrist Plate.

Motion Work.—A term applied to the various rods, levers, and links connected with an engine slide valve.

Motive Power.—The particular source of energy which is applied to actuate a prime mover or a machine. Hence motive power may be animal, steam, water, air, or gas.

Motor.—A prime mover (q.v.).

Mottled Iron.—A quality of cast iron intermediate between the grey and the white varieties, both in point of texture, crystallisation, and the state in which its carbon occurs, a portion being in the combined and a portion also being in the uncombined or graphitic state.

Mould.—The hollow matrix or enclosed space into which metal is poured to form a casting. A mould is, its cores excepted, a counterpart of its pattern (q.v.). Moulds are made in green or in dry sand.

Moulder.—An ironfounder. His work consists in moulding the reverse impressions of castings in sand, the impressions being obtained by ramming the sand around patterns of wood or other materials, which are so constructed as to withdraw therefrom without doing injury to the mould itself. Due regard must be paid to venting, to allow free egress to the liberated gases, and the metal must be selected and mixed as most suitable for the special class of work for which it is to be employed.

Moulder's Baskets.—Hand baskets made in the form of trays; used in foundries for carrying about small quantities of coke, coal, and other materials. They are made of cane or of wire.

Moulder's Bellows.—See Bellows.

Moulder's Lamp.—A cast-iron lamp whose vertical section is that of a truncated cone, and burning paraffin or benzoline. Used by moulders for throwing light into the interior portions of moulds. Candles were formerly used, but have been superseded by the lamps.

Moulder's Mallets.—See Rapping Mallets.

Moulder's Nails.—See Pipe Nails.

Moulding.—The making of moulds in sand, or loam, or plaster of Paris, with or without the aid of patterns. When patterns are used they are enclosed in sand rammed around them, from which they are subsequently withdrawn. When moulding is done without patterns the necessary shape is imparted to the mould by means of loam boards or strickles. Moulding is done either in green or in dry sand, or in loam. The moulding of work was formerly more frequently called founding.

Moulding Board.—A bottom board (q.v.).

Moulding Box.—A flask (q.v.).

Moulding Cutter.—An adjustable steel cutter for woodworking, having its edge ground and sharpened to the shape of the moulded edge which it has to form. These cutters are usually fixed in pairs on opposite sides of a disc, which is set to run with a spindle revolving over the wood to be operated on. The cutters travel at a rate of from 4,000 to 5,000 feet per minute.

Moulding Letters.—Pattern letters (q.v.).

Moulding Machine.—A machine employed for moulding patterns in a partially automatic manner, the aim being to produce moulds of precisely the same shape, and with flush joints, *i.e.* joints showing no lap. The patterns are usually made of metal, laid upon a table horizontally, and, according to the type of the machine used, the moulding box is rammed up over it by hand or by some kind of special mechanism. The plate is then usually turned over, and the box released and dropped downwards, being in this respect the reverse of ordinary moulding, where the pattern is lifted away from the box. Usually, also, there is no complete pattern, but divided patterns are put on separate plates, or on opposite sides of the same plate, the moulds made separately and only brought together for casting.

Moulding Sand.—Foundry sand (q.v.). But strictly speaking moulding sand designates the black sand which accumulates on the floor of the foundry from repeated castings, and which is only used for box filling and for the rougher class of work.

Moulding Tub.—A wooden tub of oblong form and with sloping sides, containing moulding sand, and provided with a sliding cover. It is used by brass and iron moulders who do light work, the flask lying on the cover and the tub supplying the necessary sand; the moulder therefore stands instead of kneeling at his work.

Mounted Tracing.—See Tracing.

Mounting.—The chucking of work in the lathe.

Mountings.—Commonly signifies the brass and ornamental work about an engine, or boiler, or machine, but more especially applied to boiler mountings (q.v.).

Mouth.—The opening or orifice of a pipe, or furnace, or ladle, or similar cylindrical or hollow vessels.

Mouthpiece.—(1) An attachment fitted to the mouth (q.v.) of a vessel, pipe, or tank. A door often covers the mouthpiece. (2) A bridge of wood let into the mouth of a plane to reduce the width caused by the wearing back of the iron, and of the face of the wooden stock.

Movable Expansion.—Expansion which is capable of regulation by means of a second slide valve, or other gear.

Movable Points.—(1) The movable centres of the rods in parallel motions (q.v.). (2) The various legs, pens, pencil, and points which are substituted for each other in the compass (q.v.) used for drawing.

Movable Pulley.—A pulley whose axis is movable in space.

Moving Weight.—See Rolling Load.

M Teeth.—Saw teeth shaped like the letter **M**. Used in some cross-cut saws (q.v.).

Mud Box.—A box placed in the suction pipe of a bilge pump (q.v.), to arrest dirt or foreign matters which would otherwise lodge in the valves. It is a cast-iron box fitted with a removable cover and containing a perforated plate or diaphragm placed transversely therein.

Mud Bucket.—A dredger bucket or scoop constructed of cast or sheet iron, or steel, and used for the purpose of bringing up the mud from the bed of a stream or harbour.

Mud Hole, or Mud.—An oval opening in the lower part of a boiler, through which the sediment deposited by the water is extracted. Most boilers have several mud holes, two or three near, or in the bottom, and one or two opposite every cross tube. They are closed by doors while the boiler is working. See Mud Hole Door.

Mud Hole Door, or Mud Lid.—A door or cover which closes the mud hole (q.v.) of a steam boiler when the boiler is in use. It is oval in outline, and is inserted within the hole and pulled up against its inner face by means of a bolt. The bolt is riveted into the door and passes thence through a bridge spanning the mud hole without, and is tightened against it with an ordinary nut.

Mud Lid.—See Mud Hole Door.

Mud.—A mud hole (q.v.).

Muff Coupling.—See Box Coupling.

Muffle.—A small arched vessel of fire-clay, shaped like an oven and used in the assaying of alloys of metals, the muffle being used to protect the metal from the direct action of the flame.

Multiple Boiler.—A multitubular boiler (q.v.).

Multiple Boring Machine.—A boring machine provided with three or more mandrels for simultaneous boring.

Multiple Drilling Machine.—A machine in which a number of drill spindles are arranged parallel to each other, and driven simultaneously. They are used in boiler and girder work where a large number of holes of the same size and pitch are required.

Multiple Gear.—The combination of several pinions and several wheels in train for the increase of mechanical effect. The product of all the wheels divided by the product of all the pinions represents the mechanical gain.

Multiple Threaded Screw.—A screw containing several helices winding around its body. Used to impart more rapid motion than could be obtained by one only. The various threads are parallel with each other, and of equal pitch. A worm wheel is an illustration of a multiple threaded screw, the number of teeth being equivalent to so many thread sections.

Multiplier.—When the cubic measurement of a body in any given material is known, its weight is readily deduced therefrom by multiplying the measurement by a constant number termed its multiplier. Thus cubic inches of cast iron multiplied by ·263, of wrought iron by ·28, of gun metal by ·3, of steel by ·28 give pounds avoirdupois.

Multitubular Boiler.—A boiler traversed with numerous tubes through which the hot gases pass from the fire-box on their way to the chimney, causing circulation and raising the temperature of the water in the boiler during their progress. The tubes are of brass or iron fitting into the tube plates of the fire and smoke boxes. Multitubular boilers are chiefly confined to those of the locomotive and portable and horizontal types, very few vertical boilers being provided with tubes of this character.

Muntz Metal.—An alloy of copper and zinc in the proportion of 60 of copper to 40 of zinc. It is strong, and largely used in engineering construction in the form of sheets and rods, and for the tubes of locomotives and of condensers.

Mushroom Valve.—See Lift Valve.

Myriagramme.—A French measure of weight containing ten thousand grammes (q.v.) and equivalent to 22·0462 English pounds.

Myriametre.—Ten thousand French metres (q.v.) corresponding with 393,707·9 English inches.

N.

Nail.—A strip of pointed metal provided with a head. Used for driving into timber. See Clasp Nail.

Nailing.—See Sprigging.

Name Plate.—The casting which bears the name of the manufacturers on a piece of work. The pattern is commonly made by casting the letters separately in lead or tin and cementing or tacking them upon their pattern plate of wood.

Napier's Compasses.—A drawing compass which is constructed with folding legs to be carried in the pocket. The working points are pivoted to the main legs, and are double ended and reversible on their pivots. One pivot leg carries a point and pencil at opposite ends, the other a point and a pen, so that two points, or a point and a pencil, or a point and a pen can be used at pleasure.

Naphtha.—A comparatively thin and pure rock oil, closely allied to petroleum, which is sometimes used as a solvent for shellac in the making of varnish for foundry patterns. Methylated spirit is mostly used in preference.

Narrow Gauge.—Any railway gauge in which the measurement between the rails is less than the standard 4 ft. 8½ in.

Native Copper.—Copper which is mined in the metallic state. It occurs in the district of Lake Superior, and is highly esteemed on account of its purity, which renders it suitable for electrical purposes.

Natural Draught.—See Draught.

Nature.—By the nature of a material is understood the average excellence of its qualities when unaffected by deteriorating influences.

Nave.—The boss of a wheel.

Neck.—(1) The narrowest portion of the passage in vena contracta (q.v.). (2) An entire journal is often called a neck.

Neck Ring.—The bush for the rod which is fitted into an engine cylinder or steam chest below the stuffing box, to insure durability of wear.

Needle Lubricator.—A pear-shaped or globular glass vessel containing oil, and furnished with a neck which is placed lowermost. A wire passes loosely through the plug which closes the mouth, and by its vibration and capillary action conducts the oil down to the bearing.

Needle Wire.—A vent wire (q.v.) of the smallest size or about $\frac{1}{16}$ in. in diameter.

Negative Slip.—Is applied to the speed of a ship's screw when the velocity of the vessel is greater than it should be according to the theoretical calculation based on the hypothesis that the screw works in an unyielding body. It is due probably to the energy of the current which follows the vessel's wake.

Negative Stresses.—In English practice are those which represent tension.

Negative Terms.—Algebraical or arithmetical terms preceded by the sign —.

Nest Gearing.—Gearing enclosed in a case or box, as in a capstan head, and in some forms of hoisting tackle.

Nett Indicated Horse-Power.—See Actual Horse-Power.

Neutral Axis, or **Neutral Line.**—When a beam is subjected to flexure, there is a longitudinal central line which is neither in compression nor extension, and is therefore subject to no straining action. The part where this line cuts any particular section is termed the neutral axis of the beam. Since also the tensile and compressive forces diminish as the neutral axis is approached, girders and girder-like structures are frequently lightened out in their central portions. In a beam of uniform section the neutral line corresponds with the central line of the cross section. In beams of other sections it will be the mean of the bending sections.

Neutral Equilibrium, or Indifferent Equilibrium.—This results when the centre of gravity of a body which is in equilibrium is in the central portion of the body, and cannot therefore arise into a higher or descend into a lower position. A sphere is always in neutral equilibrium.

Neutral Line.—See Neutral Axis.

Neutral Surface.—The plane of the neutral axis (q.v.).

Neutral Tint.—A purplish grey colour used to distinguish cast iron in sectional drawings.

New Sand.—Mixtures of foundry sands used for facing moulds. See Facing Sand, Old Sand, Sands.

Nickel.—Symbol, Ni. Comb. weight, 58·6. Sp. gr. 8·8. A silvery white metallic element found in the ore called kupfernickel. Used largely for electro-plating the bright portions of the best machinery.

Nicker.—The vertical cutter or lip on the circumference of a centre bit, or of a Jennings bit, which cuts into the wood the radius of the hole to be bored.

Nicking.—The cutting of a shallow vee'd groove around or across a bar or metal in order to ensure its fracture by a blow at that particular spot.

Nicking Fuller.—A tool similar to an ordinary fuller, or fullering tool (q.v.), except that instead of being straight across in the direction of its width it is hollowed, fulfilling the same purpose for round bars that the ordinary fuller does for flat bars.

Nippers.—See Cutting Nippers.

Nipping Lever.—A lever which is so constructed and adjusted, that on moving its longer arm around the fulcrum, the shorter arm on the opposite side of the fulcrum bites or nips the periphery of a smooth turned wheel; and the greater the pressure brought to bear upon the lever the greater the nipping power. Moving the lever in the contrary direction releases the bite immediately. This principle is employed in the construction of what is known as the silent feed (q.v.).

Nipple.—A short connecting piece in a union which receives the nut at one end, and screws into a socket at the other. Or a screwed stud piece used for insertion into gas piping.

Nobbing.—The same as shingling (q.v.).

Nom. H.P.—Nominal horse-power (q.v.).

Nominal Horse Power.—A vague term which represents no particular power of engine, but only manufacturers' advertised sizes. For marine engines, twenty circular inches of piston are sometimes taken as equivalent to a nominal horse-power, and for non-condensing land engines ten circular inches. The term is gradually falling into disuse excepting as a standard for purchase, indicated horse-power (q.v.) being used in preference.

Non-condensing Engine.—An engine which exhausts its steam directly into the atmosphere. See High-pressure Engine.

Non-conducting Composition.—See Boiler Coating.

Non-conductors.—In engineering, various substances are employed as non-conductors of heat. Felting is put around boilers and steam pipes, wooden handles are used for gauge glasses, and steam and hot water cocks, &c.

Normal.—The line which falls perpendicularly to the tangent of a curve at that particular point. The normals to the curves of wheel teeth should always, if the teeth are properly constructed, fall on the varying points of contact.

Normal Pitch.—The pitch of a screw wheel taken normal or perpendicular to the directions of the teeth, as distinguished from the circumferential pitch (q.v.). In screw gears, gearing together, the normal pitches must be equal.

Nose.—The front of a spindle, or generally any projecting part, as a mandrel nose.

Nose Bit.—A shell bit (q.v.) which is provided with a nose or lip at the cutting point for the withdrawal of the core from the wood.

Nose Helve.—See Helve Hammer.

Nowel.—A term sometimes used to designate the inner core of a loam or other mould, but chiefly when the mould happens to be a large one.

Nozzle.—(1) A contracted channel of, exit for effluent fluids, by which their velocity of efflux is increased. (2) Also the discharging end of any tube, as that of a tuyere, for example.

Number One Iron.—See Puddled Bar.

Number Two Iron.—See Best.

Numerical Co-efficient.—See Co-efficients.

Nut.—The loose head which tightens a bolt, usually rectangular or hexagonal in form, though sometimes circular.

Nut Brown Colour.—The colour at which most turning tools for metal, particularly for wrought iron, are tempered. It corresponds with a temperature of about 540° F.

Nut Lock.—The means adopted for securing a nut in place so that it shall not slacken back and become loose in consequence of vibration. There are many such devices, as pins, wedges, set screws, keys, &c.

Nut Machine.—A machine for cutting and punching nut blanks. Nut making is a specialty. General engineers buy their nut blanks and washers cheaper than they can make chem.

Nut Mandrel.—See Mandrel.

Nut Tapper.—A Bolt Cutter (q.v.).

Nut Wrench.—A spanner (q.v.) or a screw wrench (q.v.).

O.

ω.—The Greek letter Omega is used in mechanics to signify the angular velocity (q.v.) of a body.

Oak. (*Quercus.*)—Sp. gr. ·93. A hard, durable, very strong wood employed for a vast number of purposes by engineers. It is especially valuable in works which are exposed to the weather. There are several varieties, *Quercus pedunculata*, or the stalk-pointed oak, and *Quercus sessiliflora*, or cluster-pointed oak, being the European varieties. The latter is the more straight-grained of the two, the former the more flexible. *Quercus robur* is the English oak, and is superior to any other. The red and white oaks and the live oak come from North America. The ultimate tenacity of English oak is about 15,000 lbs. per square inch of section,

and a cubic foot weighs 58 lbs. Oak charcoal is used **for the black-**ing (q.v.) of foundry moulds.

Oak Bark.—See Tannic Acid.

Oblique Area.—The area of the face **of** a propeller blade. **See Screw** Area.

Oblique Joint Steam Chest.—See Divided Steam Chest.

Oblique Section.—A section (q.v.) taken obliquely or at an angle **across** the drawing of an object. The precise direction of an oblique **section** would be indicated by a dotted line and reference letters.

Obliquity of Connecting Rod.—This signifies the angle made by the con-necting rod of a steam engine when the crank pin is at the extreme upper and lower portions of its path respectively. The effect of the obliquity is to cause the slide valve to open the ports unequally at each end, the port being closed and opened a little earlier at one end than at the other. For this reason the connecting rod is always made as long as circumstances will permit of, in order to diminish the amount of obliquity.

Obtuse Angle.—One which is greater than a right angle.

Obtuse Angles.—See Angle Irons.

Occlusion of Gases.—Red-hot metals allow of the passage of gases through their substance; the gases are then said to be occluded. Occlusion plays an important part in the carburisation of iron. The exact method by which this is effected is doubtful, but probably the carbonic oxide is dissociated and deposits carbon in the pores.

Odd Pitch.—The pitch of a screw is said to be odd when it is either not of the same pitch, or not some aliquot part of the pitch of the leading screw of the lathe in which it is being cut. Thus with a leading screw of two threads to the inch, three, or nine, or eleven threads would be odd pitches.

Odontograph.—A scale invented by the late Professor Willis of Cam-bridge, to simplify the marking out of wheel teeth. Tables of appro-priate numbers are given upon the scale for wheels of various pitches, by which suitable radii for the teeth of those wheels are obtained, the centres of the radii being given by the setting of the slant edge of the scale against the radial lines running from the centre of the wheel through the pitch lines.

Oil Boats.—The receptacles for the waste oil at the ends of the bed slides of planing machines.

Oil Can.—A tin or brass can containing the oil used for purposes of lubri-cation. The best form of oil can is that in which the internal opening to the spout is closed by a cover when not in use, and opened only by the pressing down of a lever and spring.

Oil Cataract.—An oil cylinder (q.v.).

Oil Collector.—A vessel provided to catch superfluous oil from bearings, either to prevent the soiling of adjacent parts or for re-utilisation. It usually consists of a tray of some kind or another.

Oil Cup.—A recess or hollow formed in a casting for the reception of oil used in lubrication.

Oil Cylinder, or Oil Cataract.—A small cylinder used for controlling the amount of movement of the piston in a steam reversing cylinder (q.v.). The pressure of the oil against the piston in the oil cylinder is regulated by a cock.

Oiler.—An oil can (q.v.), or more specially an oil can of small size for bench use. Also, an Oil Cup (q.v.).

Oil Feed.—Any appliance by which oil is fed to a bearing. It may mean either an oil can, or a lubricator, or a copper tube leading from a reservoir, or a wisp of cotton wick.

Oil Fuel.—In the form of a jet is used in some boiler fires.

Oil Grooves.—Small semicircular or nearly semicircular grooves cut in the internal faces of brasses and on the sliding surfaces of machinery, for the due distribution of the oil for lubricating purposes. The oil grooves are cut diagonally across the bearing surfaces.

Oil Hardening.—The hardening of steel, effected by quenching it in oil instead of in water. The effect of oil hardening is the less rapid cooling of the steel, with a resulting greater elasticity and tensile strength, and an absence of extreme hardness.

Oil Hole.—A hole drilled down to a bearing to form a channel for the oil used for lubrication. Oil holes are countersunk at the top, the better to receive and retain the oil, and are usually covered with a pivoted disc when not in use, to keep dirt and dust from working into the bearing. The difference between an oil hole and an oil cup or a lubricator is, that the former conveys a temporary supply of oil to the bearing, the latter holds a store in reserve, so that while the former requires constant renewal, the latter is only supplied at long intervals.

Oiling.—See Lubrication.

Oil Pump.—A small force pump used to provide a constant and positive supply of oil under pressure to a bearing, as being more reliable, and therefore preferable to a lubricator, which acts by gravitation or by displacement only.

Oil Reservoir.—A vessel which contains a supply of oil for lubrication. It may be the chamber in an axle box, or in a displacement, or other lubricator, or it may be a tray only which supplies the oil feeders or oil pipes for the connecting and other rods in marine engines.

Oil Slip.—A thin bit of oil stone, whose edges are rounded in the transverse direction. Used for abrading or fretting the hollow faces of the various forms of gouges.

Oil Stone, or Hone.—A fine-grained natural stone, used for imparting the final edge to cutting tools by abrasion, oil being used to assist in the process. The principal oil stones, in their descending order of merit, are, Turkey, Charnley Forest, Arkansas, Grecian, and Washita. Oil stones are set in wooden stocks, and provided with covers to protect them from dust. They are used by wood workers, and also by iron turners and fitters, for giving a fine edge to finishing tools and scrapers.

Oil Tray.—See Oil Collector.

Oils.—The term oils is applied commonly to those fats which are liquid at ordinary temperatures, while the term fats is reserved for those which are solid at the same, notwithstanding that their essential chemical composition is the same, belonging to the class of organic compounds termed glycerides. Only the fixed oils concern the engineer; the essential, ethereal, or volatile oils belonging to the perfumer's art. It may be observed that the distinction denoted by the two terms is, that the essential oils can be distilled without undergoing change, while the fixed oils cannot be distilled without undergoing decomposition; hence the reason of the statement that oils cannot be boiled, since at their boiling points they cease to remain oils any longer. All the oils are lighter than water, and derived from the animal, vegetable, and mineral kingdoms. Their chief use to the engineer consists in their qualities as

lubricants q.v.). See also Oils, Purity of, and the various Oils noted under their special headings.

Oils, Purity of.—The purity of the oils used in lubrication is a matter of importance, and when oils ostensibly pure are purchased, the task of detecting adulteration is not difficult, since in the presence of certain test agents there are reactions which indicate the freedom from adulteration or otherwise of the oils subjected to the tests.

Old Sand.—The black sand which forms the floor of a foundry, consisting of various mixtures of sands which have been used for casting in over and over again for years. It alone is used in ordinary moulding work, but for facing, new sands are used for ramming around the pattern itself, to a thickness of an inch or two. See Facing Sand.

Olive Oil.—Once a common lubricant for machinery, used either alone or with other oils, Gallipoli being considered the best. Being a vegetable oil, and therefore containing resin, it is apt to gum. The flashing point of Gallipoli oil is 490° F.

Oliver.—(1) The small lift hammer used by smiths, but which has nearly disappeared since the introduction of the steam hammer. It consists of a horizontal shaft pivoted on end bearings and carrying a hammer at the end of its shaft. A cord passes from a treadle underneath to a lever standing out from the horizontal shaft, and thence to a spring pole overhead. The depression of the treadle therefore brings down the hammer, and its release allows the spring pole to lift it again in readiness for a fresh blow. (2) A pair of swage blocks held in their relative positions one over the other by means of a spring handle. Sometimes termed holliper.

Open Belt.—A driving belt (q.v.) which passes directly in one plane from pulley to pulley. With an open belt the driving and driven pulleys both revolve in the same direction.

Open Divided Scale.—A scale used for drawing purposes in which only the end primary divisions are divided into fractional parts, leaving the central ones open and clear without subdivisions. In reading off subdivisions therefore the measurement is commenced in the body of the scale, and the fractional portion is read off at one of the ends.

Open-end Rolls.—Mill rolls, free or unsupported with housings at one end. Used for rolling rings, tyres, &c., which must be slid on and off from the end.

Open Frame Connecting Rod.—See Bow Connecting Rod.

Open Hearth.—Has reference to steel-making furnaces built on the reverberatory type, those being termed open hearth furnaces, and their methods of manufacture open hearth processes.

Opening Out.—Enlarging the diameter of a hole by means of a reamer or broach.

Open Joint.—A term applied to the mode of jointing the broad-plated portions of foundry patterns. Since patterns are enclosed frequently for several hours in damp moulding sand, the moisture acting on the wood causes it to expand, with the result in the case of wide stuff of increasing its width, and if its width is confined by other timber attached thereto, of producing curving. Wide plates are for this reason jointed in narrow separate pieces, the edges of the joints remaining about ⅛ of an inch open, so that local extension takes place without affecting the outside dimensions to any appreciable extent. Conversely, the making of open joints prevents the plate from curving or contracting by the after drying of the stuff. Timber jointed thus is held together by cross strips or battens, or by some other portions of the pattern. Tight-fitting dowels are often inserted to assist the jointing.

Open Link.—A slot link (q.v.) in which the slot for the slide block is perfectly plain, the block bearing upon the inside faces of the slot. The term is used to distinguish it from the box link (q.v.).

Open Mouth.—A punch or punching bear is open-mouthed, when both sides and front are free for the insertion of the work, as distinguished from close mouth (q.v.). Used more especially for the punching of plates.

Open Pattern.—A skeleton pattern (q.v.).

Open Pig.—Pig-iron having a largely crystalline structure, this being a characteristic of the soft irons rich in carbon, or the Nos. 1, 2, and 3 foundry pigs.

Open Pit.—See Foundry Pit.

Open Rods—See Crossed Rods.

Open Safety Valve.—One which is not a lock-up valve.

Open Sand.—A foundry mould has in most cases a top covering of sand, and is therefore enclosed on all sides. But in very rough castings this top is often dispensed with, and the casting is then said to be made in open sand. The surface is necessarily extremely rough and uneven.

Open-topped Furnace.—The older form of blast furnace in which the waste gases were allowed to escape into the atmosphere. Open-topped furnaces now chiefly linger in Scotland and South Staffordshire.

Open-webbed Girder.—A Lattice Girder (q.v.).

Opposite Angles.—A term given to rolled iron bars whose section is that of a double angle iron, the parallel flanges being set off on opposite faces of the central web.

Order Number.—In large factories it is necessary, in order to avoid confusion in the charging of time, to give to each job or order its special number, by which alone it is known and distinguished from all others. But to avoid the inconvenience of very high numbers the custom is to class them in sets under the letters of the alphabet, in other words to give to A a thousand numbers, A1, A2, A3, &c., then to take B, B1, B2, &c., up to its thousand, and so on. These are called the order numbers, and their employment both saves time and facilitates reference at a future period.

Ordinates.—Lines drawn at right angles from either axis of a conic section to the circumscribing circle. See also Dict. of Modern Terms.

Ore.—Metals in their crude condition as found naturally associated with earthy matter or gangue. Ore is sometimes called mine.

Orifices.—Commonly applied to the openings in the sides of vessels from which liquids are permitted to issue.

Orthographic Projection.—See Projection.

Oscillating Cylinder.—A steam engine cylinder suspended upon hollow trunnions and oscillating thereon, so that the piston rod and cylinder accommodate their motion to that of the crank at all parts of the revolution. The trunnions are made hollow, and furnished with ports for the passage of the steam.

Oscillating Engine.—A marine engine furnished with oscillating cylinders (q.v.). It is a direct-acting engine and occupies but little space, and is especially adapted for paddle steamers, but for screw propulsion it has been almost entirely superseded by the more modern compound engines of the inverted cylinder type.

Oscillating Stresses, or Alternate Stresses.—Are those by which structures, or the members of structures are placed alternately in tension and compression, as for example in counterbraced structures, subject to alternate moving loads. The conclusions deduced from the experiments of Wohler in this direction show that when a bar is subject to these oscillations in stresses, the total stress on the bar is equal to their sum, that is, supposing a tensile stress of two tons, and a compressive stress of two tons, alternately applied, the equivalent is a total stress of four tons.

Oscillation, Centre of.—See Centre of Oscillation.

Out and Out.—See Over-all.

Outdoor Foreman.—A man whose duty consists in taking general supervision of repair works and the erecting of work sent out from the shops. He is expected to give estimates or to furnish the necessary data for the giving of estimates by the firm, to follow the work through, and to be generally responsible to his employers for the proper carrying through of the same.

Out of Gear.—Wheels are said to be out of gear when their teeth are disengaged from one another, either in consequence of being drawn backwards until their points clear, or endways until their flanks are no longer in contact.

Out of Truth.—A shop term signifying inaccuracy of work. A winding piece of board or metal, or a wobbling or eccentric piece of lathe work, is said to be out of truth.

Outside Caliper.—See Caliper.

Outside Crank.—An engine crank which occupies a position on the outside of the crank shaft bearing or bearings.

Outside Cylinders.—Locomotive cylinders placed outside the framing and connected to pins on the driving wheels.

Outside Fire Box.—That portion of a locomotive boiler which encloses the actual or inside fire box (q.v.) and which presents a large water surface therefore to the action of the fire.

Outside Gouge.—A firmer gouge, so called because the bevel is ground upon the outside or rounding face.

Outside Lap, or Steam Lap.—Lap (q.v.) given to a slide valve on the outside edges as distinguished from exhaust lap (q.v.). The amount of outside lap is a measure of the ratio of expansion of the steam, an early cut-off implying a high ratio of expansion.

Outside Screw Tools.—Chasers (q.v.) or comb tools, used for cutting or chasing external screws in the lathe.

Out Stroke.—The forward stroke of a gas engine piston, that is, in a direction away from the ignition chamber. It is produced by the explosion of the gaseous charge.

Outward Flow Turbine.—See Turbine.

Oval.—An egg-shaped figure, the curves of whose ends are unequal.

Oven Coke, or Hard Coke.—The hard coke produced by the distillation of coal in ovens, Of the latter there are several kinds in use, the Appolt, Coppée, Carves, and Pernolet. It differs from the gas coke of the gas works in being distilled at so high a temperature that the gaseous hydrocarbons are partially decomposed and deposit their carbon in the coke. Its value depends upon its hardness and density and richness in carbon, and it is employed for the melting of metal in the cupola, for which gas coke would be unsuitable.

Over-all.—A common term in the workshop signifying an outermost dimension embracing the utmost extent of the dimension. Out and out is an equivalent term.

Overblow.—See Afterblow.

Overflow Valve.—Any valve by which surplus liquid is allowed to run away. Overflow valves occur in injectors and tanks.

Overhanging Cylinders.—Engine cylinders which are bolted to the ends of their bed plates instead of upon their faces. The advantage claimed is the lowering of the piston rod centre and the shortening of the foundation for the bed.

Overhanging Pulley.—A pulley which is attached to the overhanging portion of an overhanging shaft (q.v.).

Overhanging Shaft.—An end portion of a shaft which projects beyond its bearing, being supported therefore in one direction only.

Overhauling.—The pulling down of the slack of a hoisting chain. To effect this it is necessary to attach a pear weight (q.v.) or a balance ball (q.v.) to the end of the chain next the hook. Where a snatch block is used, its own weight is usually sufficient to effect the overhauling of the chain.

Overhead Crank.—The crank of a vertical engine, the cylinder being lowermost, to distinguish it from an inverted cylinder engine.

Overhead Gear.—See Pit-Head Gear.

Overhead Traveller, or Overhead Travelling Crane.—Usually consists of a crab mounted on a gantry and worked either by hand or steam. When a crane is used the term gantry crane (q.v.) is more properly applied to it. When a crab is worked by steam the engine may be either mounted on the crab itself or both the crab and gantry may be driven through rope gearing from a source of power situated below, and at a distance therefrom. The advantage of overhead travellers is that they leave a clear space for working underneath, and travel up and down and across the shop without interfering with the operations on the floor. The gantry truss is carried on girders or beams at each side of the building.

Overhead Travelling Crane.—See Overhead Traveller.

Overheating.—(1) Causes iron and steel to become burnt. The overheating may be produced by a brief exposure to a white heat or by a lengthy exposure to a lower temperature. (2) The overheating of boiler plates is due to the accumulations of sludge, or of incrustation within, or to insufficient supply of water. The effect of overheating is to soften the plates, which then become bulged or fractured by the internal pressure.

Over Lap.—The overlap of plates is the amount by which one rivetted plate extends over the other. The overlap of riveted plates should be such that the distance from the edge of the rivet holes to the edge of the plate should not be less than the diameter of the rivets.

Overlap Joint.—A riveted joint of boiler or similar plate in which the edges of the plates overlap to an amount sufficient for the reception of the rivets. This is the reverse of a butt-riveted joint.

Overpoled.—See Poling.

Overpressure.—The pressure of steam in a boiler beyond that which it is designed to sustain. It is of course a relative term, depending on the capability of each boiler itself, so that while 140 or 150 pounds would not be overpressure in a new boiler of the locomotive type, 40 or 45 pounds would be overpressure in a Cornish or Lancashire boiler of bad design and improperly stayed. Overpressure follows from overloading of the safety valve, or sticking of the valve, and from the giving way of weak sections of the boiler, too weak to withstand the ordinary or the sudden strains to which they are subjected, as seams, rings, seatings, stays, &c.

Over-ramming.—Hard Ramming (q.v.).

Over-running.—Refers to the case of the unequal velocity ratio of driving and driven wheels. Thus, if the force of momentum generated by the revolution of a driven wheel exceeds that generated in the driving wheel the former will overrun the latter instead of being controlled by it. Over-running is due to extreme and rapid variations in the driving force exerted.

Over-riding.—Riding (q.v.).

Overshot Wheel.—A water wheel which is turned by the gravity or weight of the water emptying itself into buckets at the top of the periphery.

Overtime.—All time worked over the specified nominal number of weekly hours. Sometimes, in balancing up, each day's time is taken distinct from that of any other day, but commonly the whole week's account is taken, so that the advantage of overtime worked on any one day is sacrificed by time lost on another. Overtime is always paid for at a higher rate than the ordinary, usually at an advance of twenty-five per cent., or as the men say, "time and a quarter." For working all night, fifty per cent. more is commonly paid, while for very urgent work, as Sunday labour, breakdowns, &c., from fifty to a hundred per cent.

Ox Gall.—Is used by draughtsmen to impart fluidity to their colours.

Ox Hides.—The best leather belting and laces are made from ox hide tanned with oak bark, the chief supply coming from America.

Oxidation.—The chemical combination of certain of the elementary bodies with oxygen gas in the presence of water. Its effects are of vast interest to the engineer from the part which oxygen plays in the processes of the blast furnace, the processes of manufacture of wrought iron and steel, the preservation of iron work, the mixing of alloys, the making of rust joints, &c. All the metals used by engineers are directly oxidisable.

Oxide.—The direct product of oxidation (q.v.). The metallic ores are chiefly oxides. The presence of oxides prevents the making of welded and soldered joints, hence the addition of substances to unite with, and flux off the oxides. Oxides in metals in small quantities are a source of weakness.

Oxygen.—Symbol, O. Combining weight, 16. The vital element in the atmosphere. Its presence is essential to combustion and it enters into combination with the carbon in fuel to produce heat in furnaces. In union with metals it forms oxides.

Oyster Shells.—Sometimes used as a flux in iron melting in the cupola, in place of limestones.

P.

...—The Greek letter **Pi**, which is used to denote the relation of circumference to diameter $= 3\cdot14159$.

...ing.—(1) The material, hempen, metallic, or otherwise which is enclosed in a stuffing box (q.v.) for the purpose of rendering the moving rod, steam or water tight. (2) The act or process of insertion of packing material. (3) Blocking (q.v.). (4) Leathers (q.v.) are termed packings.

Packing Case.—All except the largest and roughest metal and timber work, is enclosed in packing cases before it leaves the yard for delivery. The cases are made of spruce deal, of about 6 ins. $\times$ 1 in. scantling, nailed with battens, and banded with hoop iron. For foreign transit the joints of the cases are lined with canvas cemented with marine glue. Where numerous irregular-shaped pieces are enclosed in a case, they are prevented from shifting by bars, stays, or blocks of wood placed crosswise in the box. Packing-case making is a special occupation and a department of the carpentry.

Packing Piece.—A plate of wood or metal interposed between parts to make up a dimension or to afford a base for the attachment of a bracket or other work. See Patch Piece.

Packing Ring.—(1) A piston ring (q.v.). (2) A ring of metal introduced into a groove on the back of an equilibrium slide valve. See Equilibrium Ring. (3) Metallic packing.

Pad.—The socket of an ordinary brace.

Pad Chuck.—A square hole chuck in which the ordinary brace bits are used, the square hole or socket being of the same size as that in the brace.

Paddle.—A flat tool employed by puddlers, for spreading the substances used in coating or fettling the puddling furnace, over the bottom.

Paddle Shaft.—The shaft which carries the paddle wheels of a steamer and which is driven directly by the engine cranks.

Paddle Wheel.—A wheel which propels a steamer by means of the resistance offered by the water to float boards arranged round its periphery and revolving with the wheel. A paddle wheel is hung on each end of the paddle shaft, which shaft is cranked and moved directly by the connecting or piston rods from the engine cylinders. Paddle wheels are either of the radial (q.v.) or feathering float (q.v.) types.

Pad Saw, or Keyhole Saw.—A small narrow saw which slides within a hollow handle or pad, and is secured in place by two set screws. Used for sawing curves of small radius, and holes in the central portions of work.

Paint.—All ironwork is painted to preserve it from rust. Before inspection it is only brushed over with linseed oil. Afterwards with three coats of ordinary lead or other paint. Chains are painted with hot coal tar, the chains being usually heated in a stove or over a clear fire before the application of the tar. For bright work pure white-lead paint, or tallow, or patent composition is used. See also Bituminous Paint, Iron Oxide Paint, Silicate Paint.

Painting.—The application of any coating or preservative preparation to the surface of work is designated painting. Thus foundry loam moulds are painted with black-wash, castings are painted with preservatives against corrosion, and lubricants are painted over broad surfaces. Wheel teeth are painted with anti-friction mixtures, the wafters of Roots blowers are painted to ensure smoothness of motion, &c.

Pale Oils.—See Clear Oils.

Pallets.—(1) Square or circular vessels used by draughtsmen for rubbing and mixing colours (see Square Tile, Wheel Slope, and Cabinet Nest). (2) The plates of a chain pump (q.v.).

Pane, Pean, Peen, Pein, or Pene.—The smaller or narrower end of a hammer head.   It is termed a ball pane when it is spherical in form, cross pane when in the form of a narrow round-edged ridge placed at right angles with the axis of the shaft, straight pane when a ridge of the same character runs longitudinally.

Panel.—(1) The central thinner or recessed portion of a piece of work. Generally, a plain recess.   (2) A bay (q.v.).

Panel Gauge.—A marking gauge for wood, but longer than the ordinary bench gauge, being as much as 18 ins. or 20 ins. long.   So called from its use in gauging the widths of panels.   It is used for all stuff over 8 ins. or 10 ins. in width.

Panel Machine.—A machine for planing up thin timber of small dimensions.   Also called panel planing machine.

Panel Saw.—A small hand saw 26 ins. or less in length, and containing from seven to ten teeth per inch.   It is used chiefly for cutting the shoulders of halvings and tenons, and for light work generally.

Pan-head Rivet.—A rivet the form of whose head is that of the frustrum of a cone, or of a pan inverted.

Pantagraph.—An instrument used for copying drawings either to an enlarged or to a reduced scale.   It consists of four rods jointed in pairs with knuckle joints, and a fixed point, a tracer, and a drawing point. The relative positions of the rods are arranged by means of thumbscrews set in certain holes in the two pairs, and the fixed point is set in a board or table ; the tracer is passed over the original drawing, and the drawing point, or pricker, describes the copy.

Pap.—A short pin or stud forming a projection on a casting which has to be turned, and which without it could not conveniently be chucked.   It is, therefore, a chucking or centring piece simply, and being such, is cut off after its purpose is fulfilled.

Paper Joint.—A mode of temporary jointing employed by pattern-makers. In works of a circular character, such as gear-wheels, which are built up in segments on a face chuck, the first course of segments is glued, not directly to the plate, but to an intermediate thickness of paper, the paper being laid down underneath the meeting of the end joints.   This when set dry and hard is sufficient to retain the segments in place during turning, but when the work is done the entire ring is readily lifted from the plate without splitting of the wood, the paper dividing and tearing through the centre of its substance.

Paper Scale.—A scale used by draughtsmen, made of a slip of thick paper. These are sold singly, or in sets of a dozen each.   Each slip has one edge only marked out, and that with a single scale, and fully divided. They are preferred by many to the ordinary wood scales for this reason, and also because of their flexibility, by virtue of which they can be carried round curved lines.   The objection to their use is that they become dirty, and if, on the other hand, they are varnished to preserve them clean, they shrink.

Paper Weight.—Flat paper weights are used in drawing offices for holding drawings open.

Parabola.—A section of a cone taken parallel with its side.

Parabolic Girder.—A form of bow-string girder, the outline of whose bow is that of a polygon inscribed in a parabola.   Used on bridge work.

Parabolic Governor.—A governor (q.v.) in which the points of suspension are so adjusted relatively to the length of the arms that the path described by the balls is approximately that of a parabola.   Governors of this class are extremely sensitive, and are known as crossed-arm governors.

Parallel Fence.—A fence (q.v.) which stands rigidly at right angles with the saw-table, as opposed to a canting fence, which can be altered to an angle. Most fences are made movable for both kinds of sawing.

Parallel File, or Dead Parallel File.—A file in which there is no taper whatever lengthways.

Parallel Forces.—Forces which act in parallel lines. Their resultant is equal to their algebraical sum, and the point at which their resultant acts is called the centre of forces.

Parallel Iron.—A plane iron of parallel thickness throughout. The advantage of its use is that the mouth of the plane is not increased in width by the wearing backwards of the iron, as in the ordinary taper form.

Parallel Lines.—Parallel lines are lines which are in the same plane, but which if produced to infinity would never meet.

Parallel Motion.—A series of levers employed for the purpose of adapting the curved movement of the end of the beam in a beam engine to the rectilinear motion of the piston rod. There are several such arrangements in use. See Back Link, Main Link, Parallel Point, Parallel Rod, Radius Rod.

Parallelogram.—A plane quadrangular or four-sided figure, whose opposite sides are parallel and equal.

Parallelogram of Forces.—This is a principle of the utmost importance in mechanics, and is constantly employed in calculating the strains on different members of built-up and braced structures. It may be stated thus :—If the magnitude and direction of any two forces acting upon a point be represented by two lines, and if the parallelogram of which these lines form two sides be completed by two other lines equal and parallel with the first, then the diagonal of the parallelogram, which passes through the point where these forces act, will represent the magnitude and direction of the resultant.

Parallel Point.—That point or centre in the arrangement of levers for parallel motion (q.v.) in which the motion first becomes rectilinear.

Parallel Print.—A print (q.v.) whose sides are parallel instead of being tapered in the usual manner. Prints are parallel when they are circular and mould sideways, since the curve of the circle itself then gives sufficient taper for delivery.

Parallel Rod.—The rod which connects the main and back links (q.v.), in the arrangement known as parallel motion.

Parallel Rule.—(1) A double rule employed by draughtsmen. Diagonal and parallel jointing strips of metal maintain the blades parallel with each other in all positions. Parallel lines are oftener marked without its aid by holding the stock of the T-square against an edge of a drawing board and sliding a set-square along the edge of the blade, marking lines therefrom at the distances required. (2) A common form of parallel rule is that in which a roller is inserted in the body of the rule running in bearings therein. The lower edge of the roller stands just perceptibly beyond the face of the rule, so that it carries the rule over the paper on the principle of the ordinary round ruler.

Parallel Shank.—See Shank.

Parallel Strips.—Winding strips (q.v.).

Parallel Vice.—A vice, the surfaces of whose jaws always retain a parallel position in relation to each other, no matter what the width of opening may be, which is not the case in the ordinary hinged or tail vices. Parallel vices are used chiefly in the fitting shops for the better classes of work.

Paring Chisel.—A long paring tool similar to the ordinary chisel, but about twice the length of the firmer tool. Patternmakers use these to the almost entire exclusion of the others, as being more convenient for their special work. Paring chisels are seldom driven with the mallet, but actuated by hand pressure alone. They are made in widths ranging from ¼ in. to 2 ins.

Paring Gouge.—A long paring tool differing from the ordinary firmer gouge in its increased length, and in being ground on the inner curve, hence called an inside gouge. These tools are not driven with the mallet, but with the hand only. They range from ¼ in. to 2 ins. in width, and are used by patternmakers chiefly, for cutting the various curved outlines of their work.

Paring Tools.—All tools which act by splitting, or which remove the fibres in a direction approximately parallel with their cutting faces. Chisels, gouges, planes, axes, and tools for metal turning and planing generally, may be classed as paring tools. Patternmakers' long chisels and gouges are specifically called also paring tools, to distinguish them from the short or firmer tools (q.v.).

Part.—One of the sections with which a moulding box is built. See Flask, Top, Bottom, and Middle Parts, Three-part box, Cope, Drag, &c.

Parting.—(1) The sand joint between two contiguous moulding boxes or flasks (q.v.). (2) The process of making the joint. (3) The act of separating the different box parts.

Parting Line.—The joint line of a pattern. It may be marked upon the pattern, but in most cases has no actual existence upon that, only signifying the parting or joint made by the moulder on the face of the sand itself. The parting line may be in a true plane, or its outline may be uneven, irregular, and undulatory in character, depending altogether on the outline of the pattern.

Parting Ring.—A heavy ring of cast iron by which the parting of loam moulds is effected. A joint being made as required on any portion of a loam mould, parting sand is strewn over, and the ring laid upon it. On the upper face of the ring the remaining courses of bricks are built and the loam struck. The upper portion of the mould can there- fore be lifted from the lower, by means of lugs on the ring, and after- wards replaced in the same position when ready for casting.

Parting Sand.—Sand employed in iron foundries for scattering over the parting surface in order to effect a complete separation between the joints of a mould. It consists of burnt sand scraped off the surfaces of castings, is loose, and non-adhesive in character; or baked new sand, or powdered brickdust, or similar dry material is used. By its use the sand in the top and bottom flasks is prevented from mingling, and the parting surface is unbroken. The layer of sand does not exceed $\frac{1}{16}$th in. in thickness, and it is dusted over by the hand, and blown away from the pattern edges with bellows.

Parting Surface.—The surface of a moulder's sand joint which corres- ponds with the parting line (q.v.).

Parting Tool.—A tool for metal and wood turning, used for cutting or parting off work in the lathe. It is narrow, deep, square across the end, and the width tapers slightly backwards in order that it shall clear itself in the cut, that is, shall not rub against the sides of the metal which it divides, and so set up unnecessary friction.

Passages.—The steam-ways of a cylinder, embracing both ports, and exhaust.

Passes.—The passing and re-passing of malleable iron or steel bars, through the rolls of rolling-mills. See Lost Pass.

Pasty.—The pasty condition of iron is that which is intermediate between the solid and fluid state. All irons do not pass through the pasty condition. Grey irons pass directly from the solid to the liquid state, while the white irons go through the intermediate stage. Hence the former are used for the finer class of castings, for which the latter would be unsuitable. The value of wrought iron is due chiefly to the readiness with which it can be made to assume any desired outline while in a pasty condition.

Patch.—A plate of wrought iron, or steel, or sometimes of cast-iron, riveted or bolted to broken parts for purposes of repair. Broken castings are sometimes patched in this way in preference to replacing them with new. Boiler plates are often patched, and so also is plated work generally which has been injured by corrosion or accident.

Patch Piece, or Patch Plate.—A casting bolted to wrought iron work to form the base for the attachment of another casting, the patch plate being an intermediary to facilitate the process of fitting-up and attachment. It may be considered as a packing piece.

Patch Plate.—See Patch Piece.

Patent Fuels.—These consist generally of small coal mixed with a cementing farinaceous or resinous substance, pressed into bricks and dried. They are used only where storage space is a consideration.

Path of Contact.—The path described by the faces and flanks of wheel teeth mutually engaged.

Pattern.—A model or a counterpart of a piece of work. As commonly understood, the model in wood or other material from which a founder makes his mould. This, called the pattern, in most cases constructed of wood, is rammed up in sand, and being withdrawn, leaves its exact impression behind, into which impression the metal is then poured. Cores, drawbacks, &c., have to be provided for in patterns.

Pattern Bench.—An ordinary carpenter's or joiner's bench, comprising top, legs, vice, stop, and drawer.

Pattern Letters.—These are made in cast lead, tin, or brass, and fastened upon plates of wood from which the actual name-plates (q.v.) are then cast.

Patternmaking.—That section of engineering devoted to the making of the wood patterns or models for foundry use. Patternmaking not only includes the actual construction of patterns, but the making of full-sized working drawings on boards, both for the use of the pattern shop and for foundry loam work, the marking out of cored and loam work in the foundry, the setting of cores in moulds, the gearing of mortice wheels, and much other work of a miscellaneous character, principally in connection with the foundry.

Pattern Register.—A book in which a record is kept of all patterns, and their places in the stores, to which the date of their construction and the job for which they were made, together with other useful memoranda may be added.

Pattern Shop.—That department of an engineer's works in which the foundry patterns are constructed. It contains the usual wood benches common to carpenters' shops, sawing stools, and such machines as are suited to the capacities and requirements of the shop, as circular, and band saws, panel planing machine, or general joiner, together with two or three lathes of different sizes.

Pattern Stores.—These stores contain the foundry patterns, which, having been sent back from the foundry and not being required again for immediate use, are stored away. In this the patterns are variously arranged according to their character and value. Large patterns are placed sometimes in open sheds, covered in at the top but exposed to the air at the sides, but the majority are placed in a closed building. The smaller ones are placed on shelves, the larger ones laid on the floor or suspended from the beams or rafters overhead. Standard patterns which are in frequent request are placed in such positions that they can be taken out more rapidly than those which are seldom required. In large stores an attendant is regularly employed to devote the whole of his time to their storage, and a register is also kept containing sundry useful particulars of their date of construction, purpose, place of storage, &c.

Pattern Work.—See Patternmaking.

Paul, or Pawl.—(1) The finger or click which engages with a ratchet wheel. (2) The catch which fixes a turntable. (3) The catch or dog which prevents the endlong movement of a crane spindle when made for single and double gear. See Spring Pawl.

Paul Feed, or Ratchet Feed.—The feed of a machine effected by means of a paul and ratchet or small cog wheel. The feeds of planing and shaping machines are effected by means of pauls actuating the feed screws.

Pawl.—See Paul.

Pean.—See Pane.

Pear-tree. (*Pyrus communis.*)—Sometimes, though seldom, used in the construction of patterns. It works sweetly, but is apt to warp unless very well seasoned. Used for set-squares and French curves. Sp. gr. ·73. A cubic foot weighs 45 lbs.

Pear Weight.—A weight of the shape indicated by its name, which is attached to the lifting chain of a crane just above the hook, for the purpose of overhauling or pulling it down when there is no load being lifted. The reason for its employment is that the weight of the chain itself is not sufficient to overcome its own friction on the pulley and the inertia of that portion of the chain which passes from the pulley to the barrel.

Peasemeal.—This is used in the finer and more intricate class of moulding as a fixing for the blackening, being intermediate between the blacking and the sand. In large ordinary and flat moulds the blacking is simply sleeked over without the intervention of peasemeal.

Pedestal.—A plummer block (q.v.). See Horn Plates.

Peen.—See Pane.

Peg and Cup Dowels.—Patternmakers' metallic dowels, consisting of the plain male and female portions without a plate. (See Plate Dowels.) The peg or pin is provided with a tapered shank or prolongation grooved circularly, which shank is driven into one half the pattern, and the cupped or hollow portion is also ribbed in the same manner and driven into the opposite half. The dowels hold, therefore, in the wood by the friction of their ribbed surfaces. They are made in brass.

Pegging Rammer.—A rammer with a small rounded head, used for press ing sand around the narrower portions of patterns. The end of the handle of a larger rammer is often employed as a pegging rammer.

Pein.—The pane (q.v.) of a hammer.

Pendant Bracket.—A hang-down (q.v.) bracket.

Pendulum.—The swinging ball and stem of the pendulum form of governor is called the pendulum.

Pendulum Governor.—A governor (q.v.) consisting of a pair of balls suspended from arms whose centre is at, or near, the centre line of a vertical axis around which the balls rotate. It is called the pendulum governor because the time of a revolution is affected by the length of the axis of the cone just as the time of oscillation of an ordinary pendulum is affected by the length of its rod, that is, the time of a revolution varies directly as the square root of the height of the cone.

Pendulum Hammer, or Monkey.—A heavy smith's hammer, suspended by a rope from a beam and furnished with a long tail or handle, by which it is swung to and fro for the purpose of directing blows upon the ends of large bars and forgings during the process of upsetting (q.v.).

Pene.—See Pane.

Pening.—Beating over, or smoothing over, a metallic surface with the pane or pene of a hammer.

Penstock.—A pentrough (q.v.). Also applied to the sluice cocks or hatches of large size which regulate the flow of water from the pentrough, or which close the entrance to dry docks.

Pentrough.—The reservoir above a water wheel which supplies the water directly to the wheel, the quantity being regulated by a sluice.

Percussion, Centre of.—See Centre of Percussion.

Perfect Engine.—An ideal engine which can have no existence in fact, but the theoretical performance of which it is convenient to assume as a standard for the measurement of the performances of actual engines. A perfect engine would be one in which there was no loss of heat, or one in which the indicator line would be an adiabatic curve (q.v.).

Perforated Pulley.—A wrought-iron or steel pulley, which is honey-combed with numerous small holes perforated through the plate of which it is composed, and through which the air escapes, giving the belt closer adhesion.

Periphery.—The surface or plane of the circumference of a cylinder.

Permanent Coupling.—Any coupling by means of which the ends of shafts are united to make a permanent line, being the reverse therefore of a disengaging coupling or clutch.

Permanent Load.—A load which is constant and unvarying, and a dead load, as the weight of a structure itself, or a load imposed thereon, or both taken in conjunction as distinguished from a live load (q.v.), or from a rolling load (q.v.).

Permanent Set.—That amount of deflection from which a beam or structure is unable to return to its original form, but which remains constant. The term is used in contradistinction to that bending underneath a load from which a structure recovers on the weight being removed.

Permanent Way.—Includes the sleepers, chairs, rails, fish-plates, points, crossings, &c., of a line of railway, as opposed to the temporary lines of rails and tramroads laid down by contractors.

Permanent-way Crane.—A crane used as an accident crane, and so called because made to run on the permanent way.

Perpendicular.—Synonymous with vertical; that is a line which stands at right angles with the surface of still water is strictly perpendicular. But a line is perpendicular to another line or to a plane, when the angles which it makes therewith are equal, that is right angles, no matter what position the base line or plane may occupy in relation to the horizontal.

Perpetual Screw.—An endless screw (q.v.).

Persian Drill.—See Archimedian Drill.

Perspective.—Is used only occasionally in engineering drawing, and chiefly to assist in the comprehension of a drawing which presents some difficulty. A knowledge of perspective is useful also in preparing sketches of machines for engraving, though photography has generally supplanted it.

Pet Cock.—An ordinary small plug-cock inserted in the ends of steam cylinders to allow of the escape of the water of condensation on the starting of the engine. But for this precaution the presence of water in cylinders having little clearance, would, owing to its incompressibility, be a source of danger, leading to the possible blowing out of the cylinder ends.

Petroleum Engine.—A gas engine in which petroleum oil is fed into the combustion chamber and ignited ; its action, due to the expansion of the vapour, is similar to that of coal gas.

Phœnix Column.—A built-up column made of four or more segments united by outer flanges running longitudinally.

Phosphor Bronze.—An alloy composed of copper and tin, to which a little phosphorus is added. It owes its value to its great toughness and tensile strength, its tenacity varying from twenty-two to thirty-three tons per square inch. It is now very largely employed in bearings where durability is required, and in gearing subjected to great stress and shock.

Phosphor Tin.—An alloy of tin and phosphorus which has been used with moderate success for bearings.

Phosphorus.—Symbol, P. Combining weight, 30·96. Enters largely into the calculations of the engineer by reason of its influence upon iron and brass. In wrought iron it produces cold shortness, its influence being very decided when present to the extent of ·75 per cent. It renders steel cold short and valueless for cutting purposes. In steel rails it should not exceed ·1 per cent. Its separation from the iron in the Bessemer process for steel making is accomplished during the after-blow (q.v.). In the basic process the phosphoric anhydride thus formed enters into combination with the lime of the charge to form phosphate of lime. The effect of phosphorus on pig iron is to increase its fluidity and hardness, hence phosphoric pig is used for the finer sort of castings. If present in cast iron to the extent of 1·5 per cent. it is injurious to its tensile strength. Added to gun-metal it produces a tough alloy. See Phosphor Bronze.

Phototype.—A drawing made from another drawing by a species of photography, the advantage of the process being the rapidity with which any number of copies all precisely alike can be multiplied and the time occupied in re-drawing and tracing mostly saved. There are three kinds of phototypes, the blue line, having blue lines on a white ground ; the white line, with white lines on a blue ground ; the black line, where the lines are black on a white ground. For making blue-line phototypes a printing frame, a bath containing a saturated solution of yellow prussiate of potash, a bath of hydrochloric acid composed of about one part of acid to nine parts of water, and two baths of clean water are required. The tracing or drawing to be printed is laid face downwards on the inside face of the glass in the frame, a sheet of prepared copying paper is then laid face downwards on the top of the tracing or drawing, a piece of felt is laid on that, and all are confined together with boards, which are maintained in position by means of springs and bars of wood (see Printing Frame). The frame is then swung over so that the outer face of the glass is exposed to the light. In strong sunlight a tracing will probably require from half-a-minute to two minutes, and a drawing about three minutes, but proportionally longer in dull weather. The best way to tell when the phototype is taken is to have some strips cut from the drawing or tracing paper on which the drawing or tracing is made, mark some black lines upon them, gum them in the frame, and lay some strips of prepared paper over these, and then at the end of every minute or two pull one of these test strips out and put it into the yellow prussiate bath. When the phototype is finished the blue lines will come out on a clear yellow ground colour on the test paper. When the print is complete the prepared paper is taken out of the frame, laid face downwards, and an edge of about half-an-inch in width turned up all round. It is then placed face downwards in the yellow prussiate bath. When the lines come out strong like those on the test paper it is transferred to one of the water baths and washed well by drawing it rapidly through the water, then taken to the acid bath and covered with acidulated water. A blue deposit will at once settle on the print, and on brushing this off with a soft flat brush the ground colour will be found white. It is then given a final washing with a soft brush in the last water bath, and hung up to dry. White line phototypes are put into the frame in the same way, but they take much longer than the blue, or from half-an-hour to two hours, according to the strength of the light. But instead of passing the print through the yellow and blue baths, it is only washed in water until the lines come out on the blue ground colour. Black-line phototypes are printed in the same way as the white lines. The black lines are seldom used, the white and blue coming out best.

Physic.—Small amounts of various substances added to bar iron used for the manufacture of steel in order to assist in the elimination therefrom of sulphur, phosphorus, and other deleterious ingredients. Generally speaking, also any substance added to a metal or alloy, to improve it. Burnt steel is sometimes physicked.

Picker, or Picker-out.—A fine-pointed steel wire used for withdrawing small patterns from the sand. Lifting screws (q.v.) are used for the heavier work.

Picker-out.—See Picker.

Picking-out.—The lifting of light patterns from the mould with a picker (q.v.).

Pickling.—(1) A process by which the outer hard skin is removed from castings before they are operated upon by files or cutting tools. Iron castings are pickled in sulphuric acid and water, brass castings in nitric acid and water. The practice is not much resorted to except in the case of small work. (2) The immersion of sheet iron plates used in the manufacture of tin plate in dilute sulphuric acid contained in a leaden trough, by which all oxide is dissolved off.

Piecework.—Work done by contract as distinguished from day work. The practice of piecework is adopted extensively in most departments

of engineering except in patternmaking, and there also to a limited
extent. The price for work is usually fixed by the foreman of the
department. The custom is for a chargeman to take a job for a definite
sum, to find employment for the men whom the foreman places
with him, and to divide the balance, if any, among them, reserving a
larger share for himself than for them, by reason of his increased
responsibility. The balance is not paid until the work is finished, but
each man takes his ordinary wages, together with overtime if worked,
as long as the job is going on. If there is a deficiency, the custom
varies in different shops ; in some the deficiency is made good from the
first subsequent job which shows a balance, in others each job is considered independently of every other, and though the workman may
have a deficit on one, if he makes a balance on the next he reaps the full
advantage of it. But a man rarely suffers actual loss, that is, receives
less than his usual wages.

Piecework Note, or Contract Note.—A paper given to a workman who
receives a job to be done at a definite price. It contains the description
of the job, with its order number, price, date of giving out, name of
man, and of his mate or mates ; and a counterfoil of it is kept by the
foreman or manager.

Piercer.—A vent wire (q.v.).

Pig.—(1) A bar of cast-iron of D-shaped section, and weighing about a
hundredweight. So called because when run from the smelting furnace
it passes through a large intermediate reservoir called the sow, from
which the smaller moulds or pigs diverge. See Pig Iron. (2) Lead is
cast into pigs when it leaves the improving furnace (q.v.) of about the
same weight and shape as pig-iron, but somewhat flatter. (3) Bars of
blister copper (q.v.) are termed pigs.

Pig and Ore Process.—See Siemens Process.

Pig and Scrap Process.—See Siemens-Martin Process.

Pig Bed.—The sand bed containing the moulds into which pig-iron is
run from the blast furnace.

Pig Boiling, or Wet Puddling.—That form of puddling in which the iron
is purified while in a molten condition, in a reverberatory, or a revolving
furnace. A layer of slag upon the surface protects the metal from the
action of the air, and the decarburization is effected by the oxide of
iron in the fettling (q.v.) and in the scale present.

Pig Iron.—The cast-iron of commerce. It is classed according to quality,
number one, two, three, &c. ; the greyest, softest iron being at the
beginning, and the hard white irons at the end of the list ; one, two and
three are foundry pigs specially, being most suitable for castings, while
the numbers above these are chiefly employed for conversion into
wrought iron. See Foundry Pig, and Forge Pigs.

Pig-iron Breaker.—A machine used for breaking up pig iron into short
lengths for remelting in the cupola. Blake's machine will break a ton
a minute. In the absence of a machine, pig is broken either with a
sledge hammer, or by throwing the bars down upon the angular
edge of a mass of iron embedded in or laid upon the ground.

Pig Mould, or Casting Pit.—The receptacle for the refined iron which is
tapped out from the refinery (q.v.). It consists of cast-iron blocks
rebated and luted together with fire-clay, and in communication
with the refinery through a plate. It rests upon a cistern of brickwork, or cast-iron, through which a current of cold water circulates
in order to cool the mould. A rib is often left in the bottom of the

mould to produce a line of weakness in the plate to facilitate its breaking up.

Pig Stack.—The pile of pig iron in stock in an engineering works.

Pile.—(1) The fagot of puddled bar prepared for the reheating furnace. See Fagoting. (2) Squared timber driven into the beds of streams, or into uncertain and marshy ground, to form the foundations of bridges or buildings. The point of the pile is enclosed in iron, and the head encased with an iron ring to prevent splitting of the wood. See also Screw Pile.

Pile Cap.—A beam which connects the heads of piles.

Pile Driver.—A vertical framework, provided with guides for carrying a descending weight, or monkey, which, being first of all elevated to the top of the framing, is allowed to fall by the force of gravity on the head of the pile. Pile drivers are worked either by hand or by steam, from a crab winch.

Pile Hoop.—An iron band, or bond, shrunk on the head of a pile to prevent splitting of the timber while being driven in.

Pile Screw.—A screw used for the bottoms of cast-iron foundation piles. The screw, usually of one revolution only, and cast on a centre-tapered core or cylinder, is run into the soil by worm gearing from above, and becomes the attachment for the upper columns or piles.

Pile Shoe.—An iron point with straps, fastened to the driven end of a pile to enable it to penetrate the ground.

Piling.—The placing of iron bars in pile. See Pile, Fagoting.

Pillar.—A column (q.v.).

Pillar Bolt.—Bolts shaped like gland bolts (q.v.) are sometimes termed pillar bolts.

Pillar Drill.—A form of drilling machine, which is supported by a central pillar terminating in a suitable base.

Pillar File.—A flat, thin, and narrow safe-edged file.

Pillar Iron.—Rolled malleable iron bars used for building up phœnix columns. The section of the iron is that of the quadrant of a circle with external flanges for bolting to the quadrants on each side, four such quadrants therefore forming a circle. The dimensions are given in internal diameters, and range from 6 ins. to $11\frac{1}{4}$ ins.

Pillar Pump.—A lift or force pump attached to a base plate upon which a pillar stands, the latter forming the support for a crank, fly-wheel, and handle by which the necessary movements are imparted to the rod of the bucket or piston.

Pillow.—See Pillow Block.

Pillow Block, or Pillow.—A pedestal or plummer block (q.v.), being the older name for the actual plummer block which receives the pillow bearing or " brass."

Pimple Metal.—See Fine Metal.

Pin.—(1) A small axis or spindle, as that which carries a pulley, or on which a lever oscillates. (2) A split piece of iron rod or wire which keeps a nut or collar in place, hence called a split pin.

Pin Boss.—The small boss of an engine crank which carries the crank pin (q.v.).

Pincers.—Used for drawing nails from timber. Made of iron, and steel-faced in the jaws. Pincers should be large and nearly flat across the face, for affording greater leverage.

Pincer Tongs.—A pair of smith's tongs open or globular, similar to carpenters' pincers, but Vee'd also in the jaws to grip thin rods or bolts, the head of the rod or bolt being enclosed by the globular body of the jaws.

Pinch Bar.—A shop term used to designate a crow-bar. The word indicates the method of moving trollies about the yard tramrails, by inserting the end of the bar between the wheel and the rail, and so pinching the carriage along.

Pin Drill.—A drill which is prolonged below the cutting edges into a short solid central cylinder of relatively small diameter which fits into a hole previously drilled to correspond, and becomes a concentric guide to the cutting edges for the larger hole. Used in flat countersinking.

Pine Wood, or Fir.—Is applied to timber obtained from the trees of the natural order *Coniferæ*, which order includes such trees as the yew, cedar, larch, &c. Pine wood is largely used in engineering construction. The term deals (q.v.) refers to the sizes of the timber as imported. The Baltic red or Riga deals are yielded by the Scotch fir (*Pinus sylvestris*), the white deal by the spruce fir (*Abies excelsa*) of Norway. The North American white deals are obtained from the Weymouth or white pine (*Pinus strobus*). The yellow pine is the wood of the *Pinus variabilis* or *Pinus mitis*, or *Pinus lutea*. The pitch pine is yielded by the *Pinus rigida*. The uses to which the pines are put by engineers are for foundry patterns, for which the yellow and white are employed, and for the framework of heavy structures in wood, the red and pitch pines being mostly employed for these. The sp. gr. of red deal ranges from ·48 to ·70, and the weight of a cubic foot from 29 to 43 lbs., that of spruce being about the same. The North American yellow pine, sp. gr. ·46, and weight 28 ; the pitch pine, sp. gr. ·73, and weight 45 lbs. Very wide variations, however, occur in woods of the same species but of different qualities and under different conditions of seasoning, so that these must be taken as rough average approximations only.

Pinion.—A small toothed wheel, either bevel or spur. It denotes no particular size, but in a pair the smaller of the two is the pinion.

Pinning.—The securing of the cogs of mortice wheels in their place. This is effected by driving nails or short lengths of wire into the ends of the shanks just within the rim, or sometimes through the rim itself. Wedging is often substituted for pinning.

Pinny.—(1) When wrought iron or gun metal contains numerous enclosed specks of metal harder than that in the general mass they are said to be pinny. (2) A term applied to signify the condition of a file whose teeth have become choked up with minute particles of soft abraded metal. It is cleaned out with card wire.

Pin Rammer.—A pegging rammer (q.v.).

Pipe.—(1) A cylindrical tube open at both ends and used for the conveyance of water, steam, or other liquids or gases. There are numerous kinds of pipes—cast iron, copper, rain water, steam, weldless—described under their headings. (2) A hollow spindle or quill is often termed a pipe.

Pipe Bend.—See Bend.

Pipe Bending.—When bending copper and wrought iron and lead pipes it is necessary to guard against buckling or wrinkling in the internal curve. This is prevented by pouring into the section which is to be bent a quantity of melted rosin or fusible alloy which preserves its circular shape during the bending process, and is melted out again afterwards. When wrought-iron pipes are being bent hot, sand is often used instead of an alloy, which would melt with the heat.

Pipe-clay.—Is used for making models of ornamental castings, the modelling of which demands considerable time. Clay is mixed with water or with glycerine, the latter causing it to retain its plasticity longer than the former. Moulds are taken direct from the clay model, or a metal pattern is moulded from the model and the actual mould is taken from that.

Pipe Connections.—The various parts used in making the joints of pipes, as bends, tees, unions, elbows, crosses, nipples, thimbles, &c., described under their headings.

Pipe Covering.—See Covering of Pipes.

Pipe-cutter.—A tool used for cutting off wrought-iron piping. It consists of a hook or stirrup within which the pipe is grasped, and a hardened sharp-edged steel revolving disc, whose centre is adjustable and which being rotated around the pipe by means of a lever handle cuts it off.

Pipe Flange.—When a series of pipes have to be rigidly connected in line, flanges are used in preference to sockets. These are united with bolts and the sizes of flanges and number and sizes of bolts usually bear some approximately definite relation to the diameter of the pipe. Flanges are either rough or faced ; in the former case they are jointed with lead or millboard, in the latter with red-lead alone, dependent on the purposes for which the pipes are to be used.

Pipe Joint.—Pipes are jointed in various ways, with flanges, sockets, and spigots, caulked or turned, the various hose couplings unions, &c.

Pipe Moulding.—Where cast-iron pipes are made in quantity it is the practice to mould them by special appliances, the pipe pattern being of iron, and usually moulded vertically. The cores are made on an expanding and collapsible bar (see Core Bar), so that the use of haybands is dispensed with. Pipes are either cast on a sloping bed, or else upright.

Pipe Nails.—Broad flat-headed nails of wrought or malleable iron used as chaplets (q.v.) in light cored work. Being employed largely for pipes they are thus denominated.

Pipe Ovens.—See Pipe Stoves.

Pipe Prover.—See Hydraulic Pipe Prover.

Pipe Roll.—A roll hollowed or curved longitudinally, free to turn on a bearing bracket, and used for the support of long lengths of steam and hot-water pipes, its function being to permit of the free endlong expansion and contraction of the series of pipes due to heat.

Pipe Sleekers.—Moulders' tools curved in cross sections and used for smoothing pipe moulds and hollow circular work generally.

Pipe Stand.—A bracket hollowed on its top face, its use being to afford support to pipes, and to keep them from the ground and prevent them from rusting.

Pipe Stoves, or **Pipe Ovens.**—Stoves for hot blast, in which the air, while on its way to the blast furnace, is heated by being passed through a series of cast-iron pipes arranged horizontally or vertically in a closed chamber. The air is heated sometimes by a fire, usually by the hot gases from the furnace. About one square foot of heating surface is allowed per cubic foot of blast passing through, though rather more when the waste gases are used. Pipe stoves are being supplanted by the regenerative stoves, the iron pipes in the former frequently suffering fracture, oxidation and other evils.

Pipe Tap.—A Gas Tap. See Gas Stocks and Dies.

Pipe-testing Machine.—See Hydraulic Pipe Prover.

Pipe-threader.—A machine for cutting the threads, or screws in metal tubing.

Pipe Tongs.—A pair of tongs hollowed out and finely serrated in the jaws. Used for the clasping and screwing up of wrought-iron piping, the serrations maintaining a secure hold without risk of bruising the pipes. Those of small sizes are called gas plyers.

Pipe Wrench.—A tool which fulfils the same purpose as pipe tongs, but which is provided with a single handle only, and embraces the work by means of a hinged lever which tightens with increase of pressure.

Piston.—A solid or hollow disc-like plunger moving lineally in a closely-fitting bored receptacle termed its cylinder. Engines and pumps are furnished with pistons against which steam, air, gas, or water pressure is brought to bear. Pistons are rendered close fitting by cup leathers (q.v.), junk rings (q.v.), Ramsbottom rings (q.v.), and split rings (q.v.). The revolving portions of a rotary engine or a blower are also termed pistons.

Piston Air Pump.—A marine engine air pump fitted with suction and delivery valves at both ends, and having a solid bucket.

Piston Cover.—In small engine pistons it is common to form them of a single split-ring enclosed between top and bottom discs. These discs are termed piston covers.

Piston Packing.—The packing by means of which pistons are rendered air, gas, steam or water tight. See Piston, Piston Ring.

Piston Ring.—A metallic ring either of cast iron, wrought iron or gun metal, the first being preferable, as possessing more elasticity than the others, and used as the peripheral portion of an engine piston. The ring is thin and turned slightly larger in diameter than the bore of the cylinder, usually also thicker on one side than on the other, then slotted across diagonally and forced into the cylinder between the piston covers. Its elasticity then thrusts it outwards and maintains a steam-tight or gas-tight joint with the bore.

Piston Rod.—The rod attached to the piston of an engine or pump, by which its motion is transmitted to the connecting rod or crank. The rod travels through a packed stuffing box and gland, by which leakage of the contents of the cylinder is prevented.

Piston-rod Gland.—The gland which closes the stuffing box of the piston-rod of an engine or pump cylinder. It is placed on the cover of the cylinder or barrel and renders the piston rod steam or water tight.

Piston-rod Packing.—See Packing.

Piston Speed.—The speed of an engine piston measured in feet per minute. The speed ranges ordinarily from 250 feet in condensing and heavy engines to about 800 in locomotive engines. The piston speed is one of the factors in the estimation of the engine or horse-power.

Piston Spring.—(1) A spring ring, or piston ring. (2) A spring placed within the ring to press it outwards.

Piston Valve.—A slide valve whose transverse section is circular, moving closely in a circular seating, opening and closing steam ports arranged diagonally around the seating. The advantage of these valves is that they are balanced.

Pit.—See Foundry Pit.

Pitch.—(1) The distances between the centres of wheel teeth, or of bolts, rivets, or boiler stays, or similar parts arranged equidistantly. Thus wheel teeth or rivets, &c., whose centres are 2 in. apart are of 2 in. pitch. (2) The inclination or rake of the teeth of saws, varying from those which are upright or have no pitch, as in cross cut saws, to those which are set forward very much, as in some mill saws for soft woods. (3) The angle at which a plane iron is set in its stock, the angle being measured from the back of the bedding on which the plane iron rests, to the sole of the plane. It varies sometimes for hard and soft woods, but most planes are used for both alike, and are made to an angle of 45°. Some of the planes for hard wood range to 55°. (4) The height or angle of a roof truss.

Pitch Chain.—A chain composed of built-up flat links, between whose sides the projections of a sprocket wheel engage. The centres of the link pins are pitched out with exact uniformity and correspond with the centres of the sprockets, hence the term.

Pitch Circle.—The circumference of the pitch line (q.v.).

Pitch Cone.—The imaginary cone formed by the development of the pitch plane of a bevel wheel.

Pitch Diameter.—The diameter of the pitch line of a wheel, that being the term in which the size of the wheel is given, the diameter at root and point being deduced therefrom, and from the pitch.

Pitching Out.—The marking or dividing out of the equidistant centres of wheel teeth, or rivets, or other similarly pitched work, with dividers or compasses.

Pitch Line.—The line on which the centres or the pitches of wheel teeth occur. Its position relatively to the length of the tooth is $\frac{1}{18}$ths from the point and $\frac{1}{15}$ths from the bottom, the difference in which allows the necessary bottom clearance.

Pitch Plane.—The plane of a pitch line from one end of the wheel teeth to the other. In a bevel wheel it is the frustrum of a cone.

Pit Head Gear.—The frames, rope pulleys, bearings, and bracing, erected over a pit's mouth for raising and lowering the cage.

Pitman.—A term sometimes applied to a connecting rod.

Pitman Box.—The box end of a connecting rod or pitman (q.v.).

Pit Saw.—A saw about six feet in length, having $\frac{1}{2}$-in. to $\frac{3}{4}$-in. tooth spaces, and furnished with two cross handles, one above for the top sawyer and one below for the bottom sawyer. Used for cutting planks and boards in a saw pit.

Pitting.—The corrosion of boiler and wrought-iron plates in patches, due to the action of acids, or to their inferior quality, or to electro-chemical action, or to all combined.

Pit Wheel.—A mortice wheel revolving on a horizontal axis, and which is usually the first motion wheel in a mill, so called because it revolves in a pit specially prepared for its reception.

Plan.—The appearance which a structure would have (disregarding perspective) to an observer looking at it from above. "Sectional plan" supposes the structure cut through in a horizontal plane and the observer placed as before.

Plan Angle.—The angle formed by the two edges of a double-edged metal turning tool viewed from above.

Plane.—A cutting tool which owes its value to the guidance which the wooden or iron stock imparts to the cutting iron. The stock is usually of beech wood, and the iron is held in place with a wedge or screw. Iron planes are made for various special purposes, and in most of these the iron is tightened with a screw. There are perhaps a hundred varieties of planes in use.

Plane Iron.—The cutting iron of a wood-working plane, sometimes single, sometimes double, the lower one then for cutting, the upper for imparting rigidity. The plane irons of adze blocks for machine work are always single and 12 in. or 15 in. wide. Though denominated irons they are of course always faced with best steel.

Planer.—The workman who gives his attendance to a planing machine (q.v.) for metal. He sets and fixes the work, regulates the feed, often grinds his own tools, and keeps the machine in order. His wages are those of a machinist (q.v.).

Planer Bar.—A rigid bar of iron or steel carried horizontally in, and standing out from the front of the tool box of a planing machine. Its use is in planing the interior of hollow works, the cutting tool being clamped in the end of the bar farthest from the tool box, the bar entering into narrow spaces where the tool box could not find admittance.

Planer Centres.—See Loose Centres.

Plane Surface.—A surface whose points are all on the same level. To obtain such surfaces the chisel and file, the plane, and planing and shaping machines, and scrapers are used, and to test their accuracy, winding strips, straight edges, and surface plates are employed. It is possible to obtain such accuracy that when two heavy metallic plates are placed in contact, the upper will sustain the lower, even in vacuo.

Planing.—The removal of material from plane surfaces by means of cutting tools, effected in the case of wood by hand tools, and in the case of wood in large quantities and of metal, in planing machines (q.v.).

Planing Machine.—(1) For wood. A machine in which the cutting tools are either arranged to project from the face of a disc, or around a rectangular adze block. In the former case they are augur-like in form and rotate in a plane parallel with the face of the wood ; in the latter they rotate in a plane at right angles with that face. The forms and dimensions of planing machines vary much, and there are some machines in which the cutters do not revolve at all but are set in an iron block, and the wood is literally planed as if by hand, with this difference, that while in hand planing the tool is traversed over the wood, in the type of machine here referred to the timber is passed over a fixed iron. Numerous modifications of planing machines are made. Panel-planing and thicknessing machines are used for surfacing and planing small pieces of stuff to uniform thickness, and are used in pattern making. (2) The machine for iron planing consists of bed, travelling table, standards, cross slide, tool box, and gearing ; by which metal to be operated on is carried underneath a fixed cutter, the cutter moving only after the termination of a cut by a distance equal to the amount of feed (q.v.), usually automatic, which is imparted to it. The wheel-spoke planing machine is constructed for the purpose of planing the edges of the spokes of railway wheels.

Planishing.—The smoothing and polishing of metallic surfaces by hammering or by rolling, instead of by abrasion or cutting. Shafting is very commonly planished instead of being turned.

Planishing Hammer.—A machine-driven hammer used for planishing plates of sheet metal, and capable of delivering as many as three hundred blows a minute.

Plank.—A piece of timber more than nine inches in width. Planking is used as a general term to signify timber over about two inches in thickness, and of any width over nine inches, to distinguish it from board (q.v.).

Plank Way.—When work is cut from board or plank in such a way that the cut edges stand approximately at right angles with the face of the timber it is said to be cut plank way of the grain, as opposed to end grain.

Planometer.—The older name for a surface plate (q.v.).

Plan Section, or Sectional Plan.—A section (q.v.) taken in a horizontal plane in the drawing of an object.

Plant.—Plant comprises the machines, tools, forges, furnaces, benches, cranes, &c. of a factory. Railway plant is the rolling-stock and permanent way.

Plaster of Paris.—Plaster of Paris is used for making the reverse impressions of some founders' moulds, for making the whole or portions of some intricate and ornamental patterns and core-boxes, and as bomontague (q.v.) for pattern work.

Plate.—A broad, thin sheet of metal. Plates being rolled, the fibre is developed in the longitudinal direction, in which direction therefore they are best able to resist tensile stress. There are limiting sizes and weights for plates in iron and steel. The thicknesses of plates in these materials are given in inches and fractions of an inch ; that of brass plates and of iron plates or sheets (q.v.) for tinning is given in B.W.G. Rolled plates of malleable iron, above No. 4 B.W.G., or ·238 of an inch in thickness, are called plates to distinguish them from sheets (q.v.). See also Fish Plate, Horn Plates, Plate Moulding, &c.

Plate-bending Machine.—A machine used for bending boiler and other plates to various curves. It consists of three rolls having their bearings in housings, the top roll being capable of adjustment for distance from the others, by which adjustment provision is made for the different curves to be imparted to the plates ; diminishing the centres increasing the curvature of the plates, increasing the centres diminishing the curvature.

Plate Box.—A foundry flask used for plate moulding.

Plate Dowels.—Patternmakers' metallic dowels, in which the male and female portions are carried on thin rectangular plates which are sunk flush into the jointed surfaces of the patterns, and there screwed fast. They are the best dowels because they are less liable to be shifted by rough usage than the peg and cup, or the wooden dowels. When small they are sometimes made in brass, but the majority are made in malleable cast-iron.

Plate-edge Planing Machine.—A special metal-planing machine used for truing the edges of the wrought-iron and steel plates employed in boiler and girder work. It differs from ordinary machines in this, that the tool travels while the table remains fixed, and that the box which traverses the tool over the work moves along the side of the machine. The tool box is traversed by a double-threaded screw of coarse pitch, and quick return is given by pulleys of unequal size driven by open and crossed belts.

Plate-flattening Machine.—A form of straightening machine (q.v.) for plates, which contains a larger number of rolls than the ordinary form, seven rolls being employed, four above and three beneath, by passing between which, the plates are rendered perfectly flat and true.

Plate Furnace.—A reverberatory furnace, used in boilermakers' and platers' sheds for the purpose of heating plates preparatory to flanging, bending, and welding.

Plate Gauge.—A thin, flat metal gauge, used for measuring spaces for which the ordinary cylindrical gauges would be too weak, owing to the smallness of the dimensions measured. The thicknesses of plate gauges range from about ½th to $\frac{1}{70}$th of an inch.

Plate Girders.—Girders of wrought iron built up of a central plate or web, riveted to top and bottom flanges by means of angle-irons riveted alike to the plate and to the flanges.

Plate Link Chain.—A chain formed of flat links united with pins passing through holes near the ends of the links. Suspension chains, and chains for heaving up slips, are made of flat links. The lengths of the links vary with the purpose for which they are to be employed.

Plate Metal.—See Refined Iron.

Plate Mill, or Plate Rolls.—A rolling mill similar in design to mill rolls (q.v.), the place of the grooved cylinders in the latter being taken by plain cylinders. They consist of roughing, and finishing rolls, the former being grain rolls (q.v.), the latter chilled rolls.

Plate Moulding.—A process by which the labour of ramming up patterns in sand is lessened. The opposite halves of the pattern are laid upon metal plates in such a position that when two moulding boxes are rammed, each on the separate plates and then brought together, the opposite portions of the moulds correspond exactly. Frequently the plates are manipulated by a moulding machine (q.v.).

Platen.—(1) The table of a planing machine for metal which traverses the work under the tool. So named after the platen of a printing machine. (2) The table of a moulding machine (q.v.).

Plater.—A boilermaker who devotes himself exclusively to that branch of work which consists of marking out, cutting, and punching the iron or steel plates which are used in the construction of boilers, girders, bridges, and similar structures.

Plate Rolls.—See Plate Mill.

Plate-shearing Machine.—A shearing machine furnished with specially long knives or shears for the purpose of cutting off the ragged edges of plates after they leave the rolls. The knives may average four or five feet in length.

Platform.—The planking attached to the sides of a crab (q.v.) or a gantry (q.v.), upon which the man stands to work the handles of the levers. The platform of a steam crab is usually termed a foot plate and is made in metal.

Platform Crane.—A whip crane (q.v.) which is independent of any support at the top of the post, hence termed independent whip crane. It is held firmly on a very broad cast-iron base. The rope and chain barrels are carried on a post of cast, or of wrought iron, stepped into the base.

Platform Scale.—A small weighbridge (q.v.).

Plating.—The special work of the plater (q.v.).

Play, or Slackness.—Freedom of movement of bearing or working parts, but confined within certain definite limits. The purpose of giving this freedom is to prevent jamming of parts, by heating or by oscillation or other causes. See End Play.

Pliers, or Plyers.—A pincer-like tool, having broad and flat roughened jaws, and employed by fitters, smiths, and metal workers generally for the grasping of wire or slight iron rods. Wire pliers are used for bending wire into loops and other shapes, nipping pliers are used for cutting, smiths' pliers are employed for grasping work indiscriminately.

Plomber Block.—A plummer block(q.v.)

Plomer Block.—A plummer block (q.v.).

Plommer Block.—A plummer block (q.v.).

Plotting.—The laying down of the lines of diagrams which are made with a view to the calculations of strains, &c., by the graphic method alone.

Plough.—A woodworker's plane used for grooving out the recesses for tongueing and for panels. It is guided to its work by a fence, and the depth of the cut is also controlled by a fence, adjustable with a screw.

Plug.—(1) The plug of a cock is the inner movable portion, which on being turned allows free passage to the liquid. (2) An arbor or chuck used as an intermediary attachment for small lathe drill chucks. It fits into a tapered hole in the back of the actual chuck at the one end and into the tapered mandrel nose at the other. (3) A sand plug or clay plug is a term applied to the ball of sand or clay with which the riser (q.v.) of a mould is covered while the metal is being poured at the ingate. Its purpose is to check the too rapid rushing of the air out of the mould before the advancing metal, which would be liable to result in a washing away of some portions of the sand. The plug is either floated up by the rise of the metal, or more properly the moulder removes it when he sees that the mould is nearly full.

Plug Centre Bit.—A centre bit having a cylindrical centre instead of a point. Used for the enlargement of small holes already bored in wood.

Plug Cock.—A cock in which the fluid passes through a central plug which by being made to revolve in its circular plane presents either its blank surface, or a passage through its body, to the flow of liquid. So named in contradistinction to valve cocks. The taper of plugs should not be less than one in six, or more than one in four.

Plug Gauge.—See Cylindrical Gauge.

Plug Rod.—A plug tree (q.v.).

Plug Tap.—A parallel screw tap used for taking the finishing cuts on internal screws.

Plug Tree.—A long rod suspended from the beams of old-fashioned single-acting pumping engines, and provided with tappets for moving the handles of the equilibrium and steam exhaust valves.

Plumb.—In a vertical position. The position assumed by a weighted cord at rest.

Plumbago, or Graphite.—A nearly pure form of carbon used for foundry blacking, and mixed with clay for crucibles. It is smooth and soapy to the touch, and is highly refractory.

Plumbago Blacking.—See Plumbago. Blacking.

Plumbago Crucibles.—Crucibles in which plumbago is the chief ingredient present. They contain only so much clay as is necessary to afford the requisite amount of plasticity, and are employed in preference to those made entirely of clay, where great heat or excessive alternations of temperature exist.

Plumb Bob.—A pear-shaped or globular weight suspended from the end of a plumb-line (q.v.). For engineer's work its lower end terminates in a point for the purpose of indicating the exact centre of the plumb-line.

Plumber Block.—A plummer block (q.v.).

Plumb Line.—A line or string sustaining a plumb bob, and used either alone or in conjunction with a straight-edge to ascertain the exact vertical position of portions of structures. It is employed chiefly in the erecting department.

Plummer Block.—Variously spelt plomber, plomer, plommer, plumber. The ordinary form of pedestal for carrying the journal of a shaft. It consists of a base or body, sometimes termed a pillow, and a cap enclosing a pair of brasses. The foot of the block is prolonged to receive hold-down bolts, and the bolting down of the cap secures the brasses in place. Lubrication is provided for at the top, through a hole drilled in the cap and top brass.

Plunger.—The piston or ram of a force pump. Being solid it draws the water by producing a vacuum, and delivers it by its forcing or displacing action.

Plunger Air Pump.—An air pump fitted with a solid plunger or piston.

Plunger Bucket.—A force-pump piston having no valves.

Plunger Pump.—See Force Pump.

Ply.—To bend. A fold, a twist, a single thickness of a material, as a ply of wire gauze, three ply, four ply cotton belting. meaning three and four thicknesses respectively.

Plyers.—See Pliers.

Pneumatic Lift.—See Compressed Air Lift.

Pneumatic Moulding Machine.—A moulding machine in which the pressure on the sand is imparted by means of an air bag (q.v.), instead of by a rigid platen.

Pneumatic Process.—The Bessemer process (q.v.).

Pneumatics.—Pneumatics treat of the properties of elastic fluids, as air, steam, gas.

Pocket Print.—A long or drop print used for coring holes in the sides of castings where a round print would not be available, by reason of the superimposed sand preventing it from lifting. The pocket print being therefore prolonged to the surface of the mould, the upper part of the print impression is filled up or "stopped over" with sand after the core has been dropped in.

Pockets.—(1) Curved plates inserted into, and standing out from the sides of the flue of a Galloway boiler in positions intermediate with those occupied by the Galloway tubes. Their use is to throw the flame among the tubes, and so increase the economy and efficiency of the boiler. (2) Recessed portions of a casting provided for the reception of timber or girder ends.

Podger.—See Tommy.

Point, or Switch.—A movable rail by which the direction of an engine or wagon is changed from one set of rails to another.

Point Rail.—A movable rail or switch. Also called a point simply.

Point.—This word has several significations, which are treated under their several headings, as Boiling Point, Fusing Point, &c.

Point Tool.—A diamond point (q.v.).

Poker Filing.—See Draw Filing.

Pole Lathe.—The primitive form of lathe. So called because each downward movement of the treadle was made to impart a similar movement to the end of an elastic pole fixed horizontally overhead. A cord slung from the pole was turned round the work and rotated it towards the tool. On the release of the treadle the work was rotated in the opposite direction. Hence half the time was lost in the return movement, and the tool had to be drawn away after each cut. The pole lathe is still in use in some Eastern countries.

Poling.—The stirring of molten copper with a pole of green wood, usually birch, in order to bring it to tough pitch (q.v.). The effect of poling is to remove the oxygen which is present as a suboxide of copper, and which renders the metal brittle or "dry." The oxygen combines with the gases liberated from the burning wood. If the poling is not carried on to a sufficient extent the metal is brittle, and is said to be under-poled, if it is carried too far the effect is similar, but it is then said to be overpoled.

Polishing Headstock.—A polishing lathe (q.v.).

Polishing Lathe, or Polishing Head.—A mandrel and pulley driven at a high speed and carrying small emery or buff wheels for polishing light work. It is mounted on a small head or standard, which is bolted to a bench or other suitable support.

Polishing Stick.—A couple of strips of wood used with emery for polishing work while revolving in the lathe. The ends farthest from the workman are united with hinges, those nearest him serve as handles, by which the strips are made to embrace the work tightly. The opposed faces of the sticks are hollowed out where they embrace the work, emery cloth or powder intervening. This is therefore an external lap. See Lap.

Polishing Wheel.—A buff wheel or buff (q.v.) used for finishing bright work. See Polishing Lathe.

Poll.—The larger or broad end of a hammer as distinguished from the pane (q.v.).

Polygon.—A plane figure bounded by many right lines. A regular polygon has its sides and angles equal. In an irregular polygon the reverse condition obtains. The number of sides in a polygon is indicated by the prefixes penta, hexa, &c.

Polygon of Forces.—An expansion of the triangle of forces (q.v.), and which may be thus stated:—If any number of forces be represented in magnitude and direction by the sides of a polygon taken in order they will be in equilibrium.

Polygons, Line of.—One of the sectorial scales by means of which any regular polygon can be described according to the number, four, five, six, seven, eight &c., on the line marked POL. To set out a polygon, open the legs of the sector until the divisions marked six correspond in distance apart with the radius of the circumscribing circle. Then the distance apart of four will represent the sides of a square, five of a pentagon, six of a hexagon, and so on. Conversely, having the length of the sides of a polygon given, open the legs until the distance apart of the figures representing the side of the particular polygon required corresponds with the length of the side. Then the distance between the figures six will be the radius of the circle to contain the required number of sides.

Popit.—A Poppet (q.v.).

Poplar. (*Populus.*)—A tree of the natural order *Salicaceæ*. The wood is soft, white, and light. Poplar wood is used chiefly for brake blocks (q.v.). Sp. gr. ·39. A cubic foot weighs twenty-four pounds.

Poppet, Poppit, or Popit.—The movable headstock of a lathe, upon the point of whose mandrel, work placed between centres is made to revolve.

Poppet Cylinder.—The mandrel of the poppet of a lathe. It is made cylindrical in order to allow the traversing screw to move freely through its interior.

Poppet Head.—A poppet (q.v.)

Poppit.—A poppet (q.v.).

Ports.—The passages or steamways through which the steam gains admission to the interior of an engine cylinder, and through which it also returns to the exhaust passage.

Portability.—This is carried into extensive practice in engine work and machinery. See Portable Engine, Portable Forge, &c.

Portable Crane.—A crane constructed to travel about upon tram rails. It may be worked either by hand or by steam.

Portable Engine.—Portable engines are used chiefly for agricultural purposes, but they are useful for outdoor and contractors' work, and for work of a temporary character. They are built on the locomotive type with multitubular boilers for high pressures.

Portable Forge.—A small forge supported on a light framing of wrought iron between which the bellows are situated. It is removable at pleasure, and is therefore used for outdoor and repair work.

Portable Vice.—An engineer's vice supported on a central pillar, base, and wheels.

Portable Winch.—A winch (q.v.) placed upon wheels for convenience of removal about contractors' yards, on ship board, or in other places where the work required to be done cannot be localised.

Porter.—The bar of iron which the smith holds in his hand when manipulating a forging. In fagoted work the porter is a single bar, longer than the rest and standing out from the mass. In ordinary single bars the porter is the end which the smith happens to hold in his hand.

Port Holes.—Ports (q.v.).

Portland Cement.—Cement (q.v.) made by mixing clay and chalk and burning the mixture in a kiln.

Porty.—A print (q.v.) of large diameter is often termed by pipemakers a porty.

Positive.—A positive motion or a positive stroke signifies that which is precise and exact, in opposition to that which is not so precise or exact, but only approximately so. Thus the stroke of an ordinary planing machine being accomplished by the aid of shifting bands is not precise ; but that of shaping machines, being effected by a rigid though adjustable connecting rod, is positive. Hence a positive stroke is one that is direct, a non-positive is one in which a certain amount of elasticity, or slip, or lost motion comes into play.

Positive Stroke —See Positive.

Positive Stresses.—In English practice are those which represent compression.

Positive Terms.—Algebraical or arithmetical terms preceded by the + sign ; terms preceded by no sign are also understood to be positive.

Post.—An upright timber, as a king post, a pillar, or column, a crane post, &c.

Pot.—A crucible.

Potash, Yellow Prussiate of.—The ferrocyanide of potash, symbol $K_4 FeC_6 N_6$, largely employed by smiths for the purpose of case-hardening (q.v.), and in the process of making draughtsman's phototypes (q.v.).

Potatoes.—Used to prevent incrustation of boilers.

Potential Energy.—See Energy.

Pot Lid Valve.—A hollow cup-shaped lift valve.

Pot Metal.—Used for some cheap brass goods, variously composed of copper sixteen, lead six to eight parts, the latter amount giving an inferior quality, and being called wet pot metal, because the lead partially oozes out in cooling. Tin, zinc, and antimony in small quantities are sometimes added with the view of improving the alloy.

Pot Sleeper.—A combined railway sleeper and chair, the chair being usually cast on the top of a broad dome-shaped base, which is the actual sleeper. The gauge is maintained by cross-tie rods, and the rails attached in the usual way with keys. They are used in hot countries chiefly, where timber sleepers would be liable to decay, and to suffer from the destructive agencies of insects.

Pot Valve.—A safety valve, shaped like an inverted pot. It is a lift valve, and the conical pivot of the lever drops loosely into a recess in the crown of the pot. The advantage claimed is that such valves being in a condition of unstable equilibrium are less liable to stick than the ordinary form. The lift of the valve is controlled and maintained by guides cast on the top of the seating.

Pound.—The English pound is a tenth part of a gallon of distilled water at 62° F., and 30 in. barometric pressure.

Pound Degree.—A thermal unit which denotes the quantity of heat necessary to raise a pound of water through one degree of temperature. It is a pound degree Centigrade, or a pound degree Fahrenheit, according to the thermometric scale used. Unless stated to the contrary the latter is always understood. The latter is equivalent to the common unit of heat, being 772 ft. pounds; the former is a bastard unit only derived from the other.

Pouring.—The emptying of the molten metal into a foundry mould.

Pouring Gate.—An ingate (q.v.).

Power.—(1) A term of general application signifying the actuating of motors by agencies other than those supplied by animal power. Steam, water, air, gas, &c., are power agencies. (2) The expression of the number of times by which a number is multiplied into itself.

Power Feed.—The feed (q.v.) of a lathe, planing, screwing, or other machine, which is effected automatically by the agency of the motive power which drives the machine itself.

Power Gear.—Gear, usually understood of the toothed form, which is worked by power (q.v.).

Power Lift.—A lift (q.v.) worked from any convenient source of power, as a gas or steam engine.

Power Machine.—A machine actuated by a motor (q.v.), as distinguished from a hand-worked machine.

Power of Animals.—The power of animals, as deduced by experiment, is of use in calculations connected with those machines which are worked by animals, as horse and mule gear, pumps, &c.

Power of Men.—The power of men varies greatly with the manner in which it is exerted. For practical calculations certain values are accepted.

Preadmission.—The admission of steam to an engine cylinder just previous to the termination of the stroke in the opposite direction. The amount of preadmission is governed by the lead (q.v.) it should be sufficient in quantity to allow the steam to acquire its full pressure immediately on the return stroke of the piston.

Prerelease.—The opening of a steam cylinder to exhaust just before the termination of the piston stroke, to prevent injurious and wasteful back pressure (q.v.).

Press Drill.—An old-fashioned form of drilling machine, in which the drill is pressed down to the work by a weighted lever.

Presser.—That portion of a moulding machine which imparts the necessary pressure to the moulding sand.

Press Fit.—A fitting of contiguous parts slightly tighter than a sliding fit (q v.), to allow of the sliding parts being pressed together with a hydraulic press.

Pressing.—See Stamping.

Pressure.—A compressing force which does not produce motion but imparts stress. Steam pressures are measured in lbs. per square inch, or inches of mercury. Water pressures are measured in lbs. per square inch or in the equivalent " head " (q.v.), and sometimes in atmospheres.

Pressure Blower.—This is used to denote a blower (q.v.) of rotary type, as opposed to a fan, the former producing a definite and constant amount of positive pressure, while the latter acts by displacement only.

Pressure Forging.—The practice of stamping forged works by gentle hydraulic pressure, instead of by the impact of a steam hammer. See Hydraulic Forging Press.

Pressure Gauge.—The dial gauge attached to a boiler or other vessel by which the pressure of the steam or liquid within is indicated. A curved spring within the body of the gauge actuates the pointer. Steam gauges register up to about 200 lbs. per inch, but hydraulic gauges are made for very much higher pressures, ranging to several tons per square inch.

Pressure, Lines of.—The vertical or approximately vertical lines enclosed by an indicator diagram, or a diagram of work.

Pressure Ram.—The larger of the two in a hydraulic forging press, and which imparts the necessary downward force.

Pricker.—A vent wire (q.v.).

Prime.—A number which is only divisible by itself, or by unity. See Hunting Cog and prime numbers in Appendix.

Prime Mover.—Any piece of mechanism which absorbs and gives out again the material forces of nature, whether the expansive energy of steam or gas, or the pressure of water, air, or wind, &c.

Priming.—(1) In steam boilers, is the carrying of mechanically suspended particles of water along with the steam into the steam chambers and pipes. It is due to various causes, as irregular withdrawal of the steam, the presence of oil and grease in the water, want of proper circulation, &c. Sometimes called foaming. (2) The priming of a force pump is the expulsion of the air from the water space, in order that the water shall enter into the partial vacuum thus produced. Usually it is effected by the opening of a pet cock placed in the highest portion into which air can enter; which is opened to permit of escape of air, but closed by the finger to prevent its return. This is repeated until the water fills up and begins to be ejected from the cock. (3) The fetching of a lift pump by pouring liquid into the bucket in order to produce sufficient vacuum to enable it to draw.

Priming Valve.—The valve which affords the means of escape for the water of condensation in engine cylinders. A pet cock (q.v.).

Principals.—See Rafters.

Print.—A projection put upon a foundry pattern to indicate the position of a cored hole, and to form an impression to receive the end of the core. Prints are commonly well tapered to withdraw easily, and those portions of the cores which are placed in the print impressions are filed to correspond. Prints are of various shapes, depending on the shape of their cores. At the sides of patterns, pocket prints (q.v.) and parallel prints (q.v.) are used.

Printing Frame, or Copying Frame.—A frame in which phototypes are copied. It consists of a stout oblong frame of wood, about four inches deep and from two to four feet in length, according to the sizes of paper used. A thick sheet of plate glass is laid on fillets in the bottom of the frame, through which the light penetrates to the sensitised paper. A back board or boards cover the paper and a thickness of felt, and these are kept in place by means of springs attached to cross ribs hinged to one side of the frame and folded over and clamped down on the other. The whole frame is mounted on rollers to run on rails carrying it through a window outside into the light.

Prising.—Barring round or turning round a wheel or wheels with a crowbar.

Prism.—A solid whose ends are plane figures equal and parallel, and whose sides are plane parallelograms.

Problem.—A construction which is capable of being effected by the employment of principles of construction already admitted or proved.

Prod.—The pyramidal or conical points cast on loam and core plates for the retention of the loam are termed prods.

Producer.—See Gas Producer.

Producer Gas, or Gaseous Fuel.—A mixture of combustible gases consisting chiefly of CO and hydrogen, and CO_2, which is prepared by the combustion of fuel apart from the furnaces to be heated. The gases, after passing from the gas producer furnace, are heated in regenerators and mingle and burn on their furnace hearth with atmospheric air similarly heated. Producer gas is prepared from inferior fuel, but chiefly from bituminous coal, and by its employment a saving of fuel is effected, a uniform and more regulable heat obtained, a diminished loss from oxidation of the iron and steel reheated, furnace linings saved, and a great economy generally obtained over the older methods.

Product.—The sum produced by the multiplying together of two or more factors.

Products of Combustion.—Chiefly carbonic acid and carbonic oxide and water. The former are utilised in many ways; in blast furnaces, gas engines, regenerative furnaces, &c., described under their several heads.

Profiling.—The grinding and sharpening of a cutter for working wood or metal, so that the outline of the cutting edge is that of the section which it is desired to impart to the material.

Progression.—The succession of one number after another by virtue of some definite law. Arithmetical progression is a series of numbers that increase or diminish by a common difference, as seven, ten, thirteen, sixteen, &c. In geometrical progression each term of the series is equal to that which precedes it multiplied by some factor which is constant for all the terms. Thus two, six, eighteen, fifty-four are in geometrical series.

Projection.—Signifies in a general sense the representation on a plane surface of objects as they appear to the eye of an observer, the various kinds of projection being differently named according to the position occupied by the observer; perspective, stenographic, isometric, gnomonic, and orthographic are projections, but in the working drawings of the engineer the latter alone is used. In this the eye is supposed to be at an infinite distance, and the visual rays are all parallel. Hence such drawings are suitable for direct measurement, and according as the objects are viewed the terms plan, elevation, section are employed.

Prong Chuck.—A fork-like chuck, whose prongs revolve wood set between lathe centres.

Prong Key.—A kind of spanner used for tightening up circular nuts by power applied to their faces. Two projections or prongs on the front face of the spanner fit two corresponding holes in the nut faces, the leverage being applied to a handle attached to the spanner.

Proof Bar.—The loose bar which is thrust through a hole in the trough which contains steel undergoing the process of cementation, and which is removed from time to time to enable the attendant to judge of the progress of the operation.

Proof Load, or Test Load.—A load imposed on a structure greater in amount than the working load, in order to test its capability or margin of safety. The deflection of a structure, when under its test load, is carefully noted and its capability deduced therefrom. See Proof Strain.

Proof Strain.—Any strain to which a portion of material, or a structure, or a section of structure, is subjected, in order to test its suitability for the specific purpose for which it is designed to be used. A proof strain would in all cases be short of that which would have a crippling effect, assuming of course that the material were good, but also above that which it would be expected to sustain in the ordinary conditions to which it would be subject.

Propeller, or Propeller Screw.—The helical segments used for the propulsion of steam vessels. Propellers are two, three, or four bladed, according as they form segments of two, three, or four different screws.

Propeller Blade.—A single segment of a propeller (q.v.).

Propeller Screw.—See Propeller.

Proportion.—See Ratio.

Proportional Compasses.—A compass (q.v.) for drawing purposes which is provided with two slotted and double-ended legs united by a sliding pivot and screw, by the regulation of whose position the distance apart of the extreme points may be adjusted within a very wide range. By its employment distances and drawings may be enlarged or reduced proportionally, polygons of circles and square and cube roots of numbers can be obtained rapidly without calculation.

Proportioning.—The proper proportioning of machine parts is the business of the engineer. But although a correct knowledge of the strength of materials and the nature and amount of strain and stress set up in structures is essential, there are few or no machines designed on strictly mathematical data. The art of proportioning consists in combining theory with experience, and these with an intuitive sense of fitness and beauty, in one harmonious result. So largely does experience enter into the labour of the engineer that there are few who are able, in spite of excellent theoretical knowledge, to enter successfully into the task of designing and proportioning works of a totally different class from those to which they have been accustomed.

Protracting.—The laying down of angles with or without the assistance of a protractor.

Protractor.—A drawing instrument used for laying down or for measuring angles. It is made in several forms, rectangular, semicircular, and circular, having the angles of, and divisions into degrees marked. These being of small size are most useful in the drawing offices, or for the smaller work in the shops. For large work a large wooden or cardboard instrument, or a line of chords (see Chords, Line of) is to be used in preference, as being less liable to produce errors in striking out.

Proud.—A word sometimes used by iron turners to designate a cutting

tool having a large amount of top rake (q.v.). Thus a tool for the turning of wrought iron would be too proud to turn steel or gun-metal. (2) Denotes that one part projects above another; thus, to stand proud.

Proving Machine.—A testing machine (q.v.).

Prussian Blue.—The colour used to distinguish wrought iron on sectional drawings.

Prussiate Bath.—See Phototype, Yellow Bath.

Puddle Ball.—A mass of puddled iron as it leaves the puddling furnace.

Puddled Bar, or Number One Iron.—The bar of malleable iron as it leaves the last of the grooves in the finishing series of puddling rolls (q.v.). It is not of merchantable value (see Merchant Iron), but is piled, reheated, rewelded, and rolled to make marketable qualities of iron.

Puddled Steel.—A steely iron (q.v.) produced in the puddling furnace by so regulating the process that a portion of the carbon is allowed to remain in combination.

Puddlers' Candles.—The term given to jets of carbonic oxide gas which issue from the surface of the molten metal in the pig boiling (q.v.) process. They owe their origin to the union of the oxygen in the tap cinder with the carbon in the pig.

Puddlers' Mine.—A mixture of a variety of red hæmatite and water, employed for the purpose of fettling puddling furnaces. It is smoothed over the surface of the bull-dog (q.v.) lining.

Puddling.—The series of processes by which the carbon and other foreign substances present in pig iron are extracted therefrom previous to the rolling of bars and plates for the use of the smith or plater. See Dry Puddling, Wet Puddling.

Puddling Furnace.—A reverberatory furnace (q.v.) in which cast iron is subjected to the action of the atmosphere to effect the removal of the carbon and bring it into the condition of wrought iron Some puddling furnaces are made double, two sets of men being employed, one set on each side at opposite working doors. An increase in output and economy of fuel are thus obtained.

Puddling Rolls, or Forge Train.—The first sets of rolls through which a shingled bloom is passed. The first or roughing rolls are grooved in Gothic or V-shaped channels and in diamond shape, the grooves diminishing in depth from left to right along the rolls. Those to the left are distinguished from the others as roughing rolls because there the bloom is first embraced. The surfaces of the grooves are roughened with chisel nickings to take firm hold of the blooms. To the right hand are the finishing rolls in which the grooves are rectangular in section to impart that section to the puddled bar. These grooves also diminish in depth from left to right.

Pug Mill.—A mill used for the mixing of the materials of which concrete is composed.

Pulley, or Rigger.—The wheel which carries driving belts. It is made either in cast or wrought iron and turned rounding upon the face.

Pulley Block.—A sheave pulley or series of pulleys having a hollow rim section, and enclosed between metal side cheeks in which the pulley pins have their bearings, constitute a pulley block. A hook fastened to the upper end furnishes the means of attachment to any convenient point of support. Pulley blocks are single, double, or treble according as they have one, two, or three sheaves.

Pulley Lathe.—A special kind of machine used for boring and facing pulleys. It consists of an inner ring furnished with bolts for setting the pulleys of different diameters central, and having its periphery turned truly to revolve in a bored outer ring made fast to foundations. A circle of teeth forms an integral portion of the periphery of the inner ring, which teeth are driven by a pinion, on whose shaft are the driving pulleys by which the ring is driven round. The pulleys are bored and faced, but are transferred to an ordinary lathe to be turned.

Pulling Up.—The tearing up of the sand of a foundry mould by the withdrawal of a badly made or non-tapered pattern. Pulling up from the bottom means that the whole of the sand reaching upwards from the lowermost edge of the pattern is pulled up, to distinguish it from mere breaking of the sand along the edges.

Pump.—A machine for lifting or forcing liquid either by means of a bucket, or of a piston working in a closed cylinder. See Force Pump, Suction Pump, Hydraulic Pump.

Pump Barrel.—The closed cylinder in which the bucket or the piston of a pump moves. Barrels are variously made of cast iron lined with gun-metal, or gun-metal entirely, or glass, according to the liquids which have to pass through them, or the price paid.

Pump Bit.—A bit for boring the cylinders of wooden pumps.

Pump Bob.—A bell crank, or rocking lever which converts rotary into reciprocal motion.

Pump Bucket.—The piston of a lift pump. It differs from an ordinary piston in being provided with an open central space or waterway (q.v.), which is covered with a flap valve (q.v.), which valve opens when the bucket is falling and closes when it is lifting.

Pump Case, or Pump Top.—The top of a lift pump, or that part above the working barrel (q.v.), which contains the handle, spout, and cover.

Pump Cup Leather.—See Cup Leather.

Pump Gear.—The various parts and connections of a pump.

Pump Head.—A sheet iron hood placed at the top of a chain pump to prevent any of the discharge water being thrown off by centrifugal force.

Pumping.—The term pumping is sometimes applied in the foundry to the act of feeding (q.v.), in allusion to the up and down movement of the feeding rod in a vertical direction.

Pumping Crank.—A crank or disc to which a pump rod is attached.

Pumping Engine.—An engine employed for pumping purposes. For light work, ordinary engines of horizontal or vertical types are employed, for heavy pumping, beam engines are commonly used.

Pump Leather.—See Cup Leather.

Pump Rods.—The piston rods of pumps are made of various materials, but chiefly copper and gun-metal, or Muntz metal, to prevent corrosion. Air-pump rods are commonly made of iron for strength, and sheathed with Muntz metal. The rods of deep well pumps are made of iron, and cottared together in length, or are in some cases made of wood. See Pump Spear.

Pump Spear.—A name given to the long wooden rods of deep well and mining pumps.

Pump Top.—See Pump Case.

Punch.—A shearing instrument made of steel and employed for the removal of a definite portion of metal, whose shape is the counterpart of the shape of the punch. Punches are used in many classes of work, and are actuated both by hand and by power. See also Brad. Punch and Centre Punch.

Punched Holes.—Rivet holes are commonly punched in plates for boiler, bridge, and girder, and plated work generally. Only in the best work are they drilled. The holes are either marked direct with compasses, or from a templet, and punched singly, the plates being moved by hand, or in some cases automatically. The proper spacing-out of punched holes is of the utmost importance, as preventing the injurious employment of the drift (q.v.).

Punched Plates.—Plates of wrought iron and mild steel are suitable for punching; hard steel is unsuitable. Plates which will not stand punching are too brittle for boiler and girder work. Punching involves, under the most favourable circumstances, a loss of strength, due not only to the diminution of sectional area, but to the detrusive action straining the plates beyond their limit of elasticity. Hence punched plates are sometimes annealed, or more frequently rymered out, it being considered that the stresses due to punching are only present in the immediate vicinity of the holes. The tensile strength of soft wrought-iron plates is diminished from 5 to 10 per cent. by punching, while that of steel plates is diminished from 20 to 28 per cent.

Punching.—The making of holes through plates under a punching machine. It is done in the boiler department usually by a plater (q.v.).

Punching Bear.—A portable punching machine. The punch is actuated by a screw, or in some cases by hydraulic pressure.

Punching Machine.—A machine in which the punch is actuated by power. Commonly the processes of punching and shearing are combined in one double-ended machine.

Punching Strength.—The strength necessary to punch plates of iron and steel. It has been found by experience that the resistance of a wrought-iron plate to punching is nearly the same as its resistance to tensile strain. The resistance is measured by the area of the metal separated, and the resistance increases directly as the thickness and strength of plate and diameter of hole. Holes are punched in plate thinner than the diameter of the punch, but seldom in plates whose thickness exceeds that diameter.

Punch Pliers.—Pliers in which the cutting edge takes the form of an annular ring. They are used for punching the holes in leather belting.

Pupil.—Youths of from fourteen to sixteen years of age are articled as pupils to firms of engineers for periods ranging from two to five years, during which time they pass through the various departments in turn, and obtain some practical experience in each, at the same time that they pursue theoretical studies at a technical school. The premium payment varies with different firms.

Puppet.—A poppet (q.v.).

Puppet Valve.—A lift valve (q.v.).

Purchase.—A term used in workshops, and equivalent in its meaning to leverage, more or less purchase meaning more or less leverage.

Pure Oil.—A simple oil which is not admixed with any other oil, as is the case with compound oils (q.v.).

Pure Rubber.—Applied to vulcanised india-rubber which is homogeneous throughout, to distinguish it from insertion rubber (q.v.).

Purlins.—The members which unite the trusses of roofs in longitudinal directions.

Purple.—The colour used to denote steel in engineers' drawings.

Pushing Poppet.—A poppet whose mandrel is moved in the forward direction only by the thrust of the screw, but is slid back by hand.

Putty.—Used in pattern-making for filling up the holes formed by the heads of nails and the countersunk recesses for screw-heads. Called Stopping, or Bomontague.

Pyramid Oil Can.—The small conical-shaped oil can for bench use.

Pyrometer.—See Pyrometer Gauge.

Pyrometer Gauge.—A dial gauge used for recording the heat of steam, as well as its pressure; it is therefore a combination gauge (q.v.). Another type of dial pyrometer gauge, depending for its value on the expansion of metal rods or tubes, is used for testing the heat of blast and other furnaces and ovens.

Q

Quadrant.—(1) The segment of worm wheel teeth on the upper portion of the tool box of a shaping machine, which is made to curve through an arc of a circle for hollow circular shaping. (2) A quarter circle bounded by two radii at right angles to each other and by a corresponding portion of the circumference.

Quadrant Compasses.—Wing Compasses (q.v.).

Quadrant Plate, or Wheel Plate.—The plate which carries the stud wheels in the change wheel series for screw cutting in the lathe. It is hinged on the end of the leading screw, and is provided with two parallel slits for carrying the stud or studs for the intermediate wheels. Being thus hinged, the studs with their wheels can be brought into almost any desired position in relation to the wheels on the mandrel and on the guide screw.

Quadruple Expansion Engine.—A compound engine in which steam is expanded four times; first in a high-pressure cylinder, and afterwards in three low-pressure cylinders in succession. A goodly number of engines have now been constructed on this principle, but necessarily it means high initial pressures, or from 150 lbs. to 180 lbs.

Quarter Bend.—A pipe bend which makes an arc of 90°, and is used therefore for connecting pipes at right angles with each other.

Quartering.—(1) The adjustment of cranks or crank pin holes at right angles with each other. (2) Deals cut into strips of four to the deal. Also called scantling (q.v.).

Quartering Belt.—Half crossed belting (q.v.).

Quartering Machine.—A boring machine for accurately boring out the crank pin holes in locomotive wheels, after the wheels have been fixed on their axles.

Quarter Rip Saw.—A hand saw (q.v.).

Queen Post.—A roof member which fulfils a similar function to that of the king post (q.v.), its position only being different. Queen posts occupy positions placed between the king posts and the ends of the roof truss.

Quenched.—When steel is heated for hardening or tempering, and then dipped into water or oil, it is said to be quenched.

Quenching.—The act of dipping steel rapidly into water, oil, or other hardening mixture, to impart the necessary hardness or temper thereto.

Quick.—Having a curve of lesser radius relatively to another curve.

Quick Feed.—See Slow Feed.

Quick Gear.—When a crane is lifting direct instead of through intermediate gearing it is said to be in quick gearing. A back geared lathe is in quick gear when the back gear is not in. Speed pulleys are in quick gear when they are driving from the larger to the smaller instead of under reverse conditions.

Quick Gouge.—A gouge which has the greatest amount of curvature made, though proportionate to the width of the gouge. See Flat Gouge, Middle Flat Gouge.

Quick Return.—(1) The mechanism by which the cutting tool of a machine, or the work which is traversed under the tool is ran at a quicker speed during the non-cutting than during the cutting stroke. Quick return is applied to planing, shaping, and slotting machines, and many others, and is usually effected by trains of gearing, or by link motion, that is a connecting rod sliding in a pivoted link, moving across a revolving disc, or by an arrangement of levers. (2) A lathe in which the slide rest is traversed back with a rack and pinion is said to be provided with quick return. See Quick Traverse.

Quick Sweep.—A sweep whose amount of curvature is great relatively to some other curve or curves with which it is compared.

Quick Traverse.—The hand traverse given to the slide rest of a lathe by means of a rack and pinion. After the rest has travelled down the bed, through the medium of the leading screw, the clasp nuts are disengaged, and the quick traverse or quick return is put into gear for carrying the saddle back.

Quiescent Load.—A dead load (q.v.).

Quill.—The term generally applied to a hollow shaft or spindle.

Quintal.—A French measure of weight containing one hundred kilogrammes (q.v.) or one hundred thousand grammes (q.v.), equivalent to 220·4621 English pounds, or 1·9684 hundredweight.

R

Rabbet.—A rebate (q.v.).

Rabble.—A tool used by puddlers for rabbling (q.v.) the pasty metal. It is a long bar of iron turned up at the end at right angles.

Rabbling.—The working or stirring about of the iron in a puddling furnace, removing slag from its surface, and drawing the metal into masses or balls ready for the hammer.

Race.—(1) The inner or outer ring of a ball or roller bearing. (2) A circular ring upon which travel the rollers supporting a revolving superstructure. The roller race of a revolving crane furnishes an illustration. (3) The channel by which water is conducted to a water wheel.

Racing.—Engines are said to race when their rates of speed are excessive, due to variations in the resistance which they have to overcome, or to the absence of governors, or the unsuitability of the governors provided. Racing is a source of danger because of the centrifugal and inertia stresses imposed upon moving parts.

Rack.—A straight length of toothed gearing, being a successive series of teeth pitched in a straight line instead of around a curve, as in the case of wheels. The teeth of racks are formed on the same principles as those of wheels. The term is often applied to a wheel segment or circular rack.

Rack and Pinion.—An arrangement of gear for converting rotary into linear and reciprocal motion, which is largely employed in mechanism, a pinion wheel whose centre is fixed actuating a movable rack (q.v.).

Rack Compass.—A pair of shop quadrant compasses, in which the lower edge of the quadrant is cut into minute teeth, forming a circular rack.

Rack Feed.—Feed motion imparted by a rack and pinion. The feed of some drilling machines is thus imparted, and also the longitudinal traverse of self-acting lathes, both screw-cutting and non-screw-cutting. The travelling tables of some frame saws are actuated by rack feeds.

Racking.—(1) The running of the block carriage of a crane or traveller inwards or outwards to suit the requirements of the work in hand. The racking is performed by suitable gearing, and an endless chain passing over sheave wheels. (2) The running of a slide rest along with the rack and pinion motion.

Racking Gear.—The gear which actuates the block carriage (q.v.) of a crane (see Racking). The nature of the gear will depend upon the class of crane, and whether hand or steam.

Rack Rail.—A toothed rack laid between the ordinary permanent way rails on the steep gradients in mountainous districts, into which toothed gear in rack rail locomotive engines engages, in order to move upwards by tractive force.

Rad.—The radius of a circle.

Radial.—Moving in a right ine and along the shortest distance from centre to circumference.

Radial Arm.—(1) The movable cantilever which supports the drilling saddle in a radial drilling machine (q.v.). (2) The arm which is centred on and pivots round the end of the leading screw of a screw-cutting lathe, and which carries the train of gearing connecting the screw with the wheel on the mandrel of the headstock.

Radial Axle.—See Bogie.

Radial Axle Box.—See Bogie.

Radial Drilling Machine.—A heavy drilling machine, which is so constructed that the position of the drill can be accommodated to the work without moving the latter. The whole drilling apparatus—spindle, feed, &c.—is carried on a saddle, which slides on a radial arm hinged to a post, mitre wheels actuating the spindle, and communication being maintained with the spindle in any part of the arm through a telescopic shaft. Radial drills are made both attached to a strong base or table and also for fastening to a wall.

Radial Paddle Wheel.—A paddle wheel in which the float boards are fastened directly to the arms by means of hook bolts, their faces radiating from the centre of the wheel. Hence the floats are always vertical, and consequently only develop their greatest power when in their lowest position. They enter and leave the water at an angle with the level of the water, with a consequent waste of energy. Wheels having feathering floats (q.v.) are not subject to this disadvantage.

Radiation.—(1) Emission of energy in the form of electromagnetic waves. These waves include cosmic rays, gamma rays, X-rays, ultra-violet rays, light rays, heat rays, etc. (2) Process by which heat may be transferred from a hot body to a cooler one or to the atmosphere by means of infra-red rays and heat rays.

Radius.—A radius of a circle is the straight line which extends from the centre to the circumference along the shortest possible distance.

Radius Finder.—A centre square (q.v.).

Radius Rod, or Bridle Rod.—(1) A rod in the arrangement known as parallel motion (q.v.), which is fixed on a pivot at one end and jointed to the back link at the other, and which vibrates with the motion of the beam. (2) Sometimes applied to the rod which passes from the die block of a slot link to the slide, or cut-off valve.

Rafters.—The diagonal roof ribs which extend from the king post to the tie beam in a roof truss. These are often called principals, or principal rafters, to distinguish them from common rafters (q.v.).

Rail Bender.—See Rail Straightener.

Rail Clamp.—See Rail Drill.

Rail Clips.—Clips of wrought iron attached to the front and back of a balance crane, and used for fixing the crane during the lifting of heavy loads. They are made to embrace the railway metals by bolts passing through both halves. Without rail clips, a crane would be liable to overturn with a full load.

Rail Drill, or Rail Clamp.—A portable combination tool used for drilling holes in rails without removing them from their places. It consists of a short stout clamp arching over the rail and provided with a set or tightening screw on the one side, placed in opposition to a ratchet drill upon the other.

Rail Fagot, or Rail Pile.—The fagot or pile from which the blooms for iron rails are rolled. It is composed in different ways, according to the specification, quality, and price. It is usually about $8\frac{1}{2}$ in. wide and 9 in. high. The heads, if for a double-headed rail, are formed of slabs of hard hammered iron, and the middle space is filled in with puddle bars, which may be either of the entire width or made to break joint. Steel rails are now made in preference to iron, and these are rolled from a homogeneous ingot.

Rail Gauge.—An iron bar having a projection or set-off near each end at right angles with the bar, the distance of whose outer faces apart is that of the gauge of the rails, which are laid down by direct measurement therefrom.

Rail Guards.—The curved rods or bars which extend from the front of a locomotive downwards nearly to the rails, to throw off any obstructions from the line.

Rail Ingot.—The ingot from which Bessemer steel rails are rolled, and which takes the place of the rail fagot for iron rails. These ingots are cast in moulds of cast iron, tapered to permit of ready stripping of the mould from the ingot, and are about $11\frac{1}{4}$ in. square.

Rail Mill.—A mill in which rails are rolled, the rolls being grooved out into the required sections.

Rail Pile.—See Rail Fagot.

Rail Punch.—A form of punching-bear, usually worked by hydraulic pressure, used for punching the holes in the webs of rails, to take the fish bolts.

Rails.—The bars which form the tracks for the wheels of rolling stock on tramways and permanent way. They are made in various sections, both single and double headed, and in iron and in Bessemer steel. They rest either directly on the sleepers, or in chairs spiked down to the sleepers. See details under special heads.

Rail Saw.—A saw used in rail rolling mills for cutting off the crop ends of rails after they leave the rolls.

Rail Straightener, or Rail Bender.—A powerful screw press used for straightening or bending rails and bars. Some of these machines are worked by a hand lever, or a wheel, actuating the screw, some by hydraulic pressure. The rail is sustained on two points at some distance apart, and the pressure is applied between these points. See also Squeezing Maching.

Rail Tests.—The principal test used for rails is the falling weight test (q.v.), and the usual test is a monkey of a ton weight allowed to fall from a height ranging from 15 to 20 ft. upon a rail resting on supports 3 ft. 6 in. apart. Thus with a double-headed steel rail weighing 70 lbs. to the yard, two blows from a monkey of a ton weight falling from a height of 18 ft. should not produce a deflection greater than 3 in. in amount. Under the dead weight test (q.v.), which is also applied to rails, a dead weight of 28 tons should not produce a deflection of more than ⅛ of an inch on a rail similarly supported.

Railway Axle.—Axles are made of wrought iron or mild steel. The largest part of the axle is just behind the wheel boss. The diameter is gradually reduced to the centre midway between the wheels, and again at the journals, whose lengths are about twice their diameters. The diameter of the journal is usually about 3½ in. Sharp corners at the junctions of the shouldered portions are avoided by turning a small radius.

Railway Brake.—Brakes are of various kinds—air, friction, steam, &c.— noted under their proper headings.

Railway Chair.—See Chair.

Railway Gauge.—See Rail Gauge.

Railway Wheel.—Railway wheels have their bodies either of wrought iron or wood. The different parts of a wheel are the boss or the central portion, and the body, intermediate between the boss, and the tire. The boss is usually made of cast iron cast around the free ends of the bent bars which form the arms, or of wrought iron welded and pressed into form. Tires are fastened on with rivets, bolts, or retaining rings. In wooden wheels the body is solid, built up in pieces whose shape is the sector of a circle, the grain running radially, the timber being bonded together with retaining rings (q.v.) and the boss. Wood has the advantage of elasticity; it affords continuous support to the tire, and does not beat the air or raise dust like wheels with spokes. Wheels are attached to their axles without keys, being pulled on with hydraulic pressure, or by shrinking on.

Raised Work.—Work which has been hammered into outline. See Raising.

Raising.—(1) The production of curved outlines in sheet metal by the application of blows from a hammer. (2) When a tap or die is in bad order, that is, not backed off properly, or not having a keen-cutting front, the screw thread is partly compr ssed or squeezed out of shape, and is then said to be raised instead of being cut.

Raising Hammer.—A round-faced hammer used for raising (q.v.) metal work.

Rake.—(1) A term usually applied to signify the angles of metal turning tools, as side rake, front rake, &c. (2) The amount of forward angle, or pitch of saw teeth. (3) A broad flat expansion turned down at right angles with the end of a long bar, and used for thrusting about the fuel to front or back on a furnace hearth.

Raking Bar.—A long rod of wrought iron used for raking out the cupola fire after casting.

Baking Out.—The removal of the residuary coke, slag, and metal from the bottom of a cupola furnace after the running down of the charge. This is done after every blowing, and is necessary, otherwise the semi-fused mass would form a hard agglomeration when cold, the forcible removal of which would damage the furnace.

Ram.—(1) A term sometimes applied to the monkey (q.v.) of a pile driver. (2) The arm of a shaping or slotting machine which carries the tool backwards and forwards, is termed a ram. (3) A hydraulic ram (q.v.). (4) The plunger of a hydraulic lift (q.v.). (5) A hydraulic piston (q.v.). (6) The plunger of a hydraulic press (q.v.).

Ram Leather.—A cup leather (q.v.).

Rammer.—A tool used by moulders for the purpose of ramming the sand around patterns. It consists of a head of cast iron attached to a handle of wrought iron, or sometimes of wood. In a hand rammer, or a bench rammer, the working end is made rounding, in the flat rammer it is disc shaped. Frequently the rammer is reversed in use, and the pointed end of the handle employed for ramming the sand into narrow spaces.

Ramming Blocks.—Reversed moulds (q.v.) used in some kinds of moulding work.

Ramps, or Guide Plates.—(1) Appliances for placing rolling stock upon the rails. They clip the rails, and are provided with flat helical extensions against which the wagon wheels slide up to the rail. They are made in sets of four, two right and two left. (2) Planks laid on suitable supports and forming an inclined plane, up and down which barrows of material are wheeled are sometimes termed ramps.

Ramsbottom Ring.—See Spring Ring.

Ram's Horn.—A symmetrically shaped double crane hook. Used chiefly for light weights, being inferior in strength to the single hook.

Rape-seed Oil.—See Seed Oil.

Rapping.—Rapping is the process of loosening a pattern in the foundry sand previous to its withdrawal, and is effected by inserting the pointed end of an iron bar in a hole bored in the pattern, or into a rapping plate, and striking it heavily sideways with a hammer. During the actual process of withdrawal, rapping mallets (q.v.) are employed. The process of rapping, though necessary, is often damaging to the pattern, and if carelessly performed affects the size of the mould, enlarging it to an extent which is very appreciable in small castings.

Rapping Bar.—A loosening bar (q.v.).

Rapping Hole.—A hole made in a foundry pattern for the insertion of the loosening bar used in rapping. When a single moulding, or when two or three mouldings only from a pattern are required, the rapping hole is usually bored in the wood of the pattern, but when a large number are wanted, rapping plates (q.v.) are used to prevent damage to the pattern. In large patterns several rapping holes are provided.

Rapping Mallets, or Moulders' Mallets.—Small round-faced wooden mallets used for loosening the sand from foundry patterns during the process of their withdrawal from the moulds. The faces and edges of the patterns are lightly tapped with the mallets as the pattern is gradually lifted. They are made of wood, as inflicting less injury to the pattern than iron hammers.

Rapping Plates.—Plates of malleable iron let into and screwed on the faces of foundry patterns, and provided with holes for the insertion of the rapping bar. The holes for rapping and lifting are usually inserted in the same plate, and they are then termed rapping and lifting plates.

Rasp.—Rasps are used by core makers and loam moulders for the purpose of reducing loam work to accuracy, in regard to both shape and dimensions.

Rasp Cut.—The abrasive surface formed on a file by the punching up of isolated projections, as distinguished from the ridges formed by the action of chisels.

Rastrick Boiler.—A vertical cylindrical boiler of large diameter, and having horizontal tubes. It is an antiquated form.

Ratchet.—A ratchet wheel (q.v.).

Ratchet Bar.—A straight bar serrated with teeth like those of a ratchet wheel, to receive the thrust of a paul. The function of the ratchet bar is to permit of movement in the one direction while preventing it in the direction opposite. It is attached to foot or other levers.

Ratchet Brace.—A tool used by metal workers for the drilling of holes by hand. It consists essentially of a lever which moves the drill round and feeds it forward at the same time by means of a ratchet and pawl actuating a square-threaded feed screw.

Ratchet Drill.—A drill used in conjunction with a ratchet brace (q.v.). It is furnished with a square tapered shank.

Ratchet Feed.—A paul feed (q.v.).

Ratchet Jack.—A screw jack (q.v.) rotated by means of a ratchet and click.

Ratchet Paul.—A paul or click which engages with the teeth of a ratchet or a spur wheel, as distinguished from the pauls used in the sliding shafts of cranes, and on turntables. See Spring Paul.

Ratchet Teeth.—The teeth of ratchet wheels are differently formed according to the function which they have to fulfil. When the ratchet paul has to work in opposite directions at pleasure, as when moving a feed screw to right or left alternately, the teeth are like those of ordinary spur wheels. But when the motion is always in one direction, as happens in the case of cranes, the teeth slope in the one direction only. Their outline then is roughly, though not quite, that of a right-angled triangle, the base resting on the periphery of the wheel, and the side being nearly, though not quite, radial with the centre, being slightly undercut, and the hypothenuse being the sloping face over which the paul slides. This is not, however, straight, but slightly curved outwards.

Ratchet Wheel.—A wheel provided with teeth into which a paul (q.v.) fits. The paul either moves and turns the ratchet wheel with an intermittent motion which renders it capable of feeding a machine cutter, or the wheel moves, its motion being independent of that of the paul. Owing to the shape of the teeth it turns always freely in the one direction, but when it is attempted to reverse the motion of the wheel the paul prevents it from receding. In the first case the shape of the teeth is usually symmetrical, being common wheel teeth ; in the latter they approximate roughly to that of a right-angled triangle, or to that of saw teeth for ripping. The first form is used in most machine feeds, the second in the hoisting gear of cranes for preventing the load from running down.

Rate.—See Screw Rate.

Ratio, or Proportion.—A particular relation subsisting between numbers or quantities, as shown by the division of one by the other. Thus 12 : 3 : : 8 : 2 expressing the ratios which exist between those numbers, 12 containing 3 four times, and 8 containing 2 four times. In proportional numbers the product of the first and last terms is equal to that of the second and third terms. Hence the rule to find a fourth proportional: multiply the second and third terms together and divide by the first.

Ratio of Expansion.—The proportion subsisting between the final and the initial volumes of steam in a cylinder.

Rat Tail File.—A file circular in section and tapering or bellied in the direction of its length.

Rattle Barrel.—A rumble (q.v.).

Rattling Box.—A rumble (q.v.).

Raw Hide Belting.—Belting which has not been subjected to the process of tanning, excepting on the surface.

Raw Iron.—Iron which has not been refined. See Refining.

Raw Mine.—A term applied to iron stone (q.v.).

Reaction.—When a body is at rest it presses downwards with a definite weight. The support beneath it reacts with a pressure which is equal and opposite. Hence the fundamental law, action and reaction are equal and opposite.

Reaction Wheel.—An enclosed wheel into which water enters under head or pressure, and escapes from, tangentially, the force being derived from the reaction of the weight thrown off at the periphery. See Turbine.

Reamer, Rymer, Rimer, or Rhymer.—A fluted tool used for finishing and truing cored or drilled holes. Reamers are solid when used in a socket or with a wrench, shell or hollow when bored out to fit on a mandrel. See Broach, Fluted Reamer, Rose Reamer, &c.

Rebate, or Rabbet.—A shoulder or recess on the edge of a piece of wood or metal for the reception of the edge of another similar piece.

Rebate Joint.—A joint which is made by the overlapping of the edges of material, half the thickness of the material being cut away to a little distance inwards from the matching edges.

Rebate Plane.—A plane used for the working of rebates. The cutting iron is as wide as the wood stock, and hence planes right up to the edge. Skew-mouth planes have the cutting iron at an angle with the side of the stock, and work sweeter than the square-mouthed ones. Rebating is also performed by machines furnished with revolving cutters, or by a thick circular saw.

Rebating.—The grooving or shouldering back of timber from the edge to produce a lap joint.

Reboring.—When engine cylinders have become grooved and of varying diameters through long use and wear, it is customary to bore them again and insert a larger piston. The process is termed reboring. Small cylinders are removed from their beds or foundations for the purpose, but large cylinders are rebored while in place.

Recarburisation.—The adding of a definite amount of carbon to iron which has been first completely decarburised. Steel of various grades is thus made in the Bessemer converter (q.v.) and in the cementation (q.v.) process.

Receiver.—(1) When large quantities of iron of fifteen or twenty tons weight and upwards have to be poured, exceeding the capacity of the foundry ladles, the metal is allowed to accumulate in a receiver, which is a sheet-iron vessel lined with fire bricks and clay, and from which the iron is conducted into the mould. Such vessels may be fixed or portable. (2) Generally any vessel used for the storage of liquids or

14*

gases, as for instance air in an air compressor. See also Intermediate Receiver.

Rechucking.—In all but the very plainest turned work it is necessary that it should be set a second or a third time in the lathe chuck or chucks, in order to turn some portions which could not be reached by the tool in the first chucking. This is termed rechucking or second chucking.

Reciprocal.—One quantity is the reciprocal of another when it is the result of unity divided by the other. Thus 2 and $\frac{1}{2} = 2$ and ·5 are reciprocals. Hence the product of a quantity by its reciprocal is always unity as $2 \times$ ·5 $= 1$·0.

Reciprocating.—Used in opposition to rotary. The vast majority of engines and pumps are reciprocating, that is the piston and rods move forward and backward alternately in right lines.

Reciprocating Engines.—See Reciprocating.

Reciprocating Pump.—See Reciprocating.

Reciprocating Weight.—Usually understood to signify the moving weights of an engine piston, piston rod, and half that of the connecting rod. The power which they absorb is equal to their weight multiplied by the height from which the weight must have fallen to produce their velocity.

Rectangle.—A parallelogram (q.v.) whose sides are at right angles to one another.

Red Brass.—An alloy variously composed, as: Copper, 24 lbs.; zinc, 5 lbs.; bismuth, 1 oz. Copper, 24 lbs.; zinc, 5 lbs.; lead, 8 oz. Copper, 32 lbs.; zinc, 10 lbs.; lead, 1 lb. Copper, 160 lbs.; zinc, 50 lbs.; lead, 10 lbs; antimony, 44 oz.

Red Brick Dust.—Used for parting sand (q.v.).

Red Copper Ore.—A very pure oxide of copper found in Cornwall, Cuba, and other parts.

Red Deal, or Red Pine. (*Pinus sylvestris.*)—The produce of the Scottish fir, grown in Norway and Sweden and Russia. The deals take their names from the ports whence they are shipped, as Riga, Memel, Dantzic. Riga deals are superior to the last-named.

Reddle.—A shop term for red lead, mixed with oil and used when filing and scraping surfaces to fit.

Red Hæmatite.—An ore of iron quarried largely in Cumberland, Glamorganshire, and Ireland. It is a non-hydrated peroxide of a red colour, hence its name, and furnishes excellent quality of iron.

Red Heat.—There are several grades of red heat, distinguished by the prefixes, black, bright, or low red, noticed under their headings, the distinguishing of these shades of colour being of importance in the hardening, tempering, and welding of iron.

Red Lead, or Minium.—Symbol $_2$Pb O $+$ Pb O$_2$. The red oxide of lead. It is used when mixed with boiled oil only, or with boiled oil and with white lead, in making steam joints. Also mixed in a thin semi-fluid paste, for checking the accuracy of contact of surfaces which are being fitted the one to the other, the colour being transferred from the surface which is finished to the highest portions of the surface which is undergoing the process of fitting.

Red Shortness, or Hot Shortness.—That condition of wrought iron or steel in which they are incapable of being rolled or worked at a red heat. The presence of sulphur produces red shortness.

Reducing Agent.—The agent through whose influence a chemical compound is resolved into its simple elements. The reduction of metals from their ores is due to the agency of heat acting in combination with some solid substance or substances. Thus carbon, or more properly carbon monoxide, CO, is the reducing agent in the smelting of iron ores.

Reducing Furnace.—Any furnace in which metals are separated from their ores and reduced to the metallic state. Hence blast, and some reverberatory furnaces are reducing furnaces.

Reducing Pipe.—A pipe having different dimensions, or sometimes also different shapes at its opposite ends. It is used to connect pipes of unequal sizes or of different shapes together.

Reducing Valve.—A valve constructed for the regulation of steam pressure between a boiler and its connections. A weighted lever regulates the opening of the valve.

Reduction.—The extraction of metals from their ores by depriving them of their oxygen in a reducing furnace (q.v.).

Reduction of Area.—Malleable iron and steel, when subjected to tensile stress, elongates up to the breaking strain, with a consequent reduction of area. The amount of elongation and reduction of area which it undergoes is a recognised test of its quality. The amount of reduction of area should not be less than 25 per cent. in a good specimen, and the amount of elongation 15 per cent.

Reduplication.—Reduplication refers to the gain in power obtained by the combination of pulleys in pulley blocks (q.v.). In a system of pulleys a force equal to the pull of the string comes into play at every departure of the string from a pulley.

Red Zinc Ore.—An ore of zinc of small commercial value. It occurs as an oxide, the red colour being due to contamination with the oxides of iron and manganese. It is found in New Jersey and the United States.

Reeking.—The coating over of the faces of ingot moulds for the casting of crucible steel, with a layer of carbon, to prevent the adhesion of the ingot to the mould.

Reel Hose.—See Hose Reel.

Re-evaporation.—Re-evaporation is a term used to express the influence of an unjacketed engine cylinder of long stroke, and worked to a high grade of expansion, upon the steam. The cylinder being subject to the extremes of temperature of the entering and exhausting steam, the latter is subject to initial condensation (q.v.) on its first entrance. The heat thus lost is however reimparted to it as it acquires by expansion a temperature below that of the cylinder, while towards the end of its work it acquires from the cylinder, hotter than itself, a vaporous condition, or is re-evaporated. By the use of the steam jacket these variations are prevented or minimised.

Refined Iron.—Fine metal or plate metal. It is white iron, which has passed through the refinery and been deprived of much of its carbon and silicon preparatory to dry puddling. It is rendered brittle by quick cooling with water, and is broken up into small fragments before being passed to the puddling furnace.

Refined Tin.—See Boiling.

Refinery.—The structure in which the process of refining is effected. It consists of an outer vertical framework surmounted by a low stack, and containing hearth, twyers, dam plate, tap hole, and casting pit. The hearth is about 4 ft. square by 15 to 18 in. in depth, bounded at back and sides by cast-iron water blocks. The twyers are inclined down towards the hearth at an angle of from 30° to 35°, and are five or six in number, to distribute the oxidising influence of the air equably over the area. The hearth is lined with sandstone, and coke is the fuel used, laid in alternate layers with the pig and scrap, which may amount to two tons per charge in the larger refineries, and the time of its purification three or four hours. The molten metal and slag are then tipped out into the casting pit, or pig mould (q.v.), and broken up when cooled. It is then ready for puddling. Refining is also carried on to a limited extent in the manufacture of coke and charcoal plates for tinned goods, but its commercial importance is being diminished by the substitution of steel sheets in their place.

Refining.—The process of partial decarburisation and purification of grey pig iron from silica, with its accompanying conversion into white iron. The refining process is adopted as a preliminary to dry puddling, the reason for its adoption being that white iron, when passing from the solid to the molten state, passes through a pasty condition which is favourable to its oxidation in the puddling furnace, a condition which is not the case with the grey iron, which becomes liquid without undergoing any transitional state. Refining is not practised in the open hearth or wet puddling processes. About twenty-four hundredweights of grey iron are required to produce a ton of refined iron

Reflux Valve, or **Check Valve.**—A flap valve used for the purpose of taking off the pressure of a head of water acting in a backward direction against a set of pumps.

Refractory.—A substance is said to be refractory in the degree in which it is able to resist the action of heat. Lime, silica, clay are refractory substances, and hence are used in various combinations with other substances for linings of furnaces, crucibles, &c. Substances refractory in themselves cease to be so when combined in certain proportions with bases, and due regard must also be had to the nature of the metal or alloy which is to be brought into contact with the lining or crucible.

Refrigerating Fluids.—See Hardening Mixtures.

Refrigerator.—(1) A cylindrical vessel containing a number of copper tubes and used for heating the feed water for a marine boiler above the temperature which it derives from the hot well. The waste brine from the boiler is employed to heat the feed water. (2) A machine for lowering the temperature of air, or a gas for cooling purposes, the refrigerant being made to expand.

Regenerative Furnace.—A furnace in which waste gases, gaseous fuel, and products of combustion are utilised by being taken through layers of brickwork or other regenerators, to which they give up their heat, which heat is taken up again by a similar current mingled with air turned at a certain interval in the reverse direction, which takes up the heat in the regenerators and passes into the combustion chamber of the furnace. See also Gas Producer, Pipe Stoves.

Regenerative Stoves.—See Pipe Stoves.

Regenerator.—The chequer work of a regenerative furnace. It is of glazed firebrick. See Dict. of Modern Terms.

Regulator.—Any contrivance by which motion is equalised. Hence governors, fly wheels, throttle valves, are regulators. More particularly the term is applied to the regulator valve (q.v.) of a locomotive, and the damper of a steam boiler.

Regulator Valve.—The valve which regulates the admission of steam to the cylinders of a locomotive. It is either of the double beat, or conical, or sliding type.

Regulus, or **Matt.**—Refers to different stages in the reduction and purification of certain metals; but in a general sense signifies a metal which is yet in an impure condition, that is one which has not reached the final stage of its reduction. The term is applied to a limited number only, as copper, antimony.

Regulus of Copper.—See Matt.

Reheating.—The heating a second time of puddled bar for the making of fagoted iron. It is performed in a reverberatory furnace.

Reheating Furnace.—The reverberatory furnace used for heating the fagots or piles of puddled bar preparatory to passing them under the finishing rolls.

Reins.—The handles of smiths' tongs.

Relative Strength.—It is convenient to know not only the absolute strengths of the materials of construction, but also their relative strengths when subject to uniform conditions of strain and stress.

Relative Volume.—Specific volume (q.v.).

Release.—Release signifies the opening of the steam port of an engine to allow of the escape of the exhaust steam from the cylinder.

Relief Valve.—See Cylinder Escape Valve.

Relieving.—Backing off (q.v.).

Relighting.—The ignition of the inner lighting tap of a gas engine by means of the outer or master jet (q.v.).

Remelting.—The tensile strength of cast iron is supposed to be increased by repeated remelting. This is not borne out in foundry practice. Repeated remelting tends to make the iron whiter, with a corresponding loss of tensile strength. Iron when remelted takes up impurities from the fuel. Remelting is only useful when suitable brands of pig are mixed with selected scrap, but the improved quality is not due to the remelting but to judicious mixing.

Repairs.—The proper execution of repairs to broken-down or otherwise damaged motors or machines calls for the highest skill of the working engineer. More particularly is this the case at sea, and in large factories where a temporary stoppage only, means an enormous loss in the aggregate. The best workmen are or should be entrusted with the execution of repairs. See also Auxiliary, Break Down.

Rerailing.—Placing rolling stock upon the rails. This is done with screw jacks and timber blocking, or with ramps (q.v.).

Residual Gases.—Applied to the products of combustion left in the cylinder of a gas engine after the explosion of the charge. Carbonic acid is the principal residue.

Resilience.—The amount of work involved in the resistance of a body to an impulsive load (q.v.). The total resilience of any material is the amount of work done in breaking it, and is equal to the product of its resistance into the distance through which the resistance acts. The elastic resilience of any material is the work done in straining it up to its elastic limit, and is equal to the product of the deformation by the mean load producing it.

Resin, or Rosin.—Common resin is the product of several species of pine, from which it exudes in a semi-fluid state, and is collected in vessels. It is used as a flux for soldering tinned work, and mixed with tallow and other ingredients as a dressing for belts. See Belt Dressing.

Resistance.—Resistance is the result of friction, and is a state or condition of things tending to arrest and destroy the state of motion. The study of the laws which govern the resistance of bodies under different conditions is based on direct experiments, from which coefficients of friction are deduced. A certain amount of work is necessarily absorbed in overcoming resistance, and this amount has to be estimated and provided for in calculations relating to all machines and motors. Whatever diminishes resistance becomes a clear mechanical gain. Hence the importance of giving to bearings their proper proportions of surface area, of suitable lubrication of bearing parts, the proper grinding of the edges of cutting tools, &c.

Resolution of Forces.—The process of substituting or discovering the components (q.v.) of a force from a knowledge of its resultant (q.v.). This is done either by calculations or by a graphic delineation.

Rest.—The support which takes the resistance of the tool in turning operations in the lathe. There are many kinds of rests, from the simple rests for hand-turning to the various slide rests. See Floor Rest, Hand Rest, Slide Rest, Tee Rest.

Resultant.—The sum of two or more separate forces which act upon a body in different directions (not equal and opposite), causing it to move, or producing a tendency to move in a definite direction. In other words, it may be defined as a single force which replaces two or more other forces, and which is equal to their sum.

Retaining Rings.—Rings of wrought iron employed for uniting the tyres to those railway wheels which have wooden bodies, They are annular rings, each having an internal flange on its periphery, which fits into corresponding grooves, one on each side of the ring. The rings being bolted together through the wooden body, clasp the tyres securely in the grooves.

Retaining Valve.—In pumps where the water has to be lifted from a great depth it is customary to insert in addition to the ordinary valves an additional one in the series of pipes in order to prevent much of the water from running back between the strokes. Hence the name retaining valve applied to it.

Return Block.—A snatch block (q.v.).

Return Flues.—The flues of Cornish, Lancashire, and Wagon boilers, being brought from the back of the furnace to the front, thence being carried back again to the chimney, are so termed.

Return Tubular Boiler.—A boiler of marine type in which the smoke tubes pass from the back of the boiler forwards to the smoke box or uptake. The products of combustion travel therefore first to the back through the fire box and then to the front through the tubes.

Return Valve.—A valve which allows of the return of fluid. An overflow valve.

Reverberatory Arch.—The arched roof of a reverberatory furnace.

Reverberatory Furnace.—A furnace employed in various metallurgical operations, in which the ore or metal is exposed to the action of flame, but is not in contact with the fuel. The fuel is burnt in a separate chamber, divided from the hearth by a bridge, over which the flame passes. The principle of the furnace is the deflection or concentration of the heated gases down on the metal. Reverberatory furnaces are used for puddling, and for heating iron plates, angle irons, joists, preparatory to bending; for melting iron, and for melting brass in quantity for special purposes when great purity is essential. The reverberatory furnace is an air furnace, no artificial blast being employed.

Reverse Cones.—Cones whose bases are turned away from each other. Applied to the coned bearings of lathe headstocks. See Cone Bearing.

Reversed Moulds, or Ramming Blocks.—Plaster or metal moulds used in some classes of repetition moulding work. The actual casting moulds are made from these blocks direct, instead of from a pattern, and the advantage of their employment is that a very large number of moulds can be made precisely alike without the labour of forming the parting surfaces and runners and risers at every moulding.

Reverse Jaw Chuck.—A dog chuck whose jaws can be readily reversed, end for end, for convenience of clamping work on the exterior or interior diameters. It is usually effected by running the jaw right off the screw and turning it round.

Reverse Jaws.—The jaws of a lathe chuck which are placed within the work which they clamp, for turning exterior surfaces.

Reverse Keys.—Keys (q.v.) made and used not with the object of holding machine parts together, but for the purpose of driving them asunder. They are employed in preference to the hammer, to avoid the possibility of bruising the work. They consist of two steel plates, one having a projecting slip on one edge, the other a recess of the same length. Inserted through a cottar way with the projection facing towards the larger or outer end of the strap piece, and a wedge being driven between, the projection thrusts the butt piece of the strap end outwards. They are useful with tapered piston rod ends.

Reversible Jaws.—The jaws of a reverse jaw chuck (q.v.).

Reversing.—(1) The making of a founder's mould from a clay or plaster or other model, for which the face alone is available. See Reversed Moulds. (2) The changing of the motion of an engine or machine, or machine part, into a direction directly opposite.

Reversing Countershaft.—A countershaft (q.v.) whose direction of rotation can be reversed for the purpose of driving its machine in either direction. A fast pulley, keyed on the shaft, is flanked by two loose pulleys, one carrying an open and the other a crossed belt. By shifting one or the other of the belts to the fast pulley the direction of its motion, together with that of the shaft, is immediately reversed.

Reversing Cylinder.—See Steam Reversing Cylinder.

Reversing Engine.—An engine whose motion can be changed into opposite directions. The reversal is usually effected by means of slot links, lifting links, weigh shaft, and suitable levers, actuating an eccentric, having double sheaves for forward and backward gear, from whence either port is opened to steam. A single eccentric is sometimes used instead, and reversed. See also Reversing Rolling Mill Engine.

Reversing Gear.—The gear which accomplishes the reversal of an engine or other motor, or machine. In an engine it includes the links, weigh shaft, levers. In belting, the forks, cords, levers, as the case may be. In mechanism generally, levers, pulleys, and rods of various kinds.

Reversing Handle.—The handle of a reversing lever (q.v.).

Reversing Lever.—The lever by which the reversing gear of an engine, or other motor or machine is actuated.

Reversing Link.—The slot link (q.v.) of an engine, which, through the medium of the eccentrics, alters the valve for forward or backward motion.

Reversing Mill.—A rolling mill in which the rolls are reversed after each pass of the rail, bar, or plate. Reversing mills are those with two high rolls (q.v.), and where no provision is made for reversal the bars or rails have to be brought back over the top of the rolls and a pass is thereby lost. Three high rolls (q.v.) obviate the necessity of reversal without the loss of a pass.

Reversing Motion.—Reversing motions are of various kinds. The reversing motion of a lathe is designed to enable the slide rest to traverse either up or down the bed. It is effected in back-geared self-acting lathes by an idle wheel being made to engage in turn with wheels which drive the back shaft in different directions, carrying the saddle up or down accordingly ; in screw-cutting lathes, by introducing an extra wheel into the screw-cutting train by which, the motion being reversed, left-handed screws can be cut. A reversing motion which is employed with bevel gearing is effected through the medium either of a friction or of a claw clutch. The clutch slides upon a feather sunk into a shaft upon which two bevel wheels turn freely and gear into a crown wheel common to them both. The clutch can be slid in either direction to engage with either bevel wheel, when clutch and wheel are practically fast on the shaft and revolve with it. It is easy to see that the changing of the clutch from one bevel wheel to the other effects a reversal in the direction of rotation of the crown wheel. See also Reversing Engine, Reversing Mill, &c.

Reversing Plate.—A plate or disc of metal employed for reversing the direction of travel of a single-cylinder engine, in the absence of a slot link (q.v.). It is essentially a plate keyed upon the crank shaft, and furnished either with a slot or with holes by which the eccentric proper is moved up or down for forward or backward motion as required, or for greater or less expansive working of the valve.

Reversing Rolling Mill Engine.—A type of engine specially constructed for the reversal of the rolls of rolling mills. It may be either simple or compound. It is a geared engine, the engine running faster than the rolls in the proportion of about three to one, according to circumstances. These engines are commonly in pairs connected with cranks at right angles, and the reversal is effected through the medium of a slot link, lifted or depressed for forward or backward gear by means of a small steam cylinder, or by hand only. These engines are made massive, are well fitted, and notwithstanding the rapidity of reversal, run quietly.

Reversing Rolls.—The rolls of a rolling mill whose motion is reversible to allow of passes (q.v.) both forward and backward.

Reversing Shaft.—A weigh shaft (q.v.).

Reversing Stud.—The stud or spindle which carries the idle wheel for reversing the motion of back-geared self-acting lathes. See Reversing Motion.

Reversing Valve.—The valve which directs the air and gas into one or the other of the regenerative chambers of a furnace, as they become alternately cooled and heated.

Revolution.—It is usual to give the speeds of engines, pumps, and other motors and machines, in the number of revolutions they make per minute. This is alternative with piston speed, or stroke, since one is readily calculable from the other.

Revolvers.—Rotary pistons (q.v.).

Revolving Furnace, or Rotary Furnace.—A type of furnace designed to effect the conversion of cast into malleable iron by a process of mechanical puddling, in which the furnace rotates, bringing all portions of the molten iron successively into contact with the fettling. The principal furnaces are the Danks, Spencer, Siemens, Crampton, and Pernot. They have but a limited use in England. In the first-named the hearth, or actual furnace, revolves in a vertical plane; in the Pernot, and some others, the plane is either horizontal or slightly inclined to the horizontal. The interior is lined with fettling, which is kept in place with ribs running longitudinally through the hearth. Rotary furnaces are used for the production of malleable iron direct irom the ore, as well as from the pig. The Siemens rotary furnace is of this character. It is formed of a casing of wrought-iron plates riveted together, lined with fire brick and bauxite, or magnesite bricks, and a fettling compound of hammer scale and iron ores melted together. This furnace is about 10 ft. 6 in. long by 10 ft. 6 in. diameter, having its axis horizontal. The body of the rotator revolves on friction wheels, a water jacket keeps the neck cool, the heat is supplied by gas producers and air regenerators. Bends or knees attached to the water pipes project into the body of the rotator and keep the charge continually turning over, and break it up into five or six small homogeneous balls instead of allowing it to agglomerate into one mass.

Revolving Tool Box.—A tool box employed in some planing machines for metal, which is made to turn through an angle of 180° for the purpose of cutting in both directions. It is revolved by cords moved automatically. Termed a Jim Crow box, or a Jack-in-the-box.

Rhomboid.—A parallelogram whose sides are unequal, and whose angles are not right angles.

Rhombus.—A parallelogram (q.v.) whose sides are equal in length, but whose angles are not right angles.

Rhymer.—A reamer (q.v.).

Rib.—A flange or fillet carried around the edges of or across a casting or piece of plated work, in order to strengthen and support an otherwise weak web.

Ribbon Brake.—A strap brake (q.v.).

Ribbon Saw.—A band saw (q.v.).

Riddle.—A coarse foundry sieve about half an inch in the mesh, used for sifting the coarse and old sand of the foundry floor.

Riddlings.—The residual lumpy material left in the riddle after the riddling or sifting of foundry sand. The smaller the proportion of riddlings the better the quality of sand.

Ridge.—A central horizontal foundry runner, lying longitudinally in relation to the casting or castings, and from which a number of offshoots or sprays (q.v.) proceed to convey the metal each to its own section of the mould, or to its separate casting, as the case may be.

Ridge Capping.—The covering which runs along the ridge of a roof.

Ridge Roof.—A roof whose rafters meet in an apex. Its end view is therefore that of a gable.

Riding.—When, owing to loose fitting shafts, or to bad centring of cog wheels, they slip out of gear and the points of the teeth of one come into contact with the teeth points of its fellow, they are said to ride or override.

Riffler, or Bow File.—A file curved in the longitudinal direction. Used chiefly for some kinds of brass work.

Rigger.—A pulley used for the purpose of transmitting motion through the medium of a belt or cord.

Right Angle.—When a straight line standing on another straight line makes equal angles on each side of it, those angles are right angles; a right angle is therefore 90°.

Right-hand Engine.—A horizontal engine which stands to the right of its fly wheel when viewed from the back end of the cylinder.

Right-hand Tools.—Side tools ground to an angle on the left-hand side, and which therefore cut from right to left.

Right-hand Screw.—A screw which turns from left to right. A common wood screw is right-handed.

Right Line.—A straight line, that is, one which occupies the shortest possible length in passing from one point to another, which is not the case with a curved line.

Rigid.—Strictly speaking, there are no rigid bodies existent, since absolute rigidity implies that such a body exposed to strain would suffer no stress. But conventionally speaking, the rigid bodies are the metals and harder materials.

Rigid Metals.—Metals which resist the force of impact, or of tension, or compression, with the least change of form. Cast iron and steel are the purely rigid metals from an engineer's point of view.

Rigid Wheel Base.—See Wheel Base.

Rim.—The outer portion of a circular object, as a wheel, a sheave, or a pulley.

Rimer.—See Reamer.

Ringing Engine.—A form of pile-driver worked by the combined labour of several men. The monkey slides in timber guides, and is attached to a rope passing over a pulley at the top. The rope to which the monkey is attached divides into several smaller ropes, each of which is held by one man, who lifts and lets go at a given signal. Each individual hauls about 40 lbs. of the weight of the monkey, which is lifted about three or four feet.

Rings.—Are used for many purposes, as for pistons and piston packings, the means of union between the shells and fire boxes of boilers, the jointing of boiler seams, caulking rings, retaining rings, rings of angle iron used in built-up structures, &c. Rings are cast or welded, seldom riveted.

Ring Seams.—The circumferential seams or joints of a boiler.

Ring Valve.—A lift valve (q.v.) in which the usual solid disc is replaced by a ring, in order to allow of the escape of fluid on both the outer and inner edges, and which therefore diminishes by one-half the amount of lift necessary. The valve is guided by a central block fitting within the ring. Valves having two rings are also in use, by which the amount of lift is still further lessened.

Ripping.—The sawing of timber longitudinally or with the grain, as distinguished from cross cutting (q.v.).

Rip Saw.—A hand saw from 28 in. to 30 in. long, and containing three to three and a half teeth per inch. It is used for cutting or ripping down with the grain only. It would not be called a hand saw (q.v.), but a rip saw only.

Rise.—The rise of an arch is the vertical distance between the springings (q.v.) and the centre or highest point of the intrados (q.v.).

Riser, or Air Gate, or Rising Gate.—A vertical opening through which the upper portion of a foundry mould communicates with the outer air. When the metal rises into it the founder knows the mould is full, dirt and scoriæ are carried up into it, and the first outrush of air is allowed to escape freely therefrom.

Rising and Falling Spindle.—The spindle of a circular saw, which is made to rise and fall relatively to the table, in order to cut grooves of different depths in the edges of boards. A worm and wheel are the means used to operate the saw spindle.

Rising and Falling Table.—A machine table used with drilling, panel-thicknessing, and other machines, which is made to rise and fall to suit the requirements of different kinds of work.

Rising Gate.—The riser (q.v.) of a mould as distinguished from the running gate.

Rising Main.—The first portion of a series of pipes for the delivery of water supply, that is, next the pump or pumps.

Rising Rod.—A rod which actuates, through the medium of catches, sectors, and weights, the steam and exhaust valves in a Cornish engine.

Rising Spindle Saw.—See Rising and Falling Spindle.

Rivet.—A double-headed bolt-like but solid fastening, employed for securing plates together in a permanent manner, the tail of the rivet being hammered over in place. Rivets are chiefly employed in cases where they will be subject to shearing, and not to tensile strains. They are of soft iron of superior quality to that used for plates, and are made by pressing into dies while red hot. Rivets hold partly by the contraction of the metal in cooling, nipping the plates, but also by the frictional resistance of the plates to slipping. Rivets are usually closed up while red hot, but if the length between the heads is more than four inches the head is apt to become broken off by the contraction. Over six inches rivets are seldom used, or if used are riveted up cold.

Rivet Boy.—A lad whose duty is to make rivets red hot in a furnace or on a forge, and hand them to the workmen with tongs.

Riveted Plates.—No definite rules can be laid down for the strength of riveted plates, since each example must be tested on its own merits—the mode of punching or drilling, the fairness or otherwise of the holes, the quality of the material, both in the plate and the rivet, the nature of the strains to which the plates are subject, and the forms of joints, whether lap, or butt, single or double stripped.

Riveted Stays.—Boiler stays are mostly riveted over at their ends, being first screwed into the plates, and sufficient length being left for riveting over. Sometimes a rivet is passed through a pipe or distance piece of wrought iron whose length is equal to the distance between plates.

Rivet Forge.—A portable forge (q.v.) used by boiler-makers and platers for heating their rivets alongside of the work for which they are required.

Rivet Head.—The head of a rivet is the portion which is first formed, not that which is turned over by the act of riveting, or the tail. Rivet heads are segmental, ellipsoidal, pan, snap, or countersunk, noted under their several headings.

Rivet Holes.—The holes pierced in metal plates to receive the rivets (q.v.). They are either punched, or drilled, or punched and rymered out. Punching weakens a rigid plate; rymering relieves the punched holes of some of the local tension set up therein, and leaves the plate uninjured.

Riveting.—The turning over or clenching of the heads of rivets. Large rivets are treated while red hot, small ones when cold. Two men and a boy are required for hand riveting, one to hold up, the other (often two) to hammer over the head, and the boy to fetch the rivets from the fire. The rivets are hammered sharply while red hot, the hammer beating over the tail on all sides in succession. The finish is imparted to the head by the snap, whose shape is an exact reverse of that of the rivet. See Machine Riveting.

Riveting Hammers.—Hammers of various kinds employed in riveting. They are mostly narrower than ordinary engineers' hammers and the panes are never broad.

Riveting Machine.—A machine used for closing up or forming the tails of rivets by pressure only. The essential mechanism consists of a die fixed near the end of a stout cast-iron pillar, and which receives the thrust of a movable cup or snap-head die, actuated by mechanism similar to that of a punching machine, the rivet, of course, intervening. Hydraulic and pneumatic riveting machines have come into extensive use of late years.

Riveting Set.—A steel punch hollowed upon its face either to a cup or snap-head shape, and used for closing up and forming the tails of rivets. In hand work they are used to follow after the hammer and impart a finish; in machine riveting they close at one operation.

Rivet Joint.—Riveted joints have the advantages of being quickly and readily made, and of being very reliable if properly proportioned, and hence are applied to a vast number of structures—boilers, girders, bridge work, tanks, &c. The forms of rivet joints are single and double lap, and butt, described under their headings.

Rivet Steel.—Mild steel is now used largely for the rivets of boilers. Good rivet steel should stand bending cold to a curve whose inner diameter is equal to that of the rivet. The heads of steel rivets should be hammered down quickly. Overheating of these is more injurious than in the case of iron.

Rivet Tongs.—Tongs with long narrow jaws used for grasping red-hot rivets and slipping them into their holes.

Roach Belly.—See Chambered Core.

Road Locomotive.—See Traction Engine.

Road Pen.—See Drawing Pen.

Road Sand.—The scrapings of roads in flinty districts, which, being of a silicious and refractory character, was used for mixing with sands of a more close nature, for foundry use.

Roasting.—The preparatory treatment of ores of metals, consisting in the heating them at a moderately high temperature in order to volatilise and drive off the deleterious gases, carbonic acid, sulphur dioxide, &c.

Rocking Disc.—See Wrist Plate.

Rocking Frame.—The frame which carries the gear wheels at the end of a self-acting lathe, for reversing the direction of the back shaft.

Rocking Grate Bars.—Fire bars (q.v.) to which a rocking motion is imparted in order to prevent choking of the interspaces.

Rocking Handle, or Rocking Lever.—The double-ended handle for the application of manual power to the working of double-barrelled pumps. One piston is thus at the lowest part of its stroke while the other is at its highest. Called also bob lever.

Rocking Lever.—The lever which drives the air and the circulating pumps in marine engines.

Rocking Shaft, or Rock Shaft.—A shaft which carries the double-ended lever which actuates the slide valve in the indirect acting slide valve (q.v.) type. Also any shaft or spindle which has a to-and-fro motion only, as a weigh shaft (q.v.).

Rock Sand.—The débris of abraded rock, used for foundry cores. See Foundry Sand.

Rod.—A bar of iron or other metal. The sections and sizes of rods are variable—flat, round, oval, half-round, &c.—according to the numerous purposes for which they are required.

Rod Chisel.—The smith's chisel, which is held by withy handles, to distinguish it from the anvil cutter (q.v.).

Rodding.—When, in the course of moulding a pattern, a mass of sand overhangs a flange space, the sand is supported by bent rods of iron proceeding from a drawback or other plate, and bent over the flange space to form a rigid nucleus for the binding together of the sand. But for this rodding, as it is called, the sand would probably be washed down by the rush of metal.

Rod Gauge.—See Gauge Rod.

Rolled Bars.—Bars of malleable iron or steel, the shape of whose cross section has been developed by passing through the rolls of rolling mills. All imaginable sections can be rolled thus, provided that the section is uniform and that all flanged portions stand at right angles with the main axis.

Rolled Beams.—Joists (q.v.).

Rolled Iron.—See Rolled Bars.

Roller Feed.—In machines for wood working the timber is fed up to the cutters by fluted rollers ; hence the term roller feed.

Roller Path.—A smooth surface, plane, circular, or otherwise, upon which smooth rollers, turned or chilled, travel to carry a superstructure of some kind. Rotating furnaces, revolving cranes, &c., run on roller paths.

Rolling.—The straightening, or the bending of plates of sheet or bar metal, performed between the rolls of rolling mills.

Rolling Circles.—See Rolling Curves.

Rolling Collar.—The half shrouding of gear wheels, which, terminating in each wheel on the pitch line, gives rolling contact on that line, and consequent ease and regularity of motion, in addition to strength.

Rolling Contact.—The communication of motion by means of friction wheels and toothed wheels.

Rolling Curves, or Rolling Circles.—A term given to templet curves used in striking out the shapes of the teeth of gear wheels. The fundamental circles (q.v.) are represented by arcs of circles, inner and outer respectively, laid upon the pitch lines struck on the drawing board, and the generating circles by complete discs, which are rolled upon the fundamental arcs, needle points being inserted in the generating circles, by which the cycloidal curves of the teeth are scribed upon the drawing board or paper. See Wheel Teeth.

Rolling Friction.—Friction which takes place in the contact of an axle with friction rollers (q.v.), used to distinguish it from sliding friction (q.v.)

Rolling Load, or Moving Weight.—A load which is neither concentrated, nor distributed simply, but which, either concentrated or distributed in its mass, is nevertheless in a state of motion over a structure, as in the passage of a railway train over a bridge.

Rolling Mill.—The mill in which puddled bar is converted by rolling into merchant iron. See Puddling Rolls, Forge Train, Merchant Train, Roughing Rolls, Three High Rolls, &c. All the various sections of malleable iron bars are imparted during the processes of rolling.

Rolling Mill Engine.—The term cannot be said to refer to any special type, but rather to the work which it is designed to perform. All the main types of engines are employed in rolling mills, condensing and non-condensing, horizontal and vertical, reversing and non-reversing They are massive and well made, suitable for the heavy work which they have to do.

Rolling Oil.—Oil used for lubricating the surfaces of metal passed between cold rollers.

Rolling Over.—The turning over of a foundry box after one-half has been rammed up. Strictly speaking, that process of moulding by which both sides of a pattern are directly rammed up, in contradistinction to the bedding in method, where the top side only is rammed and the bottom is bedded in the floor and packed around with sand.

Rolling Parallel Rule.—See Parallel Rule.

Rolls.—Cylinders of cast iron, ranging from a few inches to three or four feet in diameter, used in the processes of the manufacture of the various sections of malleable iron and steel. The cylinders are plain, or grooved, according to the sections which they are designed to produce. For specific terms see Puddling Rolls, Mill Rolls, Forge Train, Two High Rolls, Three High Rolls, Grain Rolls.

Roll Squeezers.—Rolls used as squeezers for consolidating puddled balls into blooms. The squeezer contains three rolls, two of them side by side and on the same level, and an upper one capable of adjustment upwards or downwards. The length of the bloom is fixed by collars on the lower rolls, and its diameter by the adjustment of the top one.

Roman Balance.—See Steelyard.

Roofing Stays.—See Roof Stays.

Roof Stays, or **Roofing Stays.**—The stays which stiffen the roof of the fire box of a locomotive or portable engine. See Girder Stays.

Roof Truss.—See Truss.

Rope Crane.—A travelling crane driven by an endless rope of cotton cord or of hemp. The source of power may be attached to or be distinct from the crane. The cord travels at a very high speed, so that a minimum of power is required to lift a heavy load. The tension of the cord is maintained, and its slack taken up by tightening pulleys having their bearings in sliding frames, which are counterbalanced by suspended balance weights.

Rope Drive.—Power is conveniently transmitted to a long distance by means of ropes running at a high speed. They are principally applicable to machinery to which the power has to be conveyed to a considerable distance, as the ropes should hang loosely on their pulleys in order to increase the extent of the arc of contact. The driving power is due principally to the weight of the ropes, and they should fit slack to prevent undue wearing of the strands in the grooves of the pulleys. The pulleys should be large—that of the smallest should not be less than 30 times the diameter of its rope. It is more convenient to use several small ropes than two or three large ones. The pulley grooves should be turned as smooth as possible to diminish undue wearing of the strands of the ropes.

Ropes.—Ropes are used for the transmission of power over long distances, as being more suitable than belt and toothed gearing. Ropes are made of Russian hemp and tarred, or of Manilla fibre, or of cotton, or steel, or iron wire. The twist of the strands of which a rope is composed diminishes its strength, but is necessary to bind it together continuously. Tarred hempen ropes are the weakest, steel wire ropes the strongest. The sizes of ropes are given in their girth or circumference, and their weight in pounds per fathom.

Rope Wheel.—A sheave pulley whose groove receives the bight of a rope for hauling purposes. The rim is veed in section, but often waved in the circumferential direction. See Wave Wheel. Rope wheels are used for overhead travelling cranes, hoists, &c.

Rose Bit.—A solid cylindrical parallel boring tool, used for making drilled holes straight and parallel and of equal diameter. This it effects by cutting only in front, or rather where the front merges into the parallel sides, the front edge being rounded slightly and serrated into a succession of small cutters so that the end is imagined to bear some rude resemblance to the whorl of a rose blossom. The rigid solid cylindrical body prevents the tool from being drawn to one side by the irregularities of the original rough drilled hole. The difference between a rose drill and a broach, or reamer, with which it is sometimes confounded, is that the latter cuts at the sides, the former at the front rounded portion only.

Rose End.—A strainer (q.v.).

Rose Jet.—The perforated strainer or nozzle at the end of a water hose employed for irrigation or other purposes.

Rose Reamer.—A broach, both fluted and having a rose end.

Rosette Copper.—A variety of copper obtained by throwing water upon the surface of the molten metal, the result being that it solidifies in thin films.

Rosin.—See Resin.

Rotary Blower.—A blower (q.v.) whose essential characteristic is the rotation of fans or pistons in a cylindrical case. The fans are usually either radial or curved. The pistons are double or treble ended, and so formed and pivoted that during rotation, escape of air is prevented by their accurate and close fitting within the cylinder, and next each other.

Rotary Cutter.—A cutter (q.v.) which is rapidly rotated on a spindle in order to impart a definite outline and a clean surface to a wood or metal blank. The teeth of gear wheels are thus formed. Milling cutters are rotary cutters; also the surfaces of timber boards and of moulded edges are produced with cutters set in a revolving adze block. The rate of revolution is variable, though very high in most cases. Cutters set in slowly revolving boring bars are not properly included under this head.

Rotary Engines.—Applied to steam engines whose pistons have a rotary movement.

Rotary Furnace.—See Revolving Furnace.

Rotary Pistons.—The pistons of rotary engines, pumps, and blowers. They are circular, or segments of circles, with or without sliding pieces or abutments moving in grooves in their bodies. The pistons of blowers are also called wafters.

Rotary Puddler.—See Revolving Furnace.

Rotary Pumps.—Pumps in which the necessary action is produced by the revolution of segmental pistons around their axes. This is also a general term applicable to all pumps whose pistons have a rotary or circular motion. See Centrifugal Pump.

Rotary Squeezer.—A form of squeezer used in the consolidation of puddled ball. It consists essentially of two cylinders, an external and internal, whose surfaces in opposition are roughened, and their axes set eccentrically one with the other. The ball introduced at the widest portion of the space between the cylinders is carried round by the rotation of the inner drum and squeezed between the narrower spaces.

Rotary Vice.—A swivel vice (q.v.).

Rotating Valve.—A cylindrical valve which turns upon its longitudinal axis, thus opening or shutting ports or passages against which its circular face abuts, its body being pierced with a hole to allow of the passage of fluid in certain positions. See Plug Valve. It is thus distinguished from disc valves, lift, and slide valves.

Rotating, or Rotational Weights.—Usually applied to the weights which tend to equalise the varying angular velocities of engine crank shafts, fly wheels, machinery, paddle wheels, propellers, the balance weights of locomotives, &c.

Rotator.—A revolving furnace (q.v.).

Rough Coat.—The first coat of loam applied to a loam pattern, mould, or core.

Rough Cut.—(1) The term used to designate the coarsest kind of file made. The degree of coarseness depends upon the length of the file. In one of 12 in. in length there would be 40 rows of cuts to the linear inch. (2) A first thickness of shaving removed from a piece of metal work in a lathe, shaping, or similar machine. The rough cut is taken to penetrate under the skin.

Rough Dimensions.—It is necessary to give certain allowances of extra thickness of metal in all work which has to be machined, whether castings or forgings, so that they are larger by this amount than the finished dimensions (q.v.). These are called the rough dimensions, and in the pieces of work, whether rough castings or rough forgings, so long as sufficient thickness is allowed for machining, very close accuracy is not looked for, nor is necessary.

Roughing.—See Roughing Down.

Roughing Down or Roughing.—The removal of the largest bulk of material, including the outer skin or scale, from a piece of work, preparatory to the more accurate and final bringing to shape with finishing cuts.

Roughing Out.—Sometimes used in the sense of roughing down (q.v.), but more properly signifies the first preparation of a forging, or a piece of wood work from the crude material. A forging would be roughed out when produced by swaging, jumping up, and welding from the original bar, while not yet finished with the fullering and flatting tools, dies, or final swagings. A pattern would be roughed out when the stuff was brought nearly, but not accurately to dimensions. A piece of bright metal work would be roughed out when it leaves the slotting or shaping machine, or when chipped rudely to shape.

Roughing Rolls.—The first rolls used for rolling puddled bar, in which the bloom is squeezed and consolidated and brought roughly to shape. See Puddling Rolls, Mill Rolls, Plate Mill.

Roughing Tool.—The ordinary tool used by iron turners and machinists for removing the outer skin, and generally for turning cast iron, wrought iron, and steel. It is either of the solid cranked form or is a short length of rod held in a cutter bar. Its angles vary with the material upon which it is used.

Round Backed Angles.—See Angle Iron.

Round Bar Iron.—See Bar, Rod.

Round Body Valve.—A pump valve whose body is of cylindrical shape.

Round File.—A file circular in section. Round files are either tapered or parallel, the tapered files of small size being termed rat tail files (q.v.). Round files are used generally for enlarging holes and shaping hollow curves. Round parallel files are also used for gulleting the teeth of large circular and pit saws.

Round Head Screws.—Button-headed screws (q.v.).

Rounding Tool.—A swage tool hollowed to a semi-cylindrical section and used by smiths for rounding and finishing iron rods.

Round Nose Tool, or Round Tool.—(1) A chisel whose cutting edge, rounding in plan, is used both in wood and metal turning for finishing curved and hollow parts. (2) A stout tool similar in shape to the foregoing, but narrower, used by fitters for cutting oil grooves in brasses and on broad bearing surfaces.

Round Plane.—A plane, the edge of whose cutting iron is an arc of a circle, and which therefore planes hollow. Used by carpenters and pattern makers. The sizes of the planes are indicated by numbers, the lowest and smallest sweep being No. 1, and the highest and largest being No. 18.

Round Tool.—See Round Nose Tool.

Router.—The lip on a centre bit or Jennings bit, which removes the wood within the circle cut by the nicker (q.v.).

Royal Drawing Paper.—Measures $24 \times 19\frac{1}{2}$.

Rubber, or Glass Paper Rubber.—(1) A flat rectangular piece of cork glued to a similarly shaped wood backing, and used in applying glass paper to woodwork. The cork is more elastic than wood alone and does not work so harshly. Rubbers are cut also into various outlines, curved or otherwise for special work. (2) India-rubber (q.v.).

Rubber File.—A coarse heavy file used only in the roughest work, and seldom by engineers.

Rubber Hose.—Hose pipe built up of vulcanised india-rubber. This is superior to canvas hose (q.v.).

Rubbing Board.—A flat board measuring about 6 in. $\times$ 3 in., held in the hand and used for sleeking over the flat faces of foundry moulds previous to the final smoothing with the trowel and the blackening.

Rubbish Pulley.—See Gin Block.

Rule.—A standard of measurement. See Arch Joint, Contraction Rule, Standard Rule, Square Joint. Engineers' rules are either two feet, two fold, that is, having one joint only; or one foot steel rules (q.v.).

Rule Measurement.—Used for all ordinary dimensions, but is not admissable for work requiring great exactness; trammeling off (q.v.) and the use of calipers, with ordinary, and micrometer gauges, &c., being employed instead.

Rumble, or Rattle Barrel, or Tumbler.—A hollow revolving cylinder, used for the purpose of cleaning off castings from the sand which clings to them from the moulds. The castings are enclosed loosely in the revolving cylinder and become cleaned by their mutual attrition.

Rumbling.—The cleaning of castings in a rumble (q.v.).

Runner.—The channel or passage through which the molten metal passes from the gate into a mould.

Runner Head.—The mass of metal which fills the ingate (q.v.) of a foundry mould. Runner heads and feeder heads (q.v.) are knocked off while red hot, and are in request in the foundry for cooling very hot metal just tapped into the ladles. They are dropped in the molten metal, and in becoming remelted, lower the temperature of that in the ladle.

Runner Pin, or Runner Stick.—A tapered pin, cylindrical or rectangular in section, used by iron moulders as a pattern from which the runner (q.v.) is formed for a mould. The sand is rammed around it in the usual way, and on the withdrawal of the pin a passage remains for the inflow of the metal.

Runner Stick.—See Runner Pin.

Running Away.—(1) Racing (q.v.). (2) The escape of metal from a mould in the act of pouring, due to bad jointing of the sand. When it occurs in quantity, a waster casting results.

Running Centre Chuck.—A driver chuck (q.v.) so called because it revolves with the lathe, as opposed to the similar points in a dead centre lathe (q.v.).

Running Down.—The melting of iron in a cupola. When a charge of iron is all melted, or completely liquid at the tap hole, it is said to be run down. A definite charge takes a definite time to run down, the time being variable with the character of the cupola, and the nature of the fuel and the blast, but constant for the same cupola.

Running Gate.—The ingate (q.v.) of a mould as distinguished from a riser (q.v.) or rising gate.

Running In.—An engine is said to run in when the valve is so set that the top of the fly-wheel rim runs towards the cylinder.

Running In Refinery.—A refinery (q.v.) which is built close to the blast furnace and into which the molten iron is run direct from the latter. See Melting Down Refinery.

Running Out.—(1) An engine is said to run out when the valve is so set that the top of the fly-wheel rim runs away from the cylinder. (2) The slipping or working of a drill to one side of the centre in which it was started, due to carelessness in centring, or to improper setting of the work, or to a badly formed drill, or to the influence of inequalities on the surface of the work itself. It is remedied by recentring with the centre punch if the hole is not entered deeply, and by chipping out if it is too deep for punching. If very deep the evil is incapable of remedy except by filing, or broaching, or boring, or drifting, according to circumstances. (3) When a piece of work is not chucked truly in a lathe, but is eccentric, it is also said to be running out.

Running Out Fire.—See Refinery.

Running Pulley.—(1) The movable pulley in a snatch block (q.v.) and in lifting tackle generally. (2) A gin pulley.

Run Out.—The escape of metal from a mould during the act of pouring, due to open joints somewhere, caused by bad fitting of the mould itself, or of cores, or through straining of the flask parts.

Rust, or Red Rust.—The hydrated peroxide of iron ($2Fe_2 O_3 3H_2 O$) formed by union of iron with the oxygen of the air, producing slow combustion. It is prevented in the workshop by brushing newly fettled castings over with a coat of boiled oil, and by painting finished work with three or four coats of oil paint.

Rusting.—The coating of bright iron patterns with rust in order that they may take the shellac varnish without risk of its peeling off. Rusting is accomplished rapidly by rubbing the iron over with a solution of sal-ammoniac in water, or of weak hydrochloric acid, and allowing it to dry.

Rust Joint.—A joint used by engineers where iron and iron have to be caulked to withstand water pressure. It is made by filling the joint with iron borings and sal-ammoniac, to which sulphur is sometimes added, and which hardens in a day or two.

Rymer.—A reamer (q.v.).

Rymering.—The practice of enlarging by means of a rymer or reamer (q.v.) holes which have been already punched or drilled. The advantages of rymering are: that in holes roughly drilled the ridges or arrisses left by the drill can be removed and the hole left truer; that holes in plates which have been punched or drilled and which are not exactly fair can be made to correspond by clamping the plates together and passing the rymer through both at once; also, that in punched holes the local tension set up in the immediate vicinity of the holes is removed entirely by rymering them about $\frac{1}{16}$ in. larger in diameter.

S

Saddle.—The base of a slide rest which lies on the lathe bed, and whose edges embrace the edges of the Vees. Similarly, the sliding plate which carries the drill spindle and gear wheels of a radial drill. The seatings or supports which carry horizontal cylindrical boilers, both land and marine, are called saddles. These are made both in wrought and in cast iron.

Saddle Back Rail, or Barlow Rail.—A rail whose sides curve rapidly outwards and downwards, in order to provide a broad base for the laying of the rail directly on the ballast without the necessity for the intervention of sleepers. Its shape is inherently weak.

Saddle Flange.—A curved flange hollowed out to fit a boiler, a pipe, or other cylindrical vessel.

Saddle Key.—A key (q.v.) whose inner face is hollowed to fit its shaft, instead of the shaft being either flattened or recessed for its reception as is the case with ordinary keys.

Saddle Tank.—See Saddle Tank Engine.

Saddle Tank Engine.—A locomotive engine in which the water tank envelops the top and sides of the boiler, thus presenting the appearance of a saddle.

Safe Edge File.—A file having one or more edges left uncut, in order that the smooth edge may serve as a guide only to the actual cutting faces of the file.

Safety, Factor of.—See Factor of Safety.

Safety Hoist.—Any hoist constructed with differential pulleys. Also an ordinary rope hoist in which an automatic catch or safety stop (q.v.) secures the rope against running down.

Safety Ladle.—The tipping of a heavy ladle for the purpose of pouring its contents into a mould was formerly effected by the power of men applied to levers fixed on the axis of its trunnions. To obviate the risks attendant upon this clumsy method, Nasmyth applied the screw and worm wheel movement thereto. One man by turning a handle affixed to a worm or screw shaft can then move several tons of metal with ease. This is called the safety ladle, and is now universally employed in foundries. Geared ladles of the largest size are often provided with a pair of mitre wheels in addition to the worm gear, by which more power is gained.

Safety Plug.—A fusible plug (q.v.).

Safety Rail.—A guard rail (q.v.).

Safety Stop.—An arrangement applied to lifts and elevators for the purpose of checking their descent in case of the accidental breaking of the rope or chain. It usually consists of levers or springs which on the release of the lift throw a paul into gear with a ratchet, one being placed upon each side of the shaft.

Safety Valve.—The escape or relief valve by which excessive and unsafe pressure in steam boilers is prevented. It consists essentially of a lift valve of some one shape or another, and contains mechanism for imparting to the valve a definite pressure. The mechanism is either a lever with an adjustable weight, or a spring whose strength is known, and whose tension may be adjusted by a movable nut; or a dead weight, or Cowburn valve (q.v.).

Safety Valve Lever.—A loaded lever, the valve to be opened by steam pressure lying between the fulcrum and the load or weight; its principle being that the weight of the total steam pressure on the valve multiplied by its distance from the fulcrum is equal to the weight on the lever multiplied into its distance from the fulcrum. This takes no account of the weight of the lever itself, or of the valve, which must also be considered.

Safety Valve Seating.—(1) In Cornish and Lancashire and marine boilers, the safety valve is not attached directly to the boiler shell, but to an intermediate casting called its seating. It is more convenient, and safer in practice to fit a seating to the curve of the boiler, and provide it with a flange for the attachment of a valve shell, than to fit the latter directly. (2) The annular opening on and within which the valve fits and through which the steam escapes, is called the seat or seating. This is usually a gun-metal casting distinct from, and driven into the safety valve shell.

Safety Valve Shell.—The cast-iron casing which carries the gun-metal seating of a safety valve.

Safe Working Load.—That which is obtained by dividing the load which would produce fracture or crippling, by a factor of safety (q.v.).

Sag.—The vertical and lateral sway of a rope or belt due to stretching and slackening.

Saggers.—Cast-iron boxes used for packing the castings and sifted red hæmatite, in readiness for the annealing oven, in the process of manufacture of malleable cast iron (q.v.).

Salamander.—See Scaffold. Also, a type of melting crucible.

Sal-ammoniac.—Symbol NH_4Cl. The chloride of ammonia obtained as a bye product in gas manufacture. Its uses are various. It is used as a flux for tinned surfaces which are to be brazed, and as an ingredient in rust cement, and for the rusting of iron patterns. It is also sometimes used to prevent or to remove incrustation in steam boilers. This it effects by the formation of chloride of lime which is soluble in water. It requires caution in its use, as the carbonate of ammonia formed will, if it becomes concentrated, attack the metal work.

Salinometer.—An instrument used for testing the degree of saltness present in the water used in marine boilers. The instrument used, is either the thermometer by which the boiling point of the water is ascertained and the amount of saturation deduced therefrom, or a hydrometer especially marked for degrees of saltness, by which the specific gravity of the water is measured.

Salting.—The accumulation of a deposit of salt on the plates of a marine boiler. A certain amount of salting is not injurious to the plates, but the density of the water should not exceed $\frac{2}{32}$, or from 8 to 10 ounces of salt to the gallon. See Two Salt Waters.

Salt Water.—See Sea Water.

Sampling.—The process of judging of the quality of pig iron or steel by the appearance of a newly fractured surface. A small button of metal is removed from the furnace and either flattened out or broken with a sledge, and by the appearance of the fracture, or by a ready chemical test, or both combined, the properties of the metal are inferred, and any necessary alteration made in the proportions of the ingredients, or in the conditions of working.

Sand.—Sand is formed by the decomposition of sandstone (q.v.) rocks, and is of very various composition and degrees of hardness. Its basis is silica, but this is so intermixed with other substances as to produce compounds varying greatly in characteristics and uses. Some are highly infusible, others very slightly refractory. The uses of sand in engineering work are multifarious and are described under their specific headings. See Old, Dry, Green, Strong, Road, Core Sand, Loam, Ganister, &c. Sand is also sprinkled by smiths over those joints which have to be welded together in order to protect them from oxidation by the atmosphere. The sand forms a glassy scale at the welding temperature. Sand and water are used for the scouring of the sheet-iron plates used in the tin manufacture.

Sand Bank.—In foundries where small pipes are cast in quantities the moulding boxes are placed, and the metal run on a bank of sand sloping at an angle of about 45° with the horizontal. The pipes are sounder when made by this method than horizontally, since the sullage rises to the upper portion as is the case when casting upright.

Sand Bed.—(1) The bed of sand in front of a blast furnace into which the sow and pigs are run. (2) Any bed of foundry sand, level or sloping, on which a moulding box or pattern is laid.

Sand Bin.—A trough or compartment in, or adjacent to foundries, used as a convenient receptacle for sand required for the use of the moulder, the different sands being kept in separate bins.

Sand Blast Sharpening.—The sharpening of files by the direction of a current of sand and water across the teeth and along the file at an angle. The mixed current is discharged by the introduction of a steam jet. The sand is as fine as flour emery. Files sharpened thus are considered superior to those cut in the ordinary manner.

Sand Box.—(1) A box containing a selected quantity of foundry sand for moulders' use. Sand boxes are kept at hand in all classes of work both hand and machine. In moulding machines the boxes are either above, or underneath in the base of the machine. (2) A box used on locomotives as a receptacle for sand, which being allowed to fall on the rails in slippery weather increases the adhesion of the wheels.

Sand Burned.—Castings become burned, or sand burnt, when the metal is poured into the mould at too high a temperature, in which case the surface of the metal combines with the silica in the sand and becomes extremely hard and rough or chilled. The use of plumbago or of blacking is to interpose a protective infusible medium between the actual sand and the intensely hot metal.

Sand Joint.—The parting or joint between the different portions of the sand of a foundry mould.

Sand Mixer.—A machine used in mixing sand for foundry use.

Sand Mixing.—The mixing of sands for foundry use is an art for which no directions can be laid down, being altogether within the range of practical experience. The proportions of the different sands necessary to produce any given result varies also with the locality from which the sands are obtained. For special notes see Strong Sand, Core Sand, Dry Sand, Loam, and other specific headings.

Sand Paper.—See Glass Paper.

Sand Papering Machines.—See Glass Papering Machines.

Sand Pit.—See Foundry Pit.

Sand Plug.—See Plug.

Sand Pump.—A pump used for drawing the wet sand out of caissons, pits, and mines, and in prospecting for gold and diamonds. It is of the centrifugal type.

Sand Rat.—A slang term commonly applied by iron moulders to members of their craft. Sand puncher is also a favorite cognomen.

Sand Sifter.—A machine made for sifting foundry sand, and worked by hand or power. For sifting sand in small quantities sieves or riddles are employed. There are different forms of machines, small sifters resting on supports, larger ones being suspended from suitable beams. The principle in each is as follows. An open sieve of rectangular form has a to-and-fro oscillatory movement imparted to it by means of three teeth, which as they are revolved by means of a winch handle within a slotted block impart to it alternately a back and forth motion, this being communicated through a rigid rod to the frame of the sieve. The movement is much more rapid than can be imparted by hand sieves, hence the advantage of using the mechanism.

Sand Sifting.—All foundry sand is subjected to a process of sifting or riddling before being mixed and used, the object of sifting being the removal of lumps and gravelly stones and foreign matters.

Sandstone.—Sandstones are of various composition and degrees of hardness. They consist essentially of silicious sand cemented together with carbonate of lime, oxide of iron, or clay, hence called calcareous, ferruginous, or argillaceous sandstones. They furnish the various foundry sands (q.v.), the linings of furnaces, are mixed with fire clays, and some fire bricks, and yield the ganister for converters, and steel moulds. They occur as rocks and as loose friable deposits, and are termed, according to their physical appearance and locality, rock sand, sea sand, pit sand, free sand, &c.

Sand Valve.—The valve by which the escape of sand from the sand box of a locomotive is regulated. It is under the control of the driver, being actuated by a foot lever.

Sap.—The circulating fluid in timber which conveys the elements of nutrition to the tree.

Sappy.—Timber is said to be sappy when it is foxey (q.v.) or when it has not been properly seasoned, so that the sap remains in large quantities.

Sap Wood.—The wood which lies immediately underneath the bark of a tree. It is of little value, and rapidly undergoes deterioration when in the plank.

Saturated Steam.—Steam is said to be saturated when it remains in contact with the water from which it has been generated. It then holds a quantity of moisture in suspension, and is not in the condition of a true gas. Also called dry steam (q.v.) and vaporous steam.

Saturation.—A vapour is said to be saturated when it is at the greatest density and pressure corresponding with its temperature, and when therefore an increase of pressure or decrease of temperature would cause some of the vapour to be condensed. It is on the point of condensation, and is therefore in a different condition from that of a true gas.

Saw.—A tool having a serrated blade, and furnished with a handle or frame. The saw plate is of steel, the temper of the steel varying with the nature of the work which the tool has to do, being hard in most cases, but soft for band saws. The saws used in the departments of engineering are the hand saw, the back, frame, and pad saws, the pit, deal frame, and circular saws for wood, and the hack saw for metal, noted under their special headings.

Saw Bench, or Saw Table.—The bench of a machine saw, upon which the timber is supported while being cut. Sometimes the word bench is used more especially when speaking of circular saws, and table in reference to band and jigger and fret saws.

Saw Dust.—This is used for cleaning the oil from bolts, and also, mixed with loam, for the same purpose as horsedung (q.v.).

Saw Fence.—See Fence.

Saw Files.—Files made specially for the use of the saw sharpener, though employed also for other work. They are known under various names and made in various classes, mostly as second cut (q.v.) and smooth (q.v.) and both single cut (q.v.) and double cut (q.v.). They are either taper files (q.v.) or blunt files (q.v.), and of great range as regards length. For saws having triangular teeth simply, the ordinary triangular single cut taper files are used, but for circular and pit, and large frame saws, the parallel, blunt, double-cut, flat-faced, round-edged files, and the parallel blunt, round double-cut files are employed.

Saw Filing Machine.—A machine used for filing band saws, in which the file is carried to and fro by an arm provided with a rectilinear motion, at the same time that the saw is traversed along a distance equal to a single tooth for each traverse of the file.

Saw Frame.—(1) A swing frame (q.v.). (2) The frame of a bow saw (q.v.).

Saw Gullet.—The curved bottom or root or interspace between contiguous saw teeth. The teeth of the larger saws alone are gulleted, the object being to secure freedom or clearance for the sawdust without choking the saw or causing it to work hard. Gulleting, called also saw gumming (q.v.), is only performed once to every two or three times of sharpening.

Saw Gumming.—The grinding out of the gullets or roots of circular saw teeth by means of emery wheels, the sections of whose edges are the counterparts of the tooth spaces.

Sawing.—The division of material by means of saw teeth. The saving of a given quantity of material and the minimum expenditure of power depends partly on the shape of the saw teeth, partly on the degree of thickness of the blade. The thinner the blade consistently with the work the saw has to perform, the easier the task and the less the waste of material. Sawing demands under any circumstances a great expenditure of power, hence saws are where practicable thinned down towards the back, which together with set, enables the plate to clear itself in the cut. Circular saws and band saws are in addition lubricated with tow soaked in tallow and oil.

Sawing Stool.—A low trestle upon which timber is laid, which has to be sawn by hand.

Saw Pit.—The pit used in country districts upon which timber is laid when being cut into planks and boards, the bottom sawyer working in the pit.

Saw Set.—A tool or appliance used for setting the teeth of saws. It is either a plyor set, or a setting hammer and round-faced block, or it is a special contrivance of an automatic character, of which latter there are many kinds.

Saw Setting.—The bending over of the teeth of a saw to right and left alternately, in order to make the kerf formed by the teeth somewhat wider than the saw blade, so that the latter shall move with the minimum of friction therein. The setting of saw teeth is not usually done at each time of sharpening, but at about every second or third sharpening.

Saw Sharpening.—The bringing of the teeth of saws to a keen edge when dulled by use. It is effected by files, mostly worked by hand. The sharpening is done from each side of the blade, every alternate tooth being filed from one side, and afterwards all the intermediate teeth from the side opposite, the teeth being filed in the same direction in which they lean, both to prevent screeching of the saw, and to give a burr on the cutting edge. For sharpening small saws a single triangular file is used, for larger saws a flat file having rounding edges for gulleting, or a flat file for the points and a round file for gulleting.

Saw Spindle.—The spindle which carries a circular saw. It consists of a turned spindle furnished with bearing necks, fast and loose pulleys, and a couple of washers, one fixed, the other movable, and tightened with a nut, the saw being clamped between the two washers, and prevented from turning by a feather fitting into a slot cut in the hole in the centre of the saw.

Saw Table.—See Saw Bench.

Saw Teeth.—The angles of saw teeth vary with the work they have to do. The angles of the faces are from 80° to 85° for hard, 65° to 70° for soft wood, and the relief angle is 65° to 70° for hard, 45° to 50° for soft wood. The spacing and the set should be greater for soft than for hard wood. For cross cutting, the teeth are set farther back than for ripping.

Scab.—An excrescence upon a casting caused by the washing out of a corresponding portion of the sand in the mould, due principally to improper ramming. The presence of the scab is not in itself very objectionable since that can be chipped off, but in most cases where scab is present the sand which has been washed away is located somewhere in the body of the casting, with the result of rendering it hollow and unsound.

Scaffold, or Salamander.—A mechanical obstruction in a blast furnace due to one or two or several causes in combination, as bad fuel, slag, bad charging, or a badly shaped furnace, and which by its presence interferes with the proper working of the furnace.

Scale.—(1) A term signifying the relative proportions of parts, and applied in mechanics to any instrument by which proportional parts are obtained. Specially an instrument for making or measuring drawings which are not drawn to full size, but to some proportional fraction of actual dimensions. Thus if an inch upon a drawing represents a foot on the actual piece of work, that is said to be drawn to a scale of an inch to a foot. That inch will be divided into twelve equal parts, and each twelfth part represents an inch of actual dimension. A fully divided scale is one each inch of whose whole length, usually of twelve inches, is subdivided thus into twelfths, so that a minute measurement can be taken at any part of its length. An open divided scale is one in which the end or terminal dimension only representing the foot, is divided into twelve parts, the parts intermediate being only partly divided. An open divided scale will thus contain more separate and distinct scales than a fully divided one, as for instance ½ in. at one end and 1½ in. or 3 in. at the other. Main divisions from which the scale takes its name, as ½ in., 1 in., 2 in., are termed the primary divisions of the scale. (2) The carbonate of lime which accumulates on the interior of boiler plates, and which is deposited by hard water.

Scaled Drawing.—A drawing made smaller than the work which it represents, but to a definite proportion which is specified on the drawing itself. See Scale. Many working drawings are made to full size (q.v.).

Scale Pan.—See Scum Trough.

Scaling.—(1) The process of removing the scale (q.v.) from the interior of boiler plates. (2) The taking of dimensions from a drawing by means of a scale (q.v.), in the absence of figured dimensions, or full sized delineations.

Scaling Hammer.—A keen-edged hammer used for the removal of the scale (q.v.) from steam boilers, which it effects by a process of chipping.

Scandinavian Cotton Belting.—A cotton belting solid woven instead of being made in plies. It can be obtained in widths ranging from one to twenty-four inches, and in light, medium, and heavy weft.

Scantling.—All quartering under five inches square. Also the transverse dimensions of a piece of timber is termed its scantling, as for example a piece having a scantling of 5 in. × 4 in.

Scarf.—See Scarf Joint.

Scarf Joint, or Scarf.—An oblique form of joint used by smiths and by carpenters, for the fastening together of rods of metal and strips of wood. The scarf joint in wood is often assisted and tightened by wedges inserted in the middle portion.

Scintillation.—Hard white iron scintillates or throws off sparks when in the ladle, so that the founder knows thereby the nature of the metal while yet in its molten condition. See Breaking, Jumper.

Scoop Wheels.—Wheels employed for lifting water and drawing it away from marshy lands. They are in shape somewhat similar to water wheels, but the place of the buckets is occupied by scoops, and the necessary power is supplied by windmills or by steam engines. These wheels are largely employed in the fen countries.

Scoria.—(1) Slag (q.v.). (2) Sullage (q.v.).

Scorification.—The melting of a small quantity of lead along with copper which is intended to be rolled into sheets, so called because of the scoria which collects on the surface.

Scotch Pig.—For ordinary foundry purposes Scotch pig is that chiefly used, both for light and ornamental castings, and mixed with other brands, for the stronger and tougher castings. No. 1 Scotch, mixed with scrap or with inferior brands, produces useful mixtures. It is smelted from the clay ironstone or black band ores, and exported from the Clyde chiefly.

Scotch Tuyere.—A form of blast furnace tuyere in which the water is made to circulate around a coil of pipe encircling the tuyere.

Scouring.—Plates of sheet iron used in the tinplate manufacture are scoured with sand and water to cleanse them, after preliminary immersion in dilute sulphuric acid.

Scouring Drum.—A rumble (q.v.).

Scrap.—Broken and worn castings, wasters, &c., clippings and ends of wrought iron and steel bars, and plates. The cast-iron scrap is remelted with an addition of new pig, and the wrought-iron scrap is piled or fagoted, and balled up or rewelded into blooms for future use.

Scraper.—The scraper for wood is a strip of old saw blade, used for smoothing over the face of planed timber whose grain is cross and crooked. It is held a few degrees over the perpendicular and pushed or dragged over the wood. It is sharpened by forcible barring up of the edge with a smooth piece of steel, such as a screw-driver or gouge. The scraper for metal is a file ground square across and sharpened at the end, and pushed with a straight or semi-rotatory movement over the metal at a few degrees only from the horizontal. It is sharpened upon an oilstone.

Scrap Heap.—A receptacle for odds and ends of waste material, which is utilised in various ways. Scrap cast iron consists to a slight extent of foundry wasters, but chiefly of old iron bought for the purpose of re-melting, while the scrap heaps of wrought iron or steel consist of the accumulation of shop refuse, and odds and ends cut to waste.

Scraping.—The bringing of surfaces to a very approximate plane condition by means of scrapers. All good wearing and sliding surfaces are scraped to a fit, to obliterate the coarse marks or grooves left by cutting tools and files, and to increase the number of minute points of contact. By scraping, the work of fitting is localised on certain spots for the time being, until at length so perfect a contact becomes possible that one plate will sustain another of equal weight on its under face. There is no mechanical operation by which such perfect results are possible of attainment as by scraping. To indicate the points of contact it is customary to rub a thin film of red lead and oil on the opposite surface to that which is being scraped, which red lead indicates by its transference to certain highest points the localised points of contact, to which points therefore the work of the scraper is directed.

Scraping Tools.—Tools which remove the particles from the materials on which they operate by abrasion, mainly or altogether, as scrapers, broaches, reamers, saws.

Scrap Iron.—See Scrap.

Scrap Steel.—Scrap steel is utilised in the open hearth processes, being mixed with pig iron and ore in the Siemens-Martin process (q.v.), and in the Siemens process (q.v.) also.

Scratch Brush.—See Wire Brush.

Screen.—A large rectangular sieve set up at an angle with the ground, and used for screening (q.v.) sand for foundry use.

Screening.—The passing of foundry sand through a screen (q.v.) in order to separate the lumpy from the finer portions, the former being pounded up afterwards and screened over again.

Screw.—A screw may be defined as a helix wound around a cylinder. The helix may be single, as in a common screw, or double or multiple, as in many forms of feed and endless screws. Fractional portions only of helices may be present, as in screw propellers. The helix, if further analysed, is found to be an inclined plane. The mechanical advantage of a screw (disregarding friction) is in the ratio of the circumference of the arm at which the power is applied to the pitch of the thread. Screws are square or angular in section, right or left-handed according to the direction in which the helix turns ; and they have different names designating their function or shape. They are cut by taps and dies, or in the lathe, and are used for almost all kinds of engineering work. The various terms used in connection with screws and screw threads will be found defined under their special headings.

Screw Area.—In a propeller is the area of the circle described by the tips of the blades.

Screw Barrel.—A chain barrel having a continuous spiral groove cut around its periphery to receive the links edgeways, the flat of the alternate links resting on the periphery of the barrel. Designed to maintain an even lap of the chain, and thus prevent overriding.

Screw Blade.—A sector of a screw thread, forming one of the arms or blades of a propeller screw.

Screw Blade Area.—The area of the oblique surfaces of the blades of a screw propeller.

Screw Bolt.—More commonly termed a bolt (q.v.), signifying therefore a fastening made by means of a nut engaging with a screw thread, as distinguished from the riveted form of fastening.

Screw Chasing.—The cutting, or smoothing and polishing of screw threads in the lathe by means of chasing tools. See Chaser.

Screw Chuck.—See Taper Screw Chuck.

Screw Clamp.—The ordinary clamp used by wood workers. Small screw clamps are sometimes called thumb screws to distinguish them from the larger timber or bar clamps.

Screw Coupling.—The ordinary coupling used for railway rolling stock. Links are attached to the opposite draw hooks, and these are united by a double screw, right and left hand respectively, by turning which in one direction or another the coupling is lengthened or shortened.

Screw Cutter's Gauge.—A centre gauge (q.v.).

Screw Cutting.—The formation of screw threads on cylinders. It is performed by taps (q.v.) and dies (q.v.) by chasers (q.v.) and in a screw cutting lathe (q.v.) sometimes with a traversing mandrel (q.v.). A good screw should have a full thread, be cut clean, and have sharp edges.

Screw Cutting Lathe.—An iron turner's lathe, in which provision is made for cutting screw threads without the aid of a die, by the use of a screw tool only. The tool is held in the slide rest and traversed in a longitudinal direction to an exact and measured distance for each revolution of the lathe, the traverse being effected by means of a guide screw (q.v.) and change wheels (q.v.). A screw cutting lathe is a self-acting lathe (q.v.) though the latter term is commonly used to designate a lathe which is non-screw-cutting.

Screw Dies.—See Die.

Screw Down Valve.—A screw valve. A valve which is raised or depressed by means of a screw. Globe valves, sluice valves, and others similar in type are familiar examples.

Screw Driver.—Screw drivers are either flat, or rectangular, or circular in section. The latter form is useful to pattern makers because of the facility which it affords for driving screws down deep into round bored holes. Breadth of handle gives increase of leverage, hence handles are properly made broad, and oval in section. Length also gives more power. For turning large screws into metal work, screw drivers with cross handles in the form of a letter $\top$ are used.

Screw Driver Bit.—A screw driver, the end of whose shank is square tapered to fit within the socket of a brace, similarly to brace bits. Being revolved with the brace its action is quicker than that of an ordinary hand-worked screw driver.

Screwed Stays.—The bar stays, or tube stays, of boilers are also termed screwed stays. Ordinary riveted stays (q.v.) are usually screwed first of all, but the term is not so commonly applied to them as to those just named.

Screw Engine.—An engine specially designed to drive the screw of a steamer. It is generally of the inverted cylinder (q.v.) type and compound (q.v.). Engines having two cylinders only, one high and one low pressure, were, until recently, mostly used, and are still for small vessels. But for large ocean-going steamers these are being replaced by triple expansion engines of three cylinders, one high and two low pressure, and to a limited extent by quadruple expansion engines (q.v.).

Screw Feed.—The feed (q.v.) of a machine which is effected through the medium of a screw and nut. Screw feeds are applied to many drilling machines, to screw cutting lathes, and other machines.

Screw Gearing.—Gearing in which the teeth are not parallel with the axes of the shafts. The axes may be placed at any angle, and the teeth may correspond with a single, or with a many threaded screw. Worm gearing and angle wheels come under this heading.

Screw Head.—The upper portion of a wood, or metal, or other screw, being the part by which it is driven into its place.

Screwing.—The cutting of screw threads by means of taps and dies. Most commonly used by workmen to denote the cutting of bolt screws in the screwing machine, or the cutting of gas threads on piping rather than of work done in the lathe, which is more often denominated screw cutting.

Screwing Flange.—A cast-iron flange, screwed internally for attachment to the similarly screwed ends of wrought-iron piping, the piping being then united by bolts passing through opposing flanges. Screwed flanges are used in preference to common tubular unions, because of the readiness with which they can be disconnected, and also for connecting pipes to other flanges, as on cast-iron work, engine cylinders, &c., into which it would be inconvenient or impossible to screw the wrought-iron tubing directly. Better steam tight joints can also be made with flanges than with screwed ends alone.

Screwing Machine.—A machine used for screwing bolts in quantity, the dies revolving.

Screwing Tackle.—A term commonly applied to the appliances used for cutting screws apart from the aid of a lathe. Stocks, dies, taps, and tap wrenches are included under this heading.

Screw Jack.—A lifting jack which is actuated by a square threaded screw worked either by a lever bar inserted in a hole at the upper end or head of the screw, or by a lever permanently attached to the head and provided with a ratchet and click similarly to a ratchet brace (q.v.), or by a worm and worm wheel, the worm wheel embracing the screw and becoming its nut. The nut not being free to move endways, the screw must of necessity traverse. The mechanical gain is obtained by multiplying the power exerted at the end of the lever into the product of its length with the pitch of the screw. In the larger jacks the means of longitudinal traverse is also provided by a horizontal screw working in a base. See Bottle Jack, Traversing Screw Jack, Hydraulic Jack.

Screw Key.—A spanner (q.v.) for turning screw heads. Usually applied to spanners which belong to machines and workshop appliances.

Screw Mandrel.—See Traversing Mandrel.

Screw Nail.—A common wood screw (q.v.) as distinguished from an ordinary nail driven in by percussion, the term being akin to the term screw bolt (q.v.).

Screw Nut.—See Nut.

Screw Pile.—A foundation pile which is not driven into the ground, but turned in by means of a tapered screw, of one or of several revolutions at its lower end. It is used in ground too hard or too uncertain for the reception of driven piles.

Screw Plate.—A steel die plate used for the cutting of small screws of about $\frac{1}{4}$ in. and under, in diameter. A number of distinct holes are screwed in a single plate, and the plates are also made in various sizes. The finest of the holes in the smallest plates used for model work would cut a thread on a very small needle. Screw plates are diminished in thickness as the holes decrease in size. Taps are supplied corresponding in size with the holes in plates.

Screw Press.—See Fly Press.

Screw Propeller.—See Propeller.

Screw Rate.—The rate of a screw signifies the number of threads per inch which it contains.

Screw Shaft.—The shaft which drives a ship's screw, and to which it is directly attached. Screw shafts are usually made in lengths and united with solid couplings. The thrust of the screw and shaft is taken by thrust collars and thrust blocks (q.v.).

Screw Slotting.—The slotting of the grooves in the heads of cheese-headed and button-headed screws. It is effected with a rotary cutter or saw.

Screw Stock.—See Die Stock.

Screw Surface.—The surface formed by the development of a helix, or of a screw blade. The latter is termed the screw blade area (q.v.).

Screw Tap.—See Tap.

Screw Threads.—Formerly each manufacturer had his special screw rates designed without reference to uniformity with any other. A sense of the advantage of a uniform series of threads has, however, resulted in the universal adoption of certain standard rates. In England the Whitworth threads are the standard, the Sellers' threads in America, and the decimal threads in France.

Screw Threaded Gauge.—An assemblage of laminal or thin steel plates, whose edges are notched to fit screws of many different pitches, and by which the pitch of any thread within the range of the gauges can be ascertained.

Screw Tool.—A tool whose edge is ground to the sectional shape of the interspace between two contiguous threads, Vee shaped, or square, as the case may be, and used for cutting screws in the screw cutting lathe. For cutting screws in the lathe by a hand feed only, a chaser (q.v.) or chasing tool is employed.

Screw Valve.—See Screw Down Valve.

Screw Worm Chuck.—See Taper Screw Chuck.

Screw Wrench.—A spanner, the width of whose jaws is rendered adjustable by a screw to an extent within the limits of the traverse of the screw. The strain put upon a spanner is so great that this traverse is necessarily very short, and hence a screw wrench will enclose a few bolt sizes only. It is convenient for use in outdoor work, saving the weight of a number of separate spanners.

Scriber.—A steel tool used for marking lines on work in wood or metal. The timber scriber is pointed at one end and knife-shaped at the other, the former for scratching, the latter for cutting. The fitter's scriber is pointed at both ends, but while one end is straight the other is usually curved for marking underneath work.

Scribing.—Before proceeding to cut a piece of wood or metal to outline or shape, whether by hand or machine, the outline is marked out on the rough material as an exact guide in the subsequent operations. The process is variously termed scribing, lining, or setting out, the former word taking its name from the scriber.

Scribing Block.—A block of metal carrying a scriber pivoted on, and pinched by a screw, for adjustment to different heights. It is used for lining off centre lines on work when laid on the surface plate. Hence called a surfacing gauge.

Scrieve.—To scribe. See Scribing.

Scroll Chuck.—A universal chuck (q.v.)

Scroll Irons.—Small brackets attached to the underside of railway wagons, to which the ends of the bearing springs are attached.

Sculls.—The lining or skin of metal and scale which forms around the interior of the ladles used in casting steel and iron. These are knocked out previous to the relining of the ladles with ganister or fire clay.

Scum Cock.—A cock placed in the side of a marine boiler for the purpose of getting rid of the dirt and scum which are carried to the surface of the water. A pipe leads from the cock through the boiler shell and terminates in a copper dish or trough immersed just below the water line, and into which the scum collects.

Scum Trough, or Scale Pan, or Sediment Collector.—A shallow trough provided in large marine and stationary boilers, for the collection of the fine sediment which is ballooned to the surface by the ebullition of the bubbles of steam. See Ballooning, and Scum Cock.

Sea Cock.—An injection cock (q.v.) which is placed upon the side of a steamer, leading from the sea to the condenser. It is supplementary to the usual injection cock.

Seam.—A lap joint made between the wrought-iron, or steel plates of which riveted and welded structures such as steam boilers are built up.

Seamless Tubes.—Solid drawn tubes (q.v.).

Seam Set.—A coppersmith's tool, which is used for closing up seams in sheet metal work.

Searing.—The practice of smoothing the surface of the rougher class of foundry patterns with a flat-faced red hot iron, as a substitute for paint or varnish. The grain is thereby smoothed and prevented from rising up when in contact with the damp sand.

Sea Sand.—The sea sands of many localities are the débris of sandstone rocks, and are suitable for core making, being free or porous, and so offering facility for the necessary venting. They are rendered adhesive by the addition of clay water.

Seasoning.—The removal of the sap from fallen timber. This is effected by allowing the moisture to dry out by free exposure to the atmosphere. Timber takes from one to three years to season in the ordinary way, but much less when artificial seasoning (q.v.) is resorted to.

Seat.—A seating (q.v.).

Seating.—(1) A casting which becomes the base or foundation for portions of work. Usually applied to the castings which are attached to boilers to carry the safety, stop, and other valves. Also called a seat.

Sea Water.—Sea water is of necessity used in marine boilers, and though a slight degree of saltness is not injurious, but rather beneficial, too much salt is detrimental, producing a rapid deposition of scale. Since the water in the boiler increases in saltness by reason of the constant evaporation going on, frequent testing by means of a salinometer is essential, and judicious blowing off of the brine at intervals

Sec.—The secant of an angle.

Secants, Line of.—See Sines, Line of.

Second Cut File.—A file having a degree of coarseness midway between that of a bastard and a smooth cut (q.v.)

Second Tap.—A tap. intermediate in size between a taper and a plug tap.

Section.—The division of a substance or the cutting of a mass in any direction. The term is used chiefly to indicate those particular views in a drawing by which the internal or central parts of a piece of work are delineated. It may therefore be defined as a drawing showing a view of an object on the assumption that the object is cut across at that particular spot. The direction in which the section is taken is indicated by the words "longitudinal," "transverse," "oblique," "vertical" or "plan,"or by lines dotted on the drawing, as "section on line A B, line C D," &c.

Sectional Area.—This has reference to the sum total of the area of the quantity of metal in a casting or forging irrespective of its diameter or linear dimensions. The term is chiefly employed in reference to columns whose strength to resist torsion varies widely with their sectional areas, for example, a hollow column whose sectional area is the same as that in a solid column, will if its inner and outer diameters are as 8 to 10, be $2\frac{1}{4}$ times stronger than the solid one, notwithstanding that there is the same amount of metal in each.

Sectional Boiler.—A steam boiler which consists of an aggregation of small independent heating tubes. The advantages of these boilers consist in the high pressures to be obtained in them, their immunity from, or at least localisation of explosions, the strength of small tubes, and rapid transmission of heat through them, and the ease with which local repairs can be effected. On the other hand there is a tendency to the accumulation of deposit in the tubes, and difficulty of clearing them out, and a tendency also to overheating. Also called water tube boiler, unit boiler. See Water Tube Boilers in Dict. of Modern Terms.

Sectional Plan.—See Plan Section.

Section Angle.—The angle between the top and front faces of a double-edged turning tool.

Section Colours.—The colours used by engineers to indicate sections of various materials. Cast-iron section is grey, wrought-iron section is blue, steel section is purple, gun-metal and brass sections generally are gamboge. Wood is brown (umber). When colours are not used, conventional varieties of shading are employed.

Sector.—(1) A universal scale by which numerous computations can be obtained by measurement simply. It consists of two legs divided out into lines, termed lines of chords, sines, polygons, &c., whose uses are described under their special names. The lines of division are termed sectorial lines. Sectors are usually six inches long when folded and twelve inches when opened out. (2) The sector of a circle is the part bounded by two radii and the arc which they enclose.

Sectorial Lines.—See Sector.

Sediment.—The muddy deposit which collects in steam boilers, consisting chiefly of carbonate, and sulphate of lime. See Incrustation, &c.

Sediment Collector.—See Scum Trough.

Seed Oil.—See Vegetable Oils.

Segment.—A segment of a circle is the portion bounded by an arc (q.v.) and its chord (q.v.).

Segmental Rack, or Segmental Wheel.—An arc of a toothed wheel, moving about its proper centre, and used for imparting reversible motion to a spindle through the medium of a wheel gearing into the rack.

Segmental Rivet.—A rivet, the vertical section of whose head is that of a segment of a sphere, being hemispherical nearly.

Segmental Wheel.—See Segmental Rack.

Seizing.—The jamming fast of bearing parts, caused by the expansion due to heat and the absence of sufficient lubrication. When bearings seize, they should be cooled by pouring hot water thereon, to be followed afterwards by applications of colder water.

Self-acting.—A machine or some section of a machine whose movements are not executed by a direct intervention on the part of the attendant, but which are derived immediately from the motions of the machine itself. The chief motions in ordinary machines are rendered self-acting through the intervention of screws, cams, gearing, levers, &c.

Self-acting Balance Crane.—A balance crane in which the balance weight is rendered automatic, with a view of preventing accidents due to the overturning of the crane, through the balance weight being run out to its full distance when not lifting a load. The ends of the tie rods terminate in looped chains near the side frame, and each of these chains is attached, one end to a lug in a derrick barrel, the other to the short end of a bell crank lever. The lower or longer limbs of the bell crank levers are attached to balance weights, and these again by means of links to the actual balance weight itself, which is a round ball rolling on the tail girders of the crane. The action of the mechanism is such that when a load is suspended from the crane the long arm of each bell crank lever with its balance weight is lifted and the actual balance run out, the distance to which it is run being proportional to the load. When the load is dropped the levers drop, and the balance ball runs inwards.

Self-acting Lathe.—A lathe furnished with a slide rest whose motions are rendered either partially or entirely self-acting. Strictly speaking a self-acting lathe will have two motions, one by rack and pinion, or by screw, or by each at pleasure, for traversing or longitudinal cutting: the other by back shaft and screw gearing, and geared spindle in the slide rest, for surfacing or cross traverse. But it is also customary to term a lathe self-acting even though the longitudinal traverse only is so rendered by means of a leading screw, while the surfacing movement is effected by hand. Almost all engineers' lathes are self-acting in both motions.

Self-acting Lubricator.—A lubricator which is self-feeding, the oil passing over from a reservoir into the actual oil hole, being drawn up in many forms through a twist of tow or gasket by capillary attraction. The difference between the steam and the exhaust pressure in steam cylinders is also made to actuate the movements of valves in some forms of lubricators, while in others the difference in the specific gravities of oil and condensed steam is utilised to cause the oil to pass through an overflow pipe into a steam cylinder. The term embraces any lubricator which is only filled at intervals, and in which the feeding goes on until the supply is exhausted. See Displacement Lubricator, Impermeator, Steam Lubricator, Syphon Lubricator.

Self-centring.—A term which signifies the automatic setting or marking of cylindrical work truly. It is effected in the lathe or similar machine by means of a chuck whose jaws are worked simultaneously to or from the centre. See Centring Chuck. The bell centre punch (q.v.) is also an example of self-centring, so also is the use of a square centre (q.v.).

Self-contained Machines.—Machines whose parts are not attached to distinct fixings, but are all so combined that one foundation or attachment is sufficient. Thus a radial drill is self-contained when the table and pillar are bolted together, instead of, as is often the case, the radial arm being attached to a wall or pillar, while the table is bolted to the stone foundation. Independent machines (q.v.) may be considered as self-contained.

Self-delivery.—The delivery of the hollow portions of patterns in the sand, due to their own taper only, as distinguished from coring out. In self-delivery the core is therefore of the same material as the remainder of the mould.

Self-fluxing Ores.—Some red hæmatites contain sufficient calcareous matter as to require no additional flux, hence the term self-fluxing. They occur in Sweden chiefly.

Sellers' Screw Threads.—The American standard engineers' screw threads in which the sides slope at an angle of 60°. The tops and bottoms of the threads are cut off square to a depth equal to one-eighth of the pitch.

Semaphore.—A signal attached to a tall pillar; as used on railways the signals consist of arms and lamps worked by levers and chains.

Semi-Beam.—A cantilever (q.v.).

Semi-Bituminous Coal.—See Bituminous Coal.

Semi-Chord.—Half the length of the chord of an arc.

Semi-Circle.—A half circle, bounded by the lines of circumference and of diameter.

Semi-Fixed Portable Engine.—A boiler and engine of the common portable type, which, instead of being mounted upon wheels, is made a fixture, the fire box resting directly upon the ground and a cast-iron standard being placed underneath the smoke box.

Semi-Girder.—A semi-beam or cantilever (q.v.).

Semi-Portable Engine.—An engine which is in the main of the portable type (see Portable Engine), but which differs from it in this respect, that instead of being mounted upon wheels it is attached to an iron base or frame, and this to a solid foundation of stone or timber. It is used generally for the same kinds of work as are portable and light horizontal engines, as sawing, pumping, winding, driving machines, and for work located in one place, and in work of a temporary character.

Semi-Steel.—A steely iron. See Puddled Steel.

Sensible Heat.—Heat which is measured by the thermometer, as opposed to latent heat (q.v.).

Service Pipe.—The small piping which conveys liquid from a main pipe to its ultimate destination.

Serving.—The covering of a rope with spun yarn to protect it from chafing against the pulley or drum over which it works.

Serving Board.—The board employed in serving or winding the spun yarn around a rope. Its shape is somewhat like that of the letter ⊤. The yarn is wound around the rope in a transverse direction, and is tightened as it is wound by being reeved round the neck of the board, while the handle affords the necessary leverage for pulling the yarn taut.

Serving Mallet.—A cylindrical block of iron used in serving (q.v.) a rope.

Set, or Sett.—(1) A narrow square nosed or round nosed chisel-like tool used by fitters and boiler makers for chipping grooves in metal. (2) Broad chisel-like tools used for cutting off hot or cold bars on the anvil, the hot set being ground thinner than the cold set. They are attached both to handles as ordinary hammers or to hazel rods. (3) A hook wrench (q.v.). (4) The deflection which a bar or structure undergoes when subjected to strain or stress is its set. See Permanent Set.

15*

Set after Fracture.—See Ultimate Set.

Set Hammer.—A form of smith's hammer whose face is rectangular in plan and its edges square, and which is laid upon the work and struck with a heavier hammer or sledge in order to form a set-off (q.v.). Also called a set, simply.

Set Off.—A term used in a general way to signify the production of an abrupt shoulder in a piece of work, or the standing out of one portion of a structure at an abrupt angle from another portion.

Set Screw.—A plain screw having a square or other shaped head, made use of for tightening purposes. It holds by friction alone, and is not suitable for machine parts subject to excessive vibrations.

Set Square.—A square whose shape is that of a right-angled triangle. the angles of the hypothenuse being usually either 45°, or 30°, and 60°. Used for laying down lines, and for checking work where the ordinary try square being of different thicknesses in blade and stock would not be suitable.

Sett.—See Set.

Setting.—(1) The fixing of engine slide valves in their proper positions relatively to the ports, valve rods, eccentrics, and cranks, is called setting of valves (q.v.). (2) The production of a set off (q.v.) in a piece of work. (3) The placing of portions of mechanism in their exact positions for the fitting of keys, the drilling of holes, and the marking of centres. See also Saw Setting.

Setting Block.—A round-edged block of iron upon which a saw is laid longitudinally during the operation of setting. The amount of set for different saws is regulated by the amount of angle given to the saws relatively to the curve of the block.

Setting Down.—The flatting or hammering down of portions of smith's work with a set hammer or flatter.

Setting Down Screws.—The top screws in the standards or housings of bending rolls, by which the position of the top roll is adjusted.

Setting Hammer.—A light cross pane hammer (see Pane) used for saw setting, purposely made narrow to permit of the striking of a saw tooth without risk of hitting its next neighbour. The teeth are struck while the saw is laid on a setting block (q.v.).

Setting of Boilers.—The placing of steam boilers permanently in place. Cornish and Lancashire boilers are set on brickwork, the flues being built in the brickwork itself. Marine boilers are set on saddles. Locomotive boilers are not set, but slung.

Setting of Valves.—The setting of engine slide valves is an operation of great nicety, and is performed in various ways, dependent upon the nature of the link motion. The aim in every case is to obtain a precise and constant amount of lead for the valve, so that the greatest angular advance of the eccentric shall coincide at each revolution with the lead. Also it is the aim, though not always practicable, to get the lead equal at both ends of the stroke, except in the case of vertical engines of large size, where the bottom lead is slightly increased over that of the top to allow of wearing down. Usually in setting valves the practice is to first place the crank on dead centres, then turn the eccentric into the position of greatest angular advance corresponding with the position of the crank, set the valve to the position open to lead, also corresponding with the position of the eccentric and crank, and then having already made the rod joints, to trammel off the precise length of the connecting rod. Then turning the crank, eccentric, and valve into the exactly opposite positions, to repeat the operation, and shut or weld up the eccentric rods to their ends. Lastly, before finishing off the rods to repeat the previous operations and make any necessary adjustments. But when the rods are already welded up before setting, it is necessary to omit the cutting of the key-way for the eccentric on the crank shaft until the setting is finished, and the length of the connecting rods may have to be adjusted by planing off some portion of the ends of the rods, or conversely by lining them up.

Setting Out.—See Lining Out.

Setting Over.—The placing of the mandrel centre of a lathe poppet out of line with the axial centre of the headstock in order to provide for the turning of taper work between centres. In many lathes the upper portion of the poppet is made movable on a lower base or foundation plate for this purpose.

Set Work.—Work which is repeated many times over or perpetually in the workshop. Work which is a speciality, and which therefore is done more cheaply and better by firms which engage in it mostly or entirely, than by those who do work of a general character. It is carried out by minute subdivision of labour, and piece work, and though cheap, tends to produce an inferior and one-sided class of workmen.

Shackle.—A loop to which the end of a chain is attached. The open ends of the loop are generally formed into eyes through which a sustaining bolt or pin is passed.

Shackle Bolt.—A bolt having a shackle formed at the end opposite to the nut.

Shading.—The colouring or lining of drawings to give to parts the appearance due to relief, or to circular, sectional, and other forms. Working drawings are shaded but slightly, elaborate and complete shading being reserved for finished and for general drawings. Conventional modes of shading are adopted for various metals.

Shaft.—A spindle which revolves in bearings and carries pulleys or gear wheels for the transmission of power. The principal strain to which shafts are subject is that of torsion (q.v.), and the force of torsion is proportional to the cubes of their diameters.

Shaft Boss.—The larger boss of an engine crank which carries the crank shaft (q.v.).

Shaft Coupling.—See Coupling.

Shafting.—A line or succession of shafts united with couplings and used for the carrying of pulleys for driving machinery.

Shafting Oil.—Oil used especially for the lubrication of shafting.

Shaft Straightener.—A machine used for straightening shafts. It acts by the uniform and adjustable pressure of three rollers divided out equally in a circle, and between which the shaft is passed on to two larger rollers.

Shaft Tunnel.—See Tunnel.

Shake.—A line of fracture in timber, the splitting being caused by the shrinking up of the fibres due to the desiccation or drying up of the moisture.

Shank.—A hand foundry ladle is sustained by a shank, or rod bent round as a hoop to envelop the body of the ladle, and prolonged at one end and there bent downwards to fall over a cross bar, and forged at the opposite end into a cross handle. Three or four men carry such a ladle, one at each end of the cross bar and one at the cross handle, or one at each end of the latter. By the leverage of the cross handle or crutch the metal is tipped into the mould. Shank ladles hold from 1½ to 3 or 6 cwt. of metal. (2) The stem or body of a cutting tool, as the shank of a drill, or the upper portion which enters into the chuck or the socket through which it receives its motion. It is tapered or parallel, round or square in section, according to the shape of the socket or chuck.

Shank Ladle.—A foundry ladle (see Ladle) suspended in the central loop of a horizontal handle or shank (q.v.).

Shaping.—The shaping of metal in a shaping machine, though analogous to that of planing, differs therefrom both in the extent of the surface operated upon, and in the outlines imparted thereto. The shaping machine only operates on small areas, seldom over 12 in. in length, commonly much less; also convex and concave cut surfaces can be shaped as readily as plain ones, the former by threading the work on an arbor under the tool, and imparting a slow rotatory movement to the arbor, the latter by using a tool holder having a worm and quadrant movement, by which the tool is caused to rotate slowly through an arc of a circle.

Shaping Machine.—A machine used for shaping metallic masses which are not sufficiently bulky to go into the planing machine. The tool is carried on a rigid arm which moves in a horizontal direction, the length of whose stroke is capable of regulation by means of a connecting rod movable in a slot in the arm. Quick return (q.v.) motion is given in the larger but not in the smaller machines. An angle plate is attached to the front of the machine and is movable vertically or transversely. Most machines have a worm and wheel segment arrangement on the tool box for circular shaping.

Sharpening.—The imparting of a fine edge to a cutting tool by abrasion or fretting on an oil stone.

Shear.—(1) To cut with shears (q.v.). (2) A bar is said to be in shear when it is subject to shearing stress (q.v.).

Shearing.—The act of cutting with shears.

Shearing Machine.—A machine in which shears or shearing knives are worked by power. Commonly the processes of punching and shearing are combined in one machine, the punching below, the shearing above, or in the larger machines at opposite ends. Shearing machines are employed in iron and steel manufactories and in boiler shops chiefly, for cutting off the rough edges of plates, or for cutting to precise dimensions. Bars and angle irons are also cut, the length and shape of the knives varying with the nature of the work. The knives are placed at an angle with each other, the cut taking place from one end towards the other like that of scissors.

Shearing Strength.—The strength necessary to shear iron is about equal to its tensile strength.

Shearing Stress.—The stress to which a body is subject when force is applied to it in a direction parallel with its section. Its mean intensity is equal to the shearing force divided by the area of the section.

Shearing Tools.—Tools which remove material by detruding action (q.v.), or violent separation. They include shears and punches.

Shear Legs, variously termed **Sheer Poles, Shears, or Sheers.**—A rude but strong form of crane, having three legs or pillars set leaning at an angle towards each other for supporting the lifting tackle. The gear wheels may be attached to one of the legs, or be fixed independently, and hand or steam power may be employed. Timber or iron are used in the construction of the legs.

Shears, or Sheers.—(1) Knives used for severing sheet metal, bars, angles, &c., by shearing, or cutting parallel with the section. They act by the opposition and pressure of the cutting edges, the faces being ground generally to an angle of from 80° to 90°. Shears are worked by hand power for the thinner sheets, but for the thicker plates shearing machines are used. (2) The term is often used to designate the bearers of a lathe bed. (3) Shear legs (q.v.).

Shear Steel.—Forged steel composed of bars of blistered steel laid side by side to form a bundle or fagot, and drawn down and welded under a hammer.

Sheave.—(1) A term applied to the eccentric body or to that part which is keyed directly on the shaft, to distinguish it from the eccentric straps (q.v.). (2) A sheave wheel.

Sheave Wheel, or Sheave, simply.—A wheel whose groove is made to receive a chain or rope. The section of the rim of the sheave may be a plain semi-circle, or veed, or a single groove, or a double groove, according to the work which the wheel has to do.

Sheer Poles.—See Shear Legs.

Sheers.—See Shears.

Sheet Brass.—Brass rolled into sheets and used for a variety of purposes, chiefly for ornament, covering steam domes, gauges, handrailing, &c. It is either hard or soft, similarly to wire.

Sheet Copper.—Used by engineers for the making of steam pipes chiefly.

Sheet Iron.—Strictly speaking, the thinnest iron which is rolled, as distinguished from plates. See Plate, Sheets.

Sheet Lead.—Used for making the joints of steam and water pipes. Used also for lining up small patterns.

Sheet Tin.—Iron sheet, coated with tin. It is used by engineers in the making of the lightest kinds of templets.

Sheet Metal Gauge.—A gauge used for measuring the thicknesses of sheet metals. It may be a micrometer caliper (q.v.) or a wire gauge (q.v.), the former measuring in terms of the inch, the second by manufacturers' numbers.

Sheet Mills.—Rolling mills of light construction, used for the rolling of sheets (q.v.).

Sheet Rubber.—Vulcanised india-rubber, insertion or plain, is made into sheets of large size for the convenience of being cut into washers, collars, &c., of any required dimensions. It may be had in sheets of from 4 to 30 yards long, by 48 in. in width, and from the $\frac{1}{75}$th of an inch upwards in thickness. Its usual thickness is from $\frac{1}{8}$ to $\frac{3}{16}$ for ordinary work.

Sheets.—All plates of malleable iron below No. 4 B.W.G. or ·238 in., are termed sheets, as distinguished from plates (q.v.).

Sheffield Files.—Sheffield is the great seat of the file industry, the majority of engineers' files being made there. For shapes and characteristics, see specific names.

Shell.—The shell of a boiler is the outer casing of all, enclosing fire box, water space, cross tubes, &c. The shell of a cock is the outer envelope or casing through which the liquid passes on the turning of the plug (q.v.).

Shellac.—Shellac is produced by the agency of members of the family Coccidæ, of the order Rhynchota or beaked insects. The true lac insect

(*Coccus lacca*) is a native of Siam, Assam, Bengal, Burmah, and Malabar. The lac is produced by the agency of the females chiefly, who, once settled upon a twig of a tree—the trees frequented being species of Butea, Ficus, and Croton—never leave it, but suck up the resinous juices, feeding on and enveloping themselves in them until they die, leaving their eggs enveloped in the secretion thus obtained, coloured with dye matter secreted from their own bodies. Generations of insects hatched from successive colonies of eggs live on the same branch until it becomes coated to the extent of half inch or more, when the twigs are collected by the natives and subjected to various processes of boiling, kneading, and squeezing, by which the resin is obtained in varying degrees of purity known as stick-lac, seed-lac, and shell-lac, the latter being the most free from colour, and most soluble in spirit, and that which is exclusively used in the mixing of varnish for foundry patterns.

Shellac Varnish.—This is made by simply dissolving shellac (q.v.) in naphtha, in methylated spirit, or in spirits of wine, the former being the worst, the latter the best solution, but the second being that most generally employed. The proportions should consist of about two pounds to the gallon, and ought to be mixed without the aid of artificial heat.

Shell Bit.—A bit used for boring wood, whose section is a convex and concave curve, roughly though not exactly of a half circular form. See Nose Bit, Gouge Bit.

Shell Plates.—The plates of which boiler shells are composed.

Shell Reamer.—See Reamer.

Shield.—(1) A covering employed to protect the bearings and spindles of emery-grinding machines from the action of the gritty dust. (2) A guard placed over or in front of band and circular saws and portions of machinery to protect the workmen from accidents. (3) The curved device or outline formed by the union of the body of a two-way cock with its branches, is often termed the shield.

Shifting Link.—A slot link (q.v.) whose concavity lies towards the crank shaft, and which is lifted or lowered relatively to the link block or pin. Originally known as Stephenson's link motion.

Shingler.—The workman whose task consists in shingling (q.v.) puddled ball.

Shingling.—The process of hammering or rolling puddled ball in order to convert it into bar or sheet iron.

Shingling Hammer.—The hammer, either tilt or steam, by which the puddled ball is converted into a bloom.

Shingling Mill.—The forge where the puddled ball is converted into malleable iron.

Shingling Tongs.—The tongs used in moving the puddled ball about under the tilt or steam hammer.

Shipping.—The putting on of a belt over its pulley. See Belt Shipper.

Shipping Measurement.—The mode of measurement adopted in the case of machinery which is received as cargo on board ship. The measurement is taken over the extreme limits of any piece of machinery. Hence projecting portions should, where practicable, be removed for shipment, and stowed into as small an area as possible.

Ship Plates.—Plates of malleable iron, or mild steel of inferior material, or material imperfectly refined. Their tensile strength is lower than that of other qualities of iron or steel, and the metal of which they are composed is not sufficiently homogeneous or reliable to render them suitable for any but the commonest work. The hulls of iron ships are largely built of these, hence the name.

Ship's Hoist.—An engine and boiler attached to a carriage or truck running on wheels, used for the general and miscellaneous work where power is required on board ship.

Ship's Pump.—A suction pump, varying in form, employed on ship board for general use, pumping bilge water, &c.

Shock.—The jerk or reflux action produced in pumps and long pipes, being due to the reversal of the water pressure. Shock is lessened by the use of an air vessel (q.v.).

Shoe.—A cap or socket which receives the end of a piece of timber, and which sustains the end pull of a truss rod (q.v.) or takes the thrust of the legs of shears, or of roof principals, or carries a pivot or bearing as in the pillars of wooden cranes, or acts in a general way as a protection to timber ends, as in pile shoes. Shoes are commonly made of cast iron, but often wrought iron is used in their construction. The base plates which support the housings of rolling mills are also called shoes.

S Hooks.—Iron hooks whose shape is roughly that of a letter §. They are suspended from the bars of moulding boxes in most cases, to aid in binding together and supporting the body of sand which is enclosed by the sides of the boxes. The adhesion of the sand is assisted by dipping the hooks into clay water.

Shooting.—The planing of the edges and ends of timber straight, with or without the aid of a shooting board (q.v.).

Shooting Board.—A board used for planing or shooting the edges of timber square and straight. It consists of two pieces of wood screwed one on top of the other, the upper one being narrower than the lower by so much as will allow a sufficient breadth of bearing for the side of the trying plane on the margin of the lower one. The stuff to be planed is laid upon the upper board, and the side of the plane being laid upon the uncovered margin of the lower one, with the cutting iron towards the edge of the stuff, a shaving is removed as the workman pushes the plane forwards. The advantage of the use of a board over that of the vice is that in the former the edges planed, assuming that both board and plane are true, are not only straight, but square also with the faces of the stuff.

Shop Tools.—Numerous tools and appliances which are not provided by the workmen in departments, but by the firm, for the common use of all. The term does not embrace machinery so much as minor articles, as standard rules, straight edges, surface plates, trestles, dies and die blocks, water cans, sieves, shovels, and scores of kindred articles used by all, but owned only by the firm.

Shop Traveller.—An overhead travelling crane employed in the workshop under cover, and used for lifting about heavy masses to the machines, various machined and finished parts for the erectors, ladles of metal or castings in the foundry, plates, girders, bars in the boiler shop, forgings in the smithy, &c. Most shops are furnished with travellers in preference to cranes. See Overhead Traveller.

Short Columns.—A term applied to columns whose length exceeds their diameter by from three to five times that amount, and which yield to rupture by simple crushing only.

Short Heat.—The making hot of a short length only of a bar of iron in a smith's fire in order that the heated portion alone shall be operated upon, without incurring the risk of distorting the adjacent portions.

The heat is localised by covering all but the portion to be heated with damp slack or small coal, or in some instances by enclosing in a matrix of damp clay.

Short Link Chain.—See Close Link Chain.

Shoulder.—That portion of a shaft, or of a stepped, or of a flanged structure where an immediate increase of diameter occurs.

Shovel.—Broad light shovels are used in foundries for filling the moulding boxes with sand.

Shrinkage.—See Contraction.

Shrinking Head.—See Feeder.

Shrinkage Hole.—A depression or slightly hollowed space left in the face or in the body of a casting, due to imperfect feeding. See Feeder. Shrinkage holes are prevented by making feeders large enough to hold a supply of hot metal whose mass is sufficient to prevent solidification taking place in the feed metal before the casting itself is set.

Shrinking On.—The clasping of wheel tires around their centres, being first expanded by heat, and then allowed to tighten on the centres by virtue of their own shrinkage in cooling down. The amount allowed for shrinkage on a wheel of about 2 ft. in diameter is barely $\frac{1}{8}$ of an inch.

Shroud.—(1) A flange cast against the ends of wheel teeth in order to increase their strength or to promote smoothness of motion. If cast up to the pitch line the wheels roll together on their pitch circles, and the term half shrouding is applied to it. If cast to the points of the teeth the purpose is to afford the maximum of strength, and this is termed full or whole shrouding. Whole shrouding is believed to increase the strength of wheel teeth by from 40 to 50 per cent. (2) The circular flanges around the edges of a water wheel, between which the buckets extend.

Shrouded Wheel.—A gear wheel provided with shrouds. A water wheel is also a shrouded wheel. See Shroud.

Shrouding.—The casting of shrouds on the ends of wheel teeth or of the buckets of water wheels. See Shroud.

Shroud Laid Rope.—An ordinary rope formed of three strands twisted together. These are stronger than cable laid ropes (q.v.) and strongest of all when untarred or white.

Shutting.—The shop term for the welding together of joints in wrought iron or steel.

Shutting Link.—The union of the free ends of chains, either to each other or to other portions of work, as rings, hooks, &c., is effected by means of a shutting link, that is a link similar to the ordinary link but about $\frac{1}{8}$ in. larger in the diameter of the iron from which it is made, and frequently also a little longer ; this increase in size is made to compensate for possible weakness due to any imperfection in the weld.

Side Chisel.—The chisel specially employed for wood turning. The cutting edge is ground obliquely in relation to the edges of the shank, and it is used purely as a cutting tool, seldom as a scraper, the cutting edge being placed approximately at a tangent to the work. The acute angled corner of the chisel is also used for parting or cutting off. The side chisel turns a remarkably clean and approximately plain surface, but is not suited for quick curves or for mouldings or ornamental work, or for work of large diameter.

Side Discharge Valve.—See Discharge Valve.

Side Discharge Wheel.—See Turbine.

Side Elevation, or **Side View.**—A view on a drawing which shows the side of a structure.

Side Frames.—The main frames, standards, or supports of cranes, engines, pumps, &c., which carry the bearings for the shafts. They are made both in cast and in wrought iron. Small side frames are often called cheeks.

Side Hooks.—The coupling hooks of railway wagons, placed at the side of the draw hook.

Side Lever Engine.—A marine engine in which motion is communicated from the side rods of the piston cross head to the connecting rods, by two levers centred nearly on a level with the base of the cylinder. Side lever engines are nearly obsolete, though in the early days of steam navigation they were much in request, and they have the advantage of maintaining the moving parts nearly in equilibrium, so that there is little difficulty in starting them. The connecting rod being long, the oblique thrust on the crank is reduced to a minimum. But they lack compactness of form and are heavy.

Side Rake.—The rake (q.v.) of a cutting tool which cuts sideways, to right or left as the case may be. Such rake may be side top rake, or side bottom rake. See Top Rake, and Bottom Rake.

Side Rods.—(1) The rods connecting the cross head with the side levers of a side lever engine (q.v.). (2) Rods fulfilling the same purpose in one of the old forms of table engine.

Side Tank.—See Tank Engine.

Side Tool.—A tool used by wood turners for finishing the inner and outer portions of work. Called a side tool because its cutting edge is ground to an angle with the axis of the tool in order to reach the sides of the work, or those which stand approximately at right angles with the tool rest.

Side Top Rake.—See Top Rake.

Side View.—See Side Elevation.

Side Winch.—A light hoisting winch furnished with a flange for attachment to any convenient support, as the side of a wall or beam.

Siemens-Martin Process.—The process of manufacture of mild steel by the fusing together of pig iron and scrap steel in a reverberatory furnace, in which the pig is decarburised, the necessary amount of carbon being added afterwards by spiegeleisen or by ferro-manganese. The cast iron is rendered fluid, before the scrap steel rendered white hot in another part of the furnace, is allowed to mingle with it. It is one of the open hearth processes, and is also termed the pig and scrap process.

Siemens Process.—The method of manufacture of steel by the fusing together of pig iron and iron ore and scrap steel, the spiegeleisen or ferro-manganese, or a mixture of both, being added afterwards. Hæmatite pig iron and steel scrap, in the proportion of about thirty of the former to seventy of the latter, are first melted, and hæmatite ore added after fusion. Spiegel, or ferro-manganese, or both, are added as the metal is being tapped into the ladle, to afford the necessary recarburisation. It is not necessary that scrap steel should be added except as a convenient mode of using up odd ends. It is one of the open hearth processes, and is also called the pig and ore process.

Sieve.—Sieves of different meshes are used in foundry work for the purpose of sifting the sands required for moulding. They are termed $\frac{1}{8}$, $\frac{1}{16}$, &c., according to the sizes of mesh. A sieve proper ranges from $\frac{1}{8}$ in. downwards, or from 8 to 18 mesh. Brass moulders' sieves range from 8 to 36 mesh. See Riddle.

Sight Feed Lubricator.—A lubricator (q.v.) in which the flowing or nonflowing of the oil is always apparent at sight, being enclosed in, or having to pass through a glass vessel,

Sight Holes.—Holes in the sides of a reverberatory, or blast furnace, or of a cupola, or of a hot blast stove, through which the attendants note the progress of melting operations.

Signs.—The arithmetical and algebraical symbols used in mathematical calculations, and expressive of the operations denoted thereby.

Silent Feed.—A piece of mechanism by which the logs are fed forward to the saw in deal and timber frames. It is used in preference to the rachet and click by reason of its freedom from noise. It consists of an adaptation of a nipping lever (q.v.) carried at the end of a radial arm, whose length can be varied to suit different amounts of feed. In Worssam's silent feed, which is that chiefly used, the nipping lever is disguised by being modified into a cam-like piece forming an arc of a circle, whose centre is set eccentrically to its periphery.

Silica.—The oxide of silicon, symbol Si O_2. A refractory substance which, in combination with ether elements, is extremely useful to the metallurgist. In combination with alumina and water it forms the fire clays of crucibles. The fire bricks of Bessemer converters contain about 90 per cent. of silica, while ganister is composed largely of it. Natural siliceous stones are used for the hearths of blast furnaces. See Silicon.

Silica Bricks.—Fire bricks whose main constituent is silica (q.v.). The rock which furnishes their constituent silica is ground, crushed, and mixed with lime and water, then moulded and burnt. Silica bricks expand in burning to the extent of $\frac{1}{4}$ inch in 9 inches. They are used for the roofs of furnaces, being set in silica cement. They are unsuitable for those hearths and other furnace linings where they would come in contact with metallic oxides. For these situations aluminous bricks (q.v.) are employed.

Silicate Cloth.—An incombustible silicious cloth used as a non-conducting coating for steam boilers.

Silicate Cotton.—See Slag Wool.

Silicate Paint.—A protective coating for iron work whose bases are silicious. It resists the action of acids and of salt water better than the ordinary paint.

Silicious Iron.—Iron which contains silicon in combination in quantity. Silicon makes iron hard and brittle, and therefore by diminishing its toughness impairs the property which renders it most valuable for the chief purposes for which it is required. But the effect of the silicon depends largely on the proportion of carbon present, a diminution in the percentage of the latter depriving the silicious iron of much of its brittleness. The influence of silicon is analogous to that of carbon. It is believed to exist both in the combined and uncombined condition. Hot blast pig contains more silicon than cold blast. It is produced by the working of poor ores and light burdens. See Blazed Pig. Silicious iron possesses both cold shortness (q.v.) and red shortness (q.v.).

Silicious Sand.—Sand composed mainly of silica, used instead of lime for setting silica bricks (q.v.), and also for mixing with fire clay for the manufacture of fire-bricks, and for the bottoms of hearths of furnaces.

Silicious Steel.—This is harder than the ordinary mild steel, but has red shortness (q.v.), and is therefore unsuitable for plates and bars, though useful for some kinds of castings.

Silicon.—Symbol Si. Comb. weight 28. Is, next to oxygen, the most abundant element in nature, occurring in combination with oxygen as silica (q.v.) in rocks and sands, and in combination with metals as metallic silicates. Pure silicon is extremely hard. Its presence in iron or steel is one of the chief factors affecting their qualities of hardness and brittleness, both during the operation of smelting or refining, and as a permanent constituent. It is present to the extent of 4 per cent. in some soft grey pig irons, but usually below 2 per cent.

Silky Fibre.—When the bending of wrought iron or steel develops a smooth homogeneous and shining fibre, it is said to be silky.

Silver Grain.—The medullary rays in timber which radiate from the heart to the bark. The position of the silver grain is of importance as affecting the shrinkage, which takes place chiefly in a direction transversely to the course of its rays.

Similar Beams.—Beams, which though differing in dimensions, are so proportioned that their relative ratios of strength are the same.

Simple Machines.—The mechanical powers in their elementary forms, as a single lever, or a single wheel and axle, &c.

Simple Train.—A train of change wheels for screw cutting in which there is only one intermediary or stud wheel used. All coarse pitches, or, say, any of less than eight to the inch, can be cut with a single train. See Compound Train.

Simply Divided Scale.—A generic term used merely to distinguish ordinary equally divided drawing scales from diagonal scales (q.v.) and vernier scales.

Sin.—The sine of an angle.

Sines, Line of.—Sectorial lines (see Sector) by which the lengths of sines and tangents, and secants to a given radius, can be obtained by simple measurement. To obtain sines, open the sector until the transverse distance of 90 and 90 on the sines is equal to the given radius, then the transverse distance measured on the numbers which correspond with the degrees and minutes will give the sines. The same operation performed on the lines of tangents opened to 45 and 45 will also give tangents. To obtain the secant, let the radius be taken at 0 and 0 at the beginning of the line of secants, and then take the transverse distance as required. Conversely, to obtain the radius from the length, make the length given a transverse distance in degrees and minutes on its proper scale. Then in the case of a sine the transverse distance of 90 and 90 on the sines will be the radius. If a tangent under 45°, 45 and 45 on the tangents; if a tangent over 45°, 45 and 45 on the upper tangents; if a secant, 0 and 0 on the secants.

Single-acting.—Signifies generally a prime mover, or machine in which the motive power is applied in one direction only; the motion in the opposite direction being effected by momentum or other extraneous form of power.

Single-acting Engine.—An engine in which the steam acts only against the under side of the piston, the weight of the atmosphere pressing it down again, against the vacuum produced by the condensation of the spent steam. Such engines are now out of date, though long used for pumping in the mines. They were also called atmospheric engines. Engines of single-acting type are used for steam hammers and also for some forms of three-cylinder engines.

Single-acting Piston.—A piston which is in contact with the fluid upon which it acts on one side only, usually termed a plunger (q.v.).

Single-acting Pump.—A pump which delivers liquid at each alternate stroke only, one stroke being spent in drawing or lifting the liquid into the pump, the other in delivering it.

Single Belting.—A belt formed of a single thickness only of leather. This thickness varies from $\frac{1}{8}$ to $\frac{1}{4}$ of an inch. Single belting is employed for light and moderately light work. For the heaviest driving, double belting (q.v.) is used.

Single Butt Strip.— A single covering strip uniting a butt-riveted joint. Joints made thus are much weaker than those having double butt strips (q.v.).

Single Cut File.—A float cut (q.v.) file.

Single Cutting Drill.— A drill which is ground to cut in one direction only, usually right-handed. The majority of drills and drill-like tools, and bits for boring wood are single cutting.

Single Cylinder Engine.—An engine having one steam cylinder only. Its action without the assistance of a fly wheel to carry it over dead centres (q.v.) would be jerky and uncertain.

Single Disc Valve.—A disc-shaped plug valve which is in contact with a seating on one face only. It is not so tight as the double disc valve (q.v.).

Single-ended Boiler.—A marine boiler fired from one end only.

Single-ended Machine.—A term applied to those punching and shearing machines where one punch or one shear only can be in operation at one time.

Single Fished Joint.—A fished joint having a single fish plate or covering strip only, as distinguished from a double fished joint (q.v.).

Single Flue Boiler.—See Cornish Boiler.

Single Gear.—Gear comprising one pinion and one wheel only in combination, for the gaining of power. See Double Gear.

Single Geared.—When the mandrel of a lathe or the spindle of a machine is driven directly by a pulley keyed upon it, and has therefore no back gear, it is said to be single geared.

Single Grease Cup.—A grease cup having one cock only, leading direct into the cylinder or steam chest.

Single Purchase.—Gear for lifting or other purposes in which one pinion and one wheel only, that is a single gear, are engaged. The velocity ratios are then inversely as the radii.

Single Rail Crane.—A light workshop crane which runs on a single rail embedded in the floor. It is maintained in a vertical position by means of wheels running on channel iron hung from or attached to the roof.

Single Reading Scale.—A scale (q.v.) used for drawing purposes in which one primary division only is subdivided.

Single Riveting.—When there is one line only of rivets in lap joints, or two in butt joints, this is called single riveting, to distinguish it from two lines in the former case, and four in the latter, or double riveting.

Singles.—Sheet iron plates used for tinning, whose thickness ranges between No. 4 (·238 in.) and No. 20 (·35 in.) B.W.G.

Single Shear.—A term applied to the strains upon rivets; a rivet which unites a lap or a butt joint with a single fish plate only, can be shorn across in but one section, and is therefore weaker than a rivet in double shear (q.v.).

Single Stroke.—One stroke of a piston, taken either backwards or forwards. Used in opposition to double stroke (q.v.) and employed in calculations relating to engine power.

Single-threaded Screw.—A screw consisting of one helix only winding around the body. Ordinary wood screws are always single threaded, as are also attachment screws, and set screws.

Single Train.—A simple train (q.v.) or single gear (q.v.).

Single-webbed Girder.—A built-up flanged girder whose flanges are connected by a single vertical web only. The term would apply both to a plated and to a lattice girder having a single web.

Sinking Engines.—Stationary or semi-portable engines, used for sinking or excavating the shafts of mines.

Sinking Pump.—A pump used for clearing the water from foundations, docks, mines, &c. The actual pump consists of two suction barrels whose pistons are worked alternately with a bob lever, either by hand or power, the barrels being sunk to the necessary depth, or about twenty feet from the surface of the water to be pumped, by means of telescope pipes. Similar pipes connect the barrels with the suction below. Another form consists of a steam pump self-contained, slung with chains and so sunk into deep wells and mines for pumping to any required depth.

Siphon Oil Cup.—See Syphon Lubricator.

Sixteenth Bend.—A pipe bend which makes an arc of $22\frac{1}{2}°$, and which therefore connects pipes whose amount of divergence is comparatively slight.

Skeleton Girder.—An open-webbed girder, or lattice girder (q.v.)

Skeleton Pattern.—An open frame pattern, that is, one which is not precisely like its casting, but whose outlines or bounding edges alone are given. In a skeleton pattern the central portions are either strickled or scraped out, or are cored or made up in some other way by the moulder. The reason of their employment is the saving of cost in the pattern making. These are only adopted in framed and plain work.

Skelp.—A strip of wrought iron which is bent preparatory to its being welded into a pipe.

Sketch.—See Hand Sketch.

Sketch Book.—A book of semi-transparent prepared tissue paper, used in drawing offices, in which is preserved a copy or print of each hand sketch (q.v.) sent out into the shops. The sketches being made in copying ink are printed on the prepared paper similarly to letters, by first damping the paper, placing the sketch between the leaves, preserving the adjacent leaves from contact with oiled sheets, and squeezing the book in the screw press for a moment or two ; in this way a permanent record is kept of all hand sketches however small, and the book is to the sketches, what the tracings are to the shop drawings.

Sketch Paper.—Foolscap paper having faint lines crossing at right angles, producing squares of 1 inch, each divided into eight equal parts. Used for proportional drawing or sketching.

Skew.—Oblique, not at a right angle. Work is said to be askew when it is out of square, or when it is atwist.

Skew Back Saw.—A hand saw whose back is curved inwards in order to lighten its weight without diminishing its stiffness.

Skew Bevels.—Bevel wheels (q.v.) whose axes are not in the same plane and which consequently would not meet if prolonged.

Skewers, or Wires.—Pieces of iron wire from 3 to 6 inches in length, pointed at one end, and turned round into a loop at the other, and employed for the temporary holding of loose pieces (q.v.) in the foundry sand. The turned round eye is for the insertion of the workman's finger, in order to effect the withdrawal of the wire from the pattern after it is partially rammed up. Wire nails or wrought brads are sometimes used, but are not so convenient as skewers with eyes. The diameter of the wire from which skewers are made varies from $\frac{1}{16}$ in. to $\frac{3}{16}$ in., $\frac{1}{8}$ in. being the most generally suitable.

Skew Rebate Plane.—A rebate plane in which the cutting iron is set obliquely to the sides of the stock, instead of at right angles. It cuts with less chatter, and sweeter than the square-mouthed plane.

Skid.—See Skidding.

Skidding.—The slipping of a wheel along its rail. A wheel skids through lack of adhesive power, or through the brake power being insufficient to overcome the momentum of the moving mass.

Skimmer.—(1) A length of rectangular iron bar which is held across the mouth of a foundry ladle while the metal is being poured, to bay back scoriæ and foreign matters which float on the top, and which if not skimmed would run into the mould. (2) The call name of the boy whose duty is to hold the skimmer across the ladle.

Skimming.—The baying back of the scoriæ or dross, which floats upon the surface of molten metal, iron, or brass, and which is thereby prevented from running into the mould to the detriment of the casting.

Skimming Chamber, or Skimming Gate.—A chamber whose function is the separation of the dross or sullage from the metal in a foundry mould. It owes its utility to the principle of centrifugal force, and usually consists of a globular, or sometimes of a discoid cavity placed in the course of the running metal between the pouring gate and the mould. The metal is run into it on one side and necessarily assumes a whirling motion which throws the scum to the centre and the heavier iron towards the circumference, where it passes through a gate or sprue (q.v.) into the mould. All work which has to be planed or turned bright all over is run with a skimming chamber, or, as it is sometimes termed, run with a ball, or run with a disc.

Skimming Gate.—See Skimming Chamber.

Skin.—A term used to denote the thin film of hard metal which exists on the surface of a casting or forging, owing to the chilling or hardening of those surfaces against the sand, or in contact with the air, or with cold iron chills. It is so hard that when turning, or planing, the first cut taken must be sufficiently deep to pierce below the skin, while if it has to be filed the skin must be first removed by grinding, or by chipping, or by using a file nearly worn out, to save the new one.

Skin Deep.—See Skin.

Skin Drying.—The drying of the surface of a green sand mould previous to closing it for the pouring in of the metal. Skin drying effects the removal of a portion of the moisture and diminishes the risk of a blown or a scabbed casting. It is effected with a devil (q.v.) or with an open tray containing burning charcoal or coke.

Skip.—A circular or rectangular shaped open vessel used for tipping rubbish, concrete, ballast, coal, &c., where desired. Skips are either turn-over skips or tipping buckets, that is, they swivel on trunnions, and throw the contents out at the top, or are drop bottom skips, in which the bottom opens downwards on the unfastening of a catch.

Skulls.—See Sculls.

Slab.—(1) A flat rectangular block of wrought iron or steel, used for piling or building up heavy forgings, or for rolling plates. Slabs of malleable iron are prepared from the puddling furnaces, or are piled from scrap, slabs of mild steel are drawn down under the steam hammer from the ingot. (2) The outer boards cut from logs of adzed timber are called slabs, and are useful for the outsides of core boxes, turn-over boards, and similar rough work.

Slabbing.—The drawing down of steel ingots into slabs under the steam hammer.

Slack.—(1) Small coal used in engine fires. See Small Coal. (2) The dust and burnt out rubbish from a smith's fire.

Slack Chain.—The loop of chain which hangs below the pulley blocks in lifting tackle, and which diminishes in amount when a load is being lowered.

Slack Fit.—A fit is said to be slack when the parts in contact have more freedom of play than is sufficient or desirable for their free and easy movement. The term is relative, because what would be a slack fit in some portions of a machine would be too tight a fit in another. See End Play, Sliding Fit. A slack fit in many cases would more properly mean a bad fit, as in the case of the fitting of shafting into wheel bosses, which must necessarily be tighter than the fitting of the same shafting into its bearings.

Slacking.—See Slacking Back.

Slacking Back.—A nut is said to slacken back when it works loose from its thread owing to the influence of constant vibration. The nuts of engine and machine parts are thus apt to slacken back, necessitating the introduction of lock nuts (q.v.).

Slacking Down.—The damping down of engine fires to prevent or diminish the generation of steam for a season.

Slacking Off.—A shop term denoting the tempering (q.v.) of steel tools, as taps, chisels, and turning tools.

Slackness.—Play (q.v.).

Slack Side.—That side of a driving belt which is lowermost. Called also the Following Half.

Slag.—Slag is a bye product of blast furnaces, cupolas, and melting furnaces generally. As relating to the iron manufacture, it consists of earthy and infusible matters present in the ore, and which are fluxed off by being brought into contact with elements for which they have a natural affinity. The slags formed in the iron furnaces are essentially silicates, that is, the basis of their composition is silica, but alumina lime, magnesia, and many other elements enter into the composition of furnace slags.

Slagging.—The tapping out of the slag from a cupola. It requires to be done several times in the course of a prolonged blow, to relieve the iron of its impurities, and to prevent the loss of heat which the presence of slag in quantity occasions. The better the quality of iron the less slagging out is required. With inferior and burnt iron the slag hole is allowed to run nearly all the time the melting is going on. By judicious slagging, one quarter to one half more metal can be melted in a given time in a cupola than could be done if it were neglected.

Slag Hole.—The hole in a blast furnace or cupola through which the slag is discharged. In a blast furnace it is situated in the lower part of the crucible. In a cupola it is situated in the breast plate (q.v.).

Slag Wool.—Used as a coating for steam boilers and steam pipes. It is prepared by directing a jet of steam against the molten slag as it issues from the furnace. Sometimes termed silicate cotton.

Slake Trough.—The vessel which contains the water attached to a smith's forge for cooling or quenching hot metals.

Slaking.—Sleeking (q.v.).

Sleaker.—A sleeker (q.v.).

Sledge.—See Sledge Hammer.

Sledge Hammer, or Sledge, simply.—The hammer swung by the smith's striker. It is a broad, flat-paned, double-ended hammer, weighing from 5 to 14 lbs., and is lifted up hand (q.v.) for the lightest blows, or about sledge (q.v.) for the heaviest.

Sleekers, or Smoothers.—Moulders' tools, made in various forms and used for smoothing and finishing the surfaces of sand moulds. See Bead, Egg, and Square-corner Sleekers. Variously spelt Slakers, Sleakers, and Slickers.

Sleeking, or Slaking, or Slicking.—Smoothing over the surface of a foundry mould with tools specially adapted for the purpose. Broad flat surfaces are sleeked with the trowel, but where that fails to enter proper sleekers (q.v.) are used. All except the roughest moulds are sleeked both before and after blackening.

Sleepers.—(1) Balks of timber of rectangular section, used as the foundations of the rails forming permanent way. Sleepers are either longitudinal, or cross sleepers, according as they run continuously with the rails, or transversely. Cross sleepers are laid underneath the longitudinal ones. The chairs are attached to the sleepers. Dantzic or Memel fir is the material commonly employed, which is usually creosoted or otherwise preserved from the weather. Iron sleepers are sometimes used in foreign countries. Timber sleepers are valuable by reason of their elasticity. Cast-iron sleepers are pot sleepers, elliptical in plan and dome-shaped in elevation, and the chair and sleeper are a single casting. Wrought iron has also been used for sleepers, both in the pot form and also having the chair and sleeper independent of one another, and held together with bolts. Iron sleepers are advantageous in hot countries infested by white ants and other destructive insects. (2) The timber imbedded in or laid on the ground to form the foundations for derrick cranes, and to which the guys are fastened.

Sleeve.—A hollow cylinder or quill (q.v.).

Sleeve Bearing.—An unjointed bearing which is unusually long in proportion to its diameter.

Sleeve Screw.—An external screw whose cylinder is hollow throughout. Examples occur in the feed screws of some drilling machines.

Sleeve Wheel.—A wheel on which is cast a long hollow boss, the boss being made hollow to fit over and slide along a shaft. Examples occur in the bevel wheels which slide along the revolving gantry shafts of overhead travelling cranes, and through which motion is conveyed to the crab in whatever part of the gantry it may happen to be situated.

Slewing.—The act of turning round, or revolving on a centre. Applied to the circular movements of cranes and other revolving machines. This is performed in hand cranes with a pair of bevel wheels actuated by a winch handle, giving motion to a small pinion gearing into a curb ring, or ring of cogs. In steam cranes the motive power is transmitted to the curb ring through movable friction crutches, generally placed on the engine shaft.

Slewing Bracket.—A bracket, or bearing, which carries the spindle of the cog wheel used for slewing cranes.

Slewing Gear.—The gear by which the slewing motion of a travelling crane is actuated, consisting usually of friction cones actuating a vertical shaft which revolves a pinion in a curb ring on the truck.

Slice, or Slice Bar.—A firing tool, being a flat thin expansion only, at the end of a long bar, and used for breaking up and separating the clinkers on a furnace hearth. See Clinkering.

Slice Bar.—A slice (q.v.).

Slicker.—A sleeker.

Slicking.—See Sleeking.

Slide.—A term which signifies in a general way any piece of mechanism which moves in a rectilinear direction over a flat or curved face, for the purpose of opening and shutting off communication with certain chambers or passages, or which becomes a guide or steady to some other piece of mechanism.

Slide Bars, or Guide Bars.—The bars by which the free end of a piston is constrained to move in a straight line through the medium of slide blocks (q.v.). The blocks move between the slide bars and are attached to the cross head (q.v.) of the piston. Slide bars have flat, or cylindrical faces. Sometimes termed motion bars.

Slide Blocks.—The blocks attached to the cross head (q.v.) of an engine or pump, which, moving between and along the slide bars (q.v.), maintain the piston rod truly rectilinear. Also termed guide blocks, slipper blocks, slippers, motion blocks.

Slide Case.—The old-fashioned steam chest affixed to an engine cylinder, within which worked the long D valve, now obsolete. The openings of the ports being near the ends of the cylinder, and the valve being single only, a case or chest, nearly equal in length to that of the cylinder itself, was employed to cover them over, hence the term.

Slide Lathe.—A slide rest lathe (q.v.).

Slide Principle.—The slide principle is embodied in that construction in which adjacent plates are so fitted one to the other with guiding strips that they can move in one direction only, over each other. It has its application in almost all kinds of modern machine tools.

Slide Rest.—A lathe rest designed for automatic or self-acting turning and screw cutting. It consists essentially of a saddle, fitted either to the vee'd edges of the lathe bed, or between the bearers, and of two slides moving on vee'd edges, one for sliding, the other for surfacing. A circular swivelling motion is also usually provided for angular turning. The slides are moved by hand screws, and the saddle is either worked by hand or power with rack and pinion movement, or through the intervention of the leading screw (q.v.).

Slide Rest Lathe, or Slide Lathe.—A lathe for metal turning furnished with a slide rest. In the larger lathes the slide rest is an integral portion of the lathe, without which no turning is done; in the smaller lathes it forms an attachment only for automatic turning, being removed for the accomplishment of hand turning also.

Slide Rod.—A slide valve spindle (q.v.).

Slide Rule.—A rule provided with logarithmic numbers arranged on a sliding scale, whose graduations are so arranged relatively to other similar graduations upon the body of the rule, that when a certain number upon the scale is made to correspond with a certain number upon the rule, a definite product or fraction is obtained without the labour of calculation.

Slides.—The edges and faces of machine parts which carry and form the guides of rectilineally moving parts. See Adjustment Strips.

Slide Valve.—Any valve whose motion is that of sliding, as opposed to rotating or lift valves. The term is usually understood to refer specially to the valves of steam and gas engines of the ordinary type (not Corliss or lift valves). These are usually of the D shape, but may be single, double, or treble ported. They are also main slide, or cut off expansion valves. They may be provided with lap or not, all of which terms are explained under their own headings. See also Equilibrium Valve, Travel, Setting of Valves.

Slide Valve Diagram.—See Valve Diagram.

Slide Valve Spindle; or Slide Rod.—The rod of a slide valve through which motion is communicated from the eccentric to the valve, with or without the intervention of a slot link. It is attached to the valve either with a brass T-headed nut, or is hooped to embrace it, or is set with nuts at each end of the valve.

Sliding.—Sliding denotes the longitudinal traverse of a slide rest, that is its motion lengthways on the bed. See Surfacing.

Sliding Blocks.—(1) The blocks which slide across the slotted cross head of a steam pump. (2) Any blocks whose motion is that of a sliding character are slide blocks.

Sliding Contact.—The communication of motion by means of cams, screws, worms, and worm wheels.

Sliding Fit.—When surfaces cylindrical or plain move one over the other freely, yet without perceptible slackness, the fit is so designated. The difference of 1-10,000th part of an inch in diameter may make the difference between a sliding and a slack fit.

Sliding Friction.—The friction which is set up between two surfaces in contact. Its amount is independent of the area, so that the larger the area the less the amount per unit of surface.

Sliding Pipe.—A suction or delivery pipe for a pump, made adjustable for length with a joint similar to an expansion joint (q.v.).

Sliding Pulley.—A guide pulley for a rope or chain which is not confined endways on its shaft, but which, while revolving, can at the same time slide from side to side along the shaft to accommodate its position to that of the rope or chain as the latter is wound or unwound towards one end or the other of its barrel. Instances occur on the jibs of cranes.

Slimes.—The deposits of tin ore left in the washing reservoirs after the process of stamping (q.v.) and subsidence is complete. The slimes are subjected to a further process of washing previous to calcination.

Sling.—A rope, or chain, or clip passed around a load for the purpose of lifting and moving it about.

Sling Chain.—An ordinary piece or a bight of chain used for encircling heavy work for the purpose of lifting about, the crane hook being inserted into a lap of the chain.

Slinging.—The securing of heavy masses of material in the ropes or chains of cranes or of pulley blocks while being hoisted about. The methods of slinging vary with the shape of the body which is being lifted.

Slinging Rings.—Rings attached to the top of a portable engine boiler by means of straps riveted to the boiler shell, and used for lifting the boiler by.

Sling Rods.—The rods or stays by which a locomotive boiler is attached to the framing.

Sling Stays.—Slings used for attaching girder stays (q.v.) to the crown

of the outer fire box shell in locomotive and portable boilers. **The** sling stays are attached to angle or T irons.

Slip.—The sliding of riveted joints one over the other to such an extent as to be visible. See also Gouge Slip, Slip of a Screw.

Slip Jaws.—Temporary facings dovetailed on the gripping portions of the jaws of lathe chucks.

Slip Knot.—A knot which allows the rope to glide through it and to become taut. It is used in slinging (q.v.).

Slip of a Screw.—The difference between the theoretical performance of a ship's screw supposing it moved in an unyielding medium, and the actual performance, the yielding nature of the water being taken into account. The slip of a propeller is always given as so much per cent. on the distance as calculated from the working of the engine.

Slipper.—A slide block (q.v.).

Slipper Block.—A slide block (q.v.).

Slipper Brake.—A form of brake which consists of iron blocks so attached to a truck by an arrangement of screws and levers that they can be thrust forcibly down on the rails when it is desired to arrest the motion of a wagon or train. It is thus the reverse of the method commonly employed, of braking the wheels themselves.

Slipper Guides.—Slide bars (q.v.).

Slippery Iron.—A mixture of cast iron specially prepared for engine cylinders, cylinder liners, slide blocks, and moving surfaces generally. It is tough and moderately hard, and its slipperiness of surface is obtained by using a manganiferous brand of iron, the manganese being present to the amount of 2 or 3 per cent.

Slipping Load.—A term sometimes used to denote the load necessary to produce slip in riveted plates. See Slip. It is stated in terms of tons per rivet subject to strain.

Slipping of Belts.—See Belt Tension, Belt Dressing.

Slit Bar.—A bar having an open slot or central clear space within which a stud is slid or tightened at pleasure. Linear and rotative motions are thereby rendered interchangeable, and the length of stroke of reciprocating rods is rendered capable of adjustment. Slit bars are used in machines of almost all descriptions, and in prime movers.

Slit Bar Motion.—The mechanism of a rod or link, the amount of whose throw is capable of adjustment within the limits of a slit or slits in the rod along which the set bolt for varied adjustment slides.

Slitting File.—A file whose section is that of a double V having two thin knife-like edges.

Slope Tile.—See Square Tile.

Slot.—An oblong hole, which may be either rounding or square at the ends. Slot holes are cut in levers to allow of adjustment for length and in machine slides to take up the wear, the adjustment being made by a small nut or set screw, or in a cottar way (q.v.) by means of a wedge or cottar (q.v.).

Slot Drilling.—The cutting of slots either for cottars or set screw adjustment, or of shallow grooves for the reception of sunk feathers, by means of a machine somewhat similar to a drill, the cutting tool being carried by a revolving spindle, though provided with no downward feed, but having a sliding table, which, while carrying the work, is fed along under the slot drill, thus cutting an oblong recess or hole of any required length and of uniform depth throughout.

Slot Hole.—See Slot

Slot Link.—The movable link, the shifting of whose position relatively to the valve rod of an engine cylinder determines the amount of cut-off, and the forward or backward motion of the engine ; called also the reversing link. These results it effects by bringing one or other of the eccentric rods in turn into a line with the valve spindle.

Slotted Cross Head.—A cross head (q.v.) slotted for the reception of the sliding block attached to the end of the connecting rod of a steam pump, and which combines rectilinear with rotational movement.

Slotting.—The vertical cutting of narrow grooves and keyways, and generally the vertical shaping of curved outlines by means of a slotting machine (q.v.).

Slotting Machine.—A machine used for shaping metals, the arm which carries the cutters moving in a vertical direction. The travel of the arm is capable of adjustment by means of a connecting rod movable in a slot in the arm ; and provision is also made for imparting feed (q.v.) motion to the table, and also motion in a circular direction. The arm also is usually supplied with provision for quick return (q.v.) motion by means of a link attached to a disc, on the same principle as that employed in shaping machines. Slotting machines occur in modified forms. What are known as curvilinear machines are constructed for slotting out the interior portions of the rims of railway wheels.

Slotting Tools.—Comprise round-edged, keyway, parting, and hook tools.

Slow Cutting Motion.—The motion of the tool box of a shaping or slotting machine during the forward or downward stroke, as opposed to the quick return (q.v.) stroke.

Slow Feed.—When provision is made for two sets of feed motions, as in many drilling machines, one is termed the slow and the other the fast or quick feed, the variation being necessary, since different work demands different rates of feed according to the coarseness or fineness of the finish required, or the nature of the metal itself.

Slow Gear.—The double or treble purchase gear of cranes and hoisting machinery, and the back gear of lathes and drilling and other machines, are termed slow gear, because with increase of power there must be corresponding and inverse diminution of velocity.

Sludge.—The muddy deposit which accumulates in a steam boiler.

Sludge Cock.—The cock at the bottom of a steam boiler which is employed for the purpose of cleaning the boiler from sludge by periodical washing through with a strong current of water.

Sludge Hole.—A mud hole (q.v.).

Sluice.—(1) A water way and gate, or a disc plug valve for the regulation of the flow of water. (2) The opening through which the water is admitted to a water wheel or turbine.

Slurry.—A term used in foundries to signify dirty water, or black wash, and in a general way any fluid used in moulding.

Small Coal.—Coal which has been screened at the pit's mouth. It is used to a limited extent in engine and smiths' fires and furnaces, but is more largely employed in the manufacture of patent fuels (q.v.).

Smelting.—The process of the reduction of metals from their ores, effected chiefly by the aid of heat, and fluxing and reducing agents.

Smelting Furnace.—A blast furnace (q.v.) or some form of reverberatory furnace (q.v.).

Smith.—An engineer's smith is a specialist, since his work relates entirely to forging of machine parts. Usually it is of a somewhat heavier nature than that of an ordinary smith, requiring the aid of the steam hammer; and in the lighter work, owing to the repetition demanded, die blocks (see Die) in which the forgings are either partly or entirely stamped, are largely used.

Smith's Brace.—A hand brace used by smiths for drilling holes of moderate diameter. Its outline is very similar to that of the ordinary carpenter's brace, and it is turned in the same way, but the requisite pressure is imparted at the top by means of a feeding screw against whose end the top of the brace, which is formed into a hardened point, is centred. The resistance is taken by a horizontal arm having a boss tapped to receive the feeding screw, the arm being capable of sliding up and down, and of being tightened at any height on a vertical pillar, whose foot is bolted to a bench or clasped in a vice. These braces are not used much in engineers' smithies, the ratchet brace and drilling machines being more convenient.

Smiths' Hammers.—Smiths' hammers embrace the sledge hammer, the ordinary hand hammer having either a straight or a ball pane at the small end, the set hammer, and the various flatters and fullers, which though not strictly speaking hammers, are nevertheless tools of a similar type. Notes on each will be found under their respective headings. See also Oliver, Tilt Hammer, &c.

Smith's Saw.—A hack saw (q.v.).

Smith's Shop.—The shed or building in which the operations of smith's work are carried on. The forges are arranged around the walls, space being left between each for the racks which carry the swages, fullers, chisels, and other similar tools. The anvils stand in front of these open spaces. A light wall crane is attached to the wall between each forge, or one to two forges. The central open space of the shop is occupied with steam hammers, forging machines, &c. A shed adjoining the shop is used as a store for bar iron.

Smiths' Tools.—Include anvils, anvil stands, forges and firing tools, hammers, tongs, flatters, fullers, swages in their various forms, sets, pliers, punches, bolt and heading tools, and drills, chisels, saws, and machines, both special and common to the fitting and erecting shops, each described in place.

Smith's Work.—The heating, handling, light hammering, and manipulation of wrought iron and steel, the direction of the heavier blows of the striker, the holding of the top swages, and other tools, the preparing of joints for welding, the hardening and tempering of steel, &c.

Smithy Fan.—A centrifugal fan used for blowing the fires in a smith's shop, branch pipes leading to each forge through tuyeres, and the amount of blast being capable of regulation by a valve.

Smoke.—Consists of a mass of minute particles of solid unburnt carbon liberated from hydro-carbons by the action of heat, and suspended in an enveloping medium of vapour of water, and unburnt gases given off by the fuel. See Smoke Prevention. Black smoke is that which contains the largest quantity of solid carbon. The so-called smoke which is of a yellowish or yellowish brown colour is not smoke (*i.e.* free carbon), but carburetted hydrogen, chiefly olefiant and marsh gases mixed with tarry and sulphurous matters.

Smoke Box.—The front part of the locomotive, or portable engine boiler, into which the tubes discharge the products of combustion on their way to the chimney.

Smoke Box Plate.—The dished plate which covers over the outer end of the smoke box of a locomotive or portable engine boiler and which is removed when the tubes require examination or cleaning. It is usually of wrought iron, but is sometimes cast.

Smoke Consuming.—The application of the principle of perfect combustion, to the smoke from furnaces, whereby the solid carbon and inflammable gases are fully oxidised. See Smoke Prevention.

Smoke Flues.—The flues of Cornish and Lancashire boilers, and externally fired boilers which are built in brick work, as distinguished from the fire flues (q.v.) of internally fired boilers.

Smokeless Coals.—Anthracite, and anthracitic coals (q.v.).

Smoke Prevention.—Without describing the various schemes that have been proposed to prevent the nuisance of smoke formation, it is as well to note the essential conditions which must obtain in order to its removal. The black smoke and yellow smoke (see Smoke) which are given off from engine fires is given off in these forms, either because there is not a sufficient quantity of oxygen present to form a chemical union therewith, or because, if present, the temperature is not sufficiently high to produce ignition. All such gases and solid carbon allowed to pass off unburnt represents a vast amount of heat wasted. Hence the adoption of measures to prevent the formation of smoke not only abates a nuisance but is an economical gain. In ordinary furnaces the remedy is obvious. The fuel should be well coked on the dead plate so that the liberated gases shall be well consumed in the combustion chamber on their way to the chimney.

Smooth Cut File.—An ordinary smooth or fine file. A smooth cut file of 12 in. in length will contain 72 lines of teeth to the linear inch.

Smoother.—A sleeker (q.v.).

Smoothing Plane.—A small plane used by wood workers for giving a smooth finish to surfaces in cases where accuracy is not very essential. It measures about 8 in. in length, and its cutting iron will range from $1\frac{3}{4}$ in. to $2\frac{1}{4}$ in. in width.

Snail.—A volute-shaped cam.

Snap.—The steel die used by riveters for finishing the heads of rivets.

Snap Head.—See Snap Head Rivet.

Snap Head Rivet, or Cup Head Rivet.—A rivet whose head is semicircular or cup-shaped in section.

Snatch Block.—The suspended block in lifting tackle, containing the pulleys which rest in the lower bight of the chain.

Snifting Valve.—A relief valve, used generally for the purpose of affording an exit for an imprisoned fluid or gas, when the pressure thereon attains a definite amount. In a condensing engine it is the valve through which the steam and air which mingle in the blowing through (q.v.) process pass out of the cylinder.

Snug.—A lug (q.v.).

Soaking.—See Soaking Pit.

Soaking Pit.—A cavity or pit in the ground or base of a steel works, lined with fire brick, and in which steel ingots are laid to soak, as it is termed, before being passed to the cogging and rolling mills; the soaking being a term applied to the uniform distribution of the temperature of the ingot. The ingot is too fluid in its central portion for rolling immediately that it leaves the mould, and if allowed to cool, the outer skin becomes set while the inner portion is yet in a semifluid condition. By cooling the ingot in a pit of fire brick, the temperature becomes uniformly distributed without much loss of heat by radiation. Hence reheating is unnecessary when soaking is resorted to, that is, so long as the ingots are not allowed to accumulate on hand.

Soap.—Soap is thrown into the melting pot containing gun-metal or brass to act as a flux. See also Soapsuds, Soft Soap.

Soapstone Gasket.—A form of packing used for hydraulic joints, made of hemp, tallow, French chalk, and other ingredients. Called soapstone because of its greasy nature.

Soapsuds.—Soap and water used in the lubrication of cutting tools, and supplied from a soap-water can (q.v.).

Soapsuds Can.—A soap-water can (q.v.).

Soap Water Can.—A cylindrical tin can, open at the top and furnished with a stop cock near the bottom. It is used to effect the lubrication of work which is being turned or otherwise machined, the stop cock being opened only just so much as will allow of the slow drip of the fluid contained in the can, upon the work. The usual lubricant is soap and water, or oil. Hence called drip oil can.

Socket.—That enlarged recessed portion of a pipe which receives the spigoted end of the pipe to which it is united. Called also the faucet. When the spigot pipe is inserted into the socket, the joint is secured by stemming or caulking with lead, or with rope gasket, or in the best pipes the joints are turned and bored.

Socket Bend.—A bend pipe furnished with a socket at one end and a spigot at the other.

Socket Chisel.—The strongest kind of chisel made, used like the mortice chisel for cutting the joints of heavy timber framing, being actuated by the mallet alone. It derives its name from the upper end terminating in a hollow tapered socket, into which a round wooden handle is driven. These chisels range from $\frac{3}{8}$ in. to $1\frac{1}{2}$ in. in width.

Socket Pipe.—A cast-iron pipe which is provided with a socket (q.v.) at one end and a spigot (q.v.) at the other. The sockets of wrought-iron pipes are unions (q.v.), and are screwed over the ends on the outside diameter.

Socket Washer.—See India-rubber Washer.

Soda.—Common soda is used for the purpose of preventing and of removing incrustation of carbonate and sulphate of lime from steam boilers. This is effected by an exchange of acids, the carbonate in the soda entering into combination with the lime, forming sulphate of lime, which is soluble, and carbonate of soda, which is deposited. The soda is either introduced in lumps or is dissolved with the feed water. Soda ash is sometimes substituted, sometimes also caustic soda. In the former case a larger quantity, in the latter a smaller amount is required. Soda is useful, neutralising the grease in greasy feed water, forming with it a soluble soap.

Soda Ash.—See Soda.

Soffit.—The intrados (q.v.) of an arch.

Soft Blast.—A blast of low velocity, or a blast insufficient in amount, its pressure being inadequate to melt the metal without an excessive expenditure of fuel.

Soft Brass.—Brass which has been annealed after drawing and rolling Used for purposes requiring ductility.

Soft Centred Steel.—See Mild Centred Steel.

Softener.—A double colour brush (q.v.) used by draughtsmen for shading.

Soft Grit.—A grindstone of soft and porous texture.

Soft Iron.—Iron which can be shaped with ordinary cutting tools or abraded readily with files. The quality is due to the amount of carbon present and the manner of its combination, and also to the mode of crystallisation. Iron which contains practically no carbon, as malleable iron, is very soft, so also is iron which contains the maximum of carbon, as foundry pigs, which may contain as much as 4 or 5 per cent. Carbon when present in the graphitical condition makes a soft iron, but a very much smaller proportion when in the combined state yields white iron, which is extremely hard. Iron allowed to cool slowly in sand is soft, while the same iron cooled rapidly against a metallic chill is hard. Soft iron is used for all ordinary castings which have to be machined; tough, slippery, and hard iron being reserved for special classes of work. Where it is necessary to machine castings of hard iron, grinding or cutting by means of an extremely slow feed is resorted to.

Soft Metal.—This is a relative term expressive of the density of the particular metal in relation to the purpose for which it is required. Metals are soft when they are more ductile, more elastic, and more easily cut than the same metals when hard. Hard brittle metals are rendered softer by annealing (q.v.), by judicious cooling, and by chemical changes affecting the relative proportions of foreign metallic or non-metallic substances, which are always present in small and variable quantites.

Soft Ramming.—See Hard Ramming.

Soft Soap.—Soft soap is used for the lubrication of the wooden patterns of pipes and columns while turning in the lathe, and supported by a steady, the soap being rubbed into the bearing of the steady to prevent heating and burning of the wood. Used also in the mixing of soap suds (q.v.).

Soft Solder.—Solder used for uniting the joints of works in lead, tin, and soft brass, and similar metals and alloys. Its composition is various, and so also are the fluxes employed with it.

Soft Water.—Water free from carbonate and sulphate of lime. The softer the water for hot water apparatus and for steam boilers the better, since it will then produce less calcareous deposit in the pipes and on the plates.

Soft Woods.—The soft woods employed by engineers are pine—both yellow, red, and white—spruce, red deal, kauri pine, &c.

Solder.—A metallic cementing material interposed between surfaces which are to be united in close and firm contact. See Hard Solders, and Soft Solder.

Soldering.—The uniting of two portions of metal by means of another metal or of an alloy. Soft soldering is effected with the aid of various alloys of lead and tin. Hard soldering applies to the alloys of copper and tin, and is called brazing (q.v.). A clean surface is essential to efficient soldering. A copper bit (q.v.) is used in soft soldering; in hard soldering the parts are held together with binding wire (q.v.), and the metal allowed to run in. See Autogenous Soldering, Brazing Metal, Sweating On.

Sole Plate.—(1) A thin base plate, differing from a bed plate in its more slender form being either unstayed or but slightly stiffened with flanges or ribs. It usually owes much of its rigidity to the timber or metal foundation upon which it is fastened. A sole plate is often a mere base for plummer blocks. Not unfrequently, however, the term is applied indiscriminately to ordinary bed plates. (2) The periphery of a water wheel from which the buckets start. (3) The bed of a reverberatory furnace (q.v.).

Soles.—The longitudinal or side timbers in the under frame of a railway truck.

Solid.—A body or substance which is able to sustain a longitudinal pressure without receiving lateral support is termed a solid.

Solid Blows.—Dead, non-elastic blows, imparted by a hammer to a sheet of metal, or similar solid body, which rests upon a firm and solid support. The effect of a solid blow is to indent or thin the metal, and spread it out over a large area. Solid blows are imparted to wedges, nails, keys, cogs, &c., by holding the hammer very firmly when striking.

Solid Box.—A solid or unjointed bearing. A dead eye (q.v.).

Solid Column.—See Hollow Column.

Solid Coupling.—A coupling forged in a solid piece with its shaft. The ordinary couplings of propeller shafts are an example of the solid form.

Solid Drawn Tubes, or Seamless Tubes.—Drawn tubes, in opposition to brazed, or welded, or riveted tubes. The copper piping for feed, bilge, blow-off, and similar purposes in connection with marine engines, the brass tubes for surface condensers, and the tubes of multi-tubular boilers, are all solid drawn. Iron hydraulic tubes also are solid drawn.

Solid Emery Wheel.—A wheel built wholly of emery, to distinguish it from a wheel in which the powder is laid on a centre of wood.

Solid Half Rounds.—See Half Round Iron.

Solidity.—The solid contents of a body. The solidity of all bodies can be obtained by calculation, and it is often necessary to be able to do so in order to obtain their mass and weight.

Solid Piston.—An engine piston which is a solid disc, instead of being built up in separate portions. The circumference of the disc is recessed for the reception of spring rings (q.v.).

Solid Reamer.—See Reamer.

Solid Rolled.—Ordinary rolled wrought-iron joists, beams, or girders, as distinguished from beams and girders which are built up with trellis work of various types.

Solid Tool.—An ordinary tool for metal turning in which the cutting point is in one with its shank, as distinguished from the smaller tool points used with a tool holder (q.v.).

Soot.—Solid unburnt carbon which accumulates in the flues of engine furnaces, and the amount of whose accumulation should be reduced as much as possible by consuming the smoke in the furnace, soot representing fuel wasted.

Soot Door.—A square iron door inserted in an iron frame, built into the front ends of the brickwork flues of Cornish and other boilers of the horizontal type, through which the periodical accumulations of soot are removed.

Sow.—See Pig.

Spacing.—The dividing or setting out of holes, rivets, bolts, &c., into their relative positions in a piece of work, the compasses or dividers or special templets being used for the purpose.

Spall.—To break off a chip. Usually applied to the accidental breaking off or out of a strip of wood when planing the end grain of timber. Spalling only occurs when the planing is clumsily or carelessly done. It is readily prevented by first chamfering the edge towards which the plane is cutting, or by planing alternately from each edge. Broken edges are said to be spalled or spalt.

Span.—(1) The horizontal distance between the springings (q.v.) of an arch. (2) The distance measured between the supports which sustain a beam or girder or girder-like structure. The strength of beams or similar section and similarly fixed will vary inversely with the span.

Spandrel.—The space enclosed between the outer curve of an arch, a horizontal line from its apex and a vertical line from its springing.

Spanish Mahogany.—See Mahogany.

Spanner.—A lever used for the tightening up of nuts. Spanners are usually double ended, that is, each end is formed to take a nut of a different size than the end opposite. The size of a spanner is reckoned by the diameter of bolt for which it is suited and not by the width of the jaws. Thus a ½-in. spanner takes a ½-in. nut, but will be an inch wide in the jaw. The angle which the spanner jaws make with the handle is not arbitrary, but is selected with the view to turning a nut in the least possible space. If the handle is straight with the jaws, a hexagon nut cannot be turned in less than 60°, or a square one in less than 90°. But if the angle of the jaws be made 15°, a hexagon nut can be turned in 30°. The length of a spanner may be from 15 to 18 times the diameter of the bolt. See also Screw Wrench.

Spare.—See Spare Gear.

Spare Gear, or Spare, or Spare Parts.—Signifies the supply of duplicate portions of machinery together with the order to which they belong. It is customary to include spare parts with work which is despatched to the colonies and with sea-going engines ; the parts consisting of those which wear out most rapidly, and therefore require frequent replacement, or those vital parts upon which the efficiency of the machine or engine mainly depends, and which in the case of a breakdown it would be difficult or impossible to replace immediately.

Spare Parts.—See Spare Gear.

Spark Arrestor.—(1) A globular cage of galvanised iron wire, or a curved plate of sheet iron, put above the chimney of a portable engine or crane to arrest and throw back the sparks. (2) A grating placed across the smoke box and beneath the blast nozzle in some locomotives to arrest the upward flight of sparks.

Spathic Ore.—Is a carbonate of iron found chiefly in Styria and Carinthia. It owes its chief value to the presence of manganese, which renders it especially suitable for the making of some qualities of steel.

Spear.—See Pump Spear.

Specification.—A document in which all particulars, both general and in detail, are stated respecting the construction of a machine or engineering work. Specifications are issued for all important works, and are binding on the contracting parties.

Specific Gravity.—The density or weight of a substance estimated relatively to that of water, if a solid or liquid, or relatively to that of air at the same temperature and pressure, if a gas. It is denoted by the initials sp. gr.

Specific Heat.—The ratio of the quantity of heat required to raise the temperature of a given weight of a substance through 1° F. to the quantity of heat required to raise the temperature of an equal weight of water at 39° F. or 4° C. through 1°. In other words, the capacity of a body for heat, relatively to that of water.

Specific Levity.—Signifies the relative tendency of those gaseous elements

and compounds which are lighter than air to rise. It is therefore the opposite to specific gravity.

Specific Volume, or Relative Volume.—The volume of a gas or vapour compared with that of the liquid from which it is generated.

Specimen Bar.—A length of bar of any material specially prepared for testing in a testing machine. Bars are either provided with shouldered ends to be held in bridles or clips, or are screwed into their bridles.

Specimen Bridle.—See Specimen Holder.

Specimen Holder.—A bit of mechanism employed for holding specimens to be tested in the testing machine (q.v.). They are made for the most part to clip the specimen, which has collars at the ends to fit thereon. Hence termed specimen bridles.

Specimen Plate.—A piece of plate of any material specially prepared for testing in a testing machine. Plate specimens are embraced at the ends, the breadth at the ends being increased to maintain the sectional area, and so compensate for the loss of strength due to the holes made for attachment. The central portion of the plate is parallel. Mr. Kirkaldy's plates are 10 in. long by 2 in. wide. The Admiralty length is 8 in., and the area of original section as nearly as possible one square inch. The French standard is 200 mm. (7·87 in.).

Specular Ore.—A non-hydrated peroxide of iron which furnishes metal of excellent quality. It is one of the hæmatites, and derives its name from the lustre of its crystals; hence also the term "iron glance" applied to it.

Speed Cones.—See Speed Pulleys.

Speeding.—The calculation of the diameters of pulleys and gear wheels in order to obtain definite velocity ratios (q.v.) between the shafts and spindles connected with each other through their medium.

Speed of Ignition.—This has reference to the rapidity with which chemical combination takes place between the air and gas in a gas engine. The speed of ignition is slow or rapid, according as the percentage of admixed air is large or small. The speed of ignition must always exceed the piston speed in order to obtain the best results due to chemical combination.

Speed Pulleys, or Speed Cones.—Pairs of pulleys, each stepped down to different diameters, but both being alike or nearly alike in dimensions. They are arranged the reverse way to each other on their shafts, the smallest diameter of the one coming in line with the largest diameter of the other. By shifting the belt to the corresponding pairs variations in speed corresponding with the variations in their respective diameters are obtainable. When speed pulleys are so placed that their centres are situated a long distance apart, the steps are equal, that is, the sums of the diameters of each pair of steps throughout are the same, but when the centres approximate, the sums are not precisely alike, the difference being due to the greater obliquity of the belt, and then the diameters may either be obtained by an elaborate calculation or by a more simple graphic delineation.

Speeds.—Speeds signifies the steps of a cone or driving pulley or drum, or the diameters around which the belt or cord laps, as three speed, four speed, &c.

Speed of Cutting Tools.—See Cutting Tools, Speed of.

Spelter.—Zinc (q.v.) cast into the ingots of commerce. Also an alloy of equal parts of zinc and copper.

Sperm Oil.—See Whale Oil.

Sp. Gr.—Specific gravity (q.v.).

Sphere.—A solid having every part of its surface equally distant from its centre. The sphere has these properties—that of all solids of a given volume it is that which has the least surface, and of all solids of a given surface it is that which has the greatest volume.

Spider Wheel.—A wheel or pulley having light arms of wrought iron or steel.

Spiegel.—An abbreviation for spiegeleisen (q.v.).

Spiegeleisen.—A variety of iron which is invaluable to the steel manufacturers, and which owes its value to the presence of definite amounts of carbon and manganese. By adding spiegeleisen in definite proportions, any desired exact quantity of these elements can be imparted to decarburised iron melted in association therewith, causing the immediate production of a definite grade of steel. See Ferro-Manganese.

Spikes.—(1) Iron fastenings used for holding railway chairs down to the sleepers. They are either cylindrical bars with circular heads, or rectangular bars with a single projecting head on one side. The latter are also termed dogs, or dog spikes. (2) Nails larger than tenpenny nails, or over three inches in length.

Spills.—The cracks and seams which occur in inferior iron bars, and in which adhesion is imperfect, owing to the presence of scales. When a bar of wrought iron or steel shows these cracks or seams in its fibre it is said by smiths to be spilly.

Spilly.—See Spills.

Spindle.—A slight metal rod used for carrying wheels and pulleys in light machinery. It is a shaft of small diameter.

Spindles.—See Breaking Pieces.

Spindle Valve.—A lift valve guided by a central axial spindle or stem.

Spindle Wheel.—The toothed wheel on the leading screw or spindle of a screw-cutting lathe, therefore being a "driven" wheel in the series of change wheels.

Spinning.—The moulding of circular articles in thin sheet metal by pressure applied during rotation in a lathe. Also termed burnishing.

S Pipe.—A pipe whose outline is roughly that of the letter S, used for connecting parallel lengths of straight piping.

Spiral Gear.—Includes helical, stepped, and worm gearing. Described under those headings.

Spiral Spring.—A helical coil of an elastic material, as steel, brass, or iron, which has the faculty of extension and compression in the direction of its length. Used in spring balances, locomotive axle springs, and as elastic cushions. A spiral spring is understood to have all its coils of the same diameter and pitch. In a helical spring the diameters of the coils increase with each revolution. A buffer spring is a good example of a helical spring. In a double helical spring the diameters of the coils increase from the centre towards the ends.

Spiral Winged Valve.—A lift valve, the wings of which instead of being at right angles to the seat, are arranged as sections of a spiral of very long pitch. The advantage is that the valve is turned round on its seating to a slight degree at each lift, thus rendering the wear uniform.

Spirit.—Methylated spirit, and spirit of wine are used for making shellac varnish (q.v.)

Spirit Level.—An instrument used for trying the horizontal and vertical accuracy of work. It consists of a glass bulb nearly but not quite full of spirit, and enclosed in a wooden casing. The bubble formed by the vacancy in the bulb indicates by its central position the accuracy of the surface under test. For testing vertical surfaces the bubble is placed transversely near one end of the stock, which is then pierced with a hole to permit of it being seen.

Spirits of Salts.—Hydrochloric acid (q.v.).

Splasher.—The guard plate or guard strap (q.v.) which covers the top of a locomotive wheel.

Splicing.—The splicing of hempen and wire rope is generally performed by some one of the engineer's labourers who has been an old sailor. The hempen ropes require to be well stretched previous to splicing, else the operation has to be repeated several times subsequently at short intervals, especially in the long ropes used for travelling and gantry cranes.

Spline.—A feather (q.v.) on a shaft.

Split Cottar Pin.—A split pin (q.v.).

Split Draught.—In an internally fired boiler (q.v.) the gases pass through the furnace tubes (q.v.) to the back of the boiler, whence they dip underneath into an external flue of brickwork, and so pass to the front end. Thence they divide or split to right and left into two distinct brick flues placed upon each side of the boiler, which conduct them to the chimney. This is called split draught to distinguish it from wheel draught (q.v.). Dampers (q.v.) are placed in these side flues at the farther ends.

Split Joint.—See Tongue Joint.

Split Pin, or Split Cottar Pin.—A split pin is formed of wire whose section is hemispherical, bent round until the flat faces meet. It is then inserted into the circular hole prepared for its reception and the free ends drawn apart to prevent the pin from falling out. It forms a secure and neat fastening for the pin of a knuckle joint, and is largely used.

Split Pulley.—A pulley made in halves and bolted together, and used in cases where it would be impossible or inconvenient to slide it on from the end of its shaft. The two halves are unscrewed, clipped over the shaft, and bolted up in place.

Split Ring.—An ordinary piston ring, or ring which is divided either diagonally or with a lapped joint to afford the elasticity necessary to render it steam tight.

Splitting.—The method of dividing castings of wheels and pulleys in the act of casting. It is done either in order to have two symmetrical halves which can be fitted to the central portion of a shaft when other wheels, pulleys, or bearings occupy the ends, or to relieve the strain on a casting due to unequal massing of metal around the boss. It is effected by introducing splitting plates (q.v.) into the mould, which usually not only divide the mould in two, but also carry the cores to form the bolt holes. When the boss of a pulley or wheel is split simply to relieve it of tension, filling-up pieces of metal are afterwards introduced into the open spaces, and the whole banded together by the shrinking on of rings of wrought iron over the ends of the bosses.

Splitting Plates.—Thin plates of wrought iron or of cast iron placed in a foundry mould to effect the division or splitting of a wheel or pulley in the act of casting. To prevent adhesion of the molten metal to their faces, they are painted over with tar, or with loam and black wash.

Split Wheel.—Cog wheels are sometimes split for the same reason as pulleys. They are usually moulded as one, but split by interposing a thin plate of wrought or cast iron between the arms, by means of which the casting is divided into two. The splitting takes place either through the arms ; or through lugs cast on the bosses, and on the inside of the rim. Sometimes the halves are moulded and cast separately, and fitted together by planing.

Spoke Machine.—A machine or lathe used for turning spokes on the copying principle. The lathe has two sets of centres, the wood to be turned revolving between one set, and an iron templet or copy of the spoke revolving at the same speed between the other set. The cutters and a roller are set in the same .lide rest : and while the latter follows the templet, being kept pressed against it, the former follows, and removes the wood, so that the latter is bound to become an exact counterpart of the templet or copy.

Spokes.—The arms of the running wheels of locomotives and rolling stock.

Spoke Shave.—A tool used by pattern makers for the working of sweeps in cases where great accuracy is not essential.

Sponge.—The ball or impure mass of iron newly reduced from the ore, in readiness for shingling, rolling, or casting.

Sponge Cloth.—A coarse cotton towel woven expressly for the use of machinists and engineers, to cleanse bright and working parts from superfluous oil and dirt. Hence termed a wiper. These have largely taken the place of waste (q.v.).

Spontaneous Combustion.—Some qualities of oils if left in saturation with cotton fibre, cotton waste, &c., for a few hours become heated and spontaneously ignite, hence a source of danger to factories which the oil manufacturers strive to diminish by careful selection and preparations of their oils. It is therefore a recommendation to an oil where, other qualities being equal, it has a high flashing point (q.v.).

Spontaneous Ignition.—See Spontaneous Combustion.

Spoon Bit.—A bit of gouge-like section (see Shell Bit) used for boring wood. Its cutting end is roughly cone or parabola-shaped, so that it does not draw its core so effectually as a nose bit (q.v.).

Spoon Gouge.—See Bent Gouge.

Spoon Tools.—Moulders' tools rudely spoon-shaped, and used for smoothing or sleeking moulds.

Spray.—A series of small or thin runners (q.v.) diverging from a common centre or ingate (q.v.), each to convey its quota of metal to a separate mould, the moulds all being in one box. Sprays are commonly employed in plate moulding when several patterns are moulded in contiguity to each other, and are used also for any light castings when it is desired to obtain sufficient area of entry for the metal without the risk of breaking the casting, which would result from the knocking off of a single large runner. See Ridge.

Spray Tuyere.—A form of blast furnace tuyere, in which the interior of the tubes of which it is composed is kept cool by a number of jets or sprays of cold water, directed from holes in supply pipes introduced into the body of the tuyere, the waste water falling down the sides and passing out through a waste pipe.

Sprig.—A cut brad. See Brad.

Sprigging.—When narrow and inherently weak and otherwise unsupported sections of sand occur in foundry moulds, it is necessary to sustain them ; and the custom is to thrust long cut brads or sprigs into the sand, reaching from the weak sections into the body of the mould. This is termed sprigging or nailing.

Spring.—(1) A contrivance in which the elasticity due to the nature of a substance or to its form is employed for the purpose of deadening shocks caused by the sudden contact of bodies, or for the exact adjustment of strains or pressures. Examples of the former are found in buffer springs (q.v.) and in the springs of railway axles, examples of the latter in the spring balance (q.v.) and in the dial gauge (q.v.). The railway, or wagon spring, consists of a bundle of strips of curved steel riveted together, the rivets having play allowed endways in slots in the springs. The springs being curved upwards sustain the pressure of loads suddenly applied to the truck or carriage to which they are attached. (2) The term spring is also applied to a wrought-iron bend pipe, the amount of whose curve is exceedingly small.

Spring Balance.—The ordinary spring balance is used chiefly for safety valves and weighing machines. The proportion for springs is that the diameter of the internal portion of the coil should be four times that of the thickness of the steel.

Spring Balance Valve.—A form of safety valve in which the lever is attached to the end of a spring balance, instead of receiving a weight. The graduations of the balance indicate the pressures.

Spring Bows, also called Bows, or Bow Compasses.—Small drawing compasses used for the most delicate and minute work, for which the ordinary form would be unsuitable. The legs, which are short, are united at their upper portions by a spring, which would maintain them wide asunder but for a milled headed screw by means of which their distances apart are adjusted as required. A set of bows comprises three instruments, points, pen, and pencil.

Spring Box.—A hollow box, containing a helical spring enclosing and confining the spring sideways. The reaction of the latter against the end of the box thrusts it against whatever is placed in line with and in contact with it. Levers, pauls, &c., are thus actuated by springs enclosed in spring boxes.

Spring Chaplet.—See Springer.

Spring Chuck.—A lathe chuck made of hard wood, generally box wood, and used for holding light and delicate work which would be injured or distorted by the application of much pressure. It is in shape like a cup chuck, and the spring is imparted to it by running a number of radial saw kerfs from the front downwards, similar to the slits in the caps of telescopes, and by clamping the sections thus formed with a ring sliding on the slightly tapered exterior diameter, so that the exact amount of pressure of the chuck upon the contained object can be regulated, and work can also be removed without using the hammer.

Spring Dividers.—Workmen's shop dividers, whose legs are united by a strong arched spring which tends always to thrust them asunder, which tendency is counteracted by a wing nut and screw pulling them together. The nut and screw furnishes the means of very minute adjustment of the legs. For pitching out wheels and centres generally, and for striking the smaller curves they are employed in preference to the wing compasses.

Springer, or Spring Chaplet.—A simple chaplet made by bending a short strip of hoop iron round to form three sides of a rectangle. It is used chiefly for assisting to steady the sides of cores at a definite distance from the sides of their moulds or from each other.

Spring Grinder. A grinder (q.v.) for emery which is made in two portions attached to opposite ends of a metal rod bent round so as to form a spring loop. Spring grinders are used for lapping out cylindrical holes, as well as the outsides of turned work.

Springings.—That part of an arch where the intrados (q.v.) meets the abutments (q.v.).

Spring Paul.—A paul which instead of pressing downwards by the action of gravity simply, in which case it must be placed over its wheel, is kept in contact with the wheel by means of a spring. The spring may be either external or internal, in the former case consisting of a strip of sheet metal fixed at one end and exerting pressure at the other, or a coiled spring enclosed within the body of the paul which by its reaction against one of three faces of a triangular pin enclosed in the paul centre maintains the paul in either one of those three positions ; though the resistance of the spring is not so great but that the paul can be turned round by manual force.

Spring Piston.—An engine piston fitted with some form of spring ring, as distinguished from a solid piston (q.v.).

Spring Pulley.—A wrought-iron pulley which is divided through the boss, and also at a point in the rim in line with the joint in the boss, in order to permit of its being opened or sprung apart sufficiently wide to pass over a central portion of a shaft. It is built up of wrought iron or steel. See also Wrought-iron Pulley.

Spring Ring.—A metallic ring by which the piston and the cylinder bore of an engine are rendered mutually close fitting. It sometimes fits between the top and bottom plates of the piston. Then it is turned in the first place somewhat larger in diameter than that of the bore of the cylinder, and slit or cut in a diagonal direction and sprung inwards by main force, and so reduced to the size of the bore. When put in place its elasticity maintains it closely fitting against the cylinder bore. Spring rings are usually a little thicker on one side than on the side opposite, in order to increase their elasticity. Ramsbottom rings, usually of steel, are thinner than ordinary spring rings, and either form a spiral of two or three coils, or two or three separate cut rings are pressed into grooves turned in the periphery of a solid piston. Spring rings are also forced outwards by set screws, wedge blocks, &c. Spring rings are usually made in cast iron, but wrought iron, steel, and gun-metal, are also employed.

Spring Safety Valve.—A spring balance valve (q.v.).

Spring Swage Tools.—Top and bottom swages (see Swage) which are connected by two parallel handles uniting to form a spring at the end farthest away from the swage blocks, roughly after the fashion of a pair of sugar tongs. They are used for hand forging chiefly.

Spring Tool.—A turning tool used for taking a finishing cut along a piece of work already roughed down nearly to size. It is curved or arched upwards and downwards just behind the cutting edge, and so possesses a slight amount of elasticity or spring, which, though not conducive to accuracy, produces a smooth and polished surface.

Sprocket.—See Sprocket Wheel.

Sprocket Wheel, or Pitch Chain Wheel.—A chain wheel which is provided with projections upon its periphery, which are so pitched out that they enter into the recesses of flat link or pitch chains, and either carry the chain along as they revolve or are themselves carried round by the moving chain. This is a method for the conveyance of motive power which has come largely into use of late years where power has to be transmitted to a moderate distance. The chain is composed of flat links whose sides are formed of plates of wrought iron or steel, riveted together, the centres of the links coinciding in distance apart with the

centres of the projections upon the sprocket wheels, but the projections themselves fall into clear interspaces midway between the centres of the pins which unite the links.

Spruce (*Abies excelsa*).—A kind of deal imported from Norway and Sweden. See Pine Wood. It is used for packing cases and the rougher class of work generally.

Sprue.—A channel or channels running from an ingate to a foundry mould. Also the small runners which are knocked off after casting.

Spun Yarn.—A loosely spun soft and flexible hempen rope. It is used in slinging bodies of light weight, for which its flexibility and the ease with which it is manipulated render it very useful. It is also used for making steam joints, and for serving (q.v.).

Spur Gearing.—Gearing composed of combinations of spur wheels (q.v.).

Spur Wheel.—A toothed wheel whose teeth are on the outer diameter, and at right angles with the wheel face.

Square.—(1) A rectangle (q.v.) having four equal sides whose angles are right angles. (2) A tool used for checking the accuracy of ends and edges of timber or metal. See Set Square, Try Square. (3) The square of a number is the product of the number multiplied by itself.

Square-bar Iron.—Malleable iron of rectangular form, rolled to various sections in rolls, and used in smiths' and platers' work generally.

Square-body Valve.—A pump-valve whose casing is of rectangular shape.

Square Centre.—A lathe centre placed in the poppet mandrel, having its end pyramidal instead of conical in form. It is used for cutting in the centres of work revolving in the lathe, the work being pressed against the sharp edges of the centre.

Square-corner Smoothers.—Moulders' tools having two faces at right angles with each other taken in cross section, and a thumb bit for the finger and thumb in the internal angle. Used for sleeking the angular portions of moulds. Both faces may be straight, longitudinally, or one may be curved in outline, the other being straight.

Squared.—Work is said to be squared when the faces referred to are brought to the exact angle of 90° with each other.

Square-edged Angles.—See Angle Iron.

Square File.—A file square in cross section, commonly tapered in length, sometimes cut on all four sides, sometimes having one safe edge. Used for cottarways and keyways, and for rectangular holes generally.

Square-hole Chuck.—A lathe chuck roughly similar to a cup chuck, but having a square tapered hole instead, in its front end. It is made in wood or in metal, and used for chucking rough pieces of wood or metal.

Square Inch.—See Inch.

Square Joint.—The joint plates of a rule which pass straight across the wood instead of being arched out. See Arch Joint.

Square-nose Tool.—An iron turner's finishing tool. So called because its cutting edge stands at right angles with the edges of the shank, and is a straight line; differing in this from the round-nose tool, which it resembles in other respects.

Square Packing.—Gasket (q.v.) made into a square sectional form. Used for water joints.

Square Root.—The square root of a number is the number which, if multiplied by itself, would have the square (q.v.) for its product.

Square-threaded Screw.—A screw the section of whose threads is rectangular in form. Its pitch is about twice that of a triangular thread, and its depth $\frac{18}{18}$ of the pitch. Square-threaded screws are used where accurate movement and good wearing surface are required, as in lathes and in engineers' machines generally. Often the points and roots are rounded and hollowed to a slight amount. Sometimes also the flanks, instead of being exactly square, slope at a slight angle.

Square Tile.—A draughtsman's rectangular palette, having three or four slopes or divisions for holding colour.

Square Trowel.—A moulder's trowel whose outline is rectangular.

Square Way.—The water passage through the body of an ordinary cock is called a square or "common" way, to distinguish it from full-water way (q.v.).

Squaring Up.—The process of bringing the faces of work into a squared (q.v.) position with reference to one another, with planes, files, or machine cutters; or the setting of faces already squared, perpendicular in relation to some other section of the work to which they belong, or to the face of a surface plate or marking-off table or bench, in readiness for the scribing off of lines and distances.

Squeezer.—(1) A machine used for the purpose of reducing a ball of puddled iron to a solid homogeneous mass, by the expulsion of the slag and cinder preparatory to rolling it into bars. The steam hammer (q.v.) is often used in preference to the squeezer. Squeezers are reciprocal, or rotational. See Alligator Squeezer, Rotary Squeezer. (2) Straightening Machines (q.v.) are also called squeezers, or squeezing machines.

Squeezing Machine.—Any machine by which bars or rails are bent or straightened. There are several forms, but the essential principle of their construction is that of a ram or presser actuated by a screw driven by a large hand wheel, or by a cam movement driven by power and pressing against the bar, whose pressure is taken by two points of support behind it. Squeezing machines form a portion of the equipment of all boiler shops and of most smithies.

Stable Equilibrium.—This results when the centre of gravity of a body which is in actual equilibrium lies in its lowest position.

Stack.—A chimney, though usually applied to a structure built of masonry. The chimney of an engine boiler is made long in stationary engines for the production of draught, its shortness in locomotives and portables being compensated for by the utilisation of the action of the blast.

Stacking Crane.—A fixed or portable crane used for stacking piles of goods or material at wharves or in quarries, and at sheds and sidings of railway stations.

Stake.—(1) A long tapering spike or bar of rectangular section, of wood or iron, used in foundry work. When a pattern is bedded in (See Bedding In) and covered with a top part (q.v.) only, the latter is guided into its proper position by several stakes driven into the sand of the foundry floor in immediate proximity to the lugs which project from the sides of the box. Being set for ramming, and reset after lifting, cleaning and coring, by means of the same stakes, the joints in the mould will correspond. (2) Small anvils having faces variously shaped and used by coppersmiths for bending and raising (q.v.) work are also called stakes.

Staking On.—A term sometimes applied to the rough method of fitting or hanging (q.v.) of wheels on shafts by the aid of keys alone without boring or other fitting. The eye of the wheel is in such cases larger than the shaft itself, and several keyways are cut in the eye, the

number of keys employed supplying the means of adjustment. It is common in millwrights' work.

Stamp.—A die for the moulding of sheet metal into desired form by means of screw or hydraulic pressure. Boiler crowns and repetition flanged work is done with stamps or dies. So also is much smith's work.

Stamping, or Pressing.—(1) The production of forged work by means of dies pressed together either under the hammer or under the hydraulic forging press. (2) The term is also applied to one of the processes in the preparation of tin ores. Tin ores are stamped in a stamp chest, and being reduced to powder are carried away by a stream of water into reservoirs where the heavier ore settles down by its greater sp. gr., the sand, &c., being carried on by the stream. The deposits of ore are termed slimes (q.v.).

Stamps.—Letters and figures cut in relief on the ends of short lengths of steel rod, and used for stamping initials and words upon patterns and machine parts, in the former case with a view to ready recognition, in the latter to indicate the contiguous parts in order to facilitate final re-erection.

Stanchion.—A light column of cast or wrought iron used for the support of a floor, or to carry hand railing and other similar light purposes.

Standard.—(1) A vertical support used for machinery, being usually a stout cast-iron structure fitted with bearings, feet, and strengthening ribs or fillets. Lathes, engines, pumps are built on or bolted to standards or upright frames. (2) Anything of fixed and exact dimensions which is used as a test or basis of final reference and appeal in a workshop. Standard rules, gauges, taps, &c., are thus preserved in workshops. (3) A wholesale unit of measurement for timber. A standard of pine timber is equal to 720 ft. of 11 in. $\times$ 3 in. cross section. Also the standard sizes of deals, as St. Petersburg, Quebec, &c.

Standard Rod, or Measuring Rod.—A rod of well-seasoned yellow pine used as a basis and reference for accurate measurement in a workshop. Such rods are usually made from 5 to 10 ft. long, the section of the wood being about 2 in. $\times$ 2 in. Each 12-in. division is marked on a brass inlaid plate. The end foot is wholly marked on brass, accurately divided out into fractional parts as required. A standard rod is either precisely the length, if required for butt measurement (q.v.), in which case its ends are protected with steel or brass. Or the length of the rod is in excess, and the requisite length is marked on brass plates let in, from which measurement is taken with trammels. Standard rods are made with metre divisions also.

Standard Rule.—A rule containing the divisions into ordinary inches and parts. Used to distinguish it from the contraction rule (q.v.).

Standing Vice.—See Tail Vice.

Stand Pipe.—(1) A safety valve seating (q.v.). (2) The pipe which is screwed on the top of a hydrant.

Staple, or Dog.—A tool used for clamping timber joints temporarily together while being glued or otherwise fastened. Its form is that of three sides of a rectangle, the free ends being pointed. It is made of wrought iron or steel rod.

Staple Vice.—The ordinary form of upright or tail vice, which is stapled to a bench and whose leg is stepped into a block on the floor.

Starting Cylinder.—The cylinder of a starting engine (q.v.).

Starting Engine.—The small engine used for starting large steam and Diesel engines.

Starting Gear.—Any arrangement used for starting engines or machines. It consists variously of starting valves and levers or wheels, barring engines, link motions, striking gears, fast and loose pulleys, &c., noted under their headings.

Starting Lever.—The lever attached to a starting valve (q.v.) when the latter is of the sliding type.

Starting of Machines.—The strains put upon machines by abrupt starting are frequently so great as to become a source of injury to them. Sudden starting is quite analogous to the sudden stopping of a mass in rapid motion, and produces jars, jerks, slipping of belts, and surging of parts. A machine therefore, especially when of a heavy character, should be allowed time to gain momentum by having its starting gear put slowly and easily into motion.

Starting Valve.—The valve employed for admitting steam from the boiler or pipes connected therewith, into the cylinder or cylinders of a steam engine. The valve may be of the sliding disc type moved by a lever, or a lift or puppet globe valve lifted by a screw, or a plug valve turned with a wheel.

Starts.—Those portions of the buckets (q.v.) of a water wheel which come next the sole boards and start therefrom.

Star Wheel.—A small wheel furnished with large projecting radial teeth, and attached to the end of a feed screw. A paul engages with the star wheel once during each revolution, and so feeds the screw round to a definite amount at each passing of the paul.

Star-wheel Motion.—A feed motion employed for surfacing in lathes and boring machines, through the agency of a star wheel (q.v.) and suitable adjuncts. The cutting tool is carried in a slide moving in a plate and actuated by the feed screw attached to the star wheel. The star wheel is moved one tooth at each revolution of the boring bar by which it is carried, through the medium of a pin attached to any convenient position on the bed, or rest, or head of the machine. When the star wheel is turned through the arc represented by the passing of a tooth, the screw being also carried with it, the slide in which the latter moves, and the cutting tool also which is fixed in the slide, must move to the precise amount which is regulated by the pitch of the wheel teeth, and by that of the screw thread.

Static Fatigue.—The fatigue of materials (q.v.), which results either from a purely dead load (q.v.) or from a variable load (q.v.).

Static Load.—A dead load (q.v.).

Static Moments.—The moments of forces which produce stresses without causing actual motion.

Statics.—The study of those forces which act upon structures without producing sensible motion, but which induce strains (q.v.) and stresses (q.v.) instead.

Stationary Engine.—An engine fixed on foundations. The reverse of a portable engine (q.v.).

Stationary Hand-wheel Feed.—A term applied to the feed of those drilling machines in which the feed screw is elevated and depressed by the revolution of a hand wheel, which is turned in a fixed horizontal plane. The interior of the hand wheel boss is cut to take the screw thread, and so forms its nut. The wheel is prevented from rising and falling by a turned collar embracing its boss.

Stationary Link.—A slot link (q.v.) which is not lifted or lowered for the purpose of reversing, but in which the pin or die block alone is moved.

The link in this case is curved in a direction away from the crank shaft.

Staunch.—Riveted joints are rendered staunch or water-tight by caulking with a drift (q.v.). Plates bolted together are rendered staunch by caulking with iron borings and sal-ammoniac.

Stave.—The projecting end or porter by which a crank shaft is held during the process of its building up and forging. It is cut off when the forging is complete.

Stay.—(1) A bar which stays or steadies a structure. (2) A boiler stay (q.v.). (3) Also applied to the bars of moulding boxes. See Bar.

Stay Bars.—Bars which stiffen the roofs of the inside fire boxes of locomotives. They rest edgeways at intervals across the roof, touching however, only through the intervention of washers at the two sections through which the bolts pass, in order to permit of the circulation of the water underneath them. See Girder Stays.

Stay Bolt.—Any bolt which stays or stiffens an otherwise intrinsically weak plate. The end flat plates of the fire boxes of locomotives are furnished with stay bolts connecting the two outer and inner shells. See also Boiler Stays. The plates of built-up cast-iron water tanks are held against the outward pressure of the water by stay bolts reaching from side to side. Side frames of pumps, crab winches, &c., are also connected with stay bolts.

Stayed Link Chain.—See Studded Chain.

Staying.—The affording of support to structures inherently weak, by means of stays of various kinds, as most suitable to the forms of the structures, and the pressures or strains to which they are subjected. Plates are stayed with ribs, chains are stayed with studs, boilers with various stays, tanks with tie rods, gantries and roofs with trusses; the arched, dished, and other forms given to many structures are with a view to proper staying.

Stay Tap.—A specially long tap, being from 15 to 20 in. in length, though sometimes reaching to 2 or 3 ft. Used for screwing the holes which receive the stay rods in the ends of locomotive and marine boilers. Its lower end is not screwed, but turned smooth, in order to maintain the tap truly concentric with the holes in both inner and outer shells.

Stay Tubes.—The smoke tubes of boilers, which are also made to stay the boiler ends by being screwed for the reception of back nuts within and without the end plates.

Steady.—A piece of mechanism used by wood and iron turners for the purpose of preventing the wobbling of long cylindrical pieces of work which are being revolved in the lathe. A portion of the cylindrical piece is turned first of all and enclosed in adjustable bearings in the steady, after which the remaining portion is turned with ease. The steady is fixed to the lathe bed by wedges or screws, or else travelled with the rest. Wood turners use a steady of wood, those of iron turners are constructed of iron. Long light shafts are also steadied by a cylindrical body or ring encircling them, set with screws and running in the bearings of a vertical steady, the latter being often simply a bit of square iron bar bent to the outline of an inverted vee over the turned portion of the cylinder, and clamped to an angle plate bolted down to the lathe bed. That form of steady which is bolted to the saddle of a slide rest is also called a following rest, or following steady, or back steady rest (q.v.), because it follows after or travels along with the work over the newly turned portions.

Steadying Roller.—A small roller employed in planing machines for wood, its function being to press the plank or board firmly down upon the table. Its position is immediately in front of the revolving cutter block.

Steady Load.—See Dead Load.

Steady Pins.—(1) The guide or set pins in the lugs of a moulding box by which the box parts are made to coincide. Dowels are sometimes termed steady pins. (2) A stud, pin, or short sunk feather, employed for the purpose of preventing a wheel or pulley from turning round upon its spindle, is also called a steady pin. The pin is made to fit a corresponding recess in the wheel or pulley.

Steam.—Water in the vaporous, or in the gaseous condition. Vapour of water is given off at all temperatures, provided the vapour already existing in the atmosphere is not of the same density. The difference between vaporous and gaseous steam is, that in the former the particles of water are in a state of mechanical suspension, and therefore visible, while the latter is in the condition of a true gas, or invisible, and its pressure and volume are strictly in inverse relation to each other. (See Boyle and Marriott, Law of.) Saturated steam is also invisible, though not absolutely in the true gaseous condition. It is usual to designate that as steam only, which is evaporated from the surface of water, at and above the boiling point (q.v.). Steam owes its value as a motive agency to its elastic force (q.v.) and as a drying or heating agent to its latent heat (q.v.). The pressure and density of steam alike rise with the temperature, and conversely the temperature rises with the pressure and density. But there is only one pressure and density for any given temperature; hence steam, generated in contact with water in a closed vessel, is at its maximum density and pressure corresponding with that temperature, and is then at its point of saturation. See Saturated Steam. If, however, the steam so generated be made to receive increments of heat in a separate vessel away from contact with water, it becomes superheated (see Superheated Steam), and approaches then to the condition of a perfect gas. A cubic inch of water vaporised at the boiling point corresponding with the ordinary atmospheric pressure is expanded 1,642 times. Steam of any pressure is capable of expansion on a reduction of pressure by virtue of its elastic force, and hence its value in the economy of engines worked expansively. See Compound Engine, Absolute Pressure, High, and Low Pressure Steam.

Steam Boiler.—See Boiler.

Steam Brake.—A railway brake in which the pressure of steam is made to actuate the brake mechanism. Such a brake is suitable for engines and tenders, but not for trains, owing to the condensation of the steam.

Steam Capstan.—A capstan (q.v.) worked by steam power, actuating a small independent pair of engines, instead of by hand.

Steam Case.—An old term for steam jacket (q.v.).

Steam Chamber.—Any reservoir or store for steam, as the steam room (q.v.) a superheating chamber, or steam dome (q.v.).

Steam Chest.—(1) The rectangular case or box attached to the side of an engine cylinder, and enclosing the slide valve (q.v.) together with a portion of the valve rod. Strips cast upon its edges afford guidance to the edges of the valve, and in large engines a facing or ring cast within its back face maintains the valve in equilibrium. See Equilibrium Valve. Steam chests are almost invariably cast separately from their cylinders and bolted thereto, for convenience of fitting up. In the larger chests the back is also made distinct as a separate cover, and bolted to a flange on the actual chest, for convenience of examination of the valve. (2) Also applied to the steam domes of locomotive and marine engines, the object of which is to dry and superheat the steam away from immediate contact with the water.

Steam Coal.—Coal used in the furnaces of steam boilers. The anthracitic coals (q.v.) are the best steam coals; being free burning, non-caking, almost smokeless, and producing comparatively little ash, they are much used for sea-going vessels. But the semi-bituminous coals (See Bituminous Coal) are those chiefly used as steam coals. They are rich in hydrocarbons, burn freely, and emit much smoke, but there is much variation in their composition and qualities.

Steam Cock.—(1) The upper of the two try cocks (q.v.) of a steam boiler, which when open should always pass steam, never water, since that would show too contracted steam room. (2) Generally, any cock by which admission of steam is regulated.

Steam Coil.—A coil of pipes through which steam is circulated for heating purposes.

Steam Corner.—The left-hand corner of an indicator diagram, or the corner which unites the exhaust with the lead line. It is not properly a corner, but becomes a small radius.

Steam Crane.—A crane to which the motive power is supplied by a steam engine, or a pair of engines forming a portion of the crane, to distinguish it from a crane worked by hand power only. Steam is applied to both fixed and portable cranes. The engines are usually of the vertical type, and attached to the side frames. See also Horizontal Crane.

Steam Cylinder.—The cylinder (q.v.) of a steam engine (q.v.).

Steam Dome.—The dome-shaped structure upon a locomotive boiler whence the supply steam is taken for the cylinders. The steam is partially superheated and is therefore hotter and drier there than it is elsewhere, hence the tendency to priming (q.v.) in the boiler is diminished.

Steam Donkey Pump.—A small pump used for supplying steam boilers with feed water. The piston rod of the steam cylinder and the ram of the pump are in line, and the connecting rod is usually of the bow or triangular form. See Bow Connecting Rod. A fly wheel carries over the dead centres (q.v.). The pump and cylinder are attached to a base plate which is bolted to the steam boiler or other convenient support. There are numerous modifications of this form of pump.

Steam Edges.—The outer edges of a slide valve by which the entering steam is cut off.

Steam Engine.—A prime mover which is actuated by the elastic force of steam urging a piston in a closed cylindrical vessel. See details under their headings.

Steam Engine Indicator.—See Indicator.

Steam Gauge.—See Pressure Gauge.

Steam Hammer.—A hammer moving vertically in guides and lifted by the force of steam, actuating the piston of an engine cylinder placed over-head on an A framing. The tup or moving mass and the hammer attached to the same, are in some hammers allowed to descend by the force due to gravity alone, in others the steam is introduced above the piston also to accelerate their descent. The valves for the admission and cut off of steam are under the control of a lever held by the attendant. The vibration of the blows is received by a massive anvil block (q.v.) embedded in the foundations. The weight of a steam hammer is expressed by the weight of the piston rod and hammer-head or tup. In a five-ton hammer, for example, five tons would be the total weight of these parts.

Steam Hammer Framing.—A term applied to the framing of ordinary inverted cylinder marine engines, on account of its resemblance to that of the steam hammer.

Steam Hammer Type.—A term applied to inverted cylinder marine engines. See Steam Hammer Framing.

Steaming.—The firing and getting up or generating of steam in an engine boiler, in preparation for testing or working.

Steam Jacket.—An open casing or envelope formed around a steam cylinder for the purpose of maintaining the temperature of the cylinder as high as that of the entering steam. It is formed by fitting an iron bush or lining ring within the bore of the main casting, which contains the casing, the piston moving within the bush. The liner, as it is commonly called, fits only for a short distance in from the ends, the steam space being formed by the intermediate recessed portion. When the variations in the temperature of the entering and exhaust steam are not excessive, the steam jacket is productive of economical working as preventing initial, or subsequent condensation and re-evaporation.

Steam Jet.—(1) A jet or blast of steam introduced either into the ash-pit or into the body of the fuel in a gas producer, to effect an economy of fuel and an increase in the volume of the gas produced. This it effects by becoming decomposed in contact with the heated fuel, the liberated oxygen then combining with the carbon to form CO, and the hydrogen mingling with the producer gas. (2) Also the jet of steam sent into the chimney of a locomotive or portable engine boiler through the blast pipe (q.v.) for the purpose of increasing the draught.

Steam Joints.—The joints made between the various flanges of steam pipes and steam connections, as cylinder covers, steam chests, stuffing boxes, &c. Yarn, copper wire, india-rubber, lead, millboard, and other substances are used in the making of steam joints.

Steam Lap.—The extension of the faces of a slide valve beyond the precise length which is necessary to cut off live steam exactly at the termination of the stroke of the piston. See Lap. Used in contradistinction to Exhaust Lap (q.v.).

Steam Lift.—A lift (q.v.) actuated by a small steam engine.

Steam Line.—The upper straight or approximately straight portion of an indicator diagram, which records the mode of entrance of the steam into an engine cylinder.

Steam Lubrication.—Engine pistons and the bores of cylinders require lubrication to prevent abrasion of the working surfaces. Saturated steam (q.v.) is in some degree its own lubricant, owing to the presence of moisture held in suspension, but superheated steam (q.v.) owing to the absence of water, abrades and scores the surfaces of cylinders, pistons, valves, and valve faces very rapidly. But all steam requires some lubrication, hence it is the custom to feed a constant supply of cylinder oil in with the steam, through the steam chest or cylinder ; this being diffused among the steam, and coming in contact with the wearing surfaces, prevents the injury due to friction.

Steam Lubricator.—A displacement lubricator (q.v.). A certain amount of steam being allowed to flow into the closed lubricator, containing oil, is condensed and floats the oil to the surface, whence it passes down through an overflow pipe to the parts requiring lubrication. See Impermeator.

Steam Nozzle.—The discharging opening of a steam blast pipe, narrowed to increase the force of the jet. Applied to the blast pipes of steam engines, and to the pipes by which steam is discharged into chimneys and furnaces, to increase the rate of combustion.

Steam Passages.—The steam and exhaust ports of an engine cylinder through which the steam obtains ingress and egress. They should bear a constant proportion to the size of the cylinder and the rate of travel of the piston, together with ample allowance for loss by friction due to bendings in the pipes, their size being such that there should not be any material difference in the pressure of steam in the cylinder and in the passages. An excess of size should always be given them.

Steam Pipes.—The pipes which form the connections between a steam boiler and an engine, or, generally, any pipes through which steam circulates for heating or other purposes. Steam pipes are made of cast iron, wrought iron, or copper. Cast-iron pipes are used for some large stationary engines and for heating apparatus, but their use necessitates the introduction of expansion joints. See Expansion Joint. Pipes of wrought iron are used for small engines, and when no very great length is required there is no necessity for the introduction of expansion joints. Copper pipes are used for marine and locomotive engines and for the best engines in other classes, and being usually subject to several bendings, expansion takes place without damage to the pipes or to their joints.

Steam Ports.—The passages through which the steam gains ingress to an engine cylinder. They enter the side of the cylinder close to the end. In small engines they take a curved course to the valve face at the centre of the cylinder; in large engines and in engines of the Corliss type they pass out direct.

Steam Pressure.—The pressure of steam is equivalent to its elastic force, and in the superheated state is in an inverse ratio to its volume. The temperature increases more rapidly than the pressure, hence the economy of using steam at high pressures. See Boiler Pressure, Expansive Working.

Steam Pump.—An ordinary force pump driven by a steam engine. Steam pumps are usually direct acting (q.v.), but are made in a very great variety of forms.

Steam Reversing Cylinder.—A small steam cylinder whose function is the reversal of the slot link on a large marine engine, for ahead or astern movements. See Oil Cylinder.

Steam Room.—The area included between the level of high water and the crown of a boiler, and which is occupied by the steam.

Steam Thermometer.—A thermometer graduated for high pressures and attached to a steam pipe.

Steam Tight.—Flanges, moving rods, pistons, and other jointed portions are rendered steam tight by various modes of packing. See Packing, Steam Joints, &c.

Steam Trap.—A piece of mechanism automatic in its action, by means of which the water of condensation which accumulates in steam pipes is periodically discharged. There are several forms, but the commonest is that in which the water on rising to a certain height lifts a float, and through its movement opens an escape or relief valve, the float falling again immediately that the trap is emptied.

Steam Valve.—A slide valve, or any valve by means of which steam supply is regulated, whether of the sliding or lifting types. A starting valve (q.v.) or a stop valve (q.v.).

Steam Whistle.—Essentially a keen edge, annular or otherwise, against which a jet of steam is projected at pleasure. Various names are given to the different whistles according to their shape and tone, as round barrel, long barrel, plug, hummer, fog, octave, treble, &c.

Steam Winch.—A winch (q.v.) driven by steam power, the engines being affixed to the outsides of the side frames which carry the gearing. The cylinders are usually placed at an angle in order to give as great a length of connecting rod as possible.

Steel.—A term of wide application, embracing cast, shear, Bessemer, and Siemens's steels, in their various qualities. It is essentially a compound of iron with carbon, upon whose proportions mainly depend the hardness or softness of the steel. The mildest steel differs little from wrought iron in respect to the amount of carbon present, while the hardest contains somewhat more. The properties which render steel so valuable are its great tensile strength, cast steel having a tenacity of fifty tons per square inch, and its capacity for receiving any degree of temper required. Details are furnished under special headings.

Steel Castings.—These have largely taken the place of cast iron in late years, for purposes where strength and durability are required. Steel castings are made both from crucible cast steel (q.v.) and from steel made by the open hearth (q.v) processes, the latter being the more recent of adoption, and also the cheaper. The castings are made from patterns as in ordinary moulding, but the sands are more silicious and more refractory, consisting of mixtures of ganister, and the moulds are always dried. Steel castings require annealing, and are seldom so clean and smooth as those made in cast iron owing to the high temperature at which the metal is poured. They are also more subject to blow holes, but are nevertheless to be preferred owing to their greater strength.

Steel Facing.—Refers to the practice of covering, or facing the working parts of many percussive and cutting tools, whose bodies are made of iron, with steel, the metals being made to adhere when at a welding heat. Smiths' anvils, tongs, &c., hammers in general, plane irons, the chisels, axes, and a multitude of similar tools are steel faced.

Steeling.—The facing of wrought-iron work with a thin layer of steel, in order to impart durability, or cutting power thereto. See Steel Facing.

Steel Pulleys.—See Wrought-iron Pulleys.

Steel Press.—A hydraulic press, used for compressing fluid steel into ingots.

Steel Ropes.—See Wire Ropes.

Steel Rule.—Engineers' rules are made of steel, usually being a foot long, and containing numerous subdivisions of the inch, ranging from eighths to sixty-fourths frequently; also English inches are upon one side and millimetres on the other. Sometimes various drawing scales are also included. Special rules are made triangular, and square in cross section.

Steel Square.—An ordinary try square (q.v.) used by a fitter or machinist. These are usually of small size, the smallest being carried in the waistcoat pocket, some being no longer than $1\frac{1}{2}$ in. in the length of blade. The blades are narrow so that they may pass through cottarways and keyways, and similar contracted spaces. Often they contain subdivisions of the inch on the blade or stock.

Steel Tyre.— See Tyre.

Steel Wire.—This is used in the manufacture of wire ropes. Its tenacity exceeds that of steel prepared in any other form and is developed in the act of drawing, since subsequent annealing weakens it. Thus steel wire unannealed was found to have a tensile strength of 54·07 tons per inch, while when annealed it gave 33·32 tons only. Some steel wires have a tensile resistance of 90 tons per inch. See Wire Ropes.

Steel Yard, or Roman Balance.—A lever in which the action of two unequal parallel forces ranged in positions inversely to their weights on opposite sides of a third force (the point of support) is taken advantage of as a means of weighing objects.

Steely Iron.—Cast iron which contains a low percentage of carbon.

Steeple Engine.—A form of marine engine now become antiquated, so called because the connecting rod passed up above the deck and was guided in massive A standards rising to a considerable height.

Steeples.—Sometimes applied to moulder's nails.

Steering.—The same as slewing (q.v.).

Steering Bracket.—A slewing bracket (q.v.).

Steering Gear.—Slewing gear (q.v.), or the gear whether hand or steam, employed for steering a ship.

Stemming.—See Lead Joint.

Stencilling.—The marking letters and figures on drawings, and on portions of wood and metal work for identification, by means of stencil plates (q.v.). Small-sized metal and pattern work is often marked with stamps (q.v.) in preference to stencilling, as being more permanent.

Stencil Plates.—Metallic plates of copper or zinc pierced with the shapes of letters or figures, narrow bridges of metal only being left to maintain the continuity of the plates. The ink or paint is used thick and rubbed on with a stumpy brush, the plate being held closely down to prevent the running in of the colour beyond the edges.

Step.—(1) See Footstep. (2) The distance from either one of the belt faces on a speed pulley to the belt face of its contiguous step measured radially. Hence twice the amount of step will give the exact difference in the diameters of the contiguous speeds. Also termed the fall, or the amount of fall.

Step Bearing.—See Footstep.

Step Gauge.—A male gauge turned to several standard diameters.

Stepped Gearing, or Hooke's Gearing.—Gearing in which the teeth are arranged as though cut into successive slices in their breadth, and these slices placed at a definite and constant distance behind one another in their plane of rotation. The arrangement is made to obtain continuous bearing contact between wheels gearing with one another, and consequent steady motion. The principle has been developed more fully in helical gear (q.v.).

Stepped Pulleys.—See Speed Pulleys, Step.

Stepped Rack.—A rack in which the teeth are stepped. See Stepped Gearing. Used chiefly for the tables of planing machines.

Stepping In.—When the lower end of a shaft or spindle is turned or formed into a step or shouldered shank to fit into a hole in a base plate or other support, it is said to be stepped in, and the process of fitting is termed stepping in.

Stepping Round.—This is often used to signify the dividing round of the rim of a toothed wheel pattern into equal parts, according to the number of teeth required. It is done with spring dividers by a process of trial and error.

Stern Tube.—The tube which carries the bearings of a propeller shaft.

Sterro Metal.—An alloy of copper, zinc, and tin with wrought iron.

Sticking of Valves.—The sticking of safety valves is a fruitful cause of boiler explosions, and is due to several causes. Too tight a fit in the first place, bending of the spindle, corrosion, bad design, are the chief causes of sticking.

Stiffeners.—See Web Stiffeners.

Stirrup.—A strap or loop which supports a rod or spindle, or any similar object, in a vertical direction. The spindles of some lever drilling machines are lifted by means of stirrups which embrace the tops of the spindles.

Stock.—The mass of hard melted caked coal which surrounds the hollow incandescent portion of a smith's fire, and within which his heat is localised at pleasure, and from which the fire is fed.

Stock Rail.—The fixed rail at a point in a permanent way, to distinguish it from the point rail (q.v.) or movable rail.

Stocks and Dies.—The common term for the ordinary dies for screwing, which are held in the stock or box provided with lever handles.

Stokehold.—The compartment in which the stoking of marine boilers is carried on. Stokeholds are open, when they receive only natural draught from above, closed, when they receive all their draught from fans. See Closed Stokehold. The term stokehole is more specially applied to land boilers.

Stoke Hole.—The area or enclosure in front of the door of a boiler furnace where the stoking is performed.

Stoker.—In the case of small land stationary boilers means one whose duties are to attend generally to a steam boiler, feed the furnace fires, maintain the water supply, keep the plates properly clean and free from scale and deposits of mud, and to see that the cocks and valves are always in working order. In large land boilers and in marine and locomotive boilers the stoker's duties are of a more restricted character, being those comprised under the head of stoking, the engineer attending to the cocks and valves.

Stoking.—The supplying of fuel to a boiler furnace, including its proper distribution over the grate area, the regulation of the draught, periodical clinkering (q.v.), and the removal of ash.

Stop.—(1) A plain cup-shaped casting used to fill up the socket end of a cast-iron water pipe when it is required to be closed. It is thrust into the socket and there caulked. (2) A term often used to denote a chaplet (q.v.) of rectangular section. (3) A bar of wood fitting and sliding closely in a hole near the front end of a bench to receive the thrust of wood which is being planed over. (4) A block of metal used under many modifications, its purpose being to cause the action of a machine to cease or to be reversed at a definite invariable position, the stop striking or being struck by some projecting portion of the mechanism.

Stop Cock.—(1) A cock used in steam boilers, whose function is the shutting off and turning on of the steam into the pipe which communicates with the engine. It is usually of a globular shape. (2) Any small plug cock (q.v.) used for making or closing communication in the small tubes of brass fittings.

Stop Drill.—A drill provided with a collar in order to prevent it from penetrating beyond a required distance into the material upon which it is being operated.

Stop Motion.—A lever arrangement in punching and shearing machines by which the punch or shear can be instantly thrown out of action without stopping the machine itself. The action of the lever is commonly to interpose or throw out a block along the line of thrust between the slide and the punch. Also termed disengaging motion.

Stopper Hole.—The hole in the side of a puddling furnace just beneath the working door through which the rabble (q.v.) is introduced for the purpose of stirring about the pasty metal.

Stoppering.—The closing of the mouth of a steel ingot mould when filled with molten metal. It is effected by strewing sand on the metal and covering it with an iron plate, which is then held down with a cross bar and wedges. Stoppering is necessary to prevent oxidation and the formation of spongy metal, and to preserve the chemical composition of the metal intact.

Stopping.—Material used for the filling up of holes in castings or in patterns; putty, plaster of Paris, chalk, shellac varnish, are used as stopping for patterns. See Beaumontague.

Stopping Off.—The device by which the shape of a foundry mould is altered without altering the pattern itself; a loose piece of the desired outline being prepared and placed in the mould, and the sand rammed around it into the new shape required. Thus pipes are stopped off shorter than their patterns, frames are made shallower than their patterns, and, in short, in an almost infinite number of ways, local parts are altered and modified. The process of filling up pocket prints is also called stopping off, or stopping over.

Stopping Over.—See Stopping Off.

Stop Valve.—Strictly speaking, a valve with a sliding face, as a sluice valve, but used indifferently to designate any cock or valve which is employed to regulate the passage of fluids. See Stop Cock.

Stop Valve Seating.—The globular stop valve of a Cornish or Lancashire boiler is fitted to a seating similar to a safety valve seating (q.v.), and employed for the same reason.

Stores.—Stores are either general or special in character. In the general stores are contained all the miscellaneous lighter articles used throughout the factory, as screws, nails, packing, tallow, oil, wire, small rivets, soap, &c., &c., all kept under the charge of a storekeeper, who gives them out as needed. The special stores comprise the larger and heavier goods. Thus there is a shed or store for bar iron, another for plates, another for angles and tees, another for stock castings, another for patterns, &c., &c., each being under the charge of a man who in large works may have his time wholly occupied in attending thereto, or in small works may divide his time among other duties.

Stourbridge Clay.—A fire clay found near Stourbridge, used for crucibles and fire bricks. The clay is mixed with ground coke to diminish its tendency to contraction when heated.

Stoves.—See Drying Stove, Pipe Stoves.

Straight Edge.—A parallel straight strip of wood or metal used for checking the linear accuracy of work. Straight edges are very difficult to make in the first instance, and are therefore constructed either in pairs or in threes, and checked one against the other in turn. Pairs of equal straight edges are termed winding strips (q v.).

Straightening Machines.—(1) Machines used for the purpose of straight ening rails and bars when they leave the rolls of the forge train. The machines are like punching machines, a flat block or squeezer taking the place of the punch. See Rail Straightener. (2) A machine in which channel, angle, and bar iron are straightened or bent, in boiler and smiths' shops, by squeezing; hence also called squeezers or squeezing machines. The bar, angle, or channel to be bent or straightened is laid against two blocks which form two points of resistance, on the same principle as bending rolls. The force necessary to bend or straighten, as the case may be, is furnished by a head situated midway between these blocks, and on the opposite side of the iron to be bent; the head is moved forward or backward by a cam, or eccentric, either driven by power, or by a screw if by hand, and being pressed against the iron, bends or straightens it according to the relative positions of the blocks and the amount of pressure brought to bear upon the iron. Since it will bend as well as straighten it is also called a bending machine.

Straight Link.—A slot link (q.v.) in which the amount of movement necessary for reversal is divided between the link and the block.

Straight Pane.—See Pane.

Straight Shank Drill.—A drill, twist or otherwise, having a parallel shank for use with self-centring chucks.

Straight Shank Reamer.—See Chucking Reamer.

Straight Taper.—A taper whose bounding line is a plane, not a curve.

Straight Way Valve.—A valve whose waterway or passage does not bend at a right or at any other angle, but passes through in one plane, or straight forward.

Strain.—The change of form produced in a structure by the action of the load. When the load is duly proportioned to the structure an elastic strain (q.v.) is the result, when the load is excessive the strain accumulates into permanent set (q.v.).

Strainer, or Wind Bore.—A bulb-like expansion at the termination of the suction pipes of a pump, perforated with numerous small holes, which while permitting of the passage of water prevent the ingress of foreign solid bodies. The collective area of the perforations should be in excess of that of the bore of the pipe, being about equal to $1\frac{1}{2}$ times the bore.

Straining.—See Strain. Specifically the tension put upon a foundry mould by the pressure of the liquid metal. It produces lifting (q.v.) in some cases, in others rupture of the mould or distortion of the casting. It is most liable to occur in deep moulds and moulds having large superficies.

Straining Beam.—A strut placed between the queen posts of a roof truss to counteract the upward thrust of the principals.

Straining Line.—A fine brass wire used by engineers for lining out shafting and similar work. It is wound on a reel when not in use, and serves the purpose of the chalk line of the wood worker.

Straits Tin.—Tin ore from Malacca and Banca. It is the purest ore known, and is valued accordingly.

Strand.—An aggregation of separate rope yarns twisted together in readiness for the laying of the actual rope. See Hempen Ropes.

Strap.—(1) A belt (q.v.). (2) An eccentric strap (q.v.).

Strap Bar.—A sliding bar which carries the belt fork or forks in the reversing motion of machines for fast or loose pulleys. It slides in guides, in hanging brackets, or other convenient supports, and cords or levers move it endways to a distance sufficient to move the belt from one pulley to another.

Strap Brake.—A brake consisting of a strap of hoop iron enveloping the periphery of a smoothly turned wheel, and operating by friction alone, the necessary friction being produced by the tightening of a lever. The hoop may or may not be lined with friction material.

Strap End.—When the end of a connecting rod (q.v.) has a loose strap or loop enclosing the brasses (See Brass) and held in place with a gib (q.v.) and cottar (q.v.), the complete fitting is called a strap end.

Strap Fork.—A belt fork (q.v.).

Strap Pulley.—See Pulley.

Straw.—Straw is mixed with loam for the same purpose as horse dung (q.v.).

Stream Tin.—Alluvial tin ore, which, as its name implies, has been washed down by torrents into valleys. It is therefore less contaminated with earthy matters than the mine tin (q.v.).

Strength.—That property by which a material or structure is enabled to resist strains and pressures. See details under specific heads.

Strength of Materials.—As it concerns us, is a measure of the value of the materials used by the engineer. These embrace iron in its various characters, copper and its alloys, timber, hemp, &c. The strains to which these materials are subjected are compressive, tensile, and shearing. See details.

Stress.—When materials and structures are subjected to straining forces, those forces are resisted by internal molecular actions called stresses. When the straining forces overpower the resisting stresses, the limit of the elastic strength (q.v.) of the material is passed and permanent set (q.v.) ensues. Stresses are tensile, compressive, transverse, shearing, or torsional, and these can be reduced to the tensile and compressive types simply.

Stress Diagram.—A diagram drawn to scale in which the strains upon the ties, struts, beams, &c., of a structure are represented in a graphic manner both in respect of direction and intensity. The diagram consists therefore of lines only, and, being drawn to scale, the exact amount of the stresses on each member can be taken by direct measurement.

Stretcher.—See Stretcher Bar.

Stretcher Bar, or Stretcher.—A long bar or bolt shouldered near each end, and used for the purpose of maintaining **A** frames and side frames at a fixed distance apart and perfectly rigid.

Stretching.—(1) The stretching of drawing paper on the drawing board in readiness for use. It is accomplished by damping evenly with a sponge all its central portions, gluing the edges for about ½ in. inwards and laying on the board. The glue will hold the paper and strain it taut as the paper shrinks with drying. (2) Driving bands and cords become permanently elongated due to the constant tension which they undergo over their pulleys. Ropes and leather bands stretch most. To compensate for the stretching of cords, tightening pulleys (q.v.) are employed. Leather bands are purposely stretched previous to use to anticipate the inevitable, and to avoid the necessity for relacing almost immediately after first using. See Belt Stretcher.

Strickle.—(1) A templet by which a founder strikes or works to shape a loam core. The strickle is usually guided in a longitudinal direction by a guide iron (q.v.). (2) Any piece of wood cut to a special shape and used to impart a special contour to a bed of foundry sand, and thus save expense in pattern making. The methods of its application are very numerous, being adapted to each special piece of work in turn.

Strickling.—The art of forming symmetrical or almost entirely symmetrical outlines in sand or loam by the aid of strickles, a practice by which the cost of much pattern making is saved. The strickling is effected by means of a strickle or striking board and a guide of some form or another, either a vertical bar, an iron guide rod, or a core plate or board, as may happen to be most convenient.

Striker.—(1) The central core or die block upon which a swage tool is moulded to shape by forging. (2) A hammerman (q.v.).

Striking Gear.—The lever and forks with their essential fittings by which a belt is shifted from the fast and loose pulleys belonging to a countershaft or a machine.

Striking Knife.—A timber scribe (q.v.).

Striking Out.—The drawing out of detailed views of work. Chiefly used in reference to the full-sized working drawings of the workshop.

Striking Screws.—The initiation or commencing of screw threads with chasing tools. See Chaser.

Striking Up.—Generally used in the same sense as strickling, but the latter is frequently employed in the more restricted sense of producing cores and patterns with strickles working against a guide iron where the longitudinal axis of the core is not in a right line, and the former of making cores or other works by the process of striking up straight though circular work on a core bar against a core board.

Strip, or Draught.—The taper of patterns for withdrawal from the sand.

Stripping.—(1) The interposing of thin strips of wood between the boards in a pile of stacked timber, in order to allow a free passage for the air between, for the purpose of effectual drying of the boards. (2) The damaging or tearing off of the teeth of wheels owing to too sudden starting or stopping. Wheels not sufficiently deep in gear are liable to stripping. It is more specially understood of the wooden cogs of mortice wheels. (3) The removal of the sand from a casting which is newly turned out of its mould.

Stripping Plate.—Stripping plates are plates whose ends are sloped to fit the several grooves of mill rolls, bevelled off chisel-shaped to lift or clear the bars from the grooves, and so prevent collaring (q.v.).

Stroke.—The distance traversed by the rod and piston of an engine or pump, or similar reciprocating part, or the length of travel of the ram of a shaping machine, or the arm of a slotting machine.

Strong Iron.—Applied usually to mixtures of iron of various brands, together with scrap iron, effected according to the judgment of the founder, and by which a definite grade of strength or toughness is obtained. Strong iron will not be soft on the one hand, nor too hard or brittle to be turned and planed on the other. When hammered with a sledge it will become largely indented before fracture. When broken, its crystals will be of moderate size, grey, and somewhat silvery, having a large proportion of bright black crystals of uncombined carbon or graphite intermixed therewith. Its tensile strength will be high, say from ten to twelve tons per square inch.

Strong Sand.—Tenacious foundry sand, containing a large proportion of loam and horse dung.

Structural Load.—The load due to a structure itself, as distinguished from the imposed load (q.v.).

Structures.—An assemblage of distinct members combined together to produce a specific piece of mechanism. In estimating the strength of structures the strains on each individual member and sectional portions are taken separately, and also the strains due to the action of certain portions on each other due to their mass and leverage, &c.

Strut Girder.—A lattice girder whose top and bottom flanges are united by vertical struts and braced by diagonal braces, or by counter bracing.

Struts.—(1) Vertical or diagonal rods which connect a beam with a truss rod, and through which the strains due to deflection of the beam are translated into those of tension in the rod. In roof trusses, those members which extend diagonally from the beam or from the foot of the king or queen posts either to the rafters or to the shoulders of queen posts. (2) Generally any member of a structure which is in a state of compression or thrust, as, for example, the bars of a lattice girder which incline downwards to the nearest point of support. The jib of a crane is also a strut.

Stub End.—A strap end (q.v.).

Stud.—(1) A projecting pin used as an attachment or centre of motion for a lever, rod, or wheel. (2) A bolt or screw having one nut only. Both ends of a stud are threaded, but while one end is tapped into a screwed hole, the other is left to receive the tightening nut. (3) A circular projection turned on a loose boss or similar loose portion of a pattern, and fitting into a corresponding recess of its fellow-piece. A stud is therefore a single dowel. (4) A form of chaplet having basal plates and disc-shaped heads.

Stud Block.—A rectangular block used for the purpose of screwing a stud (q.v.) into its hole. The block is screwed internally throughout its length; the lower screwed portion fits easily over the standard size of screw for which it is designed, and the upper part receives a set screw whose point bears on the end of the stud which is being screwed in, a spanner fitting around the outside of the block imparting to it the necessary circular motion. Since the block fits the screw loosely only, a turn of the hand alone is sufficient to run it back off the stud after the screwing in is completed. Stud blocks are kept for all the ordinary sizes of screws in workshops, and the trouble of using lock nuts for the screwing in of studs is avoided by their employment.

Studded Chain, or Stayed Link Chain.—Chain whose links are stayed in the centre by a short stud or connecting piece welded in one with them.

Stud Print.—A round print set in position with a circular stud fitting into a corresponding centre-bit hole, instead of being attached with brads or screws. Stud prints of various diameters, but whose studs are uniform in size, are therefore readily interchangeable on patterns.

Stud Wheel.—The wheel which revolves loosely on the stud or intermediate pin of a screw-cutting lathe, or the idle wheel of the train.

Stuff.—A workshop term which denotes timber in general, as dry stuff, wet stuff, cross-grained stuff, &c.

Stuffing Box.—A recessed chamber through which the rods of pumps and steam pistons pass, being surrounded therein either with hempen, metallic, asbestos, or other packing, to prevent the leakage of the fluid or the steam. The packing is retained in place by the screwing down of a gland (q.v.) thereon.

Sucker.—The bucket of a lifting or sucking pump.

Sucking Pump.—See Suction Pump.

Suction.—The drawing of air or liquid into a vacuum, the entry being due to the difference in pressure

Suction Box.—The lower or entering chamber of a suction pump, or of a series of two or three throw pumps, into which the liquid is drawn on the upward stroke of the piston.

Suction Pipes.—The series of pipes through which liquids are drawn up by pumping machinery.

Suction Pump.—A lift pump which depends on atmospheric pressure for its action. The atmosphere will sustain a column of water 32 ft. in height, and this therefore is the limit to the depth from which a suction pump will draw water. When the water is drawn into the barrel by suction, the descent of the piston allows it to pass through the bucket valve into the upper part of the pump, whence it is lifted out and delivered on the upward stroke of the bucket (see Pump Bucket). Actually a pump will not lift from a greater depth than 28 ft. owing to defects in fittings causing leakage, and 25 ft. is about the limit in practice. But less lift than this must be given if the pumping is to be continuous.

Suction Strainer.—See Strainer.

Suction Valve.—The lowermost valve in a lift pump, that is below the bucket through which the liquor is lifted by the suction of the vacuum above, or in other words by the pressure of the atmosphere beneath acting against the vacuum above.

Sudden Grip Vice.—See Instantaneous Grip Vice.

Sudden Load.—An impulsive load (q.v.).

Suet.—Used for the lubrication of engine cylinders and bearings; it is also rubbed over bright parts to preserve them from rust.

Suet Lubricator.—A tallow cup (q.v.).

Sugar Loaf.—See Thimble.

Sullage.—The scoriæ, cinder, scurf, dirt, &c., which floats on the surface of, and is skimmed off the metal, while it is being poured into a foundry mould.

Sullage Piece.—Sometime applied to a runner (q.v.) or to the head metal (q.v.) of a casting.

Sulphate of Copper.—See Margery's Fluid.

Sulphate of Lime.—Ca SO_4. A compound which ranks next to the carbonate as a destructive agent in the incrustation of steam boilers. It loses its solubility beyond 95° F., being altogether insoluble at 290°. See Soda.

Sulphur.—Symbol, S.; comb. weight, 31·98. An element which plays a most important part in the preparation of iron. Its presence, whether in wrought iron or steel or cast iron, is in most cases an evil, affecting the qualities of the metal. A very small proportion of sulphur will make cast iron harder and whiter. In excess it produces red shortness. A mere trace of sulphur in wrought iron or steel will produce red shortness. Pig iron which contains near ·03 per cent. of sulphur is not suitable for conversion into Bessemer or Siemens steel. The value of coke as a smelting agent, or for remelting iron in the cupola, is also diminished by the presence of sulphur.

Sump.—The lower part of a sinking pump (q.v.).

Sun and Planet Wheels.—A substitute invented by Watt for the engine crank. It consists of a cog wheel fixed at the end of the connecting rod, whose centre occupies the place of the ordinary crank pin. The centre of this wheel is maintained equidistant from the centre of another, or "sun," wheel keyed on the fly-wheel shaft. As the planet wheel is carried round the sun wheel, the latter is compelled to revolve, and with it also the shaft which carries the fly wheel.

Sun Copy.—A term often used to designate a phototype (q.v.).

Sunk Key.—See Feather.

Super Elevation.—The amount by which the outer rail of permanent way is elevated above the inner one when carried round a curve.

Superheated Steam.—Steam in the gaseous state, produced by imparting additional heat to saturated steam (q.v.), or else by expansion. It is brought to this condition by being passed through a superheater in which the suspended moisture is driven off by removing it from contact with the water, and by the application of heat combined, so as to render its condition very nearly that of a true gas. If superheated beyond a few degrees the steam corrodes away the metallic surfaces with which it comes in contact in a rapidly destructive manner.

Superheating.—See Superheated Steam.

Superheating Surface.—The amount of surface in a steam boiler to which steam is subject to drying or superheating (q.v.).

Super Royal Drawing Paper.—Measures 27 in. by 19½ in.

Super Salted Water.—Sea water which contains more than 3/8, or 8 oz. of salt to the gallon, and from which therefore the brine should be blown off to prevent injurious incrustation. See Two Salt Waters.

Supplement of an Angle.—Its difference from 180°.

Supply Steam.—Entering steam, or that which is about to do work in an engine cylinder, as distinguished from exhaust steam (q.v.).

Suppression.—The cut off (q.v.) of steam in a cylinder.

Surface Blow-off.—The blowing off of the scum which collects on the top of the water in a boiler.

Surface Chuck.—A face plate (q.v.).

Surface Condensation.—Condensation of steam effected in a surface condenser (q.v.) as distinguished from that effected by a jet condenser (q.v.).

Surface Condenser.—The modern form of condenser, in which the steam is brought into contact with the outside of a number of small brass tubes enclosed in a cast-iron casing, through which tubes a circulation of cold water is maintained by means of a circulating pump (q.v.). The steam is thereby condensed effectively, and the warmed water of condensation is sent to feed the boiler.

Surface Gauge.—A scribing block (q.v.).

Surface Plate.—A plate of metal which has been brought to the highest possible degree of superficial accuracy, and which is used for the purpose of testing the truth of work in course of preparation. Sir Joseph Whitworth originated the modern method of the manufacture of surface plates by the finishing operation of scraping (q.v.), previous to which they had been ground by means of emery powder. Surface plates are made in numerous sizes for workshop use. They are rendered rigid with ribs underneath. A good pair of surface plates should be separated with difficulty after being rubbed together. Professor Tyndall attributes this property of adherence to molecular attraction, due to the extreme proximity of the particles. That such is the cause, and not, as was previously supposed, the pressure of the atmosphere, there can be little doubt, since the plates will adhere if placed under an air pump vacuum.

Surfacing.—Surfacing signifies the transverse traverse of a slide rest, or its motion across the bed for face work. Hand surfacing is effected by the screw and handle which actuates the slide for cross travel, but in the larger lathes which are provided with a back shaft the surfacing is automatic, being effected by gearing or by friction cones driven from the back shaft.

Surfacing Lathe.—All ordinary engineers' lathes are surfacing lathes, but it is customary to give a lathe that appellation in specifications, to distinguish it from special lathes in which boring only, or some such single set of operations are performed.

Surging Drum.—See Warping Cone.

Suspension Links.—Two parallel flat rods or plates, by which the slot links (q.v.) of an engine are lifted or lowered for reversal, one end of the pair being loosely attached to the tail of the slot link, the other end being attached to a short lever keyed on the weigh shaft. Also called vibrating links from the manner in which they vibrate with the slot link when the engine is working. When placed in a horizontal direction they are frequently termed drag links.

Swab.—See Water Brush.

Swab Pot.—See Water Pot.

Swage.—A die used for imparting any desired shape to a piece of forging. It consists of a lower piece fitting into the anvil, and an upper piece held by the smith, between which the metal to be operated upon is wrought to shape. Called also top and bottom tools. See also Spring Swage Tools.

Swage Block.—A rectangular block of cast iron used by smiths. It is pierced through its central portions with numerous holes, both round and square in section, for the reception of work which requires shouldering, while its edges are grooved, in various sectional forms to receive pieces of forging too large for the ordinary bottom swages.

Swage Tools.—See Swage.

Swaging.—The drawing down of a piece of wrought metal to a definite form, with or without the aid of swages. The reverse of upsetting (q.v.).

Swarf.—(1) The particles of metal abraded by the cutting of screw threads. (2) See also Wheel Swarf.

Swash Plate.—A circular plate attached to a shaft, the face of the plate being at an angle with the axis of rotation of the plate. The plate therefore imparts a reciprocal movement to a rod which is in contact with its surface.

Sway.—The lateral vibration of a rope or belt due to want of accuracy in the pulleys, or to their being not in plane.

Sweating On, or Sweating.—The soldering of metallic surfaces without the aid of a copper bit. The surfaces are cleaned, heated, and covered with a film of solder; they are then brought into juxtaposition and warmed until the solder flows and unites. Sweating on is often employed for the temporary holding together of work which has to be turned or shaped, and which could not be so conveniently held by other methods. After having been turned or shaped, the separation of the parts is readily effected by the aid of heat.

Swedish Iron.—Swedish bar iron is chiefly in request for the manufacture of crucible cast steel (q.v.). It is produced from iron smelted from magnetic ores nearly free from sulphur and phosphorus, the smelting being effected by charcoal. The Dannemora brands are the Hoop L, the Double Bullet, W and Crown, Hoop F and Little S, and Gridiron brands. Swedish pig is cast in iron moulds, in rectangular slabs 16 in. long × 9 in. wide × 2¼ in. thick.

Sweep.—Any portion of work whose outline is curved is said to be sweeped. Sweeps, that is short sections of circles, are commonly used in the foundry for sweeping up entire rings, the segment being rammed up and shifted round several times in succession until the ring is complete.

Sweeping Up.—Striking up (q.v.) in loam or green sand.

Sweep Saw.—See Bow Saw, Compass Saw, Pad Saw.

Swell Piece.—Any flat piece of material which has its outer face curved. A term common in the shops.

Swing.—The swing of a lathe signifies the size of the piece of work which can be turned in it. Its dimensions are usually given in the gap, over the bed, over the rest or carriage, and between centres.

Swing Frame.—(1) The frame which carries the row of saws in a reciprocating or frame saw. It is made of cast or wrought iron, and works in vertical guides. (2) Swing frames are besides of wide application, being used in nearly all those cases in which the bearings of a piece of mechanism are required to be free to move relatively to other portions.

Swing Sledge.—See About Sledge.

Swing Table.—The table of a drilling machine which is made to swing or swivel around the central pillar, or frame work, as the case may be, in order to bring any desired portion of the work underneath the drill.

Swipe.—A term used in some districts to signify the starting lever of an engine, chiefly of the portable type.

Switch.—See Point.

Swivel.—A provision made by means of a shank and collar for circular movement. Crane hooks are thus made to swivel, the shank of the hook turning in a collar formed in the eye. Moulding boxes swivel upon shanks, or pins, cast in their ends. Swivel motions are employed in many parts of machines.

Swivel Hook.—See Swivel.

Swivel Jaw.—See Taper Vice.

Swivel Union.—A union (q.v.) which is provided with an independent movement by which it may be swivelled when the parts are screwed up.

Swivel Vice.—A bench vice which is made to swivel in a horizontal plane.

Symbols.—Letters representative of numbers, or quantities, used in mathematical calculations.

Symmetrical Beams.—Beams whose centre of gravity corresponds with the centre of their mass.

Syphon Lubricator.—A type of lubricator in which the oil is fed slowly to the bearing by cotton wick or cotton tape inserted at one end into the oil, while the other bends over and down to the bearing. The principle is of very wide application.

Syphon Oil Cup.—See Syphon Lubricator.

Systems.—Alludes to the various types of bracing and counter bracing adopted in different girder frames.

T.

θ.—The Greek letter Theta is usually applied in engineering formulæ to denote the measurement of an angle, whatever angle may happen to be under consideration in the formulæ.

Table.—Applied to the horizontal portions of machines upon which the work to be operated on is placed, as in planing, shaping, slotting, and other machines.

Table Engine.—A type of vertical engine, once very common, in which the cylinder stood on a table or plate of metal, and the crank and fly-wheel shaft worked in the top of A standards.

Table Feed.—When a shaping machine has a fixed instead of a traversing head (q.v.), the traverse is imparted to the table; hence called a table feed, to distinguish it from the feed given to the traversing head in that type of machine.

Table Moulding Machine.—A wheel moulding machine (q.v.) in which the flask is rammed on a revolving table, as distinguished from those machines which are set in the floor. The table machines are used for small work and work of an average size, the floor machines for the largest wheels.

Table Saw.—A compass saw (q.v.).

Table Vice.—A small vice for light work, attached to a table or bench by means of a screw underneath, which tightens up a triangular pointed clamp against the under side of the table.

Tackle or **Lifting Tackle.**—A term applied to the combinations of pulleys used for hoisting purposes in the erection of heavy work, and embracing blocks, chains, and similar attachments.

Tail.—(1) The hinder part of a portable crane upon which the balance box rests. (2) The vertical support of a tail vice (q.v.).

Tail Piece.—A short double-flanged casting, bolted to the bottom of the suction box of a pump for the connecting up of the actual suction pipes

Tail Race.—The channel which conducts away the tail water (q.v.) from a water wheel.

Tail Screw.—A screw situated at the hinder end of that which it actuates. Hence the back centre screw of a headstock, and the screw which actuates a poppet cylinder are termed tail screws. See Tail Stop Screw.

Tail Stock.—Another name for the poppet (q.v.) or movable head of a lathe. It is used in contradistinction to *head* stock.

Tail Stop Screw.—The back screw of the headstock of a back-geared lathe. It has a flat or slightly rounding end, which receives the thrust of the tail end of the mandrel.

Tail Vice.—The ordinary form of vice for general and rough work, which is partly attached to the bench or to a tripod stand, but whose tail or vertical support rests in a block of wood or other attachment, either on the floor or on the tripod base. It is sometimes called the standing vice.

Tail Water.—The water which flows away from a water wheel or turbine after having done its work, or the waste water. Called also tail race.

Taking Up.—(1) A shop term which signifies the making of adjustment for wear. It is nearly, though not quite, equivalent in meaning to tightening, but is rather applied to smooth bearing and moving parts than to parts tightly bolted or wedged together. The term is legitimately applied to the closing up of strap ends of connecting rods, the cone bearings of lathes, divided brass bearings, compensating collars, &c. It is accomplished by means of set screws, gibs and cottars, lock nuts, &c. See Wear. (2) Also applied to the shortening and relacing or resplicing of belts and ropes which have become stretched by use. It is then termed taking up slack.

Taking Up Slack.—See Taking Up.

Taking Up Wear.—See Taking Up.

Tallow.—Used for the lubrication of engine cylinders, as a flux for solder-ing lead, for coating bright work to prevent corrosion, and as an anti-incrustation agent in steam boilers, which it effects by uniting with the salts of lime to form an insoluble soap. Its employment for the latter purpose is, unless performed very judiciously, a practice to be strongly condemned. It is sometimes forced into a boiler by means of a syringe, in order to prevent priming (q.v.), which it does by forming a scum on the surface of the water. Tallow should not be used for the lubrication of engine cylinders unless of the very best quality, since with high-pressure steam it becomes partially converted into oleic or stearic acids, which corrode the iron away in the piston, cylinders, and steam joints, and around the bolts.

Tallow Cup.—A hollow cup-shaped vessel screwed in the upper part of an engine cylinder to hold the tallow or oil used for lubrication. It is provided with either one or two cocks.

Tallow Injector.—A syringe used for injecting melted tallow into marine and other boilers for the prevention of incrustation.

Tamping.—(1) A moulder's term which signifies the ramming up of the sand round a pattern. (2) Stopping the tap hole (q.v.) of a cupola with clay. The term is derived from the beating down a charge into a hole in blasting and mining operations.

Tan.—The Tangent of an angle or of a circle.

Tandem Engines.—Horizontal engines of the compound type in which the cylinders are placed in the same longitudinal axis, so that the piston rods are in one length.

Tang.—The shank of a cutting tool, or that portion which is driven into the handle. This portion is not hardened.

Tang Chisel.—A chisel having a tang, as distinguished from a socket chisel (q.v.).

Tangent Screw, or Tangent Wheel.—A worm or endless screw of fine pitch, consisting either of multiple or single threads, and so called because its axis stands at a tangent with the circumference of the wheel which it drives.

Tangents, Lines of.—See Sines, Line of.

Tangent Wheel.—A worm or tangent screw (q.v.).

Tanite.—A manufacturer's term given to a certain class of emery wheels.

Tank.—A receptacle for liquids, made in wrought and cast iron and of various sizes, the larger sizes being made of separate plates riveted or bolted together. Cast-iron tanks are caulked with iron borings and sal-ammoniac, sometimes from within, sometimes from without. See Tank Plates. The sides of large tanks are stayed against liquid pressure by means of stay bolts or tie rods.

Tank Base.—A water tank bed plate (q.v.).

Tank Engine.—A locomotive engine which is provided with a tank for the conveyance of its own water, thus dispensing with the necessity for a tender. Tank engines are of side tank, or saddle tank type, according as the tanks flank or arch over the boiler.

Tank Plates.—The plates of which cast-iron tanks are constructed. They are rectangular in form, with flanges cast around the edges and standing at right angles with the plate for bolt attachments. Since the difference in liquid pressure on the sides of deep tanks near the surface and near the bottom is very considerable, a corresponding difference is commonly made in the thicknesses of the upper and lower plates of deep tanks, or tanks consisting of two or more tiers. See Tank.

Tannic Acid.—Substances containing tannic acid are in request, being found useful for the prevention of boiler incrustation, the tannic acid combining with the lime to form tannate of lime. This is an insoluble scum which is blown off. Oak bark is the substance commonly employed, being suspended within the boiler.

Tap.—(1) A male templet screw used for cutting threads in holes. Taps are tapered for entering, and stump or parallel for finishing. Taps for ordinary work are cut to common Whitworth threads; gas taps have a finer thread. Taps are actuated by a tap wrench, or by screwing machines. They are grooved, and relieved. Using a tap is termed tapping or screwing. See details under these and other headings. (2) A term employed in reference to the opening of a hole in a melting furnace through which the liquid metal flows. See Tap Hole

Tap Cinder.—The slag or refuse of puddling and reheating furnaces, consisting of a ferrous silicate, or a silicate of the protoxide of iron, $2 Fe O Si O_2$, which, after a roasting process, becomes converted into $Fe_2 O_3$ and Si, the peroxide or ferric oxide, and silica, both highly infusible substances. This, under the name of bull dog (q.v.), is used for the bottom lining of puddling furnaces. It is covered with puddlers' mine (q.v.).

Tape.—See Measuring Tape, Cotton Tapes.

Taper.—(1) In a foundry pattern, is the bevelling or thinning downwards of the faces which are to draw or lift from the sand, in order to facilitate their withdrawal. In deep patterns this taper will be from $\frac{1}{8}$ in. to $\frac{1}{4}$ in., that is the bottom edge will be $\frac{1}{8}$ in. or $\frac{1}{4}$ in. thinner than the top. Also termed draught, and strip. (2) The slope of the cylindrical plug, and seating of a cock. It varies from one in four to one in six.

Taper Belt Joint.—A joint made in cotton belting by cutting off the plies at different lengths to thin the joint down, and so prevent jumping of the belt when running.

Taper File.—The commonest form of file, in which the dimensions of the cross section diminish towards the point.

Taper Plug.—See Plug.

Taper Reamer.—A reamer whose body is tapered longitudinally for cleaning out tapered holes.

Taper Screw Chuck, or Screw Worm Chuck.—A form of lathe chuck used for turning wood, in which the trouble of attaching it to the work with ordinary wood screws is saved by having a taper screw affixed to the centre of a small face plate and standing out therefrom, to which the piece of wood to be turned is screwed. These chucks are restricted in use to small work, and that also which is chiefly cut plank way of the grain, since the screw strips out end grain if much pressure is applied when turning.

Taper Shank Drill.—A drill, twist or otherwise, whose shank is tapered in its length, and circular in section, for use with the ordinary drill spindles.

Taper Shank Reamer.—A reamer (q.v.) whose shank is tapered similarly and for the same reason as a taper shank drill (q.v.).

Taper Tap.—A tap (q.v.) tapered in the direction of its length, in order to afford facility of entrance and ease of cutting when commencing the cutting of a screw thread in a drilled hole. The tap, by reason of its taper, cannot cut at its lower end, but only near the upper end or shoulder. Hence its function is to initiate a shallow thread only at the top of the hole, and so prepare the way for the plug or finishing tap, which is parallel throughout.

Taper Turning.—See Conical Turning.

Taper Vice.—A vice provided with an attachment for holding tapered work. It usually consists of a loose jaw piece rounded on the back, and movable in a corresponding hollow seating on the fixed jaw, by which means it is capable of a slewing movement to adapt it to different angles.

Tapes.—See Cotton Tapes.

Tap Grooving.—The cutting of the grooves or flutes in taps (q.v.) by means of which their cutting edges are imparted. The process is usually accomplished in a milling machine (q.v.).

Tap Holder.—The contrivance by which the taps are held in a screwing machine.

Tap Hole.—(1) The hole in a blast furnace or cupola through which the molten metal is allowed to flow into the mould or ladles. See Bott Stick, Tapping. (2) The hole in the bottom of a casting ladle (q.v.) for the discharge of the metal.

Tappet.—(1) The adjustable pin or stud bolted on the edge of a metal planing machine, and which through a series of levers and rods reverses the motion of the table. (2) The cam teeth, or wipers, on the wheel of a tilt-hammer are also called tappets. (3) The valves of internal combustion and Cornish engines are actuated by tappets.

Tappet Motion.—The feed motion derived from the action of a tappet (q.v.) as distinguished from that of a pawl, a screw, or other agency.

Tapping.—Signifies the opening of the tap hole in a cupola or melting furnace, through which the molten metal finds egress into the ladle or into the mould, as the case may be. It is closed with fire clay during the intervals of tapping. See Bott Stick. (2) The art of screwing a hole with a tap.

Tapping Bar.—The pointed bar or bott stick (q.v.) employed for tapping out the molten metal from a cupola.

Tapping Hole.—(1) A hole drilled to the same diameter as the bottom of a screw thread, or twice the depth of the thread smaller than its nominal size, thus leaving metal for the tap to remove. (2) A tap hole (q.v.).

Tap Plate.—A screw plate (q.v.).

Tap Wrench.—A double and equal armed lever used for the screwing of taps. A square hole pierced through the centre of the bar fits over the neck of the tap and gives sufficient grip to turn it by.

Tar.—Is used in foundries for imparting a hard preservative skin to the surfaces of loam patterns and sometimes of cores. See also Anhydrous Tar. Archangel and Stockholm tar are used for hemp ropes.

Tarred Ropes.—Ropes made of hemp, usually Russian, and steeped in tar, as distinguished from Manilla or white ropes.

Tar Varnish.—Bituminous paint (q.v.).

Teak.—(1) There are two woods thus designated, the one the African oak, the other the Moulmein or Indian, the true teak. The latter is the one which is chiefly in request for engineering works. It is the wood of *Tectona grandis*, found in the south-eastern parts of Asia and the East Indian Islands, Ceylon, Malabar, and Java. It is of a greasy nature, and therefore does not corrode iron in contact with it. It contains silicious matter which dulls the edge of cutting tools. It is of a light brown colour, shrinks little, is durable, straight and rather open grained, stands heat well, and is not attacked by insects. A cubic foot weighs 46 lbs. Sp. gr. ·74. (2) African oak or African teak is believed to be the product of *Oldfieldia Africana*, of the order Euphorbiaceæ. It is harder, closer grained, and heavier than the Indian teak. It is about as heavy as water, its sp. gr. being ·993.

Tedge.—An ingate (q.v.).

Tee.—See Tee Pipe.

Tee-headed Bolt.—A bolt whose head is formed of a transverse piece, so that with its body it makes the outline of a letter T. The piece which forms the upper bar fits into a recessed or undercut T slot, whose over-hanging edges furnish the necessary resistance to the force requisite for tightening up. This form of bolt is used where adjustability is neces-sary, the bolt sliding into any position within the range of the tee slots.

Tee Irons.—Rolled wrought-iron bars whose section is that of a letter T. They are universally employed in miscellaneous engineering built-up constructions, as bridge, girder, and roof work.

Tee Joint.—A welded joint employed for uniting pieces of bar iron stand-ing at right angles with each other. In forge welding the end of the vertical bar is spread out, and set down or fullered crossways, and laid upon a similar set-down and fullered portion of the face of the transverse bar. In fusion welding the base of the vertical bar is usually bevelled.

Teem, or Teeming.—A term applied to the pouring of metal, or to the appearance which it presents on being run from the furnace, or ladle, or crucible. Thus it is said to teem "dead" or "fiery," according to its sluggish or active appearance. The term is chiefly used by steel makers, who "teem," not pour, their ingots.

Teeming.—See Teem.

Teeming Hole.—One of the holes or pits in the floor of a steel melting house, in which the moulds for the reception of crucible cast steel or for ingot moulds are placed.

Tee Pipe.—A branch pipe, the branch of which comes out of the main pipe at right angles, similar to a letter T. Tee pipes are usually made with socket and spigot. Connections through flanged tees are also made.

Tee Rest.—The ordinary rest for hand-turning, so called from its re-semblance to the letter T.

Tee Slots.—Slots or grooves cast in the tables of planing, shaping, slotting, and drilling machines for carrying the heads of tee-headed bolts (q.v.). Tee slots should preferably be planed on their inner edges.

Tee Square.—A trying square formed in the shape of the letter T, the vertical portion being the blade, the transverse bar the stock. Used chiefly in the drawing office.

Teeth.—The projections on the peripheries of cog wheels and ratchet wheels. See Wheel Teeth.

Telescopic Boiler.—A boiler in which the rings forming the barrel are tapered in diameter, so that the transverse joints or seams fit with-out and within in succession. The shells of most Cornish and Lan-cashire boilers are thus made.

Telescopic Pipe.—A pipe within whose upper end the last of a series of deep-well pipes slides. It is used in deep-well pumps to permit of moderate adjustment of the length of the series. It is sometimes termed a telescopic suction slide.

Telescopic Shaft.—See Telescopic Slide.

Telescopic Slide, or Telescopic Shaft.—Is formed by the movement of hollow tubes one within the other. Such shafts are of wide applica-tion, being employed to carry a portion of a machine whose longitudinal position is always varying, as the horizontal spindles of radial drilling machines, and in blast pipes for portable forge furnaces, and in deep-well pumps.

Tell Tale.—An indicator by which the attendant of a winding engine is informed when to stop its motion, and so arrest the upward or downward movements of the cage. Tell tales are employed also on various automatic machines to indicate the precise time when a certain set of operations are terminated.

Telodynamic Transmission.—The transmission of power to long distances by means of relays of wire rope (q.v.).

Temper.—(1) The temper of a piece of steel is its elastic condition considered in reference to the purpose for which it is to be utilised. See Tempering. (2) A mixture of 16 parts of copper to 32 of tin, used to assist in the mixing of alloys of various kinds in which the proportion of copper is so small that if added alone its higher fusing point would not allow proper combination to take place. The copper and tin alloy are added as a temper. See Babbitting.

Temperature.—The temperature of a body is measured by the amount of its sensible heat, or that which can be measured by a thermometer, and is independent of its latent heat (q.v.). It is therefore that which can be communicated from one body to another, and be directly recognisable and measurable.

Tempering.—(1) The imparting of a definite degree of hardness, or elasticity to steel, by the raising or letting down (q.v.) of the metal to a certain temperature, and then rapidly cooling or quenching in water or oil, from that temperature. The temperature is indicated by the colour which the steel assumes in passing from a high to a lower degree, and is different for almost every class of cutting instruments. The steel is quenched in water, oil, soap, or other substances, the medium varying with the nature of the objects tempered and the degrees of heat. Sometimes the temperature is estimated not by colour, but by the flashing point (q.v.) of a fat. (2) Tempering also has reference to the slow drying of articles moulded in clay, as bricks and crucibles. When the paste has been moulded into form, the article is subjected to a prolonged desiccation at a gradually increasing temperature ranging from about 60° F. to 150° or 180°. The tempering may continue from two to eight months, and the longer the process is continued the longer the crucibles or bricks will last.

Template, or Templet.—Any temporary pattern, guide, or model by which work is either marked out or by which its accuracy is checked. Usually thin light plates of wood or metal cut to special outlines. Used in all the departments of engineering.

Template Maker.—See templating.

Templating.—The laying out (q.v.) of diagrams on thin sheets of metal, or thin strips of wood, from which the plates employed in boiler and sheet metal work are to be cut. Templating is usually the special work of a boiler maker or plater devoted wholly to it, hence called a template maker, whose intelligence and careful skill ensure him a rate of wages correspondingly higher than those of ordinary platers or boiler makers.

Templet.—A template (q.v.).

Temporary Centre.—A centre piece, or a centre mark, used only for a temporary purpose, either for marking out work or for machining it. A temporary centre may be a slip of wood bridging over a hole, upon which slip the centre point is marked; or it may be a mandrel filling up the hole, and having a centre point countersunk in its ends.

Temporary Mandrel.—A mandrel (q.v.) used as a temporary centre, either for supporting work which is being lined out, or for carrying work already bored, in the lathe while it is being turned on the outside diameter.

Tenacity.—That property by which a material is enabled to resist forces tending to tear it asunder. Tenacity depends mainly upon density, but is affected by conditions of cooling, drawing out, or development of fibre, temperature, and other causes. The tenacity or tensile strength of materials is usually given in the number of hundredweights or tons necessary to break a bar of the material of one square inch sectional area.

Tender.—It is the custom to prepare tenders of all large works to be contracted for, in the drawing office, in fulfilment of the terms of specifications. Before tenders can be prepared, general draw. gs and partial details at least of the machine or structure in question have to be made, in order to obtain therefrom approximate weights of all the castings, forgings, and plated work, and also to estimate the value of the labour involved. Tracings of the drawings, or sun copies are also sent with the tender or estimate. Sometimes tenders are paid for, but usually there is no compensation in the event of non-acceptance, and lost tenders amount to a very high percentage of drawing office expenses.

Tenon.—A tongue projecting from the end of a piece of timber, and which, with the mortice into which it fits, is one of the commoner joints in wood work.

Tenoning Machine.—A machine employed for cutting tenons in wood, consisting generally of four saws, two running vertically together, and capable of adjustment sideways for the different thicknesses required, and two running horizontally to cut the shoulders. Or, in more recent machines, consisting of two adze blocks (q.v.), with flat plane iron cutters revolving horizontally, and capable of vertical adjustment. The work is then bolted to a sliding table having a horizontal movement in front of the cutter blocks. Sometimes a third cutter is used to take out a central portion, and so form a double tenon.

Tenon Saw.—A saw which derives its name from the tenon form of joint. It is a thin saw ranging from 8 in. to 16 in. in length, and is supported by a back of wrought iron or brass; hence called also a back saw. It contains about ten teeth to the inch.

Tensile Strain.—A strain or pull in a longitudinal direction, and therefore the reverse of a crushing strain.

Tensile Strength.—The strength necessary to enable a bar or structure to resist a tensile strain; equivalent to tenacity. See Ultimate Strength.

Tensile Stress.—The stress to which a bar or structure is subject when in tension.

Tension.—(1) A body is in tension when it is subject to forces tending to tear or break it asunder in a direction parallel with its axis. (2) The degree of strain put upon a band saw in order to keep it taut for work. It is regulated by means of a weighted lever and rack and quadrant, or by means of a hand wheel, screw, and spring.

Tension Bar.—A bar which is being subjected to tensile stress.

Tension Bolts.—Tie bolts (q.v.).

Tension Rod.—A rod subject to tensile stress. Tie rods (q.v.) are examples of tension rods.

Terminal Pressure.—Final pressure (q.v.).

Terms.—The parts of an algebraical or arithmetical expression connected by the signs plus and minus.

Tesselated Plates.—Often applied to the thin wrought-iron chequered

plates used as foot plates, the word having reference to the form of the chequering, which is produced by the diagonal crossing of shallow ridges or elevations on the surface of the plate.

Test.—(1) Engines and machines are usually tested at the works in which they are manufactured, before delivery to their purchasers. The tests are in every case those of a recognised and standard character, and are witnessed in most instances by an appointed and experienced inspector, not the actual purchaser. (2) The tests to which the raw materials of engineers' structures are subjected are extremely numerous, and in some cases very severe.

Test Bars.—Bars employed as tests of the quality of metal in any particular set of castings. When large quantities of iron castings are ordered it is customary to have bars run from the same metal as that which is put into the castings. These bars are usually of a 2 in. × 1 in. cross section and about 3 ft. 4 in. long. They are rested upon supports 3 ft. apart, and loaded until they break, the amount of their deflection before breaking being also noted. A bad bar will break at 25 cwt., a good one at 35 or 37 cwt., 30 cwt. being a bar of fair average quality. These will deflect $\frac{1}{2}$ in. or more before fracture if the metal is tough.

Test Cocks.—Try cocks (q.v.).

Testing Machine.—A machine used for testing the elasticity and strength of materials. In its simplest form it consists of a simple lever very unequally divided, the proportions between the lengths of the arms being inversely as the weight suspended and the strain on the specimen. The best testing machines are made of a combination of two or more levers, together with sundry other attachments for the adjustment of the levers themselves, and specimen holders, and appliances also for the measurement of torsional strains.

Test Load.—A proof load (q.v.).

Test Pump.—A hydraulic force pump (q.v.) is the kind of pump used for testing purposes.

Test Specimens.—Pieces of material specially prepared for testing, either with or without the aid of a machine. See Specimen Bar, Specimen Plate, Test Bars, Test Strips, Forge Test, &c.

Test Strips.—In specifications for boiler and other works in wrought-iron and steel plates, it is usual to stipulate that a strip shall be cut off each plate to be subjected to suitable tests by the engineer or inspector; these are known by the name of test strips.

Theorem.—A truth capable of demonstration from truths previously admitted or proved.

Theoretical Quantities.—Quantities which have scarcely any existence in practice, though valuable as standards of reference when making calculations as to the useful effects given out by motive agencies. Thus the modulus of a machine represents its actual value and is a fraction only of its theoretical value.

Thermal Lines.—See Adiabatic Curve, and Isothermal Lines.

Thermal Unit.—A certain quantity of heat, or its equivalent of work; used as a standard by which to compare other quantities. See Unit of Heat.

Thermal Value.—The economic value of a combustible element or substance, estimated according to the number of heat units liberated by its perfect combustion. The thermal value is the same for any substance under similar conditions, whatever be the rate of its combustion. (See Combustion, Products of Combustion.) The thermal values of substances are higher when the products of combustion are water instead of steam, because a certain portion of the heat is absorbed in the conversion of water into steam. See Latent Heat.

Thermodynamic Engines.—Usually applied to internal combustion, gas and hot-air engines, and engines which derive their motive power from increase and diminution in temperature of the motive force. Steam engines, though thermodynamic, are not popularly classed with these.

Thermodynamics.—The science which treats of heat as a form of energy or mode of work.

Thermokinematics.—Sometimes applied to the study of the theory of the motion of heat.

Thermometer.—An instrument used for measuring temperatures. An air thermometer is graduated from absolute zero (q.v.) point upwards. Since its readings must be taken in conjunction with those of the barometer, its use is confined to scientific researches chiefly. Fahrenheit scale reckons from 32° to 212°; Centigrade, from 0° to 100°; Reaumer, 0° to 80°. To convert degrees Centigrade into degrees Fahrenheit multiply the given temperature by 9 and divide by 5 (a degree C. being to a degree F. in the ratio of 9 to 5), then add 32. Conversely, to reduce Fahr. to Cent. subtract 32; multiply by 5 and divide by 9.

Thermostatics.—Sometimes applied to the theory of the equilibrium of heat.

Thick Fire.—A furnace fire is thick when the fuel is spread to a considerable depth over the grate bars. It is more economical than a thin fire, because the heat of combustion is not so rapidly dissipated, and the fuel is more thoroughly coked.

Thickness.—Thickness in a loam pattern signifies that coat of loam which represents the pattern thickness superimposed over the core. This is stripped off after moulding, leaving the core intact, which last is then placed in the mould ready for casting.

Thicknessing Machine.—A machine used for the purpose of planing stuff to a gauged or parallel thickness, two adze blocks being used, one of which is movable for adjustment to thickness.

Thickness Pieces.—Rectangular strips of wood gauged to definite thicknesses, and used by moulders for trying the spaces or metal thicknesses in their moulds, the strips being laid between cores and the adjacent mould portions.

Thick Oil, or Heavy Oil.—A lubricating oil having much body or viscosity, as castor oil, and used for the heaviest class of machinery, being less liable than the thin oils to become squeezed out from between the bearing surfaces.

Thimble.—(1) A short cylindrical casting used for connecting the spigoted ends of pipes which lay end to end, or which have been cut off, or have become cracked and broken and have to be reunited without sockets and spigots. A thimble is of the same bore as the socket of its pipe, and is caulked similarly to a socket. (2) A tapered iron mandrel, circular in section, used in smithies for the bringing of rings and holes which have been already roughly shaped under the hammer into a truly circular form, the hole being slid over the tapered portion, and set down on the faces and struck on the outside until it assumes the circular form. Termed also a sugar loaf from the almost precise resemblance of the smaller thimbles in outline and size to a sugar loaf. Also termed a cone simply.

Thin Fire.—A boiler fire is said to be thin when the fuel is spread in a thin **stratum** over the fire bars. There is a loss of heat in using a thin fire, though in horizontal internally fired boilers its employment cannot be avoided owing to the contraction in the area of the flue and ash pits.

Thinning Out.—In boiler work where the circumferential and longitudinal joints of the plates overlap (q.v.), it is necessary that the middle plate should be bevelled or drawn down to a feather edge just at the joint, to permit of close and flush fitting. This is called thinning-out, or thinning down.

Thin Oil, or Light Oil.—A lubricating oil having little body or viscosity, as sperm oil, and used for the lightest class of machinery.

Thoroughfare Hole.—A hole carried right through a piece of work, as distinguished from a hole which is recessed only.

Thread.—The actual helix, or feather, or blade of a screw, as distinct from the solid body or bar on which it is cut and by which it is carried and supported.

Thread Gauge.—A screw-threaded gauge (q.v.).

Three-cylinder Engine.—(1) Several types of three-cylinder steam engines were made. They may be divided into two main classes—those in which the cylinders are placed at an angle of 120° to each other, and those in which they are ranged side by side. In the former or Brotherhood type, the three pistons were of the trunk form, though shallower than ordinary trunk pistons, and three piston rods pivoted loosely in the bodies of the pistons are connected together by bonding, around a single eccentric shaped crank, made solid with the crank shaft. The supply of steam took place at the end, that is to one side of the three cylinders, and was regulated by a valve common to all, which opened and closed the ports in succession. (2) In the other type, in which the cylinders were placed side by side, their pistons actuated a three-throw crank. In Willan's engine the steam was introduced only above the pistons, and the pistons acted as valves also for opening and cutting off the supply of steam to the other cylinders. The pistons of three-cylinder engines being in equilibrium no balancing of reciprocating parts is required, and they can be run at a much higher speed than engines of the ordinary reciprocating type. Hence in their modern steam or Diesel forms they are employed for driving dynamos, centrifugal pumps, circulating pumps and high-speed machinery generally.

Three-high Mill.—See Three-high Rolls.

Three-high Rolls, or Three-high Mill.—In a rolling mill this signifies the arrangement of three rolls one over the other, so that the bars or plates may be passed forward between the lower pair and backward through the upper ones, without reversing the direction of rotation of the rolls.

Three-part Box.—A moulding box or flask consisting of three parts, and used for moulding those patterns in which the middle section must be distinct from the top and bottom, as, for example, in the case of a grooved pulley.

Three-ported Cylinder.—The ordinary steam cylinder for the common slide valve, as distinguished from those cylinders having two or more openings leading into each steam passage from the valve face, in cases where double-ported valves and treble-ported valves are employed.

Three-square File.—A term commonly applied to a file triangular in section, or a three-cornered file. Used chiefly for saw-sharpening.

Three-throw Crank.—A shaft having three cranks forged upon it at angles of 120°, for driving three valves, or buckets, or pistons, and used chiefly for pumps. The cranks being so arranged, are in equilibritum in any position, and can be run smoothly at a high rate of speed, and also started in any position.

Three-throw Pump, or Treble-barrel Pump.—A pump having three working barrels arranged in line, and having their piston rods connected to three overhead cranks set at angles of 120° on a common crank shaft. Equality of motion and a continuous supply are the advantages of their use.

Three-way Cock.—A cock having three delivery branches, either sockets, or flanges, for diverting the liquid from the inlet branch into two different directions at pleasure.

Throat.—(1) The enlarged neck of a branch or offset pipe, enlarged and curved for the purpose of diminishing friction and allowing greater freedom of flow for the liquid. (2) The top or opening of an open-topped blast furnace, into which the fuel and ore are charged. (3) The inner edge of the flange of a railway wheel, or that portion which corresponds with the gauge of the rails.

Throat Core Box.—A core box used in connection with cast-iron pipe work. It forms the curved portion by which the core of a branch or Tee passes into that of the main pipe. See Throat.

Throttle Valve.—A flat, thin, disc valve, which lies diagonally within and across a round pipe and closes it partly or entirely. Used in dampers, flues, and engine steam pipes leading from the boiler, to regulate the passage of smoke and of steam.

Throttling.—The wire drawing (q.v.) of steam in a narrow passage.

Through Tubes.—The flue tubes of Lancashire and Cornish boilers, so called because they pass from end to end, being attached to the end plates.

Throw.—The total amount of eccentricity of a crank or eccentric, being equal to twice the radius.

Throw Disc.—The disc of a slotting machine which actuates the ram (q.v.) through the medium of a short connecting rod.

Throw-out Gear.—An arrangement of levers and rods by which the motion of a single eccentric is reversed in marine engines.

Thrust.—The compressive force exercised by a body transmitting pressure.

Thrust Bearing, or Thrust Block.—(1) The bearing which receives the endlong thrust or pressure of the horizontal shaft of a ship's screw. It is similar in outline to a plummer block, but its bearing portion is grooved circumferentially to receive the thrust collars (q.v.). A horseshoe thrust block is one in which the bearing grooves are not continuous around, but arched over the shaft only, so that the portions of the shaft collars which are lowermost run in a reservoir of oil, and carry it up to the actual collar bearings above. (2) The bearing or shouldered portion of a drill spindle (q.v.) where the feed screw and the revolving spindle are in contact. See Thrust Collars.

Thrust Block.—A thrust bearing (q.v.).

Thrust Collars.—(1) Collars turned on the body of the shaft of a ship's screw which transmit the endlong thrust to the thrust bearing (q.v.), the collars fitting into the circumferential grooves of the thrust bearing. (2) The compensating collars (q.v.) of drilling machine spindles, so named because they sustain the thrust due to the reaction of the work against the drill.

Thrust Screw.—A screw with or without the power of endlong adjustment, which takes the thrust of a revolving spindle. Examples of thrust screws occur at the top of the drill spindles of some drilling machines, and in the back centres of the headstocks of lathes. The bearing end of a thrust screw may be either flat, or conical in form, and is always hardened.

Thumb Plane.—A small plane about four or five inches long, and having an iron of about an inch in width, used for small work in general, where a larger plane would be inconvenient.

Thumb Screw.—A light screw having its head expanded for grasping by the finger and thumb. A wing nut (q.v.).

Ticks.—Crows' feet (q.v.).

Tie.—A bar or member introduced into a structure to resist tensile stress. Ties are used in trussed beams, in Warren and lattice girders, being the bars which incline downwards towards the centre; and to receive the pull of the load on crane jibs, and in numerous other positions besides. See Wind Tie, Tie Beam, Tie Rod.

Tie Beam.—A beam which ties together or prevents the spreading out of the lower ends of the rafters of roof trusses.

Tie Bolts.—Screw bolts, usually long, which tie or stay large flat and inherently weak surfaces. They are used in built up water tanks, as longitudinal stays in Cornish and Lancashire and marine boilers, and for similar purposes in many other structures.

Tie Rod.—(1) A long screwed bolt or stay, which passes from end to end of a Cornish or Lancashire boiler, and ties or stays the flattened plates. (2) Rods which support or tie the jib or lifting strut of a crane. These rods are subject to very severe stress. Tie rods are also termed tension bolts or tension rods.

Tightening Gear.—The arrangement by which the cotton or hemp ropes used for driving machinery are kept taut on their pulleys. The usual means of tightening is the employment of weights depending from an idle and adjustable pulley over which the rope passes. See Tightening Pulley.

Tightening Pulley.—In rope gearing, hygroscopic changes in the atmosphere, and the stretching due to straining action causes the lengths of the ropes to vary. To take up the slack, a guide pulley or pulleys are hinged on a sliding bearing plate, moving in guides so that within certain limits the extensions of the ropes are compensated. The pulleys are therefore termed tightening pulleys. In light running cords, driving lathes and other machines from overhead, similar pulleys are employed.

Tile.—The iron plate or lid which covers the mouth of a brass melting furnace. So named because a common tile was originally used.

Tilt Hammer, or Trip Hammer.—An old-fashioned form of smith's hammer, used for heavy forging before the introduction of the steam hammer. The hammer shaft swings on a pivot: and a revolving cam or wheel furnished with projecting teeth, by pressing down the end of the hammer shaft lifts the hammer, which falls by its own weight directly the cam has passed, to be again lifted for the next blow by the next cam tooth. Tilt hammers, though to a great extent superseded by the steam hammers, are still employed both in smithies and in iron, steel, and tin plate works, for forging and for slabbing and blooming down the smaller puddled balls and ingots.

Tilt Steel.—Blister steel which has been beaten out under a tilt hammer (q.v.). The name is still retained notwithstanding that the work is now done by the steam hammer.

Timber.—The various timbers employed by engineers, together with

17*

their qualities, defects, preservation, seasoning, &c., are noted under appropriate headings. See Wood.

Timber Clamp.—A bar clamp (q.v.).

Timber Measurement.—Timber is measured in various ways By the superficial foot, reckoned as one inch thick, as in boards ; by the cubic foot, as in logs or beams ; by the standard (q.v.) as board in quantity , by the load, as in wholesale dealing ; by the hundred, as in deals and battens.

Timber Rack.—A store for timber. There are two kinds of racks, one in which the wood is laid in a horizontal, the other in a vertical position. In the first case the boards are stripped (see Stripping) and laid in bays or compartments ; in the second they stand on end slightly inclined edgeways, leaning against a beam from which round rods of iron project and act as divisions between each batch of boards. The first mode is used when the boards are actually drying, the second when they have been dried and are considered ready for use. But though boards are never dried while standing on end, they are when dried, often kept in the horizontal position until required.

Timber Scriber.—See Scriber.

Timber Washers.—Washers (q.v.) of cast or wrought iron used in framed timber works to receive the pressure of the bolt heads and nuts, and so prevent bruising of the timber. When the bolts pass through the timber in an oblique direction bevelled washers are used.

Time Sheets.—Printed slips of paper with headings for dates, on which workmen write out each day's time ; that is, the number of hours which they have worked and the jobs upon which they have been engaged. These are collected and sent into the office and checked against the time as given in by the gatekeeper.

Tin.—Symbol, Sn. ; comb. weight, 117·8 ; sp. gr., 7·3. A silvery white metal of immense value in engineering work, chiefly as an alloy in the manufacture of gun-metal, brass, phosphor bronze, Muntz metal, &c., and in the mixing of solders.

Tine.—An iron or steel claw or fork used in dredging operations. The terms half tine, whole tine, refer to the length of the claws. Any fork-like projection used for piercing or digging would properly be called a tine.

Tinned Chaplets.—Moulders' chaplets coated with tin for the same reason as tinned nails (q.v.).

Tinned Nails.—Moulders' chaplet nails coated with tin to prevent rusting and the consequent formation of blow holes.

Tinning.—(1) The process of covering sheet iron plates with tin in the tinplate manufacture. It is effected by a succession of processes subsequently to the preparatory rolling and pickling. About 340 plates are immersed at a time in a bath of melted tin covered with a thick layer of tallow, and heated to nearly the flashing point of the latter. They are removed in about an hour and a half, immersed in another bath of tin, lifted out, brushed and plunged into melted tallow, and allowed to drain. They are then dipped into a hot pot to remove the rim of metal along the bottom edge, cleaned in bran, sorted and packed. (2) The coating of a soldering iron with tin previous to use.

Tin Ore.—See Tinstone.

Tin Plate.—Thin sheet iron coated with tin.

Tinstone, or Tin Ore.—The ore of tin occurring in the condition of an oxide. It is found in Cornwall, Malacca, Banca, Australia, Bohemia,

and Saxony. That from Malacca and Banca is known as Straits' tin, or Banca tin. See Stream Tin, Mine Tin.

Tipping Bucket.—See Skip.

Tire.—See Tyre.

Titan Crane.—A massive type of block-setting crane similar to a Hercules (q.v.) in lifting capacity and general form, but differing therefrom, in that it is not provided with motions for slewing. It consists of a travelling carriage built up and properly braced, spanning the pier or breakwater, and travelling on rails. A traveller and lifting crabs are mounted on this, so that there are three sets of motions available, one of the Titan framework longitudinally, another of its traveller, also longitudinally, carrying the crab, and a third of the crab transversely. Hence a considerable range in the setting of the blocks is practicable. Titans are driven by steam power, the engines being at the hinder end.

Titanic Steel.—See Titanium.

Titanium.—Symbol, Ti. ; comb. weight, 48. One of the rare metallic elements which is sometimes found associated with grey pig iron and with the slags of blast furnaces, and is also purposely alloyed with tool steel, or steel required for cutting purposes, hence called Titanic steel.

Toe.—The lower or bearing end of a vertical spindle which rests in or works in a footstep (q.v.).

Toe Step.—See Footstep.

Toggle Joint.—A form of lever knuckle joint by means of which great power is obtained. It consists essentially of two lever arms forming an angle with each other and hinged at the centre, the opposite end of one lever being hinged on a fixed pivot, the opposite end of the other being free to move. Any attempt to straighten the levers, that is to bring them into, or nearly into line, causes enormous pressure to be exerted at the ends. This form of joint is used in various presses.

Tommy.—A pointed round iron bar or lever used for insertion in the holes drilled in the circular back nuts of lathes and other machines, for the purpose of tightening them up. Also a metal rod kept for insertion in the eyes of the tightening screws of hand-rest sockets, for tightening the T rest. Sometimes called a podger.

Tommy Hole.—A small hole drilled in a circular nut or similar machine part, and into which the point of a tommy is inserted for tightening up.

Ton.—The highest unit of weight employed in English practice. It contains 20 cwt., or 2,240 lbs. The Cornish mining ton is 21 cwt.

Tongs.—A smith's tool made in diversified forms and used for holding the work which is being operated upon. See Flat Bit Tongs, Pincer Tongs, Crook Bit Tongs, Hammer Tongs, Hoop Tongs.

Tongue, or Tongue Piece.—This is a short piece of metal or wood projecting from one portion of a structure and entering into another, in order to ensure steadiness and to prevent overlapping of joints. Or a tongue may simply be of the nature of a guide piece or a check. Hence tongue and tenon are not quite synonymous, a tenon being short and narrow, and used only as a method of jointing ; while a tongue may be long and wide, and used for other purposes besides the making of joints, as for instance a projecting strip screwed on a foundry loam board to give the length of a struck-up core, or to form a groove around the core, would be termed a tongue.

Tongueing.—The jointing of the edges of boards by a groove and corrasponding slip, the tongue on the edge of one board fitting into the groove on the edge of its fellow. Tongueing is performed either by cutters set in a revolving block, or by means of a joiner's plough.

Tongue Joint, or Split Joint.—(1) A smith's joint made by inserting the tapered end or tongue of a rod into a V-shaped corresponding cleft in the end of another rod for the purpose of making a weld. (2) Also the joint formed by the tongueing of boards.

Tonne.—A French measure of weight equivalent to that of a cubic metre of distilled water at 4° C., and equal to 1,000 kilos. See Millier.

Tool.—Any implement by which mechanical operations are performed, whether by hand or machine. See Hand Tools, Machine Tools, and details under specific names.

Tool Angles.—See Angles of Cutting Tools, Angle of Relief, &c.

Tool Blanks.—Blanks (q.v.) for milling and other cutters.

Tool Board.—A board supported behind the bed of a lathe as a receptacle for the tools used in turning.

Tool Box.—(1) A piece of mechanism attached to the cross slides of planing machines, the slide rests of lathes, the arms of shaping and slotting machines, &c., to carry the cutting tools. Movement is imparted to the tool in one or more directions by automatic arrangements varying in the different kinds of machines. Tool boxes are usually made in several parts, sliding the one over the other for longitudinal or cross traverse, and often for circular movements besides. The tool itself is clamped with a screw or screws. (2) Any wooden box which is used to contain workmen's tools. Pattern-makers' tools are kept in a chest suitably fitted with trays and drawers; moulders' tools in an open rectangular box without any trays or partitions, and furnished only with a cross handle bridging over the top. Fitters and turners keep their tools in drawers or in plain boxes. For outdoor work the erectors' tools, as spanners, ratchet braces, &c., are put in strong iron-clamped packing boxes with padlocked covers.

Tool Carriage.—The sliding carriage which traverses the cutting tool in any self-acting machine. The slide rest of a lathe, and the tool box (q.v.) of a planing machine are illustrations.

Tool Handles.—The adoption of suitable forms for the handles of tools, cutting and otherwise, is not a matter of indifference, but one which is regarded as almost of exceptionable importance by workmen, since few care to use either the tools or the handles used by another man. In most cases the handles are made of wood. Thus files, chisels, gouges, and hand-turning tools have handles of wood bonded with ferrules where the tang is inserted, and the shape of the handles varies in the case of nearly every separate tool, depending upon the way in which it is held. The handles of moulders' trowels are similar in shape to those of small file handles; the handles of smiths' swages and fullering tools are of withes twisted around the body of the tool; hammer handles are nearly parallel, and oval in section, but of various lengths. Many tools, as all moulders' tools, except the trowel, have no handles, nor have cold chisels, drifts, slide-rest tools, any handles.

Tool Holder, or Cutter Bar.—A bar of iron or steel used for holding the actual cutting tools in metal turning or shaping. There are numerous tool holders in use, but they consist essentially of an arrangement by which the tool is clamped in the holder to the angle most suitable for cutting the metal in hand. Some types of cutter holders have the front portion, in which the tool is clamped, perfectly rigid with the shank ; hence in these a pair of right and left-hand holders are necessary in each size. In other types the clamping arrangement is made to swivel from one hand to the other, so that one cutter bar in each size is sufficient. These sometimes lack the rigidity of the solid bars, allowing the tool to give under heavy cutting. The advantage of the use of tool holders is that time and material are saved ; time, in drawing down and re-tempering, which are rendered unnecessary by the use of tool points, and material, because a single cutter bar will take a number of small tool points, each of which would otherwise have required its own separate shank.

Tooling.—The cutting of metals with ordinary cutting tools, as opposed to the shaping of surfaces by grinding.

Tool Post.—A circular post attached to the top of a slide rest for the clamping down of the cutting tools. It is pierced with a rectangular hole, through which the tool is slid and clamped with a set screw passing down from above. Tool posts are used only on small lathes, the lathes of larger size being provided with screwed studs and nuts and cross-bars, or with a triangular or rectangular shaped form of clamp.

Tool Rest.—See Rest, and Slide Rest.

Tool Smith.—In all large engineering firms there is at least one smith whose sole business consists in forging, hardening, and tempering the steel tools used by the fitters, turners, and machinists. The experience which this tool smith gains relative to the tempers required for different kinds of steel and different kinds of work, renders it simply a matter of economy to entrust to him alone the preparation of the workmen's tools.

Tool Stay.—A lathe stay used for the purpose of preventing drills, and double-edged cutters which are driven by the poppet centre against their work, from being rotated thereon. It is a slotted bar held in the socket of the T rest, the slot of which embraces the flattened shank of the boring tool.

Tooth.—See Wheel Teeth.

Tooth Block.—The block from which the teeth of gear wheels are moulded in a wheel moulding machine. It consists of a rectangular block, which is both long and wide enough to be embraced by the carrier of the machine, or about 9 in. × 4 in., and of the exact depth of the wheel. It carries two teeth precisely like those required for the wheel, except that their outer flanks are chamfered away to well clear the sand of the tooth mould last rammed up. The inter-tooth space alone forms the tooth mould, and the block is moved round to a distance of one tooth after each inter-space is rammed.

Toothed Gearing.—Wheel gearing (q.v.).

Toothed Segment.—Large gear wheels are seldom cast in a single ring by reason, first, of the enhanced difficulty attendant upon the making of large castings, and, second, their inconvenience as regards transit, fixing, and subsequent repair. These are therefore commonly cast in segments of six, eight, or ten to the circle, and either bolted to their arms or to something which forms a suitable base, as the shrouding of a water wheel.

Toothed Wheel.—See Wheel Gearing.

Top and Bottom Tools.—See Swage.

Top Card.—An indicator card taken from the top of a vertical, or oscillating cylinder. In the top card the lead line (q.v.) is on the left-hand side.

Top Face.—That face of a casting or mould which is uppermost in pouring. It is unsound as compared with the bottom face.

Top Iron.—The curved upper iron which is screwed to the cutting iron of a wood working plane. The function of the top iron is to impart rigidity to the cutting iron. This it seems to do simply by exerting pressure upon the cutting edge, and not by breaking the continuity of the shaving, as many suppose.

Top Part.—The upper portion of the boxes, or the top box alone, in a foundry mould ; or sometimes the upper part of the mould itself.

Top Fuller.—A fullering tool (q.v.) which is held in a hazel rod (q.v.) and struck by the sledge. See Fullering.

Topping.—(1) The reduction of the tops of saw teeth to a uniform level previous to sharpening. The topping is necessary to ensure the simultaneous cutting action of all the teeth, and is performed by passing a file over them once or twice in a longitudinal direction previous to sharpening. In circular saws it is done by holding a piece of stone to the tops of the teeth while the saws are revolving. (2) The breaking off of the unsound and hollow top portions of steel ingots which have sunk or become hollow in the centre.

Top Print.—A print (q.v.) which is placed on the top face of a pattern. It is rendered necessary when the core is too long and slender to be supported truly by a bottom print alone, or when, though it might be supported, the shape of the mould is such that it cannot be centred by measurement in the top previous to closing of the mould. The top print then guides and steadies it in correct position.

Top Rake.—In cutting tools, signifies the angle which that part of the cutting face that lies immediately behind the tool point makes with the horizontal. The top rake will vary from zero in tools cutting the hardest metals, as chilled rolls, down to 15° in wood-cutting tools. Top rake is either front top rake, when it is sloped backwards directly from the point, or side top rake, or side rake, when it simply slopes from the side or transversely to the front. In the latter case the tool is specially adapted for side cutting, which it will not effect so sweetly when front rake alone is given.

Top Steam.—The steam which enters above the piston in a double-acting steam hammer (q.v.).

Torsion.—A body is subject to torsion when two equal and opposite forces forming a couple (q.v.) act upon it in directions at right angles with its axis. It may be defined as the strain to which a bar or shaft is subject tending to twist it.

Torsional Stiffness.—That amount of rigidity of a shaft by which it is enabled to resist, not simply such strains as would produce actual twisting off, but also such excessive vibration as would prevent it from doing its work with due steadiness.

Torsional Strength.—The strength necessary to enable a bar or structure to resist a torsional strain. The strength of a shaft to resist torsion varies as the cube of its diameter.

Torsional Stress.—The stress to which a bar or structure is subject when in torsion.

Tossing.—A process of tin refinery which is sometimes substituted for boiling (q.v.). The molten metal is lifted in ladlesful to a considerable height and poured back into the bath, the effect being to oxidise out the impurities by bringing them into contact with the air during their descent.

Total Heat.—The total heat of steam, or the total heat of vaporisation, is the sum of both the sensible and latent heats, including also that which is engaged in overcoming the resistance of the atmosphere.

Total Pressure.—The total pressure of steam is that reckoned from the point of no pressure, or absolute pressure (q.v.).

Total Wheel Base.—See Wheel Base.

Touch Hole.—The aperture through which the charge of a gas engine is ignited.

Tough Brass.—Brass which is suitable for engine bearings and wearing parts. It is rather a vague term which might mean either a superior quality of brass, or gun-metal simply.

Tough Cake Copper.—See Tough Pitch.

Tough Iron.—Slippery iron (q.v.) and strong iron (q.v.) are tough irons.

Toughness.—Toughness denotes the inherent power which a material has of resisting fracture by bending. In tests (see Forge Test) it is measured by the capacity for bending through a definite angle one or more times without sustaining fracture. Toughness and strength are not quite identical, since a material may be strong, *i.e.* rigid, without possessing toughness, that is, it may be able to sustain great tensile force up to a certain point and then give way suddenly without previous warning. Thus chilled iron, and silicious iron are strong in this sense, but lack the quality of toughness. Hence for structures subject to tension, toughness is to be preferred to rigidity, while for those in compression the latter is the principal quality required.

Tough Pitch.—Blister copper after the process of refining by roasting, melting, and poling (q.v.) is said to be at "tough pitch." The process is called toughening, and the ingots cast in this condition are called "tough cake" copper. A test ingot is cast, and when hammered red hot should show no cracks at the edges.

Tow.—See Hemp.

Tracing.—The making of copies of drawings by tracing over the lines on the drawing or copy, upon a semi-transparent tracing paper (q.v.), or tracing cloth (q.v.) laid thereon. It effects a great saving of time when a number of copies are wanted, and being quite an automatic task is done by lads. A mounted tracing is a tracing pasted on thin calico to preserve it from becoming torn.

Tracing Cloth.—A specially prepared cloth glazed on one side and rough upon the other, used instead of tracing paper (q.v.) in making tracings for shop use, for which ordinary tracing paper would not be sufficiently strong.

Tracing Machines.—Wood-working copying machines used for working the tracery of panels, &c. The cutter is affixed to the end of a movable radial arm, and a collar affixed to the collar of the spindle follows the outline of a templet of the form to be cut; sharp angles and corners are of necessity finished by hand. See also Copying Machines.

Tracing Paper.—Thin semi-transparent paper used for copying drawings placed underneath. It is made by soaking tissue paper in a solution of resins, or a varnish. Spirits of turpentine and Canada balsam are the agents chiefly employed.

Tracks.—A term sometimes employed to designate the top faces of the bearers of a lathe bed, as flat tracks, V tracks.

Traction.—Locomotion by virtue of the frictional adhesion of wheels.

Traction Engine, or Road Locomotive.—An engine of the locomotive type provided with large ribbed wheels for traction on common roads.

Tractive Force.—The amount of force necessary to enable a locomotive or animal to overcome the resistance due to gravity, incline, weight, and also to travel at a definite speed. The tractive force necessary to move a train on a level line must be sufficient to overcome the friction of the wheels on the rails, the resistance of the friction of the axles, and the resistance due to the air.

Trade Marks.—See Brands.

Trailing Axle.—The hinder axle of a locomotive.

Trailing Lengths.—Those portions of the coupling rods of a fully coupled locomotive which extend backward to the trailing wheels (q.v.).

Trailing Springs.—The springs which carry the axle boxes of the trailing wheels of locomotives and railway stock, and which sustain and minimise the shocks due to concussion. They are made similarly to leading springs (q.v.).

Trailing Wheels.—The hinder wheels of a locomotive.

Train.—Parts of machines similar or identical in form arranged in series. Thus the rolls of puddling mills are termed the forge train. A number of wheels mutually dependent are termed a train of gearing, &c.

Train of Gearing.—An arrangement of wheels and pinions by which power is gained in the one direction or speed gained in the direction opposite. See Simple Train, Compound Train. In a train of gearing the power or speed of each wheel is inversely as its radius, diameter, circumference, or number of teeth, and the product of the series is in geometrical ratio.

Train of Wheels.—See Train of Gearing.

Train Oil.—See Whale Oil.

Trammel Heads.—See Trammels.

Trammelling Off.—The taking or measuring off of important lengths and centres by means of trammels. This is the only correct method recognised in the shops, rule measurement being inadmissible. Small distances and centres are similarly transferred with compasses or dividers.

Trammels.—Beam compasses, in which the heads slide along a straight bar; they are tightened by set screws, and are used to strike radii too large for the capacity of ordinary compasses. The heads are made indifferently of brass or of boxwood, the former being the better of the two. The steel points are inserted into the bodies, and frequently a socket for lead pencil is provided at the side of one of the heads.

Tram Wheel.—A flanged wheel constructed to run on rails.

Transfer of Heat.—Refers specially to the transmission of heat from a boiler furnace to the water within. (See Heating Surface.) Rapidity of circulation (q.v.) is necessary to prevent destructive overheating of the plates, which would soon happen if the heat transferred from the furnace were not carried away rapidly by the water. Incrustation (q.v.) and sooty accumulations interfere with the transmission of heat. Thin plates conduct more rapidly than thick ones, the rate of transmission being directly proportional to the thickness of the plates. Furnace area is more efficient than tube area, because of the greater concentration of heat there. The rate of transmission signifies the number of heat units transferred per hour, but the amount of heating surface will vary in different types of boilers.

Transmission of Heat.—See Transfer of Heat.

Transmission of Power.—When power is transmitted through shafts, wheels, belts, &c., the amount of horse-power that is transmitted varies as the speed. But at high speeds this general rule becomes modified.

Transoms.—The transverse distance or stretcher pieces placed between the longitudinal sleepers of permanent way.

Transverse Axis.—The right line which passes through the vertices of a curve, or through the longest diameter in the case of an oval or ellipse, or through the vertical height of a parabola.

Transverse Section.—A section (q.v.) taken across at a right angle to the longitudinal direction of a drawing.

Transverse Strain.—That kind of strain to which a bar or structure is subject when it rests upon supports, the load being placed between the supports.

Trapezium.—A quadrangular figure which has no sides parallel.

Trapezoid.—A quadrangular figure which has two sides only parallel.

Travel.—The amount of extreme linear movement of a piston or valve, or of their rods.

Traveller, Overhead.—See Overhead Traveller.

Travelling Crane.—This has reference chiefly to steam cranes which are furnished with gearing to enable them to travel along lines of rails, the term portable being applied rather to cranes travelled by hand.

Travelling Gear.—The gear by which the motions of a travelling crane are actuated, usually embracing friction clutches on the engine or intermediate shaft, whose motion is communicated by a vertical shaft to bevel wheels, one on the vertical shaft and the other on one of the wheel axles of the truck, with or without the aid of a pitch chain.

Traverser.—A low flat carriage containing a capstan, usually driven by a stationary engine fixed on the carriage, and used for the purpose of hauling railway trucks about on sidings.

Traversing.—The motion for longitudinal cutting in a self-acting lathe. It is accomplished either by means of the leading screw and clasp nut or by means of a rack and pinion. Usually in lathes furnished with both sets of gear the turning is accomplished by means of the screw, and the racking back by means of the pinion. The movement effected by the rack and pinion is called the hand traverse, to distinguish it from the traverse effected by the leading screw. The term traversing is also applied to numerous motions of machines, as the longitudinal motion of a shaping machine head, the transverse movement of a tool box effected by a traversing screw, &c.

Traversing Drill.—See Slot Drilling.

Traversing Head.—The head of a shaping machine comprising tool box and ram, which is made to slide along its bed, to distinguish it from those heads which have a motion in one direction only. See Table Feed.

Traversing Mandrel.—A headstock sliding mandrel used for cutting ornamental screw threads. It is furnished with interchangeable guide screws at the tail end, which screws work in an eccentrically fixed brass plate cut out round its circumference to corresponding pitches. Either one set of threads in guide screw and plate being placed in gear, and the plate fixed, when the lathe is started the mandrel slides forward in its bearings to the extent permitted by the brass plate, and traverses any work attached to it against a cutter set in the rest. This therefore is the reverse of the ordinary screwing process in the screw-cutting lathe, inasmuch as in this case the mandrel travels, the tool remaining fixed. It is useful chiefly for short screws and for ornamental threads, and is scarcely employed by engineers.

Traversing Screw.—Any screw employed in machine construction for the purpose of drawing along a movable block or carriage, a tool box or a saddle for instance. The screw does not traverse—that is, it has no endlong motion—but as it revolves it carries along a nut attached to the movable slide.

Traversing Screw Jack.—A jack (q.v.) which is made to travel longitudinally upon its base by means of a horizontal traversing screw turning in bearings on the base.

Tread.—(1) The upper portion of a rail upon which the wheels run. (2) That portion of the wheel which runs on the rail. (3) The bearing faces of a lathe bed.

Treadle, or Foot Board.—A strip of wood actuated by the foot and connected to the crank of a lathe, grindstone, drill, or other small machine by a crooked rod or chain, through which the vertical motion of the foot is converted into a rotative movement on the crank shaft.

Treble Barrel Pump.—A three-throw pump (q.v.).

Treble Clack Box.—A treble valve box (q.v.).

Treble Gear.—The combination of three pinions and three wheels in train for the increasing of mechanical effect.

Treble Geared.—A lathe is treble geared when it is provided with two back gear spindles (see Back Gear), as distinguished from one which is double geared (q.v.). Treble-geared drilling machines are those having three pairs of gears on two vertical spindles, so that the drill can be fed at three different speeds, the wheels on the second and third spindles being adapted for throwing into and out of gear. Treble-geared lathes are rather exceptional, being reserved for the heaviest work, and treble-geared drilling machines are chiefly hand machines where it is desired to drill holes of an inch in diameter and upwards.

Treble Ported Slide Valve.—An exhaust relief valve (q.v.) in which the steam gains access to the cylinder through two narrow ports in the body of the valve in addition to the end supply, and exhausts through two ports in the cylinder on each side of the exhaust. Its function is similar to that of the double-ported slide valve (q.v.), being only an extension of the principle.

Treble Purchase.—A combination of gear in which three pinions and three wheels are engaged, the velocity ratio being equal to the product of the radii of the wheels divided by that of the radii of the pinions.

Treble Riveting.—Three rows of rivets pitched parallel in a longitudinal direction, but whose middle row is placed intermediately with the others in the transverse direction.

Trebles, or Lattens.—Sheet iron plates prepared for tinning, whose thickness ranges between No. 25 B. W. G. ($\cdot$020 in.) and No. 27 B. W. G. ($\cdot$016 in.).

Treble Valve Box.—A pump valve casing, provided with suction and delivery valves and an intermediate check or retaining valve. The feed is then rendered continuous; used on portable or stationary engines for pumping to the boiler.

Tree Feller.—A reciprocating saw driven directly from a steam cylinder of long stroke, used for felling trees or sawing trunks and logs. The saw cuts on the backward or return stroke only, to avoid the risk of becoming bent. Several tree fellers may be worked at once from a portable boiler.

Trenails.—Wooden spikes used for holding railway chairs down on their sleepers. The timber from which they are made is compressed, and moisture excluded, so that the trenails swell in their holes and prevent the chairs from becoming slack. They are used in combination with spikes, usually two spikes and two trenails to a chair.

Tresses.—Trestles (q.v.).

Trestles, or Tresses.—Light narrow stools or supports with spreading legs, standing about 2 ft. 7 in. high, and used for laying drawing boards, patterns, foundry loam boards, core bars, and light work generally upon. They are made in wood and in iron, iron trestles being chiefly for foundry use. The trestles in common use are flat on their upper faces; those for core bars are notched to receive the journals of the bars.

Trial and Error.—A method of obtaining correct results in the production of pieces of mechanism and of tools, commonly resorted to in workshops. It simply means that when the best results are obtained possible with one set of tools, or with one method of construction, that another set of tools, or another method of construction is resorted to in order to detect the errors left uncorrected by the previous method. Thus the production of surface plates or of straight-edges is essentially one of trial and error. The centring of work in the lathe with chalk is done by a method of trial and error, and numerous other instances will occur to the mind.

Triangle.—A rectilineal or plane figure which is bounded by three lines or sides. Having all its sides equal, it is equilateral. Having two sides forming a right angle,—a right-angled triangle. With one obtuse angle,—an obtuse-angled triangle. With three acute angles,—an acute-angled triangle. A triangle is the only figure whose shape cannot be altered while the length of its sides remains constant. Hence crane frames and roof trusses are triangles.

Triangle of Forces.—It is a fundamental principle in mechanics that if three forces (equal or unequal) be represented in magnitude and direction by the three sides of a triangle taken in order, they will produce equilibrium. Hence if three forces, equal or unequal, acting upon a point, are in equilibrium, then the magnitudes of those forces may be represented in a graphic method by the lines of a triangle, the lengths of whose three sides are proportionate thereto, and drawn in the direction in which the forces act. The post, jib, and tie of a crane furnish a common example of the triangle of forces.

Triangular Compasses.—A compass (q.v.) used for drawing, provided with three movable points by means of which three centre points or dimensions can be taken off, and transferred at once from one drawing to another.

Triangular Rule.—A rule having three faces, and used with the same object as the triangular scale (q.v.).

Triangular Scale.—A drawing scale having three faces, each being differently divided. The dimensions on each side can therefore be set off directly on the paper without resorting to the use of compasses, or without having separate scales at hand.

Triangulation.—The designing of the bays in a Warren, or lattice girder. In a Warren girder the triangles are equilateral, in the lattice girder the bases are shorter than the sides, the stresses in the diagonals are proportionally reduced, and the weight is distributed more evenly over the flange; there being a greater number of apices. Triangulation is single in the Warren girder, multiple in the various forms of lattice girders.

Triblet.—(1) A smith's tool, being a round rod slightly tapered, which is used **as a** mandrel around which rings and nuts are finished **upon the anvil.** (2) The steel core upon which tubes are drawn to produce a smooth interior surface of uniform diameter.   See Tube Drawing.

Trimming.—See Fettling.

Trip Gear.—Valve gear in which the valves are opened and closed by means of cams, levers, or catches.   Trip gears include those of Corliss, Cornish, and other engines.

Trip Hammer.—See Tilt Hammer.

Triple Expansion Engine.—A compound engine in which the steam is expanded in three separate cylinders, one high pressure and two low pressure cylinders.   The triple expansion is rapidly superseding the ordinary two-cylinder compound engines for ocean service.

Trip Lever.—A lever usually of the bell crank form whose function is the rapid opening and closing of valves.   The lever being lifted at the end of one arm for a certain distance moves the valve at the end of the opposite arm, but slipping off at a definite stage allows the valve to go free.

Tripod Jack.—A screw jack (q.v.) which is supported on a tripod of three iron legs.

Tripod Vice.—See Tail Vice.

Trolly.—A low strong truck running on a tramway.   Used in engineers' yards for running castings, forgings, and heavy work generally, from one department to another.

Trolly Wheel.—A single or double-flanged wheel which is attached to the axles of a trolly (q.v.).

Troughs.—(1) The fire-clay vessels in which the bar iron which has to be converted into steel by the cementation process is placed.   Also called pots.   (2) Any open vessels containing water, hardening mixtures, sand, or other materials for smiths' use.

Trowel.—A small tool used by moulders for smoothing over the surfaces of sand, making sand joints, and generally for shaping and finishing surfaces.   See Heart, Square, Dog-tail Trowel.

Truck.—A low carriage, a trolly, a bogie, the carriage of a portable crane; or in general any low carriage running on wheels.

True Pitch.—Uniform pitch (q.v.).

True Screw.—A screw whose pitch is uniform throughout.

Trumpet Mouth.—See Bell Mouthed.

Trunk Air Pump.—A marine air pump in which the piston rod is made hollow to enclose the connecting rod, which is jointed loosely within to its lower end.   Trunk air pumps are used in cases where space is so confined that the crosshead and links cannot be employed.

Trunk Engine.—A marine engine the purpose of whose design is the production of so compact an arrangement that the engine shall be secure from the enemy's shot below the water line.   The distinguishing feature consists in the piston rod being hollow or trunk-like in form. It passes through both ends of the cylinder, being encircled by the piston at its centre, while the connecting rod is attached to its interior, the length of the piston rod being saved thereby.   Trunk engines are wasteful of steam owing to the large conducting surface of the trunk being exposed to the air, and are not constructed now.

Trunk Plunger.—A pump plunger in which the connecting rod, instead of being attached to an eye at the upper end, passes loosely through the central portion, which is hollowed out for its reception, to be attached to the lower end.   The principle is that of the trunk engine (q.v.).

Trunk Valve.—A term applied to that form of D slide valve which is made sufficiently long to govern the entrance of the steam to ports placed near the end of the cylinder. It is an antiquated form.

Trunnion.—Bearing on which a vessel or cylinder swings or oscillates. Usually cast as part of the cylinder or vessel.

Truss.—A diagonal brace employed to receive the thrust of a loaded beam, girder, or shaft. The vertical thrusts on the beam are received and transmitted through the diagonals, which are thus put into a condition of tensile stress.

Trussed Beam.—A long beam whose power of resistance to bending strains is increased by transmitting those strains through a truss rod (q.v.).

Trussed Shaft.—A long light shaft rendered rigid by truss rods arranged around it. Trussing is applied to shafts whose length is such as to require some intermediate support, which support it is not convenient to give otherwise than by trussing.

Trussing.—The affording of support to otherwise weak structures by means of truss rods (q.v.).

Truss Rod.—The rod which passes from end to end of a trussed beam and through which the strain due to downward pressure is transmitted. The strains in a truss rod are those due entirely to tension. Hence it is a tie rod. A truss rod forms an angle with the beam whose strains it receives, its distance from the beam increasing as it approaches the central portions where it receives the thrust of the strut or struts, and it forms with the beam and strut a triangle of forces (q.v.).

Truth.—A term of very common use in workshops to denote accuracy. Inaccurate work, or an inaccurate tool is said to be out of truth ; the getting of a straight-edge or grindstone or square into correct outline is termed truing up. A shaft or a spindle is true when it is straight or in line, and of equal or correct diameter. A piece of stuff is true when of equal thickness, or when not winding ; a cylindrical piece of work is true when it fits to gauge or calipers. All the tools used for measurement are employed to check the truth of work.

Try Cocks.—The cocks screwed into a boiler shell at the extreme highest and lowest water marks, which are used to check the water gauge itself in the event of the latter becoming choked up or ceasing to act; the upper one being the steam, the lower the water cock.

Trying Cocks.—Try cocks (q.v.).

Trying Plane.—A plane used for producing the utmost accuracy attainable in the surfacing of wood by hand. Its length is commonly 22 in., and the width of its iron $2\frac{1}{2}$ in. The iron is sharpened straight transversely, and this combined with the length of the plane enables very true surfaces to be produced by its use. It is used after the jack plane.

Trying Up.—The planing of the surfaces of wood true, that is, straight and out of winding and at right angles with each other; hence the terms trying plane (q.v , trying up machines (q.v.), try square (q.v.).

Trying Up Machines.—Wood working machines for planing timber, usually applied to those machines used for the heavier railway timbers for wagons, trucks, &c., in which the cutters are fixed on the face of a circular disc.

Try Square.—The ordinary square, consisting of thick stock and thin blade, used for checking the accuracy of the edges of work. The term is used to distinguish it from the set square (q.v.).

Tube.—A pipe of wrought iron or brass. It is either welded, or solid drawn. Cast-iron tube is termed pipe, or piping.

Tube Cleaner.—A stiff brush or metallic scraper screwed to the end of a rod and used for cleaning the soot out of the tubes of locomotive, portable, and marine boilers.

Tube Cutter.—An instrument by which wrought-iron tube is embraced in a curved or hollowed portion provided at the end, and cut by a smooth-edged, hardened steel roller actuated by the handle or lever of the cutter. Tube cutters of circular form are also made for cutting off boiler tubes from the inside.

Tube Drawing.—(1) The periodical taking out of the ' bes of multitubular boilers for the purpose of thorough examination. This should be done at intervals of from five to ten years. (2) The manufacture of tubes in various materials by drawing them through a draw plate, with or without the aid of a triblet (q.v.). The tubes thus made are either drawn through at a welding heat and so united, or they are welded or brazed first, and then drawn to impart the necessary finish to their exterior and interior surfaces, or they are solid drawn, that is, cast and drawn, or drawn from a semi-fluid mass only.

Tube Expander.—A tool used chiefly by boilermakers for expanding or increasing the diameter of wrought-iron or brass tubes for steam boilers in their tube plates (q.v.). There are several forms in use, but in principle they consist of a central mandrel, tapered in form, by the screwing in of which an outer tube or rollers are both turned round and increased in diameter when placed within the boiler tube and rotated.

Tube Ferrules.—The ferrules of condenser tubes. See Ferrule.

Tube Plates.—The plates into which the tubes of multitubular boilers, or of surface condensers are inserted and fastened.

Tube Plug, or Tube Stopper.—A plug of wood driven into the end of a leaky boiler tube to render it temporarily water tight, or until it is convenient to replace it with a new tube.

Tube Scraper.—See Tube Cleaner.

Tube Stopper.—See Tube Plug.

Tube Surface.—The total area of the exterior surface of the tubes in a surface condenser. The extent of this area is variable in different engines, being dependent on the efficiency of the cold water circulation, but it may be taken to average one square foot for every three to five pounds of steam condensed per hour.

Tube Vice.—A clamping apparatus fitted with a vertical screw, and used for holding wrought iron tube whilst being cut off or screwed. Also termed a pipe vice.

Tube Wrench.—A pipe wrench (q.v.).

Tubing.—(1) Tubes. (See Tube.) (2) The insertion of tubes in a boiler is called tubing the boiler. See Tube Expander.

Tubular Boiler.—A steam boiler provided with small water tubes or pipes to increase the amount of heating surface without unduly adding to the bulk of the boiler. It denotes no ne type of boiler in particular, but any form which is provided with tu s, and hence embraces marine, locomotive, portable, vertical, horizonta , and sectional boilers.

Tubular Compasses.—A pair of drawing compasses whose legs are hollow tubes, within which other tubes move telescopically for the purpose of increasing their capacity. These sliding bars also carry reversible points rotating around pivots, point and pencil on one side, point and ink on the other.

Tubular Girder.—A box girder (q.v.) of large size.

Tubulous Boiler.—A tubular boiler (q.v.).

Tucking In, or Tucking Under.—The thrusting of the moulding sand around and underneath the lower portions of patterns which are bedded in. (See Bedding In.) Tucking under is not necessary when patterns are turned or rolled over in their moulding boxes.

Tucking Under.—Tucking in (q.v.).

Tue Iron.—See Tuyere.

Tumbler.—A rumble (q.v.).

Tumbler Bearing.—A bearing used specially for gantry (q.v.). driving shafts. When the crab travels along the gantry it carries with it a sleeve bevel-wheel, having a square central hole fitting on and sliding along the square shaft, and driving it, and with it the traveller wheels at the ends. The long square driving shaft, though supported at the ends and by the bevel-wheel on the crab, requires also intermediate support, and this is supplied by three or four tumbler bearings, placed equidistantly, and pivoting on the girder sides. Their peculiarity is that they are movable, being depressed by the crab as it travels along, and returning to their positions again immediately that it has passed over. The commonest and typical form is that of two arms cast at right angles, and having their bearings at their free ends, the point of junction of the two being bossed up to swing on its pivot. Other forms more intricate in design are in use also.

Tumbling Bay.—A means of testing or ascertaining the performance of a condensing-engine. The water from the hot well is delivered into a trough and made to circulate over and under plates of metal. The water escapes from the trough through a notch of definite width cut in a thin plate of brass. This is called the tumbling-bay. The temperature of the water which tumbles over is taken by a thermometer, and the quantity passing per minute is ascertained by the difference in height of the notch and of the water in the trough itself. Knowing the quantity therefore which passes per minute and comparing the temperature of the injection water with that of the water which falls over the tumbling-bay, the amount of heat lost per indicated horse power can be ascertained.

Tungsten.—Symbol, W. Comb. weight, 184. A rather rare metal which is of interest to the steel manufacturer from the beneficial effect which it produces in combination, making, when the amount of tungsten present does not exceed from one to three per cent. a very pure silvery white steel of uniform crystallization, both hard and ductile. When present in greater amount, it produces a hard tool steel which does not require further hardening for cutting purposes. Tungsten is derived from the mineral Wolfram where it is combined with ferrous oxide.

Tunnel, or Shaft Tunnel.—The enclosed space in the stern of a steamer through which the propeller shaft passes.

Tunnel Bearings.—The propeller shaft bearings within the tunnel (q.v.) as distinguished from the thrust and main bearings.

Tunnel Head.—A chimney surmounting the charging holes of open topped blast furnaces to carry the flame clear of the charging holes.

Tup.—The mass of iron which constitutes the hammer head of a steam hammer, being attached directly to the piston-rod on one side, and receiving the hammer block itself on the other.

Turbine, Water.—A motor consisting essentially of two horizontal rings of buckets, one ring being enclosed within the other, and its buckets or chutes becoming the guides to a column of water, which having descended by gravity under a definite head, is caused to impinge on the buckets of the outer ring, and turn it by reaction. This is the principle of the outward flow or Fourneyron turbine. In the inward flow or central discharge turbine the method is reversed, the water entering from without and escaping inwards. In the Jonval or downward flow turbine the discharge is from above downwards. In the Vortex turbine the water is admitted from without and actuating the vanes passes out at the centre above and below; and the turbine may either be fixed horizontally or vertically. In the Girard turbine the water does not act by reaction at all, but impinges on the curved sides of the buckets only. This also can be used in a vertical or horizontal position. Special turbines are also constructed in which improvements or modifications or combinations of these are introduced, so that from the original simple outward flow of M. Fourneyron a vast variety of useful motors have been developed. Turbines are high pressure when running under a great head of water up to 300 or 400 feet, the rings being then of small diameter; they are low pressure when under a small head, and are then of large diameter. Heads as low as 9 in. will drive a turbine. Turbines working under a constant head have their chutes or vanes fixed; when designed for a variable head, the blades are made movable. With the exception of the Girard type, turbines will work immersed in water without loss of energy. See also Gas Turbine and Steam Turbine, Dict. of Modern Terms.

Turn Carriage.—The bogie arrangement for swivelling the axles of the front wheels of a portable pump to which the drag handle is attached.

Turned Bolts.—These are used on all good machine and engine work, their holes being drilled or bored to make a good fit. The employment of turned bolts is necessary to prevent the displacement of parts due to vibration.

Turner.—Turners are a distinct class of men in engineers' factories, seldom leaving the lathe for the vice. The skill of a turner consists chiefly in rapid chucking and in ready manipulation of tools, both in their grinding and setting.

Turnery.—See Machine Shop.

Turning.—The art of bringing work to circular form in a lathe. Turning is done by hand and machine tools, and in wood and metals; it is both plain and ornamental, and while some of its productions are of the simplest character, others are of the most complex and highly ornate kind. Pattern turning is usually of a simple character, and seldom automatic, being only so when parallel pipes or pillars are done in a slide rest lathe. Engineers' turning in the machine shop is nearly all done in automatic lathes provided with self-acting slide rests. It varies from the smallest spindles, screws, pins, &c., to that of the largest crank shafts, cylinders, and wheels.

Turning Chisel.—An obliquely-ground doubly-bevelled chisel used for turning smoothly, or finishing the surfaces of woodwork in the lathe. It is not held flat, but at an angle with the axis of the stuff. Much pattern work being of large diameter and necessitating the employment of the scraping or ordinary firmer chisel, that is also frequently though incorrectly termed a turning chisel.

Turning Gouge.—A long stiff gouge used for turning or roughing down the surfaces of wood in the lathe. Gouges range from $\frac{1}{8}$ to $1\frac{1}{2}$ in. width.

Turning Lathe.—See Lathe.

Turning Over.—The method of moulding by which both sides of a pattern are rammed up in succession. Used in opposition to bedding in (q.v.). See Rolling Over.

Turning Point.—That point of a structure around which the statical moments are supposed to act.

Turnings.—Turnings, borings, and other metallic shavings cut off by the processes of turning, boring, drilling, planing, shaping, slotting, &c., are, when of cast iron, utilised in the making of iron cement; when of wrought iron and steel are fagoted or remelted; when of brass and gun metal are remelted.

Turning Saw.—See Bow Saw.

Turning Shop.—See Machine Shop.

Turning Tools.—For pattern makers' wood turning, consist of gouges, chisels, round nose, diamond points, side tools. For metal turning they comprise roughing tools, spring tools, right and left-hand tools and gravers, the angles of each being modified for use with steel, cast iron, gun metal, and copper. See details under these specific headings.

Turn Over Board.—A board used for ramming a pattern upon; when, first, the pattern is so flimsy that it would otherwise become rammed out of truth, or when, secondly, the quantity of castings required is so large that the use of a board will save the time the moulder would otherwise occupy in making sand joints. Also called bottom board and joint board.

Turn Over Skip.—See Skip.

Turn Screw.—A screw driver (q.v.).

Turntable.—A platform or table used for changing the direction of motion of rolling stock. The table revolves on a circular line of rails, and the top of the table is furnished with rails of the same gauge and height as the permanent way. Turntables, when large, are usually built up with fish-bellied girders and distance pieces; when small, of cast iron or of a ring of H iron (q.v.), across which the plates which carry the rails are laid. Turntables commonly revolve on a live ring (q.v.).

Turret Rest.—See Capstan Tool Rest.

Tuyere.—The pipe through which a current of air or blast is conveyed to a furnace. The tuyeres for blast furnaces, cupolas, and smiths' fires are formed on different types, according to the special work which they have to do. In nearly every case they are cooled by a current of water circulating around the nozzle. Also spelt twyere. A smith's tuyere is often called a tue iron.

Twin Screws.—A pair of propeller screws, right and left handed respectively, placed on separate parallel shafts, and designed to neutralise the tendency which each alone would have to produce vibration or rolling.

Twist.—See Angle of Flexure, Torsion.

Twist Drill.—A drill grooved longitudinally and spirally for the clearing of the waste material, the swarf or waste passing spirally up the shaft or twisted grooves. A drill of this form need not be periodically withdrawn from the hole for the purpose of affording clearance to the borings, as is the case with the ordinary form, and being circular in section is not so liable to run out of truth in the hole as is a drill of the common type.

Twist Drill Grinder.—Since so much of the efficiency and sweet working of twist drills depends on the angle of their cutting lips being maintained intact, grinding machines have been devised, and are used in most large factories, by which this constant angle is maintained, the drill being held in a socket or clip against a rapidly revolving emery wheel.

Twisting Moment.—The measure of the amount of torsion set up in a bar. It is equal to the product of the twisting force into the distance from the centre of the bar at which it acts.

Twisted Belt.—Driving belts are sometimes twisted to secure their better adhesion, and so save the trouble of tightening up. But, as commonly understood, twisting refers to the turning of a belt through a definite angle, to drive pulleys whose axes are not on the same plane, or to drive machines in opposite directions. See Crossed Belt, Half-crossed Belting.

Twist Wheels.—See Screw Gearing.

Two High Mill.—See Two High Rolls.

Two High Rolls, or Two High Mill.—Two rolls only, one above the other, for rolling bar iron, or plates. In such rolls the bar must, when passed through, be either drawn back over the top, so losing a pass, or the mill must be reversed. See Three High Rolls.

Two-jawed Chuck.—A lathe chuck having two jaws only opposed to each other. They are either independent, or universal in movement; in the first case being actuated by distinct screws, in the second by a single screw spindle, whose halves are furnished with right and left-handed threads respectively.

Two Salt Waters.—Sea water contains $\frac{1}{33}$ of its weight of salt. When it is prevented from exceeding $\frac{2}{35}$ of its weight in the boilers of a marine engine, no incrustation of moment accumulates. Hence the boilers must be blown out sufficiently often to prevent the density from exceeding two salt waters.

Two-throw Crank.—An axle or shaft having two cranks forged upon it, the cranks being usually situated at right angles with each other. Used for double cylinder engines and double-barrel pumps.

Two-way Cock.—Most cocks are two-way cocks, that is, having one passage for the entrance of the fluid, the other for its exit, the valve dividing the passage and regulating the flow. In these therefore the fluid can only pass in one straight channel. See Three-way Cock, Four-way Cock.

Two-wheeled Barrow.—Barrows having two wheels extending one on each side of the front of the barrow, instead of the ordinary central one, are much used in foundries for wheeling the castings and moulding boxes about, because the risk of their overturning is minimised thereby. Used also in machine and erecting shops and yards for a similar purpose.

Twyere.—See Tuyere.

Tymp.—See Tymp Stone, Tymp Plate.

Tymp Plate, or Tymp simply.—A plate of cast iron over the hearth of a blast furnace which encloses the tymp stone.

Tymp Stone, or Tymp simply.—The stone which forms the front or top of the hearth in a blast furnace.

Typical Engine.—Signifies an ideal engine such as could not exist in fact, but which might be imagined to exist if the theoretical values of the principles involved in its construction could be obtained in practice. Also a generic term denoting any engine which might be relegated to a certain type, as a single-acting engine, a portable engine, a compound engine, &c.

Tyre.—A steel ring which forms the periphery of a truck wheel, or running wheel. It is shrunk, bolted, riveted, or otherwise fastened around a wheel centre (q.v.), or a built up centre of arms and boss, or is held by some form of continuous retaining ring (q.v.). Tyres are rolled from blooms of Bessemer steel whose section is that of the frustrum of a cone, the centre being punched out and the shape being imparted by rolling on open end rolls.

Tyre Bolt.—A bolt whose head is tapered similarly to that of a tyre rivet, but at a rather more acute angle, and used for the same purpose. The nut is fastened within the rim.

Tyre Ring.—See Retaining Rings.

Tyre Rivet.—A form of rivet specially used for fastening tyres to wheel bodies. The end which enters through the tyre is conical, the part passing through the body parallel, and the riveted end is formed within the rim. It is therefore a counter-sunk rivet in which the taper of the counter-sunk part is small.

U.

U Leather, or U Packing.—A pump leather packing whose section is that of the letter U inverted.

Ultimate Set.—The difference between the length of a specimen plate, or bar, before testing, and at the moment of fracture; and given in percentage of the length. This amount in conjunction with the reduction of area (q.v.) is a measure of the ductility of the plate or bar. Since it is measured after fracture by laying the broken ends together, it is also termed the set after fracture.

Ultimate Strength, or Ultimate Tensile Stress.—Synonymous with the load which produces actual fracture in a structure. The ultimate strength is determined by dividing the maximum load recorded before fracture, by the original cross-sectional area of the test piece.

Under Cut.—A pattern is said to be undercut when those portions of the pattern which are lowermost in the mould slope beyond the perpendicular, so that they cannot be drawn out without tearing up the sand along with them. It is the reverse of taper (q.v.). When parts must unavoidably be undercut, the only course is to core these portions, or to lift away the sand around them with a drawback (q.v.).

Under Frame.—That portion of a truck or wagon which contains the bearing springs for the axles, the buffer and drawbar springs, the axles and axle boxes, with their wheels. The under frame itself consists of soles, or side timbers, headstocks, or end timbers, cross bars, or transverse timbers, and diagonals, all formed of timber united with wrought-iron knees and bolts.

Under Poled.—See Poling.

Under-shot Wheel.—A water wheel which receives the water near the bottom of the periphery, and whose motion results from the impulse due to the head of the water. The water is confined between masonry, so that none of its impulse is wasted.

Unequal Lap.—This is caused by the obliquity of connecting rod (q.v.).

Unequal-sided Angles.—See Angle Irons.

Ungeared.—A lathe or drilling machine is said to be ungeared when it is unprovided with back gear (q.v.).

Uniformity of Section.—Signifies the proportioning of the several parts of a machine or structure so that no part shall be weaker than another.

in proportion to the strain which it has to bear. If the sections are not proportioned uniformly, an excess of stress will be put upon the weaker parts which will render them still weaker. The increase in diameter of a cottared end, and the increase in diameter of the screwed end of a rod, are illustrations of the maintenance of uniformity of section.

Uniform Load.—A load which is not a variable load (q.v.), and which therefore does not induce so great stress as the latter.

Uniform Pitch.—A screw is said to be of uniform pitch when there is an equal distance between the helices, or blades, as distinguished from increasing pitch.

Uniform Motion.—Motion, the rate of which is uniformly proportional to the time occupied therein.

Uniform Strength.—A beam or structure is said to be of uniform strength when the moment of resistance at any point is equal to the bending moment (q.v.) at the same point. In other words, when each part is in an equal condition of strain relatively to its strength.

Union.—A connection joint for water piping made in three pieces, two being nozzles for attaching to the pipes, and the third, a nut screwed internally, having also an internal shoulder which clips a corresponding shoulder upon the nozzle, whose tail end passes through the nut. A ring of india-rubber, or other impervious material is laid between the shoulders before the nut is screwed down.

Union Wrench.—A tommy-shaped wrench used for tightening up the joints of union screws. It partially embraces the circular nut, and a hole drilled in the wrench slips over one of the horns cast on the side of the nut.

Unit Boiler.—A sectional boiler (q.v.).

Unit of Heat, or Thermal Unit.—Is a quantity of heat which will produce a certain effect, and which is used as a standard by which to compare other quantities of heat. The English unit is the Joule, or 772 foot pounds, that being the amount of work necessary to raise the temperature of one pound avoirdupois of water at or near its temperature of greatest density 39·1° F. through one degree F. The French thermal unit is equal to one kilogramme of water raised one degree C. in temperature, and is equal to 3·97 English units.

Unit of Mass.—Any definite weight employed in mechanical calculations, as a pound or a ton.

Unit of Work.—A foot pound (q.v.). The French unit of work is one *kilogrammetre*, or one kilogramme raised one metre high. It is equal to 7·2331 foot pounds.

Units.—Mechanical units are either those relations of things which have their basis in natural law, or are conventional standards commonly agreed upon as being convenient for calculations. Thus the unit of heat has its foundation in a natural and interchangeable fact, but a horse-power, or a foot pound, are simply conventional rules accepted by the mutual consent and use and custom of engineers. The great primary mechanical units used by engineers are the Joule, or unit of heat (q.v.); the foot pound or ton; the inch pound, or inch ton, and the H.P. But besides these there are many scores of secondary units, as the power of men, the electrical standards, and so forth.

Unit Strength.—The ultimate unit stress (q.v.) of a material, or that which causes fracture.

Unit Stress.—The stress upon a unit of sectional area, as a square inch

or square foot, with a given amount of force. Or the ultimate stress
per square inch or square foot of section which would cause fracture.
It is represented by the letter f. In a structure whose sections are
subject to various degrees of stress the relative stresses of the several
sections are sometimes conveniently referred for comparison to a single
primary stress, and this is also termed a unit stress. Thus the unit
stress in a Warren girder is that of the bays next the end, whence in a
girder loaded uniformly or at the centre, and having the longer flange
lowermost, the stresses increase by simple proportion through successive bays towards the centre.

Universal Bend.—Two bends provided with a swivel joint, so that they
can be turned round to any desired angle in relation to each other.

Universal Chuck, or Concentric Chuck.—A jaw chuck whose jaws are
endowed with simultaneous movement for instantaneous centring of
the work. The jaws of universal chucks are actuated by gear-wheels
working in a common rack, or by a scroll, or by a combination of both,
the rack and pinions in this case moving the scroll. They are also
moved by means of a key or lever or by hand. All universal chucks
must of necessity consist of two main portions, a front and a back plate,
or an inner and outer shell enclosing the gearing between them. See
Combination Chuck.

Universal Coupling.—See Universal Joint.

Universal Joint.—In Hook's gimbal joint the ends of the shafts to be
connected are prolonged into forks which swivel on the ends of a cross,
each arm of which also swivels in an arm of the forks. Used for carrying shafting round angles of buildings. Other forms of universal joint
are used for pipes and rods, being modifications of the ordinary ball-and-
socket or globular joint, the ball being retained in place by means of a
flange slipped over it and screwed to an opposing flange.

Universal Scale.—A scale (q.v.) used for drawing purposes upon which
all proportional divisions in ordinary use are engraved in parallel rows.
The objection to universal scales is that measurements, except when
along the edges, must be taken off by dividers instead of being set off
from the edge directly on the paper.

Universal Square.—A centre square (q.v.).

Unstable Equilibrium.—Unstable equilibrium results when the centre of
gravity of a body which is in equilibrium is situated at its highest point.

Unsymmetrical Beams.—Beams whose cross section is irregular considered
with reference to the quality of their straining moments. In other
words, beams whose centre of moments does not coincide with the
centre of gravity.

Untarred Ropes.—Manilla ropes (q.v.).

U Packing.—See U Leather.

Up End, or Up Ending.—A term in frequent use by erectors to signify the
standing up of a heavy piece of work upon its end.

Up Hand.—Up hand signifies the method of using a smith's sledge hammer
for light work. The sledge is held in the left hand and lifted and
thrown down by the right; the latter slides freely up and down the
shaft.

Uphand Sledge.—Uphand sledge denotes the lifting of the sledge over the
work, in opposition to about sledge (q.v.). See Uphand.

Upright Chuck.—A wooden chuck furnished with a pair of slotted jaws,
or with a single jaw only, by which work is held which would not conveniently be held in the hollow or ordinary form.

Upset, Angle of.—See Angle of Upset.

Upsetting.—Jumping up (q.v.).

Uptake.—(1) The internal flue of a vertical boiler leading from the furnace to the outer flue or chimney. (2) The flue of a marine return flue boiler. (See Dry Uptake, and Wet Uptake.) (3) The short vertical chimney of a gas producer which forms a communication between the fire-brick chamber and the gas mains.

Uses.—Rough pieces of metal to be used in the welding (q.v.) of heavy masses of forging. Or any rough heavy mass of metal from which pieces of mechanism are to be shaped. Uses are prepared at the forge mills and delivered to the orders of customers who do not possess machinery sufficiently heavy to forge large masses of material.

V.

Vacuo, or in Vacuo.—A term used in calculations on the behaviour of falling bodies and liquids, by which their velocity is referred to that of a body falling in a vacuum, or *in vacuo*; in other words, not subject to the retarding influence of the atmosphere.

Vacuum.—A vacuum is strictly speaking a space absolutely devoid of air. In engineering it only signifies a space partially empty. Twenty-six inches, or thirteen pounds, is a very good vacuum in a condenser. See Vacuum Gauge.

Vacuum Brake.—A form of continuous brake in which the steam pressure actuating through a steam jet or an air pump is made to produce a partial vacuum against which the pressure of the atmosphere then operates.

Vacuum Engine.—An engine in which the power is generated by the explosion of a gas producing a vacuum, the air then rushing in to fill the vacuum.

Vacuum Gauge.—A gauge used to register the amount of vacuum in the condenser of a steam engine or in any vessel in which a partial vacuum is produced by the exhaustion of air. Vacuum gauges are dial gauges (q.v.) and are graduated from zero to thirty pounds.

Valve.—A movable cover or lid, by means of which the ingress and egress of fluids and gases is regulated. There are various kinds of valves. See under their heads, ball, bucket, disc, flap, lift, slide, spiral winged, equilibrium valves, and others.

Valve-box.—The box or casing, rectangular or otherwise, in which the valves of pumps are placed. Also the steam chest which contains the slide valve of an engine.

Valve-Bridle.—See Bridle.

Valve Casing.—A valve-box, or valve chamber.

Valve Chamber.—A chamber which contains a valve, and within which it works. Usually understood to denote a steam chest rather than a clack or valve box.

Valve Chest.—A valve casing or steam chest (q.v.).

Valve Circle.—The circle on a valve diagram (q.v.) whose diameter is equal to the half travel of the valve. There are two such circles on the diagram.

Valve Cock.—A cock of the ordinary outline, gland, bib, faucet, or otherwise, which opens with a lift valve, and hand wheel and screw, or by means of a sliding valve; instead of by the turning of a plug.

as in a plug cock (q.v.). A globe valve is therefore properly termed a valve cock.

Valve Diagram.—A diagram by which the position of the valve for any position of the piston, and conversely the position of the piston for any position of the valve, may be determined graphically. As a consequence the relative positions of valve and piston at the instants of opening, cut off, cushioning, release, may also be determined. There are numerous modes of constructing valve diagrams, Dr. Zeuner's being considered the simplest.

Valve Face.—The face against which a slide valve works or slides, and therefore properly distinguished from a valve seat (q.v.).

Valve Gears.—The mechanical arrangements for actuating slide valves, both for single and reversing action, embracing slot links, rods, provision for expansion, automatic or otherwise.

Valve Plates.—See Cut off Valve (q.v.).

Valve Ring.—An equilibrium ring (q.v.).

Valve Rod Gland.—The gland which closes the stuffing-box of the rod of a slide valve. It is screwed to the steam chest and maintains the rod steam tight.

Valve Seat, or Valve Seating.—The bearing surface, annular or otherwise, against which a valve fits when closing or shutting off; properly applied to the seatings of lift valves and plug valves.

Valve Sector.—(1) A slot link (q.v.), also termed a quadrant. (2) A specific name for the vertical sliding link of an oscillating cylinder engine which communicates the motion of the eccentric to the valve rod weigh shaft.

Valve Spring.—A spring of either the plated or spiral form, used to force the packing or equilibrium rings of slide valves against their working faces or to return a lift valve to its seating.

Valve Stem.—A valve spindle or rod.

Valve Yoke.—The bridle (q.v.) of a valve rod.

Vanes.—The buckets or guides of turbines, or the flat expansions of blowing fans.

Variable Expansion.—The expansion of steam in an engine whose amount of expansion is not constant, but capable of alterations under the varying conditions of the work or strain thrown upon it. It may either be automatic expansion (q.v.), or expansion effected through the medium of some type of hand gear, by which either the amount of lap of the main slide valve is altered, or by means of a back cut-off valve (q.v.).

Variable Expansion Gear.—See Automatic Expansion Gear, Variable Expansion.

Variable Load.—See Live Load, Oscillating Stresses.

Variable Motion.—When a body moves over equal spaces in unequal times its motion is said to be variable.

Varnish.—Shellac varnish (q.v.) is used to protect patterns from the action of the damp foundry sand, in which they are often embedded for several hours at a time, and likewise to furnish a glossy skin, which facilitates withdrawal from the mould. Two or three coats are laid on, the varnish being glass-papered down after each application. In spite of varnish, patterns of considerable width are found to swell in the sand. Hence the necessity for open joints (q.v.), but small patterns remain practically unaffected. See also Lacquer.

Varying Lap.—To produce variable expansion (q.v.) it is necessary that the amount of lap (q.v.) of the slide valve should be rendered capable of variation, since that governs the point of cut off.

Varying Lead.—The lead of a valve designed for the same purpose as varying travel (q.v.). It is not employed in practice.

Varying Load.—A variable load or live load (q.v.).

Varying Travel.—The difference in the lengths of travel of a slide valve, in order to furnish the different amounts of lap necessary to effect variable expansion (q.v.). Expansion gears of various kinds are employed with this object, the usual method being to set the sliding block from which the valve rod receives its motion at a point nearer to or farther from the centre of the slot link.

Vee'd Edges.—The edges of automatic metal working machines which are embraced by the sliding vee'd portions of the movable parts. The edge of a lathe bed embraced by the sliding rest, or the edge of the cross travel of a planing machine are familiar cases in point. The angles of these edges are commonly 60°.

Vees.—See Vee'd Edges, Vee Strips.

Vee Strips.—Loose strips having angles corresponding with those of the vee'd edges (q.v.), along which they slide, and by means of which the amount of slack due to wear is taken up. Hence called adjustment strips (q.v.).

Vee Threaded Screw.—The ordinary Whitworth screw of angular thread. See Screw Threads, Whitworth Thread.

Vee Welding—A mode of welding the plates of boiler flues in which there is neither butt nor lap properly so called, but in which a strip of square rod is inserted angle ways between the nearly abutting edges of the plate, so that it unites the edges upon two sides of the rod.

Vegetable Oils.—Oils expressed from the seeds of vegetables, and used in lubrication, or for paints. They are either non-drying or drying oils (q.v.), and embrace almond, beech, castor, cotton seed, nut, poppy, hemp seed, olive, palm, rapeseed, and other oils. The oil is expressed from the seeds by pressure, or by pressure and heat combined. Rape seed oil is expressed from the seeds of the rape (*Brassica napus*). It is not a good lubricant used alone, being liable to gumming and acidity. Castor oil and olive oils are good, and are noted under their headings. See also Linseed Oil.

Veins.—Irregular wavy markings on the surfaces of castings, which occur when too much blacking has been used in the mould. The metal swells the blacking up into fine ridges, hence the marks. They do not injure the casting, but look unsightly, suggesting to inexperienced eyes incipient lines of fracture.

Vein Stuff.—Gangue (q.v.).

Velocity.—The rate at which a moving body changes its position under the action of force impressed upon it from without. Velocity is commonly estimated in feet per second. It may be either uniform or variable, that is, in the former case when the body passes over equal distances in equal times, in the latter when the space moved over in consecutive seconds varies. Velocity is either linear (q.v.) or angular (q.v.). See also Virtual Velocity.

Velocity Ratio.—Signifies the proportional velocities of bodies which are mutually connected, as levers, gearing, &c.

Vena Contracta, or Contracted Vein.—The contraction in area which a jet of water, or other effluent liquid undergoes in issuing from an aperture in a thin plate. The reduction of area is in the ratio of ·6 to 1·0, the latter being the area of the aperture; the distance at which the contraction takes place from the opening being about half its diameter.

Vena Contracta Mouthpiece.—A mouthpiece adapted to the vena contracta, being made doubly conical in section, in order to prevent the loss of area due to vena contracta (q.v.).

Vent.—The calorimeter (q.v.) of a boiler multiplied by the length of the flue in feet is termed its vent.

Ventilating Bucket.—In overshot water wheels the water as it descends into the buckets would be partially driven out again by the reaction due to the compressed air within the buckets, but for the precaution of leaving a slight opening between the bucket and sole plate, through which the air escapes; hence the term.

Ventilating Fan.—An ordinary fan (q.v.), used for purposes of ventilation.

Ventilator.—The term ventilator is applied to the revolving grids, usually attached to the doors of Cornish and Lancashire boilers, through which the ingress of air to the fuel is partially regulated.

Venting.—The venting of a mould is the piercing or honeycombing of the sand of which it is composed, by a long rod of $\frac{1}{8}$ in. or $\frac{1}{4}$ in. wire, thrust in all directions to allow of the free egress of the gases generated by the decomposition of its moisture, consequent on the heat imparted thereto by the inflowing of the molten metal.

Vent Pipes.—Lengths of common piping leading up at an angle from the coke bed (q.v.) of a foundry mould, to bring off the escaping gases and carry them outside the edges of the moulding box, where they are ignited and burnt.

Vent Rope, or Vent String.—Common rope or string rammed up in certain cores and in certain sections of moulds in situations where it would be difficult, or impossible to ram up the usual vent wires, by reason of the sinuous character of the vents. After the ramming is complete the rope or string is withdrawn, leaving the vent or vents curved as required.

Vent Wire.—A rod of iron wire used for piercing a foundry mould with small holes for the escape of the gas generated in casting. Vent wires and rods will range from $\frac{1}{8}$ in. to $\frac{1}{4}$ in. diameter.

Vermilion.—A bright rich red colour, being a sulphide of mercury, Hg S. Useful for mixing with shellac varnish to impart a hard and glossy skin to the best foundry patterns.

Versin.—The versed sine of an angle.

Vertical.—Plumb, or perpendicular to the centre of the earth. A vertical must be perpendicular, but a perpendicular (q.v.) is not necessarily vertical.

Vertical Boiler.—A steam boiler whose horizontal section is circular. Boilers of this type are used only for small engines, and are more wasteful than those of horizontal type, the products of combustion passing at once into the chimney. Economy in vertical boilers is obtained by the employment of Field's tubes (q.v.), but in general such boilers consist of shell and fire box only, connected by the uptake (q.v.) leading to the chimney, and having in the larger forms one or two cross tubes (q.v.) in the fire box.

Vertical Crane.—A steam crane of the ordinary type with tall side frames, as distinguished from a horizontal crane (q.v.).

Vertical Engine.—An engine whose axis is vertical, and which therefore forms a very numerous type. The advantages of vertical engines consist in the small foundation space which they occupy, and equality of wear on the cylinder, piston, and rods. Very few vertical engines are constructed with the cylinder lowermost, and those only of the smallest sizes. Most vertical engines have the cylinders above, embracing the various steam hammer types, launch and screw engines, or inverted cylinder engines.

Verticals.—The upright members of a lattice girder.

Vertical Section.—A section (q.v.) taken in a vertical direction through a drawing of an object.

Vertical Shaping Machine.—A slotting machine (q.v.).

Vibrating Links.—See Suspension Links.

Vibration.—Vibration in engineering works is a fruitful source of deterioration of material, and if long continued produces fatigue of materials. The vibration produced by heavy caulking, hammering, &c., on rivet heads and seams is also liable to lead to fracture or starting of joints and consequent leakage. To prevent vibration in machines doing heavy sawing, planing, drilling, shaping, slotting, &c., the machine bases are bolted down to large stone or concrete foundations. Steam hammers are not only thus bolted down, but their anvils also are embedded on massive anvil blocks. The framings of heavy machines are made hollow the better to withstand vibration stresses ; ties, struts, and distance pieces, &c., are also introduced into structures for the purpose of minimising the stresses due to vibration.

Vice.—A common workshop tool employed for holding or gripping work which requires to be held firm, but which is not sufficiently heavy in itself to remain immovable under the operation of the tools. It consists of a pair of steel-faced jaws, one of which is moved by a screw or by a lever, the other jaw being rigid. In the former case the friction of the screw prevents the release of the jaws. In the latter the teeth of a ratchet fulfil the same function. Vices are made in great variety of forms and sizes, the best being those with parallel movement in the jaws. Machine vices are employed for bolting to the tables of planing, shaping, and drilling machines.

Vice Bench.—A small iron portable bench usually mounted on a tripod and made to carry the ordinary tail vice.

Vice Cheeks.—Vice jaws (q.v.).

Vice Chuck.—A machine vice (q.v.).

Vice Clamps.—See Vice Claws.

Vice Claws, or Vice Clamps.—Angle strips of lead, brass, or copper, used for insertion into vice jaws to prevent the bruising of delicate pieces of work by the serrations of the hard steel faces. They are usually made to cover the serrated jaws and to bend over the upper faces as well. For special purposes vice claws are made in one piece, the two cheeks being united at some distance below the jaws either by a spring joint or by hinges.

Vice Jaws.—Vice jaws are steel-faced, the faces being screwed to the wrought-iron backing, being first screwed and serrated while untempered, and subsequently removed for hardening, and then replaced. In the smaller vices the steel jaws are instead welded to the iron. Most jaws are thus fixed, but not invariably. See Taper Vice.

Virtual Velocity.—The law of virtual velocities is simply the expression of a fundamental principle of mechanics, that, namely, that whatever is gained in power is lost in time, with its converse axiom ; or that the power multiplied by the space through which it moves is equal to the weight multiplied by the space through which it moves. It is simply saying in other words that by no mechanical combination can we create force, but only change its mode of application. See Conservation of Energy.

Viscosity.—See Body, Thick Oil.

Visible Drop Lubricator.—A sight feed lubricator (q.v.).

Vis Inertia.—The property of inactivity inherent in matter, by virtue of which it resists external influences tending to make it change its state of rest or of motion. Its equivalent is therefore the force required to put bodies which are in a state of rest into motion, or to accelerate the speed of bodies already in motion.

Vis Viva.—See Energy.

Voltaic Action.—See Galvanic Action.

Volume.—A term of general application, signifying the cubical contents of an engine cylinder, a definite amount of steam, or of gas at a definite pressure, or a mass of liquid. The product of the sectional area of a piston and the length of its path is termed its volume. This also is equivalent to the quantity of water lifted by a pump. In the case of a steam engine the volume multiplied by the mean pressure is equivalent to the gross amount of work performed.

Volume, Lines of.—The horizontal or approximately horizontal lines in an indicator diagram of work.

Volute Spring.—A flat ribbon like spring coiled in the form of a helix, and extensible in the direction of the breadth of the ribbon. Used as a buffer spring, and also for driving springs (q.v.).

Vortex Turbine.—See Turbine.

Voussoirs.—The wedge-shaped stones of which an arch is composed.

Vulcanite.—A compound of india-rubber and sulphur. Used for draughtsmen's set squares.

Vulcanized Rubber.—See India-rubber.

W.

Wabblers.—The coupling-boxes used for connecting the breaking pieces (q.v.) with the necks of the rolls in puddling rolls.

Wabble Saw.—A drunken saw (q.v.).

Wabbling Disc.—A swash plate (q.v.).

Wafters.—The revolving discs or fans in a Root's blower. See Rotary blower.

Wagon Boiler.—An old-fashioned, externally fired form of boiler, now rapidly going out of use. It has a semi-circular, or overarching top, slightly hollowed sides, and flat or arched bottom, the flues taking the form of a wheel draught (q.v.). Its shape renders it difficult of staying, and though adapted to the low pressures of forty or fifty years ago, it is wholly unsuitable for the high pressures now commonly employed.

Wall Bearing.—A Wall Box (q.v.).

Wall Box, or Wall Bearing.—A cast-iron open frame, commonly rectangular in form, used in cases where shafting is carried through a wall, the bearing or plummer block for the shafting being bolted within the wall box. The box is built into the masonry of the wall.

Wall Bracket.—A bracket of approximately triangular outline bolted to a wall to carry a plummer block for shafting.

Wall Drill.—A drilling machine bolted to the wall of a workshop. It may be either fixed or radial, according to convenience, and the particular class of work which it is designed to perform. The advantage of wall drills is that they leave a clear floor space underneath for the manipulation of heavy work. The wall is sometimes arched out underneath the drill to afford still more room for the hauling and turning about of bulky work.

Wall Engine.—A type of small vertical steam engine, whose bed plate is bolted to a wall. Such engines are convenient for driving sectional lines of shafting, as one set of shafting can then be stopped without affecting that in another shop, or in another part of the same shop. They are also used for driving travelling cranes, foundry fans, &c. The cylinders are lowermost and the crank shaft above.

Wallow Wheel.—A bevel wheel fixed upon a vertical shaft, and having the teeth facing downwards, or the reverse of a crown wheel. It is employed in mill work.

Wall Plate.—(1) A flat cast-iron plate or angle bracket, bolted to a wall as a base for the attachment of wall brackets, bearings, &c., for the carrying of shafting, or other portions of machinery. Its function is therefore that of a base or foundation plate. (2) The plate upon which the ends of roof trusses abut, in order to distribute the strain due to their weight over a sufficiently large area.

Wall Pump.—Sometimes applied to a donkey pump when bolted to a wall.

Wall Washer.—A large flat washer against which the ties of buildings are screwed. Or a similar broad washer which receives the pressure of a bolt head or bolt heads used for fastening brackets or plates to that face of the wall which is on the side opposite to that which takes the wall washers.

Walrus Hide.—Walrus hide is used as a covering for buffs (q.v.).

Waney Boards.—The outer boards of logs whose edges are chamfered off, and irregular.

Waney Log.—A squared log or balk of timber whose angles have been adzed off. Waney log is therefore more wasteful in the sawing up than squared logs, but is cheaper in first cost.

Warehouse Crane.—A light slightly-made crane, used only for lifting weights ranging usually between 10 cwt. and 40 cwt. in warehouses, and commonly worked by hand. See also Whip Crane, Platform Crane.

Warp.—To pull a load along by the winding of a rope or chain upon a drum. See also Warping Cone.

Warping.—(1) The alteration in form, of timber planks, and boards, due to the unequal swelling or shrinking of the fibres by reason of the action of dryness or of moisture therein. (2) The act of warping a rope or chain by means of a warping cone (q.v.).

Warping Cone.—A conical or capstan-shaped drum, used to receive the coil of the rope or chain when loads are being warped along (see Warp). Sometimes called a surging drum. Also frequently termed a capstan, because its outline is that of a capstan ; sometimes, also, warping ends, when attached to the ends of horizontal shafts, as in steam winches.

Warping Ends.—See Warping Cone.

Warren Girder.—A lattice girder (q.v.) in which the struts (q.v.) and ties (q.v.) form triangles or triangular bays, the struts leaning inwards or towards the centre of the girder, and the ties leaning outwards. Warren girders are largely employed for bridges. A warren girder consists of a single system of triangles, while lattice girders contain two or more systems of triangulation.

Wash Brush.—A draughtsman's colour brush (q.v.) used for cleaning palettes and colour utensils.

Washer.—A thin ring of metal encircling the screwed end of a bolt, and receiving the thrust of the nut. These are of most value when used against rough surfaces, as affording a comparatively true face for the distribution of the pressure. Wall washers and other cast-iron and wrought-iron washers of large sizes are often made rectangular in outline.

Washing Out.—The cleansing of a steam boiler, effected by driving out the muddy deposit or sludge before a jet of water sent through a hose pipe. See also Flushing Charge.

Wash-out Plug.—A plug screwed either into the bottom plates of boilers or on the outside of a nozzle attached thereto, for the purpose of closing the opening used for washing out (q.v.).

Waste, or Cotton Waste.—The refuse of cotton mills, which being soft and of close fibre, is used by engineers and machinists to cleanse the working parts of accumulated oil and dirt, and to wipe off superfluous oil and grease from the surfaces of work in process of shaping and fitting. Sponge cloths (q.v.) are rapidly superseding cotton waste.

Waste Blocks.—Plaster moulds from which reversed moulds (q.v.) are made, the waste blocks being rammed directly on the pattern. After the reversed moulds are obtained the waste blocks are no longer of service.

Waste Heat.—See Close-topped Furnace, Hot Blast, Regenerative Furnace, &c.

Waste Pipe.—The pipe used to convey waste steam away from a bonneted safety-valve. Also a blow-off pipe. Similarly any pipe used for conveying away waste steam or water.

Waster.—(1) Waster castings are those which, owing to some defect or defects, are discarded as useless, and are broken up for the scrap heap. Wasters are due to several causes, as blow-holes, scabs, cold shuts (see Cold Short), insufficient metal, a very rough exterior, flaws arising from bad proportioning or improper cooling, and sometimes from mistakes in coring and stopping off. (2) Steel and iron plates which show blow-holes and other defects on rolling are also termed wasters.

Waster Casting.—See Waster.

Waste-water Cock.—See Pet Cock.

Waste-water Pipe.—A drain pipe (q.v.) or waste pipe (q.v.).

Water.—Symbol, H_2O. Is a compound of the two gases, hydrogen and oxygen, in the above proportions, and can be analysed into those elements. It may be built up by exploding a mixture of the gases. Its value as a motive agent is due to the pressure derived from head (q.v.), and to its capacity of vaporization (see Steam). It is the best solvent known, so that it is impossible to obtain water in a condition of purity in a natural state ; hence the incrustation (q.v.) of steam-boilers. This is especially the case with sea water (q.v.) which requires for this reason to be maintained at a certain standard of saltness by frequent blowing-off. See Two Salt Waters. The weight of distilled water is used as a standard by which to measure the specific gravity (q.v.) of bodies. Of the temperatures of water there are four being constantly employed as standards for reference : 32° F. or 0° C., the freezing-point:

39·1° F. or 4° C., the point of maximum density; 62° F. or 16·66° C., the standard of temperature as used for determining the specific gravity of bodies ; 212° F. or 100° C., the boiling point at the pressure of the atmosphere.

Water Bath.—A bath of clear water used in the development of phototypes (q.v.), its function being to check the action of the prussiate solution, which, but for this precaution, would cause the lines to break and diffuse over the paper. There are usually two baths used, one before, and the other after the acid bath, to remove all traces of the chemicals. The water bath is lined with tin.

Water Bosh.—(1) The tank which supplies the tuyere (q.v.) of a smith's shop with water. (2) A tank in a foundry which supplies the water necessary for core making and the watering of green sand moulds.

Water Bridge.—A form of bridge used for steam boilers which is made continuous with the boiler itself. It is of iron, and is hollow, thereby assisting the circulation of the water which traverses through the interior.

Water Brushes.—Soft brushes of bristles employed in foundries for the purpose of wetting or moistening the surface sand of moulds to make it cohere consistently. The water is spurted over in a fine spray, or lightly wiped around the edges, according to circumstances.

Water Can.—Ordinary gardeners' water cans are used by moulders for damping and remixing the sand on the foundry floor after the castings made over-night have been turned out.

Water Chamber.—A chamber, usually annular in form, encircling the cylinders of gas engines and air compressors to prevent the heating due to the products of combustion in the first place, and to the compression of the air in the second.

Water Cock.—The lower of the two try cocks (q.v.) of a steam boiler. So called because water should always issue from it when opened, never steam, since that would imply that the water had fallen to a dangerous level.

Water Column.—A column up which the supply water for a water crane (q.v.) passes. It is of cast-iron and encloses within its body the actual supply pipes.

Water Core.—In some large cylindrical works of considerable thickness, the central portions remain hot so long after the outer portions have cooled, that injurious internal stresses are set up between the outer and inner layers of metal, due to the unequal cooling. This is prevented, and the cooling and contraction rendered approximately uniform by the carrying of a current of cold water through the interior, by which the central portions are cooled at about the same rate as the exterior. This is called a water core.

Water Crane.—A contrivance for supplying water to the tanks of locomotives at the stations along their lines, though not properly speaking a crane at all, the term being probably derived from the appearance of the swinging supply arm. In one form the turning round of the swing arm to which the hose is attached opens a valve, and the water pressure in the reservoir forces up the liquid through the supporting column along the arm into the tender. In another form the lifting of a sluice is necessary to deliver the water supply, the water still flowing under constant head. In a third type a tank is carried on the water column and is maintained constantly full, being fitted with inlet valve and overflow pipe, and the raising of a lift valve by means of a chain is all that is necessary to allow of the descent of water from the tank by gravity.

Water Cylinder.—Sometimes applied to the pump barrel of a steam pump to distinguish it from the steam or engine cylinder.

Water Engine.—See Water Pressure Engines.

Water Float.—See Float.

Water Gauge.—A gauge affixed to a steam boiler for the purpose of registering the height of the water contained within. It consists usually of a stout glass tube fixed in brass seatings, whose lowest end is at the lowest level at which the water should be allowed to sink, and furnished with steam and water cocks. Sometimes, however, it consists of a float and wire, with a chain and register pointer attached. See Float Gauge.

Water Hardening.—The hardening of steel, effected in water, as opposed to oil hardening (q.v.). In water hardening the steel becomes more brittle and acquires a harder texture than when oil is used. Sometimes the water is medicated with salt or other substances, but pure well or spring water appears to be in most favour.

Watering.—Alludes to the practice of damping foundry sand to make it cohere properly. After castings have been turned out of the sand in the morning, a certain amount of water is sprinkled over it from the rose of a water can, or from buckets, the sand meanwhile being turned over with shovels. The amount of water should be only sufficient to render the sand perceptibly damp, without being muddy or sticky. Watering is also resorted to just previously to the withdrawal of a pattern, the edges of the sand around the margin of the pattern being damped with a swab or water brush (q.v.) to cause sufficient coherence of the sand to prevent pulling up. Broken edges of sand are also watered in the act of mending up and for the same reason.

Water Jacketing.—The casing of the cylinder of a gas engine with a water jacket. See Water Chamber.

Water Meter.—A meter used for recording the amount of feed water which passes into a steam boiler in any given time. There are many forms of water meters in use.

Water Packing.—The pistons of some air pumps are made for water packing; that is, the piston, which is solid and deep, is furnished with circularly turned grooves around its circumference, which retain a film of water of sufficient thickness to render the piston watertight.

Water Pot.—A small cast-iron pot used in foundries for holding water, for dipping the water brush or swab into. Called also a swab pot.

Water-pressure Engines.—These include generally hydraulic rams, turbines, and other motors driven by water pressure, but refer specifically to certain types of engines having cylinders and pistons, either reciprocating or rotary, whose construction is the same in principle, though differing in detail, as that of steam engines, the pressure due to a head of water being employed instead of steam.

Waterproof Cement.—Iron Cement (q.v.).

Waterproof Glue.—See Glue Cement.

Water Ram.—See Hydraulic Ram.

Water Tank.—See Tank, Tank Plates, Iron Cement.

Water-tank Bed Plate.—A form of engine bed plate which is cored out to serve as a tank for the feed water of the boiler.

Water-tube Boiler.—A sectional boiler (q.v.).

Water Tubes.—See Galloway Tubes.

Water Tuyere.—See Tuyere.

Water Wheel.—A wheel whose axis is horizontal, and whose rim is furnished with buckets or floats, against which the water acts either by impulse, or gravity, or reaction. See Breast, Overshot, and Undershot Wheels, and Turbine.

Water Way.—The full space open for the passage of fluid in the buckets of pumps, the plugs of cocks, and the area of lift of valves.

Wave Wheel.—A wheel for the reception of a rope, the bite of which is ensured by a wavy contour being given to the rim, the groove waving from side to side around the circumference.

Wax.—This is used as a material for modelling patterns of ornamental work.

Way Shaft.—A weigh shaft (q.v.).

Wear.—A term applied to the slow abrasion of surfaces in frictional contact, as the slides of machine parts, the pistons and valves of engines, the journals of shafts, lever pins, &c. Provision is made in many instances by means of adjustable pieces for the taking up of the wear, as it is called. See Taking-up.

Wear and Tear.—See Depreciation.

Weathering.—(1) Those iron ores which contain pyrites in considerable quantity, and which are not of a calcareous quality, are often weathered by exposure to the atmosphere previous to calcination, the sulphur present being slowly oxidised, and rendered soluble, and washed away by rain. (2) Foundry clay is weathered by being cut up and exposed to the frost, which facilitates the subsequent work of grinding and mixing.

Web.—The plated or central portion of a structure as distinguished from its flanges and bosses. Thus a crank web is the plate which carries the shaft and pin bosses; a girder web is the main vertical plate which becomes the connection between the top and bottom flanges. See Lattice Web.

Web Stiffeners.—Webs of $\top$ iron riveted to the sides of deep-plated girders, and continued to the inner faces of the top and bottom flanges to preserve them from wrinkling stresses.

Wedge.—A triangular prism resolvable into two inclined planes set back to back, in which the mechanical gain is greater the more acute the angle of the sides. The power of a wedge may be obtained in a graphic manner. The theoretical rule is that the pressure applied at the back of the wedge is to the resistance, as half the width of the back of the wedge is to the length of the side. But the elements necessary to the calculation are so variable that it is best not to rely much on the theoretical statement.

Wedge Gearing.—Wheels whose peripheries are grooved circumferentially; the sides of the grooves being sloped or angular in section: the projecting rings in the one wheel enter into the grooves on the other and drive by friction only. These are therefore a type of friction gearing (q.v.),

Weighbridge.—This is essentially a table carried by a system of levers comprised in a small compass, the lengths of whose arms are so proportioned that a weight of a few ounces or pounds hung upon a steelyard at one end will counterbalance a weight of several hundredweights or tons on the weighbridge table at the other end of the system.

Weighing Machines.—Light instruments made on the principle of the weighbridge (q.v.) for taking weights of a ton and under.

Weigh Shaft, or Way Shaft.—The shaft or spindle which forms the centre or motion of the lever employed for throwing the slot links (q.v.) for the reversing motion of an engine into the proper positions for forward or backward gear. Sometimes called a reversing shaft.

Weight.—The tendency of a body to move downward towards the surface of the earth, under the action of gravity. See Mass. The unit of weight employed by the engineer is the pound, cwt., or ton, dependent upon the precise work whose amount it is desired to estimate. The question of weight is of fundamental commercial importance. Castings in various metals are sold by weight at so much per lb., cwt., or ton. Rolled iron of various sections is sold by weight, and is quoted at so much weight per foot. Ropes and chains are quoted at a certain weight per fathom. Sheet metals are distinguished as of certain weights per foot; the weight of tubing is also thus quoted. In many cases, especially in plates, a slight percentage of margin is allowed either way.

Weight Case.—The casing which surrounds the accumulator (q.v.) of a hydraulic crane, and which is loaded with stones or iron to the pressure necessary for the effective working of the crane.

Weight, Distributed.—The weight of a locomotive is so distributed that the load upon leading, driving, and trailing wheels shall be as nearly as possible alike. See also Distributed Load.

Weighted Pendulum Governor.—A governor (q.v.) whose balls are attached by links to the central axis above, and to a collar below, the latter being attached to a heavy weight free to slide on the central spindle. For any given rise in the height of the balls that of the weight is doubled.

Weighting Down.—The holding down of the top part of a moulding box during casting by means of hundredweights, or half ton weights. Pins and cottars or screw bolts afford better security than weights.

Weights.—Weights are used for many purposes in engineers' workshops. For scales, weighbridges, foundry boxes, and as tests in the proving of structures. They vary from small weighbridge weights to those of half a ton or a ton each.

Weir.—The dam or barrier by which the waters of a stream are bayed back in order to afford sufficient head of water for driving a water wheel or turbine.

Weld.—A welded joint (q.v.) as a good weld, a bad weld.

Welded Joint.—A joint made between opposing faces in wrought iron and steel work by welding (q.v.). Welded joints are butt, lap, scarf, and vee, noted under those headings. They are employed both by smiths and boiler-makers.

Welded Tube.—Wrought-iron tube which is made by being brought to a circular form or skelp, raised to a welding heat, and drawn through a pair of jaws, by which the seam is closed up.

Welding.—Term covering both forge welding and fusion welding of metals. In forge welding of wrought iron and steel the joints to be united are cleaned from scale and sprinkled with a flux, usually sand or borax, to dissolve off or to prevent the formation of oxide, brought to a welding heat and hammered together. The process is commonly called shutting, or shutting up, in the shops. Fusion Welding is covered under that heading in the Dict. of Modern Terms.

Welding Heat.—The degree of heat when in smith's work the iron or steel gives out vivid sparks, and small globules melt off. It corresponds roughly with a temperature of from 2,500° F. to 2,700° F. for wrought

18*

iron, but less for steel; which requires especial care as regards the temperature, cast steel being difficult of welding, and scarcely bearing more than a cherry red. Shear steel will weld easily, and endure a white heat safely, so that in welding the judgment and experience of the smith is of first importance.

Weldless Tube.—See Solid Drawn Tubes.

Well Crane.—A fixed post crane, one-half of whose post is above ground and the other sunk in a pit, or well, dug to receive it, and encased with masonry, or with iron plates. The lower end of the post rests on a step at the bottom of the well, and the fulcrum of the post is at the ground line.

Well Plate.—A cast-iron plate put over the mouth of a well to carry the pumps, &c. See Well Stage.

Well Pump.—An ordinary lift pump of two or three throw type, affixed to staging at the required depth, and worked from the well staging by hand or cattle gear, or by power, the cranks being connected to the pump piston by a sufficient length of jointed rods to establish communication between the crank and piston. Also called deep well pump.

Well Stage.—A framing of timber erected over the mouth of a well to carry the pumps and pipe connections.

Welt.—The covering strip used in butt riveting (q.v.).

Westinghouse Brake.—An air brake used on railway rolling stock. Air is compressed by a steam pump placed underneath the engine, and is retained in a reservoir ready for use. A cylinder and piston under each carriage communicates with the reservoir by means of hose pipes and couplings. The communication between the piston rods and brakes is made through the intervention of levers.

Wet Blacking.—Black wash (q.v.).

Wet Bottom.—A puddling furnace is said to have a wet bottom when the slag is allowed to accumulate instead of being drawn off at frequent intervals.

Wet Brush.—See Wet Brushes.

Wet Puddling.—Pig boiling (q.v.).

Wet Steam.—Steam, the amount of whose moisture has been increased by mechanically admixed spray thrown off by the priming (q.v.) of the water in the boiler.

Wet Uptake.—When the uptake of a marine boiler is placed within the shell, and is therefore surrounded by water and steam, it is termed a wet uptake, to distinguish it from a dry uptake (q.v.).

Wet Wood.—Timber which has not been seasoned properly, and which therefore retains the sap in the vessels. Timber which has been seasoned may become wet by long exposure to the weather. This moisture soon dries, however, under cover, and is not so detrimental as the sap wet.

Whale Oil.—Sometimes, though improperly, termed a fish oil. There are two kinds of whale oil known, as train oil, the produce of the blubber of the Right whale, and some other species, and sperm oil, which is taken from a reservoir in the head of the sperm whale or Cacholot. The sperm oil is an excellent machinery lubricant and 5,000 or 6,000 gallons of the oil is taken from a single animal.

Wharf Crane.—A crane specially adapted for use on wharves. It is a fixed well crane of the triangular form, consisting of pillar or post, jib, and ties, and usually worked by hand.

Wheel.—A general term which denotes no wheel in particular, apart from its qualifying prefix, as sheave wheel, chain wheel, pulley wheel, toothed wheel, cog or mortice wheel, &c.

Wheel and Axle, or Windlass.—This owes its value to the principle of the lever, and has its application in trains of gearing, capstans, &c.

Wheelbarrow.—Wheelbarrows are used in foundry work for carrying sand, loam, light castings, and moulding boxes, and for yard work generally. Made in iron and wood, and with single and double wheels.

Wheelbase.—The distance apart of the centres of the leading and trailing wheels of a locomotive, or of the front and hinder wheels of a railway wagon. A rigid wheelbase denotes the distance between centres of those wheels which have rigid (*i.e.* not bogie) bearings. Total wheelbase would include the bogie wheels also.

Wheel Bossing.—The welding on of the boss pieces of railway wheels under the steam hammer.

Wheel Centre.—The central portion of a solid plated truck or wagon wheel, which corresponds with the body of a wheel with arms. For trucks and trollies it is customary to use wheels having plated centres for the attachment of steel tyres in preference to using chilled cast-iron wheels. The centres are turned and shouldered, or stepped on the circumference, and the tyres are bored to correspond. Heating of the tyres in a reverberatory furnace expands them sufficiently to clear the shallow shoulder, and in cooling they shrink over and embrace the shoulder and so remain fast. Sometimes tyre bolts are employed in addition to prevent the possibility of the working loose of the tyres.

Wheel Cutting.—The cutting of the teeth of gear wheels by means of revolving milling or gear cutters, the wheel blank being held steady while each interspace is being cut, and moved around an arc of a circle exactly equivalent to the pitch of the tooth for the next cut. The means of dividing out adopted are either a division plate (q.v.) or a tangent wheel and screw with change wheels. The practice of wheel cutting has rather to do with the making of models and the smaller portions of machines, as paul wheels and such like, than with general engineering, for which cast teeth are preferable.

Wheel Draught.—In an externally fired wagon boiler (q.v.) the gases pass under the boiler from front to back, then upwards into a flue on the right-hand side leading towards the front of the boiler, thence round into a flue on the left-hand side, and so to the back of the boiler into the chimney. This, from its circuitous motion, is called a wheel draught to distinguish it from a split draught (q.v.).

Wheel Float.— See Float, Boiler Float.

Wheel Gearing, or Toothed Gearing.—This means that type of gearing which acts by means of teeth or cogs, or modifications of the same attached to the peripheries of rings, as distinguished from gearing of other kinds. Hence it embraces spur and bevel wheels, mortice wheels, helical and worm, and internal gears, each noted under its special heading.

Wheel Lathe.—A lathe of special design, having a bed short in proportion to its size, and used for the turning of locomotive and other wheels. A duplex-wheel lathe has two heads, so that a pair of wheels can be turned at one time upon their axles.

Wheel Moulding Machines.—Machines extensively used for the moulding of toothed wheels without the making of a complete pattern. The machines, differing much in details, are in principle dividing engines.

a worm and endless screw and change wheels supplying the means of regulating the equal division into any required number of teeth. Two teeth only are required for moulding from, and the space between those teeth alone imparts the outline to the sand. When a single tooth space is rammed, the tooth block is lifted and moved round by the machine to a distance corresponding with the pitch. The arm which carries the tooth block usually revolves around a central pillar, and is adjustable for radius and also for depth. Sometimes the mould-ing table revolves, and the arm can only move radially and vertically. The arms of the wheels are made by means of cores. There are also machines which are not furnished with the tangent wheel; but a perforated drum or circular division plate instead answers the same purpose.

Wheel Plate.—A quadrant plate (q.v.).

Wheel Segment.—See Toothed Segment.

Wheel Slope.—A draughtsman's palette (q.v.), circular in plan, and having several radial slopes falling inwards.

Wheel Stamping.—The stamping out of the spokes of railway wheels under the steam hammer previous to welding up.

Wheel Swarf.—The silicious and steely mud which collects in the troughs of the grindstones of the Sheffield cutlers. It is used to form a layer within the mouths of the crucibles or pots used in the production of crucible cast steel, where it fuses and forms an air-tight glaze to protect the bars of blister steel within.

Wheel Teeth.—The teeth of gear wheels are formed on definite principles of design, the forms of their curves being either cycloidal, or involute. The cycloidal curve is that employed for nearly all wheels, the involute (q.v.) being reserved for special work. The cycloidal curves are obtained by means of the odontograph (q.v.) scale, or are formed for each separate pair or set of wheels by means of rolling curves (q.v.). The strength of wheel teeth is estimated by considering each tooth as a cantilever (q.v.), or by determining the horse power to be trans-mitted.

Wheel Tooth-cleaning Machine.—A revolving emery wheel, turned to the section of the interspaces of cog-wheel teeth, and used for grinding out the cast teeth to a smooth surface, the wheel being laid horizontally.

Wheel Valve.—A lift globe valve of the ordinary type, in which the screw which lifts and depresses the valve is worked by means of a hand wheel above.

Whelps.—The longitudinal strips arranged equidistantly around the barrel of a capstan (q.v.) or warping cone (q.v.) on which the rope coils, and by which its bite is increased.

Whiffle Trees.—Whipple trees (q.v.).

Whin.—A variety of horse gear or bullock gear (q.v.) in which the rope is wound round a barrel fastened directly to the vertical shaft which carries the cross-bar.

Whip Crane.—A slight crane used generally for warehouse purposes. It has no gearing, but is worked by a rope; hence its name. The rope is affixed to a large drum or pulley at the top of the crane, thence it passes down to a small barrel on the winch shaft. The winding up of the rope gives motion to a small barrel on the same axis as the large pulley, and the difference in the diameters of the barrels and pulley affords sufficient leverage for lifting loads as high as 30 or 40 cwt.

Independent whip cranes are those in which the top of the crane has no point of fixture for slewing, the whole structure being simply supported on a base sufficiently broad to afford the requisite stability.

Whip Gin.—See Gin Block.

Whipping Drum.—A term applied to the winding barrel of a deck winch (q.v.) or a whip crane (q.v.).

Whipple Trees, or Whiffle Trees.—The loosely swivelling bars attached to the ends of the poles in horse and cattle gear, to which the cattle are yoked. They are either single, double, or treble.

Whips.—The long arms which carry the cross-pieces and sails of a windmill.

Whistle.—See Steam Whistle.

White Brass.—See Antifriction Metal.

White Heat.—A degree of colour in smith's work, at which the scales on the surface are scarcely visible, and which corresponds roughly with a temperature of 2,370° F. It is the temperature at which most large forged work is taken from the fire.

White Iron.—A highly crystalline form of cast-iron. It is extremely hard and brittle, and contains nearly all its carbon in the combined state. Hence called a carbide of iron. White iron also results when the surface of molten grey or mottled iron is chilled against a metallic mould. See also Forge Pigs.

White Lead.—Carbonate of lead (q.v.). Used when mixed with boiled oil and red lead for making steam joints.

White Line Phototype.—See Phototype.

White Metal.—See Babbitt's Metal. These and similar mixtures are termed white metals.

Whitening.—The facing used for brass founders' moulds.

Whitening Bag.—A linen bag which contains the whitening used in brass founders' moulds, and from which it is dusted out.

White Pine (*Pinus strobus*).—A class of pine timber imported from North America. See Pine Wood.

White Rope.—See Manilla Ropes.

White Spruce.—See Pine Wood.

White's Tackle.—A system of pulley blocks, so named after the inventor, in which the diameters of the pulleys are regulated according to the rate at which they travel, in order to ensure uniformity of speed and of wear.

Whitwell Stove.—A regenerative stove used for heating the blast for smelting purposes. It is in principle like the Cowper stove, but firebrick partitions are employed therein instead of chequer work.

Whitworth Thread.—The standard engineer's screw thread of angular section. The depth of the thread, that is to the apices of the angles, is ·96 of the pitch, and the slope of the angles is 55°. The working depth of the thread is reduced by rounding off the points and roots to an extent of about $\frac{1}{6}$th each, making the working depth ·64 of the pitch.

Wholes and Halves.—A drawing compass, the lengths of whose legs are in the proportion of one to two, measured from a joint pin around which the legs play. They are used for enlarging or reducing dimensions on drawings in those proportions.

Whole Shrouding.—See Shroud.

Wide Gauge.—See Broad Gauge.

Willow (*Salix*).—A tree of the natural order *Salicaceæ*, being the order to which the poplar belongs. It is a very soft, smooth-grained, white or yellowish grey wood, easily worked, and is used by engineers **for brake** blocks. Sp. gr. ·585. A cubic foot weighs 36½ lbs.

Winch.—A fixed crane (q.v.) having no jib and no provision for slewing. It may have single, double, or treble gear, and be driven by hand or steam. See Deck Winch, Steam Winch.

Winch Handle.—The handle of a crane, crab, or other hoisting machine turned by hand power. It is a lever whose length is the measurement taken from the centre of the handle to the centre of the spindle. This length is usually 1 ft. 4 in.

Winch Shaft.—The first shaft in a train of gearing, to which the winch handle is attached.

Wind Bore.—The strainer of a suction pipe. (See Strainer.)

Wind Chest.—See Air Belt.

Wind Engine.—A windmill (q.v.).

Wind Gall.—A defect in a balk of timber caused by an old wound, subsequently covered over by a growth of wood. It becomes a centre of decay.

Winding.—The variations of a surface from a true plane, considered chiefly in relation, not so much to local inequalities of surface, as to the total superficies. Thus, if on one corner of the plate the surface were sloping $\frac{1}{8}$ below the general surface, the plate would be $\frac{1}{8}$ winding.

Winding Drum.—The barrel or cylinder which coils the wire rope for raising and lowering cages and trucks on inclines.

Winding Engine.—The engine which drives a winding drum (q.v.). It does not refer to any particular type of engine, but simply to the purpose for which it is employed.

Winding Gear.—A particular type of hoisting machine used for hauling wagons up inclines. It consists of winding drum, or clip pulley, and suitable toothed gearing, single or double purchase, with arrangements for the application of sufficient brake power, the whole being driven by a stationary, a locomotive, or portable, or semi-fixed portable engine.

Winding Sticks.—Winding strips (q.v.).

Winding Strips, or Winding Sticks.—Straight-edges of metal or of wood, made perfectly parallel, and equal in width, by means of which the inaccuracy of a supposed level surface is detected. The strips are set by their edges on those portions of the plate whose condition it is desired to test, and the amount of their divergence from parallelism is observed by the eye cast across their top edges.

Windlass.—See Wheel and Axle.

Windlass Jack.—A lifting jack provided with a double-ended lever and handles for working bevel gear wheels, by which the nut is turned around the lifting screw.

Wind Mill, or Wind Engine.—These are used to some extent for driving pumps. A windmill consists of its main support, which carries the sail frame or whip shaft. This shaft is inclined at an angle of about 8°. with the horizon on level ground, and 15° in mountainous districts, it being found that the direction of the wind is always slightly downwards. The whips or sail frames carry transverse or cross pieces, on which the actual sails are spread, the cross pieces being so hinged that their obliquity in relation to the direction of the wind is capable of variation.

Wind Ties.—Diagonal side ties of wrought iron rod, on bridges and large block setting cranes, and other structures, for steadying the structures to which they are attached against wind pressure.

Wing Compasses.—A pair of compasses used in the workshops, in which an arc-shaped wing attached to one leg passes through a slot in the other leg, to which it is set or clamped in any position by a set screw.

Wing Nut.—A form of nut which is tightened or relaxed by two thin flat wings or expansions coming out from opposite sides.

Wing Rail.—In permanent way, is a short rail laid down at a crossing near the point rail to guide the wheel flange. It fulfils the same purpose as a check rail (q.v.) on a curve.

Wing Tanks.—The tanks of a side tank locomotive engine.

Wing Valve.—A valve whose lift is guided by three or four wings, or feathers, or ribs, as they are variously termed, cast upon its under side and fitting into its cylindrical seat; called also lift, and puppet valve. Such a valve may be placed in a vertical position, as in a common safety valve, or its axis may be horizontal or otherwise, if pressed against its seating by springs.

Wiper.—(1) See Sponge Cloth. (2) The cam teeth on the wheel of a helve or tilt hammer are called wipers or wipes.

Wiper Shaft.—The shaft or axle of a wiper wheel (q.v.).

Wiper Wheel.—The wheel or disc to which the wipers (q.v.) or cams of helve hammers and stamping machines are attached.

Wipes.—The teeth or projections on the cam plate of a helve hammer.

Wire.—Iron, brass, and copper wire is used for many purposes in the factory, and in all departments, as for dowels and skewers, for binding and fastening light cores, for making steam joints, for pins and small screws, for brazing joints, for straining lines, and many others.

Wire Brush.—A brush used for the fettling off of castings. It is in shape like an ordinary hand brush, but the place of bristles is occupied by thin elastic fibres of steel, by which the sand and dirt are readily scraped off the surfaces of the castings.

Wire Card.—See Card Wire. Used by moulders for rubbing or dressing down cores.

Wire Drawing.—(1) The process of manufacture of wire by pulling it through draw plates. (2) The throttling of steam in the passages of an engine cylinder, by which some amount of expansive working is effected. This throttling takes place when the passages are so contracted that the rate of travel of the entering steam falls behind that acquired by the piston towards the termination of its stroke.

Wire Edge.—The turned over strip of metal which results when a tool is subjected to abrasive action upon a grindstone, or hone. This is removed by rubbing both faces of the tool alternately, thus thrusting the wire edge from one side to the other, until it is readily detached.

Wire Gauge.—A notched plate having a series of gauged slots, numbered according to the sizes of wire and sheet metal manufactured. The Birmingham wire gauge is that in general use.

Wire Gauze.—Fine iron wire woven into gauze, having a fine mesh. It is employed in making steam joints, the gauze being cut to the size and shape of the flanges, and smeared with red or white lead before bolting together of the flanges. Brass or copper gauze is used for filters.

Wire Nails.—Nails circular in section, parallel, and having thin circular heads. Used in general pattern work.

Wire Pliers.—Pliers (q.v.) in which a pair of smooth jaws, circular in section and tapered lengthways, are substituted for the ordinary flat and roughened jaws, their purpose being the bending of wire into small curves and loops.

Wire Rope.—Rope formed by twisting several wires around a central hemp core, to form a strand, several such strands being then twisted around a central hemp core to form the rope. The diameter of a single wire varies from $\frac{1}{80}$ to $\frac{1}{18}$ of an inch, and the twisting of the wires to form the strand is in the reverse direction to that of the strands which make up the rope. Wire ropes are formed of both iron and steel. Their strength is very great, so that a small wire rope will do the work of a much larger hempen one.

Withe Handles.—The flexible elastic handles of smith's tools. See Hazel Rods.

Wood.—Wood is employed as a fuel, and in the preparation of charcoal. It is not economical as a fuel in the green state, owing to the quantity of moisture which it contains. Wood is composed essentially of carbon and oxygen; hydrogen, nitrogen, and ash being present in small quantity. Woods are hard or soft, terms which are relative merely. The value of wood to the engineer consists in its adaptability in the form of timber boards, planks, and logs for purposes of construction.

Wood Ferrules.—Rings of wood used for making the ends of marine engine condenser tubes water-tight. They fit closely around the tube ends, and are driven tightly between them and the holes in the tube plates.

Wood, Preservation of.—Various methods have been employed for the preservation of timber from decay, the chief of which are creosoting, kyanizing, and the employment of Burnett's and Margery's fluids, described under their several heads.

Wood Screw.—A right-handed screw having a conical-shaped head provided with a slot or nick, by which the screw is turned in through the medium of a screw-driver. Formerly called a screw-nail.

Wood Spirit.—Methylated Spirit. See Shellac Varnish.

Wood Turning.—Wood turning is an essential section of pattern work, and is not performed by professional turners, but by the pattern-maker himself, who alone can know the conditions necessary to be observed in order to proper delivery from the sand. The work is seldom ornamental, and much of it is done by scraping tools chiefly.

Woolf Engine.—An early type of compound engine in which the cranks were set opposite to each other, instead of at right angles, as is now the common practice, so that when one piston was at the termination of its stroke the other was at its commencement.

Woollen Pad.—See Worsted Pad.

Work.—This is the result of resistance overcome by the action of force acting upon a body. See Energy.

Work, Diagram of.—See Diagram of Work.

Working.—The working of the frames of locomotives signifies the loosening of their joints, due to the strains communicated to them by the engines.

Working Anvil Block.—A block of steel or cast iron fitting on the top of the anvil block (q.v.) of a steam hammer. It is fitted to the latter with a dovetailed joint, to be readily removable, in the same manner as the hammer block (q.v.) is attached to the tup (q.v.).

Working Barrel.—A pump barrel proper, containing the piston and bored portion, and in some cases also the clack valves. Used to distinguish it from the pump case (q.v.).

Working Beam.—The beam of a Cornish or beam engine, having the piston rod at one end and the pump rod at the other.

Working Bed.—The lining of a puddling furnace, consisting of broken slags, tap cinder, hearth bottoms, and fettling.

Working Cylinder.—The exploding cylinder in those gas engines which are provided with a separate compressing cylinder (q.v.).

Working Depth.—In toothed gearing this is less than the total length of tooth from point to root by the amount of bottom clearance.

Working Door.—The opening through the side of a reverberatory furnace through which the charge is introduced to, and withdrawn from the hearth. The door slides in vertical grooves, is protected with fire brick, and is balanced by a chain and counterweight.

Working Drawing.—A plain, fully detailed drawing of an engine, machine, or structure, or more generally a portion only of the same, and either dimensioned, or drawn to full size, or to a uniform scale, and sent out into the workshops to be used in construction.

Working Gear.—This is applied specifically to the motion work of a locomotive or engine, namely, the piston and connecting rods, eccentrics, and rods, links, slide valves and rods, as opposed to the fixed gear, or that which has no reciprocal movement.

Working Load.—The ordinary load to which a structure is subjected, not necessarily the maximum load, but the average or mean load, as distinguished from the proof load (q.v.).

Working Nuts.—Nuts which are perpetually being tightened and relaxed by means of spanners or screw keys, as distinguished from the nuts which remain permanently fast. Working nuts are, or should be, always case-hardened (see Case Hardening) in order to prevent the wearing of their angles rounding.

Working Stress.—The safe stress which calculation and experience combined allow that structures should be subject to. It is less than the elastic strength (q.v.) by the divisor selected as the factor of safety (q.v.), which factor of safety is selected in accordance with the nature of the load imposed.

Work, Unit of.—See Unit of Work.

Worm.—A form of helical gear consisting of a continuous screw thread wrapped around a cylinder. It is used to impart a slow and equal motion to a worm wheel (q.v.).

Worm Gearing.—Gearing composed of worms and worm wheels. It is employed where great power is sought, but much of its efficiency is lost in friction, so that it is often necessary to run the worm in oil.

Worm Wheel.—Gears with a worm, and each tooth consists of a small segment of a helix. The worm commonly drives the wheel. In order that the wheel should drive the worm, the obliquity would have to be very great.

Worsted Pad, or Woollen Pad.—A pad of wool placed in the lower portions of some types of axle boxes, from which oil is led up by capillary attraction through tapes of cotton wool to the bearing. It rests on a horsehair cushion saturated with oil.

Wrench.—A spanner (q.v.) one of whose jaws is rendered adjustable by the agency of a screw, for nuts of narrow sizes. See Screw Wrench. The term is often applied to any form of spanner.

Wrinkling.—The failure of thin unstayed or improperly stayed wrought-iron plates, by wrinkling up, or the becoming corrugated under pressure.

Wrinkling Strain.—The strain necessary to produce wrinkling (q.v.).

Wrist.—Sometimes applied to a crank pin, or to any projecting pin which receives a connecting rod.

Wrist Plate, or Motion Disc, or Rocking Disc.—The plate attached to the side of the cylinder of a Corliss engine (q.v.), through which motion is transmitted by connecting rods to the valve spindle. It is worked by levers.

Wrought Iron.—Iron which has had the major portion of its carbon, as well as the foreign elements which would affect its workability, removed in the various processes of puddling, shingling, and rolling. Its value lies chiefly in its great tensile strength, averaging about twenty-three tons per square inch, in its capacity for welding, and its ready malleability. See Dict. of Modern Terms.

Wrought-Iron Castings, or Mitis Castings.—Made by a special process of melting and pouring wrought iron into moulds, by which the expense of forging is saved. The wrought iron is melted in crucibles in a petroleum furnace, and poured into a mould containing a patented mixture composed essentially of fire clay, burnt and ground, and mixed with molasses. The metal is dead melted or superheated by the addition of about 0·1 per cent. of aluminium, by which its melting point is lowered, and the entrance of deleterious gases due to the superheating of the metal alone, is prevented. It is thus rendered sufficiently fluid to take the finest impressions of the mould, while its quality as wrought iron remains unimpaired. These castings do not require to be annealed, and while cold can be bent and twisted like good malleable iron.

Wrought-Iron Plates.—Plates which have been rolled from iron prepared by puddling. See Boiler Plate, Ship Plates, Limit of Weight, &c.

Wrought-Iron Pulleys.—These are rapidly superseding the old cast-iron pulleys, from their superior lightness and reliability. The rims are framed of wrought iron or steel, and the arms are of the same material, each being made distinct from the other, and fastened together. They are put together in halves, and are, therefore, split pulleys (q.v.). There are numerous designs of these pulleys in the market. They do not break, put less strain upon the shafts, are more easily fixed, and, though costing more in the first place, are cheaper in the end than those of cast iron.

Y.

Yacht Engine.—A light engine specially designed to drive the screw of a yacht. Yacht engines are mostly of the inverted cylinder (q.v.) and compound engine (q.v.) types.

Yank.—To pull, wrench, or hammer with undue violence. A slang term.

Yard Travellers.—Ordinary overhead travellers, used for lifting heavy work about an engineer's yard, both for erecting and loading. They are worked by hand or steam with ordinary gear, or with hempen ropes.

Yarn.—Fibres of prepared hemp twisted together into small cords in readiness for laying into strands.

Yellow.—See Gamboge.

Yellow Bath, or Prussiate Bath.—A bath containing a saturated solution of the yellow prussiate of potash $K_2 Fe C_6 N_6$, and which is used for developing the lines of phototypes (q.v.) which are removed to the bath from the printing frame (q.v.). This bath is lined with gutta percha.

Yellow Brass.—A cheap kind of brass formed of an alloy of copper 70 parts, zinc 30 parts, and used for the commoner class of turned and other work, and also for name-plates and similar castings when durability and strength is not essential, or when flexibility is desired.

Yellow Metal.—See Yellow Brass.

Yellow Pine (*Pinus variabilis* or *Pinus mitis*).—The wood of an American pine of the natural order *Coniferæ*. A light faintly yellow-coloured wood, open grained, not much given to warping, and pleasant and soft to work. It is employed chiefly in the making of patterns (q.v.). Sp. gr. 448. A cubic foot weighs 28 lb.

Yield.—The quantity of iron or steel produced by a blast furnace, a puddling, or similar furnace, in a given time.

Y Lever.—The longest lever of a weighbridge, to which the rod of the actual steelyard is attached.

Z.

Zero.—(1) In mathematics, signifies 0, or nothing. (2) In thermometry, commonly denotes the temperature obtained by a mixture of salt and snow ; or the zero of Fahrenheit's scale, being 32°, corresponding with 0° on the Centigrade scale. This is the zero point used in practice ; but sometimes it is convenient to reckon from absolute zero (q.v.).

Zigzag Riveting.—Rows of rivets placed parallel longitudinally, but alternately, or in zigzag form in the transverse direction.

Zinc.—Symbol, Zn. Comb. weight, 64·9. An easily oxidizable metal, used as an alloy for brass (q.v.). The oxidization of zinc, when in the brass-melting furnace, is prevented by throwing borax, powdered glass, or powdered charcoal over the surface. It is valuable also because of its being an element electro positive to iron. In this capacity it is employed to prevent the corrosion of steam boilers ; for if strips of zinc are suspended therein they are corroded, whilst the iron plates remain unaffected.

Zinc Oxide, or Zinc White.—Used as the basis of a white paint.

Zones.—Zones signifies the various horizontal sections of areas of a blast-furnace, corresponding with certain chemical reactions which take place between the hot gases and the metal ; accompanying the formation of CO_2, and the reducing agent CO.

SELECTED BRITISH STANDARDS

RELATING TO MECHANICAL ENGINEERING

B.S.	Title
4–1962.	Structural steel sections.
15–1961.	Mild steel for general structural purposes.
18–1962.	Methods for tensile testing of metals.
21–1957.	Pipe threads.
24:Part 5–1954.	Copper.
46:Part 1–1958.	Keys and keyways.
46:Part 3–1951.	Solid and split taper pins.
51–1939.	Wrought iron for general engineering purposes.
57–1951.	B.A. screws, bolts and nuts.
61:Part 1–1947.	Copper tubes (heavy gauge) for general purposes.
61:Part 2–1946.	Screw threads for copper tubes.
84–1956.	Parallel screw threads of Whitworth form.
93–1951.	British Association (B.A.) screw threads.
122:Part 1–1953.	Milling cutters.
122:Part 2–1952.	Reamers, countersinks and counterbores.
131:Part 1–1961.	The Izod impact test on metals.
192–1954.	Open-ended spanners (not including B.A. sizes).
205–1943.	Glossary of terms used in electrical engineering.
218–1963.	Leaded brass (58 per cent copper, 2 per cent lead).
219–1959.	Soft solders.
228–1962.	Transmission roller chains, chain wheels and cutters.
240:Part 1–1962.	Brinell hardness test.
240:Part 2–1950.	Steel balls for Brinell hardness testing.
250–1960.	Wrought high tensile brass rods and sections.
265–1963.	Cold rolled brass sheet, strip and foil. Common brass.
275–1927.	Dimensions of rivets ($\frac{1}{2}$ in. to $1\frac{3}{4}$ in. diameter).
292–1958.	Dimensions of ball bearings and parallel-roller bearings.
302, 621–1957.	Steel wire ropes for cranes, excavators and general engineering purposes.
308–1953.	Engineering drawing practice.
309–1958.	Whiteheart malleable iron castings.
310–1958.	Blackheart malleable iron castings.
325–1947.	Black cup and countersunk bolts and nuts.
328A–1963.	Twist drill sizes.
405–1945.	Expanded metal (steel) for general purposes.
427:Part 1–1961.	Vickers hardness test.
450–1958.	Machine screws and machine screw nuts (B.S.W. and B.S.F. threads).
498:Part 1–1960.	Rasps and engineers' files.
499–1952.	Glossary of terms relating to the welding of metals.
620–1954.	Dimensions of grinding wheels.
641–1951.	Dimensions of small rivets for general purposes.

721–1963.	Worm gearing.
768–1958.	Slotted grub screws.
817–1957.	Cast iron surface plates for inspection and marking purposes.
818–1963.	Cast iron straight-edges (bow shaped and I-section).
863–1939.	Steel straight-edges of rectangular section.
870–1950.	External micrometers.
876–1957.	Hand hammers.
888–1950.	Slip (or block) gauges and their accessories.
891 : Part 1–1962.	Method for Rockwell hardness test.
906–1940.	Engineers' parallels (steel).
907–1954.	Dial gauges for linear measurement.
919 : Part 1–1960.	Gauges for screw threads of Unified form.
919 : Part 2–1952.	Gauges for screw threads other than those of Unified form.
939–1962.	Engineers' squares, including cylindrical and block squares.
949 : Part 1–1951.	Taps for threads other than Unified threads.
949 : Part 2–1954.	Taps for Unified threads.
957–1941.	Feeler gauges.
958–1941.	Precision levels for engineering workshops.
959–1950.	Internal micrometers (including stick micrometers).
978 : Part 3–1952.	Bevel gears.
1054–1954.	Engineers' comparators for external measurement.
1083–1951.	Precision hexagon bolts, screws and nuts.
1104–1957.	General purpose Acme screw threads.
1120–1943.	Diamond tipped boring tools.
1127–1950.	Circular screwing dies and hexagon die nuts.
1157–1953.	Tapping drill sizes.
1290–1958.	Wire rope slings and sling legs.
1342–1962.	Detail drawing paper.
1440–1962.	Endless V-belt drives for industrial purposes.
1553 : Part 1–1949.	Graphical symbols for pipes and valves.
1554–1949.	Rust, acid and heat resisting steel wire.
1574–1958.	Split cotter pins.
1580–1962.	Unified screw threads.
1591–1949.	Acid-resisting high silicon iron castings.
1609–1949.	Press tool sets.
1642–1950.	Ball and roller bearing plummer blocks.
1643–1950.	Vernier height gauges.
1657–1950.	Buttress threads.
1660 : Part 1–1950.	Machine tapers.
1685–1951.	Bevel protractors (mechanical and optical).
1709–1958.	Drawing instruments for drawing office use.
1723–1963.	Brazing.
1734–1951.	Micrometer heads.
1759–1951.	Knurling wheels.
1768–1963.	Unified precision hexagon bolts, screws and nuts.
1804–1951.	Parallel dowel pins (steel).
1916–1953.	Limits and fits for engineering.
1919–1953.	Hacksaw blades.
1937–1953.	Engineers' ratchet braces and drilling pillars.
1981–1953.	Unified machine screws and machine screwnuts.
2055–1953.	Rust, acid and heat resisting steel wire for springs.

2059–1953.	Straight-sided splines and serrations.
2061–1953.	Phosphor bronze spring washers for general engineering purposes.
2066–1953.	Balata belting.
2470–1954.	Hexagon socket screws and wrench keys.
2517–1954.	Definitions for use in mechanical engineering.
2553–1954.	Spring collets.
3468–1962.	Austenitic cast iron.
3481:Part 1–1962.	Flat lifting slings.
3555–1962.	Ring spanners.

The above-quoted Standards are obtainable from:—
The British Standards Institution (Sales Branch)
2 Park Street,
London, W.1
England

APPENDIX

TABLE OF EQUIVALENTS

12 inches = 1 foot = 30·48 centimetres = 304·8 millimetres.
3 feet = 1 yard = 914·4 millimetres = 0·9144 metre.
144 square inches = 1 square foot = 929·03 square centimetres.
9 square feet = 1 square yard = 0·836 square metre.
1728 cubic inches = 1 cubic foot = 0·028317 cubic metre.
27 cubic feet = 1 cubic yard = 0·764553 cubic metre.
16 drams = 1 ounce (*avoirdupois*) = 28·35 grammes.
16 ounces = 1 pound (*avoirdupois*) = 0·45359 kilogramme.
10 mil'imetres = 1 centimetre = 0·3937 inch.
10 centimetres = 1 decimetre = 3·9370 inches.
10 decimetres = 1 metre = 39·3701 inches = 1·0936 yards.
1 inch = 2·54 centimetres = 25·4 millimetres.
1 millimetre = 0·03937 inch.
1 kilogramme = 2·2046 pounds (*avoirdupois*).
1 pound (*avoirdupois*) = 0·4535 kilogramme.
1 Imperial gallon = 1·2 United States gallons = 4·5459 litres.
1000 kilogrammes = 1 metric tonne = 2204·61 lbs = 0·9842 ton.
$\frac{1}{64}''$ = 0·015625 inch = 0·3969 millimetres.
$\frac{1}{32}''$ = 0·03125 „ = 0·79380 „
$\frac{1}{16}''$ = 0·0625 „ = 1·5875 „
$\frac{1}{8}''$ = 0·125 „ = 3·175 „
$\frac{1}{4}''$ = 0·250 „ = 6·350 „
$\frac{3}{8}''$ = 0·375 „ = 9·525 „
$\frac{1}{2}''$ = 0·500 „ = 12·700 „
$1''$ = 1·000 „ = 25·400 „

SQUARES AND CUBES

	Squared	Cubed		Squared	Cubed
$\frac{1}{8}$	0·015	0·0019	7	49	343
$\frac{1}{4}$	0·062	0·0156	8	64	512
$\frac{3}{8}$	0·140	0·0527	9	81	729
$\frac{1}{2}$	0·250	0·1250	10	100	1000
$\frac{5}{8}$	0·390	0·244	20	400	8000
$\frac{3}{4}$	0·562	0·421	30	900	27000
$\frac{7}{8}$	0·765	0·670	40	1600	64000
1	1	1	50	2500	125000
2	4	8	60	3600	216000
3	9	27	70	4900	343000
4	16	64	80	6400	512000
5	25	125	90	8100	729000
6	36	216	100	10000	1000000

APPENDIX

GRADIENTS

Per cent	Gradient	Angle
100	1' in 1'	45 degrees
50	1' in 2'	26 deg. 34 minutes
25	1' in 4'	14 deg. 2 min.
20	1' in 5'	11 deg. 19 min.
10	1' in 10'	5 deg. 43 min.
5	1' in 20'	2 deg. 52 min.

TEMPERATURE EQUIVALENTS—CENTIGRADE, FAHRENHEIT AND RÉAUMUR

$-273°$ C. = $-459\cdot4°$ F.		212° C. = 413·6° F.	
$-273°$ F. = $-169°$ C.		212° F. = 100° C.	
0° C. = 32° F.		1500° C. = 2732° F.	
0° F. = $-17\cdot8°$ C.		1500° F. = 816° C.	
100° C. = 212° F.		0° C. = 0° R.	
100° F. = 37·8° C.		32° F. = 0° R.	
212° F. = 80° R.		100° C. = 80° R.	

CONVERSION RULES

To convert F° to C°, subtract 32 and multiply by $\frac{5}{9}$.
To convert F° to R°, subtract 32 and multiply by $\frac{4}{9}$.
To convert C° to F°, multiply by $\frac{9}{5}$ and add 32.
To convert C° to R°, multiply by $\frac{4}{5}$.
To convert R° to F°, multiply by $\frac{9}{4}$ and add 32.
To convert R° to C°, multiply by $\frac{5}{4}$.

CIRCUMFERENCE AND AREA OF CIRCLES

Diameter (in.)	Circumference (in.)	Area (sq. in.)	Diameter (in.)	Circumference (in.)	Area (sq. in.)
⅛	0·3926	0·01227	2	6·283	3·141
¼	0·7854	0·04908	3	9·424	7·068
⅜	1·178	0·1104	4	12·56	12·566
½	1·570	0·1963	5	15·708	19·635
⅝	1·963	0·3067	10	31·41	78·539
¾	2·356	0·4417	15	47·12	176·71
⅞	2·748	0·6013	20	62·83	314·16
1	3·141	0·7854	25	78·54	490·87

NOTES

NOTES

NOTES

NOTES

NOTES

NOTES

NOTES

NOTES